Grzimek's
Animal Life Encyclopedia

Second Edition

••••

Grzimek's
Animal Life Encyclopedia

Second Edition

●●●●

Volume 13
Mammals II

Devra G. Kleiman, Advisory Editor
Valerius Geist, Advisory Editor
Melissa C. McDade, Project Editor

Joseph E. Trumpey, Chief Scientific Illustrator

Michael Hutchins, Series Editor
In association with the American Zoo and Aquarium Association

GALE®

THOMSON
✳
™
GALE

Detroit • New York • San Diego • San Francisco • Cleveland • New Haven, Conn. • Waterville, Maine • London • Munich

Grzimek's Animal Life Encyclopedia, Second Edition
Volume 13: Mammals II

Project Editor
Melissa C. McDade

Editorial
Stacey Blachford, Deirdre S. Blanchfield, Madeline Harris, Christine Jeryan, Kate Kretschmann, Mark Springer, Ryan Thomason

Indexing Services
Synapse, the Knowledge Link Corporation

Permissions
Margaret Chamberlain

Imaging and Multimedia
Randy Bassett, Mary K. Grimes, Lezlie Light, Christine O'Bryan, Barbara Yarrow, Robyn V. Young

Product Design
Tracey Rowens, Jennifer Wahi

Manufacturing
Wendy Blurton, Dorothy Maki, Evi Seoud, Mary Beth Trimper

ISBN 0-7876-5362-4 (vols. 1–17 set)
      0-7876-6573-8 (vols. 12–16 set)
      0-7876-5788-3 (vol. 12)
      0-7876-5789-1 (vol. 13)
      0-7876-5790-5 (vol. 14)
      0-7876-5791-3 (vol. 15)
      0-7876-5792-1 (vol. 16)

This title is also available as an e-book.
ISBN 0-7876-7750-7 (17-vol set)
Contact your Gale sales representative for ordering information.

LIBRARY OF CONGRESS CATALOGING-IN-PUBLICATION DATA

Grzimek, Bernhard.
  [Tierleben. English]
  Grzimek's animal life encyclopedia.— 2nd ed.
    v. cm.
Includes bibliographical references.
Contents: v. 1. Lower metazoans and lesser deuterosomes / Neil Schlager, editor — v. 2. Protostomes / Neil Schlager, editor — v. 3. Insects / Neil Schlager, editor — v. 4-5. Fishes I-II / Neil Schlager, editor — v. 6. Amphibians / Neil Schlager, editor — v. 7. Reptiles / Neil Schlager, editor — v. 8-11. Birds I-IV / Donna Olendorf, editor — v. 12-16. Mammals I-V / Melissa C. McDade, editor — v. 17. Cumulative index / Melissa C. McDade, editor.
ISBN 0-7876-5362-4 (set hardcover : alk. paper)
    1. Zoology—Encyclopedias. I. Title: Animal life encyclopedia. II. Schlager, Neil, 1966- III. Olendorf, Donna IV. McDade, Melissa C. V. American Zoo and Aquarium Association. VI. Title.
QL7 .G7813 2004

                          590'.3—dc21
                          2002003351

Printed in Canada
10 9 8 7 6 5 4 3 2 1

Recommended citation: *Grzimek's Animal Life Encyclopedia*, 2nd edition. Volumes 12–16, *Mammals I–V*, edited by Michael Hutchins, Devra G. Kleiman, Valerius Geist, and Melissa C. McDade. Farmington Hills, MI: Gale Group, 2003.

Contents

Contents

Contents

Foreword

Earth is teeming with life. No one knows exactly how many distinct organisms inhabit our planet, but more than 5 million different species of animals and plants could exist, ranging from microscopic algae and bacteria to gigantic elephants, redwood trees and blue whales. Yet, throughout this wonderful tapestry of living creatures, there runs a single thread: Deoxyribonucleic acid or DNA. The existence of DNA, an elegant, twisted organic molecule that is the building block of all life, is perhaps the best evidence that all living organisms on this planet share a common ancestry. Our ancient connection to the living world may drive our curiosity, and perhaps also explain our seemingly insatiable desire for information about animals and nature. Noted zoologist, E. O. Wilson, recently coined the term "biophilia" to describe this phenomenon. The term is derived from the Greek *bios* meaning "life" and *philos* meaning "love." Wilson argues that we are human because of our innate affinity to and interest in the other organisms with which we share our planet. They are, as he says, "the matrix in which the human mind originated and is permanently rooted." To put it simply and metaphorically, our love for nature flows in our blood and is deeply engrained in both our psyche and cultural traditions.

Our own personal awakenings to the natural world are as diverse as humanity itself. I spent my early childhood in rural Iowa where nature was an integral part of my life. My father and I spent many hours collecting, identifying and studying local insects, amphibians and reptiles. These experiences had a significant impact on my early intellectual and even spiritual development. One event I can recall most vividly. I had collected a cocoon in a field near my home in early spring. The large, silky capsule was attached to a stick. I brought the cocoon back to my room and placed it in a jar on top of my dresser. I remember waking one morning and, there, perched on the tip of the stick was a large moth, slowly moving its delicate, light green wings in the early morning sunlight. It took my breath away. To my inexperienced eyes, it was one of the most beautiful things I had ever seen. I knew it was a moth, but did not know which species. Upon closer examination, I noticed two moon-like markings on the wings and also noted that the wings had long "tails", much like the ubiquitous tiger swallow-tail butterflies that visited the lilac bush in our backyard. Not wanting to suffer my ignorance any longer, I reached immediately for my *Golden Guide to North American Insects* and searched through the section on moths and butterflies. It was a luna moth! My heart was pounding with the excitement of new knowledge as I ran to share the discovery with my parents.

I consider myself very fortunate to have made a living as a professional biologist and conservationist for the past 20 years. I've traveled to over 30 countries and six continents to study and photograph wildlife or to attend related conferences and meetings. Yet, each time I encounter a new and unusual animal or habitat my heart still races with the same excitement of my youth. If this is biophilia, then I certainly possess it, and it is my hope that others will experience it too. I am therefore extremely proud to have served as the series editor for the Gale Group's rewrite of *Grzimek's Animal Life Encyclopedia*, one of the best known and widely used reference works on the animal world. *Grzimek's* is a celebration of animals, a snapshot of our current knowledge of the Earth's incredible range of biological diversity. Although many other animal encyclopedias exist, *Grzimek's Animal Life Encyclopedia* remains unparalleled in its size and in the breadth of topics and organisms it covers.

The revision of these volumes could not come at a more opportune time. In fact, there is a desperate need for a deeper understanding and appreciation of our natural world. Many species are classified as threatened or endangered, and the situation is expected to get much worse before it gets better. Species extinction has always been part of the evolutionary history of life; some organisms adapt to changing circumstances and some do not. However, the current rate of species loss is now estimated to be 1,000–10,000 times the normal "background" rate of extinction since life began on Earth some 4 billion years ago. The primary factor responsible for this decline in biological diversity is the exponential growth of human populations, combined with peoples' unsustainable appetite for natural resources, such as land, water, minerals, oil, and timber. The world's human population now exceeds 6 billion, and even though the average birth rate has begun to decline, most demographers believe that the global human population will reach 8–10 billion in the next 50 years. Much of this projected growth will occur in developing countries in Central and South America, Asia and Africa—regions that are rich in unique biological diversity.

Finding solutions to conservation challenges will not be easy in today's human-dominated world. A growing number of people live in urban settings and are becoming increasingly isolated from nature. They "hunt" in supermarkets and malls, live in apartments and houses, spend their time watching television and searching the World Wide Web. Children and adults must be taught to value biological diversity and the habitats that support it. Education is of prime importance now while we still have time to respond to the impending crisis. There still exist in many parts of the world large numbers of biological "hotspots"—places that are relatively unaffected by humans and which still contain a rich store of their original animal and plant life. These living repositories, along with selected populations of animals and plants held in professionally managed zoos, aquariums and botanical gardens, could provide the basis for restoring the planet's biological wealth and ecological health. This encyclopedia and the collective knowledge it represents can assist in educating people about animals and their ecological and cultural significance. Perhaps it will also assist others in making deeper connections to nature and spreading biophilia. Information on the conservation status, threats and efforts to preserve various species have been integrated into this revision. We have also included information on the cultural significance of animals, including their roles in art and religion.

It was over 30 years ago that Dr. Bernhard Grzimek, then director of the Frankfurt Zoo in Frankfurt, Germany, edited the first edition of *Grzimek's Animal Life Encyclopedia*. Dr. Grzimek was among the world's best known zoo directors and conservationists. He was a prolific author, publishing nine books. Among his contributions were: *Serengeti Shall Not Die*, *Rhinos Belong to Everybody* and *He and I and the Elephants*. Dr. Grzimek's career was remarkable. He was one of the first modern zoo or aquarium directors to understand the importance of zoo involvement in *in situ* conservation, that is, of their role in preserving wildlife in nature. During his tenure, Frankfurt Zoo became one of the leading western advocates and supporters of wildlife conservation in East Africa. Dr. Grzimek served as a Trustee of the National Parks Board of Uganda and Tanzania and assisted in the development of several protected areas. The film he made with his son Michael, *Serengeti Shall Not Die*, won the 1959 Oscar for best documentary.

Professor Grzimek has recently been criticized by some for his failure to consider the human element in wildlife conservation. He once wrote: "A national park must remain a primordial wilderness to be effective. No men, not even native ones, should live inside its borders." Such ideas, although considered politically incorrect by many, may in retrospect actually prove to be true. Human populations throughout Africa continue to grow exponentially, forcing wildlife into small islands of natural habitat surrounded by a sea of humanity. The illegal commercial bushmeat trade—the hunting of endangered wild animals for large scale human consumption—is pushing many species, including our closest relatives, the gorillas, bonobos and chimpanzees, to the brink of extinction. The trade is driven by widespread poverty and lack of economic alternatives. In order for some species to survive it will be necessary, as Grzimek suggested, to establish and enforce a system of protected areas where wildlife can roam free from exploitation of any kind.

While it is clear that modern conservation must take the needs of both wildlife and people into consideration, what will the quality of human life be if the collective impact of short-term economic decisions is allowed to drive wildlife populations into irreversible extinction? Many rural populations living in areas of high biodiversity are dependent on wild animals as their major source of protein. In addition, wildlife tourism is the primary source of foreign currency in many developing countries and is critical to their financial and social stability. When this source of protein and income is gone, what will become of the local people? The loss of species is not only a conservation disaster; it also has the potential to be a human tragedy of immense proportions. Protected areas, such as national parks, and regulated hunting in areas outside of parks are the only solutions. What critics do not realize is that the fate of wildlife and people in developing countries is closely intertwined. Forests and savannas emptied of wildlife will result in hungry, desperate people, and will, in the long-term lead to extreme poverty and social instability. Dr. Grzimek's early contributions to conservation should be recognized, not only as benefiting wildlife, but as benefiting local people as well.

Dr. Grzimek's hope in publishing his *Animal Life Encyclopedia* was that it would "...disseminate knowledge of the animals and love for them", so that future generations would "...have an opportunity to live together with the great diversity of these magnificent creatures." As stated above, our goals in producing this updated and revised edition are similar. However, our challenges in producing this encyclopedia were more formidable. The volume of knowledge to be summarized is certainly much greater in the twenty-first century than it was in the 1970's and 80's. Scientists, both professional and amateur, have learned and published a great deal about the animal kingdom in the past three decades, and our understanding of biological and ecological theory has also progressed. Perhaps our greatest hurdle in producing this revision was to include the new information, while at the same time retaining some of the characteristics that have made *Grzimek's Animal Life Encyclopedia* so popular. We have therefore strived to retain the series' narrative style, while giving the information more organizational structure. Unlike the original *Grzimek's*, this updated version organizes information under specific topic areas, such as reproduction, behavior, ecology and so forth. In addition, the basic organizational structure is generally consistent from one volume to the next, regardless of the animal groups covered. This should make it easier for users to locate information more quickly and efficiently. Like the original Grzimek's, we have done our best to avoid any overly technical language that would make the work difficult to understand by non-biologists. When certain technical expressions were necessary, we have included explanations or clarifications.

Considering the vast array of knowledge that such a work represents, it would be impossible for any one zoologist to have completed these volumes. We have therefore sought specialists from various disciplines to write the sections with which they are most familiar. As with the original *Grzimek's*, we have engaged the best scholars available to serve as topic editors, writers, and consultants. There were some complaints about inaccuracies in the original English version that may have been due to mistakes or misinterpretation during the complicated translation process. However, unlike the original *Grzimek's*, which was translated from German, this revision has been completely re-written by English-speaking scientists. This work was truly a cooperative endeavor, and I thank all of those dedicated individuals who have written, edited, consulted, drawn, photographed, or contributed to its production in any way. The names of the topic editors, authors, and illustrators are presented in the list of contributors in each individual volume.

The overall structure of this reference work is based on the classification of animals into naturally related groups, a discipline known as taxonomy or biosystematics. Taxonomy is the science through which various organisms are discovered, identified, described, named, classified and catalogued. It should be noted that in preparing this volume we adopted what might be termed a conservative approach, relying primarily on traditional animal classification schemes. Taxonomy has always been a volatile field, with frequent arguments over the naming of or evolutionary relationships between various organisms. The advent of DNA fingerprinting and other advanced biochemical techniques has revolutionized the field and, not unexpectedly, has produced both advances and confusion. In producing these volumes, we have consulted with specialists to obtain the most up-to-date information possible, but knowing that new findings may result in changes at any time. When scientific controversy over the classification of a particular animal or group of animals existed, we did our best to point this out in the text.

Readers should note that it was impossible to include as much detail on some animal groups as was provided on others. For example, the marine and freshwater fish, with vast numbers of orders, families, and species, did not receive as detailed a treatment as did the birds and mammals. Due to practical and financial considerations, the publishers could provide only so much space for each animal group. In such cases, it was impossible to provide more than a broad overview and to feature a few selected examples for the purposes of illustration. To help compensate, we have provided a few key bibliographic references in each section to aid those interested in learning more. This is a common limitation in all reference works, but *Grzimek's Encyclopedia of Animal Life* is still the most comprehensive work of its kind.

I am indebted to the Gale Group, Inc. and Senior Editor Donna Olendorf for selecting me as Series Editor for this project. It was an honor to follow in the footsteps of Dr. Grzimek and to play a key role in the revision that still bears his name. *Grzimek's Animal Life Encyclopedia* is being published by the Gale Group, Inc. in affiliation with my employer, the American Zoo and Aquarium Association (AZA), and I would like to thank AZA Executive Director, Sydney J. Butler; AZA Past-President Ted Beattie (John G. Shedd Aquarium, Chicago, IL); and current AZA President, John Lewis (John Ball Zoological Garden, Grand Rapids, MI), for approving my participation. I would also like to thank AZA Conservation and Science Department Program Assistant, Michael Souza, for his assistance during the project. The AZA is a professional membership association, representing 215 accredited zoological parks and aquariums in North America. As Director/William Conway Chair, AZA Department of Conservation and Science, I feel that I am a philosophical descendant of Dr. Grzimek, whose many works I have collected and read. The zoo and aquarium profession has come a long way since the 1970s, due, in part, to innovative thinkers such as Dr. Grzimek. I hope this latest revision of his work will continue his extraordinary legacy.

Silver Spring, Maryland, 2001
Michael Hutchins
Series Editor

How to use this book

Gzimek's Animal Life Encyclopedia is an internationally prominent scientific reference compilation, first published in German in the late 1960s, under the editorship of zoologist Bernhard Grzimek (1909-1987). In a cooperative effort between Gale and the American Zoo and Aquarium Association, the series is being completely revised and updated for the first time in over 30 years. Gale is expanding the series from 13 to 17 volumes, commissioning new color images, and updating the information while also making the set easier to use. The order of revisions is:

Vol 8–11: Birds I–IV
Vol 6: Amphibians
Vol 7: Reptiles
Vol 4–5: Fishes I–II
Vol 12–16: Mammals I–V
Vol 1: Lower Metazoans and Lesser Deuterostomes
Vol 2: Protostomes
Vol 3: Insects
Vol 17: Cumulative Index

Organized by taxonomy

The overall structure of this reference work is based on the classification of animals into naturally related groups, a discipline known as taxonomy—the science through which various organisms are discovered, identified, described, named, classified, and catalogued. Starting with the simplest life forms, the lower metazoans and lesser deuterostomes, in volume 1, the series progresses through the more complex animal classes, culminating with the mammals in volumes 12–16. Volume 17 is a stand-alone cumulative index.

Organization of chapters within each volume reinforces the taxonomic hierarchy. In the case of the Mammals volumes, introductory chapters describe general characteristics of all organisms in these groups, followed by taxonomic chapters dedicated to Order, Family, or Subfamily. Species accounts appear at the end of the Family and Subfamily chapters To help the reader grasp the scientific arrangement, each type of chapter has a distinctive color and symbol:

● =Order Chapter (blue background)

◓ =Monotypic Order Chapter (green background)

▲ =Family Chapter (yellow background)

△ =Subfamily Chapter (yellow background)

Introductory chapters have a loose structure, reminiscent of the first edition. While not strictly formatted, Order chapters are carefully structured to cover basic information about member families. Monotypic orders, comprised of a single family, utilize family chapter organization. Family and subfamily chapters are most tightly structured, following a prescribed format of standard rubrics that make information easy to find and understand. Family chapters typically include:

Thumbnail introduction
    Common name
    Scientific name
    Class
    Order
    Suborder
    Family
    Thumbnail description
    Size
    Number of genera, species
    Habitat
    Conservation status
Main essay
    Evolution and systematics
    Physical characteristics
    Distribution
    Habitat
    Behavior
    Feeding ecology and diet
    Reproductive biology
    Conservation status
    Significance to humans
Species accounts
    Common name
    Scientific name
    Subfamily
    Taxonomy
    Other common names
    Physical characteristics
    Distribution
    Habitat
    Behavior

Feeding ecology and diet
Reproductive biology
Conservation status
Significance to humans
Resources
Books
Periodicals
Organizations
Other

Color graphics enhance understanding

Grzimek's features approximately 3,000 color photos, including approximately 1,560 in five Mammals volumes; 3,500 total color maps, including nearly 550 in the Mammals volumes; and approximately 5,500 total color illustrations, including approximately 930 in the Mammals volumes. Each featured species of animal is accompanied by both a distribution map and an illustration.

All maps in *Grzimek's* were created specifically for the project by XNR Productions. Distribution information was provided by expert contributors and, if necessary, further researched at the University of Michigan Zoological Museum library. Maps are intended to show broad distribution, not definitive ranges.

All the color illustrations in *Grzimek's* were created specifically for the project by Michigan Science Art. Expert contributors recommended the species to be illustrated and provided feedback to the artists, who supplemented this information with authoritative references and animal skins from University of Michgan Zoological Museum library. In addition to species illustrations, *Grzimek's* features conceptual drawings that illustrate characteristic traits and behaviors.

About the contributors

The essays were written by scientists, professors, and other professionals. *Grzimek's* subject advisors reviewed the completed essays to insure consistency and accuracy.

Standards employed

In preparing these volumes, the editors adopted a conservative approach to taxonomy, relying on Wilson and Reeder's *Mammal Species of the World: a Taxonomic and Geographic Reference* (1993) as a guide. Systematics is a dynamic discipline in that new species are being discovered continuously, and new techniques (e.g., DNA sequencing) frequently result in changes in the hypothesized evolutionary relationships among various organisms. Consequently, controversy often exists regarding classification of a particular animal or group of animals; such differences are mentioned in the text.

Grzimek's has been designed with ready reference in mind and the editors have standardized information wherever feasible. For **Conservation status**, *Grzimek's* follows the IUCN Red List system, developed by its Species Survival Commission. The Red List provides the world's most comprehensive inventory of the global conservation status of plants and animals. Using a set of criteria to evaluate extinction risk, the IUCN recognizes the following categories: Extinct, Extinct in the Wild, Critically Endangered, Endangered, Vulnerable, Conservation Dependent, Near Threatened, Least Concern, and Data Deficient. For a complete explanation of each category, visit the IUCN web page at <http://www.iucn.org/>.

Advisory boards

Series advisor

Michael Hutchins, PhD
Director of Conservation and Science/William Conway
Chair
American Zoo and Aquarium Association
Silver Spring, Maryland

Subject advisors

Volume 1: Lower Metazoans and Lesser Deuterostomes

Dennis A. Thoney, PhD
Director, Marine Laboratory & Facilities
Humboldt State University
Arcata, California

Volume 2: Protostomes

Sean F. Craig, PhD
Assistant Professor, Department of Biological Sciences
Humboldt State University
Arcata, California

Dennis A. Thoney, PhD
Director, Marine Laboratory & Facilities
Humboldt State University
Arcata, California

Volume 3: Insects

Arthur V. Evans, DSc
Research Associate, Department of Entomology
Smithsonian Institution
Washington, DC

Rosser W. Garrison, PhD
Research Associate, Department of Entomology
Natural History Museum
Los Angeles, California

Volumes 4–5: Fishes I– II

Paul V. Loiselle, PhD
Curator, Freshwater Fishes

New York Aquarium
Brooklyn, New York
Dennis A. Thoney, PhD
Director, Marine Laboratory & Facilities
Humboldt State University
Arcata, California

Volume 6: Amphibians

William E. Duellman, PhD
Curator of Herpetology Emeritus
Natural History Museum and Biodiversity Research
Center
University of Kansas
Lawrence, Kansas

Volume 7: Reptiles

James B. Murphy, DSc
Smithsonian Research Associate
Department of Herpetology
National Zoological Park
Washington, DC

Volumes 8–11: Birds I–IV

Walter J. Bock, PhD
Permanent secretary, International Ornithological
Congress
Professor of Evolutionary Biology
Department of Biological Sciences,
Columbia University
New York, New York
Jerome A. Jackson, PhD
Program Director, Whitaker Center for Science, Mathe-
matics, and Technology Education
Florida Gulf Coast University
Ft. Myers, Florida

Volumes 12–16: Mammals I–V

Valerius Geist, PhD
Professor Emeritus of Environmental Science
University of Calgary
Calgary, Alberta
Canada

Contributing writers

Mammals I–V

Clarence L. Abercrombie, PhD
Wofford College
Spartanburg, South Carolina

Cleber J. R. Alho, PhD
Departamento de Ecologia (retired)
Universidade de Brasília
Brasília, Brazil

Carlos Altuna, Lic
Sección Etología
Facultad de Ciencias
Universidad de la República Oriental
del Uruguay
Montevideo, Uruguay

Anders Angerbjörn, PhD
Department of Zoology
Stockholm University
Stockholm, Sweden

William Arthur Atkins
Atkins Research and Consulting
Normal, Illinois

Adrian A. Barnett, PhD
Centre for Research in Evolutionary
Anthropology
School of Life Sciences
University of Surrey Roehampton
West Will, London
United Kingdom

Leonid Baskin, PhD
Institute of Ecology and Evolution
Moscow, Russia

Paul J. J. Bates, PhD
Harrison Institute
Sevenoaks, Kent
United Kingdom

Amy-Jane Beer, PhD
Origin Natural Science
York, United Kingdom

Cynthia Berger, MS
National Association of Science Writers

Richard E. Bodmer, PhD
Durrell Institute of Conservation and
Ecology
University of Kent
Canterbury, Kent
United Kingdom

Daryl J. Boness, PhD
National Zoological Park
Smithsonian Institution
Washington, DC

Justin S. Brashares, PhD
Centre for Biodiversity Research
University of British Columbia
Vancouver, British Columbia
Canada

Hynek Burda, PhD
Department of General Zoology Fac-
ulty of Bio- and Geosciences
University of Essen
Essen, Germany

Susan Cachel, PhD
Department of Anthropology
Rutgers University
New Brunswick, New Jersey

Alena Cervená, PhD
Department of Zoology
National Museum Prague
Czech Republic

Jaroslav Cerveny, PhD
Institute of Vertebrate Biology
Czech Academy of Sciences
Brno, Czech Republic

David J. Chivers, MA, PhD, ScD
Head, Wildlife Research Group
Department of Anatomy

University of Cambridge
Cambridge, United Kingdom

Jasmin Chua, MS
Freelance Writer

Lee Curtis, MA
Director of Promotions
Far North Queensland Wildlife Res-
cue Association
Far North Queensland, Australia

Guillermo D'Elía, PhD
Departamento de Biología Animal
Facultad de Ciencias
Universidad de la República
Montevideo, Uruguay

Tanya Dewey
University of Michigan Museum of
Zoology
Ann Arbor, Michigan

Craig C. Downer, PhD
Andean Tapir Fund
Minden, Nevada

Amy E. Dunham
Department of Ecology and Evolution
State University of New York at Stony
Brook
Stony Brook, New York

Stewart K. Eltringham, PhD
Department of Zoology
University of Cambridge
Cambridge, United Kingdom.

Melville Brockett Fenton, PhD
Department of Biology
University of Western Ontario
London, Ontario
Canada

Kevin F. Fitzgerald, BS
Freelance Science Writer
South Windsor, Connecticut

Theodore H. Fleming, PhD
Department of Biology
University of Miami
Coral Gables, Florida

Gabriel Francescoli, PhD
Sección Etología
Facultad de Ciencias
Universidad de la República Oriental
del Uruguay
Montevideo, Uruguay

Udo Gansloßer, PhD
Department of Zoology
Lehrstuhl I
University of Erlangen-Nürnberg
Fürth, Germany

Valerius Geist, PhD
Professor Emeritus of Environmental
Science
University of Calgary
Calgary, Alberta
Canada

Roger Gentry, PhD
NOAA Fisheries
Marine Mammal Division
Silver Spring, Maryland

Kenneth C. Gold, PhD
Chicago, Illinois

Steve Goodman, PhD
Field Museum of Natural History
Chicago, Illinois and
WWF Madagascar
Programme Office
Antananarivo, Madagascar

Nicole L. Gottdenker
St. Louis Zoo
University of Missouri
St. Louis, Missouri and The Charles
Darwin Research Station
Galápagos Islands, Ecuador

Brian W. Grafton, PhD
Department of Biological Sciences
Kent State University
Kent, Ohio

Joel H. Grossman
Freelance Writer
Santa Monica, California

Mark S. Hafner, PhD
Lowery Professor and Curator of
Mammals
Museum of Natural Science and De-
partment of Biological Sciences
Louisiana State University
Baton Rouge, Louisiana

Alton S. Harestad, PhD
Faculty of Science
Simon Fraser University Burnaby
Vancouver, British Columbia
Canada

Robin L. Hayes
Bat Conservation of Michigan

Kristofer M. Helgen
School of Earth and Environmental
Sciences
University of Adelaide
Adelaide, Australia

Eckhard W. Heymann, PhD
Department of Ethology and Ecology
German Primate Center
Göttingen, Germany

Hannah Hoag, MS
Science Journalist

Hendrik Hoeck, PhD
Max-Planck- Institut für Verhal-
tensphysiologie
Seewiesen, Germany

David Holzman, BA
Freelance Writer
Journal Highlights Editor
American Society for Microbiology

Rodney L. Honeycutt, PhD
Departments of Wildlife and Fisheries
Sciences and Biology and Faculty of
Genetics
Texas A&M University
College Station, Texas

Ivan Horácek, Prof. RNDr, PhD
Head of Vertebrate Zoology
Charles University Prague
Praha, Czech Republic

Brian Douglas Hoyle, PhD
President, Square Rainbow Limited
Bedford, Nova Scotia
Canada

Graciela Izquierdo, PhD
Sección Etología
Facultad de Ciencias
Universidad de la República Oriental
del Uruguay
Montevideo, Uruguay

Jennifer U. M. Jarvis, PhD
Zoology Department
University of Cape Town
Rondebosch, South Africa

Christopher Johnson, PhD
Department of Zoology and Tropical
Ecology
James Cook University
Townsville, Queensland
Australia

Menna Jones, PhD
University of Tasmania School of Zo-
ology
Hobart, Tasmania
Australia

Mike J. R. Jordan, PhD
Curator of Higher Vertebrates
North of England Zoological Society
Chester Zoo
Upton, Chester
United Kingdom

Corliss Karasov
Science Writer
Madison, Wisconsin

Tim Karels, PhD
Department of Biological Sciences
Auburn University
Auburn, Alabama

Serge Larivière, PhD
Delta Waterfowl Foundation
Manitoba, Canada

Adrian Lister
University College London
London, United Kingdom

W. J. Loughry, PhD
Department of Biology
Valdosta State University
Valdosta, Georgia

Geoff Lundie-Jenkins, PhD
Queensland Parks and Wildlife Service
Queensland, Australia

Peter W. W. Lurz, PhD
Centre for Life Sciences Modelling
School of Biology
University of Newcastle
Newcastle upon Tyne, United King-
dom

Colin D. MacLeod, PhD
School of Biological Sciences (Zool-
ogy)
University of Aberdeen
Aberdeen, United Kingdom

James Malcolm, PhD
Department of Biology
University of Redlands
Redlands, California

David P. Mallon, PhD
Glossop
Derbyshire, United Kingdom

Robert D. Martin, BA (Hons), DPhil,
DSc
Provost and Vice President
Academic Affairs
The Field Museum
Chicago, Illinois

Gary F. McCracken, PhD
Department of Ecology and Evolu-
tionary Biology
University of Tennessee
Knoxville, Tennessee

Colleen M. McDonough, PhD
Department of Biology
Valdosta State University
Valdosta, Georgia

William J. McShea, PhD
Department of Conservation Biology
Conservation and Research Center
Smithsonian National Zoological Park
Washington, DC

Rodrigo A. Medellín, PhD
Instituto de Ecología
Universidad Nacional Autónoma de
México
Mexico City, Mexico

Leslie Ann Mertz, PhD
Fish Lake Biological Program
Wayne State University
Detroit, Michigan

Gus Mills, PhD
SAN Parks/Head
Carnivore Conservation Group,
EWT
Skukuza, South Africa

Patricia D. Moehlman, PhD
IUCN Equid Specialist Group

Paula Moreno, MS
Texas A&M University at Galveston
Marine Mammal Research Program
Galveston, Texas

Virginia L. Naples, PhD
Department of Biological Sciences
Northern Illinois University
DeKalb, Illinois

Ken B. Naugher, BS
Conservation and Enrichment Pro-
grams Manager
Montgomery Zoo
Montgomery, Alabama

Derek William Niemann, BA
Royal Society for the Protection of
Birds
Sandy, Bedfordshire
United Kingdom

Carsten Niemitz, PhD
Professor of Human Biology
Department of Human Biology and
Anthropology
Freie Universität Berlin
Berlin, Germany

Daniel K. Odell, PhD
Senior Research Biologist
Hubbs-SeaWorld Research Institute
Orlando, Florida

Bart O'Gara, PhD
University of Montana (adjunct retired
professor)
Director, Conservation Force

Norman Owen-Smith, PhD
Research Professor in African Ecology
School of Animal, Plant and Environ-
mental Sciences
University of the Witwatersrand
Johannesburg, South Africa

Malcolm Pearch, PhD
Harrison Institute
Sevenoaks, Kent
United Kingdom

Kimberley A. Phillips, PhD
Hiram College
Hiram, Ohio

David M. Powell, PhD
Research Associate
Department of Conservation Biology
Conservation and Research Center
Smithsonian National Zoological Park
Washington, DC

Jan A. Randall, PhD
Department of Biology
San Francisco State University
San Francisco, California

Randall Reeves, PhD
Okapi Wildlife Associates
Hudson, Quebec
Canada

Peggy Rismiller, PhD
Visiting Research Fellow
Department of Anatomical Sciences
University of Adelaide
Adelaide, Australia

Konstantin A. Rogovin, PhD
A.N. Severtsov Institute of Ecology
and Evolution RAS
Moscow, Russia

Randolph W. Rose, PhD
School of Zoology
University of Tasmania
Hobart, Tasmania
Australia

Frank Rosell
Telemark University College
Telemark, Norway

Gretel H. Schueller
Science and Environmental Writer
Burlington, Vermont

Bruce A. Schulte, PhD
Department of Biology
Georgia Southern University
Statesboro, Georgia

John H. Seebeck, BSc, MSc, FAMS
Australia

Melody Serena, PhD
Conservation Biologist
Australian Platypus Conservancy
Whittlesea, Australia

David M. Shackleton, PhD
Faculty of Agricultural of Sciences
University of British Columbia
Vancouver, British Columbia
Canada

Robert W. Shumaker, PhD
Iowa Primate Learning Sanctuary
Des Moines, Iowa and Krasnow Insti-
tute at George Mason University
Fairfax, Virginia

Andrew T. Smith, PhD
School of Life Sciences
Arizona State University
Phoenix, Arizona

Karen B. Strier, PhD
Department of Anthropology
University of Wisconsin
Madison, Wisconsin

Karyl B. Swartz, PhD
Department of Psychology
Lehman College of The City Univer-
sity of New York
Bronx, New York

Bettina Tassino, MSc
Sección Etología

Facultad de Ciencias
Universidad de la República Oriental
del Uruguay
Montevideo, Uruguay

Barry Taylor, PhD
University of Natal
Pietermaritzburg, South Africa

Jeanette Thomas, PhD
Department of Biological Sciences
Western Illinois University-Quad
Cities
Moline, Illinois

Ann Toon
Arnside, Cumbria
United Kingdom

Stephen B. Toon
Arnside, Cumbria
United Kingdom

Hernán Torres, PhD
Santiago, Chile

Rudi van Aarde, BSc (Hons), MSc,
PhD
Director and Chair of Conservation
Ecology Research Unit
University of Pretoria
Pretoria, South Africa

Mac van der Merwe, PhD
Mammal Research Institute
University of Pretoria
Pretoria, South Africa

Christian C. Voigt, PhD
Research Group Evolutionary Ecology
Leibniz-Institute for Zoo and Wildlife
Research
Berlin, Germany

Sue Wallace
Freelance Writer
Santa Rosa, California

Lindy Weilgart, PhD
Department of Biology
Dalhousie University
Halifax, Nova Scotia
Canada

Randall S. Wells, PhD
Chicago Zoological Society
Mote Marine Laboratory
Sarasota, Florida

Nathan S. Welton
Freelance Science Writer
Santa Barbara, California

Patricia Wright, PhD
State University of New York at Stony
Brook
Stony Brook, New York

Marcus Young Owl, PhD
Department of Anthropology and
Department of Biological Sciences
California State University
Long Beach, California

Jan Zima, PhD
Institute of Vertebrate Biology
Academy of Sciences of the Czech
Republic
Brno, Czech Republic

Contributing illustrators

Drawings by Michigan Science Art

Joseph E. Trumpey, Director, AB, MFA
Science Illustration, School of Art and Design, University
of Michigan

Wendy Baker, ADN, BFA

Ryan Burkhalter, BFA, MFA

Brian Cressman, BFA, MFA

Emily S. Damstra, BFA, MFA

Maggie Dongvillo, BFA

Barbara Duperron, BFA, MFA

Jarrod Erdody, BA, MFA

Dan Erickson, BA, MS

Patricia Ferrer, AB, BFA, MFA

George Starr Hammond, BA, MS, PhD

Gillian Harris, BA

Jonathan Higgins, BFA, MFA

Amanda Humphrey, BFA

Emilia Kwiatkowski, BS, BFA

Jacqueline Mahannah, BFA, MFA

John Megahan, BA, BS, MS

Michelle L. Meneghini, BFA, MFA

Katie Nealis, BFA

Laura E. Pabst, BFA

Amanda Smith, BFA, MFA

Christina St.Clair, BFA

Bruce D. Worden, BFA

Kristen Workman, BFA, MFA

Thanks are due to the University of Michigan, Museum of Zoology, which provided specimens that served as models for the images.

Maps by XNR Productions

Paul Exner, Chief cartographer
XNR Productions, Madison, WI

Tanya Buckingham

Jon Daugherity

Laura Exner

Andy Grosvold

Cory Johnson

Paula Robbins

Peramelemorphia
(Bandicoots and bilbies)

Class Mammalia
Order Peramelemorphia
Number of families 2 or 3
Number of genera, species 8 genera; 21 species (18 extant)

Photo: An endangered western barred bandicoot (*Perameles bougainville*). (Photo by © Jiri Lochman/ Lochman Transparencies. Reproduced by permission.)

No order of marsupials has suffered so badly as a result of European settlement as the Peramelemorphia. Before the arrival of Europeans, bandicoots were plentiful, revered by the aboriginal peoples of Australia, and valued as a source of food by both the aborigines and the native peoples of New Guinea. By the twentieth century, their fortunes in Australia were in steep decline; three species became extinct and at the beginning of the twenty first century, others are still under serious threat of the same fate.

The first Europeans viewed bandicoots with some disdain, purely because of their appearance. Writing in 1805, naturalist Geoffroy wrote "their muzzle, which is much too long, gives them an air exceedingly stupid." Their rat-like shape led to the erroneous name of bandicoot—the Indian word meaning "pig-rat," originally given to the greater bandicoot rat *Bandicota indica*, of Southeast Asia. Disparaging attitudes have continued into modern times. The word "bandicoot" is still used in the Australian vernacular as a mild term of abuse.

Dismissive attitudes have traditionally been accompanied by scientific neglect. In the classic volume *Bandicoots and Bilbies* (1990), Lyne noted that of 400 references to bandicoots in scientific journals between 1797 and 1984, more than half were within the final 20 years. Knowledge of this family is still patchy, with the New Guinea species in particular woefully little understood.

Evolution and systematics

The discovery, at the beginning of this century, of an early Eocene (55 million years ago) bandicoot more than twice the age of any fossil bandicoot previously recovered, may help to shed light on an order whose evolution and taxonomy are the object of controversy. Much of the confusion centers around shared physical characteristics with other major marsupial groups.

Some scientists argue that simple dentition suggests that the perameloids evolved from the dasyurids, an order which includes quolls and phascogales. Others claim that the presence of the fused toes of the hind foot shows a closer evolutionary relationship with kangaroos, wombats, and other diprotodonts.

Bandicoots are divided into two families. Species of arid and temperate forest belong to the Peramelidae, a family consisting of four genera and 10 species. Some taxonomists treat the subfamily Thylacomyinae as a complete family. The rainforest bandicoots, found predominantly in New Guinea, sit within the family Peroryctidae, comprising four genera and 11 species.

Physical characteristics

The public perception that "all bandicoots are much the same" is understandable. In appearance, this order has a great deal of uniformity, particularly in Australia, where the species are all roughly rabbit-sized. There is more size variation in New Guinea, where the largest of the *Peroryctes* is more than

A rabbit-eared bandicoot (*Macrotis lagotis*) eating in Alice Springs, Northern Territory, Australia. (Photo by Animals Animals ©R. J. B. Goodale, OSF. Reproduced by permission.)

three times bigger than the smallest of the *Microperoryctes* mouse bandicoots.

Bandicoots have thick-set bodies with a short neck and, in most species, a long, pointed snout. The tail is short, except in the greater bilby *Macrotis lagotis*, where it is long and brush-like. In the dry-country long-nosed bandicoots *Perameles* and in bilbies, the ears are large, but the general pattern is for short ears. The front limbs are generally short. The forefeet have powerful, flat claws, used for digging. The hind limbs are longer, with powerful thigh muscles. The hind feet are elongated and, unique among marsupials, the second and third toes are syndactylous, that is, fused together. This fusion is probably an adaptation for grooming. Bandicoots generally move slowly in a bunny hop, with the front and back legs working alternately. They can, however, adopt a fast gallop and some species use their strong back legs to make sudden leaps.

Bandicoot teeth are suited to an insectivorous diet. Although small, they are sharp and the molars are slightly pointed. In common with dasyurids, bandicoots are polypro-prodont, possessing four or five pairs of upper incisors and three lower pairs.

Distribution

Bandicoots are found in Australia and New Guinea, including its surrounding islands. Only two species show a range overlap between the two countries—the northern brown (*Isoodon macrourus*) and rufous spiny bandicoot (*Echymipera rufescens*). The land masses have been connected intermittently in recent times, suggesting that distribution of species is governed by habitat rather than geographical differences.

In Australia, bandicoots are largely confined to forested coastal strips and offshore islands. The greater bilby found in central-northern parts of the dry interior is the exception. Distribution is more widespread in New Guinea, with both the forested uplands and settled lowlands occupied.

In Australia, distribution was formerly far wider. Among the four arid zone species, the greater bilby is thought to have lived over 70% of the Australian landmass. Three other species—the lesser bilby (*Macrotis leucura*), pig-footed bandicoot (*Chaeropus ecaudatus*), and desert bandicoot (*Perameles eremiana*)—all occupied large parts of the interior, but are now extinct. In 1845, Gould described the southern brown bandicoot (*Isoodon obesulus*) as "one of the very commonest of Australian mammals."

The range contraction of the northern brown bandicoot in Queensland reflects shrinking distribution of Australian bandicoots under human pressure. Once widespread in the center of the state, it has been forced out of open country by intensive livestock farming and now occupies only narrow strips of land beside rivers.

Habitat

In New Guinea, the habitats of a number of species are linked to altitude. For example, all three *Microperoryctes* mouse bandicoots are found above 3,500 ft (1,000 m). The high-altitude species are confined to primary rainforest. Among the lowland species of New Guinea and those of Australia, there is evidence that bandicoots can occupy an extremely wide range of habitats.

If there is a linking factor between species of Australia and the New Guinea lowlands, it is their preference for habitats that are temporary by nature. Areas that have been recently burned or cleared, such as light scrub or heath, generally have a wider variety of vegetation and a greater number of invertebrates than more established habitats. These temporary, often ecotonal, habitats are quickly occupied by bandicoots. Such exploitation explains the ready colonization of low-

A long-nosed bandicoot (*Perameles nasuta*) juvenile eating a bug in eastern Australia. (Photo by Animals Animals ©K. Atkinson, OSF. Reproduced by permission.)

A golden bandicoot (*Isoodon auratus*) feeds on eggs in a green seaturtle (*Chelonia mydas*) nest during egg laying. (Photo by © Jiri Lochman/Lochman Transparencies. Reproduced by permission.)

intensity farmland in New Guinea and suburban towns and cities of Australia. The downside to such opportunism is the bandicoot's reliance on a mosaic of vegetation at different stages of growth. If these become isolated or fragmented, animals are unable to disperse at the point when a particular habitat becomes unsuitable.

Behavior

These marsupials are highly solitary, coming together only to mate. Parental bonds are broken just two months after birth and juveniles show extremely high rates of dispersal. If groups of individuals are seen together, it is purely to exploit a localized food resource.

Males show clear territorial aggression towards each other. During encounters, they mark the ground or vegetation with scent from a gland behind the ear. The two males give warning puffing calls and may chase each other. Rarely, the conflict is resolved by fighting, with both combatants approaching the other raised on their hind legs.

The males have home ranges that are generally several times larger than those of the much smaller females. A single male home range can overlap that of a number of females.

The disparity in range sizes is reflected in activity patterns - females venture into the open almost exclusively to forage for food. Their time out of the nest will be limited further if they still have young in the nest which require suckling. By comparison, males spend some time each night patrolling much larger areas, partly to chase off rival males and also in search of females in estrus.

All bandicoot species studied are nocturnal. Daytime nests vary between species and habitats. Bilbies are the only bandicoots to dig burrows. Other bandicoot species living in open country make their nests among piles of rocks, down rabbit burrows, or in tree holes. They may dig a shallow hole in the ground and cover it with grasses and dead vegetation. Forest-dwelling bandicoots make use of plentiful ground cover by building a heap of grasses, twigs, and humus and hollowing out an internal chamber.

Feeding ecology and diet

Bandicoots are nocturnal, terrestrial foragers, reliant on their strong senses of smell and hearing to detect food. Species studied gain most of their food by using their powerful forelimbs to dig numerous small, conical holes vertically into the

A juvenile Raffray's bandicoot (*Peroryctes raffrayana*). (Photo by Pavel German. Reproduced by permission.)

earth or forest floor. Some food is also taken directly from the ground.

All research points to bandicoots being omnivorous. The diet includes a wide range of surface and soil invertebrates, such as ants and termites, beetles and their larvae, earthworms, moths, and spiders. Birds' eggs, small mammals, and lizards are also eaten.

Fungi and fruit are of seasonal importance to forest-dwelling bandicoots. In the few studies of New Guinea species, the large-toothed bandicoot (*Echymipera clara*) feeds on *Ficus* and pandanus fruits, while the spiny bandicoot (*Echymipera kalubu*) has been observed eating a variety of fallen fruit.

Bandicoot dentition, together with a lack of adaptations in the alimentary canal indicate that plant material is eaten selectively, with little fibrous vegetation taken. Seeds and tubers are most often eaten. Omnivorous feeding means that the teeth, sharp and better equipped for a purely insectivorous diet, become flattened with wear.

Reproductive biology

Contact between the males and females of these solitary animals is restricted to mating, when a male will follow the female until she is ready to be mounted. Mating varies between species and location; some bandicoots mate all year round, while others are limited to six or eight months of the year by factors such as day length, rainfall, and temperature. Females are polyestrous and mating is probably either polygynous or promiscuous.

Bandicoot reproduction is unusual in two major respects. They have among the shortest gestation periods of any mammals—just 12.5 days in the case of the northern brown bandicoot. Yet conversely, unique among marsupials, they have an advanced form of placentation that is more akin to that of eutherian mammals with significantly longer gestation periods.

While the embryo first develops with the aid of a yolk sac placenta as is the case with other marsupials, it is nourished in the latter stages of gestation by a chorioallantoic placenta, a more advanced physical attachment between the uterus of the mother and the embryo, that allows the exchange of nutrition, respiratory gases, and excretia. This connection is less sophisticated in the bandicoots however, since they lack villi—the finger-like projections that link the outer membrane of the embryo with the wall of the uterus. Oddly, the umbilical cord remains attached as the young leave the uterus and crawl into the backward-facing pouch. Since the attachment lasts only a matter of hours, the cord's primary purpose at this stage appears to be as a kind of safety rope.

Although the female usually has eight teats, she rarely has more than four young at a time. The young leave the pouch at 49–50 days. Weaning takes around 10 days, by which time the next litter of half inch (1 cm)-long young are ready to occupy the mother's pouch. Bandicoots become sexually mature within four months of birth, but this order's fast reproductive rate is offset by high mortality of the young. Only just over one in 10 of all baby bandicoots will survive long enough to mate. Following maturity, life expectancy is 2.5–3.0 years

Conservation

At the beginning of the twenty-first century, there are some signs that the downward trend in the fortune of bandicoots may at least be slowing. In the previous century, three species had plummeted to extinction even before scientists had gained a clear understanding of their ecology. Today, the IUCN lists (as of 2002) the golden (*Isoodon auratus*) and eastern barred bandicoot (*Perameles gunnii*) and greater bilby as Vulnerable and the western barred bandicoot (*Perameles*

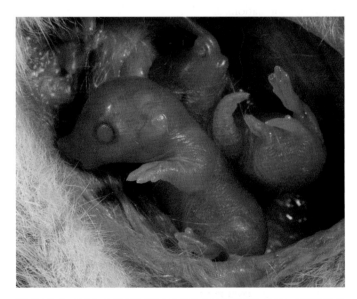

Western barred bandicoot (*Perameles bougainville*) joeys suckling in mother's pouch. (Photo by © Jiri Lochman/Lochman Transparencies. Reproduced by permission.)

A model of the extinct pig-footed bandicoot (*Chaeropus ecaudatus*). (Photo by Tom McHugh/Photo Researchers, Inc. Reproduced by permission.)

bougainville) as Endangered. Seven New Guinea species are given a Data Deficient rating.

The factors which caused catastrophic declines in Australian bandicoot populations and distribution during the nineteenth and twentieth century are still of paramount significance. Continued intensive grazing by cattle and sheep of former bandicoot habitat over much of the continent ensures that there is no realistic possibility of recolonization without major changes in land management. In Victoria, for example, the hummock grassland of kangaroo and wallaby grass was largely removed in favor of ryegrass and clover to feed grazing livestock. A lack of floristic diversity meant that there was not enough food, or shelter from predators and adverse weather to sustain the population of western barred bandicoots.

Even supposing that livestock grazing could be reduced or eliminated, an added complication is the presence of introduced predators, especially foxes and cats. The arrival of cats on Hermite Island in Western Australia caused the extinction of the golden bandicoot from that island, for example. There is also evidence that bandicoot populations are being suppressed by transmission from cats of toxoplasmosis. Introduced rabbits have also been a major cause of population declines through competition for food and habitat.

While captive breeding programs for greater bilbies and eastern barred bandicoots have proved fruitful, a prerequisite

of successful reintroductions into the wild appears to be the exclusion of predators, together with control of rabbits and kangaroos. Exclusion by use of fences can only be effective over very small areas. In some unfenced areas where bandicoots are present, conservation authorities are attempting to limit predation. In Sydney, the National Parks and Wildlife Service began a fox control program in 2001, using the presence of the southern brown bandicoot as an indicator of success.

Colonies of bilbies and western barred bandicoots are being bred at a special facility within the Francois Peron National Park in Western Australia under the Project Eden conservation program, started in 1995. These colonies are supplemented with animals from breeding programs at other agencies such as the Kanyana Wildlife Rehabilitation Centre (bilby, western barred bandicoot) and are slated for the reintroduction phase of the program.

Scientists are still exploring the relative significance of other factors in bandicoot declines. In Western Australia, for example, the extinctions of the pig-footed bandicoot and desert bandicoot are now thought to have been precipitated by the aborigines abandoning traditional burning practices starting in the 1930s. The replacing of mosaic selective burning by uncontrolled wildfires over very large areas left these less mobile species unable to escape.

A northern brown bandicoot (*Isoodon macrourus*) foraging at night. (Photo by K. Stepnell. Bruce Coleman, Inc. Reproduced by permssion.)

The conservation of some species is hindered by the fact that they do not live conveniently within protected areas. One fifth of Tasmania consists of nature preserves, yet this does not protect bandicoots, since they live largely on the periphery or outside of these sanctuary zones.

The conservation status of several bandicoot species in New Guinea remains something of a mystery. Seven species are classified by the IUCN as Data Deficient. Partly, this is a reflection of their location in remote and often inaccessible mountain rainforest habitat. But bandicoots are also notoriously difficult to trap. Their preference for natural rainforest food rather than artificial bait means that population monitoring is extremely difficult. Hunting is widespread and common in New Guinea and its islands, but without proper censuses, it is almost impossible to detect whether it is having a deleterious effect.

Significance to humans

The first aboriginal settlers of Australia venerated the bandicoot as one of the creators of life. In their spiritual "dreamtime" stories, they told of Karora, a giant bandicoot who slept in darkness under the earth, until he awoke and gave birth to the first humans from beneath his armpit!

Such reverence did not stop the aborigines from eating bandicoots—they were hunted both for their meat and for their fur. Bandicoots are still an important source of food for some native peoples of New Guinea.

Although killed incidentally by rabbit trappers, bandicoots made little impression on European settlers throughout the nineteenth and much of the twentieth century. The state of New South Wales, for example, first started giving them legal protection only in 1948, by which time three species had become extinct.

Fortunately, a growing perception in Australia of the importance of appreciating and protecting native fauna is now starting to benefit this order. Foremost in the public relations revolution is the endangered greater bilby, whose human supporters launched a sustained campaign from the 1980s onwards to substitute this rabbit-like bandicoot for the Easter bunny as an object of affection in the nation's hearts—an ironically appropriate displacement, given that the introduced rabbit is one of the primary causes of this species' decline.

In well-populated areas of southeastern Australia, where bandicoots come into contact with people, conservation organizations, and state protection departments have made attempts to promote greater tolerance of bandicoots. Residents of suburbia are urged to adopt bush-friendly backyard gardening using native plant species, and are encouraged to maintain close control of family pets at night. However, the bandicoots' habit of digging conical holes in lawns and the risk of transmitting ticks to humans does not always make them the most welcome of cohabitants.

Resources

Books

Hoser, R. *Endangered Animals of Australia.* Sydney: Pearson, 1991.

Macdonald, D. *The New Encyclopaedia of Mammals.* Oxford: Oxford University Press, 2001.

Nowak, R. M. *Walker's Mammals of the World.* Baltimore: Johns Hopkins University Press, 1995.

Seebeck, J. H. Brown, P. R. Wallis, R. L. and C. M. Kemper. *Bandicoots and Bilbies.* New South Wales: Surrey Beatty and Sons, 1990.

Strahan, R. *The Mammals of Australia.* Sydney: Australian Museum/Reed Holland, 1995.

Organizations

Arid Recovery Project. P.O. Box 150, Roxby Downs, South Australia 5725 Australia. Phone: (08) 8671 8282. Fax: (08)

Resources

8671 9151. E-mail: arid.recovery@wmc.com Web site: <http://www.aridrecovery.org.au>

Department for Environment and Heritage. GPO Box 1047, Adelaide, South Australia 5001 Australia. Phone: (8) 8204 1910. E-mail: environmentshop@saugov.sa.gov.au Web site: <http://www.environment.sa.gov.au>

Environment Australia. GPO Box 787, Canberra, Australian Capital Territory 2601 Australia. Phone: (2) 6274 1111. Web site: <http://www.ea.gov.au/>

New South Wales National Parks and Wildlife Service. 102 George Street, Sydney, New South Wales 2000 Australia.

Phone: (02) 9253 4600. Fax: (02) 9251 9192. E-mail: info@npws.nsw.gov.au Web site: <www.npws.nsw.gov.au/wildlife/factsheets/bandicoot.html>

Other

Department of Primary Industries, Water and the Environment, Tasmania. <http://www.dpiwe.tas.gov.au/inter.nsf/WebPages/BHAN-5379SX?open>

Warringah Council, New South Wales. <http://www.warringah.nsw.gov.au/bandicoot_survey.htm>

Derek William Niemann, BA

Bandicoots
(Peramelidae and Peroryctidae)

Class Mammalia

Order Peramelemorphia

Family Peramelidae and Peroryctidae

Thumbnail description
Small to medium-sized marsupials, with long, tapering snouts and short tails (most species); ears small to large, especially pronounced in rabbit-bandicoots; most species share a similar body form and are uniform in color, although some species have posterior barring or dorsal longitudinal stripes

Size
Head and body length ranges from 6.7–10.4 in (17–26.5 cm) in mouse bandicoot to 19.7–23.6 in (50–60 cm) in giant bandicoot; tail length from 4.3–4.7 in (11–12 cm) (mouse bandicoot) to 5.5–7.9 in (14–20 cm) (giant bandicoot); and weight from 4.9–6.5 oz (140–185 g) (mouse bandicoot) to 10.6 lb (4.8 kg) (giant bandicoot)

Number of genera, species
7 genera; 19 species

Habitat
Desert, grassland, woodland, forest, coastal complexes, rainforest, semi-urban

Conservation status
Peramelidae: Extinct: 3 species; Endangered: 1 species; Vulnerable: 3 species; Peroryctidae: Data Deficient: 7 species

Distribution
Peramelidae: Australia, New Guinea; Peroryctidae: New Guinea, Indonesia (Irian Jaya, Seram), Australia

Evolution and systematics

The order Peramelemorphia includes all the living bandicoots. They possess four or five pairs of blunt incisors in the upper jaw and three similar pairs in the lower jaw, and are thus polyprotodont. The hindfeet have the second and third toes joined in syndactyly. The order contains a single super-family, the Perameloidea, which is divided into two families: the Peramelidae contains all the non-spiny bandicoots and the pig-footed bandicoot (*Chaeropus ecaudatus*); the Peroryctidae includes the spiny bandicoots. While it appears that the order is intermediate between dasyuroids (polyprotodonts) and diprotodonts, the evolutionary origins of the bandicoots remain contentious and opinion varies, dependent on the significance given to dental or foot structure. Bandicoots may have evolved from dasyuroids, retaining polyprotodonty and separately evolving syndactyly, or from the diprotodonts, retaining their syndactyly and evolving polyprotodonty. But it is more plausible that they are derived from a proto-perameloid ancestor that produced two lines, one the terrestrial insectivorous/omnivorous bandicoots and the other the arboreal, herbivorous possums. Baverstock, et al. in 1990, suggested that this separation might have

occurred around 48 million years ago (mya). Such fossil bandicoots as have been described differ little from modern forms.

The two families are discriminated by skull characteristics, that of peramelids being flattened in lateral view while the skull of peroryctids is more or less cylindrical. Other skull characters have been described by Groves and Flannery in 1990. The fur in many peroryctids is harsh and spiny. Peroryctids mostly inhabit rainforest, in contrast to the relatively dry habitats used by peramelids. It is possible that the now-extinct pig-footed bandicoot should be separated from the other species within the Peramelinae, as it exhibits a number of distinguishing characters of structure and behavior.

Physical characteristics

Bandicoots are small marsupials with a long, pointed snout, and are stockily built, with short limbs and neck. The ears are generally short and rounded, although more elongate in the genus *Perameles*. The pig-footed bandicoots have long, erect ears. The tail is thin and short in most species, although it is

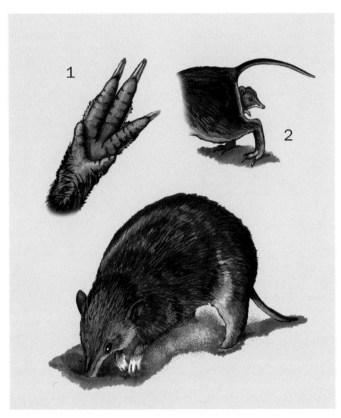

Adaptations for digging in bandicoots (*Isoodon macrourus* shown here). 1. Forepaw has long, strong claws, and digits I and V are reduced; 2. The bandicoot's pouch faces backwards. (Illustration by Gillian Harris)

long and crested in the pig-footed bandicoot. The teeth are small, relatively even in size, and sharply pointed. The dental formula is generally (I1-5/1-3 C1/1 P1-3/1-3, M1-4/1-4) but in *Echymipera* and *Rhynchomeles* there are only four pairs of upper incisors. The forefeet bear strong curved claws that are used in conjunction with the elongate muzzle to dig food items from the soil. Digits I and V are reduced in all species, and in *Chaeropus* only digits II and III are functional, giving rise to the common name. The hindfoot is elongate; digit I is reduced or absent, digits II and III are joined in syndactyly, digit IV is elongate, strong and powerful, while digit V is reduced or absent. Fur coloration is generally drab and unadorned, being darker on the dorsal surface and lighter ventrally. Exceptions exist, however. *Perameles gunnii* and *P. bougainville* have prominent pale posterior bars, *Microperoryctes longicauda*, *M. papuensis*, and *Echymipera echinista* are variously striped on the head and dorsum. The pig-footed bandicoot has a crested tail tipped with black.

Distribution

Bandicoots are confined to Australia, New Guinea, and the island of Seram. Prior to European settlement of Australia, bandicoots were widespread and at least one species was present in any given locality across almost the whole of the continent, in virtually all habitats. In the arid center of Australia, up to five species may have been found, whereas in tropical and temperate zones two or three species occurred. Some off-

shore islands are now the only places in which certain species survive. Similarly, bandicoots occur throughout New Guinea, from the coast to the central high mountain ranges, and some forms are restricted to off-shore islands. One species only occurs on Seram.

Habitat

All species are terrestrial. Australian habitats utilized by peramelids included: arid open-dense shrubland; sand plain, sand-ridge desert and spinifex grassland; temperate grasslands and grassy woodlands; wet and dry open-forest; deciduous vine thickets; heath and heathy woodlands and open-forest; savanna woodland, and shrubby grassland. Three species also use suburban gardens. In New Guinea, peroryctids occupy rainforest—lowland, primary and secondary highland; woodlands; subalpine grasslands; and are also found in gardens and regrowth forest. The Seram Island bandicoot was only found in tropical forest. Altitudinal range for bandicoots is from sea-level to 13,120 ft (4,000 m).

Behavior

Mutual avoidance is the predominant social behavior and most bandicoots are essentially solitary. During courtship and mating, male and female bandicoots associate for a limited time and several males may mate with a single female. The young may follow their mother for a short time after permanent emergence from the nest. Male-male interactions are always aggressive and in captivity will result in serious injury or death. Most species are nocturnal, some more strictly so than others, but southern brown bandicoots are often diur-

A northern brown bandicoot (*Isoodon macrourus*) foraging for insects in leaf litter in *Eucalyptus* forest. (Photo by B. G. Thomson/Photo Researchers, Inc. Reproduced by permission.)

A northern brown bandicoot (*Isoodon macrourus*) on the grass. (Photo by Tom & Pam Gardner/FLPA–Images of Nature. Reproduced by permission.)

nal. Bandicoots adopt several distinct postures when at rest, but when alert will often stand tripedally, with one foreleg raised and retracted towards the body, or stand erect on their hindfeet. Locomotion is quadrupedal and involves walking, running, galloping, and leaping. The latter is believed to be an escape mechanism. Vocalization is restricted to honks, snorts, and sneezes, which may be used to clear the nostrils after digging.

Feeding ecology and diet

Bandicoots are opportunistic and omnivorous, although the pig-footed bandicoot may have been more herbivorous. Most species obtain their food by first locating it through olfaction (and perhaps also by hearing) and then digging a conical pit to where the invertebrate or plant material is situated. The diet includes adult and larval insects (especially Coleoptera, Orthoptera, and Lepidoptera), earthworms, centipedes, seeds, bulbs, tubers, and hypogeous fungal sporocarps. Small vertebrates such as lizards and mice are occasionally eaten. In garden areas, and in tropical rainforest, fallen fruit is eaten.

Reproductive biology

The pouch opens to the rear and contains two crescentic rows of four nipples. Litters vary from one to five (average

about two) in most species. The gestation period is very short—12.5 days in *Perameles nasuta*, *P. gunnii*, and *Isoodon macrourus*. These genera are polyestrous and the estrus cycle is about 20–25 days. Growth and development is rapid and in some species sexual maturity may be reached at theee to four months of age. Bandicoots may breed throughout the year, although some degree of seasonality is shown. Such seasonality may be dependent on climatic conditions; for example, eastern barred bandicoots in Tasmania do not breed during the coldest winter months, and the same species on the mainland ceases breeding during periods of drought. Breeding may be initiated by an increase in food availability (perhaps related to rainfall events), rates of change in temperature, or photoperiod. Mating is probably either polygynous or promiscuous.

One of the most significant features of bandicoot reproduction is the presence of a functional chorioallantoic placenta in addition to the yolk-sac. The placenta has evolved independently and is probably correlated with the rapid rate of development in bandicoots.

Bandicoots are short-lived—a maximum of two to three years in the wild, although they may reach five years of age in captivity.

Conservation status

Peramelidae: Of the 10 modern species, two are extinct. These are species from the arid interior—the pig-footed bandicoot (*Chaeropus ecaudatus*) and desert bandicoot (*Perameles eremiana*). The mainland form of the eastern barred bandi-

An eastern barred bandicoot (*Perameles gunnii*) forages for earthworms in Hamilton, Victoria, Australia. (Photo by © Steve Kaufman/Corbis. Reproduced by permission.)

A southern brown bandicoot (*Isoodon obesulus*) on the ground. (Photo by Martin B. Withers/FLPA–Images of Nature. Reproduced by permission.)

coot (*P. gunnii*) is Critically Endangered; the Tasmanian form is Vulnerable. One species, the western barred bandicoot (*P. bougainville*) is Endangered—it is extinct on the main-land and occurs only on Bernier and Dorre Islands in Shark Bay, Western Australia. Four subspecies are Vulnerable—the mainland and Barrow Island forms of the golden bandicoot, Nuyts southern brown bandicoot, and the Tasmanian form of the eastern barred bandicoot. Among the other forms (mainly subspecies) at least three are Near Threatened.

Peroryctidae: Little is known about the status of most species. One species, the Seram Island bandicoot (*Rhynchomeles prattorum*), is known only from the type series collected in 1920. Several other species, including the mouse bandicoot (*Microperoryctes murina*), David's echymipera (*Echymipera davidi*), Menzies' echymipera (*E. echinista*), and Papuan bandicoot (*Microperoryctes papuensis*), are rarely encountered and may be Vulnerable.

Significance to humans

Probably all species of peramelids were used as food by native Australians. The larger peroryctids are still hunted for food in New Guinea. Minor annoyance in suburban areas is caused by bandicoots digging foraging holes in lawns.

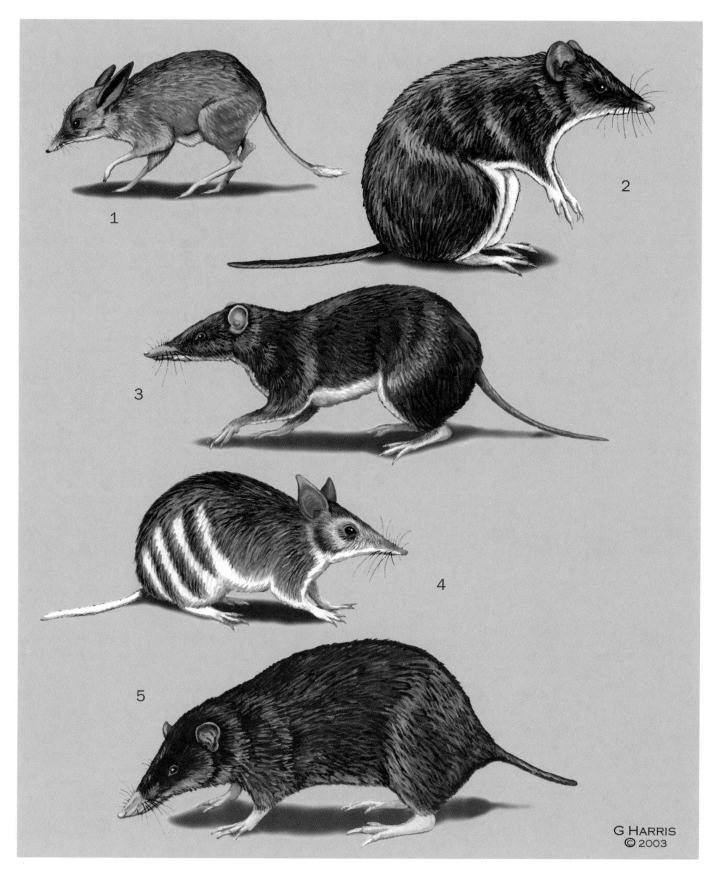

1. Pig-footed bandicoot (*Chaeropus ecaudatus*); 2. Northern brown bandicoot (*Isoodon macrourus*); 3. Raffray's bandicoot (*Peroryctes raffrayana*); 4. Eastern barred bandicoot (*Perameles gunnii*); 5. Rufous spiny bandicoot (*Echymipera rufescens*). (Illustration by Gillian Harris)

Species accounts

Eastern barred bandicoot
Perameles gunnii

SUBFAMILY
Peramelinae

TAXONOMY
Perameles gunnii Gray, 1838, Tasmania, Australia.

OTHER COMMON NAMES
English: Barred bandicoot, Tasmanian barred bandicoot, striped bandicoot, Gunn's bandicoot; German: Tasmanien-Langnasenbeutler.

PHYSICAL CHARACTERISTICS
Head and body length is 10.6–13.8 in (270–350 mm); weight is 26.5–35.3 oz (750–1,000 g). Grayish brown fur with light stripes on its hindquarters. Large ears, tapered nose, and short tail.

DISTRIBUTION
Victoria, Tasmania, and formerly South Australia, Australia.

HABITAT
Grassland and grassy woodland, pasture, also gardens in suburban areas.

BEHAVIOR
Nocturnal, solitary except when courting or mating or females with young.

FEEDING ECOLOGY AND DIET
Food is mainly obtained by digging after locating food items by smell. Small pits are dug using the forefeet and the long nose. Food is extracted and deftly manipulated in the front feet. Eats earthworms, adult and larval insects, other invertebrates, tubers, bulbs, and fallen fruit.

REPRODUCTIVE BIOLOGY
Capable of breeding year-round but may cease in colder winter months at lower latitudes (Tasmania) or during hot, rainfall-deficient summers on the mainland. Gestation period 12.5 days, polyestrous, estrus cycle about 26 days. Chorioallantoic placenta formed at about 9.5 days of gestation and is retained in the uterus after parturition. Litter size one to five, average two to 2.5. Pouch life about 55 days, weaned at 70–80 days. The nest is a grass and leaf-lined scraped depression. Growth is rapid and sexual maturity may be reached at about four months. Sequential litters may be born throughout the female's two to three year lifespan. Mating is probably promiscuous.

CONSERVATION STATUS
The mainland form is Critically Endangered. It only occurs in minuscule numbers at one site in the wild. A recovery program, involving reintroduction to protected sites of captive-bred animals has been in operation since 1989. The principal continuing threat is predation by introduced carnivores, particularly red foxes and cats, for which species continuing control is essential for the reintroduced populations to survive. The Tasmanian population appears to be declining in some parts of its range, such that it is locally threatened in its postulated focal range but has, conversely, expanded into new areas as forest has been felled and converted to pasture. The main predator in Tasmania is the cat.

SIGNIFICANCE TO HUMANS
The eastern barred bandicoot was eaten by aboriginal Australians. It is a minor annoyance to landholders in suburban areas due to foraging in lawns. ◆

Northern brown bandicoot
Isoodon macrourus

SUBFAMILY
Peramelinae

TAXONOMY
Perameles macroura (Gould, 1842), Port Essington, Northern Territory, Australia.

OTHER COMMON NAMES
English: Brindled bandicoot, large northern bandicoot; German: Grosse Kurznasenbeutler.

PHYSICAL CHARACTERISTICS
Head and body length is 11.8–18.5 in (300–470 mm); weight is 17.6–109.3 oz (500–3,100 g). Speckled brown-black fur, lighter on belly.

DISTRIBUTION
Western Australia, Northern Territory, Queensland, in which state it is coastal, New South Wales south to near Sydney, Australia. Also in southern New Guinea.

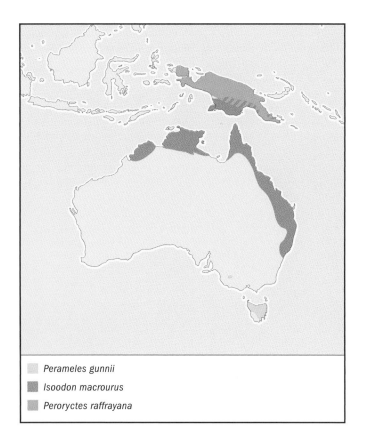

Perameles gunnii

Isoodon macrourus

Peroryctes raffrayana

HABITAT

Areas of low ground cover including tall grass and low shrubs, irrespective of tree cover. Grassland, woodland, open forest, rarely in closed forest. Gardens in settled areas. In New Guinea, found in grasslands and grassy savanna woodlands as well as gardens.

BEHAVIOR

Nocturnal, solitary except when courting or mating or females with young. Males are aggressive and use a gland behind the ear to scent-mark territory.

FEEDING ECOLOGY AND DIET

Omnivorous. Northern brown bandicoots mainly eat insects, earthworms, and other invertebrates, but also include berries, seeds, and plant fiber such as sugar cane in their diet. Food is obtained by digging conical pits with the strong forefeet after detection by smell.

REPRODUCTIVE BIOLOGY

In the southern part of its range, breeding occurs from late winter to summer; further north breeding takes place year-round. There are eight nipples and litter size ranges from one to seven, usually two to four. The gestation period is 12.5 days and young are weaned at about two months. A chorioallantoic placenta is formed and retained after parturition. Northern brown bandicoots are polyestrous and promiscuous. Growth is rapid and sexual maturity is reached well before physical maturity. The nest is a heap of ground litter covering a shallow depression, with entrances at both ends. Hollow logs are also used.

CONSERVATION STATUS

Overall, common to abundant, but has suffered local extinction due to altered habitat. Still present in urban areas in New South Wales, Queensland, and New Guinea, but such populations are at risk.

SIGNIFICANCE TO HUMANS

The northern brown bandicoot was eaten by Aboriginal Australians and is still hunted as food in New Guinea. It is considered a minor annoyance due to digging in suburban lawns. ◆

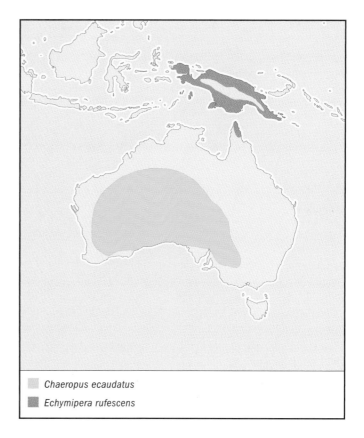

Chaeropus ecaudatus

Echymipera rufescens

mulga (*Acacia*) overstory. In the east, occupied grassy plains and open woodland with a grass and shrub understory.

BEHAVIOR

Nocturnal and presumably solitary except during mating and females with young. Gait compared with "a broken-down hack in a canter," but capable of explosive speed and agile leaps when disturbed.

FEEDING ECOLOGY AND DIET

The species' tooth structure and gut anatomy suggest that it was more herbivorous than other bandicoots. This notion is supported by observations of feeding by captives, and stomach content analysis. Termites, ants, and other insects were also eaten.

REPRODUCTIVE BIOLOGY

Breeding is speculated to have been in May and June. The pouch has eight nipples, but litters seem to have been one to two. Probably promiscuous.

CONSERVATION STATUS

Extinct. Last specimen collected in 1901 but Aboriginal testimony indicates that it probably survived in parts of its range until the 1950s.

SIGNIFICANCE TO HUMANS

None, although probably eaten by indigenous Australians. ◆

Pig-footed bandicoot
Chaeropus ecaudatus

SUBFAMILY
Peramelinae

TAXONOMY
Perameles ecaudatus (Ogilby, 1838), Murray River, New South Wales, Australia.

OTHER COMMON NAMES
German: Schweinfussnasenbeutler.

PHYSICAL CHARACTERISTICS
Head and body length was 9.1–10.2 in (230–260 mm); weight was about 7 oz (200 g); comparatively long tail, despite its vernacular, bearing a terminal crest. Forefeet digits reduced to give appearance of pig's feet or deer hooves.

DISTRIBUTION
Formerly found across much of arid Australia, including much of inland Western Australia and South Australia, the southern half of the Northern Territory and marginally in western Queensland, New South Wales, and northwestern Victoria.

HABITAT
In the central deserts, occurred on sand dunes and sand plains with hummock grassland and tussock grass, sometimes with a

---

Raffray's bandicoot
Peroryctes raffrayana

TAXONOMY
Perameles raffrayana (Milne-Edwards, 1878), Vogelkop, Irian Jaya, Indonesia.

OTHER COMMON NAMES
English: Long-legged bandicoot; German: Grossen Neuguineanasenbeutler; Spanish: Bandicut de Raffray.

PHYSICAL CHARACTERISTICS
Head and body length is 10.8–14.6 in (275–372 mm); weight is 22.9–35.3 oz (650–1,000 g). Unpatterned dark brown fur on back and on the long tail. Short, rounded ears.

DISTRIBUTION
Widespread in New Guinea, absent only from the woodlands and savanna of the south. Altitudinal range from about 130 to 13,120 ft (40 to 4,000 m), but most common at about 3,940 ft (1,200 m) along the central cordillera.

HABITAT
Lowland rainforest, lowland hill and mid-mountain oak forest, *Nothofagus* (beech) forest, mixed forest, and sub-alpine shrubs. Prefers undisturbed forest.

BEHAVIOR
Not known, but probably nocturnal.

FEEDING ECOLOGY AND DIET
These bandicoots are reported to eat fruit, particularly that of figs.

REPRODUCTIVE BIOLOGY
Females with pouch young have been captured between March and December, and the species may breed throughout the year. Litter size is one to two. It is possible that it nests communally, but that is not confirmed. One individual lived over three years in captivity. Mating is probably promiscuous.

CONSERVATION STATUS
Common.

SIGNIFICANCE TO HUMANS
Eaten by indigenous people. ◆

Rufous spiny bandicoot
Echymipera rufescens

TAXONOMY
Perameles rufescens (Peters and Doria, 1875), Kei Islands, Indonesia.

OTHER COMMON NAMES
English: Long-nosed echymipera, spiny bandicoot, rufescent bandicoot; German: Dickkopf-Stachelnasenbeutler; Spanish; Echimipera Narizona.

PHYSICAL CHARACTERISTICS
Head and body length is 11.8–16.1 in (300–410 mm); weight is 17.6–70.5 oz (500–2,000 g). Very elongate snout. Red-brown-black coarse, spiny dorsal fur, white ventrally. Short, almost naked black tail.

DISTRIBUTION
Subspecies *E. r. rufescens* is found in the lowlands of northern, eastern, and southern New Guinea, Aru Islands, and Kei Islands. *E. r. australis* is confined to Cape York, Australia.

HABITAT
In New Guinea is found only below 3,940 ft (1,200 m). Prefers rainforest but tolerates disturbed areas and grasslands. Australian subspecies occurs in closed forest including mesophyll vine forest, notophyll vine forest, and gallery forest, but also is found in eucalypt grassy woodland, coastal closed heath, and low layered open forest.

BEHAVIOR
Nocturnal. Possibly uses burrows rather than nests for daytime shelter, at least in New Guinea.

FEEDING ECOLOGY AND DIET
Omnivorous, although prefers to eat insects.

REPRODUCTIVE BIOLOGY
In New Guinea, pouch young have been recorded between March and October, but in Australia breeding may be more seasonal, with an estrus in the dry season. Possibly has lower fertility than seen in other species of the genus. Litter sizes from one to three are reported. Probably promiscuous.

CONSERVATION STATUS
Generally uncommon, but may be locally abundant in New Guinea; common in its limited range on Cape York, Australia.

SIGNIFICANCE TO HUMANS
Eaten by indigenous humans in New Guinea. ◆

Common name / Scientific name/ Other common names	Physical characteristics	Habitat and behavior	Distribution	Diet	Conservation status
Golden bandicoot *Isoodon auratus* Spanish: Bandicut de hocico corto	Mix of blackish brown and orange or yellows, underparts are yellowish gray, yellowish brown, or white. Head and body length 9.4–19.3 in (24–49 cm), tail length 3.1–7.8 in (8–20 cm), weight 9.2–23.1 oz (260–655 g).	Arid deserts and adjacent semi-arid areas and woodlands. Two to four young per litter. Generally sleeps during the day, hidden in a nest of twigs, grass, leaves, and other ground litter.	Australia.	Termites, ants, centipedes, moths, insect larvae, small reptiles, roots, and tubers.	Vulnerable
Southern brown bandicoot *Isoodon obesulus* English: Short-nosed bandicoot; Spanish: Bandicut castaño sureño	Blackish brown with hints of orange or yellow, under side is light brown, gray, or white. Pouch runs along stomach and opens backwards behind legs. Short, rounded ears, sharp claws. Head and body length 11.9–13.8 in (30–35 cm), weight 2.2 lb (1 kg).	Prefer dense ground cover, tall grass, and low shrubbery. They live near swamps and rivers as well as in thick scrub in drier areas. Generally solitary, except during breeding season. Females carry young in pouch for 50 days.	Australia.	Soil invertebrates and insects as well as fruits, seeds, fungi, and some plant fibers.	Not threatened

[continued]

Common name / Scientific name/ Other common names	Physical characteristics	Habitat and behavior	Distribution	Diet	Conservation status
Western barred bandicoot *Perameles bougainville* French: Babdicoot de Bougainville; German: Tasmanien-Langnasenbeutler; Spanish: Tejon marsupial rayado	Light brown-gray above, underparts are white. Longish snout and large erect ears. Head and body length up to 9.3 in (23.6 cm), tail is half the length of the body. Weight up to 10.1 oz (286 g).	Islands in dune scrub sys- tems. Occupies nest during day. Nocturnal.	Two islands off the northwest coast of western Australia.	Small insects, fruit, and seeds.	Endangered
Desert bandicoot *Perameles eremiana* German: Wüsten- Langnasenbeutler Spanish: Bandicut desértico	Light brown, underparts are white. Head and body length 15.7 in (40 cm), tail length 5.1 in (13 cm).	Mainly in spinifex grasslands. Generally nocturnal. Also develops pouch and holds young for 50 days.	Southern Northern Territory, northern South Australia, and southeastern Western Australia.	Small insects, fruit, and seeds.	Extinct
Long-nosed bandicoot *Perameles nasuta* German: Langnasenbeutler; Spanish: Bandicut de hocico largo	Fur is grayish brown, creamy under parts. Muzzle and ears are long and pointed. Hairy tail, no barring on rump. Pouch opens to rear. Head and body length 12–16.7 in (31–42.5 cm), tail length 4.7–6.1 in (12–15.5 cm), weight 1.8–2.5 lb (850–1,100 g).	Rainforest, wet and dry woodlands, and sometimes in more open areas with little ground cover. Mainly solitary and nocturnal.	Eastern coast of Australia.	Mainly insectivorous, although it also eats some plant material, and will occasionally eat worms, mice, and lizards.	Not threatened
Spiny bandicoot *Echymipera kaluba*	Upperparts are bright reddish brown, dark coppery brown, black mixed with yellow, or black interspersed with tawny. Underparts are buff or brown. Head and body length 7.9–19.7 in (20–50 cm), tail length 1.9–4.9 in (5–12.5 cm).	Rainforests from 0 to 5250 ft (0–1,600 m). Solitary and highly intolerant of their own kind.	New Guinea, Bismarck Archipelago, and Mysol Island.	Small insects, fruit, and seeds.	Not threatened
Large-toothed bandicoot *Echymipera clara* German: Japen- Stachelnasenbeutlers; Spanish: Echimipera de Clara	Upperparts are bright reddish brown, dark coppery brown, black mixed with yellow, or black interspersed with tawny. Underparts are buff or brown. Head and body length 7.9–19.7 in (20–50 cm), tail length 1.9–4.9 in (5–12.5 cm).	Hill forests from 985 to 5,575 ft (300–1,700 m). Generally nocturnal and solitary.	Northern Irian Jaya and Yapen.	Small insects, fruit, and seeds.	Data Deficient
Mouse bandicoot *Microperoryctes murina* German: Der Mausnasenbeutler; Spanish: Bandicut ratón	Dark gray, lighter underparts. Feet have scattered white hairs, tail is dark fuscous. Head and body length 1.9–6.9 in (15– 17.5 cm), tail length 4–4.3 in (10.5– 11 cm).	Moss forests at an altitude of 6,230 to 8,200 ft (1,900– 2,500 m). Semifossorial existence. Little known about behavior and sexual cycles.	Weyland Mountain of western New Guinea and Vogelkop Peninsula at extreme tip of the island.	Unknown.	Data Deficient
Giant bandicoot *Peroryctes broadbenti* Spanish: Bandicut gigante	Dark brown with reddish buff flanks, underparts are near white. Head and body length 6.9–7.9 in (17.5–20 cm), weight 11 lb (5 kg).	Lowland rainforest up to 6,560 ft (2,000 m) or more. Terrestrial, nocturnal, and generally solitary. Twins have been observed only once.	Southeastern New Guinea.	Consists mainly of vegetable matter.	Data Deficient
Striped bandicoot *Peroryctes longicauda* Spanish: Bandicut rayado	Reddish brown or pale brown speckled with black above, dark middorsal line, paired lateral rump stripes, and/or dark eye stripes, underparts are rufous or buff. Head and body length 9.4–12 in (23.9–30.3 cm), tail length 5.5–10.2 in (14.1–25.8 cm).	Upland forests from 3,280 to 14,763 ft (1,000–4,500 m). Terrestrial, nocturnal, and generally solitary.	Central Range of New Guinea.	Consists mainly of vegetable matter.	Not threatened
Papuan bandicoot *Peroryctes papuensis* German: Mura- Neuguineanasenbeutler; Spanish: Bandicut papá	Upperparts dark with a prominent black middorsal line, dark lateral rump stripes and eye stripes. Head and body length 6.8–7.8 in (17.5–20 cm), tail length 5.5–6.1 in (14–15.5 cm).	Highland forests. Nocturnal, terrestrial, and generally solitary.	Southeastern Papua New Guinea.	Consists mainly of vegetable matter.	Not threatened

Resources

Books

Flannery, T. *Mammals of New Guinea.* Sydney: Reed Books and Australian Museum, 1995.

Flannery, T. *Mammals of the South-West Pacific and Moluccan Islands.* Sydney: Reed Books and Australian Museum, 1995.

Gordon, G., and A. J. Hulbert. "Peramelidae." In Vol. 1B of *Fauna of Australia: Mammalia,* edited by D. W. Walton and B. J. Richardson. Canberra: Australian Government Publishing Service, 1989.

Jones, F. W. *The Mammals of South Australia.* Part 2. *The Bandicoots and the Herbivorous Marsupials.* Adelaide: Government Printer, 1924.

Mahoney, J. A., and W. D. L. Ride. "Peramelidae." In *Zoological Catalogue of Australia 5, Mammalia.* Canberra: Australian Government Publishing Service, 1988.

Menzies, J. *A Handbook of New Guinea Marsupials and Monotremes.* Madang, Papua, New Guinea: Kristen Press, Inc., 1991.

Seebeck, J. H., P. R. Brown, R. L. Wallis, and C. M. Kemper, eds. *Bandicoots and Bilbies.* Chipping Norton, Australia: Surrey Beatty & Sons, 1990.

Strahan, R., ed. *The Mammals of Australia.* Sydney: Reed Books and Australian Museum, 1995.

Stodart, E. "Breeding and Behaviour of Australian Bandicoots." In *The Biology of Marsupials,* edited by B. Stonehouse and D. Gilmore. London: Macmillan, 1977.

Periodicals

Freedman, L. "Skull and Tooth Variation in the Genus *Perameles.* Part 1. Anatomical Features." *Records of the Australian Museum* 27 (1967): 147.

Gordon, G., and B. C. Lawrie. "The Rufescent Bandicoot, *Echymipera rufescens* (Peters and Doria), on Cape York Peninsula." *Australian Wildlife Research* 5 (1977): 41.

Heinsohn, G. E. "Ecology and Reproduction of the Tasmanian Bandicoots (*Perameles gunni* and *Isoodon obesulus*)." *University of California Publications in Zoology* 80 (1966): 1.

Seebeck, John H. "*Perameles gunnii.*" *Mammalian Species* Account No. 654.

John H. Seebeck, BSc, MSc, FAMS

Bilbies

(Thylacomyinae)

Class Mammalia
Order Peramelemorphia
Family Peramelidae
Subfamily Thylacomyinae

Thumbnail description
Rabbit-sized with extremely large ears, long, thin snout, silky bluish gray fur, powerful front feet with large claws.

Size
Head and body 9–10.2 in (23–26 cm); tail 7.8–11.4 in (20–29 cm). Weight 28–88 oz (800–2,500 g).

Number of genera, species
1 genus; 2 species (one extinct)

Habitat
Arid areas of grassland and shrubs with sparse ground cover.

Conservation status
Extinct: 1 species; Vulnerable; 1 species

Distribution
Western Australia, Northern Territory, southwestern Queensland

Evolution and systematics

The only burrowing members of the Peramelemorphia, bilbies probably evolved separately during the Pleistocene period from other bandicoots.

Taxonomists disputed throughout the twentieth century whether the two species in the *Macrotis* genus, the greater bilby *Macrotis lagotis* and the lesser bilby *Macrotis leucura*, constitute a subfamily Thylacominae within the family Peramelidae, or whether they should receive full family status as Thylacomyidae. Both species show distinctive morphological features including a flattened cranium; broad braincase and narrow snout; forward-pointing rostrum; pear-shaped bullae and unique adaptations of the molar teeth. However, other characteristics, such as reproductive biology, are so similar to other species in the order that doubts continue to be expressed about separation.

The lesser bilby has been declared Extinct by the IUCN: the greater bilby is now commonly referred to as the bilby.

The taxonomy for the greater bilby is *Macrotis lagotis* (Reid, 1837), Swan River, Western Australia, Australia. Other com-

mon names include: English: Greater rabbit-eared bandicoot; French: Grand bandicoot-lapin; Spanish: Cangurito narigudo grande.

Physical characteristics

About the size of a rabbit, with huge ears that earn it the alternative name rabbit-eared bandicoot, the bilby has a very long, thin, pointed snout and an extremely long black tail with a white, crested tip. Its fur, bluish gray above and cream to white underneath, is soft and silky.

The forelimbs are strong, with three clawed and two unclawed toes used for burrowing. In common with many other marsupials, the hind feet lack a first toe.

Distribution

Last recorded alive in 1931, the lesser bilby lived in the sandhill deserts of central Australia. The greater bilby occupied a wide variety of habitats and may have lived over about

The greater bilby (*Macrotis lagotis*) uses its strong forearms to dig 3–6 ft (1–2 m) deep burrows. (Photo by Gerard Lacz/FLPA–Images of Nature. Reproduced by permission.)

The greater bilby (*Macrotis lagotis*) has very characteristic long ears and nose. (Photo by Martin Garvey; Gallo Images/Cobris. Reproduced by permission.)

70% of the Australian landmass. In the 1890s, one naturalist noted that "it was not unusual for rabbiters, even in the immediate neighborhood of Adelaide (South Australia), to take more bilbies than rabbits in their traps." Today, bilbies are only found in one fifth of their former range and are completely absent from the 386,000 mi² (1 million km²) of South Australia where they were once so common.

The species is still comparatively widespread, with fragmented populations in the Tanami Desert of the Northern Territory; the Gibson and Great Sandy Deserts and Pilbara and Kimberley regions of Western Australia; and isolated areas in southwest Queensland.

The lesser bilby (*Macrotis leucura*), also known as the rabbit bandicoot, is extinct. (Photo by Bruce Coleman, Inc. Reproduced by permission.)

Habitat

Until the arrival of European settlers, bilbies were found in a broad range of habitats. Only a small number of habitats were avoided. Rocky terrain was not used because of its unsuitability for burrowing. Places with thick ground cover were avoided too, as these marsupials need great mobility for foraging. Europeans introduced much more significant limitations. Today, bilbies are absent from areas with intensive livestock farming, as well as habitat where foxes and rabbits are present in any significant numbers.

Habitats currently occupied by bilbies fall into two types. In the south of its range, this marsupial lives on rises and ridges among sparse grasses, especially mitchell grass *Astrebla* and short shrubs. Further north, habitats are more variable. They include acacia woodland, acacia scrub with a spinifex *Triodia* understory, hummock grassland, shrub steppe, and creek beds. A critical factor in the north is the frequency of bush fires. Where fires occur at least once every 10 years, the amount of ground cover is reduced. Such fires also trigger the germination of plants such as *Yakirra australiense*, whose seeds can be an important part of the animal's diet.

Behavior

Bilbies are solitary animals. Observed groups appear to gather only in response to available food resources and show no social cohesion. Males show some territoriality by scent-marking, but there are no signs of physical aggression towards other males. Both males and females hold overlapping home

The tail of the greater bilby (*Macrotis lagotis*) is half black and half white. (Photo by Howard Hughes/Nature Focus, Australian Museum. Reproduced by permission.)

ranges, with the male ranges considerably larger. These home ranges are often temporary in nature, however, since bilbies make regular seasonal movements in response to changing food availability.

The pock-marked arid landscapes of central Australia are testimony to bilby burrowing activity. Each individual digs a number of burrows within its home range to shelter in during the day. There can be as many as 12 spiraling burrows, each up to 10 ft (3 m) long. The entrance is usually at the foot of a shrub or grass hummock, or against the base of a termite mound.

This animal can run surprisingly fast, although in an ungainly fashion, with the tail held up off the ground, the hind

feet moving together and the front feet alternately. It rarely strays more than 330 ft (100 m) from a burrow. A bilby may visit several burrows during the night, before selecting one in which to spend the next day.

Feeding ecology and diet

Emerging from its burrow about an hour after sunset, the bilby is wholly nocturnal, returning well before dawn. It searches for food by using its powerful front feet with long claws to dig numerous small conical holes in the ground up to 4 in (10 cm) deep. The long thin tongue is used to lick up much of the food—between 29 and 90% of its feces consists of earth. The senses of smell and hearing are both crucial in food detection.

This marsupial is omnivorous, with a diet that includes seeds, roots, insects, bulbs, fruit, and fungi. Research shows that individual colonies tend to favor one or two food items over all others, probably in response to their abundance within

In parts of Australia, introduced rabbits have caused a decline in the numbers of greater bilbies (*Macrotis lagotis*) due to competition for burrows. (Photo by Chris Oaten/Nature Focus, Australian Museum. Reproduced by permission.)

Greater bilby (*Macrotis lagotis*). (Illustration by Bruce Worden)

a particular habitat. Thus in the Tanami Desert, bilbies consume termites and lepidoptera larvae; in Queensland, seeds, bulbs, and acacia root-feeding grubs predominate at different locations. Bilbies do not appear to drink water; instead, they gain the moisture they need from their food.

Reproductive biology

The polyestrous females are physiologically capable of producing litters at any time of the year, although in some areas rainfall and food availability are limiting factors. Bilbies are polygynous. After mating with a socially dominant male, the female undergoes a gestation of just 14 days, then gives birth to one to three young.

No more than a centimeter in length, the newborn young crawl into the backward-facing pouch, where they will remain suckling on a choice of eight teats for the next 80 days. Even after leaving the pouch, the young will stay in the burrow for a further fortnight. The mother continues to suckle them, while at the same time making nocturnal sorties into the open for food. Although the young then leave the burrow and begin feeding on solid food, they often continue to share their mother's burrows for a short while after gaining independence. The young females attain sexual maturity at five months. Male maturity is unknown. The longevity record for a captive greater bilby *Macrotis lagotis* is seven years and two months.

Conservation status

The lesser bilby has been declared Extinct. Listed as Vulnerable under both IUCN criteria and Australian legislation, the greater bilby now exists in small, fragmented populations over about a fifth of its former range. Competition for food and nesting burrows with introduced rabbits and predation by introduced foxes are significant factors in the species' decline. Feral cats have also depleted numbers. Intensive cattle and sheep farming have limited available habitat through changes in vegetation cover and damage to the soil structure. A lack of managed burning to reduce ground cover is also implicated in localized extinctions.

The greater bilby (*Macrotis lagotis*) digs insects and larvae out of the soil for food. (Photo by Randall Hyman. Reproduced by permission.)

Despite such historical losses, a national recovery plan promises a better future for the bilby. Its key targets include managing remaining habitat and monitoring populations, as well as captive breeding and re-establishing bilbies in areas where they occurred previously. At the beginning of the twenty-first century, breeding and release schemes on predator-free islands and special enclosures within protected areas were showing signs of success.

Significance to humans

Formerly important as food and hunted for its fur by aboriginal tribes, this marsupial has gained an iconic status today as a symbol of Australia's threatened indigenous wildlife. Adopted as a mascot by the Commonwealth of Australia Endangered Species Program, the species has gained wider public awareness thanks to a campaign that began in the 1980s to replace the Easter bunny with an Easter bilby. Every Easter, thousands of chocolate bilbies are sold, often with a percentage of the profits channeled back into bilby conservation.

Resources

Books

Hoser, R. *Endangered Animals of Australia*. Sydney: Pearson, 1991.

Macdonald, D. *The New Encyclopaedia of Mammals*. Oxford: Oxford University Press, 2001.

Nowak, R. M. *Walker's Mammals of the World Online*. Baltimore: John Hopkins University Press, 1995. <http://press.jhu.edu/books/walkers_mammals_of_the _world/marsupialia.peramelidae>

Seebeck, J. H., P. R. Brown, R. L. Wallis, and C. M. Kemper, eds. *Bandicoots and Bilbies*. Chipping Norton, Australia: Surrey Beatty & Sons, 1990.

Strahan, R. *The Mammals of Australia*. Sydney: Australian Museum/Reed Holland, 1995.

Organizations

Arid Recovery Project. P.O. Box 150, Roxby Downs, South Australia 5725 Australia. Phone: (08) 8671 8282. Fax: (08) 8671 9151. E-mail: arid.recovery@wmc.com Web site: <http://www.aridrecovery.org.au>

Australian Bilby Appreciation Society. P.O. Box 2002, Rangeview, Victoria 3132 Australia. E-mail: bilbies@oze-mail.com .au Web site: <http://www.oze-mail.com.au/~bilbies>

Department for Environment and Heritage. GPO Box 1047, Adelaide, South Australia 5001 Australia. Phone: (8) 8204 1910. E-mail: environmentshop@saugov.sa.gov.au Web site: <http://www.environment.sa.gov.au>

Environment Australia. GPO Box 787, Canberra, Australian Capital Territory 2601 Australia. Phone: (2) 6274 1111. Web site: <http://www.ea.gov.au/>

Derek William Niemann, BA

▲
Notoryctemorphia
Marsupial moles
(Notoryctidae)

Class Mammalia

Order Notoryctemorphia

Family Notoryctidae

Number of families 1

Thumbnail description
Long, flexible body like a flattened cylinder a with short tail and very short stout legs; front feet bear two large spade-like claws; fur is silky and pale blond, nose has flat, callused shield and there are no visible eyes or ears; females have two teats within a backward opening pouch

Size
3.5–7 in (9–18 cm); tail about 1 in (2.5 cm); weight 1.2–2.5 oz (35–70 g)

Number of genera, species
1 genus; 2 species

Habitat
Hot arid sandy desert

Conservation status
Endangered: 2 species

Distribution
Western and central Western Australia

Evolution and systematics

In terms of appearance and habits, the marsupial moles are about as different from most other marsupials as it is possible to be. In fact, they bear an uncanny resemblance to African golden moles (*Eremitalpa* spp.). These similarities are due entirely to the convergent evolution of adaptations to a similar "sand-swimming" lifestyle. The marsupial moles have no close relatives. There is no doubt these extraordinary animals are marsupials, but even DNA analysis has been so far unable to connect them to any other living marsupial group, and it is believed they belong to a lineage that diverged from other marsupials more than 50 million years ago. The evolutionary history of the marsupial moles was made a little clearer by the discovery in 1987 of a fossil form in Tertiary rocks of the world-famous Riversleigh system in Queensland. It is now thought the ancestor of *Notoryctes* may have been a similar kind of animal, which developed its burrowing habit in order to feed in the soft litter of decomposing leaves in an ancient tropical rainforest. What started as an adaptation for rummaging through the humus layer may have evolved into the sand-swimming technique employed by marsupial moles in later, drier habitats.

Marsupial moles spend most of their time beneath the sand in one of the world's least explored regions—the deserts of central and western Australia. As a result, they are difficult to study in the wild. They have also proved impossible to keep for long in captivity and, consequently, many aspects of their biology remain a mystery.

Physical characteristics

There is no mistaking a marsupial mole for any other Australian mammal. Both species have a body shaped like a flattened cylinder, with very short legs and a short, stiff tail. The body is covered in very fine, almost iridescent golden fur, which is often stained by the red desert soil. There are no visible eyes, just dark spots marking the place where vestigial lenses lie under the skin. The ears are mere holes in the side of the head, protected by dense fur. The only distinctive fea-

A marsupial mole's (*Notoryctes typhlops*) head and shoulders. (Photo by B. G. Thomson. Reproduced by permission.)

ture of the face is a horny, hairless nose-shield, slightly bigger in the southern species, *N. typhlops*.

There are five toes on each foot. On the front feet, the first and second toes oppose the others, and the third and fourth are big with very large, spade-like claws, used for digging. The fifth front toe is small, as are the first and fifth toes of the hind feet. The tail is annulated (ringed), with a knobbly end. Females have a rearward opening pouch concealing two teats.

Distribution

The distribution of marsupial moles is difficult to map with precision, partly because of the paucity of sightings and also because most early records were not accurately mapped—locations were indicated to the nearest town or station, often 100 mi (161 km) or more away. However, it appears that the smaller species, the northern marsupial mole, *N. caurinus*, occurs from the vicinity of Eighty Mile Beach on the northwest corner of Western Australia, southeast toward the central desert. The larger species, *N. typhlops*, occurs as far south as the Nullabor Plain bordering the Great Australian Bight and east into southwestern Queensland. Where and to what extent the two species overlap is not known.

Habitat

Marsupial moles are sand-dwellers. They were originally given the generic name, *Psammoryctes*, meaning "sand dig-

ger," but this had to be changed when it was realized this name had already been given to a group of oligochaete worms. *Notoryctes* spp. live in sand dunes and flat plains, and especially in the flatlands created where seasonal rivers flood the desert. They appear to be especially active in damp sand—most surface sightings occur shortly after rain. Often these habitats are vegetated with sparse spinfex grass or mulga scrub. The moles typically dig between 4 and 8 in (9–18 cm) under the sand, but occasionally burrow as deep as 8 ft (2.5 m). They are sensitive to cold, becoming chilled at temperatures less than 59°F (15°C), and readily succumb to hypothermia.

Behavior

What little is known about the lives of marsupial moles suggests they live alone and spend almost all their time underground, moving through the sand with a swimming action. They dig incredibly fast and are capable of shoving their way into the sand in seconds. They burrow faster than most potential predators can shift sand as they attempt to dig the moles out. Marsupial moles do not leave tunnels—the sand collapses behind them as they move. On the surface, marsu-

A marsupial mole (*Notoryctes typhlops*) digging a burrow in a sand dune in Alice Springs, Northern Territory, Australia. (Photo by B. G. Thomson/Photo Researchers, Inc. Reproduced with permission.)

A marsupial mole (*Notoryctes typhlops*) in the sand in Australia. (Photo by Tom McHugh/Photo Researchers, Inc. Reproduced by permission.)

The nose of the marsupial mole (*Notoryctes typhlops*) has a horny shield. (Photo by Jim Frazier/ANTPhoto.com.au. Reproduced by permission.)

pial moles move with a sinuous gait, the body shimmying from side to side as the short legs on opposite corners move together in awkward shuffling steps, leaving a distinctive trail of wiggly lines in the sand.

Feeding ecology and diet

Stomach contents indicate that marsupial moles have a mostly insectivorous diet, comprising ants, termites, and the larvae of other insects. They also eat other invertebrates and small lizards, and occasional seeds and other plant material. Food is apparently dug out of the sand as the animal burrows.

Reproductive biology

Virtually nothing is known about breeding behavior of marsupial moles. Of the few females with young that have been found, none have more than two babies—one for each teat.

Conservation status

The IUCN lists both species of marsupial mole as Endangered. There is no hard evidence that the species' ranges have decreased, but there is anecdotal evidence for an over-decline in abundance. Normally, an animal about which

so little is known would be listed as Data Deficient, but conservationists are taking no chances with these extraordinary marsupials, which are given full legal protection. The main concerns are changes in land use, in particular burning regimes, and introduced predators such as foxes, whose dropping have occasionally been found to contain marsupial mole remains.

Significance to humans

Marsupial moles have very little practical value to humans. Indigenous Australians may occasionally have used them as food, but these days more people regard these little-known animals as rare and precious. Sightings are rare enough to make the news, and specimens are being collected at the rate of one or two a year. The last specimen to be collected alive was found in the Punmu region of Western Australia in 1999. Sadly, the animal lived for only six weeks in captivity.

Species accounts

Marsupial mole
Notoryctes typhlops

TAXONOMY
Notoryctes typhlops (Stirling, 1889), Indracowrie, Northern Territory, Australia.

OTHER COMMON NAMES
English: Greater marsupial mole; French: Taupe marsupiale; German: Großer Beutelmull; Spanish: Topo marsupial.

PHYSICAL CHARACTERISTICS
Head-body 3.5–7 in (9–18 cm); tail about 1 in (2.5 cm); weight 1.2–2.5 oz (35–70 g). Long, flattish body with short tail, very short, stout legs, and spade-like front feet; fur is silky and pale blond; no visible eyes or ears, nose has flat, callused shield; female has backward opening pouch with two teats.

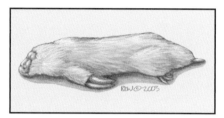

Notoryctes typhlops

DISTRIBUTION
Midwestern Australia, including central Western Australia, southern Northern Territories, and northwestern South Australia.

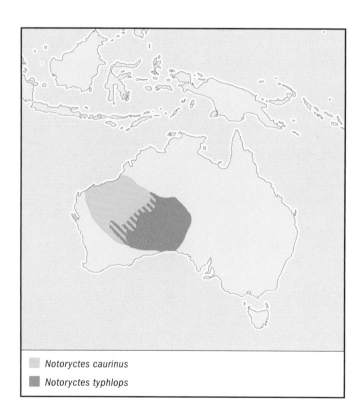

Notoryctes caurinus

Notoryctes typhlops

HABITAT
Sandy desert.

BEHAVIOR
Lives alone; burrows in loose sand using swimming action, but creates no persistent burrow system, surfaces very occasionally.

FEEDING ECOLOGY AND DIET
Mostly insect larvae.

REPRODUCTIVE BIOLOGY
Nothing is known.

CONSERVATION STATUS
Endangered.

SIGNIFICANCE TO HUMANS
None known. ◆

Northern marsupial mole
Notoryctes caurinus

TAXONOMY
Notoryctes caurinus Thomas, 1920, Ninety Mile Beach, Wallal, Western Australia, Australia. Formerly part of *N. typhlops*, but separated in 1988.

OTHER COMMON NAMES
English: Lesser marsupial mole; French: Taupe marsupiale; German: Kleiner Beutelmull; Spanish: Topo marsupial.

PHYSICAL CHARACTERISTICS
As for *N. typhlops*, but with smaller nose shield and claws and reduced lower teeth.

Notoryctes caurinus

DISTRIBUTION
Northwestern Western Australia, in the vicinity of Eighty Mile Beach.

HABITAT
Sandy desert.

BEHAVIOR
Lives alone; burrows in loose sand using swimming action, but creates no persistent burrow system, surfaces very occasionally.

FEEDING ECOLOGY AND DIET
Mostly insect larvae.

REPRODUCTIVE BIOLOGY
Nothing is known.

CONSERVATION STATUS
Endangered.

SIGNIFICANCE TO HUMANS
None known. ◆

Resources

Books

Aplin, K. P., and M. Archer. "Recent Advances in Marsupial Systematics with a New Syncretic Classification." In *Possums and Opossums: Studies in Evolution.* Sydney: Surrey Beatty & Sons and Royal Zoological Society of New South Wales, 1987.

Nowak, R. "Order Notoryctemorphia." In *Walker's Mammals of the World.* 6th ed. Vol. 1. Baltimore and London: Johns Hopkins University Press, 1999.

Strahan, R. *The Mammals of Australia.* Carlton, Australia: Reed New Holland, 1995.

Organizations

Australian Museum. 6 College Street, Sydney, New South Wales 2010 Australia. Phone: (2) 9320 6000. Web site: <http://www.amonline.net.au>

Riversleigh Fossils Centre. P.O. Box 815, Mount Isa, Queensland 4825 Australia. Phone: (7) 4749 1555. E-mail: riversleigh@mountisa.qld.gov.au Web site: <http://www.riversleigh.qld.gov.au>

Amy-Jane Beer, PhD

Diprotodontia

(Koala, wombats, possums, wallabies, and kangaroos)

Class Mammalia
Order Diprotodontia
Number of families 10
Number of genera, species 40 genera; 131 species

Photo: A mother kangaroo holding her young in Wilsons Promontory National Park, Victoria, Australia. (Photo by © Theo Allofs/Corbis. Reproduced by permission.)

Introduction

The Australasian order Diprotodontia is not particularly large (it has just 131 described living species) but is one of the most startlingly diverse of all mammal groups. Its members include animals as superficially different as the teddybear-like koala (*Phascolarctos cinereus*), the tiny feather-tailed glider (*Acrobates pygmaeus*), and the magnificent red kangaroo (*Macropus rufus*). While some diprotodonts are instantly recognizable, others are rather obscure and they include among their number some of the world's rarest animals. As a group, the diprotodonts are relatively new to science—first recognized as an order less than 150 years ago in 1866.

Evolution and systematics

The earliest known diprotodont fossils date back to the Oligocene epoch 24–35 million years ago (mya), but even these constitute a diverse assemblage of forms, so the origins of the group almost certainly go back further, to the Cretaceous period. Then, like most early mammals, the ancestors of all Australasian marsupials were probably small insect-eating animals not unlike modern bandicoots (order Peramelemorphia) or the monito del monte (order Microbiotheria).

These ancestors arrived in Australia from South America and Antarctica when all three were united in the massive tree-covered supercontinent of Gondwana. When Gondwana broke apart about 40 mya, Australia's marsupials were isolated and began a major radiation. In the absence of large numbers of placental mammals, they expanded to fill a huge range of ecological niches. No group diversified more than the diprotodonts.

There are currently 131 species of diprotodont in 40 genera, representing 10 families of koala, wombat, possum, cuscus, rat-kangaroo, kangaroo, wallaby, pygmy possum, ringtail, glider, honey possum, and feathertails. The suborder Vombatiformes, containing the living families Phascoloarctidae (koala) and Vombatidae (wombats) probably represents an early offshoot of the diprotodont lineage, characterized (at least in living members) by a reduced tail and backward-opening pouch. All other diprotodonts are placed the same group, suborder Phalangerida, which represents the main trunk of the phylogenetic tree. This trunk splits into five major branches (superfamilies) of which two, the Phalangeroidea (possums and cuscuses) and the Burramyoidea (pygmy possums) contain just one family each. The other three superfamilies each unite two families, the kangaroos and rat-kangaroos in the Macropodoidea, the ringtails and gliders in the Petauroidea, and the honey possum and feathertails in the Tarsipedoidea. This classification is broadly supported by molecular data in which the sequences of certain key molecules including DNA are compared. The diprotodonts closest relatives in other orders are probably the bandicoots and bilbies (order Peramelemorphia) and the marsupial moles (order Notoryctemorphia). However these relationships are largely speculative.

The discovery of previously undescribed mammal species is an increasingly rare event and yet this group includes among its ranks several species new to science in the last 25 years. These exciting newcomers included the long-footed potoroo, *Potorous longipes*, and the Proserpine rock wallaby, *Petrogale persephone*, both found in Australia and described in the early 1980s. Zoological expeditions to New Guinea and the Indonesian North Moluccas in the late 1980s and early 1990s yielded the previously unknown Telefomin cuscus, *Phalanger matanim*; the black tree kangaroo, *Dendrolagus scottae*; the dingiso or forbidden tree kangaroo, *Dendrolagus mbaiso*; and the Gebe cuscus, *Phalanger alexandrae*.

Physical characteristics

Diprotodont marsupials are united by two important characteristics that belie their great divergence in size and overall form. The first concerns dentition (the arrangement of teeth) in the lower jaw. Koalas, wombats, kangaroos, and possums all have only two developed incisor teeth at the front of the lower jaw. These are large and often project forwards as an adaptation to cropping vegetation. A second pair of very small incisors is present in some species, but there are no

lower canines, just a gap between the incisors and the cheek teeth. This arrangement is known as diprotodonty (literally translated this means "two first teeth")—hence the ordinal name Diprotodontia.

The second major unifying characteristic of diprotodonts is syndactyly. This means "fused toes," and refers to the structure of the hind feet, the second and third digits of which are always fused together forming a strange-looking double toe with two claws. The twin claws of the fused digits are retained and in most species serve the useful function of a grooming comb for removing caked dirt or other debris clinging to the animal's fur. Another shared feature of many diprotodonts is the arrangement of digits on the front paws. In most climbing species, the first two fingers oppose the other three, allowing the animals to maintain a firm grip of branches and stems. A notable exception to this rule is the brushtailed possum, whose forepaws are more like tiny nimble hands.

Members of the suborder Vombatiformes are set apart from other diprotodonts, being rather squat and heavy, with a pouch the opens to the rear. In burrowing wombats this prevents the pouch filling up with soil, whereas in koalas it is a rather inconvenient arrangement inherited from non-climbing ancestors. Other diprotodonts, members of the suborder Phalangerida, have a more athletic build with a long tail. In many possums the tail is prehensile. It may be thin as in potoroos or muscular as in large kangaroos, very long and brushy (as in Leadbeater's possum, *Gymnobelideus leadbeateri*) or virtually naked. Phalangerids have retained the forward opening pouch of their ancestors as a secure means of transporting young while climbing or hopping. In members of the superfamily Macropodoidea (kangaroos, wallabies, and rat-kanga-

An eastern gray kangaroo (*Macropus giganteus*) hopping over grassland. (Photo by Bill Bachman/Photo Researchers, Inc. Reproduced by permission.)

The rich color of a red kangaroo (*Macropus rufus*) male, right, compared to the smaller and more drab female at water. The males are called "boomers" and the females "blue flyers." (Photo by Wayne Lawler/Photo Researchers, Inc. Reproduced by permission.)

roos), the hind legs are larger and more powerful than those at the front, and the hind feet are very long (the group and generic names macropod and *Macropus* mean "big footed"). In most macropods the first hind toe is absent.

Diprotodonts have soft fur—that of many possums and the koala is very woolly. The majority of species are some shade of gray or brown, but a few are rather dramatically colored—for example Goodfellow's tree kangaroo (*Dendrolagus goodfellowi*), with a rich cinnamon red and gold coat and the yellow-footed rock wallaby (*Petrogale xanthopus*), whose yellow legs, feet, and ears, red and yellow-banded tail, and bold white cheek flashes contrast with gray body fur and make it one of the world's more decorative mammals.

Gliding possums are equipped with a built-in parachute, formed from a web of skin extending along each flank from the front to back legs. The precise structure of this membrane (called the patagium) varies between the three families of glider. In the pseudocheirid greater glider it stretches from elbow of the forelimb to ankle of the hind, while in the acrobatid pygmy glider it links wrist to knee. The lesser gliders (family Petauridae) have the most complete patagium, extending from wrist to ankle. None of the marsupial gliders have a tail membrane, although the tail of the pygmy glider

is modified to assist the gliding process, with a vane of hairs along each side earning the species its alternative common name feathertail glider.

Convergent evolution is a recurring theme in marsupial history, and the diprotodonts are no exception. The diverse diprotodont body forms and lifestyles show striking similarities with mammals of several other orders. For example, the head of kangaroos is deer-like, with a long muzzle, erect, mobile ears and large bulging eyes situated on the side of the head. Like deer, kangaroos have good all-round hearing and vision and long, powerful legs capable of making dramatic leaps and propelling the animal at speed—all essential for avoiding predators in open habitats. The stout, badgerlike form of the wombats is an adaptation to burrowing, with strong legs, claws and jaws all contributing to the effort of excavating what can be very hard soils. With their woolly fur, round face, large, forward-facing eyes and careful climbing technique, the cuscuses are strongly reminiscent of primates such as lorises and pottos. The pygmy possums are the marsupial equivalent of European dormouse in looks and some aspects of behavior—they climb with the aid of a prehensile tail and enter deep hibernation in cold weather. The sugar glider (*Petaurus breviceps*) and Leadbeater's possum resemble species of flying and non-flying squirrel respectively.

A red-necked wallaby (*Macropus rufogriseus banksianus*), showing the bright rufous mantle characteristic of this most common and widespread of the medium-sized macropodids of eastern Australia. (Photo by Wayne Lawler/Photo Researchers, Inc. Reproduced by permission.)

Distribution

The Diprotodontia is an exclusively Australasian group, with the bulk of its members restricted to Australia itself. New Guinea and the surrounding islands are home to about 47 endemic Indonesian diprotodonts, mainly tree kangaroos, cuscuses and forest wallabies. Only five species occur on both New Guinea and Australia but are restricted to tropical north Queensland in the latter. The most widespread species are the common brushtailed and ringtail possums, the common wallaroo and the red and gray kangaroos. With a little help from humankind, several species have made it to other parts of the world. The brushtailed possum (*Trichosurus vulpecula*) thrives in New Zealand as an undesirable alien despite sustained attempts to eradicate it. Wallabies that have escaped or been released from zoos or private collections have established colonies as far away as Great Britain. The group also includes some animals with dangerously restricted distributions—for example, several Indonesian cuscuses and possums and the Australian Gilbert's potoroo, which with an estimated

wild population of just 30 individuals living in one reserve near Albany in Western Australia, is one of the worlds most threatened animals.

Habitat

Diprotodont marsupials have successfully exploited virtually every terrestrial habitat their Australasian range has to offer. Forest diprotodonts include the cuscuses and tree kangaroos of tropical Queensland and New Guinea and Leadbeater's possum, which favors old growth highland forests of Australian mountain ash. Koalas and various wallabies and the widespread spectacled hare wallaby are animals of open woodland and bush, while rock wallabies are generally restricted to boulder slopes that less sure-footed animals find difficult to negotiate. Gray kangaroos and pademelons are grassland animals, preferring areas with regular rainfall and some tree cover, while reds make do with the sparsest vegetation in the Central Australian Desert. Honey possums (*Tarsipes rostratus*), quokkas (*Setonix brachyurus*), and various rat kangaroos are heath dwellers, while mountain pygmy possum is restricted to alpine habitats where snow covers the ground for half the year. Several species are now almost exclusively island dwellers, though not out of preference. Small wallabies such as the quokka and the rufus hare wallaby have been eradicated from much of the former range on mainland Australia by introduced predators and are only secure on offshore islands not yet colonized by foxes or cats. There are no aquatic diprotodonts, but several species of peturid, pseudochirid, and acrobatid possum readily take to the air on gliding membranes like those of flying squirrels. Not surprisingly, such species are restricted to habitats with many tall trees, from which they can launch their spectacular aerial swoops. Among the most adaptable of all marsupials, the common brushtailed possum, has adapted very well to the changes wrought by humans. It does well in suburban habitats and is a frequent visitor to gardens and outbuildings.

By no means all diprotodonts have regular nests, dens or other homes. The majority of large macropods are seminomadic and use a large home range. The red kangaroo (*Macropus rufus*) is especially wide ranging, and severe droughts or bush fires may force them to disperse 200 mi (300 km) or more in search of water and food. Being large, they have little need of a secure resting place, and most make do with patches of scant shade beneath desert shrubs such as saltbush. Smaller kangaroos and wallabies favor thickets of denser vegetation. While none of the large macropods build a nest as such, individuals may stake a claim to a particular spot. For example, the quokkas of Rottnest Island in Western Australia compete fiercely for the very limited number of shady resting places. The lack of potable water on the island in summer means that shade in which to avoid overheating is at a premium, and dominant individuals will claim the best spots as their own.

Unlike their larger cousins, many rat-kangaroos and bettongs do build a nest or den. Nest builders collect bedding material such as grass and vines, sometimes carrying them clamped between the body and the tail. These are arranged in a thicket of dense vegetation or in a shallow scrape, often

The common wombat (*Vombatus ursinus*) is an herbivore. (Photo by Animals Animals ©Robert H. Armstrong. Reproduced by permission.)

excavated in the shelter of a shrub or grass tussock. The West Australian boodie, a species of bettong, digs its own burrow or uses the abandoned tunnels of rabbits. Wombats are supreme diggers—their homes are extensive burrows in forests, grasslands or scrub.

Among the arboreal diprotodonts, most use some kind of customized resting place. Cuscuses prepare special sleeping platforms of bent twigs and leaves, while honey possums and pygmy gliders weave intricate nests of grass, shredded bark and moss. Most possums and gliders build nests of twigs and leaves and many will make use of ready-made possum or bird nests, hollow logs or naturally occurring tree holes. Both males and females build nests and the majority of possums will have several nests within a home range. Possum nests are sometimes referred to as "dreys."

Behavior

Like many marsupials, the majority of diprotodonts are primarily nocturnal or at least crepuscular, but most macropods will sometimes move about in daylight, especially under the cover of forest of scrub. A good many nocturnal species will also emerge by day to bask in the sun, especially early in the morning when the warmth helps them to digest the rewards of a night's foraging. Basking is an important part of the energy efficient lifestyle of wombats, koalas, and kangaroos. Only one species is generally thought of as completely diurnal, the musky rat kangaroo (*Hypsiprymnodon moschatus*) of tropical Queensland. However a good many macropods, in particular the tree kangaroos and forest wallabies appear to be active both night and day. The large Celebes cuscus (*Ailurops ursinus*) is also reported to be at least partly diurnal. Day active species tend to be those that live in forests or other sheltered habitats where their activity is less likely to draw unwanted attention. Some Australian pademelons (genus *Thy-logale*) tailor their activities to the time of day—grazing in open pasture by night and browsing in forest by day, pausing periodically to rest in the shelter of a dense shrub or to bask in a sunny clearing.

The annual cycles on which these various patterns of daily activity are superimposed are most apparent in animals living in temperate parts of Australia, where the seasonal climate forces some quite drastic behavioral adaptations. Surviving the winter is not simply a question of keeping warm, but of having the energy to do so. Animals that spend the warmer months feasting on seasonal foods such as fruits and certain insects must either rely on stored body fat, cache food for hard times ahead, switch to an alternative food source, or be prepared to drastically reduce energy consumption during periods of shortage. It is not surprising that the only diprotodont to use all of these strategies hails from the Southern Highlands of Australia, where the annual cycle of glut and deficiency is especially acute. The mountain pygmy possum (*Burramys parvus*) spends the spring gorging on the millions of bogong moths that visit the highlands to breed. When the moths are gone the possum switches to fruits and seeds, storing those it cannot eat immediately in a winter larder. In winter it hibernates for six months or more, the only marsupial to do so. It can also enter an energy-saving torpid sleep during periods of food shortage at any time of year—a good insurance policy again the vagaries of mountain weather. Several other diprotodonts capable of facultative torpor induced by food shortage or low temperatures—among them the honey possum and the pygmy glider (*Acrobates pygmaeus*).

Many diprotodonts are arboreal. The members of the possum families have strong grasping hands and feet. The tail is prehensile to some degree. Tails used for grasping often lack fur near the tip, especially on the underside—bare, callused skin, provides much better grip than soft fur. In species such as Leadbeater's possum the tail serves as a counterbalance, in the ringtail and dormouse possums it can be used to steady the animal in the tree, and in several the tail is strong enough to support the animal's entire weight. The koala, however, manages to climb very well with no tail to speak of. It climbs tall eucalypti by hugging the trunk, digging in with surprisingly large, hooked claws and hauling itself upwards, then reversing the process in order to descend. Koalas are also surprisingly adept at moving along the ground—this is after all the only way for them to get from one tree to another. They walk with a rolling, bowlegged gait or when pressed, proceed at a surprisingly fast, bounding gallop.

The ground is a dangerous place for most tree-dwelling animals. It is therefore not surprising that gliding as a means of traveling from tree to tree without descending to ground level has evolved several times in a variety of arboreal lineages. These include the placental flying squirrels and colugos and also the diprotodont families Pseudocheiridae, Peturidae, and Acrobatidae.

Three families of diprotodont have forsaken the trees for life at ground level—the wombats, kangaroos, and rat-kangaroos. The wombats have tackled the threat of predation by building secure underground dens and by becoming too large for most of Australia's native hunters to tackle safely. Their legs are short

Kangaroos fighting on the beach. (Photo by © Mark A. Johnson/Corbis. Reproduced by permission.)

but immensely strong as an adaptation to burrowing and they walk with a purposeful quadruped waddle.

Kangaroos and wallabies are famous for hopping. When moving at speed, they propel themselves forward with powerful leaps of the hind legs—the forelegs do not touch the ground at all, and the tail is used only for balance. Hopping is a remarkably efficient way to travel, and a large kangaroo can reach top speeds of up to 30 mph (48 kph), covering 30 ft (9 m) or more with every bound. Interestingly, swimming is the only form of locomotion in which a kangaroo moves its large back legs independently of one another.

None of the methods of locomotion described so far is especially novel—all are very similar to those employed by various groups of placental mammals elsewhere in the world, a fact reflected in much congruence of form between marsupial and placental animals. Some squirrels glide, primates and rodents climb, and several small rodents are great hoppers—some are even known as kangaroo mice, even though other group has committed quite so fully to life on the hop as the macropod diprotodonts.

However there is one form of marcopod locomotion that is entirely unique. When moving at slow speeds, large kangaroos are effectively five-legged. Their reduced forelegs are much too weak to support the entire body weight, and so the tail plays the role of sturdy prop. The animal leans forward onto its hands, and swings its hind legs forwards while supporting its rear end on the base of the tail. At times the tail can even be used to temporarily support the entire body weight, for example during combat. A fighting kangaroo may lean back onto its tail while attempting to inflict thunderous double blows on its opponent with both hind feet.

As if to prove that that no change is irreversible, some kangaroos have returned to an arboreal way of life. Tree kangaroos have evolved from ground-dwelling ancestors and thus lack many of the primary adaptations to climbing such as a prehensile tail. Nevertheless they move about the trees with great agility, using their tail as a counterbalance or brace, and grasping the branches with large, well-cushioned feet. They will leap from tree to tree and descend by jumping from branch to lower branch or shuffling backwards down the trunk. In one final phylogenetic twist, the Doria's tree kangaroo (*Dendrolagus dorianus*) has become ground dwelling once more.

Diprotodont populations exhibit a range of social structures. Among the macropods, forest-dwelling species including most of the rat kangaroos and smaller wallabies tend to live alone, except when breeding, while larger species that spend more time in open spaces are more gregarious. Kangaroo groups, known as "mobs," are rather casual associations, and members are free to come and go at will. Females are less inclined to disperse than males and often remain in the same mob as their mother. Hence the female members of a mob are often related. The size of a mob varies, with the largest groups forming when several mobs converge on a resource such as a good feeding area. Male kangaroos are socially dominant to females but they do not lead the mob—their interest in a particular group is usually confined to periods when one or more females is in or approaching breeding condition. There is no real cooperation between members of a kangaroo mob, but there is an element of safety in numbers and each individual benefits from the alertness of others.

Koalas and common wombats are solitary for most of the year. Koalas live in close proximity to one another but they require personal space. While more than one individual may use the same tree, they do so at different times. Common wombats live alone in their burrow, but they are not generally aggressive and will visit one another's homes. Hairynosed wombats are altogether more social, and up to 50 individuals have been known to share a warren of interconnected burrows.

Among the various possum species there is an inverse relationship between size and sociality. Most large species are solitary except when breeding, whereas small species such as the honey possum, the feathertails, and lesser gliders often live in pairs and may rest in quite large groups, especially during periods of cold weather when they huddle together for warmth. Leadbeater's possum is unusual in that females are socially dominant to males, and mothers drive their daughters from the territory as soon as they are old enough to fend for themselves. In this species, it tends to be males that stay put.

Diprotodonts commonly use four main types of vocalization to communicate, namely: barking, used to communicate positon to other group members; sneezing, indicative of disagreements within the group; hissing, used as a distress call; and crabbing, a sound made to convey displeasure, associated for example with being disturbed while in the nest, or while sleeping. The mountain pygmy possum makes a low guttural vocaliza-

Matschie's tree kangaroo (*Dendrolagus matschiei*) sits on a branch. (Photo by Art Wolfe/Photo Researchers, Inc. Reproduced by permission.)

A koala (*Phascolarctos cinereus*) eating eucalyptus leaves. (Photo by Daniel Zupanc. Bruce Coleman, Inc. Reproduced by permission.)

tion when distressed. Sugar gliders have an alarm call that sounds like the barking of a small dog. Squirrel gliders (*P. norfolcensis*) exhibit some unique vocal communications, they produce gurgling chatters and soft, nasal grunts, also repetitive, short gurgles. The common wombat makes a loud hissing growl when annoyed. And it has been said that the very loud hissing, crackling territorial call of the male common brushtail has a definite nightmare quality.

Feeding ecology and diet

Feeding strategies among primitive diprotodonts were aimed at a generalist herbivorous diet of leaves, fruits, and roots. The prehensile tail and opposable digits of the front feet enhanced climbing ability in many, giving them access to the branches of trees as well as some security from ground dwelling predators like the thylacine. The majority of diprotodonts are still vegetarian, but whereas early diprotodonts were mostly generalists, many now specialize in a wide range of vegetarian diets. The great radiation of ancient diprotodonts was fuelled in part by changes in the Australian climate as the continent drifted slowly north having separated from Antarctica. As Australia's forest cover dwindled, competition for certain plant foods become more intense and herbivores were forced to specialize. While opportunities for carnivorous and insectivorous lifestyle remained broadly similar, changes in vegetation cover presented a range of entirely new food types for herbivores to exploit. Large areas that were formerly covered in dense forest became drier—the lush forests giving way to more open woodland, grassland and finally to desert.

The evolution of grazing habits in Australian marsupials mirrored that which occurred in placental mammals elsewhere on the planet. Grass is a tough food. It requires a lot of chewing and very thorough digestion to unlock its nutrients. However it is abundant and easy to find, and those diprotodonts that adopted the grazing habit (mostly macropods) are now among the most widespread Australian mammals. The arrival of European settlers to the continent a little over 200 years ago has spelled disaster for many native diprotodonts, but not for large grazers, which benefited from the improvement of grassland for livestock.

Leaves and grasses are convenient dietary staples—they are easy to find and collect, and in many parts of Australia and New Guinea they are available year round. However they are not particularly nutritious, and most herbivorous animals have to eat quite a lot of them to survive. However diprotodont mammals are surprisingly fuel-efficient animals. All marsupials run their body at a slightly lower temperature than placental mammals and some are able to survive on the most meager of rations. For example, red kangaroos have a very large stomach, which when full might contain up to 15% of the animal's body weight in food. But this is processed so efficiently that the kangaroo actually eats much less than an equivalent sized sheep or other placental mammal.

The prize for most effective use of low-grade food goes to the koala. Koalas are famously lethargic animals. They sleep up to 20 hours a day and move very little when awake. In the past it was speculated that koalas were effectively drugged by toxins in the eucalyptus leaves they ate. Now it is very clear that the laid-back koala lifestyle is a highly adaptive survival strategy—by expending very little energy, koalas can survive on a food resource so low in nutritional content that no other mammal even attempts to eat it. Wombats employ a similar low-energy lifestyle, resting for long periods underground and using the warmth of the sun to reheat their body in the morning, thus avoiding the need to burn fuel.

For other diprotodonts living away from desert and dry bushland, quality food is a little easier to come by, but competition is correspondingly more intense. Rat-kangaroos and smaller macropods living on heaths and in forests eat more fruits, fungi, and starchy roots, and the active possums target sugary fruits, oily seeds, and flowers containing protein-rich pollen. Some of the most successful modern forest diprotodonts have learned to tap into the internal plumbing of trees, and feed on sugar-rich sap and gum. Early gum-eaters probably took advantage of damage to trees caused by other factors such as high winds, and fed opportunistically on the sticky secretions that leaked from damaged branches. From this it is a relatively small step to deliberately inflicting wounds on the tree in order to release sap. But sap and gum flow very slowly, even from deep wounds, and an active animal like a possum cannot afford to sit around waiting for a tree to leak enough sap to make a meal. Hence many species, in particular the petaurid gliders, have developed longer-term strategies whereby they make a cut with their sharp front teeth, and return later to harvest and eat the blob of gum that has accumulated.

Most diprotodonts consume at least some animal material, mostly insects, along with their regular diet of plants. A few species have become more actively insectivorous—for example, the mountain pygmy possum, which for several months of the year eats nothing but large moths, and the pygmy glider, which catches insects to supplement its diet of nectar and fruit.

Most divergent of all is the honey possum, which lives almost exclusively on a diet of pollen and nectar. It is one of very few mammals to do so, and has a long, specially modified tongue with a brushlike tip for collecting pollen from deep tubular florets.

Reproductive biology

Marsupial mating systems are not well known. Some species may be monogamous, but most are probably polygynous or promiscuous. The marsupials are a group defined by their reproductive biology. Female marsupials give birth to live young, but at a very early stage after a very short gestation (for example 21 days in the common wombat, 17 days in the brushtail possum, 32 days in the eastern gray kangaroo). Female marsupials have two uteri and two vaginas. Young developing in one or both uteri are born through a third opening, which only develops when the female is due to give birth for the first time. In most marsupials this birth canal is a temporary structure that seals over after the birth of every litter, but in certain diprotodonts (the kangaroos and the honey possum) it becomes permanent.

Newborn young make their own way to the mother's teats, which are usually located within a pouch (the "marsupium" for which the group is named). The young latch onto a teat and continue their development sustained by milk that changes in composition to suit their needs as they grow. The diprotodonts have some of the best developed pouches—designed to carry the young securely while the mother hops, ands burrows her way through daily life. The number of teats varies between species and gives a rough guide to maximum litter size—a female cannot rear more young than she has teats. Diprotodont litters are quite small—one or two is normal for most possums, whereas single young are the norm for kangaroos, rat kangaroos, wombats, and the koala. The mountain pygmy possum may give birth to as many as eight young, but only the four strongest will find a teat and survive.

Most mothers continue to suckle to their young when they have outgrown the pouch. Possums and wombats carry larger youngsters on their back. By accompanying their mother as she forages, young animals learn by imitation what foods are good to eat. Young kangaroos that have left the pouch may

A wallaby joey coming out of its mother's pouch. (Photo by Corbis. Reproduced by permission.)

remain close to their mother and continue to suckle for many weeks, reaching into the pouch to drink from the same teat they used as a newborn. By this stage there is often a new baby in the pouch attached to a different teat, in which case the mother produces a different kind of milk for each of her two offspring.

In many diprotodonts, including the honey possum and several species of kangaroo, the interval between one young vacating the pouch and the birth of a replacement can be as little as one day. This is thanks to a remarkable phenomenon known as embryonic diapause. Because gestation is so short, it does not interrupt the normal estrus cycle, and a female can come into season very soon after her first young is born. The new mother may mate and conceive during this "post-partum" estrous, but as long as the pouch is occupied by a suckling youngster the second embryo or litter does not develop beyond a very early stage. Instead it remains in a state of suspended animation, ready to be reactivated should the first young be lost or when it is almost ready to leave the pouch. The process is controlled by the same hormones that regulate milk production.

With the exception of the honey possum, individuals of which rarely live to see their first birthday, diprotodonts are quite long-lived animals. Possums, cuscuses, ringtails, gliders, and small macropods have a life expectancy between six and 14 years, koalas often live into their late teens, and wombats and large kangaroos well into their twenties. Even the diminutive feathertails and pygmy possums can live seven and 10 years respectively.

Conservation

Six species of diprotodont marsupial are known to have gone extinct in recent years, among them the Toolache wallaby, *Macropus greyi*. This once abundant native of South Australia was one of the swiftest of all macropods. Its agility and its very fine pelt made it a favorite quarry for "sportsmen" and it was hunted to the brink of extinction in the 1920s. The species may have hung on in some parts of its former range until as late as the 1970s, but here have been no positive records since 1924 and the species is listed as extinct. Another victim of changing times was the broad-faced potoroo—this was first described in 1844, but was extinct within 40 years.

Of the remaining diprotodont species, approximately a quarter currently appear on the IUCN Red List of Threatened Animals. In New Guinea these include the black spotted cuscus and Goodfellow's tree kangaroo and in Australia the long-nosed potoroo, the Proserpine and yellow-footed rock wallabies, the brush-tailed bettong, and the mountain pygmy possum. Two species, the Gilbert's potoroo and the Northern hairy-nosed wombat are Critically Endangered, both with less than 50 individuals left in the wild. A further quarter of all diprotodonts are regarded as Near Threatened or data deficient. This latter designation means the animals, or at least their distribution and abundance, are so little known that conservationists have been unable to assess the extent of any potential threat.

The problems facing many threatened diprotodonts are very similar. In Australia the arrival of humans and their companion animals has had a devastating effect on native wildlife, in terms of habitat modification, hunting and predation by introduced carnivores. The first aboriginal settlers arrived between 60,000 and 40,000 years ago. They introduced the semi-domesticated dingo and began managing the landscape using burning regimes to encourage fresh plant growth. However these changes were minor compared to those that came about with the arrival of European settlers in the late eighteenth century. The advance of farming, the intensification of burning regimes, and above all the deliberate introduction of foxes, cats, and rabbits, had a profoundly destabilizing effect. In New Guinea, there are two main threats to native diprotodonts, most of which are forest-dwellers. Unrestricted logging not only destroys habitat, it also opens up areas of previously inaccessible forest to hunting.

On a more positive note, several species of diprotodont that were once considered extinct have been rediscovered alive and well. While most of these are still very rare, being endangered is still very much better than being extinct. Gilbert's potoroo was presumed extinct until 1994 when a small colony was rediscovered on the coastal heathland near Albany in Western Australia. A captive breeding program is underway. Leadbeater's possum was declared extinct in the early twentieth century, but turned up again in 1961. Ironically it appears to have been saved by the enormous bush fires that devastated large areas of the Victorian Highlands in 1939—the standing dead wood left by the fire provided a sudden surplus of suitable possum nesting holes. The mountain pygmy possum had never been seen alive before 1966. The species had been described from preserved remains and

A bridled nail-tailed wallaby (*Onychogalea fraenata*) on a scientific reserve in central Queensland, Australia. This wallaby was presumed extinct in the 1960s but was rediscovered in 1973 in a small area of Queensland, Australia. (Photo by Mitch Reardon/Photo Researchers, Inc. Reproduced by permission.)

was presumed long extinct, when one turned up alive and well in a ski lodge in the winter resort of Mount Hotham. Other resurrected diprotodonts include the parma wallaby and the Mahogany glider.

There are also some conservation success stories—perhaps most notably the koala. In 1920, the species was facing extinction due to overhunting and habitat fragmentation. Thanks to a ban on hunting and the establishment of many reserves, the trend was reversed. In fact in many ways the campaign to save the koala has been too successful. Many reserves are now so overcrowded that disease has become a serious issue. While the future of the species appears secure, the outlook for many thousands of individuals living on overcrowded reserves is rather bleak. At best they face deportation to new homes (of which there are a dwindling number), at worst a slow death from infections such as *Chlamydia*. In some places, wildlife managers have had to make the unhappy decision to cull some the animals they have work so hard to save in order to improve living conditions of others.

The northern hairy-nosed wombat, *Lasiorhinus krefftii*, is now Critically Endangered. Its southern cousin *Lasiorhinus latifrons* is faring a little better, but still has a rather restricted distribution. The common wombat, *Vombatus ursinus*, while not officially threatened, is regarded as a pest in some states and is becoming a cause for conservation concern.

Significance to humans

Australia is unique among the world's nations in having such a diverse array of highly distinctive endemic mammals.

People sometimes confuse the fauna of other continents, imagining tigers in Africa and leopards in South America, but few would make the same mistake with a kangaroo or koala. As a result, these two diprotodonts in particular have achieved iconic status in the last century. They are used to market all things Australian from beer to tourism and no visit "down under" is complete without the opportunity to cuddle a koala or see a mob of kangaroos bounding across the desert.

Kangaroos are among the few marsupials to have benefited from changes in the Australian landscape since the arrival of white settlers. As pioneer farmers began developing huge areas of bushland as grazing pasture for livestock, kangaroo numbers soared. The native population benefited greatly from the improved grazing created by irrigation from artesian wells and boreholes. Being naturally adapted to arid conditions, the kangaroos fared much better than the domestic stock and when droughts hit, farmers were driven to furious despair at the sight of healthy kangaroos grazing the last dry blades of grass alongside emaciated and dying sheep and cattle. The situation now boils down to an uneasy truce in which native grazers are fenced out of core areas of pasture, where sheep and cattle are provided with feed supplements and additional water in times of drought. Kangaroos that manage to breach perimeter fences are routinely shot, but many still manage to make a good living on marginal ranchland.

Kangaroo meat is increasingly popular. It is very lean and is often marketed as a healthy alternative to beef. Large quantities are also used in the manufacture of pet foods. Kangaroo skin makes very soft leather. This is not very hard wearing but it is flexible and pleasant to touch and thus suitable for hats and other items of clothing.

Another diprotodont with a split personality—one part useful, the other a nuisance—is the common brushtailed possum. Possums are resourceful and adaptable animals and they thrive in and around human settlements, raiding garden produce and trashcans at night and generally making a nuisance of themselves. They do considerable damage to trees in plantations and carry a number of diseases. In the past some of this damage was offset by the value of the possum's fur, but this has declined of late. Possums introduced to New Zealand between 1840 and 1900 have caused untold damage to native trees and plants.

The damage caused by kangaroos or possums is as nothing compared to that inflicted by rabbits. These were introduced for food and sport in 1858 and rapidly reached plague proportions. Their depredations were as damaging to native grazers as they were to livestock and efforts at eradicating them have involved unleashing some of the most unpleasant wildlife diseases known—first myxomatosis in the 1950s and more recently rabbit hemorrhagic disease or RHD. While rabbits and kangaroos compete directly for the same foods, rabbit control measures have also affected the fate of another group of diprotodonts—the wombats. Being prolific burrowers, wombats often provide rabbits with a means to circumvent fences erected to keep them out of pastures. As a result wombats have been mercilessly persecuted.

Resources

Books

Aplin, K. P., and M. Archer. "Recent Advances in Marsupial Systematics with a New Syncretic Classification." In *Possums and Opossums: Studies in Evolution*. Sydney: Surrey Beatty & Sons and Royal Zoological Society of New South Wales, 1987

Flannery, T. *Mammals of New Guinea*. Papua, New Guinea: Robert Brown & Associates, 1995.

Flannery, T. *Possums of the World*. Chatswood, Australia: GEO Productions & the Australian Museum, 1994.

Macdonald, D. *The New Encyclopedia of Mammals*. Oxford: Oxford University Press, 2001.

Nowak, R. "Order Diprotodontia." In *Walker's Mammals of the World*. Vol. I. 6th edition. Baltimore: Johns Hopkins University Press, 1999.

Strahan, R. *The Mammals of Australia*. Carlton, Victoria: Reed New Holland, 1995.

Organizations

Austalian Conservation Foundation Inc. 340 Gore Street, Fitzroy, Victoria 3065 Australia. Phone: (3) 9416 1166. Web site: <http://www.acfonline.org.au>

Australian Museum. 6 College Street, Sydney, New South Wales 2010 Australia. Phone: (2) 9320 6000. Web site: <http://www.amonline.net.au>

IUCN—The World Conservation Union. Rue Mauverney 28, Gland, 1196 Switzerland. Phone: +41 (22) 999 0000. E-mail: mail@iucn.org Web site: <http://www.iucn.org>

South Australian Museum. North Terrace, Adelaide, South Australia 2010 Australia. Phone: (8) 8207 7500. Web site: <http://www.samuseum.sa.gov.au>

Amy-Jane Beer, PhD

Koalas

(Phascolarctidae)

Class Mammalia

Order Diprotodontia

Family Phascolarctidae

Thumbnail description
A medium-sized stocky herbivorous marsupial with a bear-like appearance, characterized by a broad face and bulbous nose, large, rounded fluffy ears, relatively long legs with large paws and powerful claws, and a short tail

Size
28–31 in (72–78 cm); 11–26 lb (5.0–11.8 kg)

Number of genera, species
1 genus; 1 species

Habitat
Subtropical/tropical dry eucalypt forest and woodland

Conservation status
Lower Risk/Near Threatened

Distribution
Eastern Australia

Evolution and systematics

The koala family, Phascolarctidae, are believed to have diverged from their nearest marsupial relatives, the wombats, around 24 million years ago (mya). At least six different members of the koala family evolved. The earliest fossil record of a koala was a browser, *Perikoala palankarinnica*, some 15 mya. More recently, a giant koala, *Phascolarctos stirtoni*, was a third as large again as our present day koala, but is believed to have died out along with other marsupial megafauna some 40,000 years ago, at around the time that aboriginal hunter-gatherers colonized Australia. Only one species, *Phascolarctos cinereus*, survives today. Phascolarctos is from the Greek words for "leather pouch" and "bear," while cinereus means "ash-colored."

There are three subspecies of koala, *Phascolarctos c. victor*, native to the state of Victoria; *Phascolarctos c. cinereus*, native to New South Wales; and *Phascolarctos c. adustus*, native to Queensland. Koalas in the north have shorter coats and are smaller than their southern cousins, and genetic studies have confirmed significant differences between the two populations, as well as suggesting that there may be a number of distinct subpopulations in the north. Southern populations appear more homogenous, probably as a result of the numerous translocation projects which have taken place.

The taxonomy for the koala is *Phascolarctos cinereus* (Goldfuss, 1817), New South Wales, Australia.

Physical characteristics

The comical, appealing, "teddy bear" appearance of the koala has made it Australia's iconic animal. Despite the misleading popular name "koala bear," koalas are not, of course, related to the omnivorous bear family, but are herbivorous marsupials. They are medium sized, with a head and body length that can be as short as 24 in (60 cm), or as long as 33 in (85 cm), but is usually in the range 28–31 in (72–78 cm). Body weight can also vary considerably, from as little as 8.8 lb (4 kg) for a northern female, to as much as 33 lb (15 kg) for a southern male, but the usual range is 11–26 lb (5.0–11.8 kg). Males are up to 50% larger than females, and there is a significant size difference between koalas in Queensland,

The Queensland koala (*Phascolarctos cinerus adustus*) is one of three subspecies of koala. (Photo by Animals Animals ©David Boyle. Reproduced by permission.)

where males average 14.3 lb (6.5 kg) and those further south, where males average 26 lb (11.8 kg).

The koala has a compact body with a broad head, large nose, and small eyes. The ears are large and rounded with white edges. Koalas have only a vestigial tail, which is of no assistance in climbing, but they have long, strong limbs, with large paws and sharp claws which are well adapted to grip smooth-barked eucalyts. Fore and hind feet have five digits, all with sharp, recurved claws, except for the first digit of the hind foot, which is short and broad. The first and second digits of the forefeet are opposable to the other three, allowing the animal to grip smaller branches and climb into the outer canopy in search of fresh leaves. The second and third toes of the hind feet are fused, with a double claw.

Koalas do not use dens nor shelters, so their fur is important for insulation. Southern koalas have dense, woolly coats, with thicker, longer fur on the back than the belly. Koalas living further north in warmer subtropical and tropical regions have shorter coats (also lighter in color), sometimes appearing almost naked. The color and pattern of coats varies considerably between individuals and with age, from gray to tawny, with white on the chin, chest, and forelimbs and whitish dappling on the rump. Males have a large chest gland that is used for scent marking trees. Females have a marsupial pouch opening to the rear and containing two teats.

Distribution

Australia's koala population is found in a broad coastal swathe down the eastern seaboard, from the Atherton tablelands in north Queensland to southwestern Victoria. Although this is an area of several hundred thousand square miles (kilometers), deforestation, habitat degradation, and historic persecution mean individual populations within this range are fragmented and often isolated.

Historically, the geographical range of koala populations was broadly similar to today, but extended into South Australia and southern parts of Western Australia. Land clearance and hunting caused populations within this range to contract, and in many cases become extinct. However, intensive koala management and reintroduction programs have reversed this decline in many places, particularly in the south, and koalas are now locally common where suitable habitat survives. A number of localized populations have been re-established in South Australia and Western Australia.

Habitat

Koalas feed almost exclusively on eucalyptus leaves, so their habitat is invariably eucalypt forest and woodland. However, they can tolerate a surprisingly diverse range of envi-

MLM © 2003

Koala (*Phascolarctos cinereus*). (Illustration by Michelle Meneghini)

A Queensland koala (*Phascolarctos cinereus adustus*) sleeps in a tree. (Photo by John Shaw. Bruce Coleman, Inc. Reproduced by permission.)

ductivity of the environment, but can be as little as 2.5 acres (1 ha) for a male in a fertile habitat, and half that for a female. Within this range an animal may live much of its life in only a dozen favored trees. A dominant adult male's range will overlap the range of up to nine females and a number of subordinate and subadult males.

Within the area of a stable social group individual koalas will "own" a number of food trees and "home range trees" marking the edge of their range. A koala trespassing into an "owned" tree may be aggressively attacked by the resident, but scent markings and scratchings usually warn animals that a tree is under possession. Other trees where ranges overlap will be shared, and it is here that limited social interaction takes place.

ronmental and climatic conditions. In the tropical north habitat is often dense thicket, with high year-round temperatures and strong seasonal rainfall. Rainfall is also high in the temperate mountain rainforests of the south, but winters can be much colder. This fertile habitat can support populations as high as three animals per acre (0.4 ha), a stark contrast to the semi-arid open woodland habitat of the west, where a single animal may require 250 acres (100 hectares) to survive.

In temperatures above 77°F (25°C) koalas use evaporative cooling in their airways to regulate temperature, by breathing rapidly, but reduce water loss by decreasing the amount of water in their urine. In cold temperatures koalas, like humans, conserve heat by reducing blood flow to extremities, and have also been seen to shiver, a means of producing heat by rapid muscle contraction.

Behavior

Koalas are essentially solitary animals, with very little social interaction other than during the breeding season. Adults occupy fixed home ranges. Range size depends on the pro-

A Queensland koala (*Phascolarctos cinereus adustus*) showing its chest gland and secretion. (Photo by Kenneth W. Fink/Photo Researchers, Inc. Reproduced by permission.)

Koalas are largely nocturnal, feeding and moving after dusk. They rarely leave the security of the trees, descending to the ground only to move to another food tree, or to consume soil, which aids digestion. Koalas walk on four legs, run with a bounding gait, and can swim if necessary.

As one adaptation to their low-energy eucalyptus leaf diet, koalas sleep up to 20 hours a day, usually wedged comfortably into the fork of a tree. Even when awake, koalas spend much time resting, and feeding occupies only 10% of their day.

Activity livens up in the summer breeding season, when dominant males will attempt to defend their territory for breeding rights with resident females. At this time of year males use their chest gland to scentmark tree trunks and can often be heard bellowing, apparently to warn off rival males and attract females. This deep, grunting bellow often provokes responses from other males in the area. At night males move around more, fighting with any competing adult males that they encounter, or mating with estrous females.

Koalas have a range of other sounds for communication. Mothers and babies make soft clicking, squeaking sounds, and

When on the ground, the koala (*Phascolarctos cinereus*) is vulnerable because it moves slowly. (Photo by R. Kopfle. Bruce Coleman, Inc. Reproduced by permission.)

Koalas (*Phascolarctos cinereus*) spend little time on the ground. (Photo by Animals Animals ©Gerard Lacz. Reproduced by permission.)

gentle murmuring, or will grunt if annoyed. All koalas are capable of a distressing, high-pitched cry like a screaming baby when afraid.

Feeding ecology and diet

Australia has some 650 species of eucalypt, but koalas are choosy eaters, and feed on only around 30 species, with just a handful, including river red gum, gray gum and manna gum, preferred. They have occasionally been known to eat non-eucalypt leaves, including acacia, mistletoe and box. Koalas eat about 1.3 to 1.8 lb (600 to 800 g) of leaves a day.

Koalas reach their food by climbing high up smooth, vertical eucalypt trunks, gripping with their sharp foreclaws and using their powerful front legs to pull themselves up, while bringing their hind legs up to their front. Although they generally move slowly and laboriously to conserve energy, they are capable of surprising agility and can leap 6 ft (2 m) from trunk to trunk.

Eucalyptus leaves are low in nutrients, contain a large proportion of indigestible cellulose and lignin, and are full of toxic chemicals. Koalas have evolved a number of ways to cope with this poor diet. They avoid the most toxic species, and vary their choice of food tree throughout the year as toxin levels vary seasonally in some species. The koala's liver is capable of detoxifying and excreting some poisonous compounds from those species they do eat. Koalas have large cheek pouches, to handle large amounts of poor quality forage, and well-developed teeth, which include a single premolar and four molars in each jaw, which grind the fibrous leaves to a fine paste. This is then digested by microbial fermentation in the animal's unusually long cecum (a blind sac in the digestive tract, between the junction of the small and large intestines). A koala's cecum can be more than 75 in (2 m) long.

The low energy yield of the koala's diet explains their slow, sedentary lifestyle. However, it is a myth that koalas are drugged by the poisonous compounds in eucalyptus leaves. What is true is that koala have one of the smallest brains of all marsupials relative to body size—only 0.2% of body weight— and this has been explained as a further response to diet, since the brain is one of the most energy-consuming of the body's organs.

Koalas obtain most of their water from leaves, but occasionally drink at streams, and in captivity often choose to drink fresh water.

Reproductive biology

Koalas are polygynous. During the summer breeding season a dominant male will attempt to mate with any estrous females he encounters in his range. Copulation lasts only a

A koala (*Phascolarctos cinereus*) with offspring. (Photo by Animals Animals ©Gerard Lacz. Reproduced by permission.)

A koala (*Phascolarctos cinereus*) feeding on eucalyptus leaves in a wildlife park in Tasmania, Australia. (Photo by Gregory G. Dimijian/ Photo Researchers, Inc. Reproduced by permission.)

couple of minutes, with the male mounting the female from behind and holding her against a branch. Females are sexually mature at two years old, but generally do not start to breed until they are older—full physical maturity is reached at about four years old in females, five years old in males.

Females have an estrous cycle of about 30 days, and usually breed once every year, between November and March. Gestation lasts about 35 days before a single young (very rarely twins) is born, weighing less than 0.02 oz (0.5g) and measuring about 2 cm long.

The tiny newborn koala crawls into the mother's large pouch and attaches itself to one of the two teats. By 13 weeks the young joey will have grown to about 2 oz (50 g), and by 22 weeks its eyes open and it begins to poke its head out of the pouch for the first time. Joeys have a pouch life of five to seven months, after which they spend most of their time out of the pouch, clinging to the mother's belly and later sitting on their back. Joeys are weaned at six to 12 months, but towards the end of their pouch life also feed regularly on soft, partially-digested leaf material passed through the mother's digestive tract. This "pap," which contains a high concentration of microorganisms, is believed to be important in intro-

A koala carrying its baby in its pouch. (Photo by © DiMaggio/Kalish/Corbis. Reproduced by permission.)

ducing to the young koala's gut the microbes it will need to digest eucalyptus leaves.

A joey will remain with its mother until about one year old, when it weighs around 4.4 lb (2 kg) and can begin to fend for itself. Juveniles disperse to find their own home range at about two years old, searching for another breeding group to join, but becoming nomadic if no area is available.

Koalas can live in excess of 10 years in the wild, and 17 years or more in captivity. Longevity is probably related to stress factors such as habitat pressure, disease, and human interference.

Conservation status

Koala conservation is a complex issue. Populations are under severe pressure from habitat loss in many parts of Australia, yet in some areas the koala is common or even over populous.

Before 1900 koalas numbered in the millions, despite regularly suffering enormous losses to bushfires and disease epidemics. But in the first decades of the twentieth century

extensive forest clearance and large scale hunting for the koala's warm, cheap, durable fur saw populations crash. The slaughter reached a peak in 1924, when over two million koala pelts were exported to Europe and America, and by the end of that year the species had been exterminated in South Australia and nearly wiped out in Victoria and New South Wales. A healthy population surviving in Queensland was next to suffer when in 1927 the state government bowed to commercial pressure and allowed an open season—600,000 more skins were exported.

Public outcry in Australia and abroad eventually resulted in legal protection, and since the 1920s intensive conservation measures, including captive breeding and translocation efforts, have allowed populations to partially recover. Today koalas are still under intense pressure in many parts of their range, but are not classified as threatened. The Australian government lists koalas as vulnerable, but has not put them on the country's endangered list. The World Conservation Union (IUCN) classifies koalas as Lower risk/Near Threatened, and lists habitat loss and degradation due to timber

A Queensland koala (*Phascolarctos cinereus adustus*) with young. (Photo by Kenneth W. Fink/Photo Researchers, Inc. Reproduced by permission.)

A koala grazes in a tree. (Photo by © Tom Brakefield/Corbis. Reproduced by permission.)

felling and urbanization, and human disturbance, particularly through fire, as the major threats.

Unfortunately, koalas inhabit precisely those eastern seaboard regions of Australia that are seeing most rapid urbanization and agricultural development. In the past 200 years an estimated one-third of Australia's eucalyptus forest has disappeared, and in semi-arid parts of Queensland, thousands of acres of woodland are still being cleared for agricultural use every year. Urbanization and tourist development along the coastal strip is further fragmenting eucalyptus woodlands. Bush fires caused by negligence or arson account for thousands of deaths each year, while an estimated 10,000 koalas are killed in road accidents. Koalas put up no resistance to attacks from domestic dogs, and do not cope well with stress, having abnormally small adrenal glands. There is also evidence that inbreeding in isolated, fragmentary populations is leading to physical abnormalities.

Despite all these pressures, there are actually locations where koala populations have thrived to such an extent that they are causing environmental damage. Koalas were translocated to a number of islands where they were not found naturally as long ago as the 1870s. Populations on Phillip and French Islands in Victoria, and Kangaroo Island in South Australia have grown so large that the unpalatable prospect of culling has been proposed. Thousands of animals have been relocated back to the mainland, causing similar overcrowding problems in some locations, and contraception is being investigated as a more publicly acceptable alternative to culling.

One possible explanation for these overpopulation problems is that the island populations are free of chlamydia, a disease that is now thought to have been endemic in koalas for many years, and may have acted as a natural population control, remaining benign when conditions were good, but killing off weaker animals during stressful times such as when habitat is reduced.

Some koala conservation work, including purchase of land for protected reserves, is being carried out by state authorities, but much koala conservation and research is in the hands of charities and privately-run welfare organizations. The koala's cute and cuddly appeal helps raise funding for such non-government organizations.

One problem in planning conservation management is the difficulty in obtaining accurate population figures. The Australian Koala Foundation suggests numbers have dropped from 400,000 in the mid-1980s to between 40,000 and 80,000 today, but this estimate can only be an educated guess. The Foundation is compiling a national atlas of surviving koala habitats, which will provide a tool to lobby for habitat conservation.

Significance to humans

The name koala is believed to have originated from Aboriginal dialect names for the animal, which include cullewine, koolewong, colo, colah, and koolah. One suggested translation for these Aboriginal names is "no water," referring to the koala's ability to largely subsist on moisture from leaves.

Australia's Aboriginal people have long hunted koalas for food—they make a slow-moving target easy to hit with a boomerang. But traditional Dreamtime stories teach Aborigines that if they fail to respect the animal they will be visited by a terrible drought. Tradition dictates that while koalas may be eaten, they must not be skinned nor their bones broken.

No such respect was accorded by early European settlers who shot koalas for "sport" and later for fur. Today koalas have a different, but equally important commercial significance, as animal ambassadors for the tourist trade. So important is the koala as a tourist draw card, particularly to the important Japanese market, that moves by state governments to ban "koala cuddling" on the grounds that it is stressful for the animals, were opposed (unsuccessfully) by tourist authorities.

Where populations are healthy wild koalas are easy to observe, but most tourists see koalas in zoos and animal sanctuaries, where petting, if not cuddling, is allowed. The Australian federal government strictly controls exports of live koalas, and this, combined with the problems of satisfying the koala's very specific dietary requirements, mean there are only a handful of zoos outside Australia, in the United States, Japan, Germany, and Taiwan, where koalas are exhibited.

Resources

Books

Australian Koala Foundation. *Proceedings of a Conference on the Status of the Koala in 2000, Incorporating the Ninth National Carers Conference—Noosa, QLD.* Brisbane: Australian Koala Foundation, 2000.

Lyons, K., A. Melzer, F. Carrick, and D. Lamb, eds. *The Research and Management of Non-urban Koala Populations.* Rockhampton: Koala Research Centre of Central Queensland, 2001.

Martin, R. W., and K. A. Handasyde. *The Koala: Natural History, Conservation and Management.* Sydney: University of New South Wales Press, 1999.

Saunders, N. R., and L. Hinds, eds. *Marsupial Biology: Recent Research, New Perspectives.* Sydney: University of New South Wales Press, 1997.

Strahan, R., ed. *The Mammals of Australia* Sydney: New Holland Publishers,1998.

Periodicals

Clark, T., N. Mazur, S. Cork, S. Dovers, and R. Harding. "Koala Conservation Policy Process: Appraisal and Recommendations." *Conservation Biology* 14, no. 3 (2000): 681–690.

Ellis, W. A, P. T. Hale, and F. Carrick. "Breeding Dynamics of Koalas in Open Woodlands." *Wildlife Research* 29 (2002): 19–25.

Martin, R. W. "Managing Overabundance in Koala Populations in South-eastern Australia: Future Options." *Australian Biologist* 10, no. 1 (1997): 57–63.

Moore, B. D., and W. J. Foley. "A Review of Feeding and Diet Selection in Koalas (*Phascolarctos cinereus*)" *Australian Journal of Zoology* 48 (2000): 317–333.

Organizations

Australian Koala Foundation. Level 1, 40 Charlotte Street, Brisbane, Queensland 4001 Australia. Phone: (7) 3229 7233. Fax: (7) 3221 0337. E-mail: akf@save the koala.com Web site: <http://www.savethekoala.com>

Stephen and Ann Toon

Wombats
(*Vombatidae*)

Class Mammalia
Order Diprotodontia
Suborder Vombatiformes
Family Vombatidae

Thumbnail description
Wombats are large burrowing herbivores, stocky with a broad massive head and short powerful limbs; the ears are small and the tail insignificant.

Size
39.4 in (1.0 m); 55–88 lb (25–40 kg)

Number of genera, species
2 genera; 3 species

Habitat
Woodlands

Conservation status
Critically Endangered: 1 species

Distribution
Southeastern Australia

Evolution and systematics

There are only three living species of wombats, but the family was more diverse in the Pleistocene (between about two million years ago [mya] and 10,000 years ago), when it was represented by a total of six genera and nine species. Some of the extinct species were much larger than the living species. *Phascolonus gigas*, for example, had a skull 16 in (40 cm) in length and may have stood about 39.4 in (1 m) high and weighed 441 lb (200 kg).

Whether these giant wombats dug burrows is unknown; they do not seem to have been as well-adapted for burrowing as their living relatives, and may only have dug short burrows for resting. The earliest fossil wombats are of early Miocene age. Wombats arose from the same stock that produced the kangaroos and possums, and their closest living relative is the koala.

Physical characteristics

The three living species of wombats are similar in size, and all have the same stocky body form. The two hairy-nosed wombats (*Lasiorhinus*) differ from the common wombat (*Vombatus ursinus*) in having a hairy covering over the rhinarium. They also have longer pointed ears and finer fur. The hairy-nosed wombats are silver-gray, but the common wombat varies in color from pale gray to rich brown. Males and females are similar in appearance.

The skeletal characters of wombats are well-suited for digging. In particular, the pectoral girdle is heavy and strong and the humerus is broad and massive. This makes the forearms very powerful, and the forepaws are broad and have strong claws.

Distribution

Wombats occur in southeastern Australia, and are reasonably widespread in New South Wales, Victoria, Tasmania, and South Australia. The northern hairy-nosed wombat (*L. krefftii*) is found just to the north of the tropic of Capricorn, and the southern hairy-nosed wombat (*L. latifrons*) has isolated populations in Western Australia.

Habitat

The two species of hairy-nosed wombats live in open woodlands, shrublands, and grasslands in semi-arid habitats, and the southern hairy-nosed wombat extends into arid regions on the Nullarbor Plain. The common wombat lives in forests and woodlands in areas of higher rainfall.

Behavior

Wombats dig by scratching with the forepaws and flinging soil behind them; the piled-up soil is then bulldozed clear of the burrow as the animal backs out of the entrance. Wombat burrows can be huge. They may consist of 98 ft (30 m) or more of tunnel length, and have several entrances as well as side tunnels and resting chambers. Warrens of the southern hairy-nosed wombat are particularly complex, and probably the same warren is used and expanded by many generations of wombats. The tunnels are wide enough to accommodate a lightly built adult human (no reasonable person would ever risk crawling down a wombat burrow, but a 15 year old boy explored many burrows of the common wom-

The common wombat (*Vombatus ursinus*) leaves its burrow at night to feed. (Photo by Mitch Reardon/Photo Researchers, Inc. Reproduced by permission.)

bat in 1960 and wrote up his observations in a now-famous article in his school magazine).

Individuals usually feed alone, but in the southern hairy-nosed wombat many animals may share the same warren. Similarly, in the northern hairy-nosed wombat burrows occur in clusters, and a group of up to 12 wombats makes common use of each cluster of burrows. However, even when two individuals use the same burrow it seems that they occupy different sections of it. There is good evidence indicating that both the female northern hairy-nosed wombat and the female common wombat are more likely to disperse from their home burrow at some stage of their lives, while the males are more philopatric. This is unusual—in most mammals dispersal is male-biased. This suggests that the groups of individuals that occupy burrow clusters in the northern hairy-nosed wombat are composed of related males and unrelated females. It is still not known at what age females disperse in the common wombat, but in the northern hairy-nosed wombat dispersal has been observed by breeding adult females.

A wombat living in the Southern Mount Lofty Ranges in southern Australia. (Photo by J. Burt. Bruce Coleman, Inc. Reproduced by permission.)

Feeding ecology and diet

Wombats are specialized grazers. They have open-rooted teeth that grow throughout life, compensating for tooth wear caused by eating abrasive grasses. The jaws are massive, and deliver powerful, short chewing strokes that reduce their fibrous food to small particles. Gut capacity is large, and the colon is expanded to house cellulose-digesting microorganisms. Food is held in the gut for long periods (70 hours or so) to maximize the breakdown of fiber.

The common or coarse-haired wombat (*Vombatus ursinus*) uses its strong claws to dig elaborate burrows. Here it scratches itself while eating. (Photo by Pavel German. Reproduced by permission.)

A common wombat (*Vombatus ursinus*) wades in a stream in Australia. (Photograph by Norman Owen Tomalin. Bruce Coleman, Inc. Reproduced by permission.)

Wombats feed mainly at night, and rest deep in their burrows during the day. Their burrows provide them with refuge from such predators as dingoes and also with protection from extreme temperatures and dry conditions. Wombats have low basal metabolic rates; this, together with the slow rate of passage of food through the gut and the efficiency with which they digest their food, means that they spend less time feeding than other grazers of their body size and they can afford to spend most of their time in their burrows. Their home ranges are small for a herbivore of their body size, typically less than 49 acres (20 ha).

Reproductive biology

The single young is born after a gestation of about 22 days, and stays in the pouch for six to nine months. It remains dependent on its mother for at least a year after leaving the pouch. Wombats have backward-opening pouches. There is no evidence of pair-bonding and there is presumably competition among males for the opportunity to mate with females, but no details of this are known.

Conservation status

The common wombat and southern hairy-nosed wombat are secure, although the ranges of both species have contracted and fragmented since European settlement. The northern hairy-nosed wombat is extremely rare. It has only been recorded in historic times from three localities, and is

extinct from two of these as of the early twentieth century. Probably, the major cause of its decline was competition for pasture from sheep and cattle. The remaining population is protected within Epping Forest National Park in central Queensland. In 2000 the size of this last population was estimated to be 116 individuals. This species is classed as endangered under Queensland State legislation and Australian Federal legislation.

Significance to humans

Wombats do not feature strongly in Aboriginal mythology. The southern hairy-nosed wombat and common wombat are sometimes regarded as pests of agriculture, because of the damage they cause to crops and fences. None of the species has commercial value. By and large, however, wombats are regarded with deep affection in Australia. They feature in many children's stories, beginning with Ruth Park's classic *Muddle-Headed Wombat* series from the 1960s. There was also a vogue for wombats in Britain in the mid-nineteenth century. The painter Dante Gabriel Rossetti regarded them as "the most beautiful of God's creatures;" when one of his two pet wombats died in 1869 he commemorated it with a touching drawing entitled *Self-portrait of the artist weeping at the wombat's tomb.*

1. Common wombat (*Vombatus ursinus*); 2. Southern hairy-nosed wombat (*Lasiorhinus latifrons*); 3. Northern hairy-nosed wombat (*Lasiorhinus krefftii*). (Illustration by Brian Cressman)

Species accounts

Common wombat
Vombatus ursinus

TAXONOMY
Didelphis ursina (Shaw, 1800), Tasmania, Australia.

OTHER COMMON NAMES
None known.

PHYSICAL CHARACTERISTICS
Head and body length 35–45 in (90–115 cm); tail length about 1 in (2.5 cm); height about 14.2 in (36 cm); weight 48.5–86 lb (22–39 kg). Coarse black or brown to gray coat; bare muzzle, short rounded ears.

DISTRIBUTION
Southeastern Australia including Flinders Island and Tasmania.

HABITAT
Temperate forests and woodlands, heaths, alpine habitats.

BEHAVIOR
Solitary, and mostly nocturnal. Burrows are dispersed, and usually simple. Each animal uses several burrows within its home range.

FEEDING ECOLOGY AND DIET
Feeds mainly on grasses, but also sedges, rushes, and the roots of shrubs and trees.

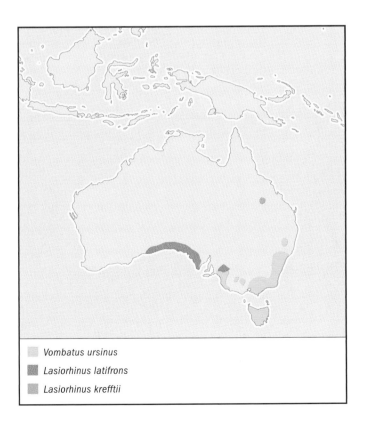

▢	*Vombatus ursinus*
▨	*Lasiorhinus latifrons*
▨	*Lasiorhinus krefftii*

REPRODUCTIVE BIOLOGY
One offspring may be born at any time of the year. Pouch life is about six months, and the young remains at heel for about another year. Sexual maturity is at two years of age. Mating system is not known.

CONSERVATION STATUS
Not threatened. Range has declined by 10–50%, but the species is common throughout large parts of its original range. *Vombatus vombatus* ursinus has gone extinct from all Bass Strait islands except Flinders Island.

SIGNIFICANCE TO HUMANS
In parts of Victoria, common wombats are considered pests because of the damage they do to rabbit-proof fences, and some local control is carried out. ◆

Southern hairy-nosed wombat
Lasiorhinus latifrons

TAXONOMY
Phascolomys latifrons (Owen, 1845), South Australia, Australia.

OTHER COMMON NAMES
None known.

PHYSICAL CHARACTERISTICS
Head and body length 30–37 in (77–94 cm); weight 42–70 lb (19–32 kg); tail and height similar to common wombat. Coat is fine, gray to brown, with lighter patches; hairy muzzle, longer pointed ears.

DISTRIBUTION
Central southern Australia.

HABITAT
Semi-arid and arid woodlands, grasslands, and shrub steppes.

BEHAVIOR
Solitary while feeding, but in many areas warrens are large and complex and used by five to ten individuals. Warrens may be connected by well-used trails, which are marked at intervals by urine splashed and dung piles. Usually nocturnal, but animals may often be seen basking outside their burrows on sunny days in winter.

FEEDING ECOLOGY AND DIET
Feeds on grasses, but also eats forbs and foliage of woody shrubs during drought.

REPRODUCTIVE BIOLOGY
A single young is born in spring or early summer and remains in the pouch for six to nine months. Weaning occurs at approximately one year, and sexual maturity at three years. Mating system is not known.

CONSERVATION STATUS
Not threatened. Range has declined by 10–50%, but the species remains common through much of its original range.

SIGNIFICANCE TO HUMANS
In some areas it damages grain crops and fences and is controlled as a pest. ◆

Northern hairy-nosed wombat
Lasiorhinus krefftii

TAXONOMY
Phascolomys krefftii (Owen, 1873), Breccia Cavern, Wellington Caves, Australia.

OTHER COMMON NAMES
French: Wombat à nez poilu de Queensland; Spanish: Oso marsupial del Río Moonie.

PHYSICAL CHARACTERISTICS
Head and body length of males 40 in (102 cm), females 42 in (107 cm); height about 16 in (40 cm); male weight 66 lb (30 kg), female weight 72 lb (32.5 kg). Coat is silky and silver gray, with dark rings around the eyes.

DISTRIBUTION
There is only one surviving population, in Epping Forest National Park near Clermont in central Queensland. Populations near St George in southern Queensland and Deniliquin in southern New South Wales went extinct early in the twentieth century.

HABITAT
Semi-arid woodland and grassland.

BEHAVIOR
Burrows are distributed in loose clusters, and up to 12 wombats make common use of the burrows in each cluster. However, individuals usually feed and rest alone, except for mothers and young. Piles of dung and urine splashes are placed outside burrow entrances and along regularly used paths that connect different burrows within a group.

FEEDING ECOLOGY AND DIET
Feeds on grasses, plus some sedges and forbs.

REPRODUCTIVE BIOLOGY
Mating associations are transient; the mating system is not known. One young born in spring or summer. Pouch life is about 10 months, weaning age unknown. On average, females breed twice every three years.

CONSERVATION STATUS
Critically Endangered; only about 116 individuals remain. Numbers have evidently increased since Epping Forest National Park was declared and cattle were excluded in 1980, when there may have been only 30 individuals in the population. Current threats include occasional predation by dingoes and (possibly) genetic decline due to isolation and inbreeding.

SIGNIFICANCE TO HUMANS
The northern hairy-nosed wombat is one of Australia's rarest species, and its rarest large mammal. ◆

Resources

Books

Long, J., M. Archer, T. Flannery, and S. Hand. *Prehistoric Mammals of Australia and New Guinea.* Sydney: University of New South Wales Press, 2002.

Wells, R. T., and P. A. Pridmore. *Wombats.* Sydney: Surrey Beatty & Sons, 1998.

Woodford, J. *The Secret Life of Wombats.* Melbourne: Text Publishing, 2001.

Periodicals

Banks, S. C., L. F. Skerratt, and A. C. Taylor. "Female dispersal and relatedness structure in common wombats (*Vombatus ursinus*)." *Journal of Zoology* 256 (2002): 389–399.

Christopher Johnson, PhD

Possums and cuscuses
(Phalangeridae)

Class Mammalia
Order Diprotodontia
Suborder Phalangerida
Family Phalangeridae

Thumbnail description
Small- to medium-sized marsupials with clawed, grasping hands and feet with five digits, a long, prehensile tail with a naked underside, and a slightly elongate muzzle

Size
Total length 23.6–47.2 in (60–120 cm); weight 2–22 lb (1–10 kg)

Number of genera, species
6 genera; 26 species

Habitat
Forest and woodlands

Conservation status
Endangered: 2 species; Vulnerable: 2 species; Lower Risk/Near Threatened: 1 species; Data Deficient: 5 species

Distribution
Australasia east of Wallace's Line, including Sulawesi, Timor, the Moluccas, New Guinea, and many closely adjacent islands, as well as Australia and Tasmania; introduced to New Zealand

Evolution and systematics

The phalangers (family Phalangeridae) are classified into six distinctive genera. The most diverse genus, *Phalanger*, comprises a group of medium-sized, soft-furred species that generally possess a dorsal stripe. *Phalanger* is primarily a New Guinean genus, but one species extends to northern Australia and several others occur throughout the Moluccan islands of Indonesia. Spotted cuscuses (genus *Spilocuscus*) are the most beautiful of the phalangers, with striking coats colored with combinations of red, white, black, brown, and yellow; they occur throughout New Guinea and on several adjacent islands, as well as in far northeastern Australia. The small-bodied species of *Strigocuscus* and the very large bear cuscuses (*Ailurops*) both occur only on Sulawesi and closely adjacent islands. Species of *Phalanger*, *Spilocuscus*, *Strigocuscus*, and *Ailurops* are all called cuscuses, whereas the remaining two genera within the family, *Trichosurus* and *Wyulda*, are called possums. Both of these latter genera are endemic to Australia. The five species of brush-tailed possums (*Trichosurus*) are (or were, until quite recently) distributed across most the Australian continent, while the rare scaly-tailed possum (*Wyulda squamicaudata*) occurs only in a small area of far northwestern Australia and is classified in a genus unique to itself. The phalanger fossil record extends back roughly 20 million years, to the middle Miocene, when three phalangerid genera (*Trichosurus*, *Wyulda*, and possibly *Strigocuscus*) occurred in northern Australia. The fossil history of phalangerids outside of mainland Australia is very poorly known. Within the family, species of *Trichosurus* and *Wyulda* are clearly closely related to one another, as are those of *Phalanger* and *Spilocuscus*. However, the exact relationships of these two groups with one another and with *Strigocuscus* and *Ailurops* remain obscure. The phylogenetic position of the genus, *Ailurops*, is particularly controversial. Anatomical evidence strongly suggests that the bear cuscuses are the most primitive phalangers—the first offshoot on the phalanger family tree. However, evidence from molecular studies does not support this hypothesis thus far. Future studies based on analysis of DNA will hopefully offer greater resolution of relationships within the family.

Physical characteristics

Phalangers are small- to medium-sized marsupials. The smallest phalanger (*Strigocuscus celebensis*) weighs 2 lb (1 kg)

A Sulawesi bear cuscus (*Ailurops ursinus*) feeds in guava tree in northern Sulawesi, Indonesia. (Photo by Janis Burger. Bruce Coleman, Inc. Reproduced by permission.)

A brushtail possum (*Trichosurus vulpecula*) mother and baby in Queensland, Australia. (Photo by Jen and Des Bartlett. Bruce Coleman, Inc. Reproduced by permission.)

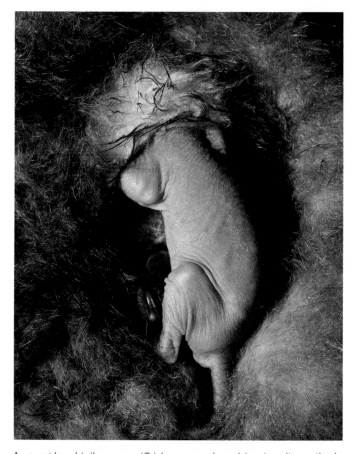

A young brushtail possum (*Trichosurus vulpecula*) enters its mother's pouch. (Photo by Joyce Photographics/Photo Researchers, Inc. Reproduced by permission.)

or less, and the largest species (*Ailurops ursinus*) weighs up to 22 lb (10 kg); but most species of the family weigh between 3 and 6 lb (1.5–3 kg). The pelage is generally soft and dense. Color varies widely, especially in the spotted cuscuses and the beautifully mottled Woodlark cuscus (*Phalanger lullulae*), but most species are brown or gray, often with a dark dorsal stripe running down the center of the head and back. In all species except those of the genus, *Trichosurus* (which has a superficially dog- or fox-like appearance), the ears are reduced and often wholly or partially hidden in the soft fur. The feet have five digits, all of which support a large claw, except the hallux (big toe). The hallux is opposable to the remaining digits of the hind foot, and the first two digits of the forefoot oppose the remaining three, allowing the feet to grasp branches firmly while climbing. The second and third digits of the hindfoot are reduced and partially united by skin to form a single functional digit with two claws (a condition called syndactyly) that is used as a hair comb. The tail is long and prehensile, with the distal part generally naked (only on the underside in the bushy-tailed species of *Trichosurus*) and (especially in

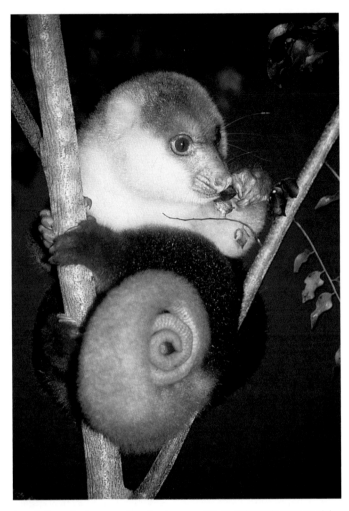

An adult spotted cuscus (*Phalanger maculatus*) nibbling leaves at night. Cuscuses are hunted for food and skins. (Photo by Zoltan Takacs. Bruce Coleman, Inc. Reproduced by permission.)

A mountain brushtail possum (*Trichosurus caninus*) foraging in Queensland, Australia. (Photo by Jan Taylor. Bruce Coleman, Inc. Reproduced by permission.)

older males) covered with small tubercles that give it a strongly rugose surface.

Distribution

Phalangers occur naturally throughout New Guinea, Australia, and Tasmania, and on a number of Indonesian islands (Sulawesi and the Moluccas). Cuscuses were prehistorically introduced from New Guinea to many nearby islands (including Timor and the Bismarck and Solomon Archipelagos), and the common brush-tailed possum was introduced to New Zealand from Tasmania and mainland Australia in the mid-nineteenth century.

Habitat

All phalangers are well adapted for climbing and are highly reliant on trees, which they use for shelter, foraging, or both. Most species are inhabitants of rainforest, although the scaly-tailed possum occurs in rocky areas in broken sandstone coun-

A possum mother carries her young on her back. (Illustration by Jarrod Erdody)

A ground cuscus (*Phalanger gymnotis*) eating seeds in the Aru Islands near New Guinea. (Photo by Rod Williams. Bruce Coleman, Inc. Reproduced by permission.)

try in northwestern Australia, and the remarkably versatile common brushtail possum (*Trichosurus vulpecula*) occupies a varied range of habitats and is common even in urban areas. Phalangers reach their greatest diversity in mid-montane rainforest in central New Guinea, where up to five species of the family coexist.

Behavior

Most phalangers are nocturnal, but the two largest species (the Sulawesi bear cuscus, *Ailurops ursinus*, and the black-spotted cuscus, *Spilocuscus rufoniger*) are often active by day. Most species are arboreal, living in tree hollows and feeding in the forest canopy. However, one species, the ground cuscus (*Phalanger gymnotis*) of New Guinea, exhibits a more terrestrial lifestyle; it lives in underground burrows and travels along the rainforest floor. However, even the ground cuscus is a very good climber, ascending into trees to feed on fruit at night. The social behavior of most phalangers has received little study to date. Several species seem to form male-female pairs (such as the small Sulawesi cuscus, *Strigocuscus celeben-*

sis, and the mountain brush-tailed possum, *Trichosurus caninus*), but the majority of species are solitary. Males of most species are aggressive toward one another and cannot be kept together in captivity.

Feeding ecology and diet

Though phalangers are almost exclusively herbivorous, the kinds of plants favored can vary greatly from species to species. Some phalangers, including the Sulawesi bear cuscus (*Ailurops ursinus*), are largely folivorous, while others such as the ground cuscus and the small Sulawesi cuscus are primarily frugivorous. Both leaves and fruit probably form high proportions of the diet of most species, especially in tropical forests; other food sources such as flowers, bark, pollen, and fungi may also be utilized in small quantities. The common brush-tailed possum is the most ecologically versatile phalanger and probably has the most generalized diet; its diet may change drastically in different local habitats, variably comprising large proportions of leaves (including some defended by highly toxic compounds), grasses and herbs, ferns and mosses, and fruits.

Reproductive biology

Although many species are monogamous, most are probably promiscuous. Female phalangers have a forward-oriented pouch with two or four teats. Adult females generally produce one or two litters per year. Up to three or four young may be born, but only one is usually reared, although in the northern common cuscus (*Phalanger orientalis*), twins are commonly raised. Gestation is 20 days or less; like other marsupials, neonates are born very small and unfurred. Infants are weaned and exit the pouch between five and eight months, after which they are carried on their mother's back.

Conservation status

Several phalanger species are in danger of extinction. Four rare species with small geographic ranges are particularly threatened: the scaly-tailed possum of northwestern Australia, the Telefomin cuscus (*Phalanger matanim*) of central New Guinea, the black-spotted cuscus of northern New Guinea, and the yellow bear cuscus (*Ailurops melanotis*) of the Sangihe and Talaud Islands in eastern Indonesia. Other species that are locally common but restricted to single islands, such as the Waigeo cuscus (*Spilocuscus papuensis*), the Gebe cuscus (*Phalanger alexandrae*), and the Woodlark cuscus, are also worthy of conservation attention. However, many other phalangers, including the northern common cuscus (*Phalanger orientalis*) and the common brushtail possum, are both geographically widespread and locally common and are presently under no threat.

Significance to humans

Cuscuses are an important source of meat for people throughout the New Guinea region, and they are widely hunted. However, in some areas of Indonesia such as the Sula Islands in the western Moluccas, cuscuses are not eaten in accordance with religious traditions. Common brushtail possums have unfortunately become a significant environmental and agricultural pest in New Zealand, where they have been introduced and are now widespread and common.

1. Sulawesi bear cuscus (*Ailurops ursinus*); 2. Common spotted cuscus (*Spilocuscus maculatus*); 3. Scaly-tailed possum (*Wyulda squamicaudata*); 4. Black-spotted cuscus (*Spilocuscus rufoniger*); 5. Small Sulawesi cuscus (*Strigocuscus celebensis*); 6. Common brushtail possum (*Trichosurus vulpecula*); 7. Ground cuscus (*Phalanger gymnotis*). (Illustration by Bruce Worden)

Species accounts

Sulawesi bear cuscus
Ailurops ursinus

TAXONOMY
Ailurops ursinus (Temminck, 1824), Sulawesi, Indonesia.

OTHER COMMON NAMES
German: Bärenkuskus.

PHYSICAL CHARACTERISTICS
A large and powerful phalanger (up to 22 lb [10 kg]), with long limbs and black fur often tipped with yellow.

DISTRIBUTION
A. u. ursinus: lowlands of Sulawesi, Peleng, and some adjacent smaller islands; and *A. u. togianus*: Togian Islands.

HABITAT
Lowland and mid-elevation rainforest.

BEHAVIOR
Generally occurs in groups of two to four individuals. It is arboreal, living and feeding in the forest canopy, and may be active at any time, day or night.

FEEDING ECOLOGY AND DIET
Largely folivorous, but flowers and fruits make up a small part of the diet.

REPRODUCTIVE BIOLOGY
Very little is known. Females give birth once or twice a year, and weaning has been reported to occur at eight months.

CONSERVATION STATUS
Possibly threatened; classified as Data Deficient.

SIGNIFICANCE TO HUMANS
As a very large phalanger, this species is a target of hunting, and is sometimes encountered as food in markets and restaurants in Sulawesi. ◆

Small Sulawesi cuscus
Strigocuscus celebensis

TAXONOMY
Strigocuscus celebensis (Gray, 1858), Sulawesi, Indonesia.

OTHER COMMON NAMES
German: Celebeskuskus.

PHYSICAL CHARACTERISTICS
This is the smallest phalanger, weighing 2 lb (1 kg) or less. The overall coloration is pale buff, a dorsal stripe is lacking, and the naked part of the tail is very sparsely haired.

DISTRIBUTION
S. c. celebensis: southern and central Sulawesi; *S. c. feileri*: north Sulawesi; and *S. c. sangirensis*: Sangihe Islands north of Sulawesi.

HABITAT
Rainforest.

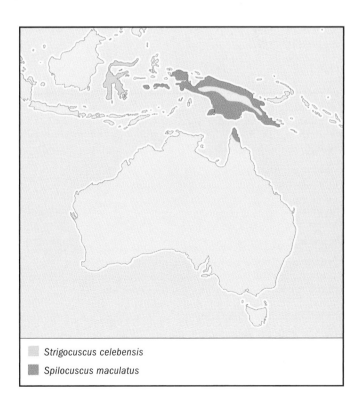

■ *Ailurops ursinus*
■ *Spilocuscus rufoniger*

■ *Strigocuscus celebensis*
■ *Spilocuscus maculatus*

BEHAVIOR
Arboreal and nocturnal, and apparently occurs in pairs.

FEEDING ECOLOGY AND DIET
Thought to be primarily frugivorous.

REPRODUCTIVE BIOLOGY
Little is known; probably monogamous.

CONSERVATION STATUS
Uncertain; classified as Data Deficient.

SIGNIFICANCE TO HUMANS
None known. ◆

Common spotted cuscus
Spilocuscus maculatus

TAXONOMY
Spilocuscus maculatus (Desmarest, 1818), Vogelkop, Irian Jaya, Indonesia.

OTHER COMMON NAMES
German: Tüpfelkuskus.

PHYSICAL CHARACTERISTICS
A relatively large phalanger (weighing up to 13.2 lb [6 kg]) with large round eyes and very small ears. The coat is colorful but very variable, and differs characteristically both between the sexes and among the four subspecies. Males may be pure white or spotted with red-orange, black, or gray, while females are often unspotted, with an unbroken black or gray "saddle" on the back.

DISTRIBUTION
S. m. maculatus: northern New Guinea; *S. m. chrysorrhous*: southern New Guinea and the central Moluccas; *S. m. goldiei*: southeastern New Guinea; and *S. m. nudicaudatus*: tropical northern Australia.

HABITAT
Lowland rainforest, from sea level to 3,900 ft (1,200 m).

BEHAVIOR
Nocturnal and arboreal, rarely descending to the ground. It has an unusually low metabolic rate, and its movement through the trees is often relatively slow.

FEEDING ECOLOGY AND DIET
Predominantly leaves and fruit.

REPRODUCTIVE BIOLOGY
Females have four mammae. Two to four young have been reported in a litter, but a single young is most common. Infants exit the pouch six to seven months after birth. It is not known whether breeding is seasonal or takes place year-round. The estrous cycle lasts four weeks; at the peak of the cycle, females are highly vocal. Mating system is not known.

CONSERVATION STATUS
Not threatened.

SIGNIFICANCE TO HUMANS
Widely hunted and often transported and sold as a pet or as food in local markets in New Guinea. ◆

Black-spotted cuscus
Spilocuscus rufoniger

TAXONOMY
Spilocuscus rufoniger (Zimara, 1937), Morobe Province, Papua New Guinea.

OTHER COMMON NAMES
German: Schwarzgeflecktkuskus.

PHYSICAL CHARACTERISTICS
A very large species, brightly colored; males with a jet black "saddle" on the lower back, females with black and white spotting on the lower back.

DISTRIBUTION
Northern New Guinea.

HABITAT
Primary (undisturbed) lowland rainforest, from sea level to 3,900 ft (1,200 m).

BEHAVIOR
Arboreal; apparently largely nocturnal, but occasionally active during the day.

FEEDING ECOLOGY AND DIET
Nothing is known.

REPRODUCTIVE BIOLOGY
Nothing is known.

CONSERVATION STATUS
This large-bodied species has undergone a widespread decline across its range in recent decades, and is listed as Endangered.

SIGNIFICANCE TO HUMANS
Hunted for food and for its beautifully colored skin. Hunting this species remains a male rite of passage in some parts of New Guinea. ◆

Ground cuscus
Phalanger gymnotis

TAXONOMY
Phalanger gymnotis (Peters and Doria, 1875), Aru Islands, Indonesia.

OTHER COMMON NAMES
German: Gleichfarbkuskus, Bodenkuskus.

PHYSICAL CHARACTERISTICS
Color light to dark gray, with a distinct dorsal stripe, prominent ears, and a coarsely tuberculated tail.

DISTRIBUTION
P. g. gymnotis: Aru Islands; and *P. g. leucippus*: mainland New Guinea.

HABITAT
Occurs in rainforest from sea level up to 8,900 ft (2,700 m), but is most common at intermediate elevations.

BEHAVIOR
This species is unique among cuscuses in that it occupies burrows in holes in the ground, often under trees, along streams,

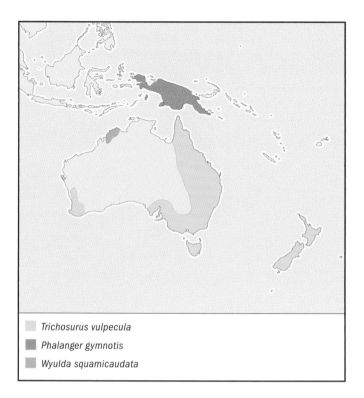

Trichosurus vulpecula

Phalanger gymnotis

Wyulda squamicaudata

reddish, brown, or white. The ears are large, with a narrowly rounded tip.

DISTRIBUTION
Occurs throughout eastern and southwestern Australia and in Tasmania and, until recently, it occupied much of central Australia. It is now common in New Zealand, where it was introduced about 150 years ago.

HABITAT
Usually forests and woodlands, but this species is extremely versatile and may occupy many different habitat types, including semiaraid areas devoid of trees and suburban and urban areas.

BEHAVIOR
Generally solitary, nocturnal, and arboreal. It most commonly nests in tree hollows, but may also nest in roofs or in burrows in the ground. In areas of low density, adults may aggressively defend discrete territories, but where population density is high, home ranges of individuals may overlap widely.

FEEDING ECOLOGY AND DIET
A wide variety of plants is eaten, and occasionally small animals and insects are taken.

REPRODUCTIVE BIOLOGY
Breeding occurs year-round, although births peak in fall and spring. Females usually produce one or, less commonly, two litters per year. After a gestation period of about 18 days, one young is usually born, which leaves the pouch after six to seven months. Probably promiscuous.

CONSERVATION STATUS
As it is very common in many areas, the brush-tailed possum is considered to be not threatened, although it has undergone a wide decline in central Australia and may be declining in southwest Australia.

SIGNIFICANCE TO HUMANS
Because it is common in developed areas such as city parks and suburban backyards, this possum has a closer interaction with people than any other Australian mammal. Pelts of common brush-tailed possums have been widely trapped and sold for the fur trade in Australia. In many areas, it is considered an agricultural pest and a potential vector of disease, and it is a pest in both crops and native forest in New Zealand, where it is non-native. ◆

or in caves. It is generally nocturnal, and forages both on the ground and in trees. New Guinea hunters claim that it suns itself outside its burrow in the morning, and that females may carry fruit to the burrow in their pouches.

FEEDING ECOLOGY AND DIET
Primarily frugivorous, but also eats leaves and, on occasion, small vertebrates and insects.

REPRODUCTIVE BIOLOGY
Both sexes are generally solitary, and breeding is continuous year-round. A single young is most common, and infants leave the pouch permanently five to seven months after birth. Mating system is not known.

CONSERVATION STATUS
Common in many parts of New Guinea and probably not threatened, but classified as Data Deficient.

SIGNIFICANCE TO HUMANS
Widely hunted with dogs throughout New Guinea, and is an important figure in tribal folklore in some areas. ◆

Scaly-tailed possum
Wyulda squamicaudata

TAXONOMY
Wyulda squamicaudata Alexander, 1918, Wyndham, Western Australia, Australia.

OTHER COMMON NAMES
German: Schuppenschwanzopossum, Schuppenschwanzkusu.

PHYSICAL CHARACTERISTICS
Medium-sized; weight 3–6.5 lb (1.4–3 kg); overall color dull gray, with a flattened head, reduced claws, and a wholly naked tail with a white tail-tip and coarse scales.

DISTRIBUTION
Known only from the Kimberley region of northwestern Australia.

Common brushtail possum
Trichosurus vulpecula

TAXONOMY
Trichosurus vulpecula (Kerr, 1792), Sydney, Australia.

OTHER COMMON NAMES
French: Phalanger-renarrd; German: Gewöhnlicher Fuchskusu.

PHYSICAL CHARACTERISTICS
Has a woolly coat and a thick, well-furred tail. Size and coloration are highly variable; individuals may be gray, black,

HABITAT
Occurs in rocky areas with trees in broken sandstone country.

BEHAVIOR
More terrestrial than most phalangers, and shows a number of specializations for moving both on rocky ground and in trees. One captive individual was reported to hoard small caches of food such as nuts, and to make chattering vocalizations.

FEEDING ECOLOGY AND DIET
Nothing is known.

REPRODUCTIVE BIOLOGY
Females breed once a year, and generally raise only a single young, which is born between March and August. Weaning occurs at about eight months. Mating system is not known.

CONSERVATION STATUS
Best considered endangered, although it is currently classified by the IUCN as Lower Risk/Near Threatened.

SIGNIFICANCE TO HUMANS
None known. ◆

Common name / Scientific name/ Other common names	Physical characteristics	Habitat and behavior	Distribution	Diet	Conservation status
Silky cuscus *Phalanger sericeus*	Compact body, small ears, and long, thick fur. Dark brown to black in coloration. Light yellowish ring around eye area.	Only occurs above 4,920 ft (1,500 m) in the central mountains of New Guinea. Found up to the tree line (around 12,800 ft [3,900 m]). Solitary, nocturnal, terrestrial.	New Guinea.	Fruits, leaves, and insects.	Not threatened
Woodlark Island cuscus *Phalanger lullulae*	Pelage is short, woolly, marbled brown, ochre, and white dorsal areas. Ventral fur is white, speckled with irregular dark spots. Black facial skin, pale ear flashes at times. Long tail. Head and body length 25–29 in (63.8–71.7 cm), weight 3.3–4.5 lb (1,500–2,050 g).	Primary and secondary lowland dry forest. Breeding season is an extended period. Solitary, completely arboreal, can be very aggressive.	New Guinea.	Two unknown types of vine.	Not threatened
Telefomin cuscus *Phalanger matanim*	Thick and woolly fur, coloration ranges from white, red, or buff to various shades of brown to light gray. Strong build, piercing eyes, naked tail. Head and body length 12.7–24 in (32.5–60 cm), tail length 9–24 in (24–61 cm).	Tropical forests and thick shrub. Terrestrial, nocturnal, and completely arboreal.	Mountains of western Papua New Guinea.	Fruits, leaves, and insects.	Endangered
Common cuscus *Phalanger orientalis* English: Northern common cuscus	Stature is heavy, powerfully built. Thick, woolly, white to medium or dark gray fur in males. Reddish brown to brownish gray in females. Tail is prehensile and naked at end. Large eyes, long snout. Head and body length 13–22 in (35–55 cm), average weight 4.6 lb (2.1 kg).	Tropical rainforests and thick scrub in the southwest Pacific. Nocturnal, solitary, females care for young.	New Guinea.	Leaves, tree seeds, fruit, buds, and flowers.	Not threatened
Moluccan cuscus *Phalanger ornatus*	Coloration is white to medium or dark gray fur in males. Reddish brown to brownish gray in females. Strong, powerful build. Tail is prehensile and naked at end. Large eyes, long snout. Large eyes, long snout. Head and body length 13–22 in (35–55 cm), average weight 4.6 lb (2.1 kg).	Tropical rainforests and thick scrub in the southwest Pacific. Nocturnal, solitary, females care for young.	Halmahera, Ternate, Tidore, Bacan, and Morotai Islands (Indonesia).	Leaves, tree seeds, fruit, buds, and flowers.	Not threatened
Northern brushtail possum *Trichosurus arnhemensis*	Coloration of coat is gray, can be reddish or brown. Tail is prehensile and covered with hair. Head and body length 13–22 in (35–55 cm), tail length 9.8–15.7 in (25–40 cm), weight 3.3–10 lb (1,500–4,500 g).	Variety of habitats, including residential areas, forests, and areas without trees that offer caves and burrows for shelter. Year-round breeding season, nocturnal, solitary.	Northern Territory of Australia, as well as in the extreme part of Western Australia.	Leaves, buds, and fruits.	Not threatened
Mountain brushtail possum *Trichosurus caninus*	Stocky, coloration is gray to dark gray. Small, rounded ears. Head and body length 29–36 in (74–92 cm), weight 5.5–10 lb (2.5–4.5 kg).	Variety of forest types in its range, although it prefers the wetter forests. Nocturnal, solitary, territorial, females care for young.	Australia, in forests of southeastern Queensland, eastern New South Wales, and eastern Victoria.	Herbivore and fruigivore, but will eat insects.	Not threatened

Resources

Books

Flannery, Timothy F. *Possums of the World: A Monograph of the Phalangeroidea*. Sydney: GEO Productions, 1994.

Long, John, Michael Archer, Timothy Flannery, and Suzanne Hand. *Prehistoric Mammals of Australia and New Guinea: One Hundred Million Years of Evolution*. Baltimore: Johns Hopkins University Press, 2002.

Kristofer M. Helgen

Musky rat-kangaroos

(Hypsiprymnodontidae)

Class Mammalia

Order Diprotodontia

Family Hypsiprymnodontidae

Thumbnail description
The one living species is a small, quadrupedally bounding animal with a sparsely haired tail, slender head, and dark chocolate-brown fur; the hindfoot has four toes; the third premolar is long and serrated ("sectorial")

Size
Head and body length, 5.9–10.6 in (15–27 cm); tail length 4.7–6.3 in (12–16 cm); weight, 12.6–24 oz (360–680 g)

Number of genera, species
1 genus; 1 species

Habitat
Tropical rainforest

Conservation status
Not threatened

Distribution
Tropical Queensland (Australia)

Evolution and systematics

The Hypsiprimnodontidae are considered to be the most basic group of Macropodoidea, on account of their simple stomach, remaining first toe, apparent inability to hop "kangaroo-like," regular twin births, etc. In skeletal morphology they share a number of characteristics with Potoroidae, i.e., large masseteric foramina, a blade-like structure of the third premolars, the presence of upper canines, and a confluence of the masseteric and inferior dental canals on their mandibles. The Hypsiprymnodontidae are separated from Potoroidae by the fact that they possess two (instead of one) lower incisors, and the fact that their squamosal bones are in broad contact with the frontals (this is similar to Macropodidae). On serological ground, all rat-kangaroos are closer to macropodids than to phalangerids, whereas their bunodont (rounded-cusp) molars resemble phalangerids more closely.

Hypsiprymnodontidae have been known in the fossil record since the Miocene. One species, *Hyspiprymnodon bartholomai*, from the early Miocene (approximately 20 million years ago), is very similar in shape and size to the living

Musky rat-kangaroo (*Hypsiprymnodon moschatus*). (Illustration by John Megahan)

A musky rat-kangaroo (*Hypsiprymnodon moschatus*) forages in the rainforest of Kuranda State Forest, Queensland, Australia. (Photo by Michael Foaden. Bruce Coleman, Inc. Reproduced by permission.)

species of musky rat-kangaroo, thus making the extant species a true living fossil. However, the most dramatic members of the family are certainly the extinct Propleopinae, the so-called giant rat-kangaroos. In the Pleistocene, some of them reached the size of a true kangaroo (up to at least 132 lb [60 kg]), and they were probably at least partially carnivorous. The size, wear, and position of their molars, their long, protruding, dagger-like incisors, and the comparatively small number of their remains in fossil deposits, all point to a position near the top of the food-chain. The smallest of these species, *Ekaltadeta sina*, reached 33–44 lb (15–20 kg), and is one of the oldest kangaroo species of all. The largest and youngest species, *Propleopus oscillans*, is a Pleistocene species whose last members probably became extinct about 10,000 years ago. Some researchers speculate that propleopines may have died out as recently as 6,000 years ago (which would mean that they survived until the dingo appeared).

The taxonomy for this species is *Hypsiprymnodon moschatus* Ramsay, 1876, Queensland, Australia.

Physical characteristics

The one extant species, *Hypsiprymnodon moschatus*, is quite easy to recognize. It has a slender head, with an almost "delicate" skull, a hairless rhinarium, and no concavity on the dorsal side. Its fur is soft and rufous-brown with gray underfur on the head. Forelimbs and hindlimbs are more equal in lengths than in potoroids (or macropodids), and the hindfoot has a claw-less, but prominent first digit, which can be opposed. They have striated pads and ceratinous scales on the palms of their hands and the soles of their feet. The animal is small, and males and females are of similar sizes. Females (18 oz/511 g) weigh slightly less than males (18.7 oz/529 g).

Distribution

The musky rat-kangaroo is found only in one small area of about 199 mi by 40 mi (320 km by 65 km), in northern Queensland. As far as is known, this range has not been any larger in historical time. All the extinct species also were rainforest-species. However, as rainforest covered more of Australia in prehistoric times, their range also was larger.

Habitat

Musky rat-kangaroos are strictly confined to tall rainforests at all altitudes. They are most regularly observed in damp areas, around lakes, or near creeks and rivers. Though basically terrestrial, they are adept climbers, and can often be found on fallen branches, trees, or logs. The striated pads and ceratinous scales on their palms and soles help them climb. In a breeding colony in Pallarenda, near Townsville, animals were observed to climb up to heights of more than 6.6 ft (2 m). Musky rat-kangaroos construct nests, similar to potoroid nests, by collecting plant material, carrying it in their tails, and forming loose piles of leaves in clumps of vines or near root-buttresses. When disturbed, the animals usually retreat into more dense, shrubby vegetation, such as near the lakeshores, returning into the tall forest after the disturbance ceases.

Behavior

Musky rat-kangaroos seem to be basically solitary in the wild, though in captivity they can be kept in pairs, or one male can be kept with two or more females. However, it has been impossible to keep two or more males together, even in enclosures of 215 ft² (200 m²). In the wild, aggregations of up to three animals are sometimes seen feeding on fallen fruit. There is no evidence of territoriality. Descriptions of social behavior are almost nonexistent, except for some descriptions of body postures (standing in front of female, pawing her head, or lateral sinuous movements of the tails in male

The musky rat-kangaroo (*Hypsiprymnodon moschatus*) is a solitary animal. (Photo by Dave Watts/Naturepl.com. Reproduced by permission.)

A musky rat-kangaroo (*Hypsiprymnodon moschatus*) in Kuranda Rainforest Park, Queensland, Australia. (Photo by © Michael Fogden/Animals Animals. Reproduced by permission.)

courtship). Musky rat-kangaroo are exclusively diurnal, spending the night and the noontime in their nests.

Feeding ecology and diet

Food is located presumably by scent and retrieved from leaf litter with the forepaws. Fruits and nuts of rainforest trees (e.g., candlenut, *Aleuntes moluccana*; king palm, *Archontophoenix alexandrae*; and kelat or watergum, *Eugenia kuranda*) are eaten, as well as insects, earthworms, and other invertebrates. Shells of small nuts or exoskeletons of insects are crushed by pushing them deep into the mouth between the sectorial blades of the third premolars. The flesh of fruit is scraped off with the lower incisors. Due to the simple, sacciform forestomach, and the rather small hindgut and caecum, musky rat-kangaroos are most probably unable to digest allozymatically (i.e., by means of bacteria and other symbiontic organisms), unlike all other macropodoids. This means that they need to forage on low-fiber, easy-to-digest foods.

Reproductive biology

Courtship and mating seems to follow typical rat-kangaroo patterns, which are polygynous. The male rapidly chases the female, or blocks her way by standing in front of her, slashing his tail laterally in undulating movements and pawing her head. If the female is not yet receptive, she throws herself onto her side and vigorously kicks at the male with her hind-limbs. When she is more receptive, both animals stand erect and touch each other's face and neck with their forepaws. These courtship phases may last a few days and normally occur between February and July. Two young, occasionally triplets, are born, and twins are normally raised in their mother's pouch. After about 21 weeks the young leave the pouch, and mostly stay in the nest for several weeks. Only then do they follow their mothers in typical young-at-heel fashion, similar to other kangaroos. Females become sexually mature at just over one year of age.

Conservation status

Though not listed by the IUCN, the 1996 Action Plan from the IUCN's Australian Marsupial and Monotreme Specialist Group suggests that *H. moschatus* be listed as Lower Risk/Least Concern due to its restricted area of distribution, and the fact that its only habitat—closed, tall rainforest—is rapidly diminishing in Australia.

Significance to humans

None known.

Resources

Books

Archer, M., S. J. Hand, and H. Godthelp. *Australia's Lost World: Prehistoric Animals of Riversleigh.* Bloomington, IN: Indiana University Press, 2001.

Johnson, P. M. "Musky Rat-kangaroo." In *The Australian Museum Complete Book of Australian Mammals,* edited by R. Strahan, 179–180. Sydney, Angus & Robertson, 1995.

Kennedy, M., ed. *Australasian Marsupials and Monotremes—An Action Plan for Their Conservation.* Gland, Switzerland: IUCN, 1992.

Seebeck, J. H., and R. Rose. "Potoroidae." In *Fauna of Australia. IB. Mammalia,* edited by D. W. Walton and B. J. Richardson, 716–739. Canberra, Australian Government Publication Service, 1989.

Vickers-Rick, P., J. M. Monaglan, R. F. Baird, T. H. Rick, eds. *Vertebrate Paleontology of Australasia.* Melbourne: Pioneer Design, 1991.

Periodicals

Johnson, P. M., and R. Strahan. "A Further Description of the Musky Rat-kangaroo With Notes on Its Biology." *Australian Zoologist* 21 (1982): 27–46.

Schürer, U. "Das Moschusrattenkänguruh, *Hypsiprymnodon moschatus*—Beobachtungen im Freiland und im Gehege." *Der Zoologische Garten N. F.* 55 (1985): 257–267.

Wroe, S. "Killer Kangaroos and Other Murderous Marsupials." *Scientific American* 280 (1999): 68–74.

Udo Gansloßer, PhD

▲
Rat-kangaroos
(Potoroidae)

Class Mammalia

Order Diprotodontia

Family Potoroidae

Thumbnail description
Small- to medium-sized marsupials that generally hop like kangaroos; they have an elongated tail; females have a pouch with four teats and usually have one young

Size
6–12 in (15–30 cm); 1.3–8 lb (500–3,000 g)

Number of genera, species
4 genera; 8 species

Habitat
Forest and open woodland

Conservation status
Extinct: 2 species; Critically Endangered: 1 species; Endangered: 2 species; Vulnerable: 1 species; Lower Risk/Conservation Dependent: 1 species; Lower Risk/Near Threatened: 1 species

Distribution
Mainly coastal Australia; absent from northern coast

Evolution and systematics

Rat-kangaroos have been traditionally classified as a subfamily of their nearest relatives, the kangaroo and wallaby family (Macropodidae). However, most taxonomists now accept the separation of two groups into families with the Potoroidae being separated from the Macropodidae on the basis of their urogenital anatomy. The two groups are often associated into a single superfamily, the Macropodoidea. These families have been separate for at least 50 million years. Within the Potoroidae there is one group: the Potoroinae, which contains the extant genera of *Potorous*, *Bettongia*, and *Aepyprymnus*. Within the Potoroinae, *Bettongia* and *Aepyprymnus* are more closely related to each other than to *Potorous*.

Physical characteristics

These animals are all smaller than a medium-sized cat. The upper fur is variously gray-brown with the belly usually much lighter. The ears are short and rounded. The face is short though the nasal region is somewhat elongated. The forelimbs are considerably shorter than the hind limbs, which are

A Tasmanian bettong (*Bettongia gaimardi*) foraging. (Photo by Tom McHugh, Melbourne Zoo, Australia/Photo Researchers, Inc. Reproduced by permission.)

elongated in a similar manner to that of the larger kangaroos. The tail is usually furred, and in the brush-tailed bettong (*B. penicillata*), there is a crest on the distal dorsal surface. In *P. tridactlus* and *B. gaimardi*, the tail tip is white; the significance of this is not known. Some bettongs have a prehensile tail that is used to transport nest material (leaves and grass) for nest building. There are three pairs of upper incisors but only one pair of lower, and young potoroids have two premolars, though near maturity these are replaced by a single sectorial molar. The molar teeth erupt gradually and the fourth molar may not erupt for three to four years. There is a lack of sexual dimorphism in the bettongs.

Distribution

The family is confined to coastal Australia in mainly the southern half, though *B. tropica* is found in northeastern Queensland. *Bettongia gaimardi* is found only on the island of Tasmania where it is still locally common. *Potorous tridactlus* is distributed from southern Queensland through New South Wales into most of Victoria as well as Tasmania. *Potorous*

Bettongs have prehensile tails with which they may carry nest materials. (Illustration by Jarrod Erdody)

longipes is only found in a very restricted area of Victoria while *P. gilberti* occurs only in Two People's Bay Nature Reserve, Western Australia. *Bettongia lesueur* is found on a few islands off the west coast of Western Australia and *B. penicillata* is found in pockets in southwest Western Australia and has been translocated to a number of islands off South Australia. *Bettongia lesueur* is restricted to only three islands of the west coast of Western Australia. *Aepyprymnus* is found only in New South Wales and Queensland.

Habitat

Most potoroids live in forest or woodland areas dominated by eucalyptus species. However, in the recent past, *B. lesueur* and *B. pencillata* lived over much of arid Australia. *Bettongia lesueur*, the burrowing bettong, lives in warrens that can be in sandy dunes. Potoroos tend to live in wetter forest areas than the bettongs, which prefer the more open forest that is better suited to their high hopping gait.

Behavior

Not a great deal of behavioral studies have been carried out on this nocturnal group. They are generally solitary, though some bettong males and females may cohabitate for short periods after mating. Males regularly check up on females living in their home range by inspecting their pouch and/or urogenital regions. This allows males to be aware of the reproductive status of the female and the imminence of estrus. Males can be aggressive towards other males, which generally takes the form of lashing out with their hind feet. Animals that are anxious often display a sinuous movement of the tail. Female bettongs not in estrous can also be aggressive towards unwelcome males, lashing out with their hind feet while lying on their side. Mating takes place usually only on the night of estrus and involves rear-entry while the male holds the female with his forelimbs around her flank. He may also hold on to her neck with his teeth. Intromission is usually of a few seconds duration but may occur several times in

A rufous bettong (*Aepyprymnus rufescens*) at night in Queensland, Australia. (Photo by Jen & Des Bartlett. Bruce Coleman, Inc. Reproduced by permission.)

a night. This group makes few vocalizations, though females make a clucking-like sound to attract their young to the pouch. Anxious animals make a vocalization with an explosive expelling of air.

Feeding ecology and diet

The bettongs and potoroos are mainly fungivores, eating a large proportion of the fruiting bodies of underground (hypogeous) fungi. In order to accomplish this, it is necessary that they have a well-developed sense of olfaction; a large proportion of their cortex is devoted to this. Animals dig with their forefeet to access their diet. The well-developed forestomach of potoroids allows them to have sufficient time to digest the fungi and the gut flora/fauna assist in providing essential amino acids that may be absent from the diet. Animals can exist almost exclusively on fungi, which provide a nourishing diet high in both protein and lipid. The fungi grow on the roots of eucalyptus and other native trees and their spores are activated during passage through the bettong's gut, thereby allowing germination at the site of defecation.

A rufous bettong (*Aepyprymnus rufescens*) standing on its hind legs. (Photo by Eric Woods/FLPA–Images of Nature. Reproduced by permission.)

The brush-tailed bettong (*Bettongia penicillata*) is also known as the woylie. (Photo by Rod Williams/Naturepl.com. Reproduced by permission.)

A brush-tailed bettong (*Bettongia penicillata*) forages on the ground. (Photo by Martin B. White/FLPA–Images of Nature. Reproduced by permission.)

Aepyprymnus rufescens with young. (Photo by L. & O. Schick/Nature Focus, Australian Museum. Reproduced by permission.)

Reproductive biology

Rat-kangaroos are probably polygynous and have a reproductive biology quite similar to that of their larger relatives, the kangaroos and wallabies. After a short gestation, usually about three weeks, the neonate crawls unaided to the pouch where it attaches to one of the four teats. On that night the mother comes into estrous, termed a postpartum estrus, and mates again. The resultant fertilized egg develops only to the blastocyst stage of about 100 cells before becoming dormant, termed embryonic diapause. It remains in the uterus until it is reactivated near the end of pouch development of the previous young. Then, usually on the same night that the large offspring finally vacates the pouch, the new offspring is born. Again, the mother will come into postpartum estrus and mate again. Thus, the rat-kangaroo can have three different generations at the same time: one young out of the pouch still suckling from her teat, one newborn young in the pouch, and one dormant embryonic stage in her uterus. All species give birth to one young at a time. The pouch contracts after the large offspring leaves the pouch, preventing its return and thereby protecting the newborn smaller young. The milk suckled by the young in the pouch is constantly changing its composition from a dilute milk low in protein and lipid but high in carbohydrate to one that becomes more concentrated and high in lipid and low in carbohydrate.

Conservation status

Bettongia tropica, Gilbert's potoroo, and the long-footed potoroo are listed as Endangered in the Environment Pro-

The long-nosed potoroo (*Potorous tridactylus*) is a nocturnal mammal about the same size as a rabbit. (Photo by Dave Watts/Nature Focus, Australian Museum. Reproduced by permission.)

tection and Biodiversity Conservation Act 1999 of the Commonwealth of Australia. *Bettongia lesueur* and the long-nosed potoroo in southeast Australia are listed as Vulnerable. The IUCN lists Gilbert's potoroo as Critically Endangered, the long-footed potoroo and northern bettong as Endangered, boodie as Vulnerable, brush-tailed bettong as Lower Risk/Conservation Dependent, and the Tasmanian bettong as Lower Risk/Near Threatened.

Significance to humans

Although rat-kangaroos are attractive animals, they have little significance in either a cultural or agricultural context.

1. Northern bettong (*Bettongia tropica*); 2. Rufous bettong (*Aepyprymnus rufescens*); 3. Long-footed potoroo (*Potorous longipes*); 4. Boodie (*Bettongia lesueur*); 5. Long-nosed potoroo (*Potorous tridactylus*); 6. Gilbert's potoroo (*Potorous gilbertii*); 7. Tasmanian bettong (*Bettongia gaimardi*); 8. Brush-tailed bettong (*Bettongia penicillata*). (Illustration by Bruce Worden)

Species accounts

Tasmanian bettong
Bettongia gaimardi

TAXONOMY
Bettongia gaimardi (Desmarest, 1822), Port Jackson, New South Wales, Australia.

OTHER COMMON NAMES
English: Bettong, rat-kangaroo, wallaby-rat; French: Kangourou-rat de Tasmanie; Spanish: Canguro-rata de Tasmania.

PHYSICAL CHARACTERISTICS
Head and body length 12–13 in (31–33 cm); weight 4–6 lb (1.5–2.2 kg). Coat color is gray-brown above, light below. Its long tail has a white tip.

DISTRIBUTION
Eastern Tasmania and Australia.

HABITAT
Open sclerophyll forest and woodland; often with a rocky substrate.

BEHAVIOR
Usually seen singly, nocturnal. Carries nest material in mildly prehensile tail. Males and females are aggressive.

FEEDING ECOLOGY AND DIET
The diet is composed of underground fungi.

REPRODUCTIVE BIOLOGY
Probably polygynous. Breeds continuously; single young born after gestation of 21 days; pouch life of 105 days and mature by one year. Mothers have embryonic diapause. Lactation lasts for approximately 22 weeks.

CONSERVATION STATUS
Extinct on mainland Australia; listed by the IUCN as Lower Risk/Near Threatened.

SIGNIFICANCE TO HUMANS
None known. ◆

Long-nosed potoroo
Potorous tridactylus

TAXONOMY
Potorous tridactylus (Kerr, 1792), Sydney, New South Wales, Australia.

OTHER COMMON NAMES
English: Potoroo, rat-kangaroo.

PHYSICAL CHARACTERISTICS
Head and body length 13.3–16 in (34–38 cm); weight females: 1.5–3 lb (660–1,350 g; males: 1.6–3.6 lb (740–1,640 g). Fur color is dark gray above, paler below. Its tapering tail has a white tip.

DISTRIBUTION
Southeast Australia and Tasmania.

HABITAT
Coastal heath and dry and wet sclerophyll forests. Often makes forays in the undergrowth.

BEHAVIOR
Lives solitarily or in pairs; males are territorial. Unobtrusive and secretive. May be monogamous or polygynous.

FEEDING ECOLOGY AND DIET
Mainly eats underground fungi.

REPRODUCTIVE BIOLOGY
Gestation lasts approximately six weeks and pouch life is 125 days. Mother has embryonic diapause and mates shortly after giving birth.

CONSERVATION STATUS
Not threatened.

SIGNIFICANCE TO HUMANS
Cell line used in genetic studies. ◆

Brush-tailed bettong
Bettongia penicillata

TAXONOMY
Bettongia penicillata Gray, 1837, New South Wales, Australia.

OTHER COMMON NAMES
English: Woylie; French: Bettongie à queue touffue, kangourou-rata à queue touffue; Spanish: Canguro-rata colipeludo.

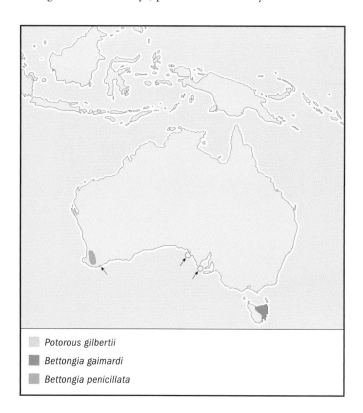

Potorous gilbertii
Bettongia gaimardi
Bettongia penicillata

PHYSICAL CHARACTERISTICS
Head and body length 11.8–16 in (30–38 cm), tail 12 in (31 cm); weight 2.6 lb (1,300 g). Coat is darker above and paler below. Similar in appearance to other bettongs, but tail has a small crest.

DISTRIBUTION
Southwestern Australia; successfully reintroduced into South Australia.

HABITAT
Open sclerophyll forest and woodland; reenters areas soon after forest fires.

BEHAVIOR
Nocturnal, solitary living in an above-ground nest.

FEEDING ECOLOGY AND DIET
Mainly feeds on the fruiting bodies of underground fungi.

REPRODUCTIVE BIOLOGY
Similar to other bettongs; gestation is 21 days and pouch life 100 days. Probably polygynous or promiscuous.

CONSERVATION STATUS
Listed as Lower Risk/Conservation Dependent. It is threatened by the fox and the cat in some areas. Increased numbers have come via translocations.

SIGNIFICANCE TO HUMANS
None known. ◆

Northern bettong
Bettongia tropica

TAXONOMY
Bettongia tropica Wakefield, 1967, Queensland, Australia.

Bettongia tropica
Aepyprymnus rufescens
Potorous longipes

OTHER COMMON NAMES
None known.

PHYSICAL CHARACTERISTICS
Head and body length 15.3 in (39 cm), tail 14 in (36 cm); weight 2.6–3.1 lb (1,200–1,400 g). Coat is darker above and paler below. Similar in appearance to other bettongs.

DISTRIBUTION
Northeastern coastal Queensland.

HABITAT
Tall wet sclerophyll forest along the edge of rainforest.

BEHAVIOR
Nocturnal, solitary living in an above-ground nest.

FEEDING ECOLOGY AND DIET
Mainly feeds on the fruiting bodies of underground fungi and cockatoo grass.

REPRODUCTIVE BIOLOGY
Similar to other bettongs; gestation is 21 days and pouch life is 106 days. Probably polygynous or promiscuous.

CONSERVATION STATUS
Endangered, and threatened by the fox and the cat in some areas. Increased numbers have come via translocations.

SIGNIFICANCE TO HUMANS
None known. ◆

Boodie
Bettongia lesueur

TAXONOMY
Bettongia lesueur (Quoy and Gaimard, 1824), Dirk Hartog Island, Western Australia, Australia.

OTHER COMMON NAMES
English: Burrowing bettong; French: Bettongie de Lesueur, kangourou-rat de Lesueur; Spanish: Canguro-rata de Lesueur.

PHYSICAL CHARACTERISTICS
Head and body length 15.7 in (40 cm), tail 11.8 in (30 cm); weight 3.3 lb (1,500 g). Coat is darker above, paler below. Similar to other bettongs, no crest on thickish tail.

DISTRIBUTION
Several islands off the west coast of Western Australia; translocated to several areas on the mainland of Western Australia and South Australia.

HABITAT
Semi-arid coastal sandy areas where it lives in burrows.

BEHAVIOR
Nocturnal, solitary, living in burrow, occasionally extends into warrens.

FEEDING ECOLOGY AND DIET
Omnivorous, eating fruit and fungi.

REPRODUCTIVE BIOLOGY
Similar to other bettongs; gestation lasts 21 days and pouch life is about 115 days. Probably polygynous or promiscuous.

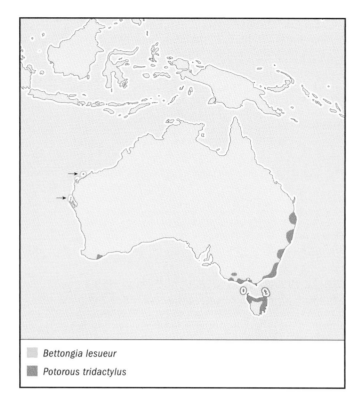

Bettongia lesueur

Potorous tridactylus

CONSERVATION STATUS
Listed as Vulnerable, due to the small area in which it lives.

SIGNIFICANCE TO HUMANS
None known. ◆

Gilbert's potoroo
Potorous gilbertii

TAXONOMY
Potorous gilbertii Gould, 1841, southwest Australia.

OTHER COMMON NAMES
None known.

PHYSICAL CHARACTERISTICS
Head and body length 13.3–15 in (34–38 cm); tail 9 in (23 cm); weight 2.2 lb (1000 g). Appearance similar to other potoroos, tail not well furred compared to body.

DISTRIBUTION
Resticted areas near Two People's Bay Nature Reserve, Western Australia.

HABITAT
Not well-described, open areas but foraging in areas with denser cover.

BEHAVIOR
No data available.

FEEDING ECOLOGY AND DIET
Feeds on underground fungi.

REPRODUCTIVE BIOLOGY
Little known, likely to be similar to long-nosed potoroo. May be monogamous.

CONSERVATION STATUS
Listed as Critically Endangered, due to its small population size.

SIGNIFICANCE TO HUMANS
None known. ◆

Long-footed potoroo
Potorous longipes

TAXONOMY
Potorous longipes Seebeck and Johnston, 1980, Victoria, Australia.

OTHER COMMON NAMES
None known.

PHYSICAL CHARACTERISTICS
Head and body length 15.7 in (40 cm), tail 12.6 in (32 cm); weight 3.3–5 lb (1.5–2.3 kg). Similar to other potoroos, but larger with hind foot longer than head.

DISTRIBUTION
Limited to two areas in northeastern Victoria and one in southeastern New South Wales.

HABITAT
Dry and wet sclerophyll forests, temperate rainforest, and montane forest.

BEHAVIOR
No data available.

FEEDING ECOLOGY AND DIET
Fungivorous, eating the fruiting bodies of hypogeous fungi.

REPRODUCTIVE BIOLOGY
May be monogamous. Gestation approximately 25–38 days, though this is not clear; pouch life 140–150 days.

CONSERVATION STATUS
Listed as Endangered, due to its fragmented populations.

SIGNIFICANCE TO HUMANS
None known. ◆

Rufous bettong
Aepyprymnus rufescens

TAXONOMY
Aepyprymnus rufescens (Gray, 1837), New South Wales, Australia.

OTHER COMMON NAMES
English: Rufous rat-kangaroo.

PHYSICAL CHARACTERISTICS
Head and body length 15 in (38 cm), tail 14 in (36 cm); weight 6.6–7.7 lb (3,000–3,500 g), females larger than males. Largest member of the family, reddish brown coloring and hairy muzzle.

DISTRIBUTION
Northeastern Queensland to northeastern New South Wales. Also found on the New South Wales-Victoria border.

HABITAT
Open sclerophyll forest and woodland.

BEHAVIOR
Nocturnal, solitary animals living in a nest by day. Males can be aggressive toward each other.

FEEDING ECOLOGY AND DIET
Mainly fungivorous.

REPRODUCTIVE BIOLOGY
Similar to other bettongs. Gestation 23 days and pouch life about 114 days. Has loose polygynous interactions.

CONSERVATION STATUS
Not threatened.

SIGNIFICANCE TO HUMANS
None known. ◆

Resources

Books

Rose, R. W. "Reproductive Biology of Rat Kangaroos." In *Kangaroos, Wallabies and Rat Kangaroos*, edited by G. Grigg, P. Jarman, and I. D. Hume. New South Wales: Surrey Beatty & Sons, 1989.

Strahan, R., ed. *Complete Book of Australian Mammals.* Sydney: Angus & Robertson Publishers, 1995.

Walton, D., ed. *Mammals of Australia.* Canberra: Bureau of Fauna and Flora, 1988.

Periodicals

Rose, R. W. "Reproductive Energetics of Two Tasmanian Rat-kangaroos (Potoroinae: Marsupialia)." *Symposium of the Zoology Society of London* 57 (1987): 149–165.

Rose, R. W., and R. B. Rose. "The Tasmanian Bettong *Bettongia gaimardi*." *Mammalian Species* 584D (1998).

Randolph W. Rose, PhD

Wallabies and kangaroos

(*Macropodidae*)

Class Mammalia
Order Diprotodontia
Family Macropodidae

Thumbnail description
Kangaroos, wallabies, and tree kangaroos are a diverse group of herbivorous terrestrial and arboreal marsupials; universally have strongly developed hind legs and long tails; all members of the family are furred and have prominent ears and thin necks

Size
Head and body length ranges from 11 to 91 in (290–2,300 mm); tail length ranges from 6 to 43 in (150–1,090 mm); weight from 3 to 187 lb (1.4–85 kg)

Number of genera, species
11 genera; 62 species

Habitat
Found in almost all habitat types, from rainforests to deserts; some degree of habitat specificity occurs within particular genera

Conservation status
Extinct: 4 species; Endangered: 7 species; Vulnerable: 9 species; Lower Risk/Near Threatened: 11 species; Data Deficient: 3 species

Distribution
Australia, New Guinea, parts of Irian Jaya, and several Indonesian islands. Introduced into Britain, Germany, Hawaii, and New Zealand

Evolution and systematics

The family Macropodidae is the largest family in the order Diprotodontia. This large and diverse order contains the two suborders, Vombatiformes, including the koala and wombats, and Phalangerida, which includes the possums, gliders, potoroos, kangaroos, wallabies, and tree kangaroos. As the order Diprotodontia sits within the subclass Marsupialia, the kangaroos, wallabies, and tree kangaroos also have strong evolutionary links to the other elements of the native Australian mammal fauna, including the carnivorous marsupials, bandicoots and bilbies. The Macropodidae have been classified into two subfamilies, the Sthenurinae and the Macropodinae. The fossil records indicate that the Sthenurinae was a successful group during the Pleistocene when it had at least 20 species. It is now represented by a single species, the banded hare-wallaby (*Lagostrophus fasciatus*). The remaining living members of the family, which includes 61 individual species from 10 separate genera, all comprise the subfamily Macropodinae. This diverse subfamily is most often treated as eight subgroups on the basis of distinctive associations within the separate genera. These subgroups are: typical kangaroos and wallabies of the genus *Macropus* (14 species); anomalous wallabies, including the two monospecific genera, *Wallabia* and *Setonix*; rock wallabies of the genus *Petrogale* (16 species); pademelons of the genus *Thylogale* (six species); nail-tailed wallabies of the genus *Onychogalea* (three species); true hare-wallabies of the genus *Lagorchestes* (four species); tree kangaroos of the genus *Dendrolagus* (10 species); and New Guinea forest wallabies, including the genera, *Dorcopsis* (four species) and *Dorcopsulus* (two species).

Physical characteristics

While there is dramatic diversity within the family, the general body shape, incorporating strongly developed hind limbs that make the forelimbs and upper body look small, a long tail, and prominent ears, is shared by all members. The family name Macropodidae is actually derived from the word *Macropus*, which means "big foot" in reference to the characteristic long hind feet that enables the kangaroos, wallabies, and tree kangaroos to adopt their characteristic hopping gait. Kangaroos are, in fact, the largest mammals to hop on both feet. Hopping is not, however, the only way that the members of this diverse family get about. In contrast to their

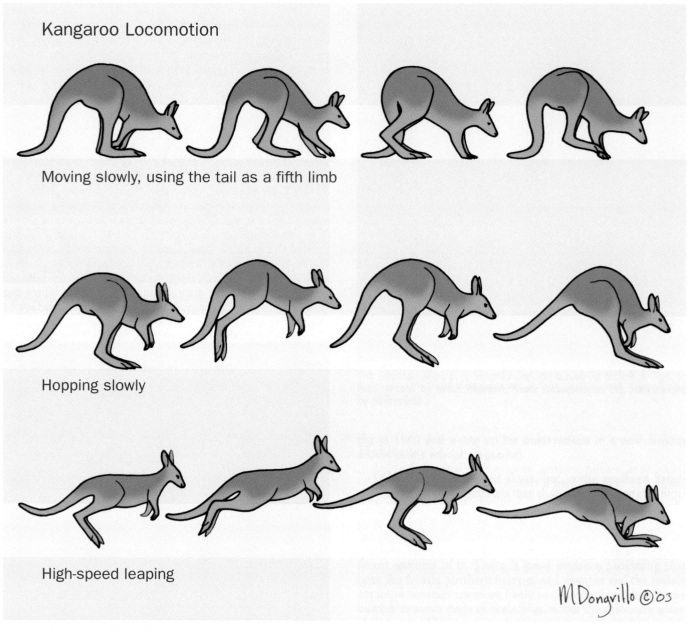

Kangaroo Locomotion

Moving slowly, using the tail as a fifth limb

Hopping slowly

High-speed leaping

Kangaroo locomotion. (Illustration by Marguette Dongvillo)

powerful hind limbs, the forelimbs of most macropodids are small and rather weakly developed. In the larger species where there is distinct size dimorphism between males and females, the forelimbs of males show a disproportionate amount of development for the purposes of display and fighting. The forelimbs of the tree kangaroos are also relatively more developed than in other genera to aid in climbing. As a consequence of the large number of species included within the family, there is an enormous size range with the group. At the smallest end of the scale are species such as the hare-wallabies that have adult head-body lengths of only 11 in (290 mm) and weigh only 3 lb (1.4 kg). At the other end are the large gray kangaroos with adult head-body lengths up to 91 in (2,300 mm) and weighing up to 187 lb (85 kg). There is also significant variation in the fur color and patterning both between and within species in the family Macropodidae. Colors range from sandy red through to black and there is a huge range of intermediate colors and mixes. Many species have distinct stripes or markings in the form of either back stripes, thigh and shoulder patches/flashes, and eye lines or patches. A few of the tree kangaroo and rock wallaby species also display distinct tail stripes.

Distribution

Representatives of the Macropodidae family are broadly distributed all over Australia and through regions of New

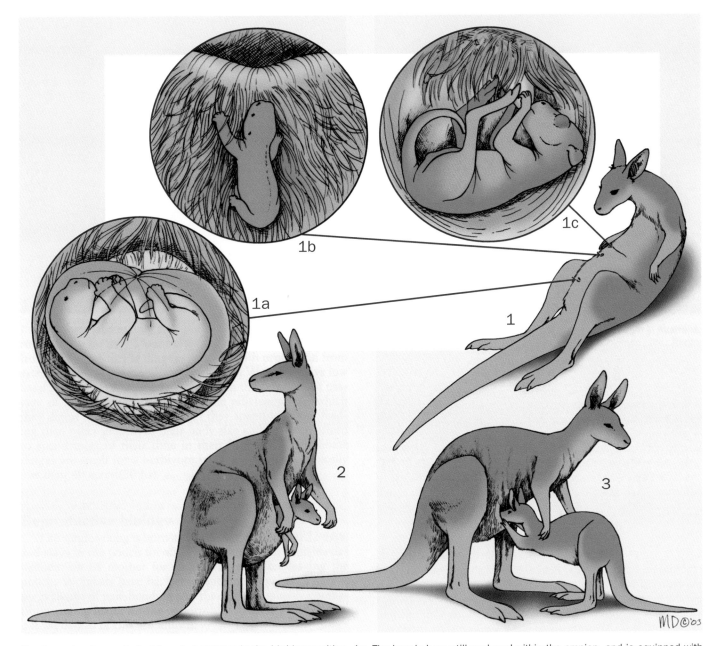

Kangaroo development. 1. A female kangaroo in the birthing position. 1a. The joey is born still enclosed within the amnion, and is equipped with sharp claws to break through it. 1b. The joey climbs to the pouch using its forelimbs. 1c. For several months the joey nurses and develops inside the pouch. 2. A joey at the stage of semi-independence; it can leave the pouch, but returns to nurse and rest. 3. A young kangaroo at the young-at-foot stage; it can no longer return to the pouch, but continues to nurse. (Illustration by Marguette Dongvillo)

Guinea, Irian Jaya, and several Indonesian islands. Of the 11 genera within the family, six are restricted to Australia: *Wallabia*, *Setonix*, *Petrogale*, *Onychogalea*, *Lagorchestes*, and *Lagostrophus*. The large and diverse *Macropus* genus is almost confined to Australia, with the agile wallaby being the only species that occurs naturally outside Australia. Two genera, *Dorcopsis* and *Dorcopsulus*, are restricted to the New Guinea/ Indonesia region. A range of other species, including eight of the 10 *Dendrolagus* tree kangaroos, and three of the six *Thylogale* pademelons, are also found only in that region. In addition to their widespread natural occurrence, several species have been introduced into other regions of the world, in-

cluding Britain, Germany, Hawaii, and New Zealand. Self-sustaining populations of brush-tailed rock wallaby continue to persist on the islands of Oahu (Hawaii) and Kawau, Rangitoto, and Motutapu (New Zealand).

Habitat

Kangaroos, wallabies, and tree kangaroos are represented in almost all habitat types in Australia and New Guinea, including alpine grasslands, high-altitude rainforests, spinifex deserts, and coastal savannahs. Many habitat types support four

The Matschie's tree kangaroo (*Dendrolagus matschiei*) is found in central and eastern New Guinea. (Photo by Animals Animals ©Michael Dick. Reproduced by permission.)

genera. The New Guinea forest wallabies (*Dorcopsis* and *Dorcopsulus*) and the tree kangaroos (*Dendrolagus*) are restricted to rainforest habitats in the northern tropics of Australia and New Guinea. While the pademelons (*Thylogale*) also have an association with moist forests, they inhabit a broader range of forest types distributed from New Guinea to Tasmania. The rock wallabies (*Petrogale*) occur in habitats ranging from the arid zone to the wet tropics, but always have a close association with rocky hillsides, boulder piles, or cliff lines. The remaining genera have less well-defined habitat associations. The hare-wallabies (*Lagorchestes*) and the nail-tailed wallabies (*Onychogalea*) are found in arid and semi-arid habitats as diverse as *Triodia* grasslands, *Acacia* shrublands, and savannas. The genus *Macropus* is undoubtedly the most cosmopolitan with respect to habitats, with members of the genus occurring in most habitat types across Australia as well as in New Guinea.

Behavior

The general patterns of dispersion and social organization within the Macropodidae share a reasonably common trend

or more species, and some woodland and forested areas along Australia's eastern Dividing Range are known to support as many as 12 different kangaroo and wallaby species. Some degree of habitat specificity occurs within particular macropodid

A Bennett's tree kangaroo (*Dendrolagus bennettianus*) surveys from the trees. (Photo by Animals Animals ©Fritz Prenzel. Reproduced by permission.)

Whiptail wallabies (*Macropus parryi*) have distinctive cheek marks. (Photo by R. Kopfle. Bruce Coleman, Inc. Reproduced by permission.)

Red-necked wallabies (*Macropus rufogriseus*) fighting in Queensland, Australia. (Photo by J & D Bartlett. Bruce Coleman, Inc. Reproduced by permission.)

terns of association of age/sex classes have been studied for a number of different species within the Macropodiae. In general, females were found to relate to the distribution of their resources in a way that maximizes their chances of successfully rearing young. Males, in comparison, overlap the distribution of females such that high status individuals gain greater access to mating opportunities. In the medium and large species, a male's status is based largely on size and this is the principal factor influencing males' mating success. This male hierarchy is extremely dynamic, and aggressive interactions between males in both solitary and gregarious species occur regularly in order to establish individual status. These aggressive interactions are quite ritualized in the larger and more sexually dimorphic species. Many of the smaller species adopt an activity pattern that is either nocturnal or crespuscular, a strategy they adopt to avoid predation and, in some cases, the harshness of hot and dry environments. Most of the larger species are active throughout the day, but their activity peaks around dawn and dusk. The social grouping displayed by larger species provides some security from predation that is absent in the more solitary species.

with like-sized bovids and cervids, whereby the small, selective feeders tend to be dispersed and solitary and the large nonselective grazers tend to be aggregated and gregarious. This pattern is most obvious in the tendency of some of the larger species within the genus *Macropus* to form groups (often called "mobs") that may contain 50 of more animals. While there is some social organization within these groups, it is extremely flexible with age/sex classes intermingling, feeding ranges not defended, and mating being promiscuous. The pat-

A Parma wallaby (*Macropus parma*) eating grass while concealed in its typical reedy habitat. (Photo by Rod Williams. Bruce Coleman, Inc. Reproduced by permission.)

An eastern gray kangaroo (*Macropus giganteus*) eating an ice cream cone at an animal farm in Australia. (Photo by Animals Animals ©Mickey Gibson. Reproduced by permission.)

Feeding ecology and diet

The kangaroos, wallabies, and tree kangaroos are predominantly herbivorous, although some of the smaller species will eat invertebrates and fungi. In general, the smaller species tend to be more selective in their feeding habits than the larger species. Smaller species such as the hare-wallabies and nail-tailed wallabies preferentially seek out scattered, high-quality food items such as seeds, fruits, and actively growing plants. In contrast, the larger kangaroos are better able to process lower-quality food items and can incorporate a wide range of plants. An enlarged fore stomach in macropodids performs in a similar way to the rumen in eutherian grazing species and enables kangaroos and wallabies to digest low-quality food. Some of the large plains kangaroos rely almost entirely on grasses. Food availability is clearly influenced by habitat productivity and seasonal conditions. The diets of species that occupy moist forest habitats, therefore, generally include more fruits and dicot leaves compared to the diets of species occupying more arid habitats. The dramatic effects of season ensure that most species adopt an opportunistic feeding strategy that takes advantage of particular feed resources that may be available in the environment for only limited periods.

Reproductive biology

The mating system is promiscuous. The age at which kangaroos, wallabies, and tree kangaroos reach sexual maturity

A red kangaroo (*Macropus rufus*) with nursing joey. (Photo by Animals Animals ©Frutz Prenzel. Reproduced by permission.)

is strongly correlated with adult body size. Females of the larger kangaroo species can breed at two to three years, whereas in some of the smaller species, females can conceive at or slightly before the time of weaning at four to five months. In most cases, males reach sexual maturity some

Agile wallabies (*Macropus agilis*) congregating at a waterhole in Irian Jaya. (Photo by Harald Schütz. Reproduced by permission.)

Tiny baby red kangaroo (*Macropus rufus*) in its mother's pouch. (Photo by Animals Animals ©A. Root, OSF. Reproduced by permission.)

Conservation status

The conservation status of the Macropodidae has altered greatly in modern times, principally in response to human disturbances associated with European colonization and development of Australia and New Guinea. Of the 62 modern species described for the family, four are now Extinct, one species is Critically Endangered, seven species are Endangered, and a further 18 species are considered to be Vulnerable or Near Threatened. Smaller species from the more arid regions have faired worse and account for three of the four Extinct species. The primary causes for these declines include loss of habitat, competition from herbivores such as the introduced European rabbit, sheep, and goat, and predation by introduced carnivores such as the red fox and cat. The impacts of habitat loss have been particularly severe for several of the New Guinea tree kangaroo species, as rainforest areas are cleared for forestry and agriculture. In contrast to the situation with the smaller species, European settlement has had little impact on the larger kangaroos and wallabies. A number of the larger species may in fact have increased in range and numbers in response to both the clearing of native vegetation for the establishment of pastures and the provision of artificial watering points for domestic livestock. Comparisons between historical accounts and recent surveys suggest that species such as the red kangaroo, eastern gray kangaroo, western gray kangaroo, and common wallaroo may now be more abundant than at the time of European settlement of Australia.

Significance to humans

Rock art in Australia's Arnhem Land plateau clearly depicts a number of kangaroo and wallaby species that were used

time after the females, and in some species their participation in mating is further delayed due to social dynamics within cohorts. The larger species exhibit significant sexual size dimporphism, as larger males are able to get the majority of mating opportunities. This dimorphism is less prominent in the smaller species where males and females reach the same adult size. All the macropodids produce only a single young at a birth. Most species exhibit the reproductive phenomenon of embryonic dipause, in which the development of a new embryo is halted at an unimplanted blastocyst stage until about a month before an existing pouch-young permanently vacates the mother pouch. This reproductive adaptation effectively reduces the time interval between births and enables macropodids to respond quickly to both favorable seasonal conditions and to the loss of an existing young. While all species appear to be capable of producing young at all times throughout the year, some species have distinct breeding seasons, most probably in response to seasonal conditions. Like other members of the subclass Marsupialia, kangaroos, wallabies, and tree kangaroos give birth to tiny (0.2–0.6 in [0.5–1.5 cm]), unfurred young in which the tail, hind limbs, and eyes are not fully developed. At birth, this macropodid neonate uses its strong forelimbs to clamber from its mother's cloacal opening to a teat inside her forward-opening pouch. The neonate attaches to the mother until it is relatively well developed. The duration of the pouch life varies between species from 180 to 320 days. Having vacated the pouch, the young-at-foot will continue to suckle from the mother for between one to six months. This young-at-foot stage is longer in the larger species, and may be virtually absent in some of the smaller species.

A tammar wallaby (*Macropus eugenii*) mother with joey, Kangaroo Island, South Australia. (Photo by Daniel Zupanc. Bruce Coleman, Inc. Reproduced by permission.)

by Australian Aborigines. Characteristics of species were generally emphasized in these drawings, which date to 20,000 years before present. These very early pictorial records provide useful information on past distributions of some macropod species and give an indication of the long history of human exploitation. The Aborigines considered the unique Australian wildlife both as a food source and as "partners" in the land. Macropods were the principal animal group consistently exploited by Aborigines. In addition to their use as a food resource, kangaroos and wallabies also feature prominently in traditional dreamtime stories and were culturally significant. Tim Flannery provides detailed accounts of traditional uses of kangaroos and wallabies by New Guinea natives. Early European settlers and pastoralists in both Australia and New Guinea initially valued kangaroos and wallabies as sources of food and hides. It was not long, however, before sheep ranch-

ers regarded macropods as competitors for stock fodder. A trade in macropod skins developed in the mid-nineteenth century. By this time, the large macropods were considered a pest by pastoralists in New South Wales and Queensland, and killing was required by legislation. This legal harvesting was the precursor to the current commercial harvesting of kangaroos that sees some two to four million red kangaroos, eastern and western gray kangaroos, and common wallaroos shot every year in Australia. This modern harvest is strictly monitored and regulated and, for the most part, humane. Population monitoring and a sustainable quota system are used to avoid overexploitation. Kangaroos are a quintessential part Australia's national and international identity. They feature prominently in coats of arms, flags, and corporate logos, and are the principal wildlife experience sought by inbound tourists to Australia.

1. Agile wallaby (*Macropus agilis*); 2. Parma wallaby (*Macropus parma*); 3. Bennett's tree kangaroo (*Dendrolagus bennettianus*); 4. Rufous hare-wallaby (*Lagorchestes hirsutus*); 5. Matschie's tree kangaroo (*Dendrolagus matschiei*); 6. Banded hare-wallaby (*Lagostrophus fasciatus*); 7. Gray dorcopsis (*Dorcopsis luctuosa*); 8. Papuan forest wallaby (*Dorcopsulus macleayi*); 9. Eastern gray kangaroo (*Macropus giganteus*); 10. Red kangaroo (*Macropus rufus*). (Illustration by Marguette Dongvillo)

1. Bridled nail-tailed wallaby (*Onychogalea fraenata*); 2. Brush-tailed rock wallaby (*Petrogale penicillata*); 3. Swamp wallaby (*Wallabia bicolor*); 4. Nabarlek (*Petrogale concinna*); 5. Red-legged pademelon (*Thylogale stigmatica*); 6. Yellow-footed rock wallaby (*Petrogale xanthopus*); 7. Quokka (*Setonix brachyurus*). (Illustration by Marguette Donvillo)

Species accounts

Banded hare-wallaby

Lagostrophus fasciatus

SUBFAMILY
Sthenurinae

TAXONOMY
Lagostrophus fasciatus (Perón and Lesueur, 1807), Bernier Island, Australia. Two subspecies.

OTHER COMMON NAMES
English: Munning; French: Wallaby-lièvre rayé, wallaby-lièvre à bandes; Spanish: Canguro-liebre rayado.

PHYSICAL CHARACTERISTICS
Head and body length 16–18 in (400–450 mm); tail length 14–16 in (350–400 mm); weight 3–5 lb (1.3–2.1 kg). Small wallaby with dark grizzled gray coat on back and sides featuring distinctive transverse bands from the mid-back to the base of the tail.

DISTRIBUTION
Extinct on Australian mainland; remaining only on Bernier and Dorre Islands.

HABITAT
Extant populations associated with thick low scrub and *Triodia* grasslands.

BEHAVIOR
Solitary, nocturnal species that sits beneath thick cover during daylight.

FEEDING ECOLOGY AND DIET
Dicotyledonous plants, including malvaceous and leguminous shrubs, represent main food items. Grasses account for less than half the dietary intake.

REPRODUCTIVE BIOLOGY
Females reach sexual maturity at 12 months; pouch life 180 days. Probably polygynous or promiscuous.

CONSERVATION STATUS
Lagostrophus fasciatus albipilis: Extinct; *Lagostrophus fasciatus fasciatus*: Vulnerable.

SIGNIFICANCE TO HUMANS
Significant to Aborigines as food source and included in dreamtime stories. ◆

Agile wallaby

Macropus agilis

SUBFAMILY
Macropodinae

TAXONOMY
Macropus agilis (Gould, 1842), Port Essington, Northern Territory, Australia.

■ *Wallabia bicolor*
■ *Setonix brachyurus*
■ *Lagostrophus fasciatus*

■ *Macropus agilis*
■ *Macropus rufus*

OTHER COMMON NAMES
English: Sandy wallaby, jungle wallaby.

PHYSICAL CHARACTERISTICS
Head and body length 23–33 in (593–850 mm); tail length
23–33 in (593–850 mm); weight 20–60 lb (9–27 kg). Sandy-
brown color on back and sides, with whitish underside. Distinctive dark head stripe and light thigh stripe.

DISTRIBUTION
Tropical coastal areas of northern Australia and southern New
Guinea.

HABITAT
Monsoon woodlands and grasslands.

BEHAVIOR
Gregarious, forms groups of up to 10 individuals or more
where food is abundant.

FEEDING ECOLOGY AND DIET
Feeds on native grasses and known to excavate roots of some
species. Also known to feed on fruits of Leichhardt tree and
native figs.

REPRODUCTIVE BIOLOGY
Females reach sexual maturity at 12 months; gestation period
29 days; pouch life 219 days. Probably promiscuous.

CONSERVATION STATUS
Not threatened.

SIGNIFICANCE TO HUMANS
Important food source to Australian Aborigines and New Guinea
natives. Agricultural pest to crops and pastures in some areas. ◆

☐ *Macropus parma*

■ *Macropus giganteus*

▨ *Dorcopsis luctuosa*

Eastern gray kangaroo
Macropus giganteus

SUBFAMILY
Macropodinae

TAXONOMY
Macropus giganteus Shaw, 1790, Queensland, Australia. Two
subspecies.

OTHER COMMON NAMES
English: Great gray kangaroo, forester.

PHYSICAL CHARACTERISTICS
Head and body length 38–91 in (958–2,302 mm); tail length
18–43 in (446–1,090 mm); weight 8–146 lb (3.5–66 kg). Males
and females are uniformly gray-brown with paler underside.
Distinguished from other species by hairy muzzle.

DISTRIBUTION
Eastern Australia, including majority of Queensland, New
South Wales, and Victoria.

HABITAT
Common throughout distribution in a range of habitats, including grassy woodlands, forest, and open grasslands.

BEHAVIOR
Gregarious and known to occur in large mobs.

FEEDING ECOLOGY AND DIET
Feed on grasses and forbs in the early morning and late afternoon.

REPRODUCTIVE BIOLOGY
Females reach sexual maturity at 18 months; gestation period
36 days; pouch life 320 days. Probably promiscuous.

CONSERVATION STATUS
Not threatened. Mainland subspecies is highly abundant across
distribution.

SIGNIFICANCE TO HUMANS
Subject to commercial harvesting for meal and skins. Presumably was an important food species for Aborigines. ◆

Parma wallaby
Macropus parma

SUBFAMILY
Macropodinae

TAXONOMY
Macropus parma Waterhouse, 1845, New South Wales, Australia.

OTHER COMMON NAMES
English: White-throated wallaby.

PHYSICAL CHARACTERISTICS
Head and body length 18–21 in (447–528 mm); tail length
16–21 in (405–544 mm): weight 7–13 lb (3.2–5.9 kg). Grayish
brown back and shoulders with characteristic white throat and
chest. White stripe on upper cheek and dark dorsal stripe ending mid-back.

DISTRIBUTION
Eastern Australia on Great Dividing Range between the Gibraltar Range and the Watagan Mountains. Introduced to Kawau Island (New Zealand).

HABITAT
Wet and dry sclerophyll forests and occasionally rainforest.

BEHAVIOR
Solitary and nocturnal.

FEEDING ECOLOGY AND DIET
Feed on grasses and herbs.

REPRODUCTIVE BIOLOGY
Females reach sexual maturity at 16 months; gestation period 34 days; pouch life 212 days. May be promiscuous or polygynous.

CONSERVATION STATUS
Lower Risk/Near Threatened. Considered at one time to have been driven to extinction, but rediscovered in 1967 and subsequently detected at a number of sites across range.

SIGNIFICANCE TO HUMANS
Considered a pest to forestry operations on Kawau Island. ◆

Red kangaroo
Macropus rufus

SUBFAMILY
Macropodinae

TAXONOMY
Macropus rufus (Desmarest, 1822), Blue Mountains, New South Wales, Australia.

OTHER COMMON NAMES
English: Plains kangaroo, blue flier.

PHYSICAL CHARACTERISTICS
Head and body length 29–55 in (745–1,400 mm); tail length 25–39 in (645–1,000 mm); weight 37–187 lb (17–85 kg). Red-brown to blue-gray above and distinctly white underneath.

DISTRIBUTION
Near continental distribution across arid and semi-arid Australia. Absent from coastal and subcoastal regions of eastern, southern, and northern Australia.

HABITAT
Semi-arid plains, shrublands, grasslands, woodlands, and open forest area.

BEHAVIOR
Crepuscular. Gregarious in small groups, but will form larger groups in response to resource availability.

FEEDING ECOLOGY AND DIET
Grazer, feeding almost exclusively on grasses.

REPRODUCTIVE BIOLOGY
Females reach sexual maturity at 14–20 months; gestation period 33 days; pouch life 235 days. Estrous cycles in females and sperm production is males show responses to environmental conditions. May be promiscuous or polygynous.

CONSERVATION STATUS
Not threatened; they have expanded their distribution and population numbers in some areas in response to development of pastures and establishment of artificial water points.

SIGNIFICANCE TO HUMANS
As one of the largest land mammals in Australia, it is an important cultural symbol for both European and Aborigine Australians. The animal features extensively in Aborigine dreamtime stories. The species is commercially harvested for meat and skins in four states of Australia. ◆

Swamp wallaby
Wallabia bicolor

SUBFAMILY
Macropodinae

TAXONOMY
Wallabia bicolor (Desmarest, 1804), locality unknown.

OTHER COMMON NAMES
English: Black wallaby, stinker.

PHYSICAL CHARACTERISTICS
Head and body length 26–33 in (665–847 mm); tail length 25–34 in (640–862 mm); weight 23–45 lb (10.3–20.3 kg). Dark chocolate brown to black above, grading to a strong red-orange color below.

DISTRIBUTION
Eastern Australia from Cape York to southeast South Australia, extending inland up to 250 mi (400 km) from the coast.

HABITAT
Occupies areas of forest, woodland, and health with a dense understory.

BEHAVIOR
More diurnal than most macropods. Solitary.

FEEDING ECOLOGY AND DIET
Specialized browser, feeding on foliage of shrubs, ferns, sedges, and some grasses.

REPRODUCTIVE BIOLOGY
Females reach sexual maturity at 15 months; gestation period 35 days; pouch life 256 days. Probably promiscuous.

CONSERVATION STATUS
Not threatened.

SIGNIFICANCE TO HUMANS
Unattractive to commercial shooters due to small size and coarse fur. Considered a pest to forestry operations in some areas. ◆

Quokka
Setonix brachyurus

SUBFAMILY
Macropodinae

TAXONOMY
Setonix brachyurus (Quoy and Gaimard, 1830), King George Sound (Albany), Western Australia, Australia.

OTHER COMMON NAMES
English: Short-tailed wallaby.

PHYSICAL CHARACTERISTICS
Head and body length 16–21 in (400–540 mm); tail length
10–12 in (245–310 mm); weight 6–9 lb (2.7–4.2 kg). Generally
grizzled gray-brown with reddish tinge. Fur long and thick,
which gives coat a shaggy appearance.

DISTRIBUTION
Southwestern Western Australia, including Rottnest Island.

HABITAT
Densely vegetated areas of moist forest, heath, and swampy
flats.

BEHAVIOR
Nocturnal. Males known to aggressively defend resting sites.
Populations living in areas distant from free water may form
groups of 25–150 individuals.

FEEDING ECOLOGY AND DIET
Browses and grazes, feeding on grasses, sedges, succulents, and
foliage of shrubs.

REPRODUCTIVE BIOLOGY
Females reach sexual maturity at eight to nine months; gestation period 27 days; pouch life 190 days. Mainland populations
breed year-round, but Rottnest Island population has only a
brief breeding season. Probably promiscuous.

CONSERVATION STATUS
Vulnerable. Declined significantly on mainland during the
twentieth century, but has recently recovered in the moister
parts of southwest.

SIGNIFICANCE TO HUMANS
There is major tourist interest in the Rottnest Island quokka
population. ◆

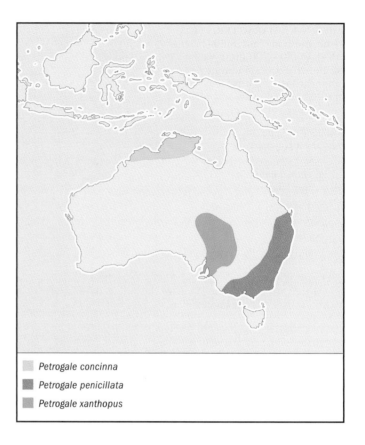

☐ Petrogale concinna

■ Petrogale penicillata

▨ Petrogale xanthopus

Narbalek
Petrogale concinna

SUBFAMILY
Macropodinae

TAXONOMY
Petrogale concinna Gould, 1842, Wyndham, Western Australia,
Australia. Two subspecies.

OTHER COMMON NAMES
English: Little rock wallaby.

PHYSICAL CHARACTERISTICS
Head and body length 11–14 in (290–350 mm); tail length
9–12 in (220–310 mm); weight 2–3 lb (1–1.5 kg). Back and
shoulders colored a dull reddish brown, marbled with light
gray and black. Tail has a definite black brush tip.

DISTRIBUTION
Northern Australia in Mary and Victoria Rivers district, eastern Arnhem Land, and northwest Kimberley.

HABITAT
Low rocky hills, cliffs, and gorges in savanna grasslands.

BEHAVIOR
Partly diurnal.

FEEDING ECOLOGY AND DIET
Feeds on grasses, sedges, and ferns.

REPRODUCTIVE BIOLOGY
Females reach sexual maturity at 12–24 months; gestation period 30–32 days; pouch life 180 days. Breeding probably occurs
throughout the year. May be promiscuous.

CONSERVATION STATUS
Lower Risk/Near Threatened.

SIGNIFICANCE TO HUMANS
None known. ◆

Brush-tailed rock wallaby
Petrogale penicillata

SUBFAMILY
Macropodinae

TAXONOMY
Petrogale penicillata (Gray, 1827), Sydney, New South Wales,
Australia.

OTHER COMMON NAMES
English: Western rock wallaby.

PHYSICAL CHARACTERISTICS
Head and body length 20–23 in (510–586 mm); tail length 20–28 in (500–700 mm); weight 11–24 lb (4.9–10.9 kg). Blackish brown above, grading to a reddish chocolate brown on hindquarters; much paler on chest and underside. Black dorsal head stripe and white to buff cheek stripe. Prominent and distinctive dark brushy end.

DISTRIBUTION
Eastern Australia on Great Dividing Range from Grampians in Victoria to Nanago region in Queensland. Successfully introduced onto Oahu Island (Hawaii) and Kawau, Rangitoto, and Motutapu Islands (New Zealand), where in all cases it continues to persist.

HABITAT
Suitable rocky areas in a range of habitats, including rainforest, wet and dry sclerophyll forests, and open woodland.

BEHAVIOR
Gregarious, and shelter by day in caves and deep fissures. Mostly nocturnal, but known to bask in sun on rocky platforms in winter.

FEEDING ECOLOGY AND DIET
The preferred food is grass but, depending on available habitat and seasonal conditions, significant portions of herbs, browse, and selected fruits are also eaten.

REPRODUCTIVE BIOLOGY
Females reach sexual maturity at 18 months; gestation period 31 days; pouch life 204 days. In the north of its range, there is a narrow hybrid zone in which there is interbreeding between *Petrogale penicillata* and the neighboring species, *Petrogale herbertii*. Probably promiscuous.

CONSERVATION STATUS
Vulnerable. The species has declined significantly in Victoria and southern New South Wales, and is considered regionally endangered in both areas. It remains reasonably common in northern New South Wales and Queensland.

SIGNIFICANCE TO HUMANS
Many thousands shot annually for skin trade between 1884 and 1914. In 1908, a single dealer traded a total of 92,590 skins in Sydney. No information available on their significance to Aborigines. ◆

Yellow-footed rock wallaby
Petrogale xanthopus

SUBFAMILY
Macropodinae

TAXONOMY
Petrogale xanthopus Gray, 1855, Flinders Range, South Australia, Australia. Two subspecies.

OTHER COMMON NAMES
English: Ring-tailed rock wallaby.

PHYSICAL CHARACTERISTICS
Head and body length 19–26 in (480–650 mm); tail length 22–28 in (565–700 mm); weight 13–26 lb (6–12 kg). One of

the most strikingly colored and patterned members of the family. Generally a fawn-gray coloration above and white below; ears, forearms, and hind legs distinctly orange' white cheek stripe and white hip stripe. Tail is orange with distinct regular dark bands. *Petrogale xanthopus celeris* generally paler in general color, and tail markings less distinct than *Petrogale xanthopus xanthopus*.

DISTRIBUTION
Both subspecies occupy spatially distinct distributions: *Petrogale xanthopus xanthopus* occurs in western New South Wales and central-eastern ranges of South Australia; *Petrogale xanthopus celeris* occurs in the ranges of southwestern Queensland.

HABITAT
Semi-arid rangelands, principally in rocky habitats supporting *Acacia* shrublands.

BEHAVIOR
Gregarious in isolated colonies. Due to extreme temperatures in parts of distribution, they are strictly nocturnal in hotter months. Known to bask in sun on rock platforms during winter.

FEEDING ECOLOGY AND DIET
Grass, forbs, and browse.

REPRODUCTIVE BIOLOGY
Gestation period 31–32 days; pouch life 194 days. Polygynous.

CONSERVATION STATUS
Lower Risk/Near Threatened.

SIGNIFICANCE TO HUMANS
No specific details known, but presumed to have been significant to inland Aborigines both as a food source and as a component of their dreamtime mythology. ◆

Red-legged pademelon
Thylogale stigmatica

SUBFAMILY
Macropodinae

TAXONOMY
Thylogale stigmatica (Gould, 1860), Point Cooper, Queensland, Australia. Four subspecies.

OTHER COMMON NAMES
English: Pademelon, northern red-legged pademelon.

PHYSICAL CHARACTERISTICS
Head and body length 15–21 in (386–536 mm); tail length 12–19 in (301–473 mm); weight 6–15 lb (2.5–6.8 kg). Grizzled gray-brown, with cheeks, shoulders, forearms, and inside of hind legs reddish brown. Underside cream to pale-gray. Some variation in color between subspecies with rainforest forms generally darker.

DISTRIBUTION
Distributed from northeastern New South Wales to far north Queensland and in southern New Guinea.

HABITAT
Rainforest is preferred habitat, but also recorded in wet sclerophyll forests and deciduous vine-thickets.

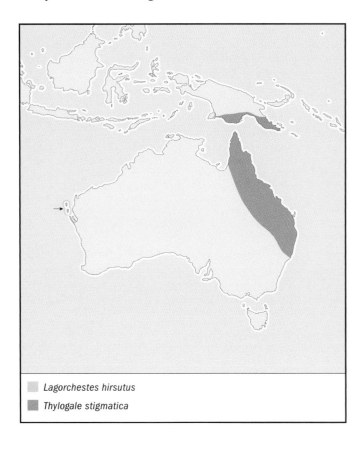

▢ *Lagorchestes hirsutus*

▨ *Thylogale stigmatica*

BEHAVIOR
Generally a solitary species, but may form feeding aggregations in areas where food resources are abundant. They are diurnally active within dense rainforest cover. Nocturnal activity concentrated on feeding areas that may be in more open habitat.

FEEDING ECOLOGY AND DIET
Browsers, feeding on leaves and fruits of a range of rainforest plants. Known also to graze on grasses and commercial crops at forest edges. There is considerable latitudinal variation in the diet, most likely associated with habitat differences due to climate.

REPRODUCTIVE BIOLOGY
Females reach sexual maturity at 12 months; gestation period 28–30 days; pouch life 180 days. May be promiscuous or polygynous.

CONSERVATION STATUS
Not threatened.

SIGNIFICANCE TO HUMANS
None known. ◆

Bridled nail-tailed wallaby
Onychogalea fraenata

SUBFAMILY
Macropodinae

TAXONOMY
Onychogalea fraenata (Gould, 1841), New South Wales, Australia.

OTHER COMMON NAMES
English: Flashjack, merrin; French: Onychogale bridé; Spanish: Canguro rabipelado oriental.

PHYSICAL CHARACTERISTICS
Head and body length 18–28 in (450–700 mm); tail length 15–21 in (380–540 mm); weight 9–18 lb (4–8 kg). General body coloration is grizzled ashy gray above and pale gray below. Characteristic white bridled stripe occurs from center of neck to behind the forearms on both sides. There is a horny spur on the end of the tail.

DISTRIBUTION
Formerly widespread through central Queensland, central-western New South Wales, and western Victoria. Range extended from Charters Towers in the north to Lake Hindmarsh in the south. Following a dramatic decline, it is now known only from a single extant population near Dingo in central Queensland, from an reintroduced population on Idalia National Park and Avocet Nature refuge, and from sanctuary populations at Scotia Sanctuary, Genaren Hill Sanctuary, and Western Plains Zoo.

HABITAT
Acacia-dominated woodlands and shrublands.

BEHAVIOR
Nocturnally active, resting during the day under dense shelter. Typically solitary animals, but feeding aggregations of up to six to eight animals are not uncommon.

FEEDING ECOLOGY AND DIET
Generally feed on mixed forbs, grasses, and browse on the edge of dense vegetation used for shelter. Chenopod forbs (plant family Chenopodiaceae) are particularly favored as well as soft-leaved grasses and some malvaceous species.

▢ *Onychogalea fraenata*

▨ *Dendrolagus matschiei*

REPRODUCTIVE BIOLOGY
Breeding occurs throughout the year. Females reach sexual maturity at nine months; gestation period 23–26 days; pouch life 119–126 days. Promiscuous species.

CONSERVATION STATUS
Endangered. Subject to major recovery program involving captive breeding and reintroduction to parts of former range. Reintroductions to Idalia National Park and Avocet Nature Refuge have significantly improved the species' conservation status.

SIGNIFICANCE TO HUMANS
Mentioned in the diaries of numerous early explorers in western New South Wales as one of the most abundant kangaroo species they encountered. No specific details of its use by humans, but presumed to have been significant food source and also utilized for its distinctive skin. Recent conservation actions to recovery for this endangered species have served to raise its public profile. ◆

Rufous hare-wallaby
Lagorchestes hirsutus

SUBFAMILY
Macropodinae

TAXONOMY
Lagorchestes hirsutus Gould, 1844, York district, Western Australia, Australia.

OTHER COMMON NAMES
English: Mala, ormala, western hare-wallaby, wurrup; French: Wallaby-lièvre de l'ouest, wallaby-lièvre roux; Spanish: Canguro-liebre peludo.

PHYSICAL CHARACTERISTICS
Head and body length 12–15 in (310–390 mm); tail length 10–12 in (245–305 mm); weight 2–4 lb (0.9–1.8 kg). Uniform sandy-red color; fur on back and hindquarters is long, giving the animal a shaggy appearance.

DISTRIBUTION
Previously extant across central and western desert areas of Australia. Now confined to Bernier and Dorre Islands and experimental reintroductions to large enclosures in Northern Territory and Western Australia, and to Trimouille Island off the Western Australian coast.

HABITAT
Spinifex grasslands.

BEHAVIOR
Solitary and nocturnal.

FEEDING ECOLOGY AND DIET
Feeds selectively on grasses, herbs, succulent shrubs, and seeds.

REPRODUCTIVE BIOLOGY
Females reach sexual maturity at five months; pouch life 124 days. Capable of producing three offspring per year under good conditions. Probably promiscuous.

CONSERVATION STATUS
Critically Endangered. A recovery plan is currently being implemented for the species and includes significant emphasis of captive breeding and reintroduction.

SIGNIFICANCE TO HUMANS
Features prominently in Aborigine dreamtime stories associated with Ayers Rock (Uluru). Also hunted by Aborigines using fire. ◆

Bennett's tree kangaroo
Dendrolagus bennettianus

SUBFAMILY
Macropodinae

TAXONOMY
Dendrolagus bennettianus De Vis, 1887, Daintree River, Queensland, Australia.

OTHER COMMON NAMES
English: Gray tree kangaroo, dusty tree kangaroo, tcharibbeena; French: Dendrolague de Bennett; Spanish: Canguro arborícola de Bennett.

PHYSICAL CHARACTERISTICS
Head and body length 27–30 in (690–750 mm); tail length 29–33 in (730–840 mm); weight 18–30 lb (8–14 kg). Color is dark brown, forehead and snout with grayish tinge; rusty brown coloration on shoulders, neck, and back of head.

DISTRIBUTION
Eastern Cape York from the Daintree River to Mt. Amos, and extending west to the Mount Windsor Tablelands.

◻ *Dorcopsulus macleayi*

◼ *Dendrolagus bennettianus*

HABITAT
Tropical rainforests, vine, and gallery forests.

BEHAVIOR
One of the few macropods to defend a discrete territory. Adult males principally solitary, but have home range that overlaps with numerous females.

FEEDING ECOLOGY AND DIET
Feeds mainly on leaves, but takes some fruit when available.

REPRODUCTIVE BIOLOGY
Females breed annually and exhibit embryonic dipause. Pouch life is approximately 270 days. The young may accompany the mother for up to two years. Probably polygynous.

CONSERVATION STATUS
Lower Risk/Near Threatened.

SIGNIFICANCE TO HUMANS
Heavily hunted by Aborigines in lowland forests; highland areas were not often visited by natives due to taboos. ◆

Matschie's tree kangaroo
Dendrolagus matschiei

SUBFAMILY
Macropodinae

TAXONOMY
Dendrolagus matschiei Forster and Rothschild, 1907, Rawlinson Mountains, Morobe Province, Papua New Guinea.

OTHER COMMON NAMES
English: Huon tree kangaroo.

PHYSICAL CHARACTERISTICS
Head and body length 16–25 in (412–625 mm); tail length 16–27 in (408–685 mm). Generally wood-brown in color with gold on the tail and limbs. Further distinguished by shorter tail and lack of golden back stripes.

DISTRIBUTION
In Papua New Guinea, confined to the Huon Peninsula and Umboi Island.

HABITAT
Dense mountain forests between 3,300–10,800 ft (1,000–3,290 m).

BEHAVIOR
Several studies of captive animals have been completed detailing courtship and mating behavior. Considered to be solitary and crepuscular in wild.

FEEDING ECOLOGY AND DIET
Nothing known.

REPRODUCTIVE BIOLOGY
Breeding can occur at any time of the year in captivity. Gestation period 44 days; pouch life 280 days. Probably promiscuous.

CONSERVATION STATUS
Endangered.

SIGNIFICANCE TO HUMANS
Hunted by New Guinea natives with the aid of dogs, which chase the tree kangaroos out of the canopy. ◆

Gray dorcopsis
Dorcopsis luctuosa

SUBFAMILY
Macropodinae

TAXONOMY
Dorcopsis luctuosa (D'Albertis, 1874), southeast of New Guinea.

OTHER COMMON NAMES
None known.

PHYSICAL CHARACTERISTICS
Head and body length 21–38 in (525–970 mm); tail length 12–15 in (310–388 mm); weight 8–25 lb (3.6–11.5 kg). Generally black/gray coat with a prominent yellow patch around the pouch and cloaca.

DISTRIBUTION
Southern New Guinea from the Mreauke area to Milne Bay. Thought to be locally abundant in the Moresby region.

HABITAT
Lowland rainforest.

BEHAVIOR
Crepuscular. Observations of captive populations suggest they are a social species, living in loosely knit groups that can contain several adult males and females.

FEEDING ECOLOGY AND DIET
Known to feed on a wide range of vegetable matter.

REPRODUCTIVE BIOLOGY
Nothing is known. May be promiscuous.

CONSERVATION STATUS
Not threatened.

SIGNIFICANCE TO HUMANS
No specific details known, but presumed to have been significant food source for New Guinea natives. ◆

Papuan forest wallaby
Dorcopsulus macleayi

SUBFAMILY
Macropodinae

TAXONOMY
Dorcopsulus macleayi (Mikluho-Maclay, 1885), Papua New Guinea.

OTHER COMMON NAMES
English: Macleay's dorcopsis.

PHYSICAL CHARACTERISTICS
Head and body length 17–18 in (435–460 mm); tail length 12–14 in (315–346 mm); weight 6–7 lb (2.5–3.4 kg). Smaller and more densely furred than species of *Dorcopsis*. Tail furred for two-thirds to three-quarters of its length.

DISTRIBUTION
Eastern New Guinea, south of Central Cordillera and east of Mount Karimui.

HABITAT
Mid-montane rainforests between 3,300–5,900 ft (1,798 m).

BEHAVIOR
Nothing is known.

FEEDING ECOLOGY AND DIET
Reported to favor the fruit and leaves of *Ficus* spp., *Pangium edule*, and *Syzgium* sp., as well as other trees.

REPRODUCTIVE BIOLOGY
Nothing is known.

CONSERVATION STATUS
Listed as Vulnerable because of the small area it occupies.

SIGNIFICANCE TO HUMANS
Presumed to have been a significant food source for New Guinea natives. ◆

Common name / Scientific name/ Other common names	Physical characteristics	Habitat and behavior	Distribution	Diet	Conservation status
Antilopine wallaroo *Macropus antilopinus* English: Antilopine kangaroo	Head and body length 31–47 in (77.8–120 cm); tail length 27–35 in (67.9–89 cm); weight 35–108 lb (16–49 kg). Males reddish-tan, females can be either pale gray or reddish tan, both noticeably paler on underside.	Monsoonal forests and woodlands. Gregarious, occuring in groups of 3–8. Births occur throughout the year. Gestation period 34 days. Pouch life 269 days.	Tropical northern Australia.	Feeds almost entirely on grass.	Not threatened
Black wallaroo *Macropus bernardus* English: Black kangaroo, Bernard's wallaroo	Head and body length 25–29 in (64.6–72.5 cm); tail length 23–25 in (57.5–64 cm); weight 29–49 lb (13–22 kg) Males dark-brown to black. Females gray to gray-brown.	Woodland and grassland habitats on steep rocky escarpments and plateau tops. Solitary, no more than 3 adults ever seen together. Nocturnal.	Northern Territory, western and central Arnhem Land in Australia.	Grazing animal. No details of preferred species.	Lower Risk/Near Threatened
Black-striped wallaby *Macropus dorsalis* English: Scrub wallaby	Head and body length 21–32 in (53–82 cm); tail length 29–33 in (74–83 cm); weight 13–44 lb (6–20 kg). Pelage is medium brown above with a distinctive mid-back stripe from the neck to the base of the tail. Generally paler on sides and white underside.	Forested country with a dense shrub layer including the margins of vinescrubs, rainforests, and brigalow scrubs. Social species sheltering by day in groups of up to 20 animals. Nocturnal. Females reach sexual maturity at 14 months. Gestation period 33–35 days. Pouch life 210 days.	Eastern Australia bounded approximately by Dubbo, Blackall, and Charters Towers.	Feeds predominantly on native and introduced pastures.	Not threatened
Tammar wallaby *Macropus eugenii* English: Dama wallaby	Head and body length 20–27 in (52–68 cm); tail length 13–18 in (33–45 cm); weight 9–22 lb (4–10 kg). Grizzled dark gray-brown above tending to red-brown on the sides. Paler gray-brown underside.	Low dense vegetation for shelter including coastal scrub, heath, dry sclerophyll forests, and mallee thickets. No social grouping observed. Nocturnal. Females reach sexual maturity at 8 months. Gestation period 29 days. Pouch life 250 days.	Coastal areas in south-west Western Australia and southern South Australia. Also from islands including Abrolhos, Garden, St Peter, Flinders, and Kangaroo Islands.	Mixed herbivorous diet including browsed and grazed species. Able to drink saltwater.	Two subspecies Lower Risk/Near Threatened; one subspecies Extinct in the Wild
Western gray kangaroo *Macropus fuliginosus* English: Black-faced kangaroo, mallee kangaroo, stinker	Head and body length 38–88 in (97.1–222.5 cm); tail length 17–39 in (44.3–100 cm); weight 10–118 lb (4.5–53.5 kg).	Common throughout distribution in a range of habitats including grassy woodlands, forest and open grasslands. Gregarious and known to occur in large "mobs." One of femacropod species confirmed as not employing embyonic diapause. Females reach sexual maturity at 14 months. Gestation period 31 days. Pouch life 310 days.	Southern Australia including southern parts of Western Australia and South Australia and western portions of Victoria and New South Wales. Northeastern distribution extends slightly into Queensland.	Feeds predominantly on grass species.	Kangaroo Island subspecies is Lower Risk/Near Threatened; mainland subspecies is not threatened
Whiptail wallaby *Macropus parryi* English: Prettyface wallaby	Head and body length 30–36 in (75.5–92.4 cm); tail length 29–41 in (72.8–104.5 cm); weight 15–57 lb (7–26 kg). General body color light gray to brownish gray. Distinctive facial markings including dark brown coloration of forehead and bases of ears, white ear tips, and white cheek stripe.	Open forests and grassland in undulating and hilly areas. Detailed studies completed of social ethology. Social species living in groups of up to 50 individuals. Females reach sexual maturity at 24 months. Gestation period 36 days. Pouch life 275 days.	Eastern Australia from south of Cooktown in Queensland to Dorrigo in northern New South Wales.	Feeds predominantly on grasses and other herbaceous plants, including ferns.	Not threatened
Common wallaroo *Macropus robustus* English: Euro, hill wallaroo	Head and body length 44–78 in (110.7–198.6 cm); tail length 21–35 in (53.4–90.1 cm); weight 14–103 lb (6.3–46.5 kg). Significant color and size differences across distribution. Varies from dark gray to reddish above, paler below. Fur is distinctly shaggy in appearance.	Diverse habitats across distribution but usually features steep escarpments, stony rises, or rocky hills. Mostly solitary and crepuscular. Females reach sexual maturity at 15 months. Gestation period 33 days. Pouch life 231–270 days.	Almost continental distribution in Australia, excluding extreme south of Western Australia, Victoria, Tasmania, and western Cape York.	Grazes primarily on grasses and shrubs. Well adapted to an arid environment and can survive on low protein food and without free-flowing water.	One subspecies is Vulnerable; three subspecies are not threatened

[continued]

Common name / Scientific name / Other common names	Physical characteristics	Habitat and behavior	Distribution	Diet	Conservation status
Red-necked wallaby *Macropus rufogriseus* English (Tasmania): Bennett's wallaby	Head and body length 26–36 in (65.9–92.3 cm); tail length 25–34 in (62.3–87.6 cm); weight 24–59 lb (11–26.8 kg). Grizzled gray to reddish coat above which is noticeably paler in females. Neck and shoulders with pronounced reddish brown coloration. Tasmanian subspecies overall darker in color.	Eucalypt forests. Solitary, but feeding aggregations of up to 30 animals can be seen grazing together at night. Females reach sexual maturity at 11–21 months. Gestation period 29 days. Pouch life 270 days. Tasmanian subspecies has a distinctly seasonal pattern of births.	Southeastern Australia from Mt. Gambier in the south following the coast line to slightly north of Rockhampton. Subspecies occur on Tasmanian and Bass Strait Islands.	Essentially a grazing animal, feeding on grasses and herbs.	Not threatened
Allied rock-wallaby *Petrogale assimilis* English: Torrens creek rock-wallaby	Head and body length 18–23 in (44.5–59 cm); tail length 16–22 in (40.9–55 cm); weight 9–10 lb (4.3–4.7 kg). Mostly gray-brown above and paler underneath but subject to variation according to type of rock on which population exists.	Rocky clifflines and escarpments with open forest. Detailed behavioral studies suggest complex hierarchies exist in local populations. Some evidence of pair-bonding from observational studies. Females reach sexual maturity at 18 months. Gestation period 30–32 days. Pouch life 180–210 days.	Northeastern Queensland, Australia, bounded by Home Hill, Croydon, and Hughenden. Also occurs on Magnetic and Palm Islands.	Forbs and browse form major part of the diet. Grasses, fruits, seeds, and flowers are also eaten.	Not threatened
Doria's tree kangaroo *Dendrolagus dorianus* English: Unicolored tree kangaroo	Head and body length 20–31 in (51.5–78 cm); tail length 18–26 in (44.5–66 cm); weight 10–32 lb (4.5–14.5 kg). Somber brown coat and a paler, shorter tail (77–80% of head and body length) distinguishes it from other tree kangaroo species.	Montane rainforest between 2,000 ft and 10,800 ft (610–3,290 m). Social organization has been studied in captivity suggesting that *Dendrolagus dorianus* is a highly social species, living in one-male groups in which the male is dominant. This study also reported details of vocalizations, social playing, and mother-young interactions.	Central highlands and southeast mountains of Irian Jaya/New Guinea.	Observed to feed on the leaves of *Asplenium*-like epiphytic ferns at a sancuary in New Guinea.	Vulnerable
Lumholtz's tree kangaroo *Dendrolagus lumholtzi*	Head and body length 20–26 in (52–65 cm); tail length 26–29 in (65.5–73.6 cm); weight 11–19 lb (5.1–19 kg).	Montane rainforests. Nocturnal. Generally a solitary animal but feeding aggregations of up to four animals have been observed. Detailed observations of behavior have been recorded from captive animals. Pouch life 230 days.	Southeastern Cape York, Australia, between Kirrama and Mount Spurgeon.	Primarily a leaf eater. Known to feed on ribbonwood, wild tobacco (introduced), and some rainforest fruits.	Lower Risk/Near Threatened
White-striped dorcopsis *Dorcopsis hageni*	Head and body length 17–24 in (42.5–60 cm); tail length 12–15 in (31.5–37.8 cm); weight 11–13 lb (5–6 kg). Similar in coloration to *D. veterum*. Distinguished by the presence of a single white dorsal stripe running from the rump to the crown.	Mixed alluvial forest. Partly diurnal. No other details known.	Restricted to lowlands of northern New Guinea between Mamberamo and Lae.	Reported by indigenous hunters to feed on cockroaches and other invertebrates, which it sources from under rocks on river banks.	Not threatened
Brown dorcopsis *Dorcopsis veterum*	Head and body length 24–30 in (60–77 cm); tail length 18–21 in (46.5–53.5 cm); weight 11 lb (5 kg). Light to dark brown in color and lacking the distinctive dorsal stripe characteristic of *D. hageni*	Low altitude rainforests. No specific details known about behavior.	Lowlands of Irian Jaya, as far east as Danau Biru in the north, and the Setakwa/Mimika Rivers district in the south. Offshore islands of Misool, Salawatti, and Yapen.	Diet consists of roots, leaves, and grasses.	Not threatened
Lesser forest wallaby *Dorcopsulus vanheurni* English: Little dorcopsis	Head and body length 12–18 in (31.5–44.6 cm); tail length 9–16 in (22.5–40.2 cm); weight 3–5 lb (1.5–2.3 kg). Smallest of the New Guinea macropods distinguished from nearest congenor, *Dorcopsulus macleayi* by a greater portion of the tail being naked.	Montane rainforests. No specific details known about behavior.	Central Cordillera and the Huon Peninsula on Irian Jaya/New Guinea.	The regrowth plant *Rungia klossii* is reported as a favored food.	Not threatened

Resources

Books

Dawson, T. J. *Kangaroos: The Biology of the Largest Marsupials.* Kensington, Australia: University of New South Wales Press/Ithaca, 2002.

Flannery, T. F. *Mammals of New Guinea.* 2nd ed. Chatswood, Australia: Reed Books, 1995.

Grigg, G. C., P. J. Jarman, and I. D. Hume. *Kangaroos, Wallabies and Rat-kangaroos.* Chipping Norton, Australia: Surrey Beatty and Sons, 1989.

Maxwell, S., A. A. Burbidge, and K. D. Morris. *The 1996 Action Plan for Australian Marsupials and Monotremes.* Gland, Switzerland: IUCN/SCC Australasian Marsupial and Monotreme Specialist Group, Wildlife Australia, 1996.

Menkhorst, P. W. *A Field Guide to the Mammals of Australia.* Melbourne, Australia: Oxford University Press, 2001.

Ramono, W. S., and S. V. Nash. "Conservation of Marsupials and Monotremes in Indonesia." In *Australasian Marsupials and Monotremes, An Action Plan for Their Conservation,* edited by Michael Kennedy. Gland, Switzerland: IUCN/SSC Australasian Marsupials and Monotreme Specialist Group, 1992.

Strahan, R. *The Mammals of Australia.* 2nd ed. Sydney: Reed New Holland, 2002.

Periodicals

Burbidge, A. A., K. A. Johnson, P. F. Fuller, and R. I. Southgate. "Aboriginal Knowledge of the Mammals of the Central Deserts of Australia." *Australian Wildlife Research,* 15 (1988): 9–39.

Organizations

Environment Australia. John Gorton Building, King Edward Terrace, Parkes, Australian Capital Territory 2600 Australia. Phone: (2) 6274 1111. Fax: 61 2 6274 1666. E-mail: info@ea.gov.au Web site: <http://www.ea.gov.au>

Other

Kangaroos—Faces in the Mob, Videotape. Green Cape Pty. Ltd., Sydney, New South Wales, 1993.

Geoff Lundie-Jenkins, PhD

Pygmy possums
(Burramyidae)

Class Mammalia
Order Diprotodontia
Family Burramyidae

Thumbnail description
Small omnivores/insectivores, characterized by
low-crowned molars with rounded cusps,
reduced premolars, and reduced digital pads

Size
Head and body length 2.7–5 in (7–13 cm),
tail 1.5–6.3 in (4–16 cm); weight 0.2–1.4 oz
(6–40 g)

Number of genera, species
2 genera; 5 species

Habitat
Forests, woodlands, and subalpine meadows

Conservation status
Endangered: 1 species

Distribution
Australia and New Guinea

Evolution and systematics

Fossils of pygmy possums have been found in central Australia, northern Queensland, and western Victoria, since the
beginning of the Miocene (i.e., approximately 20–25 million
years ago). They were mostly associated with rainforest habitats. The genus *Burramys* seems to have remained mostly unchanged since these early forms, thus forming a true living
fossil. The family is most closely related to Acrobatidae. The
burramyid family contains two genera, *Burramys* with only
one species (*B. parvus*) and possibly two living subspecies, and
Cercartetus with four species, three of which have two subspecies each. *Burramys parvus*, the mountain pygmy possum,
certainly is one of the most exciting discovery stories of all
mammals. It was first described in 1895, based on several
fragments of skulls and jaws from a cave in New South
Wales. Its long, blade-like (sectorial) premolar immediately
caught the zoologists' attention, as similar teeth had also
been described for several rat-kangaroos (not to mention a
truly fossil order, Multituberculata). For several decades,
it was undecided whether this mysterious fossil belonged
to the rat-kangaroos or to the possum group. In 1956, Dr.
David Ride firmly stated that *Burramys* was supposed to

A long-tailed pygmy possum's tail (*Cercartetus caudatus*) can be more than
6 in (15.1 cm) long. (Photo by Pavel German. Reproduced by permission.)

The eastern pygmy possum (*Cercartetus nanus*) feeds on nectar and pollen with its brush-tipped tongue. (Photo by Pavel German/Nature Focus, Australian Museum. Reproduced by permission.)

be an extinct possum. However, 11 years later, a live specimen was caught in a skiing hut on Mount Hotham in Victoria, far above the snow line.

Physical characteristics

The pygmy possums closely resemble small dormice of the Northern Hemisphere, such as the hazel mouse (*Muscardinus avellanarius*). Species range in size from 2.3–2.7 in (6–7 cm) and 0.2 oz (7 g) in the smallest species, *Cercartetus lepidus*, to the largest *Burramys parvus* with 4.3 in (11 cm) and 1.4 oz (40 g). Coat colors are mostly brownish on the upper, cream-colored to grayish on the underside. Some species store fat in the tails.

Distribution

The range of the family as a whole covers parts of Australia (the southeast, including Tasmania, the southern west, and a small part of the Queensland wet tropics) and the central mountain range of New Guinea. *Burramys* has been

found, in fossil records, in several parts of eastern and central Australia.

Habitat

Pygmy possums generally occur in wet sclerophyll or eucalypt forests. *Cercartetus caudatus*, the long-tailed pygmy possum, the only tropical species, lives in the mountain rainforests of New Guinea's Central Cordillera above 4,900 ft (1,500 m), in a small area near Townsville in tropical Queensland, and also in mountainous areas above 985 ft (300 m). In New Guinea, specimens are often trapped on the ground in subalpine grasslands above the tree line. Extended torpor (up to six days duration in *C. lepidus*) has been described in all species, either during winter or as daily torpor.

Behavior

All species are nocturnal, and all routinely construct nests of leaves, bark, and other plant material, in tree hollows or as independent structures. *Burramys parvus* is terrestrial, living in alpine meadows and boulder fields. All the other species are arboreal, using their long prehensile tails as a support or for balance when climbing small branches. It is appropriate to distinguish between the two genera, as they differ in many aspects. In fact, they are only united in one family based on anatomical and morphological characteristics. Mountain

A long-tailed pygmy possum (*Cercartetus caudatus*) showing the dark rings around its eyes. (Photo by Pavel German/Nature Focus, Australian Museum. Reproduced by permission.)

The western pygmy possum (*Cercartetus concinnus*) is mostly arboreal. (Photo by Robert Valentic/Nature Focus, Australian Museum. Reproduced by permission.)

pygmy possums are only found in two small areas between 4,265 and 7,300 ft (1,300 and 2,230 m) on the peak of Mt. Kosciusko, the highest mountain in Australia. Its habitat mostly is subalpine, shrubland, and meadows. *Burramys* has to cope with at least three months of snow cover, during which time it tends to live under the snow, climbing within and between rock crevices, or climbing into bushes to collect seeds and berries. *Burramys* also stores fat under its skin, and develops a thick fur in autumn. The heaviest animal ever found in autumn weighed 3 oz (82 g). Adults tend to enter hibernation earlier than juveniles, and can remain torpid for periods of up to 20 days. Another means of energy conservation is communal nesting. These nests are normally of either all males or all females, and can be found throughout the year, except when females breed. The social organization of *Burramys* is more complex than expected in such a small mammal. Up to 10 females (probably related kin such as mother with daughters, granddaughters, etc.) occupy communal, overlapping home ranges. Female ranges are higher up in the mountains, in more productive areas. Males visit these female ranges only briefly for mating, and emigrate again after breeding season, returning to their own ranges in lower less productive areas of their habitat. Males are quite tolerant of each other, both in captivity and even during breeding times. It is not clear whether females evict males, or whether they emigrate on their own, but the resulting social/mating system is one of female communal resource defense and male polygamy. Data on population structure and longevity are in accordance to trapping data, which is highly female-biased, and show that females live for up to 11 years, while males live only up to four years.

Feeding ecology and diet

Cercartetus caudatus is primarily insectivorous, but also can be found eating flowers and possibly plant exudates. Some species, particularly *C. nanus*, regularly visit flowering plants and feed on nectar and pollen predominantly. Others are more insectivorous or even kill small lizards. Mountain pygmy possums feed on seeds, fruit, insects, and other small invertebrates. In summer, the Bogong moth (*Agrotis infusa*) is of particular dietary importance. *Burramys* stores only hard seeds, nuts, etc., in the fall, while soft berries and fruits are eaten at once. All species are prey for owls, carnivorous marsupials, snakes, and feral cats.

Reproductive biology

Females have four teats. The mountain pygmy possum mother rears four young on her own, once per year. Most

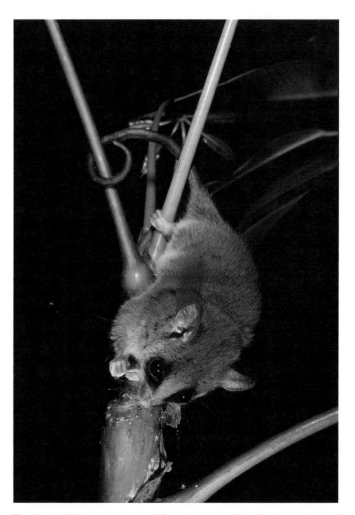

The long-tailed pygmy possum (*Cercartetus caudatus*) eats nectar and insects. (Photo by Dick Whitford/Nature Focus, Australian Museum. Reproduced by permission.)

An eastern pygmy possum (*Cercartetus nanus*) holds on to a plant with its hand-like paws. (Photo by Pavel German. Reproduced by permission.)

possums are found in communal nests of up to five individuals. Females carry one to three young, and these remain in the pouch until they are 0.2 oz (7 g), which is remarkable considering the fact that the mothers themselves weigh only 0.5–0.8 oz (15–24 g). The young appear to become independent at a weight of above 0.35 oz (10 g). Breeding in this polygamous species occurs twice per year in Australia (January–February and August–November), indicated not only by the distribution of pouch-gravid females but also by a regression in size of testes in males.

Conservation status

Due to its very limited range, the mountain pygmy possum is classified as Endangered. Habitat fragmentation and disturbance are the most imminent dangers for its survival. However, two long-term threats must also be considered: Global warming could easily change its alpine habitat into less productive ones, and the population of Bogong moths depends on both the amount of rainfall as well as the preservation of their own habitats, which could be compromised by agricultural activities.

All the other species belonging to the genus *Cercartetus* are currently not threatened in status with regard to IUCN listings. However, *C. lepidus* is considered to be Vulnerable due to its contracted range.

Significance to humans

None known.

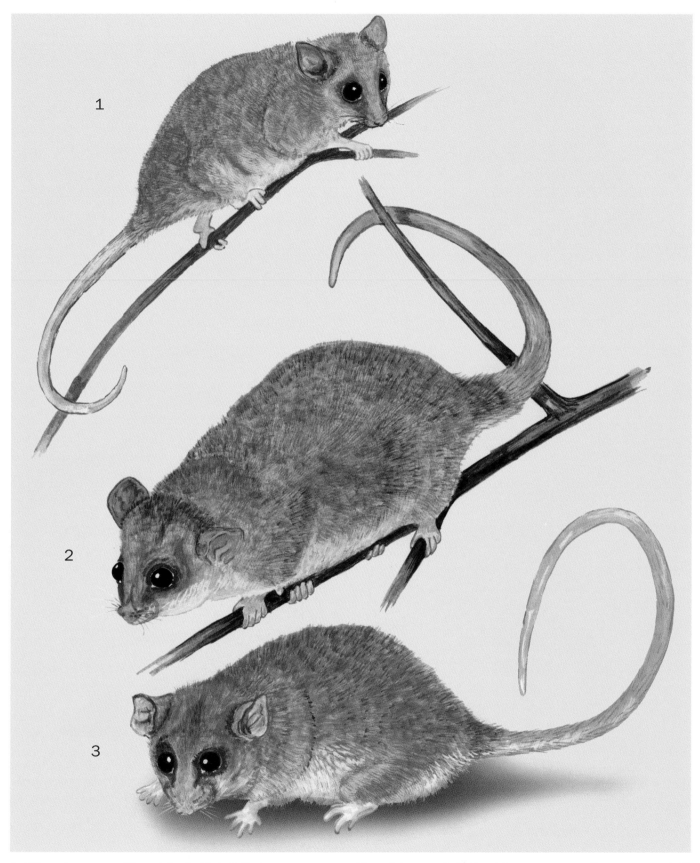

1. Little pygmy possum (*Cercartetus lepidus*); 2. Eastern pygmy possum (*Cercartetus nanus*); 3. Mountain pygmy possum (*Burramys parvus*). (Illustration by Amanda Smith)

Species accounts

Eastern pygmy possum
Cercartetus nanus

TAXONOMY
Cercartetus nanus (Desmarest, 1818), Tasmania, Australia.
Two subspecies.

OTHER COMMON NAMES
French: Phalanger-loir de l'Est; German: Östlicher Bilchbeutler.

PHYSICAL CHARACTERISTICS
Length 7–7.8 in (18–20 cm). Brownish fur with gray belly.
Base of tail thickened with storage fat. Tongue long, brush-
tipped.

DISTRIBUTION
East and southeast of Australia, and Tasmania.

HABITAT
Heathland and dry and wet sclerophyll forest.

BEHAVIOR
Mostly solitary. Constructs spherical nests of leaves and bark in
tree-hollows.

FEEDING ECOLOGY AND DIET
Feeds on nectar, pollen, and insects.

REPRODUCTIVE BIOLOGY
One to three young, twice per year. May be promiscuous or
polygamous.

CONSERVATION STATUS
Not threatened.

SIGNIFICANCE TO HUMANS
None known. ◆

Little pygmy possum
Cercartetus lepidus

TAXONOMY
Cercartetus lepidus (Thomas, 1888), Tasmania, Australia.

OTHER COMMON NAMES
French: Phalanger-loir mineur; German: Zwergbilchbeutler.

PHYSICAL CHARACTERISTICS
Length 5 in (13 cm); 0.2–0.3 oz (6–9 g). Grayish brown with
belly almost white.

DISTRIBUTION
Tasmania, Kangaroo Island, and northwestern Victoria.

HABITAT
Dry sclerophyll forests and mallee shrubland.

BEHAVIOR
Constructs nests in tree-hollows, in wall cavities, under turf, or
in birds' nests.

FEEDING ECOLOGY AND DIET
Insectivorous and carnivorous; preys even on lizards.

Cercartetus nanus

Cercartetus lepidus
Burramys parvus

REPRODUCTIVE BIOLOGY
Nothing is known.

CONSERVATION STATUS
Not threatened, but rare throughout its range.

SIGNIFICANCE TO HUMANS
None known. ◆

Mountain pygmy possum
Burramys parvus

TAXONOMY
Burramys parvus Broome, 1896, New South Wales, Australia.

OTHER COMMON NAMES
French: Souris-oppossum de montagne; German: Bergbilch-beutler.

PHYSICAL CHARACTERISTICS
Length 9–11.4 in (23–29 cm); 1–2.1 oz (30–60 g), up to 2.8 oz (80 g) when fat is stored.

DISTRIBUTION
Two small, isolated areas in Victoria, Australia.

HABITAT
Alpine meadows and rocky, boulder-strewn areas with heath-land vegetation.

BEHAVIOR
Nocturnal, spending day in nests in crevices, under rocks, etc. Periods of torpor in winter of up to 20 days or more; female social system is gregarious, possibly kin-related; males also possibly non-solitary, but in different habitat.

FEEDING ECOLOGY AND DIET
Omnivore, with heavy reliance on one moth species and seeds.

REPRODUCTIVE BIOLOGY
Four young born after 12–16 days of gestation, leave the pouch after about three weeks, and are adult sized in three to four months. Probably polygamous.

CONSERVATION STATUS
Listed as Endangered due to its fragmented populations.

SIGNIFICANCE TO HUMANS
None known. ◆

Resources

Books

Flannery, Tim. *Possums of the World.* Chatswood, Australia: Geo Productions, 1994.

Kennedy, Michael, ed. *Australasian Marsupials and Monotremes—An Action Plan for Their Conservation.* Gland, Switzerland: IUCN Publishing, 1991.

Mansergh, Ian, and Linda Broom. *The Mountain Pygmy-possum of the Australian Alps.* Kensington, Australia: New South Wales University Press, 1994.

Strahan, Ronald, ed. *Complete Book of Australian Mammals.* Sydney: Australian Museum, 1994.

Udo Gansloßer, PhD

Ringtail and greater gliding possums

(Pseudocheiridae)

Class Mammalia

Order Diprotodontia

Family Pseudocheiridae

Thumbnail description
Medium-sized possums, most species rather slow-moving, with short limbs; their teeth, particularly molars, have selenodent (i.e., half-moon shaped) crests, ideal for cutting and grinding leaves; ears are small and furred, fur color is mostly brown or gray with the last part of the (prehensile) tail more or less hairless

Size
12.6–37.4 in (320–950 mm), 4.1–79.4 (115–2250 g)

Number of genera, species
6 genera; 16 species

Habitat
Forests and woodlands, suburban areas

Conservation status
Vulnerable: 3 species; Lower Risk/Near Threatened: 4 species; Data Deficient: 3 species

Distribution
New Guinea, Australia including Tasmania

Evolution and systematics

Pseudocheiridae are most closely related to Petauridae, and together they form a superfamily Petauroidea. Even though this superfamily is named after petaurids, the pseudocheirids are obviously the more primitive and generalized family. This is evident both from morphology and from paleontological data. Pseudocheirids have been found in very old Tertiary deposits. The oldest undoubtedly pseudocheirid known so far comes from the late Oligocene-middle Miocene, and those genera are already rather distinct from each other. Oligocene as well as Quaternary species have been found in several central Australian and Queensland records, which points to a wide-ranging distribution already there. These paleontological data, together with the fact that today only two genera (nine species) live in New Guinea, but six genera (seven species) live in Australia, suggest that the family had its origin in Australia and immigrated into New Guinea quite recently. Dental morphology of pseudocheirids is also considered rather primitive: The W-shaped molars are considered (Archer) as more primitive than the rounded cusps of petaurids. Even though koalas also have similar molar crowns, this is being regarded as a convergence due to leaf-eating. Both cytogenetic studies and al-

bumin microcomplement fixation techniques have yielded quite clear-cut results recently concerning taxonomy within the family. No subfamily has been erected, but *Hemibelideus* and *Petauroides* are closely related to each other, and only distantly related to the rest. *Petropseudes* and *Pseudochirops* similarly are separate from the rest, but also with a certain distance between the two genera. *Pseudochirulus*, which now contains several species previously assigned to *Pseudocheirus*, also is an isolated genus. Some of the species (e.g., *P. peregrinus*, *Pseudochirulus canescens*) are subdivided into up to six subspecies, which is further evidence for both long isolation and old radiations.

Physical characteristics

The greater glider *Petauroides volans* is quite distinct from the rest of the family. It possesses a gliding membrane from elbows to lower legs similar but convergent to *Petaurus* spp., and a bushy tail. Its ears are large, the fur is long and woolly, and varies from white or gray to dark brown, sometimes with tail and body being in different colors. Its size puts it apart, with up to 37.4 in (95 cm) head-to-tailtip, and a weight of 32.8–42.3 oz (930–1,200 g). The rest of the family, in exter-

A common ringtail (*Pseudocheirus peregrinus*) foraging in the trees at night. (Photo by E. R. Dessinger. Bruce Coleman, Inc. Reproduced by permission.)

nal appearance is rather similar. All have short, stocky limbs, short round ears, and tails that are bare at the lower side at least about the final third of their lengths. Many species have dorsal stripes, though these are not always clearly seen when the fur itself is rather dark. Fur colors differ from light gray/cream, to orange, to dark brown. One species is greenish due to a mix of yellow, black, and white hairs. Internally, all species (including *Petauroides*) have a large cecum for fermenting their leaf-based diet, and their cheek-teeth are formed in the shape of cutting edges, scissorlike, to both cut and grind xerophytic leaves (e.g., Eucalyptus, which at least some species can detoxify).

Distribution

Pseudocheirids obviously already were part of the Australian fauna in times when the Australian continent still rested in wetter climatic zones. Species existing then, similar to other families, were larger than today's representatives, reaching up to 33 lb (15 kg). With increasing aridification of the Australian continent, pseudocheirids withdrew to the still wetter areas, and today are restricted to coastal, or coast-near areas in the east, northeast, southwest and northwest of the continent, as well as Tasmania. New Guinea seems to have been invaded rather late, and the species there are mostly restricted to mountainous, i.e., central areas.

Habitat

Nearly all of the New Guinean species, except for two rather poorly known and obviously rare species (*Pseudochirulus canescens*, *Pseudocheirus caroli*), inhabit mountain forests, where they are rather abundant. All of them prefer primary forests except *Pseudocheirus forbesi*, who seem to be more adapted to secondary, i.e. disturbed forest areas. There are up to four sympatric species in certain New Guinean places, and in those communities there are clear niche-separations in size, ranging from 5.3 to 70.5 oz (150 to 2,000 g) and nesting habits (from tree-hollows to dreys to sleeping exposed on branches). In Australia, one species (*Petropseudes dahli*) is semi-terrestrial, preferring rocky areas in Northern Australia, two species are found in sclerophyll forests (*P. volans* and *Pseudocheirus pere-*

A green ringtail (*Pseudochirops archeri*) in the trees of Australia. (Photo by C. B. & D. W. Frith. Bruce Coleman, Inc. Reproduced by permission.)

A lowland ringtail (*Pseudochirulus canescens*) foraging at night in the Uplands Tropical Rainforest, northeast Queensland, Australia. (Photo by Frithfoto. Bruce Coleman, Inc. Reproduced by permission.)

A greater glider (*Petauroides volans*) on a branch at night. (Photo by Tom & Pam Gardner/FLPA–Images of Nature. Reproduced by permission.)

grinus), and the rest are living in higher altitude rainforests of the Atherton Tableland. Contrary to New Guinea, these sympatric species (which also are in part sympatric to some phalangerid species) are all in the same size class, namely, in the 31.7–42.3 oz (900–1,200 g) range, and all except one, rest in tree-hollows. Supposedly, niche-differentiation must be along some other axes, one of them possibly activity—each species has characteristic times of emerging and returning from and to nests, the aspect of the nesting-tree, or size of the hollows. Among the sclereophyll forest-dwellers, *Pseudocheirus peregrinus* has been able to adapt to suburban habitats, feeding on rosebuds and rose leaves in front- and backyards, and day-nesting under roofs.

Behavior

Pseudocheirids are among the most diverse of all possums in term of social organization. The rock ringtail (*Petropseudes dahli*) seems to live in long-term pairs with overlapping young, as they have been observed as pairs with a young-at-heel, and another smaller one riding on back. Lemuroid ringtails are long-term monogamous, perhaps with a potential to an expansion similar to *P. dahli*, *P. peregrinus* is reported as either pair- or family-living, with paternal care such as carrying grooming and defending young. Sleeping groups of adults of varying sexual composition have also been recorded. Most of the rainforest species and *P. volans* are solitary both while foraging and resting.

Deposit of feces on specific sites has been found in *P. dahli*, and a sternal gland is used by sternal rubbing in several species of *Pseudocheirus*. Acoustic communication seems to be of medium importance, with distress calls uttered by young in trouble, soft contact calls by family members of *P. peregrinus*, loud contact calls, possibly for spacing, as a "dusk" chorus of

P. forbesi, and vocal recall of young from dangerous situations in *P. peregrinus*. There is no evidence of territorial defense in any species, even the solitary ones, except a certain spatial distribution pattern of *P. volans* that is more evenly distributed than statistically expected from randomness. This suggests at least active avoidance, or perhaps some sort of defense. All

A coppery ringtail (*Pseudochirops cupreus*) foraging. (Photo by C. B. Frith. Bruce Coleman, Inc. Reproduced by permission.)

The lemuroid ringtail (*Hemibelideus lemuroides*) has long claws to help it cling to trees. (Photo by E. R. Degginger. Bruce Coleman, Inc. Reproduced by permission.)

species are mostly nocturnual. *Pseudochirops archeri* sometimes can be seen active during the day.

Feeding ecology and diet

Petauroides volans is a highly specialized folivore, who feeds about 90% of its diet on eucalyptus leaves. It relies on high nutrient species or growth stages, and seems to be able to actively select these high-quality leaves based on tree species and leaf age. All the remaining species, so-called ringtails, also are folivores, and at least some species, such as *Pseudocheirus peregrinus*, are also capable of eating high amounts of eucalyptus leaves. However, they mostly do not rely so heavily on eucalypts alone, eating buds, flowers and fruits of other plants, in the understory of eucalypt forests as well. All leaf-eating pseudocheirids are hindgut-fermenters, which means that in their large colon and cecum, a specialized community of bacteria and other microorganisms is kept to digest cellulose (plant cell walls cannot be digested by any animal). They are all capable, in their hindgut, to separate the contents into large particles, which are defecated rather quickly, and small particles, solubles and bacteria themselves. This mechanism serves to retain the more digestible parts and nitrogen-carrying bacteria. Ringtail possums also are caecotrophic; the colon contents are defecated separately, and ingested again to gain proteins and amino acids by foregut digestion. Eucalypt-feeding ringtails often show strong preferences for certain trees, and dislike for other trees of the same species. Subsequent

chemical analysis revealed a higher proportion of toxic or unpalatable substances in the rejected individuals. They are, however, when not given a choice, capable of detoxifying the toxic plant secondary metabolites, though at a cost of higher energy and nitrogen requirements.

Reproductive biology

Mating systems of pseudocheirids, as already suggested by their diverse social organization, are rather variable. Several species are long-term monogamous, which may also include paternal care. Others, the solitary ones included, are polygamous, with evidence for shifts from monogamy to polygyny between years for the greater glider, depending on the food supply. Breeding in *Petauroides* is seasonal, one young per year being born in autumn to winter. The young stays in the pouch for 90–120 days, and is carried on the mother's back for another 90 days. Young become independent around 10 months of age, and sexually mature during their second year. The tropical ringtails, particularly those from New Guinea, seem to have no reproductive season. Consort relationships, an accompanying by a male for an estrous female for a few days, have been described for some species, nothing else is known about mating systems of the solitary ones. Number of young per litter is one or two, depending on the species, and it seems that, as far as known, the age of 120 days for leaving the pouch characterizes them all. Weaning occurs, again known only for

The green ringtail (*Pseudochirops archeri*) gets its greenish coloring from black, white, and yellow banding. (Photo by © Martin Harvey/Gallo Images/Corbis. Reproduced by permission.)

some species at 150–160 days (*Pseudochirulus herbertensis*) to 180 days (*Pseudocheirus peregrinus*). Females, at least in *P. peregrinus*, are capable of producing two litters per year. Longevity in the wild is approximately four to five years.

Conservation status

Several species of pseudocheirids are either widespread, and ecologically adaptable (e.g., *P. peregrinus*), or quite abundant in their current habitat, which is not under immediate threat. Others, however, are vulnerable due to a restricted area of distribution (e.g., some of New Guinean species, or the rock ringtail). Among the New Guinean species, several are unknown in status and biology. None of the potentially vulnerable species is currently protected under Indonesian or Papua law, and the system of proposed or established natural reserves may be inadequate to protect them, also because the reserves have been proposed without considering the needs of local people or animals themselves. *Petauroides volans* is considered vulnerable due to its dependence on old, hollow trees

for nesting, and due to its patchy distribution even in prime habitat. Over 50% of the forest, even in areas where it is common, is never used, and over 60% of individuals used 9% of the available forest in one study. Logging of such areas has tremendous negative effects.

A different kind of threat is expected to arise over the next few years or decades. Due to global warming and increasing carbon dioxide (CO_2) levels, eucalypts are expected to increase the level of toxins and antinutrients in their leaves (because these secondary metabolites are a common way of storing chemical waste in plants). It has been calculated that, given the current amount of increasing CO_2 levels, within the next 50 years most of Australian's eucalypt forests will be uninhabitable for leaf-eating species, such as greater gliders.

Significance to humans

Some of the larger New Guinean species are hunted for meat.

1. Greater glider (*Petauroides volans*); 2. Rock ringtail (*Petropseudes dahli*); 3. Common ringtail (*Pseudocheirus peregrinus*); 4. Pygmy ringtail (*Pseudochirulus mayeri*); 5. Lowland ringtail (*Pseudochirulus canescens*); 6. Green ringtail (*Pseudochirops archeri*); 7. Lemuroid ringtail (*Hemibelideus lemuroides*); 8. Herbert River ringtail (*Pseudochirulus herbertensis*). (Illustration by Marguette Dongvillo)

Species accounts

Lemuroid ringtail
Hemibelideus lemuroides

TAXONOMY
Hemibelideus lemuroides (Collett, 1884), north Queensland, Australia.

OTHER COMMON NAMES
English: Lemur possum; French: Ringtail lemurien; German: Lemur-Ringbeutler.

PHYSICAL CHARACTERISTICS
Length 2.6–2.8 in (65–72 mm), weight 28.2–44.8 oz (800–1,270 g). Woolly coat is grey on back, yellow on belly, and brown on shoulders. Bushy tail has bare tip.

DISTRIBUTION
Atherton and Mount Carbine Tablelands, above 1,575 ft (480 m), in North Queensland.

HABITAT
Mountain rainforest.

BEHAVIOR
Arboreal, nocturnal, leaping vividly from tree to tree; monogamous pairs are formed and mostly observed in close proximity.

FEEDING ECOLOGY AND DIET
Leaf-eater (at least 37 species), but supplements its diet with flowers and fruits.

REPRODUCTIVE BIOLOGY
Monogamous. One young per year, from August to November.

CONSERVATION STATUS
Not threatened.

SIGNIFICANCE TO HUMANS
None known. ◆

Greater glider
Petauroides volans

TAXONOMY
Petauroides volans (Kerr, 1792), Sydney, Australia. Two subspecies.

OTHER COMMON NAMES
French: Phalanger volant geant; German: Riesengleitbeutler.

PHYSICAL CHARACTERISTICS
Length 35.4–41.3 in (90–105 cm), weight 31.7–60.0 oz (900–1,700 g). Dark brown body with white underside. Gliding membrane from elbow to ankles. Long bushy tail.

DISTRIBUTION
Eastern Australia.

HABITAT
Sclerophyll forests, dry and wet, but no rainforest.

BEHAVIOR
Gliding for up to 330 ft (100 m), capable of mid-glide turns (by using tail as a rudder) of up to 90°; as gliding membrane is only

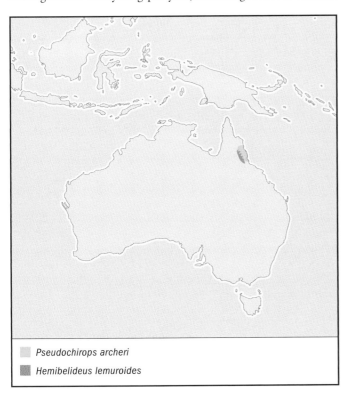

Pseudochirops archeri
Hemibelideus lemuroides

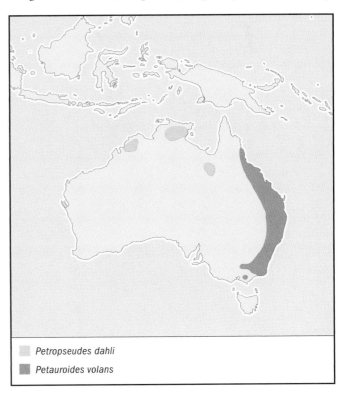

Petropseudes dahli
Petauroides volans

from elbow, wrists are not extended but tucked under chin in flight. Solitary home range of males and females may overlap.

FEEDING ECOLOGY AND DIET
Folivore, only very limited amount of other plant parts.

REPRODUCTIVE BIOLOGY
One young per year is born between March and June, stays in the pouch for 120 days, rides on mother's back for another three months. Usually monogamous.

CONSERVATION STATUS
Vulnerable.

SIGNIFICANCE TO HUMANS
None known. ◆

Green ringtail
Pseudochirops archeri

TAXONOMY
Pseudochirops archeri (Collett, 1884), north Queensland, Australia.

OTHER COMMON NAMES
French: Ringtail vert, ringtail au pelage raye; German: Grüner Ringbeutler.

PHYSICAL CHARACTERISTICS
Length 25.6–28.0 in (65–71 cm), weight 37.9–47.6 oz (1,075–1,350 g). Fur appears green due to thin bands of black, yellow, and white fur.

DISTRIBUTION
Upland rainforest in tropical northeastern Queensland coastal areas.

HABITAT
Rainforest.

BEHAVIOR
Mostly nocturnal, arboreal animal probably solitary. Sometimes seen active during the day, rests on branches instead of in tree hollows. Adults seem not to vocalize at all. Young are carried on mother's back for longer than other ringtails, and follow her around at-heel afterwards.

FEEDING ECOLOGY AND DIET
Green ringtails eat leaves of rainforest trees, also of rather toxic species. This may be the reason why the young follows the mother for an extended period of time.

REPRODUCTIVE BIOLOGY
Almost nothing is known except that normally only one young is around at a time.

CONSERVATION STATUS
Not threatened.

SIGNIFICANCE TO HUMANS
None known. ◆

Rock ringtail
Petropseudes dahli

TAXONOMY
Petropseudes dahli (Collett, 1895), Northern Territory, Australia.

OTHER COMMON NAMES
English: Rock-haunting possum, rock possum; French: Ringtail des rochers; German: Felsenringbeutler.

PHYSICAL CHARACTERISTICS
Length 20.9–25.2 in (53–64 cm), weight 45.2–70.6 oz (1,280–2,000 g).

DISTRIBUTION
Three isolated populations in northwestern Queensland, Northern Territory, and northern Western Australia.

HABITAT
Rocky areas.

BEHAVIOR
Partly terrestrial, hiding and resting in crevices and under rocks during the day. During the night they emerge and climb trees to feed. Rock ringtails appear to live in extended family groups, up to seven animals have been seen together, including young of several developmental stages.

FEEDING ECOLOGY AND DIET
Leaves, flowers of trees and fruits are eaten of a variety of species.

REPRODUCTIVE BIOLOGY
Normally one young per time is observed, and reproduction seems to be non-seasonal. Usually monogamous.

CONSERVATION STATUS
Though not listed by the IUCN, considered vulnerable by some authors due to a contraction of range since discovery in 1890s.

SIGNIFICANCE TO HUMANS
None known. ◆

Common ringtail
Pseudocheirus peregrinus

TAXONOMY
Pseudocheirus peregrinus (Boddaert, 1785), Queensland, Australia. Five subspecies.

OTHER COMMON NAMES
English: Tasmanian ringtail, western ringtail, rufous ringtail; French: Ringtail commun; German: Gewöhnlicher Ringbeutler.

PHYSICAL CHARACTERISTICS
Length 23.6–27.6 in (60–70 cm), weight 24.7–38.8 oz (700–1,100 g). Gray-brown fur with light underside. Tail is long and thin; end half is pale color.

DISTRIBUTION
Southwestern Australia coastal areas along eastern coast from Cape York to Southern Australia; Tasmania and Bass Strait islands.

HABITAT
All types of vegetation with dense shrubs and understory, from heathland to rainforest.

BEHAVIOR
Nocturnal (mostly before midnight), building large nests from twigs and leaves (dreys), often several members of a family group building nests immediately next to each other, or nesting together. Living in pairs or groups. Males take part in care of the young.

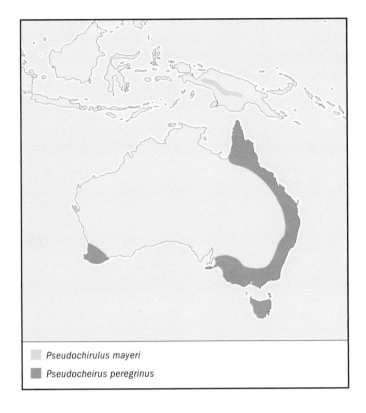

Pseudochirulus mayeri
Pseudocheirus peregrinus

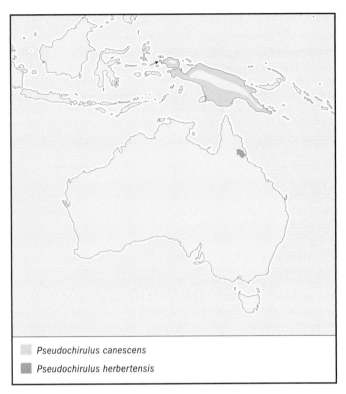

Pseudochirulus canescens
Pseudochirulus herbertensis

FEEDING ECOLOGY AND DIET
Leaves, fruits and flowers are eaten. Common ringtails can detoxify eucalypts. Coprophagy of cecum contents is common.

REPRODUCTIVE BIOLOGY
Normally two young are born, between April and November. They leave the pouch within 120 days, are weaned at about 180 days and sexually mature within 12 months. Mating system is not known.

CONSERVATION STATUS
Not threatened.

SIGNIFICANCE TO HUMANS
Suburban individuals can be a minor nuisance to gardeners by eating flowers, buds etc of ornamental plants (e.g., roses). ◆

Lowland ringtail
Pseudochirulus canescens

TAXONOMY
Pseudochirulus canescens (Waterhouse, 1846), Irian Jaya, Indonesia. Five subspecies.

OTHER COMMON NAMES
French: Ringtail à plaine; German: Tiefland-Ringbeutler.

PHYSICAL CHARACTERISTICS
Length 15.0–17.0 in (38–43 cm), weight about 12.3 oz (350 g).

DISTRIBUTION
Lowland areas in all New Guinea.

HABITAT
Lowland and foothill rainforest, which is an exception for New Guinean species of this family.

BEHAVIOR
Probably solitary. Noctural, arboreal.

FEEDING ECOLOGY AND DIET
Herbivore, folivore.

REPRODUCTIVE BIOLOGY
Nothing is known.

CONSERVATION STATUS
Not threatened, but rare all over its range.

SIGNIFICANCE TO HUMANS
None known. ◆

Herbert River ringtail
Pseudochirulus herbertensis

TAXONOMY
Pseudochirulus herbertensis (Collett, 1884), Queensland, Australia.

OTHER COMMON NAMES
French: Ringtail d`Herbert; German: Herbert Ringbeutler.

PHYSICAL CHARACTERISTICS
Length 24.4–30.3 in (62–77 cm), weight 24.7–51.1 oz (700–1,450 g). Dark fur with white belly. Long thin tail with dark hairs.

DISTRIBUTION
Limited range in tropical Queensland mountains.

HABITAT
Upland rainforest.

BEHAVIOR
Noctural and solitary animals, normally resting in tree hollows, but also capable of building tree nests. Young ride on mother's back only for a short time, and are then left in the nest.

FEEDING ECOLOGY AND DIET
Leaves, fruits, and flowers are eaten.

REPRODUCTIVE BIOLOGY
Juvenile females have four teats, adults only two. Normally two young are born, mostly between May and June. They leave the pouch after about 120 days and are weaned within 150 to 160 days. Sexual maturity is reached by about 12 months. Mating system is not known.

CONSERVATION STATUS
Not threatened.

SIGNIFICANCE TO HUMANS
None known. ◆

Pygmy ringtail
Pseudochirulus mayeri

TAXONOMY
Pseudochirulus mayeri (Rothschild and Dollman, 1932), Irian Jaya, Indonesia.

OTHER COMMON NAMES
French: Ringtail mineur; German: Zwergringbeutler.

PHYSICAL CHARACTERISTICS
Length 13.4–15.0 in (34–38 cm), weight 3.7–7.3 oz (105–206 g).

DISTRIBUTION
Central Cordillera of New Guinea, at higher than 4,920 ft (1,500 m).

HABITAT
Moss forests, most abundant at 6,560 ft (2,000 m) and above.

BEHAVIOR
Nocturnal, during the day they seem to be partly torpid, as they are rather sluggish and timid when caught.

FEEDING ECOLOGY AND DIET
Possibly too small to be an effective folivore; local people report it eats moss and lichen. Captive individual refused leaves, but took sugar water.

REPRODUCTIVE BIOLOGY
Nothing is known, except one young is seen at a time.

CONSERVATION STATUS
Not threatened.

SIGNIFICANCE TO HUMANS
None known. ◆

Common name / Scientific name	Physical characteristics	Habitat and behavior	Distribution	Diet	Conservation status
Weyland ringtail *Pseudocheirus caroli*	Fur is dense, soft, and woolly. Upperparts gray or brown, often very dark, underparts are white, yellowish, or almost as dark as back. Mottlings of white and black ventral surface may exist. Tail is usually curled into a ring at the end. Head and body length 6.4–12.6 in (16.3–32 cm), tail length 6.7–15.8 in (17–40 cm).	Rainforest, sclerophyll forest, woodland, or brush. Scansorial, nocturnal, movement is slow and quiet.	West-central New Guinea.	Variety of leaves, fruits, flowers, bark, and sap.	Not threatened
Moss-forest ringtail *Pseudocheirus forbesi*	Fur is dense, soft, and woolly. Upperparts gray or brown, often very dark, underparts are white, yellowish, or almost as dark as back. Mottlings of white and black ventral surface may exist. Tail is usually curled into a ring at the end. Head and body length 6.4–12.6 in (16.3–32 cm), tail length 6.7–15.8 in (17–40 cm).	Rainforest, sclerophyll forest, woodland, or brush. Scansorial, nocturnal, movement is slow and quiet.	Interior eastern New Guinea.	Variety of leaves, fruits, flowers, bark, and sap.	Not threatened
Arfak ringtail *Pseudocheirus schlegeli*	Facial mask of orange, black, and white fur. Upperparts gray or brown, often very dark, underparts are white, yellowish, or almost as dark as back. Mottlings of white and black ventral surface may exist. Tail is usually curled into a ring at the end. Head and body length 6.4–12.6 in (16.3–32 cm), tail length 6.7–15.8 in (17–40 cm).	Mid-montane forests of New Guinea, between 1,640 and 9,190 ft (500–2,800 m). Scansorial, nocturnal, movement is slow and quiet.	Extreme northwestern New Guinea.	Consists mainly of leaves.	Not threatened
d'Albertis's ringtail *Pseudochirops albertisii*	Fur is dense, soft, and woolly. Upperparts gray or brown, often very dark, underparts are white, yellowish, or almost as dark as back. Mottlings of white and black ventral surface may exist. Tail is usually curled into a ring at the end. Head and body length 12.6–18 in (32–46 cm), tail length 6.7–15.8 in (17–40 cm).	Rainforest, sclerophyll forest, woodland, or brush. Scansorial, nocturnal, movement is slow and quiet.	Northern and western New Guinea, including Yapen Islands (Indonesia).	Variety of leaves, fruits, flowers, bark, and sap.	Vulnerable

[continued]

Common name / Scientific name	Physical characteristics	Habitat and behavior	Distribution	Diet	Conservation status
Golden ringtail *Pseudochirops corinnae*	Upperparts gray or brown, often very dark, underparts are white, yellowish, or almost as dark as back. Mottlings of white and black ventral surface may exist. Tail is usually curled into a ring at the end. Head and body length 12.6–15.8 in (32–46 cm), tail length 6.7–15.8 in (17–40 cm).	Rainforest, sclerophyll forest, woodland, or brush. Scansorial, nocturnal, movement is slow and quiet.	Interior New Guinea.	Variety of leaves, fruits, flowers, bark, and sap.	Vulnerable
Coppery ringtail *Pseudochirops cupreus*	Upperparts gray or brown, often very dark, underparts are white, yellowish, or almost as dark as back. Mottlings of white and black ventral surface may exist. Tail is usually curled into a ring at the end. Head and body length 12.6–15.8 in (32–46 cm), tail length 6.7–15.8 in (17–40 cm). Observed female weight 3 lb (1.4 kg).	Rainforest, sclerophyll forest, woodland, or brush. Scansorial, nocturnal, movement is slow and quiet.	Interior New Guinea.	Variety of leaves, fruits, flowers, bark, and sap.	Not threatened

Resources

Books

Cork, Stephen J., and William J. Foley. "Digestive and Metabolic Adaptations of Arboreal Marsupials for Dealing with Plant Antinutrients and Toxins." In *Marsupial Biology—Recent Research, New Perspectives*, edited by Norman Saunders and Lyn Hinds. Sydney: UNSW Press, 1997.

Flannery, Tim F. *Possums of the World.* Chatswood, Australia: Geo Productions, 1995.

Goldingay, Ross L. "Gliding Mammals of the World. Diversity and Ecological Requirements." In *Biology of Gliding Mammals*, edited by Ross L. Goldingay and John S. Scheibe. Fürth: Filander Verlag, 2000.

Kennedy, Michael, ed. *Australasian Marsupials and Monotremes—An Action Plan for their Conservation.* Gland, Switzerland: IUCN Publication Department, 1992.

Strahan, Ron, ed. *The Australian Museum Complete Book of Australian Mammals.* Sydney: Australian Mammals; Sydney: Australian Museum, 1995.

Winter, John W. "Australasian Possums and Madagascan Lemurs." In *Comparison of Marsupial and Placental Behaviour*, edited by David B. Croft and Udo Ganslosser. Fürth, Germany: Filander Verlag, 1996.

Udo Gansloßer, PhD

Gliding and striped possums
(Petauridae)

Class Mammalia

Order Diprotodontia

Family Petauridae

Thumbnail description
Medium-sized, squirrel-like, active possums, all with marked dorsal facial stripes, four rounded cusps per molar

Size
1–2.5 ft (320–780 mm); 3.3–25.3 oz (95–720 g)

Number of genera, species
3 genera; 12 species

Habitat
Open forest, woodland, closed forest, and rainforest

Conservation status
Endangered: 3 species; Vulnerable: 4 species; Data Deficient: 1 species

Distribution
New Guinea, Australia, including Tasmania

Evolution and systematics

Petaurids are the closest living relatives of pseudocheirids, based on cranial morphology, serology, and microcomplement fixation. Both families together are placed into a superfamily Pseudocheiroidea. There are some late Oligocene to middle Miocene and some early Pliocene records of species most likely belonging to *Petaurus*. Petauridae is subdivided into two subfamilies, Petaurinae and Dactylopsilinae, the latter being restricted to tropical northern Australia and New Guinea. The subfamily distinction is again based primarily on serology.

Physical characteristics

Aside from their distinct facial and dorsal stripes and the bunodent molars (four low, rounded cusps per molar crown), the two subfamilies are quite different in appearance. Petauridae are between 12.5 and 30.7 in (320 and 780 mm) long, weigh 3.3–25.3 oz (95–720 g), and are gray, brown, or cream-colored. They possess distinct frontal and sternal glands, particularly the males. All species of *Petaurus* are characterized by the gliding membrane, spanning from wrist to ankle, and

a bushy tail, primarily used for steering in flight, but with a slightly prehensile tip. Dactylopsilinae are black and white with a very prominent longitudinal stripe pattern. Their fourth fingers are more or less elongated, and they have a very strong, almost skunk-like body odor.

Distribution

Dactylopsilinae are found in New Guinea, and one species occurs in the rainforest areas of northeastern Australia. The absolute lack of any fossil records suggests a recent immigration from New Guinea to Australia. Petaurinae are more widely distributed from New Guinea, through northern and eastern Australia to Tasmania.

Habitat

All species are arboreal, living in anything from open woodland to closed rainforests. They are nocturnal, spending the days in nests, normally in tree hollows. They are active climbers, running up and down thick tree trunks (also descending head first) as well as along horizontal or inclined

A striped possum (*Dactylopsila trivirgata*) uses its long fourth finger to extract larvae from wood. (Illustration by Jarrod Erdody)

The Leadbeater's possum (*Gymnobelideus leadbeateri*) is found only in a remote region in Victoria, Australia. (Photo by Rod Williams. Bruce Coleman, Inc. Reproduced by permission.)

branches and boughs. *P. breviceps*, as found in both field and laboratory studies, regularly experience daily torpor in winter, usually after several days of food withdrawal or restriction. Body temperatures drop to about 50°F (10°C) in these periods.

Behavior

Social organization markedly differs between the subfamilies. Dactylopsilinae are solitary (though no detailed field study has yet been conducted). There are some observations of vicious fighting between males over an estrous female, and the strong odor, as well as the loud, raucous calling during copulation, suggests a possible territoriality. Petaurinae all live in family groups of different sizes, with several adults of both sexes often sharing one nest. The larger species, particularly *Petaurus australis*, are described as monogamous or polygynous, and the smaller ones as polygynous. Hierarchies were found in *P. breviceps*, both for males and females. Subordinate females are harassed by the dominant one, and often lose their young during the first weeks after birth. Males often form co-dominance relationships with a brother or son, both being reproductively active. Dominant males perform paternal care such as baby-sitting, huddling, and grooming the young while the female is out of the nest. Home-range sizes vary between 1.2 acres (0.5 ha) (some *P. breviceps*) and 210 acres (85 ha) (*P. australis*). In *P. australis* there seems to be exclusive use and active defense, thus justifying the term territory. All species of *Petaurus* regularly use vocalization; *P. australis* utters loud,

The striped possum (*Dactylopsila trivirgata*) emits a strong odor from its glands. (Photo by © Roland Seitre/Peter Arnold, Inc. Reproduced by permission.)

long-ranging calls that most likely serve as territorial advertising. All species use a variety of odors that seem to serve as nest and group odors, which emit from pedal, gular, frontal, sternal, and other glands. Feeding sites of *Petaurus* groups are defended against intruders. In *P. breviceps*, there is evidence for male philopatry and female (forced) dispersal. Not much is known about social dynamics in other species. Leadbeater's possum also lives in family groups, and seems to be territorial, which might be related to its feeding behavior.

Feeding ecology and diet

Dactylopsila is insectivorous, using its strong incisors as chisels to pry insects such as beetle larvae from their hollows in trees, or extracting them with their long fourth finger. The latter is also used in a tapping movement to detect hollows of larvae under the bark or in dead wood. Petaurinae are om-

The squirrel glider (*Petaurus norfolcensis*) is similar to the sugar glider (*Petaurus breviceps*). (Photo by Pavel German. Reproduced by permission.)

The yellow-bellied glider (*Petaurus australis*) uses its incisors to break the bark of eucalyptus trees and drink the sap. (Photo by Dr. Robert Thomas and Margaret Orr © California Academy of Sciences. Reproduced by permission.)

The striped possum (*Dactylopsila trivirgata*) has three dark dorsal stripes. (Photo by R. & D. Keller/ANTPhoto.com.au. Reproduced by permission.)

A sugar glider (*Petaurus breviceps*) pair, nuzzling. (Photo by Alan & Sandy Carey/Photo Researchers, Inc. Reproduced by permission.)

A squirrel glider (*Petaurus norfolcensis*) gliding at night, showing its marsupial pouch opening. (Photo by © Hans & Judy Beste/Lochman Transparencies. Reproduced by permission.)

nivorous, but with a heavy emphasis on sap feeding, nectar, and blossoms. All species of *Petaurus* are able to bite wounds into tree trunks to start sap flow, whereas *Gymnobelideus* is reportedly unable to actively prepare feeding sites. This is discussed as a reason for *Gymnobelideus* defending a territory and for *Petaurus* only defending feeding trees. Apart from sap feeding, nectar and pollen are important parts of *Petaurus* diet, while several species are known to be pollinators.

Reproductive biology

Apart from some anecdotal observations on copulation, which takes place on branches and is reported to be accompanied by loud screeches and calls similar to domestic cats, there is nothing known about *Dactylopsila*. Two young are born, and there are two teats in the pouch. In New Guinea, pouch young were found in January and October, while in Australia, reproduction takes place from February to August. Young are carried on the females' back once they leave the pouch. *Petaurus* young are also carried on the females' back after leaving the nest. The largest species (*P. australis*) is usually monogamous and carries only one young once per year, while the other species have one to two young once or twice per year, with births taking place April to September (as far as records from field data are available). In *P. australis* and *P. breviceps*, parental care is also provided by the resident male(s). Gestation in *P. breviceps* is 16 days, and the young remain in the pouch for 60 days. They first leave the nest at an age of 110 days, and are weaned at 110–120 days. Smaller species are typically polygynous.

Conservation status

Several species are currently regarded as not threatened as a result of their widespread distribution. Three species are listed as Endangered: *Dactylopsila tatei* lives on Fergusson Island and possibly on Goodenough Island, a small island west of New Guinea; *P. gracilis* lives only in a small area in north Queensland, and relies on sap of grass trees (*Xanthorrhoea*) part of the year; and Leadbeater's possum, which lives in a very small and diminishing area in Victoria. Almost all the known populations of *G. leadbeateri* live in a eucalypt area destined for rotary clear-felling every 50–80 years. Some other species are considered Vulnerable due to habitat loss. A captive breeding program and recovery plan has been established for *Gymnobelideus*. For *P. australis* and *P. norfolcensis*, recovery plans are being implemented to identify habitat threats, protect key habitats, etc.

Significance to humans

Petaurus breviceps are currently becoming more and more popular as pet species, with a varying degree of adequacy in husbandry techniques. There is some welfare concern about this, and imports from New Guinea might be illegal and of future conservation concern.

A squirrel glider (*Petaurus norfolcensis*) foraging at night in eastern Australia. (Photo by Tom McHugh/Photo Researchers, Inc. Reproduced by permission.)

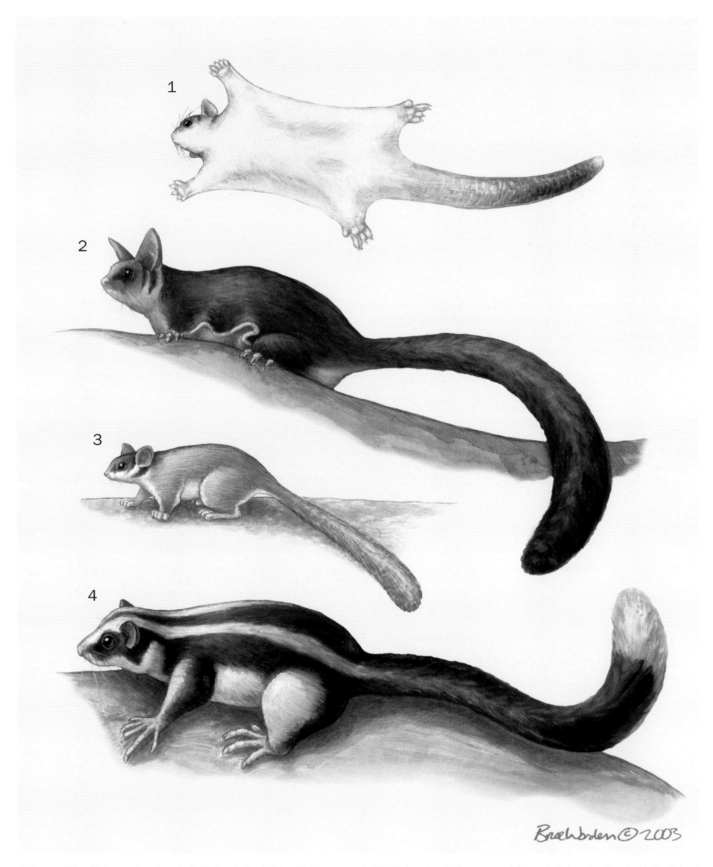

1. Sugar glider (*Petaurus breviceps*); 2. Yellow-bellied glider (*Petaurus australis*); 3. Leadbeater's possum (*Gymnobelideus leadbeateri*); 4. Striped possum (*Dactylopsila trivirgata*). (Illustration by Bruce Worden)

Grzimek's Animal Life Encyclopedia

Species accounts

Sugar glider
Petaurus breviceps

SUBFAMILY
Petaurinae

TAXONOMY
Petaurus breviceps Waterhouse, 1839, New South Wales, Australia. Four subspecies.

OTHER COMMON NAMES
French: Phalanger volant; German: Kurzkopf-Gleitbeutler.

PHYSICAL CHARACTERISTICS
Length 1–1.3 ft (325–420 mm); weight 3.3–5.6 oz (95–160 g). Blue-gray coat with light underside. Black stripe running length of body. Two smaller stripes on sides of face. Gliding membrane extends from fore foot to ankle.

DISTRIBUTION
Tasmania, eastern and northern Australia, New Guinea, and some Moluccan/New Britain islands.

HABITAT
Forests and woodlands.

BEHAVIOR
Nocturnal, arboreal animal, living in family groups. Spends the day in nests, mostly in tree hollows. Capable of gliding flight, for at least 230 ft (70 m).

FEEDING ECOLOGY AND DIET
Sap-feeder and omnivore; feeds also on nectar, pollen, insects, and even small vertebrates.

REPRODUCTIVE BIOLOGY
One and two young, once or twice per year, gestation 16 days, pouch-gravidity 60 days, weaning age 120 days. Monogamous.

CONSERVATION STATUS
Not threatened.

SIGNIFICANCE TO HUMANS
Increasingly kept as pets. ◆

Yellow-bellied glider
Petaurus australis

SUBFAMILY
Petaurinae

TAXONOMY
Petaurus australis Shaw, 1791, Sydney, Australia. Two subspecies.

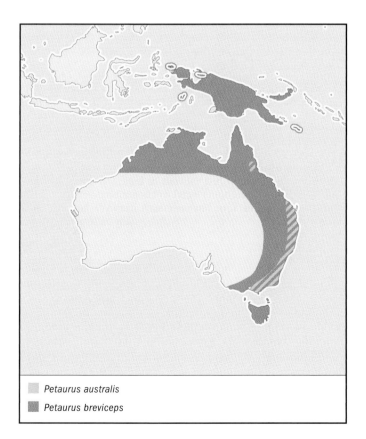

Petaurus australis
Petaurus breviceps

Dactylopsila trivirgata
Gymnobelideus leadbeateri

OTHER COMMON NAMES
English: Fluffy glider; French: Grand phalanger volant; German: Riesenbeutelflughürnchen.

PHYSICAL CHARACTERISTICS
Length 2.2–2.5 ft (690–780 mm); weight 15.8–25 oz (450–710 g). Gray-brown silky fur with light underside and black feet. Dark stripe on thigh. Unique compartmental pouch.

DISTRIBUTION
A continuous range exists along the Australian east coast and several isolated populations in three areas in Victoria. Additionally, there is *P. a. reginae* in tropical Queensland.

HABITAT
Tall, mature eucalypt forests in temperate to tropical Australia.

BEHAVIOR
Nocturnal, arboreal glider with gliding leaps up to 328–393 ft (100–120 m); living in family groups of one pair, or up to one male, three females, plus dependant offspring. Extensive marking behavior with scent glands, including loud, long-ranging calls. Probably territorial.

FEEDING ECOLOGY AND DIET
Sap-feeder and omnivore; feeds also on nectar, pollen, insects, and even small vertebrates.

REPRODUCTIVE BIOLOGY
One young per year, pouch life lasts up to 100 days, another 60 days spent in nest; juveniles disperse not before 18–24 months old. May be monogamous or polygynous.

CONSERVATION STATUS
Considered Lower Risk/Near Threatened due to its reliance on tall, mature eucalypt trees for feeding and nesting, which are threatened by logging, even when done selectively.

SIGNIFICANCE TO HUMANS
None known. ◆

Leadbeater's possum
Gymnobelideus leadbeateri

SUBFAMILY
Petaurinae

TAXONOMY
Gymnobelideus leadbeateri McCoy, 1867, Bass River, Victoria, Australia.

OTHER COMMON NAMES
French: Phalanger de Leadbeater; German: Leadbeater Hörnchenbeutler.

PHYSICAL CHARACTERISTICS
Length 0.4–1.1 ft (129–350 mm); weight 3.5–5.8 oz (100–166 g). Gray-brown coat with light underside. Dark stripe runs along the back. Lacks gliding membrane.

DISTRIBUTION
Several scattered, isolated populations in southeastern Victoria, near Melbourne.

HABITAT
Woodland and eucalypt forests.

BEHAVIOR
Nocturnal, arboreal, lacks a gliding membrane but is capable of quite extensive leaps. Lives in family groups of up to eight animals. Mating system monogamous, females force female offspring to disperse. Territories of 2.4–2.9 ac (1–2 ha) size. No scent glands.

FEEDING ECOLOGY AND DIET
Sap-feeder, insectivore.

REPRODUCTIVE BIOLOGY
One to two young, births peak in May–June and October–November, but occur in all months except January–February. Monogamous.

CONSERVATION STATUS
Listed as Endangered due to its small, declining population.

SIGNIFICANCE TO HUMANS
None known. ◆

Striped possum
Dactylopsila trivirgata

SUBFAMILY
Dactylopsilinae

TAXONOMY
Dactylopsila trivirgata Gray, 1858, Aru Islands, Indonesia. Four subspecies.

OTHER COMMON NAMES
English: Striped phalanger; French: Grand phalanger à pelage raye; German: Streifenbeutler.

PHYSICAL CHARACTERISTICS
Length 1.8–2.0 ft (560–610 mm); weight 8.4–15.5 oz (240–440 g). Coarse gray-white fur with three black stripes along the back. Small black spot on chin. Tail is bushy with white tip.

DISTRIBUTION
All of New Guinea, including several small islands, tropical northeastern Australia.

HABITAT
Upland and medium mountainous forests, up to 4,593 ft (1,400 m), most common between 656–3,609 ft (200–1,100 m); common in secondary forest.

BEHAVIOR
Nocturnal, arboreal, mostly solitary (rare records of two animals sharing one nest); nests from plant material in tree hollows. Walks very fast and in strange, rowing limb action.

FEEDING ECOLOGY AND DIET
Insectivorous; chisel-like teeth and elongated fourth finger are used to pry open or extract insect larvae from wood; also eats leaves, small vertebrates, fruit, and honey.

REPRODUCTIVE BIOLOGY
Mating takes place from February to August; two young per litter are born. Probably polygynous.

Common name / Scientific name	Physical characteristics	Habitat and behavior	Distribution	Diet	Conservation status
Great-tailed triok *Dactylopsila megalura*	Thick, close, woolly fur, dark stripes on back. Basal color is white or gray. Black spot on chin. Head and body length 6.6–12.5 in (17–32 cm), tail length 7.4–15.7 in (19–40 cm).	Rainforest or sclerophyll forest. Nocturnal, arboreal, can give off unpleasant odor.	Interior New Guinea.	Insects, fruits, and leaves.	Vulnerable
Long-fingered triok *Dactylopsila palpator*	Coloration is smoky brown on a basal color of grayish tawny. Fur is thick, close, and woolly. Slender body. Head and body length 6.6–12.5 in (17–32 cm), tail length 7.4–15.7 in (19–40 cm).	Rainforest or sclerophyll forest. Nocturnal, arboreal, can give off unpleasant odor.	Interior New Guinea.	Insects, fruits, and leaves.	Not threatened
Tate's triok *Dactylopsila tatei*	Thick, close, woolly, harsh, fur. Three parallel, dark stripes on back over basal color of white or gray. Slender body. Head and body length 6.6–12.5 in (17–32 cm), tail length 7.4–15.7 in (19–40 cm).	Rainforest or sclerophyll forest. Nocturnal, arboreal, can give off unpleasant odor.	Fergusson Island; Papua New Guinea.	Insects, fruits, and leaves.	Endangered
Northern glider *Petaurus abidi*	Upperparts are generally grayish, under parts are paler. Fur is fine and silky. Dark dorsal stripe runs from nose to rump. Stripes on each side of face through the eye to the ear. Head and body length 4.7–12.5 in (12–32 cm), tail length 5.9–18.9 in (15–48 cm).	Wooded areas. Prefer open forest. Arboreal and nocturnal.	North-central New Guinea.	Insects and the sap and gum of eucalyptus and acacias.	Vulnerable
Mahogany glider *Petaurus gracilis*	Coloration of upperparts varies from mahogany brown to smoky gray with patches of yellow-brown on shoulders, flanks, and rump. Blackish midline from between eyes to lower back. Long tail, usually honey-gray or blackish in color. Head and body length 8.5–10.5 in (21.8–26.5 cm), tail length 11.8–15 in (30–38 cm), weight 8.9–14.5 oz (255–410 g).	Found in a mosaic of habitats dominated by medium to low eucalyptus-acacia woodland on swampy coastal plains and extensive beach ridges at an altitude between sea level and 390 ft (120 m). Nocturnal, arboreal.	Known only from the region of Barrett's Lagoon, near Tully, Queensland, Australia.	Insects and the sap and gum of eucalyptus and acacias.	Endangered
Squirrel glider *Petaurus norfolcensis*	Coloration of fur is pale gray on dorsal surface, dark brown or black stripe down middle. Prehensile tail, long gliding membrane that extends from outside of forefoot to ankle. Weight 6.7–10.6 oz (190–300 g).	Dry sclerophyll forest and woodlands. Nocturnal, arboreal, live in small family groups. Breeding season is during June and July.	Eastern Queensland, eastern New South Wales, and eastern Victoria, Australia	Insects (mainly beetles and caterpillars), acacia gum, sap from certain eucalyptus, nectar, pollen, and the green seeds of the Golden Wattle.	Lower Risk/Near Threatened

CONSERVATION STATUS

Not threatened.

SIGNIFICANCE TO HUMANS

None known. ◆

Resources

Books

Flannery, Tim F. *Possums of the World.* Chatswood, Australia: Geo Production, 1994.

Goldingay, Ross L., and Rodney P. Kavanagh. "The Yellow-bellied Glider: A Review of Its Ecology and Management Considerations." In *Conservation of Australian Forest Fauna,* edited by D. Lunney. Mossman, Australia: Royal Zoological Society of NSW, 1991.

Goldingay, Ross L., and John H. Scheibe, eds. *Biology of Gliding Mammals.* Fürth, Germany: Filander Verlag, 2000.

Kennedy, Michael, ed. *Australasian Marsupials and Monotremes. An Action Plan for their Conservation.* Gland, Switzerland: IUCN Publication Department, 1992.

Salamon, Birgitt. *The Consequences of Social Dominance in the Sugar Glider* (Petaurus breviceps). University of Erlangen, Nürnberg PhD dissertation, 1998.

Strahan, Ronald, ed. *The Australian Museum Complete Book of Australian Mammals.* Sydney: Australian Museum, 1995.

Udo Gansloßer, PhD

Honey possums
(*Tarsipedidae*)

Class Mammalia
Order Diprotodontia
Family Tarsipedidae

Thumbnail description
Very small mouselike possum with long, tapering snout and very long, partially prehensile tail; eyes large and round, ears large, rounded, and sparsely haired; fingers and toes are long with rounded tips and small nails; fur is grayish brown with three dark dorsal stripes

Size
Head and body length 1.6–3.7 in (4–9.5 cm); tail length 1.8–4.3 in (4.5–11 cm); weight 0.2–0.6 oz (7–16 g)

Number of genera, species
1 genus, 1 species

Habitat
Arboreal in flowering trees and shrubs, mainly in heathland

Conservation status
Not listed as threatened, but may be at risk from habitat loss

Distribution
Southwestern Australia

Evolution and systematics

The Australian honey possum, *Tarsipes rostratus*, was formerly classified along with other Australian possums in the family Phalangeridae. It was deemed unusual enough to deserve its own subfamily (Tarsipedinae), but it was not until the mid-1970s that the true extent of its uniqueness became apparent. Today, this tiny possum is known to be one of the most divergent marsupials, easily worthy of its own family, Tarsipedidae. It is the sole survivor of an otherwise extinct marsupial group that diverged from the diprotodont lineage during the Pleistocene, about two million years ago. In this relatively short geological time span, the lineage has evolved rapidly and developed one of the most specialized lifestyles of any mammal group. Its closest relatives appear to be the feather-tailed possums and gliders of the family Acrobatidae.

The taxonomy for this species is *Tarsipes rostratus* Gervais and Verreaux, 1842, Western Australia, Australia.

Physical characteristics

The honey possum is a dainty animal, adapted for climbing. Its hands and feet bear long, grasping, monkeylike fin-

The honey possum (*Tarsipes rostratus*) is also known as the noolbender. (Photo by C. H. Tyndale-Biscoe/Mammal Images Library of the American Society of Mammalogists.)

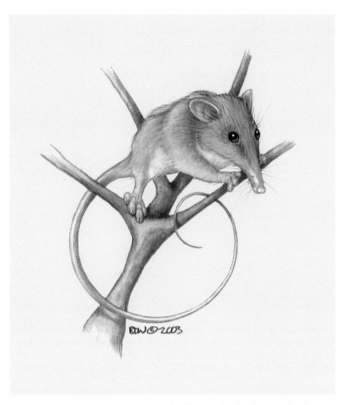

Honey possum (*Tarsipes rostratus*). (Illustration by Bruce Worden)

gers and toes. As in all diprotodont marsupials, the second and third toes of the hind feet are fused, or syndactylous. Only the fused toes of the hind foot have claws, which are used for grooming. The other digits are equipped instead with a small, round nail. The first toes are opposable, like those of a climbing primate. The honey possum's agility is greatly enhanced by its long tail, the tip of which is prehensile. The tail is used as a counter-balance, a safety line, and a fifth limb. It is very sparsely haired, and naked on the underside toward the tip. The bare skin provides excellent grip when the tail is curled around branches and stems.

The honey possum's head is distinctive, with a long, shrew-like snout bearing long whiskers. The tongue is longer still and can be extended the length of the head beyond the tip of the snout. The teeth are reduced. Only the lower front incisors are at all developed, while the others are little more than weak pegs. The fur is coarse and grayish brown, with very long guard hairs. The honey possum's eyes are large, as befits a nocturnal animal, and situated so that they face both forward and upward. This arrangement combines the advantages of binocular vision, which they need to move confidently around a three-dimensional environment, as well as be alert for airborne predators such as owls.

Distribution

Honey possums have a very limited distribution. They live in the extreme southwestern corner of Australia, hemmed in to the south and west by oceans, to the east by the vast desert of the Nullabor and to the north by high ground.

Habitat

Within its limited range, the honey possum lives only on the sandy and coastal heaths, where conditions suit nectar-rich myrtles, proteas, banksias, and other flowering plants that, between them, bloom throughout the year.

Behavior

Honey possums are nocturnal. Their foraging behavior is a cross between that of a monkey and a honeybee—they scurry nimbly about the branches of flowering trees and shrubs, pausing often to thrust their snout into every floret. Being small, the honey possum uses a lot of energy keeping warm and, when resting in cool conditions, it curls up to reduce its surface area. For most of the year, the animals live alone and rest in tree holes or abandoned bird nests. In periods of food scarcity, the honey possum runs low on energy and is sometimes forced into periods of inactivity during which it enters a deep torpid sleep; its heart rate is reduced and its body temperature drops. Under these circumstances, several animals

The honey possum (*Tarsipes rostratus*) uses its tail to steady itself while feeding on a flower. (Photo by Fredy Mercay/ANTPhoto.com.au. Reproduced by permission.)

A honey possum (*Tarsipes rostratus*) on coral gum. (Photo by Animals Animals ©G. Thompson, OSF. Reproduced by permission.)

The honey possum's (*Tarsipes rostratus*) tongue can reach about 1 in (2.4 cm) past its nose, enabling it to forage deep into flora. (Photo by Martin B. Withers/FLPA–Images of Nature. Reproduced by permission.)

may cluster together because, by sharing small amounts of body heat, they can save more energy still. They wake periodically and when things have improved, they resume diligent solitary feeding to replenish their depleted reserves.

Feeding ecology and diet

The teeth of honey possums are few and underdeveloped—little more than short pegs incapable of biting or chewing. Despite their common name, honey possums rarely eat honey. In fact, they feed more or less exclusively on pollen and nectar, a diet that makes them highly unusual among mammals other than bats. In a dramatic example of convergent evolution, these tiny marsupials are equipped with a long tongue, like that of nectar-feeding insects or hummingbirds, with which they can reach into the nectaries of deep flowers. Unlike an insect proboscis, however, the tongue is not tubular. Instead, it bears at its tip a highly specialized arrangement of bristles, like a tiny brush. This collects pollen grains and sticky nectar and delivers them into the possum's mouth where they are sucked off or scraped onto ridges on the roof of the mouth. Even in very small doses, this diet provides ample sugar and protein to fuel and maintain a small body, so long as there is a more or less continuous supply.

Reproductive biology

Female honey possums are larger than males and socially dominant. They can come into breeding condition at any time of the year, providing food is plentiful, though there is often a lull during mid-summer when nectar and pollen are less abundant. Males gather around an estrous female and com-

The honey possum (*Tarsipes rostratus*) has a dark stripe along its back to the end of its tail. (Photo by Dick Whitford/Nature Focus, Australian Museum. Reproduced by permission.)

A honey possum (*Tarsipes rostratus*) feeding on a *Banksia* flower. (Photo by © Jiri Lochman/Lochman Transparencies. Reproduced by permission.)

sums give birth to the smallest mammalian babies, each one weighing no more than 0.002 oz (5 mg). Most females will mate again almost immediately after giving birth. This second litter is an insurance policy—the fertilized eggs do not develop beyond a very early stage until the previous litter has left the pouch or died. Hence, while honey possum gestation requires only 3 weeks, pregnancies can last anywhere up to 13 weeks. Litters never contain more than four young; this is the maximum number a female can suckle on the four teats in her pouch. The litter spends about eight weeks in the pouch, during which time each baby grows to about 0.1 oz (2.5 g), which is a remarkable 500 times its birth weight. After that, the young are too big to fit in the pouch and the female deposits them in a spherical woven nest until they are weaned at about 10 weeks of age. A healthy female will normally rear two litters in a year, but will rarely live long enough to breed a third time.

Conservation status

This species remains common in areas of suitable habitat, despite predation by introduced carnivores such as foxes and cats. The biggest threat comes from habitat loss due to urbanization and land development, but conservation authorities are monitoring the situation closely to ensure that this unique marsupial lineage continues.

Significance to humans

Honey possums are popular animals, though their secretive, nocturnal lifestyles make them difficult to see. They do no harm to human interests and perform a valuable service in helping to pollinate many species of heathland flower.

pete for the right to mate, before moving quickly on. Amazingly for such small animals, the sperm produced by males are very large—at 0.01 in (0.3 mm), they are the longest of any mammal. As if to redress the balance, female honey pos-

---

Resources

Books

Aplin, K. P., and M. Archer. "Recent Advances in Marsupial Systematics with a New Syncretic Classification." In *Possums and Opossums: Studies in Evolution.* Sydney: Surrey Beatty and Sons and Royal Zoological Society of New South Wales, 1987.

Morris, P., and A. J. Beer. "Honey Possum." In *World of Animals: Mammals.* Vol. 10, *Marsupials.* Danbury, CT: Grolier, 2002.

Nowak, R. "Honey Possum (Diprotodontia; Family Tarsipedidae)." In *Walker's Mammals of the World,* 6th ed.,

Vol I. Baltimore and London: Johns Hopkins University Press, 1999.

Strahan, R. *The Mammals of Australia, Revised Edition.* Sydney: New Holland, 1998.

Organizations

CALM (Department of Conservation and Land Management) South Coast Office. 120 Albany Highway, Albany, Western Australia 6330 Australia. Phone: (08) 9841 7133. Fax: (08) 9842 4500. E-mail: albanydistrict@calm.wa.gov.au Web site: <http://www.calm.wa.gov.au>

Amy-Jane Beer, PhD

Feather-tailed possums
(*Acrobatidae*)

Class Mammalia
Order Diprotodontia
Family Acrobatidae

Thumbnail description
Small possums with a very long feather-like tail, fringed on two sides with straight hairs; the body fur is predominantly pale gray, white on the underside, with white and dark face and body markings that vary with species

Size
Head and body length 2.4–4.7 in (6–12 cm); tail 2.6–6 in (6.5–15 cm); weight 0.35–2.1 oz (10–60 g)

Number of genera, species
2 genera; 2 species

Habitat
Forests, woodland, scrub, and gardens

Conservation status
Not threatened

Distribution
Eastern Australia and New Guinea

Evolution and systematics

This small family of just two living species has created more than its share of taxonomic debate. Feather-tailed possums were once thought to be true possums of the family Phalangeridae. In the 1970s, they were moved to the pygmy possum family, Burramyidae. Soon after that, they were given full family status of their own and became the Acrobatidae, allied first to the pygmy possums in the superfamily Burramyoidea, then with gliding possums in the Petauroidea. In 1987, the family was repositioned again, this time alongside the honey possum in the superfamily Tarsipedoidea.

Acrobatids differ from other possums in having six pads on their feet instead of five (an adaptation to enhance grip when climbing) and a tail with rows of long stiff hairs along each side, forming a feather-like structure. Undoubtedly, this is an adaptation to gliding. In the Australian pygmy glider (*Acrobates pygmaeus*), the tail is used as a rudder to give the animal increased control as it glides though the air on membranes of skin stretched between its front and back legs. But the other species, New Guinea's feather-tailed possum (*Distoechurus pennatus*), does not have such a membrane and cannot glide.

A pygmy glider (*Acrobates pygmaeus*) in mid-leap. (Photo by C. & S. Pollitt/ANTPhoto.com.au. Reproduced by permission.)

The tail of the pygmy glider (*Acrobates pygmaeus*) is its most notable characteristic. (Photo by Pavel German. Reproduced by permission.)

The featherlike tail is a relic that shows that this species has abandoned the gliding lifestyle its ancestors spent several million years evolving.

Physical characteristics

Feather-tailed possums possess a variety of adaptations to life in the trees, as well as a number of physical characteristics that set them well apart from other possums. *Acrobates* is tiny—the smallest gliding marsupial, weighing less than 0.5 oz (14 g). The gliding membrane is a narrow strip of furry skin. The fingers and toes end in expanded round pads with minutely serrated surfaces and sharp claws, thus ensuring good grip on rough and smooth surfaces alike. *Distoechurus* is longer and three to four times heavier than its gliding cousin and its fingers and toes lack expanded tips. Its face bears distinctive markings in the form of black and white stripes. Both species have large eyes and rounded, sparsely haired ears with complex fleshy nodules at the opening to the ear canal.

Distribution

Both species of feather-tailed possum are doubtless descended from a common ancestor that lived throughout what is now eastern Australia and New Guinea before rising sea levels flooded the Torres Straits. *Distoechurus* is endemic to New Guinea and *Acrobates* is found from the extreme northeastern tip of Australia at Cape York all the way down to southernmost Victoria and South Australia. It does not occur on Tasmania.

A pygmy glider (*Acrobates pygmaeus*) on a *Banksia* flower. (Photo by Pavel German. Reproduced by permission.)

At 0.4–0.5 oz (10–14 g), the pygmy glider (*Acrobates pygmaeus*) is the world's smallest glider. (Photo by Zoological Society of San Diego. Reproduced by permission.)

A pygmy glider (*Acrobates pygmaeus*) hunts along a slender branch for insects in Toowoomba, Queensland, Australia. (Photo by Bruce G. Thomson/Photo Researchers, Inc. Reproduced by permission.)

Habitat

Acrobatids are strictly arboreal. Individuals rarely, if ever, venture to ground level, where they would be vulnerable to predators, including other mammals and reptiles. The pygmy glider lives in a diverse range of forest habitat from tropical rainforest of northern Queensland to the montane eucalyptus woodlands of the south.

Behavior

Feather-tailed possums are nocturnal, shy, and elusive. They live in tall trees, often in remote forests, and *Acrobates* has the ability to swoop from one tree to another. All this makes close observation extremely difficult; many details of their everyday lives remain unknown. Even pygmy gliders kept in captivity have proved tricky subjects for study—gliders kept in sparse enclosures or in pairs rather than groups do not survive well or breed. *Acrobates* is highly social. In the wild, groups of up to 20 or more animals of both sexes and various ages may build and share a series of nests within a joint home range. Aggression is apparently rare, and animals of all ages rest huddled close together, especially during periods of cold weather

or food shortage when they enter a deep torpid sleep. In contrast, *Distoechurus* spends more time alone or in pairs. Being relatively large, it is less dependent on shared body heat to conserve energy.

Both species of acrobatid possum are superb climbers, but only *Acrobates* has retained the ability to glide. A glide is an extended leap, during which the animal splays its four limbs to open a narrow web of furry skin called the "patagium." This serves as a parachute, slowing the rate of descent and giving the possum some control over its trajectory and allowing it to travel 65 ft (20 m) or more through the air. Just before landing, the possum swings its hind feet forward, thus reducing airspeed, and bracing all four legs for landing on a vertical surface such as a tree trunk. The featherlike tail is used as a rudder and a brake to help control the short "flight."

Feeding ecology and diet

Feather-tailed possums specialize in high-energy, high-protein foods such as nectar, flowers, and insects. Most feeding occurs at night, although nursing mothers are sometimes forced to forage during the day to meet their family's demands for milk.

A pygmy glider (*Acrobates pygmaeus*) leaps from tree to tree. (Photo by C. & S. Pollitt/ANTPhoto.com.au. Reproduced by permission.)

Reproductive biology

Breeding can happen at any time of year in the tropics, but births in both species have a seasonal peak in spring. Female pygmy gliders breed in their first year and may share a nest with other adults, including other breeding females. A post-partum estrus and embryonic diapause mean that a second litter may be born within a day or two of the first litter being weaned. Very little is known about the reproductive biology of *Distoechurus*. Both species might be polygynous.

Conservation status

Both *Distoechurus* and *Acrobates* are thought to be common in suitable habitat, though detailed population information is strictly limited.

Significance to humans

None known.

1. Feather-tailed possum (*Distoechurus pennatus*); 2. Pygmy glider (*Acrobates pygmaeus*). (Illustration by Joseph E. Trumpey)

Species accounts

Feather-tailed possum
Distoechurus pennatus

TAXONOMY
Distoechurus pennatus (Peters, 1874), Irian Jaya, Indonesia.

OTHER COMMON NAMES
French: Possum plumée de Nouvelle Guinée; German: Feder-schwanzbeutler; Spanish: Opósum de cola plumose.

PHYSICAL CHARACTERISTICS
Head and body length 4–5 in (10–13 cm), tail 5–6 in (12–15 cm) long; fur mainly gray, with black and white facial stripes. No gliding membrane.

DISTRIBUTION
New Guinea.

HABITAT
Disturbed and secondary forest, rainforest, scrubland, and gardens.

BEHAVIOR
Nocturnal, arboreal, solitary most of the year.

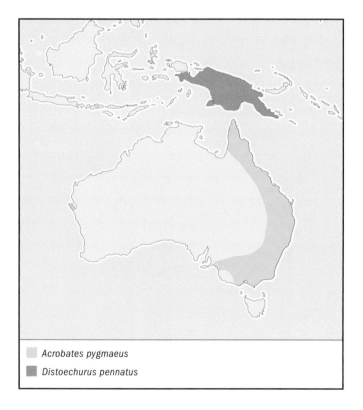

Acrobates pygmaeus
Distoechurus pennatus

FEEDING ECOLOGY AND DIET
Flowers, fruit, and insects.

REPRODUCTIVE BIOLOGY
Probably polygynous. One or two young born in spring.

CONSERVATION STATUS
Not threatened; apparently abundant.

SIGNIFICANCE TO HUMANS
None known. ◆

Pygmy glider
Acrobates pygmaeus

TAXONOMY
Acrobates pygmaeus (Shaw, 1793), Sydney, Australia.

OTHER COMMON NAMES
English: Feathertail glider, pygmy gliding possum, pygmy phalanger, flying mouse; French: L'acrobat pygmée; German: Zwerggleitbeutler; Spanish: Acróbata pigmeo.

PHYSICAL CHARACTERISTICS
Head and body length 2.5–3 in (6.5–8 cm); tail 2.5–3 in (6.5–8 cm) long; soft gray fur, white on underside, dark eyerings and variable body markings; loose skin along flanks forms gliding membrane when limbs are spread wide.

DISTRIBUTION
Eastern Australia.

HABITAT
Eucalyptus forest and woodland.

BEHAVIOR
Nocturnal, arboreal, highly social, glides between trees.

FEEDING ECOLOGY AND DIET
Insects and nectar.

REPRODUCTIVE BIOLOGY
Probably polygynous. Litters of three to four young born at any time of year; spend 65 days in the pouch, then suckled in the nest for another month.

CONSERVATION STATUS
Not threatened, and presumed secure.

SIGNIFICANCE TO HUMANS
None known. ◆

Resources

Books

Flannery, T. *Mammals of New Guinea*. Carina, Australia: Robert Brown and Associates, 1990.

Morris, P., and A.-J. Beer. *World of Animals: Mammals*. Vol 10, *Marsupials*. Danbury, CT: Grolier, 2002.

Nowak, R. "Feather-tailed Possums (Diprotodontia; Family Acrobatidae)." In *Walker's Mammals of the World*. Vol I, 6th edition. Baltimore and London: Johns Hopkins University Press, 1999.

Strahan, R. *The Mammals of Australia, Revised Edition*. Sydney: New Holland, 1998.

Amy-Jane Beer, PhD

Xenarthra

(Sloths, anteaters, and armadillos)

Class Mammalia

Order Xenarthra

Number of families 4

Number of genera, species 13 genera; 30 species

Photo: A three-toed sloth (*Bradypus infuscatus*) hangs from a tree limb with its baby. (Photo by M. Fogden. Bruce Coleman, Inc. Reproduced by permission.)

Evolution and systematics

Evolved in South America, this diverse order first appears in the fossil record in the Paleocene, about 60 million years ago (mya). It has two main groups, the Pilosa and the Cingulata. The Pilosa contains sloths and anteaters, also known as the hairy xenarthrans, and the Cingulata includes the extinct glyptodonts and armadillos, the animals with bony carapaces. The group name "Xenarthra" refers to the additional articulations between the lumbar vertebrae, called xenarthrous processes. These extra articulations provide an unusually large amount of stability to the pelvic region, and have been hypothesized to confer advantages on sloths and arboreal anteaters in reaching out from one support to another with the body held horizontally, and to the armadillos for digging with great strength and speed. The hands in xenarthrans usually have two or three digits larger than the others, although the number of digits in tree sloth hands is reduced to two or three. All fingers have sharp claws, which are often long and laterally compressed. Other than the tree sloths, which have three toes on the hind foot, most xenarthrans have five toes, each bearing a sharp claw. Sloths and anteaters have a tendency toward supination of the forearm and hands. Freedom to rotate the wrist is useful for climbing in trees as well as allowing anteaters to turn the sharp claws upward and inward for protection while walking terrestrially. The hind foot

in some extinct sloths was also capable of supination, and the animals walked on the outsides of the feet, again to protect long, sharp claws. The radius and ulna are separate, which also contributes to supination ability. The scapulae are large and have a second spinous process posterior to and parallel to the first. This characteristic is particularly important for increasing the surface area for attachment of the muscles used in retracting the forelimb as would be necessary in digging, and is most striking in the armadillos and anteaters although also present in the sloths, contributing to their ability to climb.

The earliest xenarthrans were small, resembling primitive armadillos more than sloths or anteaters. Presently, no direct ancestral fossil lineage leading to the xenarthran groups is known. By the time sloths, anteaters, armadillos, and glyptodonts occur in the fossil record, the distinctions among lineages as well as those within the lineages are clear. Past authors considered Paleanodonts to be xenarthran ancestors, but they are now recognized instead as ancestors to the Pholidota, a group that includes the living pangolins. Even though the Paleanodonts were not of the direct lineage leading to the Xenarthra, they probably were of a body form similar to xenarthran ancestors and may be the equivalent of a sister group. These animals were small, armadillo-like, but lacked the bony armor seen in the earliest xenarthrans. They had reduced dentitions, lacking enamel, as was probably a primitive characteristic of

The southern two-toed sloth (*Choloepus didactylus*) uses all four limbs to move along a tree branch. (Photo by Kate McDonald. Bruce Coleman, Inc. Reproduced by permission.)

the ancestral xenarthran. Both groups arose from Insectivores in the late Cretaceous, over 70 mya.

The earliest sloths appear in the fossil record in the Oligocene. These are the mylodonts, the family Mylodontidae, the group that retains more primitive characters than the other two, including dermal ossicles. The other two lineages, the family Megatheriidae and the family Megalonychidae, are first recorded in the fossil record in the Miocene. Although some extinct sloths were very large, all three early lineages were small to moderate and increased size through time, culminating in the giant Pleistocene megatheres and eremotheres. Although the largest sloths went extinct at the end of the Pleistocene, some may have persisted to less than 13,000 years ago. Speculations that extinct large sloths coexisted with humans and may have been driven to extinction by hunting activities are refuted by differences in the level of the strata in which sloth remains and human artifacts are found in caves. These differences indicate that the sloths used the caves at some time prior to their occupation by humans. To date, there is no conclusive evidence of human and sloth interaction.

The fossil record of the anteaters is the most fragmentary of the xenarthrans, and the earliest species do not significantly differ from the more recent ones. No fossil anteater had teeth, and even the earliest had elongate heads and, presumably, long tongues. Together with a body form reminiscent of the living anteaters, it is likely that by the time they appear in the fossil record, they were restricted to eating insects.

The Cingulata, or the armored xenarthrans, are closer in body form to the earliest of the xenarthrans. Of the three main groups, the chlamytheres or giant armadillos, the glyptodonts, and the armadillos, the first two are entirely extinct. The chlamytheres were similar to living armadillos in

body form, although some were much larger. The cingulates that differed the most from the other lineages were the glyptodonts. As did the sloths, these animals increased in body size through time, and the largest survived into the Pleistocene. Not only did the glyptodonts have solid armor on their bodies, but their tails were encased in bony plates and some had solid bone club-like expansions at the ends. They probably were used very effectively for defense.

The dentition in xenarthrans is typically reduced in tooth types and numbers and all lack enamel. There is no milk dentition, and the teeth are ever-growing. No xenarthran has identifiable incisors. Sloths are the only xenarthrans with canine-shaped teeth, and in these animals they occlude upper in front of lower, opposite from the pattern in other mammals, making their relationships to true canine teeth uncertain. Therefore, in sloths those teeth are called "caniniform." Likewise, neither premolars nor molars can be distinguished in sloths, armadillos or glyptodonts, and the cheek teeth are all similar in appearance and all called "molariform." The anteaters are the only edentulous xenarthrans, although the

Almost all of the Hoffmann's two-toed sloth (*Choloepus hoffmanni*) typical behavior happens while the animal is upside down. (Photo by Tom Brakefield. Bruce Coleman, Inc. Reproduced by permission.)

A pichi armadillo (*Zaedyus pichiy*) is also known as pygmy armadillo. (Photo by Tom McHugh/Photo Researchers, Inc. Reproduced by permission.)

ture of having additional vertebral articular surfaces that allow them to stretch themselves horizontally from a vertical support while holding on only by the hind limbs.

Anteaters are recognized by their long, tapered snouts, remarkably long and thin tongue, and large, powerful foreclaws. The foreclaws are used both for defense and for the purpose of tearing open ant and termite mounds. The pelage is long and thick enough to temporarily protect the animals from invading insects. All but one species has a grasping prehensile tail.

Armadillos are the only living mammals with a protective bony skin armor. Unlike reptile shells, an armadillo's mail is interrupted by several folds of skin to assist with agility. The skin's surface is gray or brown, and quite soft. They are stocky, medium-sized mammals that walk low to the ground. Head and ear shape varies among species, and powerful limbs bear enlarged claws for digging burrows and gathering food.

group was previously known as the Edentata. The teeth in sloths erupt as simple cones, and acquire the cusp pattern characteristic of each species through wear caused by movements of the masticatory muscles. The generation of tooth wear patterns in other xenarthrans has not been studied.

Physical characteristics

Xenarthrans living today range in size from the smallest, *Chlamyphorus*, the fairy armadillo, at about 5 in (12.5 cm) head and body length, to *Myrmecophaga*, the giant anteater, at about 47 in (120 cm) head and body length. Some extinct forms were larger; the extinct glyptodonts were over 6.5 ft (2 m) in head and body length, and the largest of the extinct sloths probably exceeded 10 ft (3 m) and were as heavy as modern elephants. It is probable that all early xenarthrans had some form of bony armor. Dermal ossicles occur in some extinct sloths, although living sloths and anteaters have entirely lost this tendency.

Tree sloths have pear-shaped bodies with large abdomens allowing a large cecum and long, slender limbs ending in elongated curved claws. Their fingers are bound together by skin, so the claws are often mistaken for their fingers or toes. All tree sloths have three claws on their hind feet, but the common names of "two-toed" and "three-toed" are based on the number of digits present on their front feet. Their outer pelage is long and coarse, and there is a short, soft, dense undercoat. Their heads are small and rounded. The eyes in *Bradypus* are small and dark; both sloths show only pinhole opening for the pupil and are not believed to see well. Tree sloths have external ears, but these are typically hidden in the elongate guard hair of the outer fur coat. Sloths are extremely unusual among mammals in having a variable number of cervical vertebrae (six in *Choloepus* and eight or nine in *Bradypus*). These additional vertebrae cause the neck to be longer in *Bradypus* and may contribute to the ability of this sloth to turn its head around to a greater distance than is typical for a mammal. The sloths also share the unusual xenarthran fea-

A female brown-throated three-toed sloth (*Bradypus variegatus*) with green algal growth on her hair. (Photo by Michael P. L. Fogden. Bruce Coleman, Inc. Reproduced by permission.)

A female three-toed sloth (*Bradypus tridactylus*) sleeping. (Photo by Michael P. L. Fogden. Bruce Coleman, Inc. Reproduced by permission.)

Distribution

Xenarthra is found strictly in the New World and evolved in the early Cenozoic of South America, with some species migrating to Central America, North America, and the Carribbean in the late Pliocene and in several different later waves as land bridges appeared. All living families now occur in Central and South America, with one species of armadillo ranging into the United States. Xenarthrans were more common in North America in the past, with extinct large sloths, giant armadillos, and glyptodonts being important elements of the fauna, and ranging as far north as Alaska, where some extinct sloth remains have been discovered.

Habitat

Xenarthrans are found in most habitats of Central and South America. The living species of sloth are the most restricted, both genera living mainly in the lowland tropical rainforests and two-toed sloths inhabiting cloud forests to an altitude of about 6,560 ft (2,000 m). Anteaters and armadillos share these habitats, but anteaters, particularly the giant anteater, *Myrmecophaga*, also occupy savannas and pampas and more open areas. The armadillos are even more cosmopolitan, with some species occurring in the driest of South American desert habitats to savannas and tropical rainforests.

Behavior

Xenarthrans are solitary, although mothers keep their young with them, in some cases for up to as long as a year. Sloth and anteater mothers carry their single babies with them until long after weaning, including when the young have achieved two-thirds the size of their mothers. Armadillos do not carry their babies but do stay with them, foraging together in a family group. Some armadillos may form loose social groups of adults, but more commonly are solitary.

Although xenarthrans do not typically associate with others of their own species, they do maintain territories, which include feeding locations and favorite resting places. Sloths eat the leaves of about 60 species of trees and vines, although they have preferences for a more limited number of food trees that they learned by accepting partially digested leaves from the mouths of their mothers as early as a few weeks after birth. Anteater young learn the locations of the nests their mothers frequent, although after weaning, they may move to a different or partially different territory. Armadillo young disperse into areas adjacent to those occupied by their mothers.

Tree sloths are well known for the slowness of their movements, and their propensity to sleep for many hours a day. Common names include "peresozo," which translates to "lazy

A southern tamandua (*Tamandua tetradactyla*) climbing. (Photo by Animals Animals ©Partridge, OSF. Reproduced by permission.)

man," and the name "sloth" symbolizes one of the seven deadly vices. However, the slowness with which the animals move is determined by their low metabolic rate, necessitated by a leaf diet that provides poor quality nutrition. Another correlate of their low metabolic rate is that sloths are poor at regulating their body temperature, and become even more inactive in cool weather. The diet of one extinct sloth of the genus *Nothrotheriops* has been determined from the analysis of dung balls preserved in a cave in Arizona. These animals ate a variety of plant materials that include some species present in the area today. With a similar quality of nutrition to that of living sloths, it is reasonable to assume that these extinct sloths were also slow moving. Anteaters also have a low metabolic rate although they do not sleep as much or move as slowly as do sloths, and armadillos are as active as most other mammals.

Feeding ecology and diet

Although they are descended from insectivorous ancestors, the sloths have become herbivores, feeding primarily on leaves. It is commonly believed that they subsist entirely on the leaves of *Cecropia* trees, although they actually consume leaves, buds, and flowers from more than 60 species of trees and lianas. The myth about sloths spending their entire lives in a single tree probably arose because they move slowly, sleep for approximately 20 hours a day and are most easily seen in *Cecropia*

trees, which have a single slender trunk and a compact canopy, making sloths more visible when in them. They prefer new, young growth, and so move frequently to take advantage of plants in their home ranges that are putting out new growth. All sloths probably ingest insects upon occasion, and *Choloepus* will actively seek bird eggs and will take nestlings if it can capture them. Sloths are cecal, or hindgut, fermenters, and process their leafy diet slowly, eliminating solid and a small amount of liquid waste once every five or so days.

Anteaters feed almost entirely on ants and termites, and in short bouts. They typically dig a small hole in the nest with the largest claw on one forefoot and lick up the insects that come to investigate the problem and to repair the nest. Within a short time soldier ants or termites become alerted to the breach and swarm out to defend their home. Anteaters therefore feed in short bouts, and can often be seen brushing at their eyes and ears after leaving the nest. Although they have thick, tough skin, evidently ant and termite soldier mandibles are able to pinch hard enough to annoy them. In a single day an anteater will visit many nests, feeding for only a minute or two at each. The three living anteater genera differ greatly in size and partition the insect resources similarly. The smallest genus, *Cyclopes*, feeds on the smallest insects, medium-sized *Tamandua* feeds on medium-sized prey, and the giant anteater takes the largest ants and termites.

The three-toed sloth (*Bradypus infuscatus*) is extremely slow when moving on the ground. (Photo by © Michael Fogden. Bruce Coleman, Inc. Reproduced by permission.)

In contrast to herbivorous sloths and insectivorous anteaters, the armadillos are omnivores, feeding on insects, small invertebrates, and plant materials as they occur in their habitat. They may take different foods at different times of the year and according to seasonal availability. This is probably the ancestral diet for the group. The extinct giant armadillos probably had food habits similar to those of living armadillos. It has been assumed, based primarily on the dentition that emphasizes the ability to grind plant foods, that the glyptodonts were herbivores and may have specialized on grasses.

Reproductive biology

Few courtship behaviors have been observed in sloths, and males and females remain together only for the length of time required to mate several times. Females have a simplex uterus and give birth to a single young, born fully furred, and with eyes open or soon open, that can hang onto it's mother's fur shortly after birth. The mother carries the baby continuously for six months, nursing it but also allowing it to feed on leaves by which it learns the mother's feeding preferences prior to weaning. During their time together, the baby infrequently moves far enough from the mother to lose physical contact. Even when exploring or feeding, the young sloth maintains contact with the mother by using at least one foot. Sloth babies as young as two weeks have been observed to feed on some of the leaves the mother is eating, and it has been reported that mother sloths may regurgitate partially digested leaves for the baby to lick from her lips. Whether this is true or not, young sloths frequently are seen licking at the lips of their mothers, and this is evidently how they learn her food preferences. At weaning, the mother leaves the territory they occupied together and moves to a different part of her range. She will remain away from the part of her territory where she left her offspring for anywhere from two weeks to a few months.

Anteaters, which are probably polygynous, generally produce only a single offspring; twins are quite rare. Offspring nurse from a single pair of mammae. Because the mother is unable to grasp the newborn due to her enlarged foreclaws, a young anteater must climb up the fur to the mammae. In most species, the young are still carried on the mother's back for a great deal of time after weaning.

For most armadillo species, mating occurs in the summer and is probably polygynous. A majority of species have a litter of one to three young per year, although *Dasypus* females exhibit obligate polyembryony, giving birth to two to 12 genetically identical young. Gestation is 140 days, and newborns are blind and naked. Their soft, leathery skin hardens into armor within a few days. The young are nursed for 2–2.5 months, begin to walk around after a week, and open their eyes after 3–4 weeks.

Conservation status

Of the 30 total Xenarthra species, three are ranked as Endangered, five as Vulnerable, two as Lower Risk, and six as Data Deficient—12 of these are armadillo species. Primary threats are human encroachment in the form of habitat de-

The teeth of the two-toed tree sloth (*Choloepus didactylus*) never stop growing. (Photo by Zoological Society of San Diego. Reproduced by permission.)

The three-toed sloth (*Bradypus infuscatus*) is an excellent swimmer. (Photo by Wolfgang Bayer. Bruce Coleman, Inc. Reproduced by permission.)

the only member of the family Myrmecophagidae recognized by the IUCN—more detailed study is necessary to determine the status of the remaining anteater species.

Sloths are a CITES I endangered species, primarily as a result of habitat destruction. In areas where the rainforest survives and the traditional degree of diversity is maintained, sloths still can do well. For example, on Barro Colorado Island in Panama, a protected habitat, there were up to 8.5 three-toed sloths per 2.5 acres (hectare). This high density is supported because sloths learn tree species preferences from their mothers and inherit part of her territory at social weaning. Neighboring sloths prefer different tree species combinations, and so pass on different preferences to their young, thereby reducing the competition for the same trees in any area.

Significance to humans

Sloths are infrequently used by humans as a food source because they have a small muscle mass and are reputed to be tough chewing. They are renowned as the namesake of the vice "slothfulness" because of their slow-moving habits, although the general public is not very familiar with their appearance or the details of their habits. *Choloepus* does well in zoological parks, and can be tamed and kept as a pet, but the more docile *Bradypus*, although hardly in need of taming, does not survive in captivity away from its home rainforests, probably because of specialized dietary preferences. Three-toed sloths are referred to as "ai's," named after one of the soft calls they make. The common name for two-toed sloths is "unau," although both genera may be referred to as "peresozo," or "lazy man," because of the slowness of their movement.

Anteaters are hunted as a food source as well as for their skin—some species are killed for the thick tendons in their tails, which can be used to make rope. Native Amazonian tribes have also been known to use anteaters to rid their homes of ants and termites. Armadillos are known to be valuable for medical research on leprosy, typhus, trichinosis, and birth defects. They are also used as a food source and as a material for crafts such as musical instruments, decorations, and charms. Pichi armadillos (*Zaedyus pichiy*) are occasionally taken in as pets.

struction and exploitation for food. Accurate census data on anteaters, however, is difficult to obtain due to their solitary behavior and large ranges. It is also unclear how well they adjust to changing habitat. The giant anteater (Vulnerable) is

Resources

Books

Carroll, R. L. *Vertebrate Paleontology and Evolution.* New York: W. H. Freeman and Co., 1998.

Martin, P. S. *Pleistocene Overkill.* New York: Columbia University Press, 1984.

McKenna, M. C., and S. K. Bell. *Classification of Mammals above the Species Level.* New York: Columbia University Press, 1997.

McNab, B. K. "Energetics, population biology and distribution of xenarthrans, living and extinct." In *The Evolution and Ecology of the Armadillos, Sloths and Vermilinguas,* edited by

G. G. Montgomery, 219–232. Washington, DC: Smithsonian Institution Press, 1985.

Nowak, R. M. *Walker's Mammals of the World.* Baltimore and London: The John Hopkins University Press, 1999.

Webb, S. D. "Late Cenozoic mammal dispersal between the Americas: The great American biotic interchange." In *Topics in Geobiology 4,* edited by F. G. Stehli and S. D. Webb. New York: Plenum Press, 1985.

Periodicals

Hanson, R. M. "Shasta ground sloth food habits, Ramparts Cave, southwestern Arizona." *Paleobiology* 4 (1978): 302–19.

Resources

McNab, B. K. "Food habits, energetics and the population biology of mammals." *American Naturalist* 116 (1980): 106–124.

Naples, V. L. "Cranial osteology and function in the tree sloths *Choloepus* and *Bradypus*." *American Museum Novitates* 2739 (1982): 1–41

————. 1990 "Morphological changes in the facial region and a model of dental growth and wear pattern development in *Nothrotheriops shastensis*." *Journal of Vertebrate Paleontology* 10, no. 3 (1990): 372–389.

Virginia L. Naples, PhD

West Indian sloths and two-toed tree sloths

(Megalonychidae)

Class Mammalia

Order Xenarthra

Family Megalonychidae

Thumbnail description
Small- to medium-sized sloths, from the size of a large domestic cat to a small bear; all species show large caniniform teeth, several molariform teeth, long limbs and strong, curved claws; they have pear-shaped bodies and long, shaggy coats ranging in color from gray to brown, often with lighter fur around the heads and faces

Size
Sloths of the family Megalonychidae range in weight from about the size of a large domestic cat (20 lb [9 kg] for *Choloepus*) to as large as a large bear; however, none of the West Indian forms grew larger than a black bear

Number of genera, species
1 genera; 2 species; extinct forms grouped with *Choloepus* include as many as 11 genera, each containing a single species

Habitat
Tropical rainforests and cloud forests

Conservation status
Extinct: all species of West Indian sloths; Data Deficient: 2 species

Distribution
Choloepus occurs in Central America and into South America as far as Brazil. The West Indian sloths may have shared part of that range as well as the islands of the West Indies

Evolution and systematics

Evolved in South America, this diverse order first appears in the fossil record in the Paleocene. It contains sloths and anteaters, the Pilosa, or hairy xenarthrans, and the Cingulata, or plated xenarthrans, which includes the extinct glyptodonts and armadillos. The group is named for the additional articulations between their vertebrae, called xenarthrous processes. The family Megalonychidae includes a single living genus, *Choloepus*, with two species, and at least 11 extinct genera from the West Indies as well as one from Curaçao. The Megalonychidae appeared first in the fossil record as distinct from the other two families of sloths (the Mylodontidae and the Megatheriidae) in the early Miocene. The early genera were small and differed from the earliest sloth family, the Mylodontidae, in lacking dermal armor. Their long claws were laterally compressed and curved, and the forelimbs were almost as long as the hind limbs. They had caniniform teeth and peg-like molariforms. It has been suggested that their body forms resemble the modern *Choloepus*, and that they might have been arboreal.

Physical characteristics

West Indian sloths are small to medium sized, and the living *Choloepus* is the larger of the two tree sloth genera, being the size of a large domestic cat. These animals have slender limbs, pear-shaped bodies, and long claws. Claws in *Choloepus* are more hook-like than those of the other genera, but all are recurved to some degree. *Choloepus* and the West Indian sloths have enlarged anterior caniniform teeth, followed by a diastema and a series of molariform teeth separated by small spaces in the jaws. The teeth in sloths emerge as simple rounded cones. The distinctive pattern of cusps and basins that characterize genera individually are formed entirely by wear caused by slight differences in the genus specific pattern of masticatory movements.

Distribution

Choloepus occurs in tropical rainforests and cloud forests up to an altitude of 6,000 ft (1,830 m) in Central America, from Nicaragua south to Brazil in South America. Extinct mega-

A juvenile two-toed sloth (*Choloepus didactylus*) feeding. (Photo by Byron Jorjorian. Bruce Coleman, Inc. Reproduced by permission.)

A close up of the Hoffmann's two-toed sloth's (*Choloepus hoffmanni*) limbs and claws. (Photo by Anup Shah/Naturepl.com. Reproduced by permission.)

A Hoffmann's two-toed sloth (*Choloepus hoffmanni*) in Costa Rica. (Photo by Tom Brakefield. Bruce Coleman, Inc. Reproduced by permission.)

lonychids have been found on most of the islands of the West Indies and Curaçao.

Habitat

The living genus, *Choloepus*, primarily occurs in the tropical rainforests of Central and South America, although some animals survive in Central American cloud forests at altitudes as high as 6,000 ft (1,830 m). The Pleistocene and recent habitats of the West Indian sloths were probably similar to those in which *Choloepus* lives today, and the present habitats may not have altered much from the time when the sloths were alive.

Behavior

Choloepus is solitary and arboreal. Adults maintain a territory that encompasses a wide variety of tree and vine species. Sloths prefer trees in their home range; they are chosen primarily because of a heavy concentration of lianas in the crowns. Sloths are nocturnal, and use vine-covered trees for resting and sleeping during the day, because predators cannot approach without alerting the sloth. *Choloepus* is slow moving (as possibly were the extinct sloths), a behavior partially attributable to the low amounts of energy obtainable from the animals' diet as well as a strategy that assists the animals to avoid predators by remaining cryptic. *Choloepus* has grooves in the outer guard hairs of its coat that house two species of blue green algae that turn the sloths a greenish color, especially during the wet season. It would be reasonable to assume that the fur of some West Indian sloths also housed algae that enhanced their ability to blend into their environment.

Feeding ecology and diet

All sloths are herbivores, and anatomical similarities between *Choloepus* and West Indian sloths indicate that the latter were probably also folivores, although the leaf species eaten probably differed. *Choloepus* may also feed on fruits.

Reproductive biology

Living sloths are solitary once weaned, and are polygynous, meeting only to mate. Males do not assist with rearing the young. References suggest a gestation period for *Choloepus* of 11 months and/or an ability to store sperm; for larger bodied West Indian sloths it was probably longer.

A southern two-toed sloth (*Choloepus didactylus*) hanging by its hind legs. (Photo by Erwin & Peggy Bauer. Bruce Coleman, Inc. Reproduced by permission.)

A Hoffmann's two-toed sloth (*Choloepus hoffmanni*) climbs a vine in a rainforest in central Costa Rica. (Photo by Janis Burger. Bruce Coleman, Inc. Reproduced by permission.)

The southern two-toed sloth (*Choloepus didactylus*) sleeps for about 15 hours during the day. (Photo by © J. C. Carton. Bruce Coleman, Inc. Reproduced by permission.)

A Hoffmann's two-toed sloth (*Choloepus hoffmanni*) climbs a tree. (Photo by Tom Brakefield/OKAPIA/Photo Researchers, Inc. Reproduced by permission.)

Conservation status

Choloepus is listed as Endangered by the IUCN, mostly due to loss or degradation of the rainforest habitat. Roads also cause mortality because slow-moving sloths are often unable to cross quickly enough to avoid vehicles.

Significance to humans

Choloepus has been used occasionally as food by humans. West Indian sloths disappeared shortly after humans invaded their islands less than 2,000 years ago. They may have also been a source of pelts. Tree sloth pelage is used in a few human societies but, in general, never achieved high fashion status. The claws are sometimes incorporated into jewelry.

Species accounts

Lesser Haitian ground sloth

Synocnus comes

SUBFAMILY
Megalonychinae

TAXONOMY
Synocnus comes Paula Couto, 1967

OTHER COMMON NAMES
None known.

PHYSICAL CHARACTERISTICS
The size of a medium-sized dog, weighing about 50 lb (23 kg). The animal is known only from skeletons from Haiti, but bones recovered indicate that it showed typical sloth body proportions,

Synocnus comes

with a broad trunk, slender limbs, and long claws. In contrast to tree sloths, the ground sloth had a tail long enough to touch the ground. The caudal vertebrae were broad, and this morphology is associated in other extinct sloths with a tail robust enough to serve as a tripodal support to allow the animal to stand bipedally in a fashion similar to tamanduas. Ground sloth has large, triangular caniniform teeth, separated by a diastema from molariforms with sharp cusps and basins. The skull was deep, with a large sagittal crest and the deep mandible allowed large masticatory muscles.

DISTRIBUTION
Known from Haitian cave deposits.

HABITAT
Resembled *Choloepus*, with similar habits, but was semi-arboreal.

BEHAVIOR
Nothing known, although a semi-arboreal ground sloth may have behaved in a manner similar to *Choloepus*.

FEEDING ECOLOGY AND DIET
As do their living relatives, ground sloths probably fed on leaves.

REPRODUCTIVE BIOLOGY
Nothing is known. Probably polygynous.

CONSERVATION STATUS
Extinct.

SIGNIFICANCE TO HUMANS
When living, these animals may have been killed for food and pelts. Now they are of interest to students of evolutionary history and ecology. ◆

Synocnus comes

Resources

Books

Carroll, R. L. *Vertebrate Paleontology and Evolution*, Ch. XXI. New York: W. H. Freeman and Co., 1998.

Hall, E. R., and H. Kelson. *The Mammals of North America*. Princeton: Princeton University Press, 1981.

Martin, P. S. *Pleistocene Overkill*. New York: Columbia University Press, 1984.

McKenna, M. C., and S. K. Bell. *Classification of Mammals above the Species Level*. New York: Columbia University Press, 1997.

Nowak, R. M. *Walker's Mammals of the World*. New York: Columbia University Press, 1999.

Webb, S. D. "Late Cenozoic Mammal Dispersal between the Americas." In *The Great American Biotic Interchange Topics in Geobiology 4*, edited by F. G. Stehli, and S. D. Webb. New York: Plenum Press, 1985.

Virginia L. Naples, PhD

Three-toed tree sloths

(Bradypodidae)

Class Mammalia

Order Xenarthra

Family Bradypodidae

Thumbnail description
Arboreal with long coarse shaggy fur and long limbs that appear thick and powerful; both sets of limbs end in three stout hooked claws; small eyes and ears, short snout, peg-like teeth, and a small, stumpy tail

Size
Head and body length 15.8–30.3 in (40–77 cm); tail 1.9–3.5 in (4.7.6–9.0 cm); weight 5.1–12.1 lb (2.3–5.5 kg)

Number of genera, species
1 genus; 4 species

Habitat
Tropical forests

Conservation status
Endangered: 1 species

Distribution
Central and South America

Evolution and systematics

Though members of the order Xenartha have been recorded from Eocene Europe and ground sloths reached North America across the Central American isthmus, there are no fossil records of tree sloths further north than southern Mexico. *Bradypus* sloths are only distantly related to two-toed *Choloepus* sloths.

Physical characteristics

Digits on fore- and hindlimbs fused to a mitten-like structure from which only the 3.2–3.9 in (8–10 cm) long claws protrude. These allow branches to be gripped without expending muscular force. The number of digits on the forelimb distinguish *Bradypus* from *Choloepus* sloths. They would be better named, three- and two-fingered sloths, since both have three digits on the hindlimbs. There are eight or nine neck vertebrae (most species of mammal, even giraffes, have seven). This allows the head to be turned with a considerable range, an important advantage for an animal with otherwise rather limited flexibility. The testes are internal. There are no incisors or canine teeth and the simple, peg-like incisors lack enamel. Sloths have poor hearing, but fairly good eyesight and smell.

A three-toed sloth (*Bradypus variegatus*) climbs a tree. (Photo by Peter Oxford/Naturepl.com. Reproduced by permission.)

To accommodate a largely suspended, upside-down lifestyle, the fur hangs down from the belly to the back. The underfur is short and fine. The coarse very thick outer fur is grooved along its length, providing attachment for two species of blue-green algae. Along with sebaceous secretions from the sloth, excretions from the algae are food for adults of *Cryptoses choloepi*, a species of pyralid moth. Adult moths reach densities of up to 132 per sloth. Larval *Cryptoses choloepi* feed on sloth dung, as do the larvae of three types of beetle (*Trichilium* spp.) and at least three types of mite (*Amblyomma varium* and two species of *Boophilus*). Experimentally, decolonized sloths have been recolonized after 40 days. The algal growth on the hairs is rarely dense enough to make the sloth appear distinctly green. There is no size difference between the sexes.

Distribution

Three-toed sloths live from Honduras to southern Argentina.

Habitat

Various forest types from primary closed canopy to highly disturbed secondary and seasonally dry forests.

A three-toed sloth (*Bradypus infuscatus*) hangs from a tree in a rainforest in Panama. (Photo by M. Fogden. Bruce Coleman, Inc. Reproduced by permission.)

A three-toed sloth with *Bradipodicola* moths. The moths lay their eggs in the sloth's feces. (Illustration by Patricia Ferrer)

A pale-throated three-toed sloth (*Bradypus tridactylus*) clings to a tree in South America. (Photo by John Giustina. Bruce Coleman, Inc. Reproduced by permission.)

A brown-throated three-toed sloth (*Bradypus variegatus*) climbs through the trees in Ecuador. (Photo by Harald Schütz. Reproduced by permission.)

Behavior

Feeding, mating and birth all occur in trees. Defecation and urination, however, occurs on the ground and sloths make their way to the ground once or twice a week to eliminate in a hole that is dug by the tail while the sloth clings with its forelegs to the tree trunk or vine above. It is during this process that females of the various moths, beetles and mites that live on the sloth fur will temporarily leave their shaggy host to deposit eggs on its dung. Sloth dung consists of hard rounded pellets about 0.3 in (8 mm) in diameter. About a cupfull are deposited on each occasion. Stereotyped movements of the tail and/or hindlimbs ensure that the hole is covered with leaf litter on completion. The entire process usually takes less than 30 minutes, but many jaguar kills of sloths are reported to occur during this period. Locomotion in trees generally proceeds with the claws used as hooks both in vertical and horizontal progression. Terrestrial movement is a slow flailing crawl with the animal preferring to hook objects with its claws and pull itself forwards. Progress under such conditions has been clocked at 0.25 mph (0.4 km/h). Swimming appears to be much easier, and sloths are frequently encountered crossing rivers. Though movement is generally slow, sloths can move quite quickly if threatened. Despite their general immobility, they are preyed on by large eagles (especially the harpy eagle, *Harpia harpyja*) and by jaguars. Males advertise presence by wiping pungent smelling secretions from anal glands onto branches. Dung middens also smell strongly and may serve as trysting locations.

A brown-throated three-toed sloth (*Bradypus variegatus*), less than 1 ft (30 cm) long, found abandoned by a remote trail in lowland tropical rainforest of La Selva Biological Station, Costa Rica. (Photo by Gregory G. Dimijian/Photo Researchers, Inc. Reproduced by permission.)

Feeding ecology and diet

In all species, the predominant diet is shoots and leaves of forest trees. Sloths feed on *Cecropia*, the most abundant tree of the Amazonia forest. Regenerating agricultural land river margins and natural gaps may sometimes be important (despite the biting ants that swarm in the tree's hollow stems), but it is never the sole food source. The idea that *Bradypus* sloths feed only on *Cecropia* probably arises because an open growth form makes a sloth in a *Cecropia* easier to see than in almost any other kind of rainforest tree. When feeding, the forelimbs are used to pull leaves slowly towards the mouth. Sloths are highly specialized for an existence that centers around squeezing as much energy as possible out of a low-intake rate of highly indigestible food. This is because leaves eaten by *Bradypus* sloths are, though energy-rich, also rich in tannins and fiber. Digestion must therefore be a simultaneous process of detoxification and energy extraction. The gut is extensive, making up 30% of the total body weight. Digestion, by bacterial fermentation in a complex multi-chambered stomach, occurs over an extended period to permit maximum absorption of scarce resources. Passage of food through the intestine is also very slow, providing maximum opportunity for the uptake of nutrients and energy. This

A male pale-throated three-toed sloth (*Bradypus tridactylus*) eats a cercropia leaf. (Photo by Dan Guravich/Photo Reseachers, Inc. Reproduced by permission.)

A brown-throated three-toed sloth (*Bradypus variegatus*) hangs by two limbs off of a tree on Bocas del Island, Panama. (Photo by Art Wolfe/Photo Researchers, Inc. Reproduced by permission.)

A three-toed sloth (*Bradypus tridactylus*) forages in Costa Rica. (Photo by J-C Carton. Bruce Coleman, Inc. Reproduced by permission.)

A three-toed sloth (*Bradypus infuscatus*) carries its young in the trees of Panama. (Photo by M. P. L. Fogden. Bruce Coleman, Inc. Reproduced by permission.)

means that sloths have little energy to spare and hence move slowly. Their metabolic rates are about half of what would be expected for an animal of their size (sloth adaptations provide plenty of opportunities to contemplate the chicken–and–egg nature of complex physiological and behavioral adaptations). The three-toed sloth has struck a compromise between being large enough to move about efficiently between the canopies of tall trees, the weight of the long gut and big stomach needed for food processing and the need to be light enough to avoid breaking the limbs from which it feeds. In minimizing weight, while maximizing mass, the sloth has compromised on muscles and has the lowest ratio of muscles to skeleton of any comparably sized ground-dwelling mammal. Its muscles are thin and ribbon-like. Much of the sloth's apparent volume comes from its long hair. The compromise works: sloths are the most abundant large mammal in neotropical rainforests (up to 70% of the arboreal mammalian biomass), and cropping some 2% of the forest's annual leaf production. It has been suggested that the blue-green algae may also provide some nutrition, being licked directly from the hair or having nutrients absorbed via the skin. Despite great similarities, two species of *Bradypus* sloth coexist in certain parts of their range. *Bradypus* and *Choloepus* sloths can also coexist.

Reproductive biology

Sexual maturity is reached at about three years. One young is produced per year. Nursing requires 6–8 weeks. Weaning occurs as infants first lick leaf fragments from their mother's fur and lips and later sharing the leaves being eaten by the mother. To save energy, sloths barely regulate their body temperature, however pregnant females do invest energy to keep their bodies a few degrees above ambient temperature, the better to develop their embryo. Mating takes 3–5 minutes and

may be conducted with the two animals face to face and hanging by their front legs from a branch, or with the female suspended by all four limbs and the male on her back. Sloths may live 10–20 years in the wild and are presumably polygynous.

Conservation status

One species, *B. torquatus*, is classified as Endangered by the IUCN and the U.S. Department of the Interior, due to its habitat being depleted by lumber extraction and argricultural activities (plantations, cattle pasture).

Significance to humans

Several tropical rainforest species, including sloths, are slowly being recognized for their potential to further human medicine. Sloths are known for their ability to heal quickly, avoid infection, and survive the most severe injuries. Researchers are investigating the basis of this healing response so as to develop improved drugs or treatment methods for severe wounds.

A pale-throated three-toed sloth (*Bradypus tridactylus*) juvenile. (Photo by Jany Sauvanet/Photo Researchers, Inc. Reproduced by permission.)

1. Pale-throated three-toed sloth (*Bradypus tridactylus*); 2. Monk sloth (*Bradypus pygmaeus*); 3. Brown-throated three-toed sloth (*Bradypus variegatus*); 4. Maned sloth (*Bradypus torquatus*). (Illustration by Amanda Humphrey)

Species accounts

Brown-throated three-toed sloth
Bradypus variegatus

TAXONOMY

Bradypus variegatus Schinz, 1825, Brazil. The genus name *Bradypus* comes from the Greek *bradus* for "slow" and *pous*, for "foot," *podos*.

OTHER COMMON NAMES

Portuguese: Preguica-de-bentinho (Brazil); Spanish: Perezoso de tres dedos (European Spanish), perico (Bolivia, Colombia, Ecuador), pelejo (Peru).

PHYSICAL CHARACTERISTICS

Color varies over wide geographical range, long coarse body fur brownish gray with white patches on hindlegs and lower part of back. Extent of white patches highly variable, some populations are nearly all white. Others are a deep foxy red-brown. Fur may have a greenish tinge due to algae. Fur on head shorter and denser. Head with a black "mask" across the eyes extending back to the ear region (ears are hidden in fur and not easily visible). Throat and chest brown. Adult males have a "speculum," a patch of short deep orange fur between the shoulder blades that is bisected by one of more deep brown-black horizontal lines. Distinguished from the two-toed sloth by the number of forefoot claws and a shorter dark muzzle (bigger, paler and more pig-like in *Choloepus*).

Bradypus torquatus

Bradypus variegatus

DISTRIBUTION

Southern Mexico to northern Argentina, to elevations of at least 3,610 ft (1,100 m).

HABITAT

Evergreen and seasonally dry forests. A natural tolerance of disturbance and secondary growth also allows them to survive in isolated trees in deforested pastures, and even in city parks.

BEHAVIOR

Active at any time of day, though generally more active at night. Drops its body temperature each night (an energy-conserving strategy) and must warm up each morning by basking. This is the time when harpy eagle predation most frequently occurs. When not basking or feeding, likely to be sleeping curled up in a ball in the crook of a tree. Hard to see under such circumstances. Spends up to 18 hours a day asleep to conserve energy. Adults are solitary but home ranges may overlap. Neighbors rarely feed in the same tree, and males may fight each other. Individuals may spend many days in the same tree, and can pass their entire 20- to 30-year life span in home ranges of less than 4.9 acres (2 ha). One of the most common animals in the South American rainforest, *Bradypus* sloths can occur at densities of six or seven per 2.5 acre (1 ha). When not hanging suspended, may rest in a fork of a branch with head between forelimbs. May be very difficult to see under such conditions. Vocalizations, a shrill whistle and a low reptilian hiss, are given only under duress. The shrill "ai, ai" sounds whistled through the nostrils, are the basis for the name for this animal in the indigenous Guarani language.

FEEDING ECOLOGY AND DIET

Within home range, a sloth may use up to fifty trees of thirty different species. To avoid toxification by the tannins, phenols and other chemicals in the leaves it ingests, sloths change trees (and tree species) on average once every 1.5 days. Passage of food through the gut is measured in days, rather than the hours usual for most mammals. This is necessary to extract all possible energy from the low-quality forage. Diet preferences are inherited from mother during several months of a "social weaning" process. Since these differ considerably, several sloths may coexist in the same area, but not compete.

REPRODUCTIVE BIOLOGY

Breeding occurs throughout the year. A single young is born, though twins have been reported once or twice. Gestation is 5 to 6 months. Young weigh 0.44–0.55 lb (0.20–0.25 kg) at birth. Young are weaned within 4 weeks, but are carried by the mother for another five months. Babies are carried resting on their mother's abdomen, graduating to dorsal carriage as they get older and larger. Once the young has learned the location of the trees in the maternal patch, the female leaves, bequeathing the young one all or part of her foraging area. This highly unusual arrangement is thought to minimize energetically wasteful conflict between individuals. Probably polygynous.

CONSERVATION STATUS

CITES Appendix II. Not threatened.

SIGNIFICANCE TO HUMANS

Hunted for meat in certain parts of their range. ◆

Pale-throated three-toed sloth

Bradypus tridactylus

TAXONOMY

Bradypus tridactylus Linnaeus, 1758, Suriname.

OTHER COMMON NAMES

French: Mouton parasseux (French Guiana); Surinamese: Dri-teenluiaard.

PHYSICAL CHARACTERISTICS

Back darker, buff to dark brown, with contrasting pale or dark grizzling. Belly paler, off-white to very deep cream. Back and rump with variably sized irregular roundels of cream or dirty orange. Facial area, a cream colored mask extending back to the ears and onto the throat. No black contrasting "mask" as in *B. variegatus*, though some dark patterning round eyes may occur. Adult males posses a speculum like that of *B. variegatus*.

DISTRIBUTION

Replaces *B. variegatus* in eastern Venezuela, the Guyanas, and northeastern Brazil. The two species may coexist in the lower Amazon.

HABITAT

Lowland rainforest. Less flexible than *B. variegatus* and rarely recorded from seasonally dry forests or highly disturbed areas.

BEHAVIOR

Active at any time during day or night. Ecology believed to be very similar to that of *B. variagatus*. Occurs together with *Choloepus didactylus*, the two-toed sloth. Resources are partitioned between the two by differing diets, activity patterns and use of different forest strata.

Bradypus tridactylus

Bradypus pygmaeus

FEEDING ECOLOGY AND DIET

Believed to be similar to that of *B. variegatus*.

REPRODUCTIVE BIOLOGY

Gestation lasts 106 days. In Guyana, births occur only in the rainy season, but elsewhere, reproduction seems flexible and dependant on local conditions. This may be due to the female's ability to halt an embryo's development until conditions are favorable. Female may be sexually receptive while still nursing and can also be both nursing and pregnant at the same time. An interval of seven months between births has been reported under good conditions. Probably polygynous.

CONSERVATION STATUS

Not threatened.

SIGNIFICANCE TO HUMANS

Sometimes hunted for meat. ◆

Monk sloth

Bradypus pygmaeus

TAXONOMY

Bradypus pygmaeus Anderson and Handley, 2001. Known only from Isla Escudo de Veraguas, an island of the Bocas del Toro, off the Caribbean coast of Panama. This recently described species provides a fascinating example of evolution in action. Sloths from the mainland have colonized five of these islands at least four times in the past few thousand years. Each time, they have changed in their isolation, becoming smaller and adapted to their new island homes. The oldest island, Escudo de Varaguas, is 8,900 years old. Only here has the population differentiated enough to be called a new species. Populations on the other islands (1,000 to 5,200 years old) are still many generations away from this.

OTHER COMMON NAMES

English: Dwarf sloth, pygmy three-toed sloth.

PHYSICAL CHARACTERISTICS

Small (20% less in all measurements) in comparison to other sloths. The speculum is pure orange. The face is tan with a distinctive dark band across the forehead, and a dark throat and an orange wash to the face. Long hair hangs forward from the forehead, giving the impression of a hood. Back with a strong spinal stripe.

DISTRIBUTION

Only found on Escudo de Veraguas island, off the Caribbean coast of western Panama.

HABITAT

Found only in red mangroves at sea level.

BEHAVIOR

Not yet studied. Lives only in coastal mangroves.

FEEDING ECOLOGY AND DIET

Believed to consist entirely of the leaves of red mangroves (*Rhizophora mangle*).

REPRODUCTIVE BIOLOGY

Nothing is known. May be polygynous.

CONSERVATION STATUS

Nothing known.

SIGNIFICANCE TO HUMANS
Of great interest to evolutionary biologists.

Maned sloth
Bradypus torquatus

TAXONOMY
Bradypus torquatus Illiger, 1811, Brazil.

OTHER COMMON NAMES
Portuguese: Preguica preta, Preguica de coleira (Brazil).

PHYSICAL CHARACTERISTICS
Shares the body shape of other sloths. But both head and body are the same color, grizzled tan brown all over. Long black hairs, to 5.9 in (15 cm), fall from the nape, over the neck and shoulders to form the characteristic mane. Speculum absent. In infants and juveniles, the fur is very pale, whitish to pale brown and lacks a mane.

DISTRIBUTION
Forests of the Atlantic Coast of Brazil ("Mata Atlantica"). Distribution very patchy and populations now highly isolated in highly fragmented forests. Few known populations coincide with location of existing protected areas.

HABITAT
Atlantic Coastal forests of Brazil. Able to survive in secondary forest but prefers vine-rich primary ones.

BEHAVIOR
May be active during day or night, but are most active during the day. Home ranges average 4.9 acres (2 ha). A new range is often colonized each rainy season.

FEEDING ECOLOGY AND DIET
Unlike other *Bradypus* species, *B. torquatus* adds liana and vine leaves (16%) to a diet of tree leaves. It prefers young leaves (68%), whereas in other three-toed sloths mature leaves dominate. Twenty-one species have been recorded in their diet, with individuals showing strong preferences and eating between seven and twelve species. The chosen trees are actively sought and are not simply the most abundant ones in the forest. Trees of the Moraceae family are highly sought after. Unlike most other tropical trees, Moraceae trees have a continuous production of young leaves, the preferred food of this sloth. The small number of species in the diet may allow the individual sloth's physiology to adapt to detoxifying them. Shares the mechanism of maternal transmission of foraging preferences and foraging home range.

REPRODUCTIVE BIOLOGY
Single young, weighing 10 oz (300 g) at birth. Reproduction is non-seasonal. Presumably polygynous.

CONSERVATION STATUS
Classified by the IUCN as Endangered.

SIGNIFICANCE TO HUMANS
Formerly hunted for food. This species is now one of several Atlantic Coastal Forest endemics being used in public awareness campaigns to promote conservation of this highly threatened ecosystem. ◆

Resources

Books
Eisenberg, J. F., and K. H. Redford. *Mammals of the Neotropics.* Vol. 3, *The Central Tropics: Ecuador, Peru, Bolivia, Brazil.* Chicago: University of Chicago Press, 1999.

Janzen, D. H. *Costa Rican Natural History.* Chicago: University of Chicago Press, 1983.

Reid, F. A. *A Field Guide to the Mammals of Central America and Southeast Mexico.* Oxford: Oxford University Press, 1997.

Periodicals
Anderson, R. P. and C. O. Handley, Jr. "A new species of three-toed sloth (Mammalia: Xenartha) from Panama, with a review of the genus *Bradypus.*" *Proceedings of the Biological Society of Washington* 114 (2001): 1–33.

Chiarello, A. G. "Diet of the Atlantic forest maned sloth *Bradypus torquatus* (Xenartha: Bradypodidae)." *Journal of Zoology (London)* 246 (1998): 11–19.

Richard-Hansen, C., and E. Tuabe. "Note on the reproductive behavior of the three–toed sloth, *Bradypus tridactylus*, in French Guiana." *Mammalia* 246 (1997): 259–263.

Waage, J. K., and G. G. Montgomery. "*Cryptoses choloepi*: A coprophagous moth that lives on a sloth." *Science* 193 (1976): 157–158.

Adrian A. Barnett

▲
Anteaters
(Myrmecophagidae)

Class Mammalia

Order Xenarthra

Family Myrmecophagidae

Thumbnail description
Small to large functional insectivores, characterized by a very elongated tapered and tubular snout, teeth are absent, wormlike tongue that is capable of extending beyond head length, large powerful curved claws, all but one species has grasping prehensile tail

Size
12–110 in (0.32–2.8 m); 0.5–86 lb (0.15–39 kg)

Number of genera, species
3 genera; 4 species

Habitat
Neotropical forests, savannas, and grasslands

Conservation status
Vulnerable: 1 species

Distribution
Belize, Mexico, Central and South America

Evolution and systematics

Fossil evidence indicates that the family Myrmecophagidae was present during the early Miocene period (25 million years ago [mya]) in South America. However, the fossil record is poor and it is possible that the family is much older. Myrmecophagidae is the only member of the infraorder Vermilingua, which means worm-tongue. There are three genera and four species. Anteaters were once thought to be closely related to sloths. However, analysis of albumin samples indicates that the members of the order Xenarthra diverged about 75–80 mya and are very distinct.

Physical characteristics

All members of this group have elongated snouts and a thin tongue that is capable of extending outward to a length greater than the length of the head. They have a tubular mouth with lips but they do not have teeth. They also have large curved foreclaws that are used to tear open ant and termite mounds. The powerful foreclaws can also be used as lethal weapons for defense. All but one species has a grasping prehensile tail. The

Northern anteaters (*Tamandua mexicana*) live in trees. (Photo by Michael Fogden. Bruce Coleman, Inc. Reproduced by permission.)

A southern tamandua (*Tamandua tetradactyla*) with its tongue extended. (Photo by Animals Animals ©Alan G. Nelson. Reproduced by permission.)

fur is long and thick to protect them briefly from the attack of ants as well as termites.

Distribution

Two members of this group are found as far northward as Southeastern Mexico, the other two members begin their northernmost range in Cental America. The ranges of three members of this group overlap to eastern Brazil. Two species extend southward to Uruguay.

A silky anteater (*Cyclopes didactylus*) hangs from a tree in a defensive posture in the Caroni Swamp, Trinidad. (Photo by © Kevin Schafer/Corbis. Reproduced by permission.)

A silky anteater (*Cyclopes didactylus*) living in the rainforests of Panama. (Photo by Michael Fogden. Bruce Coleman, Inc. Reproduced by permission.)

Habitat

Anteaters can be found in tropical dry forests, rainforests, grasslands and savannas. The silky anteater (*Cyclopes didactylus*) is an arboreal specialist that is most commonly found in rainforest. The tamandua are arboreal and terrestrial opportunists in regards to terrain and food resources. They are most commonly found in dry forests near streams and lakes. The giant anteater (*Myrmecophaga tridactyla*) is almost entirely terrestrial and usually found in grasslands and savannas.

Behavior

Anteaters are thought to be mostly solitary. The limited number of field studies done indicate that all members of this group will defend their 1–1.5 mi^2 (2.6–3.9 km^2) territories. Males often enter the territories of associated females but do not enter the territories of other males. Likewise, females do not enter the territories of other females. If a territorial dispute occurs, they will vocalize, swat with the foreclaws and sometimes sit on and even ride the back of a subordinate animal.

A giant anteater (*Myrmecophaga tridactyla*) mother and baby in South America. (Photo by Animals Animals ©John Chellman. Reproduced by permission.)

Observations indicate that anteaters have a poor sense of sight, being able to see better the closer they get to the subject. However, their sense of smell is exceptional. Most species depend on smell for direction, foraging, feeding and defense. They are thought to be able to hear very well.

Anteaters are unique in the fact that they have the lowest body temperature of any mammal. Their normal body temperature is also more variable and they can safely tolerate more fluctuation in body temperature than most mammals. The body temperature of a giant anteater fluctuates between 90 and 95°F (33–36°C). There is evidence from field studies that anteaters operate at the minimal energetic requirements for a mammal feeding on insects. In other words, energy consumed from food is only slightly greater than the energy used in everyday activities. It is suspected that anteaters coordinate their body temperatures with respect to activity and energy requirements. They can conserve energy with a lower body temperature during periods of rest; conversely when increased activity levels are needed during periods of foraging or hunting for example, the body temperature increases as a result of this energy requirement.

Feeding ecology and diet

Anteaters are specialized to feed on formidable insects. Very few mammals would consider feeding on invertebrates

that are capable of defending themselves with powerful jaws, a potent sting, prickly armor, and in some species the ability to shoot acids and toxins at the enemy. The feeding strategy of anteaters is to lick up as many ants and termites as possible, as quickly as possible. The insect attack quickly becomes unbearable. In fact, an anteater spends only about a

A giant anteater (*Myrmecophaga tridactyla*) looks for insects in a termite hill. (Photo by F. Brian Erize. Bruce Coleman, Inc. Reproduced by permission.)

A giant anteater (*Mymecophaga tridactyla*) extending its tongue. (Photo by John H. Hoffman. Bruce Coleman, Inc. Reproduced by permission.)

minute at a typical nest before it must move on to another. These animals must feed on thousands of ants and termites each day to satisfy their caloric requirements. For example, a giant anteater may visit close to 200 ant and termite nests a day in order to get enough food. Anteaters feed on many different species of ants and termites, both terrestrial and arboreal; however, the bulk of their diet is composed of only a few species. Each anteater species has specific insect preferences, depending on the locality in which it is found. Anteaters also practice resource partitioning. The species of insects upon which they feed depends on ant and termite nest construction and the location of that nest. Some ant and termite species are arboreal and only found on small branches, making them inaccessible to large anteater species. Others have a hard covering on the nest making it impenetrable to smaller anteater species.

An anteater's tongue is darted in and out very quickly and is covered with thousands of tiny hooks known as filiform papillae. These hooks help to grasp insects. Large quantities of saliva also help to hold the insects until they reach the mouth. Slight side to side movements of the jaws aid in moving the tongue and swallowing. The stomach of an anteater is specialized to aid in digestion much like the gizzard of a bird. The stomach has hardened folds and uses strong contractions as well as small bits of ingested sand and dirt to grind the insects.

Reproductive biology

Visual sex determination of anteaters is sometimes possible. Adult males tend to be slightly larger overall and have a wider

Silky anteaters (*Cyclopes didactylus*) grooming, in Ecuador. (Photo by Animals Animals ©Juan Manuel Renjifo. Reproduced by permission.)

Male anteaters have internal testes, making it difficult to tell the difference between males and females without closer examination. (Illustration by Jarrod Erdody)

grasp the newborn offspring due to the enlarged foreclaws; therefore, newborn youngsters must climb up the long extended fur to the mammae. In most species, the young are transported on the mothers back until they are self-sufficient.

Conservation status

Accurate census numbers on these animals have been difficult to obtain. They are solitary, have a low reproductive rate, are difficult to find, and seem to have large home ranges; these factors make population studies very challenging. As a result, their natural history is poorly understood and their conservation status is difficult to assess. They are found in a wide range of habitats. However, much of their range is suffering from the pressures of habitat alteration, destruction and human encroachment. It has yet to be determined how well they can adjust and survive in disturbed habitat. Their survival is also linked to the availability and health of ant and termite populations. This group is in desperate need of detailed study.

Significance to humans

Members of this group are hunted for sport and for their skin. Rope is made from the tendons of the tail from the species that have prehensile tails. They are often kept as pets and used by native peoples in their homes for ant and termite control.

head and neck as well as a more muscular build. The penis and testes are located internally in the abdominal cavity between the rectum and urinary bladder. As a result, the only sure way to determine the sex of an anteater visually is to observe the shape and size differences of the urogenital opening. In males, the opening is more ventral and greatly reduced. Thus, the distance from the anus to the urogenital opening is greater in males. Females have a shorter ano-genital distance, the urogenital opening is long and has a mound-like shape. This can sometimes be seen at a distance with the larger species when the tail is raised, otherwise the animal must be captured for close identification. There is no intromission during breeding, fertilization occurs as a result of contact transfer similar to that observed in some species of lizards. Usually only a single offspring is produced from polygynous mating, twins are possible but very rare. Females have a single pair of mammae near the armpits from which the offspring nurse. Mothers are unable to

A silky anteater (*Cyclopes didactylus*) baby clings tightly to its mother. (Photo by Wolfgang Bayer. Bruce Coleman, Inc. Reproduced by permission.)

1. Silky anteater (*Cyclopes didactylus*); 2. Northern tamandua (*Tamandua mexicana*); 3. Southern tamandua (*Tamandua tetradactyla*); 4. Giant anteater (*Myrmecophaga tridactyla*). (Illustration by Joseph E. Trumpey)

Species accounts

Silky anteater
Cyclopes didactylus

TAXONOMY

Cyclopes didactylus (Linnaeus, 1758), Suriname.

OTHER COMMON NAMES

English: Pygmy, golden or two-toed anteater; French: Flor de balsa; German: serafin; Spanish: Angelito, tapacara, gato balsa.

PHYSICAL CHARACTERISTICS

Total length 12–21 in (32–52 cm); weight 6–13 oz (175–357 g); tail length 6–12 in (16–30 cm). Small arboreal mammal with long, wavy, soft and silky fur. Silvery gray to golden yellow in color with a brown mid-dorsal stripe. Small tubular mouth with a pink nose. Tail is highly prehensile. There are two toes on each forefoot, each with large curved and sharp claws. Four toes on each hindfoot, each with small claws. The hindfoot is highly modified to grasp small branches.

DISTRIBUTION

Mexico, Central America, Brazil, Peru and Bolivia.

HABITAT

Lives among the trees and lianas of moist tropical forests, rarely descending to the ground. The silky anteater shows a preference for the crown of the silk-cotton tree of the genus *Ceiba*, it is concealed very well among the golden fibrous seed pods produced by this tree.

BEHAVIOR

Nocturnal, slow-moving and inoffensive; however, it will defend itself with quick, forceful slashes of the powerful claws. Silky anteaters rarely spend more than one day in the same tree. Their principal predators are harpy eagles (*Harpia harpyja*), eagle-hawks, and spectacled owls (*Pulsatrix perspicillata*).

FEEDING ECOLOGY AND DIET

Forages about the canopies of trees in search of arboreal insects, predominantly ants. Its small size and specialized hind foot allow the silky anteater to use higher and smaller branches and associated ant colonies that larger insectivores cannot physically access. Adults typically consume about 5,000 ants per day.

REPRODUCTIVE BIOLOGY

Usually a single young is born after a gestation period of 120–150 days. Depressions or holes in trees that are partially filled with dry leaves are often used as nests. May be polygynous.

CONSERVATION STATUS

Not threatened. However, this is a very secretive and solitary species, which makes a census very difficult. Also, this species has a poor husbandry record in captivity, seldom surviving for more than 30 days. The longevity record for this species in captivity is two years and four months. At the time of this writing this species was not represented in captivity.

SIGNIFICANCE TO HUMANS

Occasionally hunted for food. ◆

Cyclopes didactylus

Myrmecophaga tridactyla

Southern tamandua
Tamandua tetradactyla

TAXONOMY

Tamandua tetradactyla (Linnaeus, 1758), Pernambuco, Brazil.

OTHER COMMON NAMES

English: Collared anteater, lesser anteater; French: Tamandua-colete; German: Termieteneter; Spanish: Oso colmenero.

PHYSICAL CHARACTERISTICS

Total length 37–58 in (93–147 cm); weight 7–16 lb (3–7 kg); tail length 16–26 in (40–67 cm). Pale golden yellow with a black "vest" over the shoulders, chest, belly and lower back. The vest is highly variable and may cover the entire body or be totally absent. Prehensile tail, head long and curved downward, long coarse hair. Forefeet with four long, powerful curved claws. Hindfeet with five smaller claws. Tamanduas walk on the outside of the hand with the claws turned inward.

DISTRIBUTION

East of the Andes from Venezuela to Argentina and Uruguay.

HABITAT
Savanna, thorn scrub and dry forests, rainforest.

BEHAVIOR
Nocturnal, crepuscular, or diurnal. Tamanduas are opportunistically terrestrial and arboreal depending on terrain and available resources. When threatened it may hiss and release a fowl odor from the anal gland. As a last resort they will defend themselves with the powerful foreclaws and often grab and hold the opponent, keeping it away from the body. Tamanduas seek shelter in hollow trees or holes in the ground.

FEEDING ECOLOGY AND DIET
Opportunist feeding on terrestrial as well as arboreal ants and termites. It can open arboreal nests too tough for silky anteaters. They occasionally feed on bees and honey.

REPRODUCTIVE BIOLOGY
May be polygynous. Mating takes place in the fall, usually a single young is born in the spring after a gestation period of 130–150 days. The offspring is carried on the back of the mother for about a year, gradually becoming self sufficient. Tamanduas commonly reproduce in captivity and have survived for more than 9 years.

CONSERVATION STATUS
Not threatened.

SIGNIFICANCE TO HUMANS
Sometimes used by Amazonian Indians to rid their homes of ants and termites. ◆

Tamandua tetradactyla
Tamandua mexicana

Northern tamandua
Tamandua mexicana

TAXONOMY
Tamandua mexicana (Saussure, 1860), Mexico.

OTHER COMMON NAMES
English: Collared anteater, lesser anteater; French: Tamanduacolete; German: Termieteneter; Spanish: Oso hormiguero comun, brazo fuerte.

PHYSICAL CHARACTERISTICS
Total length 40–51 in (102–130 cm); weight 7–12 lb (3–5 kg); tail length 16–26 in (40–67 cm). Coarse, dense fur is colored fawn to brownish with a black "vest" down its back. Tongue can extend 1.3 ft. (0.4m). Tail is prehensite; fore feet have four claws, and hind feet have five claws.

DISTRIBUTION
Southeast Mexico, Central America, South America west of the Andes from northwestern Venezuela to northwestern Peru.

HABITAT
Savanna, thorn scrub and dry forests, rainforest.

BEHAVIOR
Diurnal and nocturnal, arboreal and terrestrial. Tamanduas are often accompanied by thick clouds of flies and mosquitoes in the rainforest. A population density of 0.05 animals per 2.5 acres (1 ha) and a home range size of 61.8 acres (25 ha) has been recorded on Barro Colorado Island, Panama.

FEEDING ECOLOGY AND DIET
Feeds on terrestrial as well as arboreal ants and termites. Montgomery (1985) reports about 9,000 ants per day consumed by animals on Barro Colorado Island.

REPRODUCTIVE BIOLOGY
A single young is born in spring after a gestation of 130–150 days. Probably polygynous.

CONSERVATION STATUS
Not threatened at this time, however, habitat destruction is a threat to this species in much of the range. At the time of this writing this species had only one known representative in captivity.

SIGNIFICANCE TO HUMANS
The tendons of the tail are used to make rope. ◆

Giant anteater
Myrmecophaga tridactyla

TAXONOMY
Myrmecophaga tridactyla Linnaeus, 1758, Pernambuco, Brazil.

OTHER COMMON NAMES
English: Ant bear; French: Tamandua bandeira; Spanish: Oso caballo.

PHYSICAL CHARACTERISTICS

Total length 68–110 in (174–280 cm); weight 48–88 lb (22–39 kg); tail length 25–35 in (64–90 cm). Long, thick, coarse and stiff fur colored with black, brown, white and gray bands. Forelegs are white, wrists are crossed by a black band. The head is greatly elongated and narrow. The eyes are small and the ears are rounded. The worm-like tongue can extend more than 2 ft (0.6 m) outward. The tail is long, bushy and it is not prehensile. Forefeet with four sickle-shaped claws, the center two being greatly enlarged and powerful, the hindfeet have five short claws.

DISTRIBUTION

Guatemala, Panama, South America, east of the Andes to Northern Uruguay and west of the Andes to Northern Argentina.

HABITAT

Dry forest, rainforest, savanna, and grasslands. Primarily terrestrial but capable of climbing. Can also swim across large bodies of water.

BEHAVIOR

Nocturnal and diurnal. A large terrestrial animal known to cover an average of 7 miles (11 km) a day while foraging in its home range of 22,240 acres (9,000 ha). Giant anteaters walk on their knuckles due to the extreme size of the foreclaws. They can be functionally bipedal while searching, digging, feeding and during times of defense. Depressions in the ground are used as shelter and the animal covers itself with the tail. Adults make a roaring sound when disturbed. Giant anteaters are capable of flicking their tongue at speeds of 160 times per minute. In captivity, they have been known to live for over thirty years.

FEEDING ECOLOGY AND DIET

Feed primarily on ants and termites. Occasionally feed on beetle larva, soft fruits and carrion. Giant anteaters often consume more than 30,000 insects per day.

REPRODUCTIVE BIOLOGY

Probably polygynous. Sexual maturity has been achieved at 18 months of age in captivity. Females give birth to a single young after about 190 days gestation. The young are patterned identically to the mother; while being carried on the mother's back the bands of the two animals visually unite and provide camouflage. The young nurse for approximately two months, then begin taking insects. Neonates often ride on the mothers back for six to nine months before they become self-sufficient. They grasp the long hairs of their mother and produce a shrill call if isolated.

CONSERVATION STATUS

Vulnerable; threatened in much of the range due to habitat loss and hunting. There are also many losses due to wild fires, as the fur is quick to ignite from stray sparks.

SIGNIFICANCE TO HUMANS

Hunted for trophies, food and skin. ◆

Resources

Books

Emmons, Louise H. *Neotropical Rainforest Mammals, A Field Guide.* Chicago: University of Chicago Press, 1990.

Engelmann, G. F. "The Phylogeny of the Xenarthra." In *The Evolution and Ecology of Armadillos, Sloths, and Vermilinguas,* edited by G. Gene Montgomery. Washington, DC: Smithsonian Institution Press, 1985.

Flint, Mike. *North American Regional Studbook for the Giant Anteater.* Tucson: Reid Park Zoo, 2001.

Montgomery, G. G. "*Cyclopes didactylus.*" In *Costa Rican Natural History,* edited by D. H. Janzen. Chicago: University of Chicago Press, 1983.

Montgomery, G. G. "Impact of Vermilinguas On Arboreal Ant Populations." In *The Evolution and Ecology of Armadillos, Sloths, and Vermilinguas,* edited by G. Gene Montgomery. Washington, DC: Smithsonian Institution Press, 1985.

Periodicals

Best, R. C., and Y. Harada. "Food Habits of the Silky Anteater (*Cyclopes didactylus*) in the Central Amazon." *Journal of Mammalogy* 66 (1985): 780–781.

Ken B. Naugher, BS

Armadillos

(Dasypodidae)

Class Mammalia

Order Xenarthra

Family Dasypodidae

Thumbnail description
Small to medium-sized omnivore or insectivore
with homodont teeth and a tough carapace
covering portions of the back, face, tail and
legs; sparse hairs cover soft underside;
elongated or shovel-shaped head has small
eyes and ears and powerful legs have large
claws for digging

Size
Head-to-tail length: 6–59 in (0.15–1.5 m);
weight: 0.18–66 lb (0.08–30 kg)

Number of genera, species
8 genera; 20 species

Habitat
Forest (deciduous, cloud, and rain), savanna,
pampas, desert, and scrub

Conservation status
Endangered: 1 species; Vulnerable: 5 species;
Lower Risk/Near Threatened: 2 species; Data
Deficient: 4 species

Distribution
Latin America with one species ranging into southern North America

Evolution and systematics

Fossil armadillos found in South America are considered
the oldest members of this order. During the Tertiary Period, they diversified in South America when it was isolated
from the north. After the land bridge formed, armadillos invaded North America but became extinct there only 10,000-
15,000 years ago. These invaders included giant herbivorous
armadillos (pampatheres) and those resembling armadillos living today. Also present were rhinoceros-sized glyptodonts
with solidly sutured carapaces without movable banding that
may have had limited locomotion. Spines or club-like structures on the end of their tails were used for deterring predators or fighting other glyptodonts. These clubs could exert a
force that could have caused fractures observed on fossil carapaces. Since the mid-nineteenth century, the nine-banded
armadillo (*Dasypus novemcinctus*) has reinvaded North America, expanding its range into the southeastern United States.
Similarities observed between living armadillos and their
ancestors led Charles Darwin to develop the law of the
succession (i.e., the same types of organisms replace one another within the same area), an adjunct to descent with modification.

The giant armadillo (*Priodontes maximus*) eats primarily termites.
(Photo by Paul Crum/Photo Researchers, Inc. Reproduced by permission.)

Some phylogenies based on molecular data have used armadillos to place xenarthrans as a sister taxon to ferungulates (carnivores and hoofed mammals), whereas others place them with Afrotheria (elephants, hyraxes, and aardvarks). The armadillo suborder, Cingulata, split from anteaters and sloths around the Cretaceous/Tertiary boundary. Cingulata has only one living family, Dasypodidae, which has three main clades represented by the subfamilies: Dasypodinae, the long-nosed genus that arose first, Tolypeutinae (giant, naked-tailed, and three-banded armadillos), and the closely related Euphractinae (hairy, yellow, and pichi armadillos). Phylogeny within the subfamilies remains unclear. The genus, *Chlamyphorus*, does not appear in the analysis because populations were rare and not sampled.

Physical characteristics

The Spanish word armadillo means "little armored one." Armadillos are quite unique in that they are the last mammals to have a shell, that is, an ossification of the corium interrupted by several folds of the skin. The surface of the skin is gray or brown, quite soft and feels like leather. Thanks to the skin folds, the animal is very agile. Surprisingly stocky and low to the ground, armadillos are medium-sized mammals with extra articulating structures in the vertebral column that presumably improve digging. Armadillos are named for the tough bony carapace that covers the pelvic and pectoral girdles as well as areas on the head, limbs, and portions of the tail. Made of ossified dermal tissue covered with a horny epidermis, the carapace, ranging in color from dark to yellow-white, provides protection from thorny vegetation, conspecifics and predators. When harassed, armadillos will tuck their eyes under the shoulder shield and coil slightly to minimize the amount of exposed flesh. Three-banded armadillos (*Tolypeutes*) take this to the extreme and bend completely into a ball, exposing only thick carapace. Girdle-like armor banding (3–13) separated by

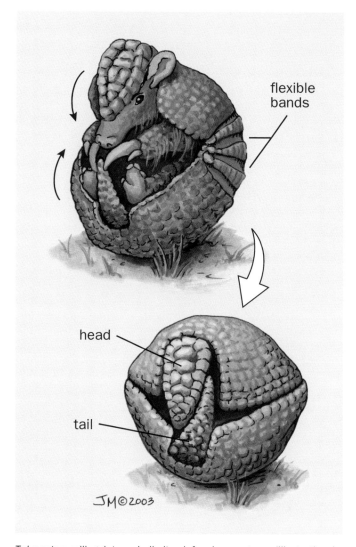

flexible bands

head

tail

JM©2003

Tolypeutes rolling into a ball, its defensive posture. (Illustration by Jacqueline Mahannah)

folds of skin provides flexibility and agility in locomotion. Black to white-colored hairs may be interspersed on the carapace and cover their soft underbellies.

The head varies from shovel-shaped to elongated and narrow. The ears vary in length as well and can be pointed or rounded. Powerful limbs bear formidable claws for digging burrows and gathering food. Hind limbs always have five digits while the number of forelimb digits varies (three to five) depending on the species. Naked-tailed (*Cabassous*) and giant armadillos (*Priodontes maximus*) possess an enlarged claw on the forelimb used to tear into termite and ant mounds. Wielding this large claw and rotating their carapaces back and forth enable these armadillos to escape predators by burying themselves within minutes.

Distribution

Most armadillos are restricted to South America, specifically east of the Andes to the Atlantic coast and south to the

A nine-banded armadillo (*Dasypus novemcinctus*) drinks from a shallow stream in Texas, USA. (Photo by A. Blank. Bruce Coleman, Inc. Reproduced by permission.)

Dasypus moves backwards as it gathers nesting material between its belly and front paws. (Illustration by Jacqueline Mahannah)

Strait of Magellan. Exceptions include the northern naked-tailed (*C. centralis*) and the nine-banded armadillos that have expanded their ranges into southern Mexico and the southeastern United States, respectively.

Habitat

Armadillos are terrestrial to fossorial, exploiting habitats ranging from rainforests to desert, including deciduous and cloud forests, grasslands, llanos, savanna, and thorny scrub. In tropical rainforests armadillos are second only to sloths in mammalian biomass. Adult armadillos are prey to jaguars, pumas, jaguarundis, wild dogs, maned wolves, black bears, and alligators. Probably due to their smaller size, nine-banded armadillo juveniles have twice the mortality of adults, falling prey to these and other animals including raptors.

Burrows, built for sleeping, nesting, escaping predators, or creating an insect reservoir, are dug in well-drained soils or into ant and termite mounds. Naked-tailed armadillos dig a fresh burrow nightly while nine-banded armadillos may reuse 20 burrows per year. In some species, adults and juveniles build nests in burrows by gathering and tucking grass or dead leaves between body and forelimbs, hopping backwards to the burrow, and depositing material by kicking their back feet. Female larger hairy (*Chaetophractus villosus*), nine-banded and Brazilian seven-banded (*D. septemcinctus*) armadillos construct above ground grass shelters prior to parturition.

Behavior

Armadillos are primarily solitary though young siblings and consorting pairs forage together. Llanos long-nosed armadillos (*D. sabanicola*) have been seen feeding in groups in elevated areas in floodplains. Social organization for most species is unknown with the exception of nine-banded armadillos. In this species, males and females have overlapping

A three-banded armadillo (*Tolypeutes matacus*) rolls into a ball for protection. (Photo by George B. Schaller. Bruce Coleman, Inc. Reproduced by permission.)

A hairy armadillo (*Chaetophractus villosus*) near Puerto Piramide, Argentina. (Photo by © Steve Kaufman/Corbis. Reproduced by permission.)

A seven-banded armadillo (*Dasypus septemcinctus*) foraging in Bolivia. (Photo by Harald Schütz. Reproduced by permission.)

home ranges. Females may or may not share an area with each-other depending on density. Although males do overlap in home range, breeding males may use more exclusive areas. These breeding "territories" are maintained by aggression directed at non-breeding males. Female aggression in nine-banded, yellow (*Euphractus sexcinctus*), and larger hairy armadillos is associated with lactation. Defense of space was also seen in northern long-nosed armadillos during the breeding season.

These animals communicate through scent and sound. Secretions from glands in the anal region, on the feet, ears, and pelvic shield function for marking of habitat, identifying individuals, and advertising sexual receptivity. Chemical composition of anal sac glands changes during estrus in nine-banded armadillos and paired females conspicuously wag their tails after male solicitation. The position of large smelly glands on the yellow armadillos pelvic shield suggests a burrow-marking function. Armadillos give off a snuffling sound while foraging and some make a growling sound or a scream when captured. Both sexes in nine-banded armadillos softly chuck during courtship. A buzzing sound may be heard between mother and young. Strangely, armadillos seem unaffected by human voice. Eyesight is so poor that they may run into objects in their path.

Activity is mostly crepuscular and/or nocturnal although yellow, three-banded, northern long-nosed, and pichi (*Zaedyus pichiy*) armadillos forage during the day. Many species shift activity periods seasonally, becoming more diurnal as temperatures drop. Only Andean hairy (*C. nationi*) and pichi armadillos hibernate. Young nine-banded armadillos have a morning and an evening peak of activity. When active, armadillos mostly forage. When disturbed, many balance on back feet and tail and sniff to monitor for predators or conspecifics. If suddenly surprised, nine-banded armadillos will leap into the air and land running, startling a predator. This escape strategy is used unsuccessfully with automobiles, resulting in road kills. Armadillos are champion sleepers, spending upwards of 16 hours snoozing per day.

Feeding ecology and diet

Armadillos are primarily insectivores, feeding on adult and larval forms of beetles, ants, and termites. Some species are myrmecophagic while others opportunistically forage on invertebrates, small vertebrates, carrion, and plant material. Some ingest fruit seasonally. Armadillos ingest large amounts of dirt for mammals. It is unknown whether dirt is required for proper digestion or trace minerals or enters the gut incidentally with food.

Armadillos root around in leaf litter and pause periodically perhaps to sense soil-dwelling prey. Once prey is detected, armadillos use their formidable claws to dig rapidly, excavating small conical pits or tearing into ant and termite mounds. Their sticky tongues effectively lap up the scurrying insects. One stomach had more than 40,000 ants present. Armadillos have a very low metabolic rate, which means that they do not waste a lot of energy producing heat. This also means that they are not good at living in cold areas, because they are not efficient at keeping warm. They do not have any fat reserves, so they must forage for food on a daily basis. A few consecutive cold days can be deadly to the animals.

The southern naked-tailed armadillo (*Cabassous unicinctus*) is smaller than the greater naked-tailed armadillo (*Cabassous tatouay*) but shares the same nocturnal habits. (Photo by Jany Sauvanet/Photo Researchers, Inc. Reproduced by permission.)

A nine-banded armadillo (*Dasypus novemcinctus*) with young. (Photo by Jeff Foott. Bruce Coleman, Inc. Reproduced by permission.)

Reproductive biology

Mating for most species appears to be polygynous and occurs in the summer but some species breed year round in captivity. During courtship, a male follows the female and, for nine-banded armadillos, they forage together for several days. The male checks receptivity by soliciting the female to lift her tail by lightly touching her back. Fertilization occurs but implantation of the embryo is delayed for four months in some species. Most species have a litter of one to three young per year. Unique to mammals, *Dasypus* females exhibit obligate polyembryony, thereby giving birth to genetically identical young (two to 12 depending on the species). Parental care is solely the job of the female. Mating systems are unknown for most species except nine-banded armadillos. Gestation is 140 days, newborn are blind and naked with soft leathery skin, that hardens into armor within a few days. The young are nursed for 2–2.5 months, and start to walk around after a week and open their eyes after 3–4 weeks.

Conservation status

Of the 20 species of armadillos, 12 are listed as Vulnerable, Endangered, Near Threatened, or Data Deficient. Exploitation for food and loss of habitat are the main reasons for decline. Many populations are fossorial and have not been studied thoroughly and so their current status is unclear. The hairy long-nosed armadillo (*D. pilosus*) is known only from a few skins from mountains in Peru.

Significance to humans

Armadillos are exploited throughout Latin America for food. They are considered so tasty that one Mexican society circumvented food taboos by calling them turkeys. In the

United States during the Great Depression in the thirties, armadillos were readily consumed and given the names Texas turkey and Hoover hog. Souvenir purses and baskets with tail handles are formed from hollowed-out carapaces. Stuffed specimens on tip toes still line shop shelves in Mexican border towns. Armadillos are unwanted guests in suburban settings and agricultural fields. Ranchers have also targeted armadillos for extermination because their burrows reportedly lead to broken limbs of livestock and horses. Many change their minds upon hearing that armadillos are the only known predator of fire ants in the United States. Armadillos also are used as research models in the study of leprosy and development of a vaccine, because they are the only animals that can transmit leprosy.

The pink fairy armadillo (*Chlamyphorus truncatus*) is the smallest of all armadillos. (Photo by N. Smythe/Photo Researchers, Inc. Reproduced by permission.)

1. Yellow armadillo (*Euphractus sexcinctus*); 2. Pink fairy armadillo (*Chlamyphorus truncatus*); 3. Small hairy armadillo (*Chaetophractus vellerosus*); 4. Greater naked-tailed armadillo (*Cabassous tatouay*); 5. Giant armadillo (*Priodontes maximus*); 6. Nine-banded armadillo (*Dasypus novemcinctus*); 7. Southern three-banded armadillo (*Tolypeutes matacus*). (Illustration by Jacqueline Mahannah)

Species accounts

Nine-banded armadillo

Dasypus novemcinctus

SUBFAMILY
Dasypodinae

TAXONOMY
Dasypus novemcinctus Linnaeus, 1758, Brazil.

OTHER COMMON NAMES
English: Common long-nosed armadillo; French: Tatou à neuf bandes; German: Neunbinden-Gürteltier; Spanish: Mulita.

PHYSICAL CHARACTERISTICS
Length 25.4 in (64.6 cm); weight 9.9 lb (4.5 kg). Dentition: 7–9/7–9. Has 7–9 bands, a long banded tail, an elongated face, and large ears held close together.

DISTRIBUTION
Latin America, southern North America.

HABITAT
Forested areas preferred.

BEHAVIOR
Crepuscular and nocturnal but more diurnal during the winter. Solitary. Prescribed home ranges maintained. Male breeding territories suggested. Polygynous. Can go without oxygen for short periods while foraging in soil. Walks across the bottom of small streams but gulps air and dogpaddles across larger bodies of water. Armadillos are fond of water; under arid, dry climatic conditions, they concentrate in the vicinity of streams and water holes. Tracks in the mud around small ponds give evidence that the armadillos visit them not only for purposes of drinking and feeding, but also to take mud baths. Armadillos are timid animals. They are almost constantly active when foraging and probing into crevices and under litter for food. They continuously grunt while foraging and do not seem to be particularly attentive to their surroundings. They communicate with each other by low-volume sounds.

FEEDING ECOLOGY AND DIET
Eats beetles, beetle larvae, ant larvae, other insects and invertebrates, small vertebrates, and fruit seasonally. Moves noisily through leaf litter stopping periodically to probe the soil.

REPRODUCTIVE BIOLOGY
Polygynous mating occurs in summer months, June though August in the United States. Courtship may last several days. Implantation of embryo is delayed for four months or as long as two years. Gestation lasts four months with births occurring about 65 days after implantation. Females exhibit polyembryony, giving birth to four genetically identical young. Not all individuals breed in a given year. In one population, genetic studies showed only one third of adults were parents over a four-year period. Ovulation is inhibited during drought conditions.

CONSERVATION STATUS
Not threatened.

SIGNIFICANCE TO HUMANS
Used as food and as an animal model for penile erection and leprosy studies. ◆

Dasypus novemcinctus

Small hairy armadillo

Chaetophractus vellerosus

SUBFAMILY
Euphractinae

TAXONOMY
Dasypus vellerosus (Gray, 1865), Bolivia.

OTHER COMMON NAMES
English: Small screaming armadillo; French: Petit tatou velu; German: Weisshaar-Gürteltier; Spanish: Quirquincho chico.

PHYSICAL CHARACTERISTICS
Length 14.6 in (37 cm); weight 1.9 lb (850 g). Dentition: 9/10. Smallest of the hairy armadillos, it has a broad head shield with widely spaced ears and 18 bands. Silky hairs sparsely cover the body. Kidneys concentrate fluids and are thus adapted to arid climates.

DISTRIBUTION
Bolivia, Paraguay, and Argentina.

HABITAT
Sandy desert habitat that is not rocky.

Euphractus sexcinctus

Chaetophractus vellerosus

OTHER COMMON NAMES
English: Six-banded armadillo; French: Tatou à six bandes;
German: Sechsbinden-Gürteltier; Spanish: Gualacate.

PHYSICAL CHARACTERISTICS
Length 24.3 in (61.6 cm); weight 9.9 lb (4.5 kg). Dentition:
9/10 or 8/9. Has a yellowish carapace with long light hairs, a
broad face shield with small separated ears, and a pelvic shield
with 2-4 holes secreting scent.

DISTRIBUTION
Suriname, Brazil, Uruguay, Paraguay, and parts of Argentina.

HABITAT
Inhabits savanna, steppe and forest edge.

BEHAVIOR
Diurnal. Reuses burrows probably marking them with pelvic-
shield glands. Unusual in that they bite when handled.

FEEDING ECOLOGY AND DIET
Omnivorous, feeding on plant material, invertebrates and vertebrates including carrion. Four rodents were found in stomach
of one road kill.

REPRODUCTIVE BIOLOGY
Females give birth to one to three young of mixed sex per litter. Does not exhibit polyembryony. In captivity gestation
length is 60–65 days. May be polygynous.

CONSERVATION STATUS
Not threatened.

SIGNIFICANCE TO HUMANS
Used as food. ◆

Southern three-banded armadillo
Tolypeutes matacus

SUBFAMILY
Tolypeutinae

TAXONOMY
Loricatus matacus (Desmarest, 1804), Argentina.

OTHER COMMON NAMES
French: Tatou à trios bandes du Sud; German: Kugel
Gürteltier; Spanish: Bolita.

PHYSICAL CHARACTERISTICS
Length 12.4 in (31.4 cm); weight 2.4 lb (1.1 kg). Dentition:
9/9. Three bands. Thick carapace. Short tail. Walks on tips of
claws on forelimb.

DISTRIBUTION
Argentina, Bolivia, Brazil, and Paraguay.

HABITAT
Grassland and open plains.

BEHAVIOR
When threatened, rolls up into a ball, exposing only its carapace and tail and head shields. Diurnal. Does not dig burrows
but uses those built by other species. Individuals found sleeping together in winter (also the breeding season).

FEEDING ECOLOGY AND DIET
Feeds on ants and termites obtained by digging shallow pits.

BEHAVIOR
Fossorial and nocturnal to avoid the day's heat but become
more diurnal in the winter. Spend most of their time foraging
near vegetation in prescribed home ranges (4.7 ha on average).
In captivity, forages systematically, spiraling inwardly in a
patch. Multiple burrows, found in sand dunes or near vegetation, are used for resting, shelter and foraging. Gives off an
eerie scream when handled.

FEEDING ECOLOGY AND DIET
Omnivorous diet consists of invertebrates, vertebrates and
plant material. Summer diet includes rodents (20%). Diet
adapted to desert life where insects are scarce.

REPRODUCTIVE BIOLOGY
Captive animals give birth to one to two young and have a gestation period of 65 days. Probably polygynous.

CONSERVATION STATUS
Not threatened.

SIGNIFICANCE TO HUMANS
None known. ◆

Yellow armadillo
Euphractus sexcinctus

SUBFAMILY
Euphractinae

TAXONOMY
Dasypus sexcinctus (Linnaeus, 1758), Brazil.

Tolypeutes matacus
Priodontes maximus

Cabassous tatouay
Chlamyphorus truncatus

REPRODUCTIVE BIOLOGY
Pairing occurs during the breeding season. Captive males solicit females by gently touching their dorsal side. Gives birth to one offspring per year in spring or summer but in captivity birthing occurs year round.

CONSERVATION STATUS
Lower Risk/Near Threatened. Another species, *Tolypeutes tricinctus*, thought to be extinct in wild, was rediscovered in Brazil.

SIGNIFICANCE TO HUMANS
Used for food. ◆

Greater naked-tailed armadillo
Cabassous tatouay

SUBFAMILY
Tolypeutinae

TAXONOMY
Loricatus tatouay (Desmarest, 1804), Paraguay.

OTHER COMMON NAMES
French: Grand tatou à queue nue; German: Grosses Nacktschwanz-Gürteltier; Spanish: Tatú-ai mayor.

PHYSICAL CHARACTERISTICS
Length 25 in (63.7 cm); weight 11.8 lb (5.35 kg). Dentition: 9/8. They resemble small versions of giant armadillos except no armor on their tails.

DISTRIBUTION
Brazil, Uruguay, Paraguay, and Argentina.

HABITAT
Found along rivers but also reported in grassland communities.

BEHAVIOR
Nocturnal forager but may continue to feed near its burrow shortly after sunrise. When harassed they quickly bury themselves within minutes. When handled they vocalize like a grunting pig.

FEEDING ECOLOGY AND DIET
Primarily myrmecophagous but also feeds on incidental invertebrates present in the mounds. Holes are dug directly into the mounds or where insects are foraging.

REPRODUCTIVE BIOLOGY
Females give birth to one offspring per year. No information available on seasonality or mating behavior.

CONSERVATION STATUS
Not threatened.

SIGNIFICANCE TO HUMANS
None known. ◆

Giant armadillo
Priodontes maximus

SUBFAMILY
Tolypeutinae

TAXONOMY
Dasypus maximus (Kerr, 1792), French Guiana.

OTHER COMMON NAMES
French: Tatou géant; German: Riesengürteltier; Spanish: Tatú carrera.

PHYSICAL CHARACTERISTICS
Length 4.9 ft (1.5 m); weight 66 lb (30 kg). Dentition 18/19. Largest of the living armadillos. Colored darkly with many narrow bands and an armored tail.

DISTRIBUTION
Venezuela, Colombia, Brazil, and Argentina.

HABITAT
Tropical forest and open savanna.

BEHAVIOR
Fossorial and nocturnal. Digs its burrows in open fields and termite mounds, destroying the mound in the process.

FEEDING ECOLOGY AND DIET
Myrmecophagous (feeds on ants).

REPRODUCTIVE BIOLOGY
Females give birth to one to two young per litter per year. Polygynous.

CONSERVATION STATUS
Listed as Endangered because of overexploitation and loss of habitat.

SIGNIFICANCE TO HUMANS
Hunted for food within its range. ◆

Pink fairy armadillo
Chlamyphorus truncatus

SUBFAMILY
Chlamyphorinae

TAXONOMY
Chlamyphorus truncatus Harlan, 1825, Argentina.

OTHER COMMON NAMES
English: Lesser pink fairy armadillo; French: Petit pichiciego; German: Kleiner Gürteltier; Spanish: Pichiciego menor.

PHYSICAL CHARACTERISTICS
Length 5.9 in (15 cm); weight 4.2 oz (120 g). Has reduced eyes and ears. Carapace is attached along backbone. Long silky hair found ventrally, extending up under the carapace. Face and rear shields are flexible.

DISTRIBUTION
Argentina.

HABITAT
Sandy to mixed soil types with limited thorny, scrubby vegetation.

BEHAVIOR
Fossorial and nocturnal. Captive animals are active periodically throughout the day. They burrow under obstacles rather than going around.

FEEDING ECOLOGY AND DIET
Omnivorous diet of insects, other invertebrates, plant material, and occasionally carrion.

REPRODUCTIVE BIOLOGY
Gives birth to one young per litter per year. Probably polygynous.

CONSERVATION STATUS
Listed as Endangered.

SIGNIFICANCE TO HUMANS
None known. ◆

Common name / Scientific name/ Other common names	Physical characteristics	Habitat and behavior	Distribution	Diet	Conservation status
Northern naked-tailed armadillo *Cabassous centralis* French: Tatou; German: Weisskopf-Zweizehenfaultier; Spanish: Armado de zapilots	Dark brown to almost black with yellow lateral areas, underparts are yellow-gray. Head is broad, snout is short and wide, ears are well-separated. Head and body length 12–28 in (30–71 cm), tail length 4–7 in (10–18 cm).	Grasslands and wooded areas. Solitary and nocturnal.	Mexico (Chiapas) to northern Colombia.	Mainly insects, including larvae and adult scarab beetles, termites, and ants. Also consume earthworms, bird eggs, and small reptiles and amphibians.	Data Deficient
Chacoan naked-tailed armadillo *Cabassous chacoensis*	Upperparts dark brown to black, lateral edges of carapace are yellow, underparts are dull yellowish gray. Five large claws on forefeet. Snout is short and broad, head is broad, ears are widely separated. Head and body length 11.8–19.2 in (30–49 cm), tail length 3.5–7.9 in (9–20 cm).	Grasslands, semiarid and moist lowlands, upland areas, and riversides. Solitary and nocturnal.	Gran Chaco of western Paraguay and north-western Argentina.	Termites and ants.	Data Deficient
Southern naked-tailed armadillo *Cabassous unicinctus*	Dark brown to almost black with yellow lateral areas, underparts are yellow-gray. Head is broad, snout is short and wide, ears are well-separated. Head and body length 12–28 in (30–71 cm), tail length 4–7 in (10–18 cm), weight 4.8–10.6 lb (2.2–4.8 kg).	Grasslands and wooded areas. Solitary and nocturnal.	South America east of the Andes from Colombia to Mato Grosso do Sul, Brazil.	Termites and ants.	Not threatened

[continued]

Common name / Scientific name/ Other common names	Physical characteristics	Habitat and behavior	Distribution	Diet	Conservation status
Andean hairy armadillo *Chaetophractus nationi* Spanish: Quirquincho de la puna	Varies from yellowish to light brown. Eighteen dorsal bands, 8 are moveable. Has hair between scales and a head shield 2.3 in long by 2.3 in wide (6 by 6 cm), head and body length 8.6–15.7 in (22–40 cm), tail length 3.5–6.9 in (9–17.5 cm).	Burrows in steep slopes. Nocturnal and solitary. Body temperature is regulated ectothermically.	Cochabamba, Oruro, and La Paz, Bolivia.	Some small vertebrates, many insects, and some vegetation.	Vulnerable
Large hairy armadillo *Chaetophractus villosus* Spanish: Quirquincho grande	Skin is brown to pinkish and hair is grayish brown to white. Double layer of horn and bone covers majority of dorsal side. Small shield on head. Carapace consists of 18 bands, 7 to 8 of which are moveable. Ventral area covered by soft skin. Head and body length 8.6–15.7 in (2.2–4.0 cm), average weight 4.4 lb (2 kg).	Open, semi-desert environments. Diurnal, year-round breeding, maximum life span is 30 years.	Gran Chaco of Bolivia, Paraguay, and Argentina south to Santa Cruz, Argentina, and Magallanes, Chile.	Insects, invertebrates, small vertebrates, plants, and carrion.	Not threatened
Chacoan fairy armadillo *Chlamyphorus retusus* German: Gürtelmull; Spanish: Pichiciego chaqueño	Characteristic pelvic armor firmly attached to spine and pelvic bones. Twenty-four mobile, dorsal bands. Head shield, white hair sparse on dorsal surface, curved claws on hands and feet. Head and body length 5.5–6.9 in (14.0–17.5 cm), tail length 1.4 in (3.5 cm).	Dry grasslands of Argentina, Paraguay, and Bolivia. In underground burrows in warm and dry soils. Rare and nocturnal, have been observed making baby-like crying sounds.	Argentina, Paraguay, and Bolivia.	Insects, insect larvae, worms, snails, roots, and small seeds.	Vulnerable
Southern long-nosed armadillo *Dasypus hybridus* Spanish: Mulita orejuda	Very little hair on upper part of body, sparsely scattered and pale yellow on undersides. Carapace ranges from mottled brown to yellow in coloration. Long, pointed nose, short legs. Head and body length 9.4–22.6 in (24.0–57.3 cm), tail length 5–19 in (12.5–48.3 cm), weight 2.2–22 lb (1–10 kg).	Dense, shady cover and limestone formation from sea level to 9,840 ft (3,000 m). Build large nests of grass or leaves, share burrows with other armadillos. Primarily nocturnal.	Argentina, Paraguay, and southern Brazil south to Río Negro, Argentina.	Insects, spiders, and small amphibians.	Not threatened
Great long-nosed armadillo *Dasypus kappleri* Spanish: Armadillo narigón mayor	Sparse, pale yellow hair scattered on underparts. Carapace coloration varies from brown to yellowish white in color. Long, pointed nose, and short legs. Head and body length 9.4–22.5 in (24.0–57.3 cm), tail length 5–19 in (12.5–48.3 cm), weight 2.2–22 lb (1–10 kg).	Dense, shady cover and limestone formation from sea level to 9,840 ft (3,000 m). Build large nests of grass or leaves, share burrows with other armadillos. Primarily nocturnal.	Colombia (east of the Andes), Venezuela (south of the Orinoco), Guyana, Suriname, and south through the Amazon Basin of Brazil, Ecuador, and Peru.	Insects, spiders, and small amphibians.	Not threatened
Hairy long-nosed armadillo *Dasypus pilosus*	Almost no hair on head. Long white and pale yellow hair on shell and underparts, giving furry appearance. Carapace coloration varies from mottled brown to yellowish white. Long, pointed nose and short legs. Eleven moveable bands on shell. Head and body length 9.4–22.5 in (24.0–57.3 cm), tail length 5–19 in (12.5–48.3 cm), weight 2.2–22 lb (1–10 kg).	Dense, shady cover and limestone formations from sea level to 9,840 ft (3,000 m) in elevation. Dig deep burrows and share them with other armadillos. Primarily nocturnal.	Only from the Peruvian Andes in San Martín, La Libertad, Huánuco, and Junín.	Mainly insects, spiders, and small amphibians.	Vulnerable
Llanos long-nosed armadillo *Dasypus sabanicola*	Almost no hair on head. Carapace coloration varies from mottled brown to yellowish white. Long, pointed nose and short legs. Head and body length 9.4–22.5 in (24.0–57.3 cm), tail length 5–19 in (12.5–48.3 cm), weight 2.2–22 lb (1–10 kg).	Dense, shady cover and limestone formations from sea level to 9,840 ft (3,000 m) in elevation. Dig deep burrows and share them with other armadillos. Primarily nocturnal.	Llanos of Venezuela and Colombia.	Mainly insects, spiders, and small amphibians.	Data Deficient
Seven-banded armadillo *Dasypus septemcinctus* Spanish: Armadillo narigón de siete bandas	Coloration of carapace varies from mottled brown to yellowish white. Sparse, white, long hair on underparts. Head and body length 9.4–22.5 in (24.0–57.3 cm), tail length 5–19 in (12.5–48.3 cm), weight 2.2 –22 lb (1–10 kg).	Dense, shady cover and limestone formations from sea level to 9,840 ft (3,000 m) in elevation. Dig deep burrows and share them with other armadillos. Primarily nocturnal.	Lower Amazon Basin of Brazil to the Gran Chaco of Bolivia, Paraguay, and northern Argentina.	Mainly insects, spiders, and small amphibians.	Not threatened

[continued]

Common name / Scientific name/ Other common names	Physical characteristics	Habitat and behavior	Distribution	Diet	Conservation status
Brazilian three-banded armadillo *Tolypeutes tricinctus*	Coloration is blackish brown. Most have moveable bands. Can completely enclose themselves in their own shell by rolling into a ball. Tail is short and thick. Head and body length 8.5–10.7 in (21.8–27.3 cm), tail length 2.3–3.2 in (6.0–8.0 cm), weight 2.2–3.5 lbs (1–1.6 kg).	Tropical deciduous forest on elevated portions of the Brazilian plateau. Forages with powerful claws.	Brazilian states of Bahia, Ceará, and Pernambuco.	Termites, invertebrates, and fruit.	Vulnerable
Pichi *Zaedyus pichiy* English: Dwarf armadillo; Spanish: Piche de Patagonia	Small ears, developed claws, and dark brown armor. Armor has white to yellow edges and hairs sticking up between them. Hair coloration ranges from yellow to white. Tail is usually yellow. Head and body length 10–13.2 in (26.0–33.5 cm), tail length 3.9–5.5 in (10.0–14.0 cm).	Grasslands and arid regions of southern South America. Usually resides in areas with sandy soils. Primarily nocturnal.	Mendoza, San Luis, and Buenos Aires, Argentina, south through Argentina and eastern Chile to the Straits of Magellan.	Insects, worms, some plant matter (like tubers), carrion, and other animal matter.	Data Deficient

Resources

Books

Eisenberg, John F. *Mammals of the Neotropics.* Vol. 1, *The Northern Neotropics: Panama, Colombia, Venezuela, Guyana, Suriname, French Guiana.* Chicago: The University of Chicago Press, 1989.

———. *The Mammalian Radiations: An Analysis of Trends in Evolution, Adapation, and Behavior.* Chicago: The University of Chicago Press, 1981.

Eisenberg, John F., and Kent H. Redford. *Mammals of the Neotropics.* Vol. 3, *The Central Neotropics: Ecuador, Peru, Bolivia, Brazil.* Chicago: The University of Chicago Press, 1999.

Emmons, Louise H. *Neotropical Rainforest Mammals: A Field Guide.* Chicago: The University of Chicago Press, 1990.

Montgomery, G. Gene, ed. *The Evolution and Ecology of Armadillos, Sloths, and Vermilinguas.* Washington: Smithsonian Institution Press, 1985.

Redford, Kent H., and John F. Eisenberg. *Mammals of the Neotropics.* Vol. 2, *The Southern Cone: Chile Argentina, Uruguay, Paraguay.* Chicago: The University of Chicago Press, 1992.

Periodicals

Carter, T. S., and C. D. Encarnaçao. "Characteristics and Use of Burrows by Four Species of Armadillos in Brazil." *Journal of Mammalogy* 64 (1983): 103–108.

Delsuc, Frederic, et al. "The Evolution of Armadillos, Anteaters, and Sloths Depicted by Nuclear and Mitochondrial Phylogenies: Implications for the Status of the Enigmatic Fossil Eurotamandua." *Proceedings of the Royal Society of London, Series B* 268 (2001): 1605–1615.

Greegor, D. H. "Preliminary Study of Movements and Home Range of the Armadillo, *Chaetophractus vellerosus.*" *Journal of Mammalogy* 61 (1980): 334–335.

McDonough, Colleen M. "Social Organization of Nine-banded Armadillos (*Dasypus novemcinctus*) in a Riparian Habitat." *American Midland Naturalist* 144 (2000): 139–151.

Merritt, Dennis A., Jr. "The La Plata Three-banded Armadillo in Captivity." *International Zoo Yearbook* 16 (1976): 153–156.

Pacheco, J., and C. J. Naranjo. "Field Ecology of *Dasypus sabanicola* in the Flood Savanna of Venezuela." In *The Armadillo as an Experimental Model in Biomedical Research.* Washington: Pan American Health Organization No. 366 (1978): 13–15.

Shaller, G. B. "Mammals and Their Biomass on a Brazilian Ranch." *Arquivos de Zoologia, S. Paulo* 31, no. 1 (1983): 1–36.

Colleen M. McDonough, PhD
W. J. Loughry, PhD

Insectivora

(Insectivores)

Class Mammalia

Order Insectivora

Number of families 7

Number of genera, species 67 genera; 426 species

Photo: A lesser hedgehog tenrec (*Echinops telfairi*) mother and babies foraging. (Photo by Harald Schütz. Reproduced by permission.)

Introduction

In 1758, when Linnaeus published his taxonomy masterpiece, *Systema Naturae*—the bible for all past, present and undoubtedly future taxonomists—the organisms were classified according to their visible physical characteristics. The order Insectivora was reserved for organisms observed eating insects. Linnaeus listed three families in this category: Talpidae (Old World moles), Erinaceidae (hedgehogs), and Soricidae (long-tailed shrews).

Today, the order Insectivora consists of seven families: the Erinaceidae, with seven genera and 21 species of gymnures and hedgehogs; the Chrysochloridae includes seven genera and 18 species of golden moles; the Tenrecidae features ten genera and 24 species of tenrecs; the Solenodontidae consists of one genera and three species of solenodons; the Nesophontidae has one genus, *Nesophontes*, and six species of extinct West Indian shrews; the Soricidae includes 24 genera and 311 species of shrews; and the Talpidae features 17 genera and 41 species of moles, shrew moles, and desmans.

The order Insectivora has 426 species of small mammals, three-fourths of which are shrews, quite possibly the smallest of all mammals. The Insectivora is the third largest order of mammals. The Rodentia is the largest with over 2,000 species and the Chiroptera (bats) is the second with over 900 species.

Evolution and systematics

The timing of the origin of extant Insectivora species has been and continues to be a topic of great debate among sci-entists. It is widely accepted, however, that Insectivora are the most primitive of the true placental mammals existing today and the ones from which present day mammals evolved. Most of the primitive eutherian (placental) mammals were insectivores. To date, common classification practice has been to include some of the primitive eutherians with all recent Insectivora members based on their similar dentition. The earliest insectivore fossils are believed to be those of *Batodon* and *Paranyctoides*. These remains date from the mid to late Cretaceous period, approximately 100 million years ago (mya). Remains of small insect eaters from Asia, zalambdalestids, along with kennalestid remains from Central America, are believed to date back to the Late Cretaceous.

Erinaceidae fossils date back to the Paleocene to early Pliocene in North America. Other fossils discovered in Africa date from the early Miocene to Recent, from late Paleocene to Recent in Europe, and from Eocene to Recent in Asia. The Solenodontidae dates from the late Mesozoic and early Cenozoic in North America and the Caribbean to Recent in Cuba and Hispaniola. Eight species of Nesophontidae are believed to have survived the Pleistocene period on into the 1900s. Lack of physical evidence of their ongoing survival has led scientists to conclude that they are now extinct. Most of the Tenrecidae began to evolve in Madagascar from the Pleistocene to Recent. Other records show that they existed during the middle Eocene to middle Oligocene in North America and from the Miocene to Recent in Africa. Chrysochloridae fossils dating back to the Miocene (Kenya) and Pleistocene (South Africa) and resembling Recent species were discovered in Africa. There are varying opinions regarding the evolution

A smoky shrew (*Sorex fumeus*) foraging in leaf litter. (Photo by Gary Meszaros. Bruce Coleman, Inc. Reproduced by permission.)

of the Soricidae. This is due to the rarity of soricid fossils. Some scientists believe that the Eocene period in Africa, Eurasia, northern South America, and North America marks the onset of this family's existence. Others believe that the Soricidae existed only in Europe and North America during this period and from Miocene to Recent in Africa and Asia, Pleistocene to Recent in northwest South America, and early Oligocene to Recent in Europe and Asia. We know of recent Talpidae fossils dating back to Miocene in Europe and recent protein studies show that they were present in North America during that period as well. Otherwise their geological range is also believed to be early Oligocene to Recent in North America, and late Miocene to Recent in Asia.

For a long time, Insectivora was the order into which scientists put species of questionable lineage or those mammals that were characterized by the lack of distinctive features possessed by other mammals. As phylogenetics (identifying and understanding relationships between life forms) and other scientific dating methods improve, and as additional fossils are discovered, Insectivora is becoming less of a catchall order but not necessarily less controversial.

There are many examples of the difficulties in classification surrounding Insectirora. Prior to 1972, for example, elephant shrews were part of Insectivora. Then one group of scientists decided that because these mammals possessed certain physical characteristics not typical of most other insectivores—moles, shrews, hedgehogs, and tenrecs—such as an auditory bulla, an entotympanic bone, a complete zygomatic arch and a large jugal bone, elephant shrews should be placed in a separate order, namely Macroscelidea. Other taxonomists believed that Insectivora should be divided into two suborders: Menotyphla consisting of Macroscelididae (elephant shrews), Tupaiidae (tree shrews), and Lipotyphla with the living families Erinaceidae (gymnures and hedgehogs), Chrysochloridae

(golden moles), Tenrecidae (tenrecs), Solenodontidae (solenodons), Soricidae (shrews), and Talpidae (moles, shrew moles, and desmans). Other scientists felt that tree shrews belonged with the primates and yet others believed that they should be placed in a separate order altogether, Scandentia. There are also those who believe that they have molecular evidence for the origins of Insectivora and for a new order, Afrosoricida, of endemic African mammals to include the families Chrysochloridae (golden moles) and Tenrecidae (tenrecs). Reallotment of these two families would decrease the number of Insectivora species by 51. The elephant shrews (*Macroscelididae*) would be included in this new order as well. The Cynocephalidae (colugos) has been moved around over the years between Insectivora and Dermoptera, where they now reside. With Insectivora, it would appear that as scientific techniques evolve, so families and species fragment, multiply, and/or relocate.

Physical characteristics

The size of living insectivores ranges from a small mouse to a large house cat. Savi's pygmy shrew (*Suncus etruscus*), is believed by many scientists to be the world's smallest living mammal weighing 0.04–0.10 oz (1.2–2.7 g) and measuring 1.4–2.1 in (36–53 mm) without the tail. The 1998 discovery of a 65 million year old fossil jaw measuring 0.3 in (8 mm) and belonging to the extinct Batonodoides, suggests that some prehistoric insectivore mammals were even tinier than some Savi's pygmy shrew.

The moonrat, (*Echinosorex gymnura*) is the largest living insectivore, weighing as much as 4.4 lb (2 kg) with a body measuring up to 16 in (40.6 cm) and an 8 in (20.3 cm) long tail. Found on the Malay Peninsula, Sumatra, and Borneo this invertebrate and fruit-eating mammal keeps predators at bay by releasing a foul odor. The largest extinct member of Insectivora is thought to be *Deinogalerix* which means "terrible hedgehog." This 2-ft-long (61-cm-long) hairy mammal with elongated snout, sharp teeth, and short legs lived in Europe during the middle Miocene period about 15 mya.

Many members of Insectivora share the following characteristics: five-clawed digits on each limb (pentadactyly); short legs; annular (ring-shaped) tympanic bone; long, flat, small skull; flat cranium (brain case); a small, smooth brain with hemispheres that do not extend backward over the cerebellum; an incomplete zygomatic arch (cheek bone); largely interorbital, well developed and sharply demarcated olfactory bulbs; small ears (exterior ones nonexistent in Talpidae) often lacking ossified auditory bulla (the bony covering of the middle ear cavity); no intestinal caecum; pollex (thumbs) and hallux (big toes) are not opposable; tibia and fibula are often fused near the the ankle whereas the radius and ulna are separate. The otter shrew (*Potamogale*) is the only Insectivora genus that does not have clavicles. A cloaca, the common posterior chamber into which the digestive, urinary, and reproductive tracts all discharge, which is uncommon in most placental mammals, is frequently found in Insectivora species. Male testes are abdominal, inguinal, or borne in a sac in front of the penis; a baculum (*os penis*) is present in some species. All members of this order have a chorioallantoic placenta that allows the young to develop fully in the womb.

A short-tailed shrew (*Blarina brevicauda*) eats a worm. (Photo by Dwight Kuhn. Bruce Coleman, Inc. Reproduced by permission.)

Tiny eyes and poor eyesight are typical of insectivores. Moles and desmans (talpids) are almost blind and in many mature species the eyes are entirely covered with fur. Most insectivores have small external ears, some concealed under fur. Moles have no external ears. Asiatic water shrews (*Chimarrogale*) have tiny ears featuring a valvular flap that seals the opening when they submerge themselves in water. The cone-shaped snouts characteristic of insectivores come in many different sizes and perform a wide variety of tasks. The ethmoturbinal bones (bony plates that support nasal membranes) are large. Coiled scrolls of bone make up the nasal chamber that is covered with olfactory epithelium, the receptors of which are stimulated by airborne chemical molecules. The insectivore olfactory sense is very keen. Some snouts are long and thin with flexible tips for prodding in muddy areas, others are short and stubby with leathery pads on the tip. Solenodons have a small round bone—*os proboscis*—located at the tip of their hairless nose that supports the snout cartilage.

Vibrissae—sensory hairs located on many insectivoran tails, snouts, behind ears, and even on feet—are large in diameter. They are relatively rigid so that they do not bend much, but instead act as levers, transmitting the applied force to their base. They are embedded in a fluid filled sack, which allows them to move about under the skin. Water shrews, like many other insectivores, have vibrissae on their snouts believed to help them locate prey. It is thought that unless these shrews actually touch their prey with their vibrissae, no matter how close they are, they cannot capture it. Long-tailed shrews (*Sorex dispar*) rotate their long, slender, vibrissae-adorned snouts constantly. Moles and desmans (Talpidae) depend primarily on their olfactory (smell) and tactile (touch) senses. The talpid snout is usually long, narrow, very mobile and extends beyond the end of the upper jaw. The tip is hairless, apart from a few vibrillae, and features Eimer's organs (minute sensory receptors). The star-nosed mole (*Condylura*

cristata) has a snout tip featuring 22 fleshy pink appendages or tentacles consisting of 25,000 sensory receptors used to help it locate food. When the star-nosed mole eats, the tentacles curl up out of the way. The snout of the desman is such that it can be used as a snorkel or periscope that continously monitors the air for prey or predators.

Many variations on the dental formulae can be found among Insectivora but one of the most frequent is (I3/3 C1/1 P 4/4 M3-4/3-4) $\times$ 2 = 44, 46, or 48. All insectivores have rooted, primitive teeth. Deciduous teeth, the first set that develops in mammals, are shed early on and rarely serve a purpose. Insectivores have unspecialized sharp teeth and often a crown pattern typical of primitive placentals. Some have front teeth modified by specialized or at times enlarged incisors and canines with a varying morphology, sometimes shaped like incisors or premolars. Shrews and moles often have dilambdodont upper molars (W-shaped crest pattern). Tenrecs, golden moles, and solenodons have zalambdodont upper molars (V-shaped crest pattern). The upper molars in hedgehogs and gymnures are quadrate meaning they have four main cusps. The Haitian solenodon (solenodon means "grooved tooth") and European water shrews—both omnivorous but preferring an animal diet—have unusual dentition: a large upper incisor points slightly backward and a deeply grooved channel in the lower incisor at the base of which is a duct that transports venomous saliva.

The majority of insectivores have plantigrade feet meaning that, like humans, they place the full length of their foot on the ground during each stride. Fossorial insectivore limbs are specialized for digging tunnels and burrows. The forelimbs are short, powerful, and shovel-like, rotated so that the elbows face upwards with strong large-clawed paws facing

An eastern mole (*Scalopus aquaticus*) emerging from its hole. (Photo by L. L. Rue III. Bruce Coleman, Inc. Reproduced by permission.)

backwards. Moles (Talpidae) have five claws and a falciform bone, which expands the palms and supports the digits. It is sometimes referred to as a "sixth digit." Tenrecs and golden moles have four claws. Talpids also have an elongated olecranon process (projection on the proximal, or elbow, end of the ulna). The aquatic desmans (Desmaninae) do not dig much and have small forelimbs with webbed feet. Moles, but not desmans, use their weaker, less developed posterior limbs for propulsion only. Shrew moles (*Uropsilus*) seldom dig as they spend most of their time under leaf litter. As a result, their forelimbs and claws are not as well adapted for digging. Several species of semi-aquatic insectivores have long stiff hairs (fibrillae) on their feet to help propel them in the water like paddles. Several aquatic forms of family Soricidae (shrews) have webbed feet with hairy fringes that increase the surface area of the foot. This allows air bubbles to be trapped making it possible for the shrew to run on the water's surface for up to several seconds at a time. No insectivore forelimbs are adapted for jumping or leaping.

Insectivore tail length and texture vary like most other physical characteristics belonging to this order. Most talpids and hedgehogs have relatively short tails, some covered with vibrissae. Several of the semi-aquatics, the desman for example, have stiff flat rudder-like tails. Golden moles (Chlrysochloridae) have no external sign of a tail. The star-nosed mole has a tail that is almost as long as its body. Mole tails brushing against tunnel walls and roofs are able to pick up a variety of information including ground vibrations. The tail of the Madagascar hedgehog (*Microgale longicaudata*) is 1.5 to 2.6 times longer than its body and has 47 vertebrae. It is one of the longest mammal tails, second only to the pangolin

The Pacific mole (*Scapanus orarius*) has claws that are useful for tunneling. (Photo by R. Wayne Van Devender. Reproduced by permission.)

(scaly anteater). *Microgale* also have a tail that is modified for prehension (grasping). The lesser hedgehog tenrec (*Echinops telfairi*), on the other hand, has no external tail whatsoever. Several shrew names denote the nature of their tails—short-tailed, pen-tailed—the other 300 plus have tails that range from short to long, thick to thin, hairy to sparse, and tubular to flat depending on the species. The giant otter shrew (*Potamogale velox*) has an eel-like, lateral tail that allows it to swim like a fish while its keeps its hind feet glued to its flanks.

One of the mammals' most unusual skeletal structures is found among Insectivora. The West African armored, or hero, shrew (*Scutisorex somereni*), is believed to possess amazing strength. It has vertebrae with many interlocking spines as well as dorsal and ventral spines. The spines appear to allow it to bend considerably as well as bear phenomenal amounts of weight. The way in which the muscles are attached is complex and it seems that the extra joints allow additional flexibility.

Insectivores have fur or, in the case of some tenrecs (Tenrecinae) and hedgehogs (Erinacidae), a spiny coat, or even a combination of hairs and blunt spines that serve as protection against predators. When threatened, these mammals activate a set of muscles that cause their spines to become erect. Hedgehogs can protect themselves further by rolling up into a tight ball by activating their *panniculus carnosus* (a powerful orbicular muscle). Most aquatic insectivores have water-repellent pelage. Fur colors go from light to black, covering all the earth tones in between. Insectivores can be mono-, bi-, or multi-colored. Many feature lighter shades on their undersides. Fur texture can be glossy on golden moles to short, dense and velvety in least shrews and thick and woolly in hero shrews. Water shrews have been reported to have iridescent fur when exiting the water. Other species with iridescent fur

A greater white-toothed shrew (*Crocidura russula*) feeds on an earthworm. (Photo by Jose Luis Gomez de Francisco/Naturepl.com. Reproduced by permission.)

include the ground-dwelling Cape golden moles (*Chrysochloris*). Some of them are reported to display greenish, violet, or purplish tinges.

Distribution

Insectivores are found everywhere on Earth but Australia, Antarctica, and most of South America. The shrews (Soricidae) are the most widespread of all Insectivora families covering the entire planet except the poles, Australia, and the greater part of South America. Shrew mice (Soricidae) and moles (Talpidae) are the only families found in North America and a few wanderers have found their way from there to the northern edges of the South American continent. Hedgehogs and gymnures (Erinaceidae) are found in Africa, Eurasia, Southeast Asia, and Borneo. Hedgehogs were introduced to New Zealand where they are now flourishing. Solenodons, originally found in Hispaniola, Puerto Rico, Haiti, and Cuba are now Endangered and found only in Haiti and Cuba. The West Indian shrews (Nesophontidae) are believed to have existed in the West Indies until the Spaniards settled there at the beginning of the sixteenth century. The 1930 discovery of West Indies shrew remains in Haitian barn owl pellets ignited hopes that they still existed but no further evidence of their presence has come to light. Moles (Talpidae) are not found in Africa. Tenrecs (Tenrecidae) inhabit Madagascar and are found, to a lesser extent, on the Comoro Islands and in western central Africa. The golden moles (Chrysochloridae) exist on the southern half of the African continent. Recent discoveries of tiny placental mammal-like fossils in Australia, are leading scientists to believe that insectivores may have been on the Australian section of Gondwana when it separated 115 mya.

Habitat

The Insectivora are primarily terrestrial mammals, living either on or under the ground. Many insectivores are fossorial and a few are aquatic. Insectivores have been recorded to inhabit altitudes ranging from sea level up to 14,760 ft (4,500 m) in the mountains of Nepal. Approximately 30 species of the Talpidae are fossorial (burrow diggers) to some extent. A number of tenrecs and soricids as well as all the chrysochlorids are fossorial as well. Golden moles dig burrows in sandy areas, plains, cultivated areas, and forests. Gymnures find shelter under brush piles, tree roots, deserted burrows, or even termite mounds. Solenondons prefer forests and rocky areas whereas tenrecs like rainforest and brushlands. Shrews, the most widespread members of the order, have been recorded living in a multitude of different habitats and altitudes ranging from the dry hot desert to high snowy mountains and everything in between.

Most shrew species prefer moist habitats. They live in shallow runways that they dig themselves or that are made by other animals, under decomposed leaves and twigs, and several species construct nests in hollowed out logs, under rocks, and in tunnels. Solenodons tend to build nests only during the mating season. The short-tailed shrews (*Blarina brevicauda*) build two types of leaf or grass nests in tunnels or

A Grant's golden mole (*Eremitalpa granti*) consumes a locust in the sand dunes of the Namib Desert, Nambia. (Photo by Michael Fogden. Bruce Coleman, Inc. Reproduced by permission.)

under logs and rocks: a small resting nest and a large mating nest.

The aquatic moles, desmans, shrews, and moonrats live right next to bogs, swamps, springs, rivers, and streams. Some, Eurasian water shrews (*Neomys fodiens*) for example, even dig tunnels that open into bodies of water. These mouse-sized mammals make sure the diameter of their tunnels is small enough to squeeze the water from their fur as they exit the water. Although most arboreal species appear in other orders, several insectivore species are known to seek refuge in trees when attacked or to forage (*Sylvisorex*, the forest musk shrews). Some shrews, such as the short-tailed shrew, are good climbers and have been observed scampering up several feet (meters) up trees to pilfer suet from a bird feeder.

Behavior

Insectivores are notoriously shy, secretive, and active mammals. Their vision is not highly developed, therefore they rely mostly on their other senses, hearing, touch, and smell. Some species of moles and shrews depend on echolocation, the use of sound to navigate. Insectivores have various means of protection, the most obvious being nocturnal, subterranean, or aquatic habitats. Most seek cover in deep forests, holes, and tunnels deserted by other animals, under leaves, branches, stems, and roots of plants. Hedgehogs and certain tenrecs have spiny armor for protection.

Insectivores communicate in a wide variety of ways. Insectivore communication can be both interspecific (between dif-

East African hedgehog (*Erinaceus albiventris*) in Nairobi, Kenya. (Photo by Animals Animals ©Bruce Davidson. Reproduced by permission.)

ferent species) and intraspecific (among members of the same species). Adult hedgehogs make hissing snorts when threatened, and young make birdlike whistles and quacks while in the nest. Solenodons emit a high frequency clicking sound similar to that of many Soricidae, which may serve an echolocation function. White-toothed shrews (*Crocidura leucodon*) emit metallic squeaks. *Suncus murinus* chirps and buzzes. Among many tenrecs, communication is tactile, with a few audible signals. Some insectivores mark territory with bodily secretions. Scent glands are used for communication between golden moles, especially between mothers and their offspring and between sexes during the mating season. When marking their territory, moonrats (*Echinosorex gymnura*) produce a foul smell reminiscent of rotten onions or ammonia from small anal glands. *Suncus murinus*, the musk shrew, emits a pungent odor from a gland located between its last rib and hip bone. The Haitian solenodon (*Solenodon paradoxus*), has scent glands located in its groin and armpits that secrete a goatlike odor.

The Haitian solenodon produces a venom in a gland located in the mandible (lower jaw) then squirts it through a channeled bottom incisor. The southern short-tailed shrew (*Blarina carolinensis*), the short-tailed shrew, and Elliot's short-tailed shrew (*Blarina hylophaga*) have submaxillary glands that secrete venom as well. The toxicity of their saliva allows these tiny mammals to attack prey larger than themselves as their bites immobilize their victims. It is also thought that shrews use their venom to immobilize their prey for consumption at a later time.

Most of the species belonging to Insectivora are nocturnal (active between dusk and dawn). Sclater's golden moles, however, are among those who are active day and night. They must keep moving in order to maintain their body temperature and they do so by digging almost non-stop. When they

do sleep, they keep their bodies warm with involuntary muscle twitches. Many shrew species are active day and night as well, because their metabolism requires them to eat continuously. Hedgehogs located in cooler climates hibernate during the winter and some desert-dwellers estivate (go into torpor by lowering their body temperature and slowing their metabolism) during very hot weather. In the months preceding their hibernation, hedgehogs build up substantial fat reserves to see them through their "down" time. Tenrecs and hedgehogs for example, are known to go into a state of torpor when food gets scarce. The tenrec's heartbeat can drop to one beat every three minutes accompanied by a drop in body temperature.

In the wild, insectivores are, for the most part, solitary animals whose social life is limited, for the most part, to mating and rearing of offspring. When hedgehogs gather during the breeding season; the animals establish a hierarchy. When solenodons meet, there may be some initial scuffling but they eventually tolerate each other. Solenodons are among the more social species with the young remaining for long periods with their mother. The nomadic shrew moles travel in groups of up to 11. Small-eared shrews (*Cryptotis*), are known to be sociable, with several adults sharing a nest at the same time. Outside of mating season, many Insectivora species do not tolerate each other in the wild and become extremely aggressive and violent. When in captivity however, these same animals initially avoid each other or exhibit aggressive behavior but eventually learn to live together quite peacefully.

Fossorial insectivores—moles, desmans, and a few species of tenrecs and shrews—spend much of their time digging burrows and tunnels for food and shelter. Eastern moles (*Scalopus aquaticus*) found in Central and North America, dig up to 102 ft (31 m) per day. The short-tailed shrew digs tunnels through leaves, plant debris and snow. The yellow golden mole (*Calcochloris obtusirostris*) is nicknamed "sand swimmer" because of the way it tunnels through the sand at an impressive speed.

Feeding ecology and diet

Insects are the mainstay of many insectivores. Others such as solenodons, hedgehogs, tenrecs, and some shrews prefer an animal-based diet that may include snails, reptiles, worms, and ground-nesting birds' eggs. The aquatic species subsist on crustaceans, mollusks, fish, and frogs. Plant matter and even fruits, vegetables, seeds, and nuts are other sources of insectivore nutrition.

Many moles and shrews have a very high metabolic rate and a voracious appetite causing some species to eat up to several times their own body weight in a 24-hour period. The long-tailed shrew (*Sorex dispar*) eats almost continuously, day and night. Insectivores forage for food in many different ways. Ground dwellers forage for insects under leaves, grasses, dirt, branches and among rocks. The fossorial insect eaters dig for food in sand or dirt or hunt in their own or borrowed burrows and tunnels. Desmans have flexible snouts that allow them to poke around in the dirt at the bottom of streams, lakes, and ponds for food. The aquatic moles, desmans, shrews, and moonrats or gymnures, are good swimmers and/or divers able to catch small fish and invertebrates in the water. Otter shrews are fond of freshwater crab. The life span of insectivores is relatively short and ranges from less than one year (shrews) to 11 years (solenodon in captivity).

Reproductive biology

Little is known about the reproductive biology of most members of Insectivora. Those with short life spans, the majority of the shrew species for example, produce several litters a year while longer living insectivores, such as tenrecs, mate only once a year. Insectivores produce as few as one offspring per litter but some tenrecs can have as many as 32 young at one time (the maximum offspring produced by mammals). Gestation periods vary from two weeks to two months. The same is true for the length of time the young nurse. The number of teats ranges from two to 24 and their location on the body depends on the species. Many male insectivores have testes in the abdominal cavity lying close to the opening of the perineum. Moles and tenrecs have a baculum (penis bone). Courtship and mating practices, which have seldom been observed but are probably polygynous, tend to be short and to the point lasting as little as a few seconds in some species, but lasting hours in other species. Some young, such as certain tenrecs, are born with hair, able to run around immediately following birth and are totally independent within weeks. On the other hand, moles are born completely naked save for a few vibrissae and guard hairs. The latter are helpless and highly dependent on the mother for several months.

There are some unusual reproductive and parenting habits among insectivorans. The female lesser hedgehog tenrec (*Echinops telfairi*), for example, emits odors during the mating season that cause males to secrete a milky white substance from glands near the eyes. Their mating ritual can last several hours, as it does among certain species of Erinaceinae (spiny hedgehogs). The male hedgehog tenrec stays with his mate up to a few hours before young are born, then he leaves and stays away as long as the young are dependent on their mother. The male spiny hedgehog must be persistent in his courtship as he often ends up chasing after his mate for several hours before she allows him to mount. The female spiny hedgehog flattens its spines during mating, if she is receptive. Because the spines are slippery, the male must hold on to his mate's shoulder with his teeth. Shrews are known for their antagonistic conduct towards one another, even during mating. Among many shrews, exposure to a conspecific (member of the same species) immediately triggers aggressive behavior. The sexual odors emitted by both parties during courtship eventually overpower the tendency towards violence, allowing copulation to take place.

The offspring of the white-toothed shrew (*Crocidura leucodon*), and musk shrews (*Suncus murinus*) or pygmy shrews (*Sorex hoyi*)—anywhere from three to seven at a time—are led by their mother in line formation, also known as caravaning. The first young shrew grabs a mouthful of hair right above its mother's tail with its teeth. The siblings follow the example with each other down the line.

A yellow streaked tenrec (*Hemicentetes semispinosus*) showing defensive behavior. (Photo by Harald Schütz. Reproduced by permission.)

Conservation

The World Conservation Union (IUCN) carried out species status assessments in 1996. Over one third (39%) of Insectivora species were placed on the Red List. Of those, 21% were Critically Endangered; 26.5% were Endangered; 40.5% were Vulnerable; 3% were Lower Risk/Near Threatened; 3.5% are Extinct; and 5% were Data Deficient. The majority of these Red-Listed species are from sub-Saharan Africa, south and Southeast Asia, and east Asia.

The reason why so many species worldwide are under threat is simple: human population growth. The major causes for the decline of whole species and genera, according to the IUCN and most conservation groups worldwide, are believed to be human-induced habitat loss, degradation, and fragmentation. Land clearing, logging, slash and burn, increased agriculture, and its widespread use of toxic fertilizers, pesticides and herbicides—all of these practices take a toll on the above ground and fossorial insectivores. Pollution of fresh water resources—springs, steams, rivers, ponds, and lakes—as well as clearing of banks, the construction of hydroelectric dams and canals, and the drainage of wetlands are greatly contributing to the reduction in numbers of aquatic species. Introduction of non-native plant and animal species are also a contributing factor to the decline of certain insectivores.

Because so little is known about the great majority of insectivores—their behavior, reproductive biology, social organization, migratory movements, and general ecology—it is difficult to identify strategies that could help predict potential threats, so as to design effective conservation plans. There are many specialist groups, however, the most active of which work under the auspices of the IUCN, dedicated to reversing the negative trend affecting the survival of mammals. The Internet allows the members of these groups to share information

A hedgehog (*Erinaceus europaeus*) baby learning to forage. (Photo by Animals Animals ©Robert Maier. Reproduced by permission.)

easily with each other and with conservation groups that are in a position to apply the knowledge obtained to studies and projects underway worldwide.

Significance to humans

Insectivores are secretive, nocturnal, shy, and for most humans fall into neither the cute and cuddly nor the big and scary attention-generating categories. A few individual insectivore species, however, have been singled out for attention. The greater hedgehog tenrec (*Setifer setosus*), for example, is probably the most recognizable of the insectivores. This creature is easy to keep as a pet and has a large enthusiastic following. Hedgehogs are also important in research on diseases such as foot and mouth disease, yellow fever, and influenza. Old home remedies sometimes called for hedgehog blood, entrails or ashes. Mixed with pitch or resin, the burned hindquarters supposedly helped cure baldness. The hedgehog was used to predict the weather and if buried under a building was thought to bring good luck. The Romans kept hedgehogs as pets during the fourth century B.C. and they remain domesticated to this day. Other uses over the years for this spiny insectivore include: consumption of its meat, use of its spiny coat to card wool, deterring horses prone to leaning, and as a tool to help wean calves.

Shrews and moles are the next insectivores most familiar to humans. The armored shrew is important to some Africans who believe that any part of the animal will act as a talisman to save them from peril. Mummified shrews dating back to the twenty-seventh dynasty (ca. 400 B.C.) were discovered in a ceremonial animal tomb near Thebes, Egypt. According to archaelogists, shrews played an important part in the religion of ancient Egyptians as representatives of nocturnal creatures those who "came from the dark." In the West, the shrew's biggest contributions to culture are most likely linguistic and literary as in Shakespeare's *The Taming of the Shrew*. The term "shrew" is used to describe a quarrelsome, nagging, irritable woman because the little mammal was believed to have a poisonous bite. Beatrix Potter's *The Tale of Mrs. Tiggy-Winkle* and Kenneth Grahame's *The Wind in the Willows* have immortalized the hedgehog and the mole, respectively, in their children's stories. John Le Carre, among others, has made sure that moles remain infamous in the memory of all spy story buffs. In the East, shrews are called "chien shu" in Mandarin and "chin chih" in Taiwanese. Both terms mean "money" and "mouse" or "rat." The noise that shrews make sounds like the Chinese pronunciation of money. In Taiwan, a shrew in one's home is not killed as the presence of the small animal means money is coming. If the shrew is killed the source of the fortune is eliminated.

The economic benefits to humans offered by insectivores are few. Mole skin from the European mole (*Talpa europaea*) was a popular trim for women's cloaks during some time. The skin of the Russian desman (*Desmana moschata*) was also used for trimmings and perfume manufacturers used the strong musk from the glands located under the desman's tail. The common tenrec has been an important food source for the human inhabitants of Madagascar for thousands of years.

Very few insectivores are labeled as pests. Poultry farmers are not fond of hedgehogs that readily eat eggs and chickens. Ticks, mites, and fleas thrive on hedgehogs and these mammals are also known to carry and transmit ringworm, influenza, yellow fever, *Salmonella eteritidis*, leptospirosis, and foot and mouth disease. While some insectivores have rodent-like habits when it comes to humans, i.e., noisy, smelly, intrusive, and destructive, they make up for it by destroying insects and even the real rodent pests. Several mole species are not very popular with gardeners because of the damage they can cause with their tunnels.

Insectivores make meaningful contributions to the environment. A majority of the species help keep insects under control. The short-tailed shrew, for example, keeps a check on insect crop pests, especially the larch sawfly, with its voracious appetite. This shrew, along with other insectivores, including Sclater's golden mole (*Chlorotalpa sclateri*), destroy snails and mice that damage crops and are pests to humans. Fossorial insectivores do a good job aerating the soil. The burrowing and digging done by desert-dwelling fossorial insectivores in desert areas helps the vegetation in harsh sandy areas to flourish. The hedgehog controls vermin and is a helpful scavenger. And let us not forget the insectivores' invaluable contribution to the food chain.

Resources

Books

Butler, P. M. *Studies in Vertebrate Evolution.* Edinburgh: Oliver and Boyd, 1972.

Kingdon, Jonathan. *East African Mammals: An Atlas of Evolution in Africa.* Part A, *Insectivores and Bats.* Chicago: University of Chicago Press, 1984.

Nowak, R. *Walker's Mammals of the World,* 6th ed. Baltimore: Johns Hopkins University Press 1999.

Vaughan, T. A., J. M. Ryan, and N. J. Czaplewski. *Mammalogy,* 4th ed. Philadelphia: Saunders, 2000.

Yates, T. L. *Order and Families of Recent Mammals.* New York: John Wiley and Sons, 1984.

Periodicals

Anderson, I. "Uprooting our family tree—Ancient Australian teeth are upsetting cherished ideas about the evolution of mammals." *New Scientist* 156, 2110 (1997): 4.

Catania, K. C. "Epidermal sensory organs of moles, shrew-moles, and desmans: a study of the family Talpidae with comments on the function and evolution of Eimer's organ." *Brain, Behavior and Evolution* 56 (2000): 146–174.

Hecht, J. "Small bite makes big impression." *New Scientist* 160, 2155 (1998): 15.

Leszek, R., and E. Jancewicz. "Prey size, prey nutrition, and food handling by shrews of different body sizes." *Behavioral Ecology* 13, no. 2 (2000): 216–223.

Mouchaty, S. K., A. Gullberg, A. Janke, and U. Arnason. "The phylogenetic position of the Talpidae within Eutheria based on analysis of complete mitochondrial sequences." *Molecular Biology and Evolution* 17, no. 1 (2000): 60–67.

———. "Phylogenetic position of the tenrecs (Mammalia: Tenrecidae) of Madagascar based on analysis of the complete mitochondrial genome sequence of *Echinops telfairi.*" *Zoologica Scripta* 29, no. 4 (2000): 307–317.

Other

IUCN 2002. *2002 IUCN Red List of Threatened Species.* <http://www .redlist.org>.

Maddison, D. *Tree of Life Web Project.* <http://tolweb.org/tree?group =Insectivora&contgroup=Eutheria>.

Museum of Zoology, University of Michigan. *Mammmalia.* <http://animaldiversity.ummz.umich.edu/chordata/mammalia. html>.

Kupitz, David. 2001. Tenrec Resources and Information. <http://www.tenrec.org/>.

The Shrew (-ist's) Site. <http://members.vienna.at/shrew>.

Sam Houston State University. *Order Insectivora.* <http://www.shsu .edu/~bio_mlt/Insectivores.html>

Lee Curtis, MA

Gymnures and hedgehogs

(Erinaceidae)

Class Mammalia
Order Insectivora
Family Erinaceidae

Thumbnail description
Small, short-legged animals, five toes on each foot, gait plantigrade; muzzle generally pointed, moderately to greatly elongate, eyes small; pelage may be spiny (hedgehogs) or soft (gymnures)

Size
Head and body length: 4–12 in (10–30 cm); tail: 0.4–12 in (1–30 cm); weight: 0.5–38 oz (15–1,100 g)

Number of genera, species
8 genera; 21 species

Habitat
Varies from woodland to grassland, deserts and urban parks for hedgehogs; gymnures restricted to humid forests and meadows

Conservation status
Critically Endangered: 1 species; Endangered: 3 species; Vulnerable: 2 species; Lower Risk/Near Threatened: 1 species

Distribution
Europe, Asia, and Africa

Evolution and systematics

Erinaceidae is a small family, containing just 21 species. The fossil record is rather sparse, with a further 21 extinct species described so far. The family appears to have its origins in North America in the mid Palaeocene, about 60 million years ago (mya). American erinaceids went extinct about 5 mya, but not before further lineages had been established in Europe, Asia, and Africa (about 58, 55, and 23 mya respectively).

Modern erinaceids are split into two quite distinct subgroups, the spiny hedgehogs (subfamily Erinaceinae) and the soft-furred gymnures (known most correctly as subfamily Hylomyinae, but also often referred to as Echinosorinae or Galericinae). Within the hedgehogs there are 14 species in five well-described genera. The taxonomy of the gymnures is rather less clear-cut. There are seven species, including the long-eared lesser gymnure (*Hylomys megalotis*), which was described in 2002. Prior to 1991, there were five recognized genera, but *Neohylomys* and *Neotetracus* have since been reclassified as subgenera of *Hylomys*. As of 2002, *Hylomys* is the largest genus in the family with five species, but these are highly variable and the expectation is that further revisions will come.

Physical characteristics

The characteristics that describe members of the family Erinaceidae are generally considered primitive. They are typical insectivores—small animals, with short legs and large feet. Feet have five digits, except in some of the African hedgehogs (genus *Atelerix*) in which the hallux or big toe is reduced or vestigial. All erinaceids walk with a flat-footed "plantigrade" gait. The two bones of the lower hind leg, the tibia and fibula, are fused into one. The tail is hairy and variable in length, the muzzle is elongated—more so in the gymnures than the hedgehogs. The eyes are small, though better developed than those of most other insectivores. The skulls of hedgehogs and gymnures vary quite considerably, from long and narrow to short and broad. All have a small braincase.

As a general rule, hedgehogs are more derived than gymnures, which have retained many characteristics of their early insectivore ancestors. The most obvious difference between the two subfamilies is the coat. While the gymnures and moonrats are covered in pelage of soft fur, the hedgehogs sport a dense coat of narrow spines, starting on the head and covering the back and flanks. The color varies between and within species, but is usually some shade of yellow- or grayish brown to black.

Hedgehogs and gymnures have similar dentition. The dental formula for the family is (I2–3/3 C1/1 P3–4/2–4 M3/3) × 2 = 63–44. The first incisors are large. In gymnures, there are three pairs of incisors in each jaw, while hedgehogs have lost the third lower pair. The muzzle or rostrum of hedgehogs is shorter than in the gymnures, which have retained a narrow, shrew-like snout.

A Malayan moonrat (*Echinosorex gymnura*) foraging for insects in Southeast Asia. (Photo by N. Smythe/Photo Researchers, Inc. Reproduced by permission.)

Distribution

The Erinaceidae is now an exclusively Old World family, with representatives throughout Europe, Asia, and Africa. There is also a pronounced geographical demarcation between the two subfamilies—the Erinaceinae (hedgehogs) are widespread, but except for the Chinese hedgehog (*Erinaceus amurensis*), they are not found with the Hylomyinae (gymnures and moonrats). The latter are restricted to Southeast Asia. Hedgehogs fare well in cool temperate to tropical climates between about 66°N (*Erinaceus europaeus*) and 34°S (*Atelerix frontalis*). The gymnures are essentially a tropical and subtropical group, but they live in a wide range of altitudes, from sea level to 11,000 ft (3,400 m) (*Hylomys suillus*).

Habitat

As a group of small, "primitive" animals, the erinaceids, and especially the hedgehogs, exploit a surprising diversity of habitats, from mangrove forest to stony desert, urban parks to alpine meadows. In all cases, the main limiting factors are

the availability of suitable daytime shelters and food—especially of invertebrate prey. A hedgehog has a fairly small home range with an approximate 120-yd (110-m) radius from its nest. The nest is built in dry litter under tangles of hedge or bush, rock crevices, termite mounds or underbuildings. The hedgehog chatters, snorts or softly growls if its range is invaded by another animal. The gymnures live predominantly in humid forests, but this is more indicative of the availability of suitable food in such places than of their intolerance of dry habitats.

Behavior

For the most part, erinaceids are nocturnal; although the Malayan moonrat (*Echinosorex gymnura*) and lesser gymnure (*Hylomys suillus*) may also forage by day. Species living in temperate zones may be forced to begin foraging before sunset in mid-summer. As a general rule, erinaceids are terrestrial, living and feeding at ground level. Most are competent climbers and swim well—*E. gymnura* may even be least partially aquatic. Digging ability varies, but some species, such as the Indian and the long-eared hedgehogs (*Paraechinus micropus* and *Hemiechinus auritus*) are very good burrowers.

Most members of the Erinaceidae are able to enter torpor and thus tolerate bad weather and seasonal food shortages by hibernating in winter or estivating during droughts. Hibernation of the European hedgehog, (*Erinaceus europaeus*), in colder parts of its range may last six or seven months, during which the heart rate drops from about 188 beats per minute to around 22, and body temperature may fall to just 1°C (34°F). Prolonged hibernation puts enormous strain on the animals reserves of stored fat and those that do not put on enough weight in the fall will not survive.

All the well-studied species are essentially solitary and territorial as adults. Small groups of three or four have been reported for some species, but these probably represent females with subadult young of the year. Fights are seemingly rare, but individuals may react aggressively to threats, giving rapid hissing snorts. Interestingly, an almost identical sound is pro-

A western European hedgehog (*Erinaceus europaeus*) on a lawn in Surrey, England. (Photo by Stephen Dalton/Photo Researchers, Inc. Reproduced by permission.)

A central African hedgehog (*Atelerix albiventris*) peering out from its log den. (Photo by © Joe McDonald/Corbis. Reproduced by permission.)

A western European hedgehog (*Erinaceus europaeus*) crossing a small brook. (Photo by Animals Animals ©Robert Maier. Reproduced by permission.)

duced by female hedgehogs during courtship, presumably in response to close approach of another individual, the male.

All erinaceids use scent to mark their home range, but olfactory cues appear to be particularly important for the gymnures, several of which have a strong musky odor that is obvious even to the human nose. Among the other senses, hearing is apparently most sensitive, while eyesight is not particularly acute. One of the most distinctive and unusual aspects of erinaceid behavior is "self-anointing," an activity performed by all species of hedgehog. Apparently triggered by olfactory cues such as strong-smelling or noxious plants or chemicals, this behavior starts with the animals licking or chewing the source of the smell, and producing copious amounts of frothy saliva. The saliva (and presumably with it the chemical trigger, whatever it may be) is then spread over the spines with quick flicks of the tongue and distinctive jerking movements of the head. Presumably the spines provide a surface area from which scent can be dispersed, but so far there is no really satisfactory explanation for the function of self-anointing. Theories range from attracting mates to deterring predators or repelling parasites, but none stand up to rigorous investigation or satisfactorily account for the diver-

sity of trigger substances or the wide range of situations in which animals will suddenly devote all their attention to this odd behavior.

Gymnures are relatively secretive animals, while hedgehogs, with the benefit of their spiny defenses, are more bold. When a hedgehog is threatened, it curls itself into a tight ball by means of well-developed abdominal muscles that act like the cord on a drawstring bag. The head, feet and tail are all tucked away and the spines are erected, presenting a potential predator with nothing but a puzzling spiky ball. Some predators, including badgers and foxes, learn the art of unrolling hedgehogs and may become specialist hedgehog eaters.

Western European hedgehog (*Erinaceus europaeus*) babies sleeping in their nest. (Photo by Animals Animals ©Robert Maier. Reproduced by permission.)

A southern African hedgehog (*Atelerix frontalis*) devouring a lizard. (Photo by Animals Animals ©Anthony Bannister. Reproduced by permission.)

Feeding ecology and diet

All hedgehogs and gymnures feed primarily on insects and other invertebrates including worms, spiders, and terrestrial mollusks. Being larger than most other members of the order Insectivora, some are also able to take some larger prey, including reptiles, amphibians, and in some cases small mammals. Hedgehogs are also rather notoriously fond of birds' eggs—the introduction of the European hedgehog to islands including New Zealand has proved disastrous for populations of ground nesting birds. Most species will also consume non-animal matter such as fruit, seeds, and fungi.

Several species of hedgehog sometimes eat venomous animals such as vipers, scorpions, and bees, as well as toxic beetles and spiders, with no apparent ill effects. Their resistance to adder venom can be up to 40 times that of laboratory mice, and they can consume the beetle toxin cantharidin in quantities equivalent to 3,000 times the dose toxic to humans.

Feeding takes place predominantly at night, when the animals may travel some distance along regular pathways, shoving their noses into nooks and crannies and rummaging about in the leaf litter, or sometimes digging into the top layer of soil. Most food is eaten as and when the animal finds it but the Indian hedgehog (*Paraechinus micropus*) is known to hide excess food in its burrow as an insurance policy against future shortages.

Reproductive biology

Female erinaceids have between two and five pairs of mammae (a maximum of four pairs in the gymnures). Thus the theoretical maximum litter size is 10, but in reality they average between two and five young per litter. Gestation periods range between 30 and 40 days and the young are born blind, helpless, and virtually naked. In the hedgehogs, the first

A central African hedgehog (*Atelerix albiventris*) mother and babies. (Photo by Animals Animals ©Jim Tuten. Reproduced by permission.)

A central African hedgehog (*Atelerix albiventris*) foraging in Africa. (Photo by Animals Animals ©Breck P. Kent. Reproduced by permission.)

Long-eared hedgehogs (*Hemiechinus auritus*) crossing the desert in Egypt. (Photo by Animals Animals ©Ashod Francis. Reproduced by permission.)

spines may erupt before birth but they are very soft and unpigmented. Adult-type spines usually begin to grow within the first week of life, and the curling response develops within a few weeks. Care of the young is always the sole responsibility of the mother—males play no part beyond courtship and mating, which is polygynous. The suckling period in hedgehogs is between five and seven weeks, and longevity is up to seven years.

Conservation status

Of the 22 species in the family Erinaceidae recognized here, three are officially threatened. Hugh's hedgehog (*Mesechinus hughi*) is listed as Vulnerable by the IUCN, the Hainan gymnure (*Hylomys hainanensis*) and both species of Philippine gymnure (*Podogymnura truei* and *P. aureospinula*) are Endangered. The IUCN regards *Hylomys suillus parvus* as a full species and lists it as Critically Endangered, as well as listing *Hemiechinus nudiventris* as Vulnerable and *Hylomys sinensis* as Lower Risk/Near Threatened. The chief threats to these species are habitat fragmentation and modification, for example, logging and agricultural development. The southern African hedgehog (*Atelerix frontalis*) is considered Rare and is listed on Appendix 2 of the CITES in order to prevent uncontrolled collection for the pet trade.

Significance to humans

Erinaceids have little economic importance. As a group, the gymnures are not well known and have little use to humans save occasional use in laboratory experiments. Hedgehogs, on the other hand, have been used to a limited extent by humans throughout history. The meat of hedgehogs is good, but while most species are sometimes eaten as bushmeat or feature in tra-

ditional country dishes they are not bred in captivity for this (or any other) purpose. There is a strong association between hedgehogs and European gypsies, who not only eat the animal's meat, but also regard it as an ally against malign *mochadi* entities such as cats and non-gypsy people.

Hedgehog body parts have been used by many cultures in traditional healing, magic and witch doctoring. The meat has been purported to have cleansing properties and various body parts have apparently been used in the treatment of ailments including leprosy, boils, colic and baldness.

Hedgehogs also appear widely in folklore. They are mentioned in the writings of Pliny and Shakespeare and famous hedgehogs include Mrs. Tiggywinkle, the bustling, petticoated washerwoman in the story by Beatrix Potter and Sonic the Hedgehog—the manic blue-spined hero of the computer game by Sega.

Hedgehogs are among the few wild mammals that have adapted to life alongside people in towns. They are kept as pets and are welcome visitors to gardens, where they help out by eating invertebrate pests. Many people put out food specially for hedgehogs and enjoy watching their spiny visitors dining on bread and milk or canned dog food from a saucer on the back lawn.

1. Malayan moonrat (*Echinosorex gymnura*); 2. Lesser gymnure (*Hylomys suillus*); 3. Mindanao gymnure (*Podogymnura truei*); 4. Long-eared hedgehog (*Hemiechinus auritus*); 5. Indian desert hedgehog (*Paraechinus micropus*); 6. Western European hedgehog (*Erinaceus europaeus*); 7. Southern African hedgehog (*Atelerix frontalis*); 8. Daurian hedgehog (*Mesechinus dauuricus*). (Illustrated by Marguette Dongvillo)

Species accounts

Western European hedgehog

Erinaceus europaeus

SUBFAMILY
Erinaceinae

TAXONOMY
Erinaceus europaeus Linnaeus, 1758, Sweden. Formerly included *E. concolor* and *E. amurensis*.

OTHER COMMON NAMES
French: Hérisson de l'ouest; German: Westeuropäisch Igel; Spanish: Erizo europeo.

PHYSICAL CHARACTERISTICS
Head and body length: 9–11 in (23–28 cm); tail: 0.5–1.2 in (1.5–3 cm); weight: 14–42 oz (400–1,200 g). Large brown hedgehog with short tail. Spines are banded yellow and brown, with pale tips.

DISTRIBUTION
Widespread throughout Western and Central Europe including Great Britain, Ireland, southern Scandinavia and northwestern Russia.

HABITAT
Deciduous woodland and scrub, grassland and pasture, suburban parks and gardens. Alpine regions below the treeline.

BEHAVIOR
Nocturnal and solitary outside breeding season, non-territorial but intolerant of conspecifics; where winters are cold, enters deep hibernation from October–December to March–April; rolls up into spiny ball when threatened; engages in self anointing with frothy saliva.

Hemiechinus auritus
Atelerix frontalis
Erinaceus europaeus

FEEDING ECOLOGY AND DIET
Omnivorous, but eats mostly invertebrates such as beetles, larvae, slugs, worms and spiders; also small vertebrate prey, eggs, fruit and fungi. Apparently relishes bread and milk, dog food and kitchen scraps provided by people.

REPRODUCTIVE BIOLOGY
Probably polygynous. Mates from April to August, the later pairings generally occur if first litter is lost or aborted. Gestation lasts 31–35 days, litter size varies from 2 to 10, and 4 to 6 is normal for most parts of range. Young are born in nest of leaves, blind and virtually naked. First soft, white spines begin to be replaced by adult-type spines between 2 and 7 days old. Young are tended by female only and weaned at 4 to 6 weeks. Sexually mature at 12 months, maximum longevity 7 years (10 in captivity), usually much less.

CONSERVATION STATUS
Not threatened; common and widespread. Road traffic and winter starvation are major factors in mortality.

SIGNIFICANCE TO HUMANS
Generally a well-known animal, welcomed into gardens where it eats many invertebrate pests; a popular figure in folklore and children's stories; may be eaten, though the practice of hunting for meat is increasingly uncommon. ◆

Southern African hedgehog

Atelerix frontalis

SUBFAMILY
Erinaceinae

TAXONOMY
Atelerix frontalis (Smith, 1831), Cape Colony, Cape Province, South Africa. Formerly in *Erinaceus*.

OTHER COMMON NAMES
French: Hérisson Sud-Africain; German: Kap-Igel; Spanish: Erizo enano Africano.

PHYSICAL CHARACTERISTICS
Head and body length: 7–8.6 in (18–22 cm); tail: 0.8–1 in (2–2.5 cm); weight: 10.5–17.5 oz (300–500 g). Gray-brown to grizzled black fur and spines, with prominent white band from forehead to flanks, narrow center parting in spines on forehead.

DISTRIBUTION
South and east from southern Angola, through northern Namibia, eastern Botswana and western Zimbabwe to eastern South Africa.

HABITAT
Scrub and savanna grassland, also in suburban parks and gardens.

BEHAVIOR
Nocturnal and solitary. Hibernation lasts from June to September, slightly less in northern parts of range, will rouse during warm weather. Curls into spiny ball when threatened. Engages in self-anointing behavior.

FEEDING ECOLOGY AND DIET
Omnivorous, eats invertebrates and small vertebrate prey including beetles, worms, frogs, small reptiles, rodents and birds' eggs, also takes fruit and fungi.

REPRODUCTIVE BIOLOGY
Polygynous. Females can produce several litters a year, but births peak from October to March (the wet season); gestation lasts 35 days, litters contain 4 to 5 (maximum 10) blind, helpless young. Weaned at 6 weeks, sexually mature at 9 to 10 weeks.

CONSERVATION STATUS
Though not officially listed as threatened, it is considered Rare in South Africa by the IUCN and listed on CITES Appendix II. Has suffered from localized hunting/collecting and habitat loss due to agricultural development.

SIGNIFICANCE TO HUMANS
Kept as a pet in some places and hunted for food in others. Generally considered beneficial as a predator of invertebrate pests. ◆

Indian desert hedgehog
Paraechinus micropus

SUBFAMILY
Erinaceinae

TAXONOMY
Hemiechinus micropus (Blyth, 1846), Bahawalpur, Punjab, Pakistan. Formerly also in *Erinaceus*. Includes *Paraechinus nudiventris*, considered by some as a separate species.

OTHER COMMON NAMES
English: Pale hedgehog; French: Hérisson du desert Indien; German: Indisch Wüstenigel; Spanish: Erizo del desierto.

Hylomys suillus
Paraechinus micropus

PHYSICAL CHARACTERISTICS
Head and body length: 5.5–9 in (14–23 cm); tail: 0.4–1.5 in (1–4 cm); weight: 10.5–17.5 oz (300–600 g). Pale gray-beige except for dark facial mask.

DISTRIBUTION
Pakistan, western and southern India.

HABITAT
Desert and semidesert.

BEHAVIOR
Nocturnal and solitary but generally non-aggressive; digs own burrow and caches food; does not hibernate; engages in self-anointing behavior.

FEEDING ECOLOGY AND DIET
Hunts insects and small vertebrates, especially toads. Apparently stores excess food.

REPRODUCTIVE BIOLOGY
Breeds from April to September, litters of 1 to 3 (maximum 6). Probably polygynous.

CONSERVATION STATUS
Not threatened. Widespread and common but local populations are increasingly isolated due to desertification and agricultural development. *P. m. nudiventris* is Vulnerable.

SIGNIFICANCE TO HUMANS
None known. ◆

Long-eared hedgehog
Hemiechinus auritus

SUBFAMILY
Erinaceinae

TAXONOMY
Hemiechinus auritus (Gmelin, 1770), Astrakhan, south Russia. Formerly in *Erinaceus*. Includes *H. a. megalotis*, regarded by some as a separate species.

OTHER COMMON NAMES
French: Hérisson à longues oreilles; German: Großohrigel; Spanish: Erizo de orejas largas.

PHYSICAL CHARACTERISTICS
Head and body length: 6.6–12 in (17–30 cm), tail: 0.6–2 in (1.5–5.5 cm); weight: 8.4–17.5 oz (240–500 g). Long legs and large, mobile ears. Spines are grooved.

DISTRIBUTION
Northeastern Africa, the Middle East and central Asia from northern Libya to the Gobi desert of northern China and Mongolia and including steppe and desert regions of Egypt, Jordan, Israel, Syria, Iraq, Iran, Afghanistan, western states of the former Soviet Union and Pakistan. The Gobi population is apparently isolated from the west by the Altai mountains.

HABITAT
Semidesert, arid grassland and montane steppe up to 8,200 ft (2,500 m), also frequents parks and gardens in suburban areas.

BEHAVIOR
Nocturnal and solitary, generally non-aggressive; will dig own burrow where alternative ready-made dens are not available.

Hibernates in parts of range where winters are cold, elsewhere may enter facultative torpor during periods of drought or food shortage; self-anoints when encountering certain trigger substances.

FEEDING ECOLOGY AND DIET
Insects and other invertebrates, also some small vertebrate prey, fruit and seeds. Can survive prolonged periods without eating or drinking.

REPRODUCTIVE BIOLOGY
Polygynous. Breeding occurs in spring and summer, hibernating individuals mate soon after waking up. Gestation lasts 35–42 days, litter size varies from 1 to 6 (usually 2 to 3 in the Gobi, 5 to 6 in western temperate areas). Weans at 5 weeks, may be sexually mature at 6 weeks. Maximum recorded longevity 6 years.

CONSERVATION STATUS
Not threatened; common and widespread.

SIGNIFICANCE TO HUMANS
Occasionally eaten as bushmeat, no real significance. ◆

Daurian hedgehog
Mesechinus dauuricus

SUBFAMILY
Erinaceinae

TAXONOMY
Hemiechinus dauuricus (Sundevall, 1842), Dauryia, Transbaikalia, Russia. At one time also considered to belong to *Erinaceus*.

OTHER COMMON NAMES
French: Herisson Daurian; German: Daurisch Igel; Spanish: Erizo el Dauryia.

Podogymnura truei
Echinosorex gymnura
Mesechinus dauuricus

PHYSICAL CHARACTERISTICS
Records are scarce. Head and body length about 9 in (24 cm); tail about 1 in (2.5 cm); weight probably similar to *H. auritus.* 8.4–17.5 oz (240–500 g). Longish ears, fur slightly coarse.

DISTRIBUTION
Gobi Desert regions of Eastern Mongolia and northern China.

HABITAT
Steppe and steppe woodland.

BEHAVIOR
Nothing is known. May be comparable to *Hemiechinus.*

FEEDING ECOLOGY AND DIET
Nothing is known.

REPRODUCTIVE BIOLOGY
Nothing is known. May be comparable to *Hemiechinus.*

CONSERVATION STATUS
Not threatened.

SIGNIFICANCE TO HUMANS
None known. ◆

Lesser gymnure
Hylomys suillus

SUBFAMILY
Hylomyinae

TAXONOMY
Hylomys suillus Müller, 1840, Indonesia. Includes very rare *H. s. parvus*, regarded by some to be a separate species.

OTHER COMMON NAMES
English: Short-tailed gymnure; French: Petite gymnure; German: Kleinegymnure; Spanish: Gymnure pequeño.

PHYSICAL CHARACTERISTICS
Head and body length: 3.5–5.7 in (9–14.5 cm); tail: 0.4–1.2 in (1–3 cm); weight: 0.5–2.8 oz (15–80 g).

DISTRIBUTION
Yunnan Province, China, Myanmar (Burma), Indochina, peninsular Thailand and Malaysia, Sumatra, Borneo, and Java.

HABITAT
Dense, damp forest from sea level to 11,000 ft (3,400 m).

BEHAVIOR
Active day and night, moves quickly and nimbly; does not hibernate. Probably solitary, though records of small groups exist—these may represent females with large young. Builds nests of leaves in rock crevices and hollows among tree roots.

FEEDING ECOLOGY AND DIET
Insects, worms, and other invertebrates found in leaf litter, will also eat fruit.

REPRODUCTIVE BIOLOGY
Litters of 2 to 3 young born at any time of year. Polygynous.

CONSERVATION STATUS
H. s. parvus is Critically Endangered.

SIGNIFICANCE TO HUMANS
None known. ◆

Mindanao gymnure
Podogymnura truei

SUBFAMILY
Hylomyinae

TAXONOMY
Podogymnura truei Mearns, 1905, Mt. Apo, Mindanao.

OTHER COMMON NAMES
English: Mindanao moonrat, Mindanao wood-shrew, Philippine wood shrew; French: Gymnure de Mindanao; German: Mindanao Gymnure; Spanish: Gymnure de Mindanao.

PHYSICAL CHARACTERISTICS
Head and body length: 5–9 in (13–15 cm); tail: 1.5–3 in (4–7 cm). Soft gray-brown fur with stiffer guard hairs.

DISTRIBUTION
Endemic to the Philippine island of Mindanao.

HABITAT
Damp forest at altitudes up to 7,500 ft (2,300 m).

BEHAVIOR
Little known, probably nocturnal with habits comparable to true shrews (family Soricidae).

FEEDING ECOLOGY AND DIET
Insects, worms and probably other invertebrates; has been known to take carrion.

REPRODUCTIVE BIOLOGY
Nothing is known. Probably polygynous.

CONSERVATION STATUS
Listed as Endangered by the IUCN because of its small population and declining habitat.

SIGNIFICANCE TO HUMANS
None known. ◆

Malayan moonrat
Echinosorex gymnura

SUBFAMILY
Hylomyinae

TAXONOMY
Viverra gymnura (Raffles, 1822), Sumatra.

OTHER COMMON NAMES
English: Malayan gymnure; French: Gymnure de Malaysie; German: Malaysische Gymnure; Spanish: Gymnure de Malasia.

PHYSICAL CHARACTERISTICS
Head and body length: 10–18 in (26–46 cm); tail: 6.5–12 in (16.5–30 cm). Coarse, shaggy fur, black on body, pale on face and neck, tail sparsely furred, white towards the tip.

DISTRIBUTION
Myanmar (Burma), Peninsular Thailand and Malaysia, Borneo, and Sumatra.

HABITAT
Lowland forest and mangroves.

BEHAVIOR
Nocturnal and solitary. May be partially aquatic. Aggressive towards other moonrats, uses powerful scent from anal glands to mark home range.

FEEDING ECOLOGY AND DIET
Worms, insects, crabs, shrimps, snails and other invertebrates caught on land and at the edge of water.

REPRODUCTIVE BIOLOGY
Breeding is apparently aseasonal. Females rear one or two litters a year with an average two young per litter; gestation is 35–40 days. Details of life history are not known. Maximum recorded longevity four years seven months. Presumably polygynous.

CONSERVATION STATUS
Not threatened.

SIGNIFICANCE TO HUMANS
None known. ◆

Common name / Scientific name/ Other common names	Physical characteristics	Habitat and behavior	Distribution	Diet	Conservation status
Eastern European hedgehog *Erinaceus concolor* English: White-breasted hedgehog; French: Hérisson de l'est; German: Östlicheuropäisch Igel; Spanish: Erizo d'Europa oriental	Brown with a white chest patch. Length 8–12 in (20–30 cm).	Deciduous woodland and scrub, grassland and pasture, suburban parks and gardens. Alpine regions below the treeline. Nocturnal and solitary outside breeding season.	Europe east of 15°E (Czech Republic and the former Yugoslavia) as far as 80°E, central Russia.	Invertebrates, small vertebrates, some plant matter and fungi.	Not threatened
Chinese hedgehog *Erinaceus amurensis* French: Hérisson de Chine; German: Chinesisch Igel; Spanish: Erizo el chino	Large brown hedgehog with short tail. Spines are banded yellow and brown, with pale tips. Length 8–12 in (20–30 cm).	Deciduous woodland and scrub, grassland and pasture, suburban parks and gardens. Alpine regions below the treeline. Nocturnal and solitary outside breeding season.	Southeastern Siberia northeastern China, Manchuria and Korea.	Eats mostly invertebrates such as beetles, larvae, slugs, worms and spiders; also small vertebrate prey, eggs, fruit and fungi.	Not threatened
Algerian hedgehog *Atelerix algirus* French: Hérisson d'Algérie; German: Algerisch Igel; Spanish: Erizo argelino	Slim with brown fur and short spines and a noticeable center parting on the forehead. Length 8–10 in (20–25 cm).	Lives in scrub, nocturnal and solitary.	Northwest Africa, southeastern Spain, extreme southwestern France, the Balearic and Canary Islands.	Invertebrate and small vertebrate prey, fruit, fungi.	Not threatened
[continued]					

Common name / Scientific name/ Other common names	Physical characteristics	Habitat and behavior	Distribution	Diet	Conservation status
Central African hedgehog *Atelerix albiventris* English: White-bellied hedgehog, four-toed hedgehog; French: Hérisson d'afrique centrale; German: Zentralafricanisch Igel; Spanish: Erizo africano central	Gray-brown to grizzled black fur and spines, with prominent white band from forehead to flanks, narrow center parting in spines on forehead. All-white belly and vestigial big toe. Length 5.5–8.5 in (12–21 cm).	Lives in scrub, nocturnal and solitary.	Senegal to the Sudan and south to Zambia.	Invertebrate and small vertebrate prey, fruit, and fungi.	Not threatened
Somalian hedgehog *Aterlerix sclateri* French: Hérisson de Somalie; German: Somalisch Igel; Spanish: Erizo somalí	Gray-brown to grizzled black fur and spines, with prominent white band from forehead to flanks, narrow center parting in spines on forehead. White belly with dark fur on lower abdomen. Big toe is present. Length 5.5–8 in (14–20 cm).	Lives in scrub, nocturnal and solitary.	Northern Somalia.	Invertebrate and small vertebrate prey, fruit, and fungi.	Not threatened
Brandt's hedgehog *Paraechinus hypomelas* English: Long-spined hedgehog; French: Hérisson longues épines; German: Wüstenigel; Spanish: Erizo d'aguja large	Large and dark-colored, with long spines. Length 8–11 in (21–29 cm).	Dry rocky habitats, will also venture onto cultivated land.	South of the Aral Sea in Kazakhstan, Uzbekistan, Turkmenistan, Pakistan, Iran, southern Afghanistan and northern India, also southern Arabian Peninsula.	Insects and other arthropods, small vertebrates, eggs and fruit.	Not threatened
Ethiopian hedgehog *Paraechinus aethiopicus* French: Hérisson d'Éthiopie; German: Äthiopien Igel; Spanish: Erizo d'Etiopía	Long ears, dark fur and a wide parting of spines on the forehead. Length 5.5–9 in (14–23 cm).	Desert and semidesert.	North Africa and the Arabian Peninsula.	Insects and other arthropods, small vertebrates and eggs.	Not threatened
Collared hedgehog *Hemiechius collaris* English: Hardwicke's hedgehog; French: Hérisson longes- oreilles; German: Großohrigel; Spanish: Erizo de orejas largas	Dark brown spines and fur, paler on face and chin. Length 5.5–7 in (14–18 cm).	Desert and plains, behavior similar to *H. auritus*.	Eastern Pakistan and northwestern India.	Insects and other invertebrates, small vertebrates, carrion and fruit.	Not threatened
Hugh's hedgehog *Mesechinus hughi* English: Shanxi hedgehog; French: Hérisson de Shanxi; German: Zentralchinesisch Igel; Spanish: Erizo el chino central	Similar to *M. dauuricus*, with shorter ears and brown fur. Length 8 in (20 cm).	Subalpine coniferous forest.	Central China.	Not known.	Vulnerable
Hainan gymnure *Hylomys hainanensis* French: Gymnure de Chine; German: Chinesische Gymnure; Spanish: Gymnure el chino	Vole like, with dark brown fur, dark dorsal stripe and long, finely haired tail. Length 4–5 in (10–12.5 cm).	Damp forest habitat, solitary and nocturnal.	South-central China.	Invertebrates and plant material.	Endangered
Shrew gymnure *Hylomys sinensis* English: Hainan gymnure; French: Gymnure d'Hainan; German: Hainanische Gymnure; Spanish: Gymnure el Hainan	Gray-brown fur, pointed snout, short sparsely haired tail. Length 5–6 in (12– 15 cm).	Rainforest habitat, spends much of time in underground burrow.	Hainan Island, China.	Invertebrates.	Lower Risk/Near Threatened
Dinagat gymnure *Podogymnura aurospinula* English: Dinagat moonrat; French: Gymnure de Dinagat; German: Philippinische Gymnure; Spanish: Gymnure dinagato	Spiny brown fur, short tail. Length 7.5– 8.5 in (19–21 cm).	Nocturnal forest dweller.	Dinagat Island, Philippines.	Invertebrates.	Endangered

Resources

Books

McDonald, D. *Collins Field Guide: Mammals of Britain and Europe*. London: Harper Collins, 1993.

Nowak, R. "Hedgehogs and Gymnures (Insectivora; Family Erinaceidae)." In *Walker's Mammals of the World*, 6th ed. Vol. 1, Baltimore and London: Johns Hopkins University Press, 1999.

Reeve, N. *Hedgehogs*. London: Poyser Natural History, 1994.

Periodicals

Corbet, G. B. "The Family Erinaceidae: A Synthesis of its Taxonomy, Phylogeny, Ecology and Zoogeography." *Mammal Review* 18 (1988): 117–172.

Jenkins, P. D., and M. F. Robinson. "Another Variation on the Gymnure Theme: Description of a New Species of *Hylomys* (Lipotyphla, Erinacaidae, Galericinae)." *Bulletin of the Natural History Museum, Zoology Series* 68 (2002): 1–11.

Organizations

IUCN Species Survival Commission, Insectivore Specialist Group, Dr. Werner Haberl, Chair. Hamburgerstrasse 11, Vienna, A-1050 Austria. E-mail: shrewbib@sorex.vienna.at Web site: <http://members.vienna.at/shrew/itses>

Other

Hedgehog Valley. March 31, 2003. <http://hedgehogvalley.com>.

The International Hedgehog Association. *The International Hedgehog Club*. January 1, 2003. <http://hedgehogclub.com>.

Stone, David R. "Family Erinaceidae: The Hedgehogs, Moonrats and Gymnures." In *Eurasian Insectivores and Tree Shrews: Status Survey and Conservation Action Plan*. 1995. *ITSES Specialist Group, IUCN*. <http://members.vienna.at/shrew/itsesAP95-erinaceidae.html>.

Amy-Jane Beer, PhD

Golden moles
(Chrysochloridae)

Class Mammalia

Order Insectivora

Family Chrysochloridae

Thumbnail description
Small ears hidden by their fur, short tails, not visible externally; large leathery pads on their noses help them to burrow through the ground; thick, glossy fur, triangular-shaped head with skin covering eyes and leathery pad over nostrils; short, powerful forearms and claws

Size
Head and body length 2.7–9 in (7.0–23.5 cm); weight 0.5–17.6 oz (16–500 g)

Number of genera, species
9 genera; 21 species

Habitat
Burrows beneath a wide range of habitats from sand dunes to tropical forest

Conservation status
Critically Endangered: 4 species; Endangered: 1 species; Vulnerable: 6 species

Distribution
Sub-Saharan Africa

Evolution and systematics

Fossil records for this family date back to the late Eocene (40 million years ago). Traditionally, both golden and "true" moles (Talpidae) have been placed within the order Insectivora. Golden moles have been considered part of a suborder Tenrecomorpha, together with tenrecs.

Molecular studies of phylogenetic relationships now conclude that golden moles fit within a superordinal group called Afrotheria that evolved separately from primitive placental mammals of northern continents. The similarities between moles and golden moles are thus a case of parallel adaptation. Scientists propose that golden moles and tenrecs be placed in an order Afrosoricida—the "African shrew-like mammals."

The family Chrysochloridae is divided into two subfamilies. Chrysochlorinae consists of six genera: *Eremitalpa* has a single species, Grant's golden mole *Eremitalpa granti*; *Chrysospalax*, the giant golden moles, has two species; *Chrysochloris*, the Cape golden moles, has three species; *Cryptochloris*, the cryptic golden moles, has two species; *Carpitalpa* has a single species, Arend's golden mole *Carpitalpa arendsi*; *Chlorotalpa* has two species. The subfamily Amblysominae has three genera: *Amblysomus*, the

A Grant's desert golden mole (*Eremitalpa granti*) catches a locust in the sand dunes of the Namib Desert, Namibia. (Photo by Michael Fogden. Bruce Coleman, Inc. Reproduced by permission.)

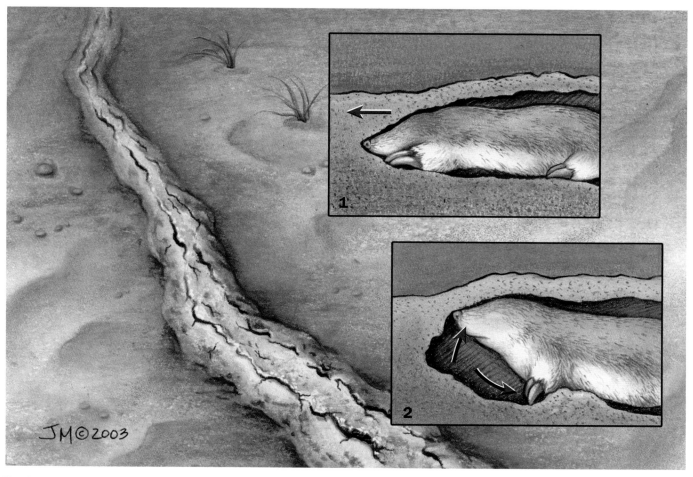

The Grant's desert golden mole (*Eremitalpa granti*) uses its powerful forelimbs to burrow through the sands of the Namib Desert in southern Africa. The golden mole moves forward (1), and enlarges the tunnel by pushing dirt up with its head and back with its claws (2). (Illustration by Jacqueline Mahannah)

narrow-headed golden moles, has five species; *Neamblysomus* has two species; *Calcochloris* has three species.

Physical characteristics

Golden moles get their descriptive name from the iridescent golden, blue, purple, or bronze sheen on their brown fur. The skin is tough and loosely attached to the body and the fur has a woolly insulating undercoat.

In common with true moles, golden moles have big shoulders and short, extremely powerful forelimbs, with curved claws suitable for digging. The hind feet have webbed toes, enabling the animals to kick soil backwards effectively. The head, conical in shape, is designed to protect the moles from the substrate through which they push. The nostrils are concealed behind a leathery pad and the vestigial eyes of these blind animals are covered by a thick layer of skin.

Distribution

This is the only mammal family with a southern African center of diversity. Eighteen of the 21 species occur only in

southern Africa, with 15 endemic to South Africa. The remaining three species are found in central and east Africa, including Somalia, with *A. julianae* in southern and eastern Transvaal. Golden moles show a scattered distribution within their range, probably as a result of climatic changes during their evolution which are no longer evident.

Habitat

Although golden moles are adapted for an almost exclusively burrowing lifestyle, they occur in a wide range of habitats. These include sandy deserts, forests, swamps, grassland, and mountains up to 13,000 ft (4,000 m) in altitude.

Behavior

Golden moles are solitary and fight to defend their burrows against intruders of either sex, especially in winter and in less fertile areas where food is in short supply. Dominant individuals may subsume a rival's tunnel within their home range. These animals are, however, tolerant of herbivorous rodents such as mole-rats *Bathyergus* in their burrows.

Burrowing activity is most intense in summer, when the ground is moist. The moles include chambers for nesting and defecation within their tunnels. The network also features several spiraling bolt-holes, so that the moles can return to the tunnels quickly when surface foraging. These animals have an uncanny ability to head unerringly to a bolt-hole if disturbed, despite being completely blind.

Golden moles have six or seven periods of activity over 24 hours, interspersed with longer spells of rest. Over a considerable temperature range of between 73° and 91°F (23–33°C), they do not regulate their internal body temperature. This thermo-neutrality, together with an unusually low metabolic rate, and an ability to sustain periods of torpidity lasting up to three days, enables golden moles to survive temperature extremes. When torpid, golden moles halve their energy consumption.

Feeding ecology and diet

Golden moles rely on touch, smell and sensitivity to vibration to hunt their food both below and, more rarely, above the soil. Sand-burrowing species, including Grant's golden mole—captured memorably on David Attenborough's *Life of Mammals* series "swimming" through a subsurface tunnel—are able to detect vibrations from dune grass in their middle ear. This enables them to head "blind" to vegetation where ants and termites are regular prey. De Winton's golden mole *Cryptochloris wintoni* surfaces to catch and kill legless lizards (*Typhlosaurus*) and invertebrates with its front claws.

Foraging behavior is determined by rainfall in other areas. The hottentot golden mole *Amblysomus hottentotus* eats mainly earthworms and soil insect larvae, taken from burrows near the surface in moist soil, but digs much further underground in dry conditions. Heavy rain and waterlogged burrows force some species to become temporarily terrestrial foragers. Under such conditions, Cape golden moles and giant golden moles root around on the surface for worms and insects.

Reproductive biology

Golden moles are believed to be polygynous and breed during the winter months of April to July. Males and females exchange chirping and squealing calls; the male shakes his head, stamps his feet and pursues the female. Mating generally occurs in the spring, and gestation is believed to last from 4 to 6 weeks. The female makes a grass-lined nest deep in a chamber of her burrow and gives birth to two or three young. She raises the young alone and suckles them for two to three months, then evicts them from her burrow.

Conservation status

Eleven out of 21 species were under threat of extinction, according to the year 2002 IUCN Red List. Destruction of habitat includes poor forestry practices, agricultural develop-

A Grant's desert golden mole (*Eremitalpa granti*) hunting at night in the Namib Desert, Namibia. (Photo by Michael Fogden. Bruce Coleman, Inc. Reproduced by permission.)

ment, livestock overgrazing, and removal of dunes for diamond mining. As human settlements move into golden mole habitats, the animals fall prey to domestic dogs and cats.

Scientists do not fully understand the distribution, status and ecology of this family, because golden moles are cryptic and trap-shy.

A Grant's desert golden mole (*Eremitalpa granti*) feeds on a palmato gecko in Namibia. (Photo by Animals Animals ©Austin J. Stevens. Reproduced by permission.)

Significance to humans

In South Africa, the burrowing activities of golden moles through vegetable plots and grass lawns make them unpopular with some farmers and householders. Many are killed as pests and are sometimes skinned for their glossy fur. Conversely, other farmers appear to favor their presence because of their insect pest diet. Bakiga tribesmen of Uganda use Stuhlmann golden mole *Chrysochloris stuhlmanni* skins as lucky charms.

1. Hottentot golden mole (*Amblysomus hottentotus*); 2. Grant's desert golden mole (*Eremitalpa granti*); 3. Stuhlmann's golden mole (*Chrysochloris stuhlmanni*); 4. Large golden mole (*Chrysospalax trevelyani*). (Illustration by Jacqueline Mahannah)

Species accounts

Hottentot golden mole
Amblysomus hottentotus

SUBFAMILY
Amblysominae

TAXONOMY
Amblysomus hottentotus (Smith, 1829), east Cape Province, South Africa.

OTHER COMMON NAMES
English: Narrow-headed golden mole.

PHYSICAL CHARACTERISTICS
Head and body length 4–5.5 in (10–14 cm); weight 1.4–3.5 oz (40–100 g). Dark red-brown fur, white hairs around ear cavities and vestigial eyes. Four-clawed toes, but mainly digs with only two.

DISTRIBUTION
Southern and eastern South Africa, Lesotho, southern Mozambique.

HABITAT
Grassland and woodland from sea level to 10,800 ft (3,300 m).

BEHAVIOR
Some golden moles inhabit areas of peaty soil in mountain or forest regions; others live in escarpment forests, and certain species are favor open areas where there is good grass cover. Some golden moles burrow just below the surface of the ground, while others burrow deeper depending on local soil conditions.

In fertile areas, they dig 13 ft (4 m) per day, but in poorer soil they tunnel 40 ft (12.4 m).

FEEDING ECOLOGY AND DIET
Mainly earthworms and larvae, but also slugs, snails, crickets, and spiders.

REPRODUCTIVE BIOLOGY
May breed all year round. Known to raise three young per litter.

CONSERVATION STATUS
Not threatened.

SIGNIFICANCE TO HUMANS
None known. ◆

Grant's desert golden mole
Eremitalpa granti

SUBFAMILY
Chrysochlorinae

TAXONOMY
Eremitalpa granti (Broom, 1907), Cape Province, South Africa.

OTHER COMMON NAMES
English: Desert golden mole.

PHYSICAL CHARACTERISTICS
Head and body length 3.0–3.3 oz (7.6–8.6 cm); weight 0.5–1.0 oz (16–32 g). Pale gray-yellow fur covers body and eyes; belly is more yellow.

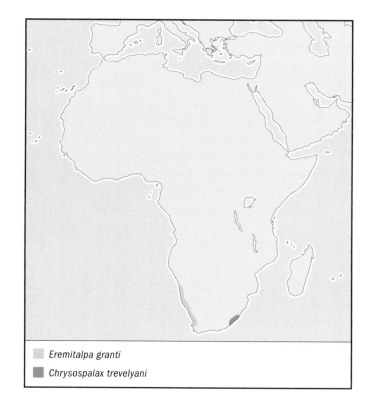

Amblysomus hottentotus
Chrysochloris stuhlmanni

Eremitalpa granti
Chrysospalax trevelyani

DISTRIBUTION
Coastal dunes in Cape Province, Little Namaqualand in South Africa, and the Namib Desert.

HABITAT
Sand dunes.

BEHAVIOR
Nocturnal. Burrows under sand just below the surface during the day.

FEEDING ECOLOGY AND DIET
Nocturnal forager in search of termites, ants and spiders. Also eats geckos, beetles, and legless lizards.

REPRODUCTIVE BIOLOGY
Digs deeper tunnels in which the female raises at least one young. Presumably polygynous.

CONSERVATION STATUS
Vulnerable according to IUCN criteria.

SIGNIFICANCE TO HUMANS
None known. ◆

Large golden mole
Chrysospalax trevelyani

SUBFAMILY
Chrysochlorinae

TAXONOMY
Chrysospalax trevelyani (Günther, 1875), Cape Province, South Africa.

OTHER COMMON NAMES
English: Giant golden mole.

PHYSICAL CHARACTERISTICS
Head and body length 5.0–6.8 in (12.5–17.5 cm); weight 3.8–5.0 oz (108–142 g). Red-brown fur, lighter on belly. Eyes are covered by hairy skin. Nose is tapered and pink.

DISTRIBUTION
Eastern Cape Province, South Africa.

HABITAT
Flat or gently sloping ground in damp forests.

BEHAVIOR
Makes small mounds above the entrances to tunnels. The only species to show complex social behavior, because several animals may share burrows in winter.

FEEDING ECOLOGY AND DIET
Earthworms and other forest invertebrates taken underground. Feeds above ground when tunnels are flooded.

REPRODUCTIVE BIOLOGY
Probably polygynous or promiscuous. Female raises one or two young in summer.

CONSERVATION STATUS
Endangered according to IUCN criteria.

SIGNIFICANCE TO HUMANS
None known. ◆

Stuhlmann's golden mole
Chrysochloris stuhlmanni

SUBFAMILY
Chrysochlorinae

TAXONOMY
Chrysochloris stuhlmanni Matschie, 1894, Ruwenzori region, Uganda.

OTHER COMMON NAMES
None known.

PHYSICAL CHARACTERISTICS
Head and body length 3.5–5.5 in (9–14 cm); weight 2.6 oz (75 g). Dark brown-green fur, appears iridescent. Front feet have long thick claws, hind feet are webbed.

DISTRIBUTION
Tanzania, Uganda, Kenya, Democratic Republic of the Congo, Cameroon.

HABITAT
Can live in highland forests up to 9,180 ft (2,800 m).

BEHAVIOR
Burrows just below surface. May make overland movements at night.

FEEDING ECOLOGY AND DIET
Invertebrates, especially beetles, larvae, and worms. Forages on surface when soil is waterlogged.

REPRODUCTIVE BIOLOGY
Two young, raised during rainy season of April–July. Presumably polygynous.

CONSERVATION STATUS
Not threatened.

SIGNIFICANCE TO HUMANS
Skins kept as lucky charms in Uganda. ◆

Common name / Scientific name/ Other common names	Physical characteristics	Habitat and behavior	Distribution	Diet	Conservation status
Gunning's golden mole *Amblysomus gunningi* French: Taupe dorée de Gunning	Coloration is very dark, smoky brown with a greenish sheen. Head and body length 3.7–5.7 in (9.5–14.5 cm).	Areas of peaty soil in the sheltered ravines of mountains or in forests, escarpment forests, or open areas with good grass cover. Solitary, no particular breeding season.	South Africa, Transvaal, Woodbush.	Worms, larvae, pupae, and insects.	Vulnerable
Zulu golden mole *Amblysomus iris* French: Taupe dorée zouloue	Coloration of upperparts is dark reddish brown with a bronze sheen. Head and body length 3.7–5.7 in (9.5–14.5 cm).	Areas of peaty soil in the sheltered ravines of mountains or in forests, escarpment forests, or open areas with good grass cover. Solitary, no particular breeding season.	South Africa, Zululand, Umfolozi Station.	Worms, larvae, pupae, and insects.	Not threatened
Juliana's golden mole *Amblysomus julianae* French: Taupe dorée de Juliana	Coat is shiny, fur is dense, has streamline, formless appearance. No visible eyes or ears. Head and body length 3.7–5.7 in (9.5–14.5 cm), average weight 0.7–2.6 oz (21–75 g).	Drier uplands in open country with sandy soils. Blind. Usually two young are born, sometimes one. As long as the mole is awake, it keeps moving and this keeps body temperature normal. Sleep may be detrimental, as the mole cools off.	Pretoria, Nylstroom/ Nylsvley, and Kruger National Park, Transvaal, South Africa.	Invertebrates, such as crickets, grasshoppers, locusts, cockroaches, earthworms, and snails.	Critically Endangered
Yellow golden mole *Calcochloris obtusirostris* French: Taupe dorée jaune	Upperparts vary from glossy brown to bright golden yellow. Broad yellow or buffy band across top of snout. Sides of face are yellow. Head and body length 3.8–4.3 in (9.7–10.8 cm), weight 0.07– 1 oz (20–30 g).	Sandy soils, sandy alluvium, and coastal sand dunes. Stays in burrow systems most of its life. Will burrow deeper if it hears noise at ground level.	Zululand and eastern Transvaal, South Africa; southern Zimbabwe; and southern Mozambique.	Earthworms and insects.	Not threatened
Congo golden mole *Chlorotalpa duthieae* French: Taupe dorée de Duthie	Upperparts are very dark brown with a green sheen. Underparts are lighter and sides of face are yellowish. Head and body length 3.7–4.4 in (9.5–11.1 cm).	Alluvial sand and sandy loam in coastal areas. Little is known of behavioral and reproductive patterns.	Southern Cape Province in South Africa.	Mainly earthworms and insects.	Vulnerable
Cape golden mole *Chrysochloris asiatica* German: Kapgoldmull	Shiny coat of dense fur, giving stream- lined, formless appearance. No eyes or ears visible. Head and body length 3.5–5.5 in (9–14 cm), weight 0.8–1.3 oz (25–35 g).	Species is known only from a single specimen collected at Gouna, 87 km (54 mi) east of Calvinia, Cape Province, South Africa. Behavior likely similar to *Amblysomus julianae*	Western Cape Province and Robben Island, South Africa; perhaps Damaraland, Namibia.	Earthworms, insects, and snails.	Not threatened
Rough-haired golden mole *Chrysospalax villosus* French: Taupe dorée ébouriffée; German: Riesengoldmulle	Rough hair, streamlined body. Light brown to buff. No eyes or ears visible. Weight 3.8–5 oz (108–142 g).	Dry grassy habitats, particularly meadow-like ground bordering wet marshes. Spends majority of life in chambers and passages in mounds, connected by a system of tunnels.	Transvaal and Natal, South Africa.	Mainly worms and insects.	Vulnerable
De Winton's golden mole *Cryptochloris wintoni* French: Taupe dorée de De Winton; German: Der De- Winton-Goldmull	Fur is short, soft, and dense. Upperparts drab lead color with violet iridescence. Underparts are lead gray. Head and body length 3.1–3.5 in (8.0–9.0 cm).	White, coastal sand dunes. Generally burrows below surface.	Little Namaqualand, Cape Province, South Africa.	Legless lizards and various invertebrates.	Vulnerable
Van Zyl's golden mole *Cryptochloris zyli* French: Taupe dorée de Van Zyl	Flat, round body shape. Fur is short, soft, and dense. Upperparts drab lead color with violet iridescence. White facial markings. Underparts are lead gray. Head and body length 3.1–3.5 in (8.0–9.0 cm).	White, coastal sand dunes. Generally burrows below surface.	Northwestern Cape Province, South Africa.	Legless lizards and various invertebrates.	Critically Endangered

Resources

Books

Apps, P. *Smithers' Mammals of Southern Africa.* Cape Town, South Africa: Struik Publishers, 2000.

Kingdon, J. *The Kingdon Field Guide to African Mammals.* San Diego, CA: Academic Press, 1997.

Macdonald, D. *The New Encyclopaedia of Mammals.* Oxford: Oxford University Press, 2001.

Periodicals

Kuyper, M. A. "The ecology of the golden mole *Amblysomus hottentotus.*" *Mammal Review* 15, no. 1 (1985): 3–11.

Mason, M. J., and P. M. Narins. "Seismic sensitivity in the desert golden mole *Eremitalpa granti*: A review." *Journal of Comparative Psychology* 116, no. 2 (2002): 158–163.

Stanhope, M. J., et al. "Molecular evidence for multiple origins of insectivores and for a new order of endemic African insectivore mammals." *Proceedings of the National Academy of Sciences USA* 95 (1999): 188–193.

Organizations

IUCN Species Survival Commission, Afrotheria Specialist Group. Web site: <http://www.calacademy.org/research/bmammals/afrotheria/ASG.html>

Other

IUCN Red List of Threatened Species—Species information. <http://www.redlist.org>.

The Life of Mammals, (episode one). BBC Television, 2002

Nowak, R.M. *Walker's Mammals of the World Online.* Baltimore: John Hopkins University Press, 1995. <http://press.jhu.edu/books/walkers_mammals_of_the_world/insectivora.chrysochloridae.htm.>

Derek William Niemann, BA

Tenrecs
(Tenrecidae)

Class Mammalia
Order Insectivora
Family Tenrecidae

Thumbnail description
Very small to small mammals showing a considerable range in external morphology with regards to, for example, tail length (very short to three times body length), fur color and texture, and foot structure; some species have dorsal fur modified into spines and others with elongated trunk-like bodies; four to five fingers and five toes; tooth count ranges from 32 to 40, depending on genus

Size
Head and body length 2–14 in (5.5–35.7 cm); weight 0.1–70 oz (3.5 g–2 kg)

Number of genera, species
10 genera; 27 species

Habitat
Forest, pseudosteppe, scrublands, and spiny bush

Conservation status
Critically Endangered: 1 species; Endangered: 6 species; Vulnerable: 3 species

Distribution
Most restricted to Madagascar; otter shrews, mainland Africa

Evolution and systematics

Four subfamilies and 10 genera of Tenrecidae are generally recognized: the Tenrecinae or spiny tenrecs (*Tenrec*, *Echinops*, *Setifer*, and *Hemicentetes*); the Oryzorictinae or shrew tenrecs (*Microgale* and *Limnogale*) and the rice tenrecs (*Oryzorictes*); the Geogalinae or large-eared tenrec (*Geogale*); and the Potamogalinae or otter shrews (*Micropotamogale* and *Potamogale*).

Recent studies indicate that the tenrecids form a monophyletic group, with the Potamogalinae or African otter shrews as the sister group. This indicates that the Tenrecidae are the result of a single colonization event and subsequently speciated into one of the most extraordinary adaptive radiations of mammals in the world. When and how this event took place is an open question. The mammalian fossil record on Madagascar is rather fragmentary and no deposits are known from the Late Cretaceous to the Late Pleistocene, a span of about 65 million years. This is presumably the period tenrecids colonized and radiated on Madagascar. Their arrival on the island is probably the result of the proto-Tenrecidae

rafting on vegetation across the Mozambique Channel from Africa. Fossils from the African continent from the Miocene have been attributed by certain paleontologists to the Tenrecidae, but these designations are in need of reevaluation.

Recent field inventories and a reevaluation of existing museum specimens have resulted in considerable taxonomic changes to the Tenrecidae, most notably amongst *Microgale*. Several new species to science have been named in recent years and numerous others remain to be described. Further, molecular studies are uncovering several previously unrecognized cryptic species. It is certain that the next decade will see a considerable augmentation in the number of recognized species and refinement of the evolutionary history of Tenrecidae.

Physical characteristics

Tenrecidae show an assortment of morphological characters. They maintain certain characteristics that have been considered primitive among living mammals. These include, for

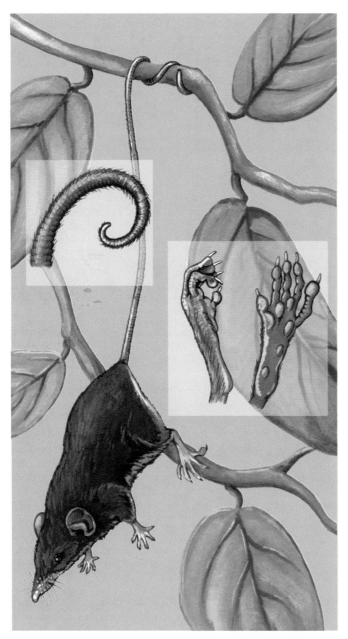

A long-tailed tenrec uses its extremely long prehensile tail. Left inset shows that the tip of the tail is naked underneath, and right inset shows the enlarged fifth digit of the hind foot. (Illustration by Gillian Harris)

A rice tenrec (*Oryzorictes hova*) foraging. (Photo by Harald Schütz. Reproduced by permission.)

example, small body size, common urogenital opening, and abdominal testes in reproductive males. Tenrecidae range from small- to medium-sized mammals and span a remarkable morphological gamut from the large (up to 44 lb; 2 kg) spiny tenrec, the hedgehog-like *Echinops* and *Setifer*, the desman-like *Limnogale*, the mouse-like *Geogale*, the molelike *Oryzorictes*, and the shrew-like *Microgale* (with adults of some species weighing less than 0.14 oz; 4 g). The tails of Tenrecinae are very short or at least not discernable, while among most other species in the other two subfamilies, the tail varies in length and is covered with short fine fur. Members of this family, particularly *Setifer* and *Echinops*, can roll themselves

into tight balls when disturbed. No species, with the exception of *Tenrec ecaudatus*, display any measurement or phenotypic feature that differs between the sexes.

Even within the genus *Microgale* there is considerable variation in body morphology. For example, *M. gymnorhyncha* and *M. gracilis* are probably semi-fossorial and have dense velvety pelage, well developed forelimbs and associated digging claws, and reduced ears and eyes, while the partially arboreal or scansorial *M. principula* and *M. longicaudata* have notably long hind limbs and prehensile tails measuring more than twice as long as their body length.

Otter shrews in the genus *Micropotamogale* are small, brownish gray animals with unwebbed feet and a round tail. The giant otter shrew (*Potamogale velox*), as its name implies, is considerably larger (about twice as long and much heavier) and has glossy chocolate-brown fur and a thick, rudderlike tail.

Distribution

The family is confined to Madagascar with the exception of the otter shrews, which are found in Africa. *Tenrec ecaudatus* has been introduced to the western Indian Ocean islands of the Comoros, Seychelles, and Mascarenes as a food resource.

Habitat

Most Tenrecidae are forest dwelling animals, particularly in the humid forests of eastern Madagascar. The major exceptions are the genera *Tenrec*, *Setifer*, *Hemicentetes*, and *Echinops* that can also be found in human-created grasslands or pseudosteppe covering significant portions of the island, and the first three genera also occur in agricultural areas and

The greater hedgehog tenrec (*Setifer setosus*) emerges from its burrow to forage. (Photo by Tom McHugh/Photo Researchers, Inc. Reproduced by permission.)

The common tenrec (*Tenrec ecaudatus*) is the largest of all tenrecs. (Photo by R. Mittermeier. Bruce Coleman, Inc. Reproduced by permission.)

Behavior

Given that most species of Tenrecidae are small nocturnal and forest-dwelling animals it is not surprising that few details are available about their life history.

Two genera of Tenrecidae have a dense grouping of spiny hairs on their back that have been modified into quill-like structures and form a stridulating organ. In young *Tenrec*, this organ is well developed and is used for communication within family groups, and with increasing age the quill-like struc-

near human habitations. Further, *Tenrec* and *Setifer* occur in all of the natural forest formations on the island. Most Tenrecidae species are terrestrial, although certain *Microgale* are best considered scansorial and two genera of spiny tenrecs (*Echinops* and *Setifer*) climb in vegetation several meters off the ground.

Microgale, with its currently recognized 18 species, is largely confined to the humid forest formations of the island. One member of this genus, *Microgale brevicaudata*, also occurs in the dry deciduous forests and spiny bush of the western and southern portion of the island. Two genera (*Echinops* and *Geogale*), each with a single species, are also confined to these drier native vegetal formations. *Limnogale* is confined to streams within native forest within the eastern humid forests and is the only largely aquatic mammal on the island. *Microgale pusilla* and *Oryzorictes hova* can be found in forested areas with wet soils or in marshlands, including those converted to rice paddy, outside of forest formations. *Micropotamogale* are found in upland forest streams of western and central Africa, while *Potamogale velox* occurs in rivers, streams, and swamps in the rainforests of tropical Africa.

At certain sites in the eastern humid forest, the density and diversity of Tenrecidae are notably high. For example at one forested site near the village of Tsinjoarivo in the eastern central portion of the island, 17 species of Lipotyphla (including 16 tenrecids) were found in sympatry. This high density in a relatively small block of forest is perhaps without parallel anywhere else in the world.

The yellow streaked tenrec (*Hemicentetes semispinosus*) uses its long snout to find earthworms. (Photo by H. Uible/Photo Researchers, Inc. Reproduced by permission.)

The lesser hedgehog tenrec (*Echinops telfairi*) is nocturnal. (Photo by H. Uible/Photo Researchers, Inc. Reproduced by permission.)

A yellow streaked tenrec (*Hemicentetes semispinosus*) forages for insects. (Photo by Harald Schütz. Reproduced by permission.)

tures are replaced with standard spines and the organ no longer functions. In *Hemicentetes* adults possess a mid-dorsal stridulating organ, also formed of modified spines. These quills, when vibrated by muscle action, produce an ultrasonic sound that can be detected by nearby conspecifics.

The spines of Tenrecinae are used in a defensive fashion against predators. In *Hemicentetes*, the easily detachable spines

A long-tailed shrew tenrec (*Microgale principula*) in front of its burrow. (Photo by Harald Schütz. Reproduced by permission.)

of the back with their very fine fishhook tips are raised when a potential predator approaches. If the predator makes direct contact, it can receive a face full of spines. Although this must be a formidable deterrent, spiny tenrecs are frequent prey for native predators.

During the austral winter, several species of Tenrecidae show a shift in activity patterns behavior, ranging from a reduction in daily movements to complete hibernation. These shifts are correlated with changes in ambient temperature, photoperiod, and food availability. In certain species, such as *Hemicentetes nigriceps* or *Tenrec ecaudatus*, animals enter a profound torpor during the period from approximately May to October. A further variation within the family is found in *Geogale*, which is heterothermic and body temperature follows the ambient temperature. This species has a notably low resting metabolic rate, even for a Tenrecidae.

Feeding ecology and diet

Most Tenrecidae feed on insects and a variety of other soil invertebrates—few quantitative studies are available. In the western dry deciduous forests, *Tenrec* feed extensively on soil-dwelling beetle (Coleoptera) larvae of the families Scarabaeidae and Alleculidae, and to a lesser extent on ants (Hymenoptera), scolopenders (Chilopoda), and butterfly (Lepidoptera) larvae. They are known to also feed on fruits and vertebrates.

A study conducted in the humid forests of eastern Madagascar on the diets of *Microgale* spp., based on stomach contents, found that various types of Orthoptera were the most frequently consumed prey, followed by Hymenoptera and Coleoptera. Remains of other animals were also identified, including Dermaptera, Annelida, Arachnida, and Amphipoda. At the level of determination, no clear dietary differences were found between sympatrically occurring *Microgale* subspecies.

The most common prey of *Limnogale*, an aquatic tenrec with a flattened tail and webbed feet that hunts while swim-

A large-eared tenrec (*Geogale aurita*) foraging in Madagascar. (Photo by Harald Schütz. Reproduced by permission.)

The burrows of common tenrecs (*Tenrec ecaudatus*) are usually found near water. (Photo by Hugh Clark/FLPA–Images of Nature. Reproduced by permission.)

ming, is Ephemeroptera nymphs, followed by larvae of Odonata, Trichoptera, and Coleoptera. A few other types of aquatic invertebrates (crabs and crayfish) and vertebrates (frogs) are also taken. In addition to worms and insects, otter shrews also prey upon crabs, fishes, frogs, and mollusks found in their aquatic habitat.

It is assumed that most terrestrial Tenrecidae actively hunt their prey amongst leaf and rotten wood litter, fallen branches, and root tangles of standing trees or scansorial species on vegetation in the forest understory to mid-canopy. In some cases, particularly for genera with well developed digging claws, prey is excavated from the ground and may be detected by scent and acoustic signals. For most Tenrecidae prey are pinned down with the forelimbs and then seized with the mouth.

Reproductive biology

The reproductive biology of most Tenrecidae is poorly known—a few species have been studied in the wild or in captivity and certain details are available. Some species build nests out of vegetation that are placed in concealed places or exca-

A white streaked tenrec (*Hemicentetes nigriceps*) probes the bark of a tree for insects. (Photo by Harald Schütz. Reproduced by permission.)

vate burrows ending in a nest chamber. The majority of species have three to five pairs of mammae.

Among *Microgale* living in humid forests, reproduction starts in September to October, coinciding with the start of the rainy season. Litters, generally varying from one to four, are born in November and early December. Neonates are naked with non-functional ears and eyes. Young start to make foraging bouts away from the nest after their eyes open at 22 to 27 days old. Most females probably have one or maximum two litters per year. Among certain species, there is evidence that they are reproductively mature before they have their full adult dentition. In captivity, certain species seem to form male-female pairs.

Tenrec has one of the highest breeding potentials of any mammal species in the world. Litters of up to 32 individuals have been documented, and more than 20 individuals is not uncommon. Females often have 32 to 36 mammae. After being born in late November to January, young emerge from the burrow at between 18–20 days old and commence actively foraging with their mother, and become independent at about 35 days.

Within the genus *Geogale*, there is a post-partum estrus and females are able to nurse one litter while another is developing in the uterus. This reproductive strategy is unknown in any other Tenrecidae, and may be an adaptation to the extremely dry conditions where this animal lives, allowing it to maximize reproductive output when optimal conditions prevail.

Conservation status

The main threat to members of the Tenrecidae, as with all forest-dwelling organisms on Madagascar, is the continued destruction of forest habitat for slash-and-burn agriculture, charcoal, and cattle pasture. The IUCN Red List indicates that several species are threatened with extinction. However, the information used to compile this list is out of date, as is the taxonomy. A meeting held in Madagascar in 2001 and attended by numerous specialists proposed new conservation status designations for Malagasy vertebrates. Among the Tenrecidae one species is classified as Endangered, one as Vulnerable, none as Near Threatened, 22 as Least Concern, and three as Data Deficient.

The African otter shrews face similar threats as the members of the Tenrecidae inhabiting Madagascar. Habitat destruction and degradation is the primary concern. The IUCN

Red List classifies all three species of otter shrews as Endangered.

Significance to humans

Throughout much of Madagascar, spiny tenrecs, most notably *Tenrec* and to a lesser extent *Echinops* and *Setifer*, are widely hunted by people as a supplementary source of meat. In fact, in some areas, these animals are sufficiently relished that their flesh is more expensive per unit weight than beef. In certain regions, the consumption of Tenrecinae meat is considered taboo (fady) and these animals are not hunted locally.

1. Large-eared tenrec (*Geogale aurita*); 2. Common tenrec (*Tenrec ecaudatus*); 3. Nasolo's shrew tenrec (*Microgale nasoloi*); 4. Dobson's shrew tenrec (*Microgale dobsoni*). (Illustration by Gillian Harris)

Species accounts

Common tenrec
Tenrec ecaudatus

SUBFAMILY
Tenrecinae

TAXONOMY
Erinaceus ecaudatus (Schreber, 1777), Madagascar.

OTHER COMMON NAMES
None known.

PHYSICAL CHARACTERISTICS
Head and body length 10.5–15.3 in (26–39 cm); weight 42–70 oz (1.2–2 kg). Fur is grey-brown or red-brown and has long stiff hairs along the back. Front legs are longer than hind legs.

DISTRIBUTION
Across much of Madagascar. It has been introduced to the western Indian Ocean islands of Comoros, Seychelles, and the Mascarenes.

HABITAT
Occurs in a wide variety of habitats from humid forests, to deciduous forest, gallery forest, and spiny bush. It also is found in human modified habitats including savanna, pseudosteppe, agricultural zones, and near habitations.

BEHAVIOR
Usually seen singly and generally nocturnal, although sometimes crepuscular. During the breeding season females can be seen accompanied by a bevy of young even during daylight hours.

FEEDING ECOLOGY AND DIET
Terrestrial. The diet includes mostly invertebrates, but fruits and small vertebrates have also been recorded.

REPRODUCTIVE BIOLOGY
Very high reproductive potential with litters of up to 32 young. Females often have 32 to 36 mammae. May be polygamous or promiscuous.

CONSERVATION STATUS
Not threatened. Widely distributed on the island and even in zones with considerable human hunting pressure this species seems to be maintaining viable populations. Its ability to live in human-modified habitats is an important factor in this regard.

SIGNIFICANCE TO HUMANS
Widely relished for its meat. ◆

Large-eared tenrec
Geogale aurita

SUBFAMILY
Geogalinae

TAXONOMY
Geogale aurita Milne-Edwards and Grandidier, 1872, Morondava, Madagascar. *G. a. orientalis* is known from one specimen collected along the east coast.

OTHER COMMON NAMES
None known.

PHYSICAL CHARACTERISTICS
Head and body length about 2.8 in (7 cm); weight 0.2–0.3 oz (5–8.5 g). Fur is red-brown, yellow on underside. Tail has fine hair covering.

DISTRIBUTION
Restricted to western and southwestern portion of Madagascar, except for the specimen noted above.

HABITAT
Occurs in dry deciduous, gallery, and spiny bush forest. Unknown outside of forested zones. Often associated with fallen and rotten tree trunks.

BEHAVIOR
Few details available about this nocturnal species. Usually seen singly, although apparent male-female pairs can be found in close proximity to one another. Heterothermic with body temperature corresponding to the ambient temperature; enters a daily state of torpor.

FEEDING ECOLOGY AND DIET
Terrestrial. The diet includes mostly invertebrates, particularly termites.

REPRODUCTIVE BIOLOGY
May be monogamous. Mating in late September to March. Gestation 54–69 days. Litter size 1–5 and neonates naked with nonfunctional ears and eyes. Eyes open between 21 and 33 days old and the young are soon thereafter weaned.

CONSERVATION STATUS
Within its range this species can be notably common. Its continued existence depends on the maintenance of natural forest.

Geogale aurita

Tenrec ecaudatus

SIGNIFICANCE TO HUMANS
None known. ◆

Nasolo's shrew tenrec
Microgale nasoloi

SUBFAMILY
Oryzorictinae

TAXONOMY
Microgale nasoloi Jenkins and Goodman, 1999, Vohibasia Forest, Madagascar.

OTHER COMMON NAMES
None known.

PHYSICAL CHARACTERISTICS
Head and body length about 3 in (8 cm); weight 0.5 oz (7 g).

DISTRIBUTION
Only known from southwestern Madagascar.

HABITAT
Occurs in transitional dry and spiny bush forest. Unknown to occur outside of forested zones.

BEHAVIOR
Nothing is known.

FEEDING ECOLOGY AND DIET
Scansorial. The diet is presumably mostly invertebrates.

▢ *Microgale nasoloi*
■ *Microgale dobsoni*

REPRODUCTIVE BIOLOGY
Nothing is known.

CONSERVATION STATUS
Only recently described from two specimens taken at two nearby sites—one in a national park and the other in a forest protected by local traditions. This species probably has a very limited distribution.

SIGNIFICANCE TO HUMANS
None known. ◆

Dobson's shrew tenrec
Microgale dobsoni

SUBFAMILY
Oryzorictinae

TAXONOMY
Microgale dobsoni Thomas, 1884, Fianarantsoa, Madagascar.

OTHER COMMON NAMES
None known.

PHYSICAL CHARACTERISTICS
Head and body length 3.9–4.5 in (10–11.5 cm); weight 0.6–0.9 oz (18.0–25.5 g). Fur is gray-brown with gray belly. Tail is long, usually length of body.

DISTRIBUTION
Across much of eastern and northern Madagascar.

HABITAT
Found in lowland and montane humid forests. Occurs in some secondary forest or other human-modified habitats.

BEHAVIOR
Nocturnal and generally terrestrial. Towards the end of the rainy season this species accumulates body fat, particularly in the tail, and is known to aestivate.

FEEDING ECOLOGY AND DIET
Largely terrestrial, but occasionally scansorial. Well developed canines and molariform teeth indicate that it is a formidable predator. Prey located by visual, olfactory, and visual means, and then subdued with forelimbs or seized directly with the mouth. The diet consists mostly of invertebrates, however it is known to predate on small land vertebrates.

REPRODUCTIVE BIOLOGY
In captivity mating in September to October. Gestation 62 days. Litter size one to five and neonates naked with non-functional ears and eyes. Eyes open between 22 to 27 days old. One litter per year. Presumably polygynous.

CONSERVATION STATUS
A widely distributed species and known to occur in several protected areas. As with other forest-dwelling organisms on Madagascar, the continued existence of this species depends on the maintenance of forest habitat.

SIGNIFICANCE TO HUMANS
None known. ◆

Common name / Scientific name/ Other common names	Physical characteristics	Habitat and behavior	Distribution	Diet	Conservation status
Aquatic tenrec *Limnogale mergulus* French: Tenrec aquatique; German: Der Wassertenrek	Hind feet are webbed, tail is flat. Pelage is dense and soft, no spines. Upperparts and head are brownish with red and black guard hairs poking through. Head and body length 4.7–6.7 in (12–17 cm), tail length 12–6.3 (12–16 cm), weight 2.1–3.2 oz (60–90 g).	Freshwater rivers. Nocturnal, sleeps in burrows beside rivers.	Eastern Madagascar.	Mostly frogs, small fish, crustaceans, and aquatic insect larvae.	Endangered
Mole-like rice tenrec *Oryzorictes talpoides* German: Maulwurfartigen Reistenrek	Fur is velvety and gray brown or dark brown in color. Underparts are gray or buffy brown. Tail is bicolored. Head and body length 3.3–5.1 in (8.5–13.0 cm), tail length 1.2–2.0 in (3.0–5.0 cm).	Marshy areas, especially the moist banks of rice fields. Nocturnal. Spends most of its time in tunnel system.	Northwestern Madagascar.	Mainly insects and other invertebrates, such as mollusks.	Not threatened
Nimba otter shrew *Micropotamogale lamottei* German: Ottersptizmäuse	Coloration is brownish gray above and gray below. Unwebbed feet and round tail. Head and body length 4.7–7.9 in (12.0–20.0 cm), tail length 3.9–5.9 in (10.0–15.0 cm), weight 4.8 oz (135 g).	Upland forest streams. Nocturnal and semiaquatic.	Guinea, Mt. Nimba, Ziela.	Worms, insects and their larvae, crabs, fish, and small frogs.	Endangered
Ruwenzori otter shrew *Micropotamogale ruwenzorii*	Brownish gray above and gray below. Unwebbed feet and round tail. Head and body length 4.7–7.9 in (12.0–20.0 cm), tail length 3.9–5.9 in (10.0–15.0 cm), weight 4.8 oz (135 g).	Upland forest streams. Nocturnal and semiaquatic.	Ruwenzori region (Uganda, Democratic Republic of the Congo [DRC]), and west of Lake Edward and Lake Kivu (DRC).	Worms, insects and their larvae, crabs, fish, and small frogs.	Endangered
Giant otter shrew *Potamogale velox* French: Potamogale; German: Die Große Otterspitzmaus; Spanish: Potamogalo	Long, tapered body. Thick, tapered, rudder-like tail, flattened muzzle with numerous whiskers. Dense, soft fur, glossy chocolate in colorations. Underparts are white. Head and body length 11.5–14 in (29.2–35.6 cm), weight 2.2 lb (997 g).	Occurs in rivers, streams, and swamps in rainforests, as well as montane torrents. Two offspring in each litter.	Tropical Africa; from Nigeria to Angola and east to the Rift Valley.	Mainly frogs and mollusks.	Endangered
Lesser hedgehog tenrec *Echinops telfairi* French: Petit tanrec hérisson; German: Kleiner Igeltanrek	Dorsal side is covered in spines. Spines vary from pale whitish to black. Paws and ventral sides covered with fine hairs.	Arid regions of Madagascar, such as dry forests, scrub, cultivated areas, dry coastal regions, and semidesert. Nocturnal and communication is tactile.	Southern Madagascar.	Invertebrates, but occasionally baby mice.	Not threatened
Yellow streaked tenrec *Hemicentetes semispinosus* French: Tenrec zébré des terres-basses	Pelage is spiny, sharply pointed, black in color. Distinct whitish markings. Hair on underside is very spiny and chestnut brown in color. Head and body length 6.3–7.5 in (16.0–19.0 cm).	Live within burrow systems in rainforests and brushland habitats. Social groups are variable.	Madagascar.	Mainly earthworms.	Not threatened
Greater hedgehog tenrec *Setifer setosus* French: Grand hérisson; German: Großer Igeltanrek	Upperparts covered with short, stiff spines. Spines begin at forehead and extent to hindquarters. Head and body length 5.9–8.7 in (15.0–22.0 cm), tail length 0.59–0.63 in (1.5–1.6 cm).	Dry forest and agricultural areas. Diurnal, solitary, active throughout year.	Madagascar, central plateau.	Worms, insects, ground meat, and the carcasses of dead mice.	Not threatened

Resources

Books

Eisenberg, J. F., and E. Gould. "The Insectivores." In *Madagascar*. Edited by A. Jolly, P. Oberlé, and R. Albignac. Oxford: Pergamon, 1984.

Garbutt, N. *Mammals of Madagascar*. New Haven: Yale University Press, 1999.

Goodman, S. M., and J. P. Benstead, eds. *The Natural History of Madagascar*. Chicago: The University of Chicago Press, 2003.

Genest, H., and F. Petter. Part 1.1. "Family Tenrecidae." *The Mammals of Africa: An Identification Manual*. Edited by J.

Meester and H. W. Setzer. Washington, DC: Smithsonian Institution Press, 1975.

Heim de Balsac, H. "Insectivores." In *Biogeography and Ecology in Madagascar*, edited by R. Battisitini and G. Richard-Vindard. The Hague: W. Junk, 1972.

Nicoll, M. E., and G. B. Rathbun. "African Insectivora and Elephant-shrews: An Action Plan for their Conservation." *IUCN/SSC Insectivore, Tree-Shrew and Elephant-Shrew Specialist Group*. Gland, Switzerland: IUCN, 1990.

Racey, P. A., and P. J. Stephenson. "Reproductive and Energetic Differentiation of the Tenrecidae of

Resources

Madagascar." *Biogéographie de Madagascar*, edited by W. R. Lourenço. Paris: ORSTOM, 1996.

Periodicals

Benstead, J. P., K. H. Barnes, and C. M. Pringle. "Diet, activity patterns, foraging movement and responses to deforestation of the aquatic tenrec *Limnogale mergulus* (Lipotyphla: Tenrecidae) in eastern Madagascar." *Journal of Zoology* 254 (2001): 119–129.

Butler, P. M., and A. T. Hopwood. "Insectivora and Chiroptera from the Miocene rocks of Kenya Colony." *Fossil Mammals of Africa* 13 (1957): 1–35.

Eisenberg, J. F., and E. Gould. "The tenrecs: A study in mammalian behavior and evolution." *Smithsonian Contributions to Zoology* 27 (1970): 1–137.

Ganzhorn, J. U., A. W. Ganzhorn, J. P. Abraham, Andriamanarivo, and A. Ramananjatovo. "The impact of selective logging on forest structure and tenrec populations in western Madagascar." *Oecologia* 84 (1990): 126–33.

Goodman, S. M., and D. Rakotondravony. "The effects of forest fragmentation and isolation on insectivorous small mammals (Lipotyphla) on the Central High Plateau of Madagascar." *Journal of Zoology* 250 (2000): 193–200.

Goodman, S. M., et al. "Inventaire biologique de la forêt de Tsinjoarivo, Ambatolampy." *Akon'ny Ala* 27 (2000): 18–35.

Gould, E. "Evidence for echolocation in the Tenrecidae of Madagascar." *Proceedings of the American Philosophical Society* 109 (1965): 352–60.

Gould, E., and J. F. Eisenberg. "Notes on the biology of the Tenrecidae." *Journal of Mammalogy* 47 (1966): 660–86.

Jacobs, L. J., W. Anyonge, and J. C. Barry. "A giant tenrecid from the Miocene of Kenya." *Journal of Mammalogy* 68 (1987): 10–16.

MacPhee, R. D. E. "The shrew tenrecs of Madagascar: Systematic revision and Holocene distribution of *Microgale* (Tenrecidae, Insectivora)." *American Museum Novitates* 2889 (1987): 1–45.

Nicoll, M. E. "Mechanisms and consequences of large litter production in *Tenrec ecaudatus* (Insectivora: Tenrecidae)." *Annales de la Musée Royale de l'Afrique Centrale, Sciences Zoologiques* 237 (1983): 219–26.

Nicoll, M. E. "Responses to Seychelles tropical forest seasons by a litter-foraging mammalian insectivore, *Tenrec ecaudatus*, native to Madagascar." *Journal of Animal Ecology* 54 (1985): 71–88.

Olson, L. "Phylogeny of the Tenrecidae (Mammalia, Lipotyphla): morphological support for a single invasion of Madagascar." *Zoological Journal of the Linnean Society* (2003) in press.

Petter, J. J., and A. Petter-Rousseaux. "Notes biologiques sur les Centetinae." *La Terre et la Vie* 1963 (1963): 66–80.

Poduschka, W., and C. Poduschka. "Zur Frage des Gattungsnamens von "*Geogale aletris*" Butler und Hopwood, 1957 (Mammalia: Insectivora) aus dem Miozän Ostafrikas." *Zeitschrift für Säugetierkunde* 50 (1985): 129–40.

Stephenson, P. J. "Reproductive biology of the large-eared tenrec, *Geogale aurita* (Insectivora: Tenrecidae)." *Mammalia* 57 (1993): 553–63.

Stephenson, P. J. "Notes on the biology of the fossorial tenrec, *Oryzorictes hova* (Insectivora: Tenrecidae)." *Mammalia* 58 (1994): 312–15.

Stephenson, P. J. "Taxonomy of shrew-tenrecs (*Microgale* spp.) from eastern and central Madagascar." *Journal of Zoology* 235 (1995): 339–50.

Steven Goodman, PhD

Solenodons

(Solenodontidae)

Class Mammalia

Order Insectivora

Family Solenodontidae

Thumbnail description
A rabbit-sized insectivore with shrew-like features

Size
Up to 2.4 lb (1,100 g)

Number of genera, species
1 genus; 2 extant species

Habitat
Tropical moist montane forests

Conservation status
Endangered: 2 species

Distribution
Caribbean/Greater Antilles islands of Hispaniola and Cuba

Evolution and systematics

The relationships among the numerous insectivore species is far from settled. Evolutionary biologists have not settled the question of whether the families listed within the order Insectivora are monophyletic (sharing a common ancestor species) or polyphyletic (deriving from several separate origins). Presently, the solenodontids appear to be most closely related to several species of extinct shrews, the Nesophontidae, native to various islands of the Antilles, while both families share similarities among the family Tenrecidae (the tenrecs of Madagascar and mainland Africa) and the subfamily Potamogalinae (otter shrews of mainland Africa).

The Solenodontidae show an array of primitive mammalian features, but whether they represent a relict family from the mainlands, little changed, or are the derived descendants of smaller colonizing insectivores that evolved to "giant" size on the islands, is not clear. These ancestors most likely rafted on vegetation from the mainlands of the Americas to Cuba and Hispaniola as long ago as the Mesozoic and early Cenozoic eras. The Solenodontidae survived on the islands without competition from more advanced mammal types.

Solonodontidae contains the living genus *Solenodon*; two living species, the Hispaniolan or Haitian solenodon, *S. paradoxus*, in Hispaniola, and the Cuban solenodon or "almiqui," *S. cubanus*, in Cuba. Also included in this family are two recently extinct (late Pleistocene to present) species, Marcano's solenodon, *S. marcanoi*, in Hispaniola, and Arredondo's solenodon, *S. arredondoi*, in Cuba.

Physical characteristics

At first glance, a typical solenodon resembles a rabbit-sized shrew, with a shrew's characteristic long, conical, whisker-studded snout, even more pronounced in the solenodon. Cuban solenodons are slightly smaller, on average, than Hispaniolan solenodons. An adult Hispaniolan solenodon can weigh up to 2.4 lb (1,100 g), its combined head and body length can reach 15 in (39 cm) and the tail length, 8 in (21 cm). An adult Cuban solenodon's maximum weight is 1.7 lb (800 g), and its maximum head and body length is 14 in (36 cm). The extinct *Solenodon arredondoi* was much larger. Its weight, based on fossil remains, has been roughly estimated at 3.3–4.4 lb (1,500–2,000 g), and its head and body length at

The Hispaniolan solenodon (*Solenodon paradoxus*) is considered Endangered. (Photo by N. Smythe/Photo Researchers, Inc. Reproduced by permission.)

18–22 in (45–55 cm). There is no sexual dimorphism in the living species of solenodon (i.e., no differences between the sexes in size, shape, or coloration).

The solenodons show a puzzling mix of primitive and derived traits. Among the primitive characters are a poisonous bite and the ability to echolocate. Solenodon sight is poor, but hearing, olfactory, and tactile senses are acute. Derived

The Cuban solenodon (*Solenodon cubanus*) was thought to be extinct until the mid 1970s. The Burmese mongoose and feral cats are believed to be the cause of the reduction of the Cuban solenodon. (Photo by J. A. Hancock/Photo Researchers, Inc. Reproduced by permission.)

traits include longevity, low birth frequency, low number of young per litter, and the os proboscis bone in the Hispaniolan solenodon.

The eyes are tiny and shrew-like. The large ears partially protrude from the fur. The legs are relatively long, and well-muscled. Each of all four paws carries five digits with large, strong claws.

The fur is dense, coarser in the Hispaniolan solenodon, finer and softer in the Cuban solenodon. Fur color in *S. paradoxus* ranges from brown through reddish brown to yellowish brown on the upper, dorsal body, with a lighter underside. The fur of *S. cubanus* varies from dark brown to black, with a white or yellowish face, snout, and shoulders. The forehead in both species is sparsely covered with hair, while the ears, legs, snout, and scaly, rat-like tail are nearly hairless. The snout sprouts long vibrissae, or sensory whiskers, along its length.

Individuals of both species carry scent glands in the armpits and groin. In females, the two mammae are located in the inguinal (groin) region. In males, the penis, except during mating, and the testes are carried within the abdominal cavity.

In all individuals of both species, the lower second incisors are grooved to channel poison. The name solenodon, meaning "slotted tooth," is based on this feature. The grooves connect with a duct that supplies a poisonous neurotoxin from glands below the incisors. Some species of shrews carry similar modified teeth and poison.

Distribution

Caribbean/Greater Antilles islands of Hispaniola and Cuba.

Habitat

Tropical montane forest; also brush country and near plantations.

Behavior

Both species are nocturnal, hiding and sleeping during the day in hollows or burrows. Adults are solitary in their foraging, although several individuals may shelter together.

Solenodons are easily startled, especially by sharp or high-pitched sounds, and are easily provoked into scrappy rages against other solenodons or other animal species. Given its jaw full of sharp teeth, modeled according to the shrew pattern with some modification, an enraged solenodon can deliver a severe bite.

Solenodons have long life spans and low reproductive rates, unusual in so small an animal. One captive Hispaniolan solenodon lived for eleven years, while a captive Cuban solenodon lived for five years. Insectivores rarely live longer than a couple of years.

Individuals walk in a winding course, holding the tail level. When alarmed, a solenodon can run in a straight line, with some speed, but will more likely run off in an erratic pattern. Solenodons can climb, but seldom do so. When sitting up or scratching itself, a solenodon uses its tail as a prop.

Solenodons produce a variety of vocalizations, including wheezing, snuffling, grunting, squeaking, and shrieking.

Feeding ecology and diet

Solenodons are generalist feeders with some emphasis on insects and spiders. When foraging, a soledonon takes a winding course using its flexible snout to root in dirt and poke into leaf litter and cracks. A soledonon uses smell and tactile senses to find earthworms, arthropods, and snails, digging them out of the soil or snagging them in leaf litter or tight spaces in rock and wood. It also uses its powerful forelimbs and claws to tear into soft, rotten wood in search of resident edibles.

Reproductive biology

A female gives birth to a litter of one to three young, and can birth two litters within a year. The mother bears only two teats, situated ventrally near the groin. The young are weaned after 75 days, although they may remain with the mother while subsequent litters are born and raised.

Apparently, males are always able to breed, since they will attempt copulating with females at any time, while females go in and out of estrous according to no strict pattern. Courting behavior is similar to insectivores in general, with scent-marking and mutual sniffing. May be polygamous or promiscuous.

Conservation status

Both solenodon species are listed as Endangered by the IUCN. Both species were written off as extinct in the early twentieth century, but were subsequently rediscovered. Ironically, solenodons are among a mere handful of surviving mammal species among a large and varied array of recently

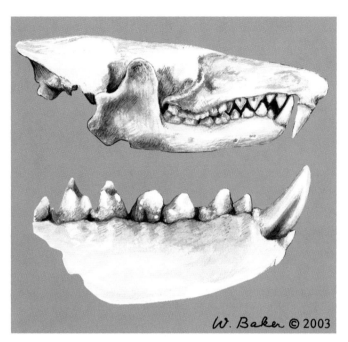

Side view of a solenodon skull (top). Side view of the inner lower jaw showing the groove along which venomous saliva flows. (Illustration by Wendy Baker)

extinct native mammals in the Antilles, including rodents, New World monkeys, sloths, and other insectivores.

Solenodons are threatened by deforestation, predation by introduced animals, and killing by humans who consider them pests.

Significance to humans

Rural people in Hispaniola frequently kill solenodons, blaming then for eating crops, although most likely the solenodons do not eat crops but damage them while grubbing for insects. In Cuba, the remaining solenodons are too remote from settled humanity to pose any sort of threat. Solenodons are not hunted for food in Hispaniola or Cuba.

Species accounts

Hispaniolan solenodon

Solenodon paradoxus

TAXONOMY

Solenodon paradoxus Brandt, 1833, Dominican Republic.

OTHER COMMON NAMES

English: Haitian solenodon; French: Solénodonte d'Haiti; German: Dominikanischer Schlitzrüssler, Haiti-Schlitzrüssler; Haitian French: Nez longue; Spanish: Solenodonte haitiano.

PHYSICAL CHARACTERISTICS

Individuals resemble very large shrews. An adult weighs up to 2.4 lb (1,100 g). Its combined head and body length ranges 11–12 in (28–32.5 cm), and the tail length is 7–10 in (17.5–25.5 cm).

Solenodon paradoxus

There are five fingers and five toes, for, respectively, the forepaws and hind-paws, and all digits are equipped with large, strong claws, although the forelimbs are more powerfully muscled than the hindlimbs, and the forepaws are bigger. The underfur is short and dense, overlain by a coarser secondary coat. The forehead is sparsely covered with hair, while the ears, legs, snout, and scaly, rat-like tail are nearly hairless. The snout sprouts long vibrissae, or sensory whiskers, along its length.

Solenodon paradoxus

Fur color ranges from brown through reddish brown to yellowish brown on the upper, dorsal body, with a lighter shade on the underside. A light-colored, roughly rectangular spot in the fur of the nape changes slightly in size and shape throughout the animal's life.

A signature characteristic of *Solenodon paradoxus* is the *os proboscis*, a small, round bone, located at the tip of the rostrum (bones of the nasal region), and which holds the proximal end of the cartilaginous support for the long snout. The os proboscis articulates with the rostrum in a ball-and-socket joint, a most unusual feature among the vertebrates. The Cuban solenodon lacks this feature.

DISTRIBUTION

The Caribbean island of Hispaniola; in the Dominican Republic, in several forested areas; in Haiti, only in the extreme southwest.

HABITAT

Montane tropical rainforest. The animals may sometimes settle in scrub, or near plantations.

BEHAVIOR

The Hispaniolan solenodon forages nocturnally, hiding and sleeping during the day in spaces between rocks, in hollow trees, or in networks of burrows that the animals excavate. Usually, several individuals rest together in a hollow or burrow. Captive Hispaniolan solenodons will sleep piled in heaps within their shelters.

Wild male adults, and female adults without young, are solitary when foraging, even if they share burrow space.

Field researchers have often found collections of snail shells in the burrows of wild Hispaniolan solenodons. Captive solenodons have been observed collecting similarly small, hard objects, such as coarse chunks of peat moss and beechnut husks, and dragging them into their sleeping enclosures.

Captive solenodons show nesting activity, lining their sleeping shelters with hay, dried leaves, peat moss, and similar materials, pushing it into little heaps with their forepaws, then dragging it, backwards, with forepaws or jaws, into the nest enclosure. A pregnant female will prepare a litter nest.

Hispaniolan solenodons, in the wild or in captivity, produce a variety of vocalizations: wheezing and snuffling like hedgehogs, grunting like pigs, squeaking like guinea pigs, twittering like mice, and whimpering like young kittens. An excited or alarmed solenodon will sound off with penetrating shrieks. Mothers and their young, or mates, make bird-like contact calls, while courting individuals make repeated "piff" noises.

Hispaniolan solenodons appear to echolocate. An individual will produce a series of high-pitched clicks when checking out a new area or encountering a strange animal. Several species of shrew (Soricidae) and tenrecs use echolocation.

Sudden high, shrill, or sharp noises will panic the animals and send them fleeing. If threatened by a rival solenodon or another species, the solenodon will first take an upright, defensive posture and warn its opponent with a loud "chirp." If fleeing the opponent is the chosen option, a Hispaniolan solenodon first runs in an erratic pattern, then crouches motionless and hides its head. This behavior may have given it a survival edge against predators, particularly birds of prey, before human settlement of Cuba and Hispaniola, but those traits work against it when dealing with introduced predators like dogs, cats, and mongooses.

FEEDING ECOLOGY AND DIET

Hispaniolan solenodons are omnivorous, generalist feeders with some emphasis on insects and spiders, while varying their fare with worms, snails, small reptiles, roots, fruits, and leaves.

A Hispaniolan soledonon forages along an erratic, winding course, probably in accordance with a food-search pattern, while using its flexible snout to root in dirt and poke into leaf litter, using smell and tactile senses to find earthworms, arthropods, or land snails, digging them out of the soil or snagging them in leaf litter. A solenodon will carefully clean the dirt from a snagged earthworm with its front paws before eating it. The animal also uses its powerful forelimbs and claws to tear into soft, rotten wood in search of resident edibles.

Hispaniolan solenodons drink water by lapping it up with their tongues, while holding the snout bent upward, then swallow it while throwing their heads back.

REPRODUCTIVE BIOLOGY

Mating behavior is common to insectivores, presumably polygamous or promiscuous. In captivity, a male introduced into a female's cage explores, and scent-marks by rubbing his anogenital area over objects and spots in the cage, often in areas already scent-marked by the female. As the two move closer to each other, they repeatedly make reassuring "piff" noises. Physical contact begins with nose-to-nose touching, followed by the two animals sniffing different parts of each other's body. As the male mounts, he secures his and the female's position by a neck grip on the female.

Both sexes of the Hispaniolan solenodon exude a greenish, oily fluid from scent glands in the armpits and groin. However, the amount of a female's excretions may indicate the level of arousal, while the male that can mate at any time excretes a similar fluid at a constant rate.

A female gives birth to a litter of one, two, or even three young in a nesting burrow, and she can birth two litters within a year. There is no set mating season. A mature Hispaniolan solenodon female's estrous lasts about a day, with intervals of nine to thirteen days between estrous periods. The gestation period is unknown. Individual young weigh 1.4–1.9 oz (40-55 g)

at birth, and are blind and nearly hairless, although they've grown a complete coat of fur by two weeks after birth.

The female has two teats, situated near the groin. Out of a litter of three, one will probably die, since the mother can nurse only two at a time. When seven weeks old, the nursing youngsters accompany the mother outside in her foraging by hanging fast to her teats with their jaws, being dragged along wherever she goes. This behavior, known as "teat transport," is well known among rodent species, but among the insectivores, known only in solenodons. The young are weaned after 75 days, a long time for insectivores, while the young may sometimes remain with the mother while she births and raises subsequent litters. Males do not share in parenting.

CONSERVATION STATUS

The Hispaniolan solenodon is listed as Endangered by the IUCN. The species is beset by deforestation, killing by humans who consider them pests, and by introduced house cats, dogs, and mongooses, against which the animals have little defense.

Hispaniolan solenodons sometimes nest and burrow near plantations, where the soil, made softer and more pliable by farming, makes for easier burrowing, while the plantations themselves are abundant with insects. Farmers kill solenodons, accusing them of attacking crops, when most likely the animals were foraging for insects in the fields, while the proximity to settled areas renders solenodons more vulnerable to harm from humans and introduced animals.

By 1907, the Hispaniolan solenodon had been written off as extinct, but several isolated populations have been rediscovered in Hispaniola. Of the two Hispaniolan nations, Haiti and the Dominican Republic, a 1996 survey found *S. paradoxus* in both countries, although its survival isn't likely in Haiti, which has been all but completely deforested.

SIGNIFICANCE TO HUMANS

Negatively, solenodons are considered only minor nuisances when they disturb agricultural land. There are not many of them, and they are not hunted for food. They are of no direct positive significance to humanity, but more abstractly, they are conservation symbols and living studies in the evolution of primitive placental mammals and in adaptive evolution on islands. ◆

Resources

Books

Eisenberg, John F. "Tenrecs and Solenodons in Captivity." In *International Zoo Yearbook 15.* London: Zoological Society of London, 1975.

Ottenwalder, Jose. "Systematics and Biogeography of the West Indian Genus Solenodon." In *Biogeography of the West Indies: Past, Present, and Future,* edited by Charles Woods. Gainesville, FL: Sandhill Crane Press, 1989.

Woods, C. A., and F. Sergile, eds. *Biogeography of the West Indies: Patterns and Perspectives,* 2nd ed. New York: CRC Press, 2001.

Periodicals

Eisenberg, John F., and Edwin Gould. "The Behavior of *Solenodon paradoxus* in Captivity with Comments on the Behavior of other Insectivora." *Zoologica* 51 (1966): 49–57.

MacFadden, B.J. "Rafting Mammals or Drifting Islands?: Biogeography of the Greater Antillean Insectivores, *Nesophontes* and *Solenodon.*" *Journal of Biogeography* 7 (1980): 11–22.

McDowell, S. B., Jr. "The Greater Antillean Insectivores." *Bulletin of the American Museum of Natural History* 115 (1958): 113–214.

Morgan, Gary S., and Jose A. Ottenwalder. "A New Extinct Species of *Solenodon* (Mammalia, Insectivora, Solenodontidae) from the Late Quaternary of Cuba." *Annals of the Carnegie Museum* 62 (1993): 151–16.

Patterson B. D. "An Extinct Solenodontid Insectivore from Hispaniola." *Breviora* 165 (1962): 1–11.

Woods, Charles A. "The Last Endemic Mammals in Hispaniola." *Oryx* 16 (1981): 146–152.

Resources

Organizations

IUCN—The World Conservation Union. Rue Mauverney 28, Gland, 1196 Switzerland. E-mail: mail@hq.iucn.org Web site: <http://www.iucn.org>

The Nature Conservancy of the Dominican Republic. 4245 North Fairfax Drive, Arlington, VA 22203-1606 USA. Phone: (703) 841-4878 or (800) 628-6860. E-mail: comment@tnc.org Web site: <http://nature.org/wherewework/caribbean/dominicanrepublic/>

Other

Animal Diversity Web. [17 March, 2003]. <http://animaldiversity.ummz.umich.edu>.

The Shrew Site. [17 March 2003]. <http://members.vienna.at/shrew>

2002 IUCN Red List of Threatened Species. [17 March 2003]. <http://www.redlist.org>.

Walker's Mammals of the World (Online). Nowak, R. M. "Solenodons (Solenodontidae)." 1999 [17 March 2003]. <http://www.press.jhu.edu/books/walkers_mammals_of_the_world/insectivora.solenodontidae.solenodon.html>.

World Wildlife Fund Global 200 Ecoregions. "Greater Antillean Moist Forests." [17 March 2003]. <http://www.panda.org/resources/programmes/global200/pages/regions/region037.htm>.

World Wildlife Fund Global 200 Ecoregions. "Hispaniolan moist forests." [17 March 2003]. <http://www.worldwildlife.org/wildworld/profiles/terrestrial_nt.html>.

Kevin F. Fitzgerald, BS

Extinct West Indian shrews
(Nesophontidae)

Class Mammalia

Order Insectivora

Family Nesophontidae

Thumbnail description
Extinct shrew-like insectivores, known only from subfossil remains

Size
Species ranged from mouse-sized to chipmunk-sized

Number of genera, species
1 genus, at least 8 species

Habitat
Varied

Conservation status
All Nesophontidae species are considered Extinct

Distribution
Several islands of the Greater and Lesser Antilles

Evolution and systematics

The Nesophontid shrews comprise one family, Nesophontidae, one genus, *Nesophontes*, and eight species, namely: Puerto Rican nesophontes (*Nesophontes edithae*); slender Cuban nesophontes (*Nesophontes longirostris*); greater Cuban nesophontes (*Nesophontes major*); lesser Cuban nesophontes (*Nesophontes submicrus*); western Cuban nesophontes (*Nesophontes micrus*); Atalaye nesophontes (*Nesophontes hypomicrus*); Haitian nesophontes (*Nesophontes zamicrus*); and St. Michel nesophontes (*Nesophontes paramicrus*).

Present scientific knowledge places family Nesophontidae closest to the solenodons (family Solenodontidae), of which two species survive in Cuba and Hispaniola. Nesophontidae are also considered to be closely related to the more generalized shrew species of the family Soricidae, which they most physically resemble.

The scientific jury is still out on the exact origins of the Nesophontidae. One or more founder species may have rafted on vegetation to the Antilles from the mainlands of North or Central America, or they may have been carried along on dry land as plate tectonic movements separated the proto-Antilles

Islands from Central America (vicariance). In either scenario, there may have been secondary colonization after the Antilles became an isolated archipelago; Nesophontidae species already on the Antilles may have rafted among some of these islands, colonized them, and produced new species.

Physical characteristics

Judging from skeletal remains, Nesophontidae closely resembled the more generalized species of the soricid shrews in appearance and morphology, with some minor variation due to isolation and adaptive evolution. Structurally, they departed little from the standard shrew body plan, except for size variations. All had typically long, narrow skulls, long and mobile snouts, perhaps more moveable than those of soricid shrews, small eyes, and a tail about as long as the body. Each of the four feet carried five fingers and five toes, each toe bearing a claw.

The exact body dimensions and weights of the Nesophontidae species can only be estimated, since their remains are merely skulls and isolated postcranial (non-skull) bones. Judging from these remains, *Nesophontes edithae*, the Puerto

Rican nesophontes, was the largest species, with a skull length of up to 1.7 in (44 mm), a femur length of 1.0–1.1 in (27–28 mm), a head-body length of 6.3–7.5 in (160–190 mm) and an estimated living weight of 6.4–7.1 oz (180–200 g). It was about the size of a chipmunk or laboratory rat, although it probably had a more lithe build. *N. zamicrus*, the Haitian nesophontes, was the smallest species.

Hispaniola was home to three native species of Nesophontidae, large, medium-sized, and small. The largest species was about two-thirds the size of *N. edithae*, while the smallest was about the size of a large soricid shrew. There was a similar, size-ranked array of three nesophontid species on Cuba, the sizes about the same as those of the Hispaniolan species.

Evidence of sexual dimorphism was observed in skeletal remains of *Nesophontes edithae*, male skulls being larger than female, but size disparities could also be due to age differences. Nevertheless, the case for sexual dimorphism among the nesophontids has recently been reopened.

Distribution

The Caribbean/Antillean islands of Cuba, the Isle of Pines, Hispaniola, Ile de Gonave, Puerto Rico, plus the Grand Cayman Islands, home of two as yet unnamed species.

Habitat

Lowland and montane tropical forests.

Behavior

Behavior can only be surmised from skeletal remains, but is likely to have been similar to that of soricid shrews (Soricidae).

Feeding ecology and diet

Food habits and feeding habits are unknown, but judging by the teeth and morphology, the nesophontids were likely omnivorous, with some emphasis on animal food, mostly invertebrates, especially insects, as with soricid shrews. None of the Nesophontidae species developed any specialized or derived dentition as adaptations to new diets.

Reproductive biology

Nothing is known.

Conservation status

All known species of Nesophontidae are considered extinct. Most or all species of Nesophontidae were probably contemporaneous with man, including Native Americans and

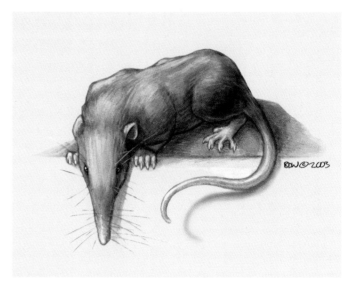

Nesophontes sp. (Illustration by Bruce Worden)

the early European colonizers of the Antilles, the latter from A.D. 1500 onward.

There are no documented records of living nesophontid species. All that zoologists know about the genus and species is from skulls and scatted postcranial bones found in owl pellets. Some such pellets are found alongside other owl pellets containing bones of Old World rats and mice, indicating that some of the species were still alive when Europeans began colonizing the islands. Predation by and competition with Old World rats, along with extensive deforestation, probably brought about the extinction of the Nesophontidae.

There is a small possibility that a few nesophontid species may still survive. Fresh remains, in barn owl pellets, were found as recently as the 1930s in Haiti. Some of the bones retained bits of dried, yet fresh-looking, tissue. It was accordingly suggested that some species of *Nesophontes* might yet be extant. Nevertheless, more recent searches for living nesophontids by researchers in Hispaniola and Puerto Rico found no convincing evidence of survivors. More recently, researchers recovered some *Nesophontes* material from Cueva (cave) Jurg, Parque Nacional Sierra de Baoruco, Dominican Republic, that included a few hairs and tiny bits of dried tissue, but radiocarbon dating of the bone collagen estimates their age at about 700 years old.

Significance to humans

The Nesophontidae probably had little if any direct significance to either Native Americans or European colonists. There is no compelling evidence that Native Americans (or Europeans) used any of the nesophontid species for food. Like shrews in general, Nesophontidae lived secretive existences.

Resources

Books

Morgan, G. S. "Late Quaternary Fossil Vertebrates from the Cayman Islands." In *The Cayman Islands: Natural History and Biogeography*, edited by M. A. Brunt and J. E. Davies. New York: Kluwer, 1994.

Nowak, R. M., ed. "Family Nesophontidae (Extinct West Indian Shrews)." In *Walkers Mammals of the World*, 6th ed. Baltimore: Johns Hopkins University Press, 1999.

Whidden, H. P., and R. J. Asher. "The Origin of the Greater Antillean Insectivores." In *Biogeography of the West Indies: New Patterns and Perspectives*, edited by C. A. Woods and F. E. Sergile. Boca Raton, FL: CRC Press, 2001.

Periodicals

Alcover, J. A., A. A. Sans, and M. Palmer. "The extent of extinctions of mammals on islands." *Journal of Biogeography* 25 (1998): 913–918.

Anthony, H. E. "Mammals of Puerto Rico, living and extinct—Chiroptera and Insectivora." *Scientific Survey of Puerto Rico and the Virgin Islands, New York Academy of Sciences* 9, no. 1 (1925).

Choate, J. R., and E. C. Birney. "Subrecent Insectivora and Chiroptera from Puerto Rico." *Journal of Mammalogy* 49, no. 3 (1968): 400–412.

MacPhee, Ross, et al. "Radiometric dating and 'last occurrence' of the Antillean insectivore *Nesophontes*." *American Museum Novitates* 3264 (1999): 1–19.

McFadden, Bruce J. "Rafting mammals or drifting islands? Biogeography of the Greater Antillean insectivores *Nesophontes* and *Solenodon*." *Journal of Biogeography* 7 (1980): 11–22.

McFarlane, Donald A. "A note on dimorphism in *Nesophontes edithae* (Mammalia: Insectivora), an extinct island-shrew from Puerto Rico." *Caribbean Journal of Science* 35 (1999): 142–143.

Miller, G. S. "Three small collections of mammals from Hispaniola." *Smithsonian Miscellaneous Collections* 82, no. 15 (1930): 1–10.

Morgan, G. S., and C. A. Woods. "Extinction and zoogeography of West Indian land mammals." *Biological Journal of the Linnaean Society* 28 (1986): 167–203.

Whidden, H. P., and C. A. Woods. "Assessment of sexual dimorphism in the Antillean insectivoran *Nesophontes*." *American Zoologist* 40, 6 (2000): 1257.

Woods, Charles A., Jose A. Ottenwalder, and W. R. Oliver. "Lost mammals of the Greater Antilles." *Dodo: Journal of the Jersey Wildlife Preservation Trust* 22 (1985): 23–42.

Other

"Extinct Mammals of the West Indies." <http://www.jsd .claremont.edu/bio/extinct/extinctmammals>

"Family Nesophontidae." *Mammal Species of the World*. National Museum of Natural History. <http://www.nmnh.si.edu/msw>

Kevin F. Fitzgerald, BS

Shrews I

Red-toothed shrews (Soricinae)

Class Mammalia
Order Insectivora
Family Soricidae
Subfamily Soricinae

Thumbnail description
Small mouse-like mammals with long pointed
snouts, short legs, and small eyes

Size
1.5–5.3 in (4.0–13.5 cm); 0.06–1.5 oz (2–40 g)

Number of genera, species
11 genera; 122 species

Habitat
Cold and wet environments, forests, banks of
rivers and streams, exceptionally also warm and
arid habitats

Conservation status
Critically Endangered: 5 species; Endangered: 8
species; Vulnerable: 7 species; Lower Risk/Near
Threatened: 3 species; Data Deficient: 1 species

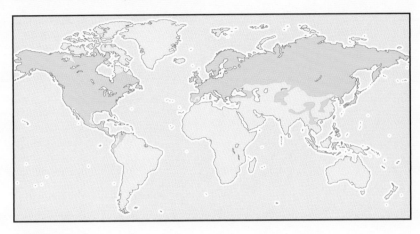

Distribution
Temperate zone in the Northern Hemisphere, Indo-Malaysia, Central America, and
the northernmost parts of South America

Evolution and systematics

Shrews (Soricidae) are small mammals with rather unspecialized body plans, retained almost unchanged since they evolved about 45 million years ago (mya). The family is usually divided into a number of extinct and extant subfamilies. The two extant subfamilies are the Crocidurinae and the Soricinae. The members of the subfamily Soricinae, red-toothed shrews, are further subdivided into seven tribes, one of them extinct (Anourosoricini, Beremendiini, Blarinellini, Blarinini, Neomyini, Notiosoricini, and Soricini). The delimitation of some tribes of the subfamily Soricinae is still open to discussion. The genetic distances among different genera of soricines based on allozyme variation generally support a subdivision of the living species into the tribes. Species of soricine shrews may show considerable chromosomal and genetic differentiation. Karyotypic and protein electrophoresis studies have been very helpful in identifying many sibling and cryptic species that were difficult to recognize by conventional methods.

According to the number of the tribes and genera included, the Soricinae constitute the most diverse subfamily among shrews. Their fossil record comes from Eurasia, Africa, and North America. The fossil remains are still insufficiently studied, and many gaps exist in our knowledge. The oldest soricines have been reported from the early Miocene of North America about 24 mya (*Antesorex*), and the Holarctic region in the Northern Hemisphere is considered the area of origin of this lineage. The species richness in the fossil record remained low in the Tertiary and during the Early and Middle Pleistocene. In the Late Pleistocene a plethora of forms appeared in Europe, Asia, and North America. The large number of living taxa unrepresented by late Pleistocene records suggests that we have only a minute sample of the total number of species that have occurred in the past. From North America, the soricine shrews spread into the extreme northern parts of South America, and from Europe and Asia to tropical regions of Indo-Malaysia. Soricines are almost completely absent from Africa, except for one extinct species, *Asoriculus maghrebensis*.

Shrews have played an important role in reconstructing faunal, floral, and climatic changes within various continents. Occurrence of specifically identifiable fossil remains allows mapping of environmental changes through time. This is particularly striking in regard to Pleistocene faunas, where massive rearrangements of the geographical distribution of shrews can be followed and allow reconstruction of vegetation and climatic changes.

Physical characteristics

Soricine shrews are small mouse-sized insectivores with a pointed snout, short legs, and usually a long tail. The smallest soricines (*Sorex minutissimus*, *Sorex hoyi*) have a body mass of 0.07–0.1 oz (2–3 g), and are among the smallest known mammals. The size of the largest soricines slightly exceeds that of the house mouse. The largest representatives of the subfamily, water shrews of the genus *Chimarrogale*, weigh up to 1.5 oz (40 g), their head and body length ranges up to 5.3 in (13.5 cm), while tail length can be up to 5 in (12.5 cm). They walk on the soles of their feet, which have five clawed

A desert shrew (*Notiosorex crawfordi*) eating a centipede. (Photo by Jack Couffer. Bruce Coleman, Inc. Reproduced by permission.)

digits. The eyes are extremely small and their sight is poor, but the sense of smell is keen as indicated by long, mobile snouts. The external ears are usually hidden in their fur. Pelage is usually dark, with brown, grayish, or black coat color. The skull is narrow and elongated, and the brain is small and smooth, dominated by large olfactory bulbs.

Certain skeletal and dental characters are diagnostic for the subfamily: the articulation of the mandibular condyle, the position of the mental foramen, and the morphology of the lower premolar. The dentition is highly specialized and similar from species to species. They have continuous rows of teeth classified as incisors, antemolars, premolars, and molars. Homologies of the antemolars are difficult to determine and thus, the dental formula is expressed using different terms than in other mammals. The number of antemolars is the only difference seen between living species in the dental formulae. Part of dental variation in soricines is clearly correlated with ecological adaptations.

In most of the Soricinae, a reddish, iron-containing pigmentation on cusps of their teeth is present. The function of tooth pigmentation is not yet clearly understood, but it is supposed that pigmented enamel should be harder than unpigmented and should provide a protection against abrasion. The intensity of pigmentation varies. Some species (*Blarina*, *Blarinella*, *Sorex daphaenodon*) have a very strong dark pigmentation, other species (*Anourosorex*, *Chimarrogale*, *Nectogale*) appear to have reduced red enamel or even unpigmented teeth. The absence of tooth pigment may also be explained by ecological, particularly dietary, factors.

Distribution

The subfamily Soricinae seems to be of Holarctic origin and likely evolved in a temperate climate. This is reflected in the pattern of present day distribution. Soricines inhabit most of the temperate and arctic parts of the Northern Hemisphere in Europe, Asia, and North America. In the New World, they occur also in Mexico and Central America and a single genus (*Cryptotis*) inhabits the mountain ranges of the Andes in northern South America. In the Old World the only soricines occurring in the tropics are members of several genera endemic to Southeast Asia (*Anourosorex*, *Chimarrogale*, and *Soriculus*).

Habitat

Soricines usually inhabit cool and moist areas in forests of various types. They are mainly terrestrial, but some take to water freely, and others burrow a little. They are usually abundant wherever there is sufficient ground vegetation to provide cover. They occur over a great altitudinal range and in many kinds of plant communities. They prefer moist habitats, but some species are found also in arid regions (*Notiosorex*). Other species of soricines inhabit banks of rivers, streams, and lakes, and are modified for an aquatic or semiaquatic life. Soricine shrews live mainly in areas with distinct seasonal changes in weather and habitat, often in areas with extreme low temperatures in winter. An adequate snow cover ensures stability and a relatively mild subnivean (under snow) microclimate.

Behavior

Shrews live hidden under cover and mostly lead nocturnal lives. Their prey does not require group hunting and

Common shrews (*Sorex araneus*) use their long pointed snouts to unearth insects and other food. (Photo by Stephen Dalton/Photo Researchers, Inc. Reproduced by permission.)

A diving shrew diving after prey. (Illustration by Emily Damstra)

solitary foraging prevails. Predation avoidance in soricines does not depend on a long-distance escape but rather on finding shelter. Soricines live solitarily and they are territorial almost all their life. Their strict territoriality is promoted by exploitation of scarce and evenly distributed resources. Shrews establish large territories within which most foraging and nesting, as well as courtship and mating, take place. In autumn and winter, territories are maintained to maximize survivorship, and in spring and summer, to maximize reproduction. The maintenance and defense of territories is based on acoustic and olfactory communication, but direct aggression involving combat and biting is also frequent. With respect to the distribution of food resources and their predictability, two social systems of territorial behavior have been described in soricine shrews—stable and shifting territories.

The shrews with stable territories do not nest communally in order to conserve heat by huddling and they may be categorized as winter-solitary species. In spring, the territories enlarge considerably and the territories of opposite sexes may partly overlap. Different males may have different territorial behavior and mating patterns and strategies at the same time and within the same habitat. Long-distance wandering males have a lower reproductive success and they probably suffer higher predation rates. However, their movements are very important because they facilitate gene flow within the species. Courtship and mating last only a short period. Nevertheless, a female can copulate with several males during this short time and multiple paternity of pups within one litter may occur. This female strategy can reduce inbreeding. The mating system is thus rather promiscuous.

The system of shifting territories is typical for semi-aquatic shrews (*Chimarrogale, Nectogale, Neomys, Sorex palustris, Sorex bendirii*). In these species, the exploitation of food resources is clumped in space and undergoes unpredictable changes that

A Eurasian water shrew (*Neomys fodiens*) pulls a frog from a stream. (Photo by Dwight Kuhn. Bruce Coleman, Inc. Reproduced by permission.)

Northern short-tailed shrews (*Blarina brevicauda*) demonstrating offensive behavior in a territorial dispute. (Photo by Dwight R. Kuhn. Bruce Coleman, Inc. Reproduced by permission.)

A Eurasian water shrew (*Neomys fodiens*) diving. (Photo by Stephen Dalton/Photo Researchers, Inc. Reproduced by permission.)

do not favor strong and stable territoriality. Territories are not maintained for the whole life, and semi-aquatic shrews usually shift along banks of flowing waters and water reservoirs, so that their territories are changed every few weeks or months. The mating system of shrews with shifting territories is promiscuous, and the breeding females are the most aggressive members of populations. The dissimilarity of social systems of certain species (*Notiosorex, Cryptotis*) to those of other soricines can be related to the fact that they inhabit regions with relatively warm climates and/or with poor food resources.

Reactions of soricines with stable territories to shrews of other species, being potential competitors, are also agonistic, including the larger shrews preying upon the smaller species. In interspecific competition, behavior leading to mutual avoidance of the individuals plays a more important role than direct aggression.

Feeding ecology and diet

Shrews feed mainly on soil invertebrates and as a whole may be seen as opportunistic and generalized invertebrate feeders, although some partitioning may occur among syn-

A masked shrew (*Sorex cinereus*) foraging. (Photo by Lynn Stone. Bruce Coleman, Inc. Reproduced by permission.)

topical species. No extreme specialist occurs, even though some shrews are more or less aquatic, which also is reflected in their diet. The fossil species *Arctosorex polaris* from the late Neogene of Ellesmere Island might be the only shrew specialized for frugivory.

Sorex shrews form a well-defined guild of insectivorous mammals with opportunistic feeding habits and largely overlapping resource requirements. In a guild, up to six species may occur together in the same time and place. Competition is likely to be common in such a case. Major differences in dietary composition and foraging mode reflect body size in shrews. Small species are often epigeal, feed mainly on arthropods, and have a relatively narrow niche breadth. Large species are usually hypogeal, and feed on earthworms and other soil-dwelling invertebrates. Body size thus plays an important role in ecological separation in multi-species assemblages of terrestrial soricine shrews. Assemblages of shrews in boreal forests show a shift from the dominance of small species in unproductive habitats to the dominance of large species in productive habitats.

Shrews have two features in common that influence their whole energetic design: a small body size and an insectivorous food habit. They do not hibernate in winter and, with few exceptions, they are not able to enter torpor. The soricines have very high basal rates of metabolism and they maintain a high and precisely regulated body temperature. The very high metabolic rates in shrews are markedly higher than would be expected in mammals of their body size (up to 315% of the expected value). Extremely high metabolic rates are characteristic especially for the genus *Sorex*. The high

basal rates of metabolism were attributed to their origin in a temperate climate and to their large litter size. High energy costs of reproduction are apparently associated with high metabolic rates. The high metabolic rates of most soricines make them susceptible to food shortage and result in the requirement of a constant food supply. Starvation time for the genus *Sorex* varies between five and 10 hours.

Soricine shrews in northern temperate regions undergo a decline in body mass and body size during winter (known as the Dehnel's effect). This is interpreted as an adaptation permitting reduction in food requirements. Winter weight loss in *Sorex* usually amounts to 25–40%, and it is accompanied by a reduction in the size of the skull and most organs.

Reproductive biology

Seasonal regimes of resource availability favor high fecundity during the flush of resources occurring during summer. Therefore, the reproduction of soricine shrews is strongly seasonal, occurring when food is highly available and weather conditions are optimal. Soricines produce significantly larger litters than crocidurines, on average 5.1 young per litter (5.9 in *Sorex*), but more than 10 young can also be born in a litter. One female can produce one to four litters per year. Increased reproductive output may be a major advantage derived from the high rate of metabolism in many soricines.

Evidence from the field and laboratory studies suggests that female shrews of a variety of species show a tendency to mate with many different males. Multiple paternity in the litter of wild-caught females has been demonstrated in the promiscuous common shrew, with up to six different fathers per litter. Most soricines have a short gestation period (on

An ornate shrew (*Sorex ornatus*) eating a lizard. (Photo by Alan Blank. Bruce Coleman, Inc. Reproduced by permission.)

A southern short-tailed shrew (*Blarina carolinensis*) eating an insect. (Photo by R. Wayne Van Devender. Reproduced by permission.)

average 21 days) and a longer weaning period. Young are born in a very early stage of development and are among the most altricial placental mammals. The pups develop quickly, and usually they are weaned within three to four weeks. After the final break of mother-offspring bond, the young leave the nest and their dispersal begins. Juvenile shrews usually do not reproduce during their first summer because generally they mature in the next spring. High social intolerance and aggressiveness with increased population density are important factors inhibiting the maturation and reproduction of juveniles. After reproduction, and as the autumn approaches, the parental population quickly dies. The average life-span of a reproducing soricine shrew is 14.7 months.

Conservation status

The soricine shrews included in the 2002 IUCN Red List often belong to very rare and endemic species with restricted distribution and low population densities. Some of the threatened species are only known from the locality of their original description. However, the major threat for the soricine shrews is environmental disturbance and pollution. Loss of forest habitats has a considerable effect on forest-dwelling species. The rate of deforestation taking place in south and Southeast Asia is an apparent reason for including several soricine species from this particular geographic region among the category of Critically Endangered. Similar threats may arise in other regions from human interference with wetlands and from freshwater pollution. Shrews are exceptionally susceptible to accumulation of heavy metals in their tissues through their diet consisting of large amounts of earthworms. Long-term persistence of DDT has also been reported in soricine shrews.

Conservation of soricine shrews is only possible if it is fully integrated with the broader issues of environmental management and sustainable development. Furthermore, there is an urgent need for considerable research because important information is still lacking for many species.

Significance to humans

Soricine shrews are an important component of natural communities and ecosystems and they have an important ecological role. The impact of soricines upon the natural environment, whether through the large amount of invertebrates consumed, or their role as prey species for many predators, is considerable. Shrews have a significant impact on litter decay, and they diminish the population density of soil invertebrates thus increasing their productivity. In this way, soricine shrews in temperate habitats accelerate decomposition of forest litter and organic matter in the upper soil layer. They can play a role in natural control of various insect pests.

Accumulation of heavy metals in insectivorous mammals through their diet containing earthworms is extremely high, and the burden in tissues of shrews may be used as a useful indicator of environmental contamination. It is unknown whether shrews have a greater resistance to heavy metals than other mammals. Strong effects of rapid environmental change on developmental stability were also recorded in free-living populations of soricine shrews and provided a tool to monitor changes in natural environments. Shrews are becoming favorable model species in research of various issues of biomedicine and evolutionary biology.

An American pygmy shrew (*Sorex hoyi*) emerging from its burrow. (Photo by E. R. Degginger. Bruce Coleman, Inc. Reproduced by permission.)

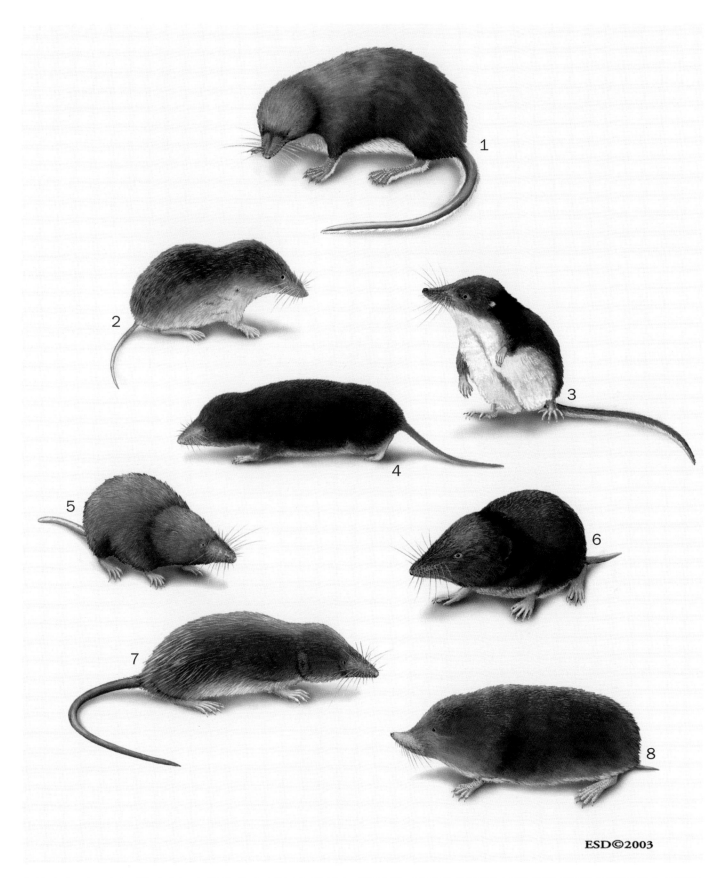

1. Elegant water shrew (*Nectogale elegans*); 2. American least shrew (*Cryptotis parva*); 3. Eurasian water shrew (*Neomys fodiens*); 4. Chinese short-tailed shrew (*Blarinella quadraticauda*); 5. Mérida small-eared shrew (*Cryptotis meridensis*); 6. Northern short-tailed shrew (*Blarina brevicauda*); 7. Himalayan water shrew (*Chimarrogale himalayica*); 8. Mole-shrew (*Anourosorex squamipes*). (Illustration by Emily Damstra)

1. Alpine shrew (*Sorex alpinus*); 2. American water shrew (*Sorex palustris*); 3. American pygmy shrew (*Sorex hoyi*); 4. Common shrew (*Sorex araneus*); 5. Giant shrew (*Sorex mirabilis*); 6. Desert shrew (*Notiosorex crawfordi*); 7. Hodgson's brown-toothed shrew (*Soriculus caudatus*); 8. Eurasian pygmy shrew (*Sorex minutus*). (Illustration by Emily Damstra)

Species accounts

Mole-shrew
Anourosorex squamipes

TAXONOMY

Anourosorex squamipes Milne-Edwards, 1872, Sichuan, China. A number of subspecies have been proposed but it is doubtful if these justify recognition.

OTHER COMMON NAMES

French: Musaraigne taupe, musaraigne à queue moignon; German: Stummelschwanzspitzmaus.

PHYSICAL CHARACTERISTICS

Head and body length 3.4–4.3 in (8.5–10.8 cm); tail 0.4–0.7 in (0.9–1.7 cm); weight 0.5–1.1 oz (15–31 g). It looks more like a mole than a shrew. Short feet with long claws.

DISTRIBUTION

The only extant species of the tribe Anourosoricini that lives in Southeast Asia. In the past, the tribe was very diverse and also inhabited North America and Europe. The present range includes Bhutan, Assam, Burma, Thailand, Laos, northern Vietnam, southern China, and Taiwan.

HABITAT

Forest-dwelling species, lives in elevations of 1,000–11,000 ft (300–3,300 m). It is associated with moist habitats and areas along streams in broadleaf and conifer forests.

BEHAVIOR

Several mole-shrews may share the same burrow system. It is reported to be semifossorial, but no signs of its underground runways or burrows were observed in Taiwan.

FEEDING ECOLOGY AND DIET

It is a burrowing shrew, but it also searches for food on the forest floor. The diet consists mainly of insect larvae and lumbricids, with probably only a small amount of hard-shelled forms like coleopterans.

REPRODUCTIVE BIOLOGY

Promiscuous. It breeds primarily in the wet season in Taiwan. Litter size averages 2.5, range 2–4. Local populations contain subgroups of related animals as shown by DNA analysis.

CONSERVATION STATUS

Not listed by the IUCN, but included in the Conservation Action Plan for the Eurasian Insectivores.

SIGNIFICANCE TO HUMANS

None known. ◆

Chinese short-tailed shrew
Blarinella quadraticauda

TAXONOMY

Blarinella quadraticauda (Milne-Edwards, 1872), Sichuan, China.

OTHER COMMON NAMES

French: Musaraigne à queue courte de Chine; German: Asiatische Kurzschwanzspitzmaus.

PHYSICAL CHARACTERISTICS

Head and body length 2.4–3.2 in (6.0–8.2 cm); tail 1.2–2.4 in (3–6 cm). Fur is brown-gray with lighter underside. Large claws; short, thin tail; and no visible ears.

Sorex mirabilis

Anourosorex squamipes

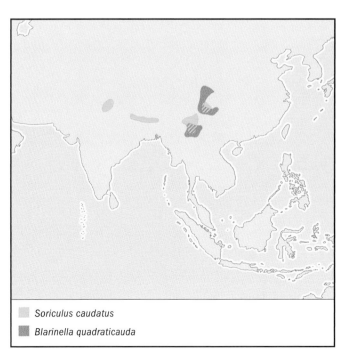

Soriculus caudatus

Blarinella quadraticauda

DISTRIBUTION
The tribe Blarinini is at present confined in distribution to East Asia, but it also has an extensive fossil record from Europe and North America. At present, endemic to central and southern China.

HABITAT
It is only found in montane taiga forest.

BEHAVIOR
A fossorial species with adaptations for burrowing.

FEEDING ECOLOGY AND DIET
Nothing is known.

REPRODUCTIVE BIOLOGY
Nothing is known.

CONSERVATION STATUS
Not threatened, though restricted endemic distribution. Included in the Conservation Action Plan for the Eurasian Insectivores (IUCN).

SIGNIFICANCE TO HUMANS
None known. ◆

Northern short-tailed shrew
Blarina brevicauda

TAXONOMY
Blarina brevicauda (Say, 1823), Nebraska, United States. Phylogeographic analysis showed a significant partitioning between eastern and western populations on either side of the Mississippi Basin.

OTHER COMMON NAMES
French: Grande musaraigne à queue courte; German: Kurzschwanzspitzmaus.

PHYSICAL CHARACTERISTICS
Head and body length 3–4.1 in (7.5–10.5 cm); tail 0.7–1.2 in (1.7–3 cm); weight 0.5–1.1 oz (15–30 g). Soft gray fur. Short snout and tail.

DISTRIBUTION
Representatives of this genus live in North America today, but its fossil members have also been found in Europe and Asia. The present range in North America includes central and southern Saskatchewan to southeastern Canada, and southward to Nebraska and northern Virginia in the United States.

HABITAT
Tall, dense grass and/or deep woods with thick ground litter. May shelter in logs, stumps, or crevices of building foundations.

BEHAVIOR
The most fossorial of American shrews, but they are effective climbers. These shrews are active all year and are seen by day and night. In captivity, individuals seem to live together peacefully if provided enough room. In the wild it is a solitary territorial species. Populations contain both resident and nomadic components. Residents mark their ranges with scent and threaten and fight intruders. A variety of sounds and postures are employed during intraspecific threat situations; a clicking sound accompanies courtship behavior.

FEEDING ECOLOGY AND DIET
This shrew has an opportunistic way of feeding, with a large portion of predaceous habits included. The diet consists of invertebrates, small vertebrates (salamanders and anurans), and plant material. Toxic saliva enables the shrew to deal with rather

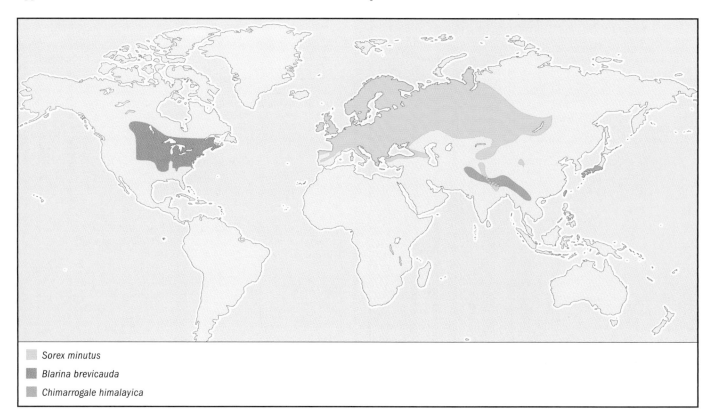

☐ Sorex minutus
■ Blarina brevicauda
▨ Chimarrogale himalayica

large prey that the saliva immobilizes. It stores snails, beetles, seeds, and other edibles where they can be retrieved later.

REPRODUCTIVE BIOLOGY
Promiscuous. The breeding season usually extends from early spring to early fall. There may be up to three litters per year. Litter size is 3–10, usually 5–7. Population density is about 1–12 individuals per acre (3–30 per hectare), and home range is about 0.5–2 acre (0.2–0.8 hectare). Few wild individuals survive for more than one year.

CONSERVATION STATUS
Not threatened.

SIGNIFICANCE TO HUMANS
It often serves as an important check on larch sawflies and other destructive insects. The poison produced by the submaxillary glands can cause pain for several days in humans. ◆

American least shrew
Cryptotis parva

TAXONOMY
Cryptotis parva (Say, 1823), Nebraska, United States. The mosaic occurrence of relic Pleistocene populations has recently been indicated in an allozyme study from the southwestern United States.

OTHER COMMON NAMES
French: Petite musaraigne à queue courte; German: Nordamerikanische Kleinohrspitzmaus.

PHYSICAL CHARACTERISTICS
Head and body length 2.2–3.1 in (5.5–7.8 cm); tail 0.4–1.1 in (1–2.7cm); weight 0.1–0.3 oz (4–8 g). Soft brown-black fur with white underside.

Cryptotis parva
Cryptotis meridensis

DISTRIBUTION
Southeastern Canada through the east-central and southwestern United States and Mexico to Panama and Costa Rica. The only species of the genus found north of Mexico.

HABITAT
A habitat specialist among shrews, living in open grassy habitats and along streams in otherwise dry habitats. Also found in cold, wet forest. Occurs from the lowlands to 7,900 ft (2,400 m). In Central America usually above 2,600 ft (800 m).

BEHAVIOR
Somewhat gregarious and colonial. May form quite large colonies of up to 30 adult individuals. Group members share one nest, rest in huddling, cooperate in burrow building, and share a common home range. They emit a variety of sounds that seem to correspond to friendly social behavior. All these facts suggest that lack of territoriality and group living is characteristic of the social system of this species. It is active throughout the year and at all hours of the day, with greatest activity during the night. This species uses runways and burrows of other small mammals, but also makes its own tunnels in loose, soft soils. Ultrasonic clicks are also given, which may be used for echolocation.

FEEDING ECOLOGY AND DIET
Feeds on invertebrates (insect larvae, centipedes, and earthworms), small lizards and frogs, and carrion.

REPRODUCTIVE BIOLOGY
This promiscuous species probably breeds throughout the year in the southern parts of its geographic range, and the greatest reproduction activity was recorded in winter and spring. The gestation period is 21–22 days, and the mean litter size is 4–5 young (range 1–9). Population density of about 12–13 individuals per acre (31 per ha) was recorded. Average life expectancy in the laboratory was eight months, with a maximum longevity of 31 months.

CONSERVATION STATUS
Not threatened.

SIGNIFICANCE TO HUMANS
Laboratory breeding colonies of least shrews have been established. The species has become an important biomedical research model. ◆

Mérida small-eared shrew
Cryptotis meridensis

TAXONOMY
Cryptotis meridensis Thomas, 1898, Mérida, Venezuela. This species was commonly included in *C. thomasi* but it is much larger and has a more robust dentition.

OTHER COMMON NAMES
Spanish: Musaraña de Mérida.

PHYSICAL CHARACTERISTICS
Head and body length 4.6–5.2 in (11.7–13.1 cm); tail 1.4–1.6 in (3.5–3.9 cm); weight 0.4 oz (11.5 g). Grayish brown fur with paler underside.

DISTRIBUTION
A restricted range in the Cordillera de Mérida and mountains near Caracas in Venezuela, South America.

HABITAT

Humid environments with constant moderate temperature in cloud forests and paramos in altitudes from 6,500 to 9,200 ft (2,000 to 2,800 m). An important feature of the cloud forest habitat is the presence of tree ferns. The shrew burrows in the leaf litter, under logs and bases of hollow trees, and throughout moss and lichen mats.

BEHAVIOR

A burrowing species living underground. Activity on the ground may be particularly reduced at the time of low precipitation.

FEEDING ECOLOGY AND DIET

A wide range feeder with its main diet consisting of hypogeal and epigeal invertebrates (earthworms, snails, spiders, isopods, insects). Prevalence of the hypogeal prey suggests primarily underground foraging.

REPRODUCTIVE BIOLOGY

The abundance of this species seems to be correlated with the rainfall pattern, and it is supposed that breeding takes place in the wet season from April to December. Presumably promiscuous

CONSERVATION STATUS

Not immediately threatened, but with restricted endemic distribution.

SIGNIFICANCE TO HUMANS

None known. ◆

Eurasian water shrew
Neomys fodiens

TAXONOMY

Sorex fodiens (Pennant, 1771), Berlin, Germany. A survey of allozymic variation in Europe suggests that populations may show extensive differentiation within short distances.

◻ *Sorex alpinus*

◼ *Nectogale elegans*

◼ *Neomys fodiens*

OTHER COMMON NAMES

English: Northern water shrew; French: Musaraigne aquatique; German: Große Wasserspitzmaus; Spanish: Musgaño patiblanco.

PHYSICAL CHARACTERISTICS

Head and body length 2.8–3.8 in (7.0–9.6 cm); tail 1.8–3 in (4.7–7.7 cm); weight 0.4–0.7 oz (12–20 g). This species has a keel of stiff hairs on the underside of the tail, and fringes of stiff hairs on the hands and feet that aid in swimming.

DISTRIBUTION

Distributed in Europe and Asia. The range includes most of Europe except Iceland, Ireland, and most of Iberia. Farther east, it ranges through western Siberia as far as the Yenisei River and Lake Baikal, and south to Tien Shan and northwestern Mongolia. In east Siberia it has apparently disjunct populations from northeastern China, North Korea, Sakhalin Island, and the Far East.

HABITAT

It is particularly fond of streams with densely overgrown banks that provide ample hiding places among roots and stones. Inhabits clear, fast-flowing rivers and streams, lakes with abundant riparian vegetation, and ponds, marshes, watercress beds, and boulder-strewn sea shores. Only occasionally found far from water. Up to 8,200 ft (2,500 m) in the Alps.

BEHAVIOR

A semi-aquatic species that moves quickly in swimming and diving. It is a marvelous experience to watch this animal hunting in water. Typically, it dives for 3–25 seconds in the field. No special adaptations have evolved to permit long-lasting diving, because this would invariably lead to hypothermia. Active by day and night, but it has an apparent preference for darkness. It is essentially solitary outside of the breeding season. Individuals are aggressive toward one another. Exhibits a marked territorial behavior with shifting territories. Home ranges extend from 210 to 320 ft² (20–30 m²) on land, and from 650 to 860 ft² (60–80 m²) along brooks.

FEEDING ECOLOGY AND DIET

It mainly feeds on macrobenthic invertebrates (crustaceans and insect larvae) in running water. Foraging is conducted in the water and on land surface, feeding on a wide range of invertebrates, as well as frogs, newts, and small fishes. The species paralyses large prey with venomous saliva.

REPRODUCTIVE BIOLOGY

Probably polygamous or promiscuous. The breeding season extends from April to September, reaching a peak in May and June. Two litters are generally produced each breeding season. Litter size is 3–8 (average about 6). A life span of 14–19 months has been reported.

CONSERVATION STATUS

Not listed by the IUCN, but included in the Conservation Action Plan for the Eurasian Insectivores.

SIGNIFICANCE TO HUMANS

Sometimes it does damage to fish fry. ◆

Himalayan water shrew
Chimarrogale himalayica

TAXONOMY

Crossopus himalayicus (Gray, 1842), Punjab, India.

OTHER COMMON NAMES
French: Chimarrogale de l'Himalaya; German: Himala-jawasserspitzmaus.

PHYSICAL CHARACTERISTICS
Head and body length 3.1–5.3 in (8–13.5 cm); tail 2.4–5 in (6–12.6 cm); weight 0.9–1.4 oz (25–40 g). Dark brown fur with light underside and feet.

DISTRIBUTION
The range extends in Asia from the Himalayan region west to Kashmir (Pakistan and India) through Southeast Asia to Indochina (Laos, north Vietnam), central and southern China, and Taiwan.

HABITAT
The banks of clear high-gradient mountain streams and the basins of waterfalls at altitudes of 2,600–4,900 ft (800–1500 m).

BEHAVIOR
This species is modified for an aquatic life. It is apparently able to swim well under water and it is occasionally caught in fish traps. As in other semi-aquatic insectivores, its dense fur is water repellent and considerable time must be spent grooming to ensure that the fur is maintained in good condition.

FEEDING ECOLOGY AND DIET
The diet consists of insects, aquatic larvae, crustaceans, and possibly small fishes.

REPRODUCTIVE BIOLOGY
Nothing is known.

CONSERVATION STATUS
Restricted endemic range. Not listed by the IUCN, but included in the Conservation Action Plan for the Eurasian Insectivores.

SIGNIFICANCE TO HUMANS
None known. ◆

Elegant water shrew
Nectogale elegans

TAXONOMY
Nectogale elegans Milne-Edwards, 1870, Sichuan, China.

OTHER COMMON NAMES
English: Tibetan water shrew; French: Nectogale élégant; German: Gebirgsbachspitzmaus.

PHYSICAL CHARACTERISTICS
Head and body length 3.5–5 in (9–12.5 cm); tail 3.5–4.5 in (8.9–11 cm). The only member of the subfamily with webbed feet. Long dark tail is fringed with white hairs.

DISTRIBUTION
An Asian range in Tibet, Yunnan, Sichuan, and Shaanxi (China), the Himalayas west to Sikkim, Nepal, Bhutan, and Burma.

HABITAT
It frequents high altitude, clean, mountain streams and torrents at altitudes of 2,950–7,500 ft (900–2,270 m).

BEHAVIOR
This species swims and dives remarkably well and shelters in burrows in stream banks.

FEEDING ECOLOGY AND DIET
Aquatic invertebrates and small fishes. Secondary cuspids on certain teeth may be an adaptation to a diet of fishes, providing a better hold on prey.

REPRODUCTIVE BIOLOGY
Nothing is known.

CONSERVATION STATUS
Restricted endemic range. Not listed by the IUCN, but included in the Conservation Action Plan for the Eurasian Insectivores.

SIGNIFICANCE TO HUMANS
None known. ◆

Hodgson's brown-toothed shrew
Soriculus caudatus

TAXONOMY
Sorex caudatus (Horsfield, 1851), Sikkim.

OTHER COMMON NAMES
French: Musaraigne à longue queue d'Hodgson.

PHYSICAL CHARACTERISTICS
Head and body length 2.4–2.7 in (6.1–6.9 cm); tail 1.9–2.2 in (4.8–5.6 cm).

DISTRIBUTION
Kashmir (Pakistan and India) to northern Myanmar and southwestern China.

HABITAT
Damp habitats at edges of rhododendron and coniferous forests; also found in alpine meadows at altitudes of 3,300–13,800 ft (1,000–4,200 m).

BEHAVIOR
Nothing is known.

FEEDING ECOLOGY AND DIET
It feeds on earthworms, insects, and other invertebrates and has also been known to eat small mammals caught in traps.

REPRODUCTIVE BIOLOGY
Breeding peaks in April–May and August, with one or two litters per year and an average litter size of 5 (3–6).

CONSERVATION STATUS
Restricted endemic range. Not listed on the IUCN Red List, but included in the Conservation Action Plan for the Eurasian Insectivores.

SIGNIFICANCE TO HUMANS
None known. ◆

Desert shrew
Notiosorex crawfordi

TAXONOMY
Notiosorex crawfordi (Coues, 1877), Texas, United States. Specimens from Tamaulipas, Mexico, were shown to represent a separate species described as *Notiosorex villai*. Another Mexican subspecies, *N. c. evotis*, was also elevated to specific status.

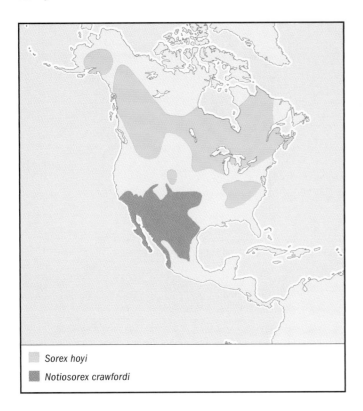

■ Sorex hoyi
■ Notiosorex crawfordi

OTHER COMMON NAMES
French: Musaraigne du désert; German: Graue Wüstenspitz-maus.

PHYSICAL CHARACTERISTICS
Head and body length 1.9–2.6 in (4.8–6.5 cm); tail 0.9–1.2 in (2.2–3.1 cm); weight 0.1–0.2 oz (3–5 g). Dark gray fur, lighter underneath. Long tail, visible ears. Reddish nose, hairless feet.

DISTRIBUTION
North and Central America from the southwestern United States east to central Texas and western Arkansas, Baja California, and northern and central Mexico.

HABITAT
An arid-adapted shrew occurring in a variety of xeric habitats. The most common occupied community is a semi-desert scrub association, with such vegetation as mesquite, agave, and scrub oak. Food and water resources are scarce, and amplitudes of daily temperatures are high in these habitats. Nests are usually constructed on the surface of the ground. The species has also been found in beehives.

BEHAVIOR
This species seems to be more social than other shrews. Captives are able to live together with little antagonism, and they live in high densities on small areas in the wild. It is active at night, twitching its snout and vibrissae while foraging. The known vocalization is a high-pitched squeak, emitted during occasional bouts of fighting or disturbance.

FEEDING ECOLOGY AND DIET
The diet consists of a variety of insects and other invertebrates, and dead vertebrates. It has a greater ability to cool itself by evaporation than do other shrews and the energy metabolism is lower than in other shrew genera. The desert shrew is able to enter shallow torpor that is believed to be an adaptation for coping with heat, aridity, and a fluctuating food supply.

REPRODUCTIVE BIOLOGY
Probably promiscuous. Pregnant females have been taken from April to November. Known litter size is 3–5. Its moderate basal rate of metabolism is apparently associated with a moderate litter size.

CONSERVATION STATUS
The loss of native coastal scrub flora and the increasing presence of the Argentine ant colonies may significantly effect the distribution and abundance of the species.

SIGNIFICANCE TO HUMANS
None known. ◆

Alpine shrew
Sorex alpinus

TAXONOMY
Sorex alpinus Schinz, 1837, Canton Uri, Switzerland. Three karyotypic forms known from the Western Alps, Swiss Jura, and Central Europe.

OTHER COMMON NAMES
French: Musaraigne alpine; German: Alpen-Spitzmaus; Spanish: Musaraña alpina.

PHYSICAL CHARACTERISTICS
Head and body length 2.3–3 in (6.0–7.7 cm); tail 2.1–3 in (5.4–7.5 cm); weight 0.2–0.4 oz (5.5–11.5 g). Nearly black fur, feet are hairless.

DISTRIBUTION
Endemic to Europe. Disjunct populations occur in the Alps, Balkans, Carpathians, and several isolated mountain ranges in Central Europe.

HABITAT
Moist and shady habitats in the altitudes from 660 to 8,200 ft (200–2,500 m). Often found in rocky habitats under mossy rocks and logs, and in the dense weed growths at the banks of mountain streams and torrents. At lower elevations confined to cool and humid environments in deep valleys and ravines.

BEHAVIOR
A nocturnal species with good climbing abilities.

FEEDING ECOLOGY AND DIET
Hypogeal feeder. Eats snails, earthworms, spiders, isopods, chilopods, insects, and insect larvae.

REPRODUCTIVE BIOLOGY
Breeding from April to October. There are two or three litters annually. Litter size 3–9 (average 5–7). The young grow very quickly and are probably sexually mature in the first year of their life. Probably a promiscuous species.

CONSERVATION STATUS
Restricted endemic range. Small peripheral populations threatened by habitat loss. Probably extinct in certain mountain ranges (e.g., Pyrenees, Harz). Listed in the Bern Convention, Annex III. Included in the Conservation Action Plan for the Eurasian Insectivores, and in the Red Data Books of several European countries.

SIGNIFICANCE TO HUMANS
None known. ◆

Common shrew
Sorex araneus

TAXONOMY
Sorex araneus Linnaeus, 1758, Uppsala, Sweden. The common shrew displays phenomenal chromosomal variation. About 70 karyotypic races have been described throughout the distribution range.

OTHER COMMON NAMES
French: Musaraigne commune; German: Waldspitzmaus; Spanish: Musaraña colicuadrada.

PHYSICAL CHARACTERISTICS
Head and body length 2.6–3.4 in (6.5–8.5 cm); tail 1.3–1.9 in (3.2–4.7 cm); weight 0.2–0.5 oz (5–14 g). Gray-brown fur with light underside. Feet are hairless.

DISTRIBUTION
Europe and Asia. The European range includes Great Britain and the Pyrenees, but the species is absent from Iberia, most of France, and Ireland. It extends eastwards as far as Lake Baikal.

HABITAT
It is the most abundant species of the European shrews, and it can be found everywhere in sites with enough humidity, soft soil layers, and some undergrowth. It occurs in a wide range of habitats including woodlands, grassland, hedgerows, heath, dunes, and scree. May live up to the limits of the summer snow line.

BEHAVIOR
Solitary and aggressive. It is active during day and night with about 10 periods of almost continuous activity. The common shrew makes its own surface runways through the ground vegetation, but it may also use the subterranean burrows of voles and moles.

FEEDING ECOLOGY AND DIET
It is an opportunistic feeder, preying on a wide range of insects, spiders, small mollusks, earthworms, and wood lice.

REPRODUCTIVE BIOLOGY
The female gives birth to the young from April to October, three or four times a year. The average litter size 5–7, a maximum of 11 young. Young disperse shortly after weaning and individuals of both sexes establish their own home range, varying in size from 0.09–0.16 acre (0.04–0.06 ha). Population densities are highly variable and may range from 17 to 110 individuals per acre (42–270 per ha) in summer, and from 2 to 11 individuals per acre (5–27 per ha) in winter. Probably promiscuous.

CONSERVATION STATUS
Not listed by the IUCN, but included in the Conservation Action Plan for the Eurasian Insectivores. The species is especially vulnerable during the juvenile period of dispersal. Because of the composition of its diet with a considerable proportion of earthworms, it very intensively accumulates heavy metals in polluted areas.

SIGNIFICANCE TO HUMANS
An important model species in evolutionary studies. ◆

Eurasian pygmy shrew
Sorex minutus

TAXONOMY
Sorex minutus Linnaeus, 1766, Siberia, Russia. Status of the populations from Central Asia is uncertain.

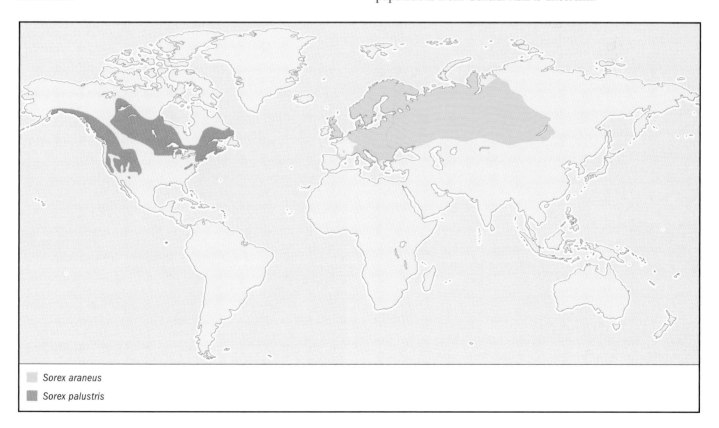

☐ *Sorex araneus*
■ *Sorex palustris*

OTHER COMMON NAMES
French: Musaraigne minuscule; German: Zwergspitzmaus; Spanish: Musaraña enana.

PHYSICAL CHARACTERISTICS
Head and body length 1.6–2.4 in (4–6 cm); tail 1.3–1.8 in (3.2–4.6 cm); weight 0.08–0.2 oz (2.4–6.1 g). Fur is dark gray-brown, lighter white on belly. Tail may have white tip.

DISTRIBUTION
Most of Europe to Yenisei River and Lake Baikal (Russia), and southwards to the Altai and Tien Shan Mountains in Asia

HABITAT
The lesser shrew is often found in the same habitats as the common shrew, but it is able to tolerate sparser ground cover. It prefers heaths, grasslands, sand dunes, and woodland edges. It seems to be adapted to cold and dampness better than the common shrew and can be found even on moorlands at higher altitudes.

BEHAVIOR
It is solitary and aggressive towards others of the same species. Territories of immature animals are largely mutually exclusive but the strict territoriality is abandoned at sexual maturity, particularly by males as they search for mates. Home range size varies from 0.1 to 0.5 acre (0.04–0.2 ha). Pygmy shrews are active during day and night, spending more time on the surface, unlike other shrews. They do not burrow, but use runways of other species.

FEEDING ECOLOGY AND DIET
Diet includes invertebrates in leaf litter—mostly beetles, spiders, and wood lice—but not earthworms. In captivity the food intake is about 100% body weight per day. Like other shrews, it is very vulnerable to starvation and will die quickly if its food supply runs out.

REPRODUCTIVE BIOLOGY
The breeding season lasts from April to October, with probably one or two litters per year. Litter size is 4–8. Pygmy shrews generally overwinter as immature animals, maturing the following spring. Population density ranges from one to nine individuals per acre (4–22 per hectare), depending on season and habitat. Species is probably promiscuous.

CONSERVATION STATUS
Not listed by the IUCN, but included in the Conservation Action Plan for the Eurasian Insectivores.

SIGNIFICANCE TO HUMANS
None known. ◆

Giant shrew
Sorex mirabilis

TAXONOMY
Sorex mirabilis Ognev, 1937, Russia.

OTHER COMMON NAMES
French: Musaraigne géante.

PHYSICAL CHARACTERISTICS
Head and body length 2.9–3.8 in (7.4–9.7 cm); tail 2.5–2.8 in (6.3–7.2 cm); weight 0.4–0.5 oz (11–14.2 g). The largest of the Eurasian shrews of the genus *Sorex*.

DISTRIBUTION
The range is restricted to eastern Asia in the Ussuri Region (Russia), northeastern China, and North Korea.

HABITAT
Humid environments in lowland and montane forests.

BEHAVIOR
Nothing is known.

FEEDING ECOLOGY AND DIET
The most important food items are earthworms; also feeds on centipedes, beetles, and insect larvae.

REPRODUCTIVE BIOLOGY
The species reproduces regularly once a year; exceptionally there may be two litters annually under favorable conditions. May be promiscuous.

CONSERVATION STATUS
Restricted endemic range. Not listed by the IUCN, but included in the Conservation Action Plan for the Eurasian Insectivores (IUCN/SSC) and in the Red Data Book of the former Soviet Union.

SIGNIFICANCE TO HUMANS
None known. ◆

American pygmy shrew
Sorex hoyi

TAXONOMY
Sorex hoyi Baird, 1857, Racine, Wisconsin, United States. Subspecies differ in weight by a factor of three, and some of them may eventually be shown to represent distinct species.

OTHER COMMON NAMES
English: Pygmy shrew; French: Musaraigne pygmée d'Amérique.

PHYSICAL CHARACTERISTICS
Head and body length 1.6–2.6 in (4.1–6.7 cm); tail 0.8–1.5 in (2.1–3.9 cm); weight 0.07–0.3 oz (2.1–7.3 g). Soft, dense fur, gray-brown in color.

DISTRIBUTION
A North American species occurring in the northern taiga zone, with southern outliers in montane forests of the Rocky and Appalachian Mountains.

HABITAT
Pygmy shrews exhibit wide tolerance for wet, dry, cold, and warm environments. They occupy forests, swamps, marshes, disturbed habitats, wet-dry soils, and grassy and herbaceous understory. The southern subspecies inhabits relatively moist, cool microhabitats. The den may be a burrow or a shelter under a log or in the roots of old stumps.

BEHAVIOR
They often stand on their hind limbs in kangaroo fashion and run quickly with the extended tail slightly curved. They climb with agility and bound as high as 4 in (10 cm). Their calls are sharp squeaks, low purrs, or high-pitched whistling and whispering.

FEEDING ECOLOGY AND DIET
It feeds on small arthropods (larchfly larvae, lepidopteran larvae, grasshoppers, crane flies, beetles, and probably spiders). Occasionally it eats carrion.

REPRODUCTIVE BIOLOGY

An extended promiscuous breeding season was proposed, with birth every month of the year. Actual collection of pregnant animals, however, suggests a more restricted breeding season, especially in northern parts of the range. Females likely produce more than one litter per year in favorable areas, and litter size ranges from two to eight.

CONSERVATION STATUS

Not listed by the IUCN, but the subspecies *S. h. winnemana* is considered rare or threatened.

SIGNIFICANCE TO HUMANS

None known. ◆

American water shrew

Sorex palustris

TAXONOMY

Sorex palustris Richardson, 1828, Canada. *Sorex alaskanus* is considered a subspecies of the water shrew, although some studies have suggested it might be a distinct species.

OTHER COMMON NAMES

English: Water shrew; French: Musaraigne palustre; German: Nordische Wasserspitzmaus.

PHYSICAL CHARACTERISTICS

Head and body length 2.5–3.2 in (6.3–8.1 cm); tail 2.2–3.5 in (5.7–8.9 cm); weight 0.3–0.6 oz (8–18 g). Fur is dark gray black, silvery on underside. Long tail is dark on top, light underneath. Feet have stiff, sparse hairs.

DISTRIBUTION

Montane and boreal areas from Alaska to Sierra Nevada, Rocky and Appalachian Mountains in North America.

HABITAT

Water shrews are typical animals of northern and montane forests. They are almost invariably found in the vicinity of streams or other bodies of water. Heavy vegetation cover and plentiful logs, rocks, crevices, or other sources of shelter that offer high humidity and overhead protection are common habitat attributes. In Virginia, water shrews inhabit canopied streams with a high gradient. They prefer fully vegetated channel banks with extensive undercut areas and many crevices.

BEHAVIOR

It uses its aquatic habitat to find food and to escape from predators. Water shrews readily dive to stream bottoms, paddling furiously to keep from bobbing to the surface. Their fur, full of trapped air, makes them buoyant. The shrews appear even more at home on the top of the water, skittering across the surface like water striders. Food frequently is cached for retrieval at a later date. Shrews are active throughout the year, and they are active primarily at night.

FEEDING ECOLOGY AND DIET

Insect larvae, adult aquatic invertebrates, and even small fishes. Much food is gathered from the near-water larder, including terrestrial insects, snails, earthworms, and even appreciable amounts of fungi and green plant materials. The daily intake seems to be about 5–10% of body weight.

REPRODUCTIVE BIOLOGY

Births occur in spring or summer, and males usually attain sexual maturity the following winter. Rarely, females may be reproductively active during their first summer. They may produce two or three litters of 3–10 offspring (average about 6). Gestation is estimated at about three weeks. Presumably promiscuous.

CONSERVATION STATUS

Not threatened.

SIGNIFICANCE TO HUMANS

None known. ◆

Common name / Scientific name/ Other common names	Physical characteristics	Habitat and behavior	Distribution	Diet	Conservation status
Sunda water shrew *Chimarrogale phaeura*	Pelage is usually dark, with brown, grayish, or black coat color. Head and body 3.1–5.3 in (8.0–13.5 cm), tail 2.4–5.0 in (6.0–12.6 cm), 0.9–1.4 oz (25–40 g).	Semi-aquatic species occurring along mountain streams.	Malay Peninsula, Sumatra, and Borneo.	Aquatic invertebrates.	Endangered
Mexican giant shrew *Megasorex gigas* English: Merriam's desert shrew; French: Grande musaraigne du désert; German: Große Wüstenspitzmaus	Pelage is usually dark, with brown, grayish, or black coat color. Head and body 3.3-3.6 in (8.3-9.0 cm), tail 1.5-2.0 in (3.9-5.0 cm), weight 0.3-0.4 oz (9.5-11.7 g).	Moist areas along streams in tropical forests, in deciduous forests, and in semiarid areas. The known altitudinal range is from near sea level to about 5,600 ft (1,700 m).	Southwestern Mexico in the states of Nayarit, Jalisco, Colima, Michoacán, Guerrero, and Oaxaca.	Insects and other invertebrates.	Not threatened, though restricted endemic range
Tundra shrew *Sorex tundrensis* French: Musaraigne de la toundra	Pelage is usually dark, with brown, grayish, or black coat color. Head and body 2.0-2.8 in (5.1-7.0 cm), tail 0.8-1.9 in (2.1-4.7 cm), weight 0.1-0.3 oz (3-9 g).	An ecologically plastic species, with wide niche. It occurs in mixed ground vegetation in well-drained patches of broad-leaved and taiga forest and tundra. It may be found also in forest-steppe and steppe habitats in Siberia, usually in the vicinity of rivers and water reservoirs.	Asia and North America, restricted to a northern distribution, the limits of which are not known.	Insects, earthworms and floral parts of small grasses have been identified from digestive tracts of specimens from Alaska. In Siberia, beetles and grasshoppers prevail in the food, and it feeds also on snails.	Not threatened

Resources

Books

Corbet, G. B., and J. E. Hill. *The Mammals of the Indomalayan Region: A Systematic Review.* Oxford: Oxford University Press, 1992.

Eurasian Insectivores and Tree Shrews—Status Survey and Conservation Action Plan. Compiled by David Stone and the IUCN/SSC Insectivore, Tree Shrew and Elephant Shrew Specialist Group. Gland, Switzerland: IUCN, 1995.

Mitchell-Jones, A. J., et al. *The Atlas of European Mammals.* London: Poyser Natural History-Academic Press, 1999.

Wilson, D. E., and S. Ruff, eds. *The Smithsonian Book of North American Mammals.* Washington, DC: Smithsonian Institution Press, 1999.

Wójcik, J. M., and M. Wolsan, eds. *Evolution of Shrews.* Bialowieza, Poland: Mammal Research Institute, Polish Academy of Sciences, 1998.

Organizations

IUCN Species Survival Commission, Insectivore Specialist Group, Dr. Werner Haberl, Chair. Hamburgerstrasse 11, Vienna, A-1050 Austria. Phone: +4315861094. Fax: +4315861094. E-mail: shrewbib@sorex.vienna.at Web site: <http://members.vienna.at/shrew/itses.html>

Jan Zima, PhD

Shrews II
White-toothed shrews (Crocidurinae)

Class Mammalia
Order Insectivora
Family Soricidae
Subfamily Crocidurinae

Thumbnail description
Small mouse-like mammals with long pointed snouts, white teeth, short legs, and small eyes

Size
1.3–5.3 in (3.5–13.5 cm); 0.06–1.5 oz (2–40 g)

Number of genera, species
11 genera; 212 species

Habitat
Typically in moist habitats that are rich in invertebrate prey; can also be found in habitats ranging from arid regions to tropical forests

Conservation status
Critically Endangered: 16 species; Endangered: 16 species; Vulnerable: 36 species; Lower Risk/Near Threatened: 7 species; Data Deficient 6 species

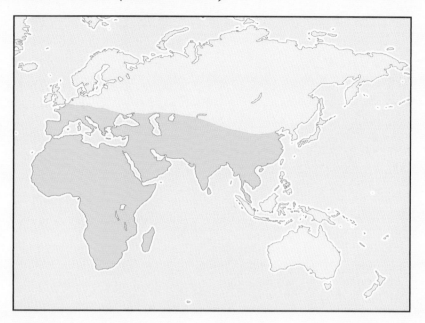

Distribution
Throughout the Old World, found in many African countries, much of continental Europe and southeastern Asia

Evolution and systematics

Among the most ancient of all living animals, shrews have remained almost unchanged over the past 45 million years. This ancient family diverged from other insectivores before the Eocene period, with modern genera (subfamilies Crocidurinae and Soricinae) first appearing in the Miocene in Europe. Soricids later migrated to Africa and North America. The taxonomic status of modern shrews (Soricidae) is currently undergoing a significant overhaul due to considerable recent taxonomic research on the genetics, morphology, ecology, and behavior of shrews. At present, the shrew family is divided into two living subfamilies, the Crocidurinae and the Soricinae. The members of the subfamily Crocidurinae, white-toothed shrews, are further subdivided into 11 genera (*Crocidura, Diplomesodon, Feroculus, Myosorex, Paracrocidura, Ruwenzorisorex, Scutisorex, Solisorex, Suncus, Surdisorex,* and *Sylvisorex*). Current generic boundaries are based on very few characters. The delimitation of some genera of the subfamily Crocidurinae is still the focus of considerable genetic and morphologic research.

The genetic distances between different genera of Crocidinurae based on allozyme variation generally support a subdivision of the living species into these genera. Those species of Crocidurinae that have been studied to date show considerable chromosomal and genetic differentiation. Karyotypic and protein electrophoresis studies have been very helpful in

identifying many sibling and cryptic species that were difficult to recognize by conventional methods.

Crocidurinae includes the genus *Crocidura*, which has 151 species—the largest number of living species of any placental mammal genus.

Physical characteristics

Shrews are the smallest of the insectivores. All shrews have short legs, five claws on each foot, short dense fur, small external ears, an elongated, pointed snout with long tactile hairs (vibrissae), and most have relatively long tails. All have extremely small eyes (often hidden in the fur) and relatively poor eyesight, but the sense of smell is keen, as suggested by their long, mobile snouts. The external ears are reduced in some species and usually hidden in their fur. Hearing is acute. The fur is short, dense, and usually some shade of brown or gray. The skull is long and narrow and has no zygomatic arch. The shrew has one of the most primitive brains of all placental mammals; the brain is small and smooth, dominated by large olfactory bulbs. The dentition is unlike any other family. The very large upper and lower incisors slant forward and meet like forceps. The external genitals of some species are enclosed in a fold of skin. Some species have a venomous saliva. Shrews have skin glands and genital or marking glands that secrete a substance with an unpleasant musky odor. The foot is not specialized, except in some aquatic species.

A common European white-toothed shrew (*Crocidura russula*). (Photo by © Peter Baumann/Animals Animals. Reproduced by permission.)

Members of the white-toothed shrew (Crocidurinae) subfamily are initially distinguished from their red-toothed (Soricinae) counterparts by the color of the enamel of the tips of their incisors. Red-toothed shrews have red teeth due to red pigmentations, and white-toothed shrews lack pigmentation. The two exceptions are *Chimarrogale* and *Nectogale*, two white-toothed aquatic species placed in the red-toothed subfamily.

Crocidurinae are also characterized by the retention of primitive dental characteristics. Modern forms differ from those of the later Miocene and from one another by the loss of one, two, or three upper and lower antemolars; reduction in the talonid of the lower third molar and greater emargination of the posterior basal outline of the upper premolar and upper first molar. The first set of teeth is shed in the embryonic stage, so that the teeth at birth are the permanent set. Crocidurinines have 26–32 teeth, normally six on each side of the lower jaw.

Differences in physiology, brain anatomy, and morphology are also diagnostic for this subfamily, including the articulation of the mandibular condyle and the position of the mental foramen. The two bones of the lower leg are fused.

The cerebrum is less highly developed in the white-toothed shrews than the red-toothed shrews.

There is considerable anatomical variation among genera of Crocidurinae. *Suncus etruscus* of southern Europe has a body mass of approximately 0.07 oz (2 g), making it one of the smallest mammals known. One of the largest true shrew representatives of the subfamily Crocidurinae is *Suncus murinus*, with a head and body length of 6 in (15 cm) and weight of 1.0 oz (30 g) for males and 0.7 oz (20 g) for females.

Myosorex is considered the most ancient of the living genera because it has the largest number of teeth. *Suncus* differ from *Crocidura* by retention of a fourth upper antemolar. *Sylvisorex* differ from both *Suncus* and *Crocidura* by a lack of tail bristles, and differ even more from *Crocidura* by the retention of a fourth upper antemolar. In most genera, the genital and urinary systems have a common opening though the skin. *Myosorex* has an independent urinary tract system.

There are also significant differences in brain development between genera within the Crocidurinae subfamily. Members of *Crocidura* show more brain development than members of the thick-tailed shrews of the genus *Suncus*.

Distribution

The subfamily Crocidurinae is Paleotropical in origin; the white-toothed shrews originated in the Old World tropics and radiated across Africa, Europe, and Asia. This is reflected in the pattern of present day Crocidurinae distribution. Some species of Crocidurinae live in arid regions and semideserts.

Habitat

Crocidurinae shrews usually inhabit damp and dry forests, grassland, cultivated areas, and occasionally human settlements and buildings. Some white-toothed shrews in the

A bicolored shrew (*Crocidura leucodon*) climbing among the rocks. (Photo by Rudolf Höfels/OKAPIA/Photo Researchers, Inc. Reproduced by permission.)

A lesser white-toothed shrew (*Crocidura suaveolens*) foraging in northern France. (Photo by J-C Carton. Bruce Coleman, Inc. Reproduced by permission.)

Diplomesodon and *Crocidura* genera live in arid areas and semi-deserts. They are mainly terrestrial, but some take to water freely, and others burrow a little. They are usually abundant wherever there is sufficient ground vegetation to provide cover. They occur over a great altitudinal range and in many kinds of plant communities. Most species prefer moist habitats, but some species are also found in arid regions.

Behavior

Shrews are extremely active, nervous animals. Some are active day and night, others only at night. When frightened, their heart may beat at a rate of 1,200 times per minute. Shrews have been known to die from loud noises, even thunder. Experiments with oxygen consumption suggest that shrews have a higher metabolic rate than mice of similar size. They are active throughout the year, though some go into torpor.

Some genera of Crocidurinae are notably aggressive and voracious. Shrews have been known to fight to the death. It is widely believed that most of these cases occurred during famine, when shrews may have become cannibalistic. The chirping and buzzing vocalizations sometimes heard are thought to be aggressive signals, and shrews may emit a squeak when alarmed or threatened.

Most shrews are solitary, except when they pair off during breeding. Members of *Suncus* and *Crocidura* can be tolerant of other shrews of the opposite gender during mating. After a period of initial fighting, a pair may live together. Once the pair bond has formed, the parent shrews may share a nest. Pairs from some species, such as *Crocidura russula*, have been observed building nests together.

Shrews are often territorial and have relatively small home ranges (0.1–0.25 acres [0.04–0.1 ha]) that vary in size with seasons of the year and with the timing of the animal's mating

season. Shrews mark their territories with their scent glands. Data on population densities are limited to a few species. The population densities of shrews varies considerably between species. The recorded population density of the lesser white-tailed shrew on Corsica is 1.9–5.3 animals per hectare.

White-toothed shrews can dig their own burrow but often use those of other animals such as woodmice or molerats. They can tunnel through loose humus and leaf mold and often are active under fallen trees and heaps of brushwood or stone.

Feeding ecology and diet

Shrews are primarily insectivorous and carnivorous, but will eat some plant materials such as seeds and nuts. Considering their high energy needs, it is likely that shrews need to adapt their diet according to food availability throughout the year. Their diets may include frogs, toads, and lizards. In captivity, many species will consume all but the skin, tail, and parts of the limbs of a small mammal; the brain is always eaten first. If no other food is available, some species resort to cannibalism.

The common belief that shrews cannot survive without eating every few hours is not a rule. In some Crocidurinae species, animals are able to lower their body temperatures in response to food scarcity. This is an example of food deprivation triggering torpor, a state of reduced activity level. This energy-saving state permits the animal to survive difficult conditions temporarily. For example, Savi's pygmy shrew (*Suncus etruscus*) loses 10–15% of its body weight by day, then increases food consumption at night to make up the loss. If food is scarce, the shrew may enter torpor.

Some species of shrews have salivary glands that secrete a poisonous substance that usually immobilizes the prey. There are records of shrew bites to humans that caused great pain. There is some question over whether the saliva of Savi's pygmy shrew is poisonous. These diminutive shrews need to bite their prey only once to subdue them.

The smallest mammal in the world is the Savi's pygmy shrew (*Suncus etruscus*). (Photo by Harald Schütz. Reproduced by permission.)

A common European white-toothed shrew (*Crocidura russula*) eating a worm. (Photo by B. B. Casals/FLPA–Images of Nature. Reproduced by permission.)

Reproductive biology

In the tropics, shrews form monogamous pairs and breed throughout the year, while those living in northern temperate zones usually breed from March to November. The known gestation periods are 17–28 days. There are one or several litters a year with 2–10 young in each litter. The young are born naked and blind in a nest of dried grass or leaves placed under a shelter or in a ground cavity. Weaning appears to occur at 2–4 weeks in most forms. Some zoologists estimate the life span in the wild to be 12–18 months, possibly longer. The oldest known Crocidurinae lived approximately four years.

Among Crocidurinae, the young are hairless for the first week and fully haired at 16 days. The auditory canals of Crocidurinae species open between the fifth and ninth days of life and the eyes open between the thirteenth and fifteenth days. Crocidurinae species begin to wean their young after 20–22 days. The young are practically adults and are sexually mature by 2–3 months.

Within the *Crocidura*, gestation is 27–31 days; litter size varies from one to 10. Larger litter sizes are associated with higher energy demands on the mother. The young weigh about 0.04 oz (1 g) at birth and are weaned at around 20 days. In *Crocidura russula*, young females can conceive at approximately 30 days of age.

Some Crocidurinae species and at least one *Sorex* species, display an unusual caravan behavior. The mother initally carries the young infants in her mouth. Starting on the sixth to tenth day, depending on the species, the mother and her young move in a caravan. At the slightest suspicious sound, one of the baby shrews grasps its mother's fur near the base of the tail. A second shrew grabs onto the first, a third onto

the second and so on until the whole litter is lined up behind the female who then pulls her train of offspring.

Conservation status

The greatest threat for the Crocidurinae shrews is environmental disturbance and pollution. Loss of forest habitats has a considerable effect on forest-dwelling species.

Data on the conservation status of most Crocidurinae shrew species are limited due to a paucity of population research. The Crocidurinae shrews included in the 2002 IUCN Red List often belong to very rare and endemic species with restricted distribution and low population densities. Some of the threatened species are only known from the locality of their original description. There is an urgent need for considerable research on most shrew species.

Significance to humans

Shrews may be valuable to agricultural communities as a form of natural control of insect pests due to their consumption of large numbers of insects and insect larvae. They also play a significant role in increasing the rate of litter decay.

Shrews are frequently used for neurological research and in other areas of biomedicine and evolutionary biology. They can also serve as useful indicators of environmental contamination such as heavy metals. Shrews feeding on earthworms in areas with heavy metals in the soil can accumulate significant levels of heavy metals in their tissues.

A lesser white-toothed shrew (*Crocidura suaveolens*) with young. (Photo by Andrew Cooper/Naturepl.com. Reproduced by permission.)

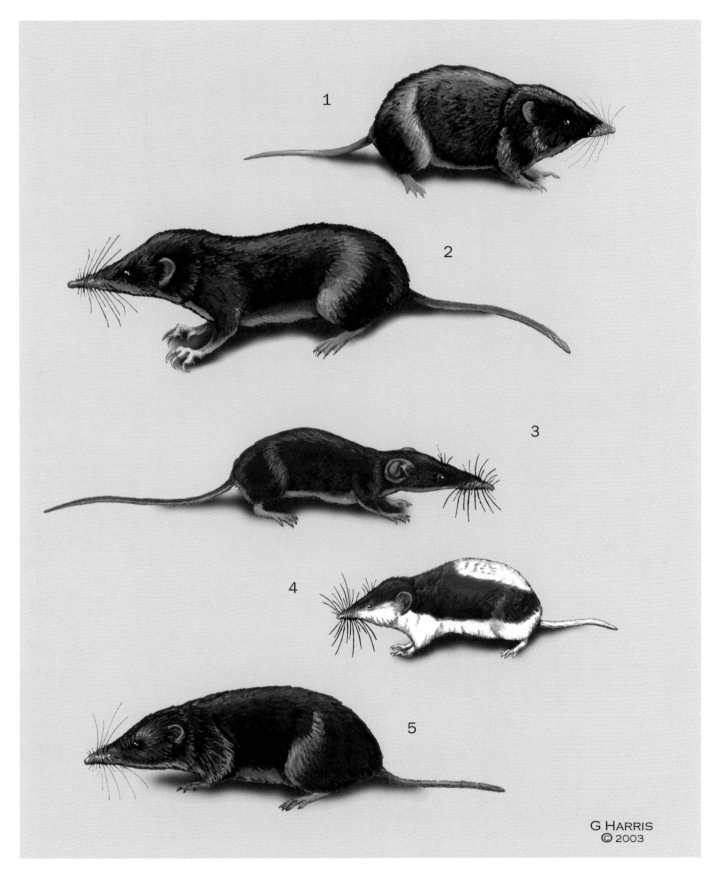

1. Ruwenzori shrew (*Ruwenzorisorex suncoides*); 2. Kelaart's long-clawed shrew (*Feroculus feroculus*); 3. Forest musk shrew (*Sylvisorex megalura*); 4. Piebald shrew (*Diplomesodon pulchellum*); 5. Schouteden's shrew (*Paracrocidura schoutedeni*). (Illustration by Gillian Harris)

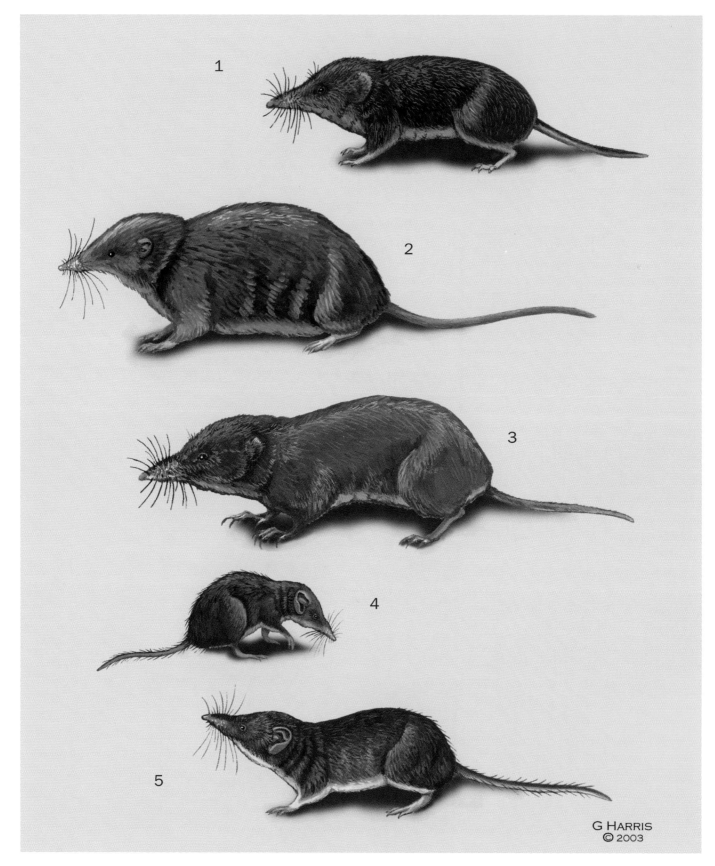

1. Forest shrew (*Myosorex varius*); 2. Armored shrew (*Scutisorex somereni*); 3. Pearson's long-clawed shrew (*Solisorex pearsoni*); 4. Savi's pygmy shrew (*Suncus etruscus*); 5. Common European white-toothed shrew (*Crocidura russula*). (Illustration by Gillian Harris)

Species accounts

Common European white-toothed shrew
Crocidura russula

TAXONOMY
Crocidura russula (Hermann, 1780), Bas Rhin, France.

OTHER COMMON NAMES
French: Musaraigne musette; German: Hausspitzmaus.

PHYSICAL CHARACTERISTICS
Head and body length 2.4–3.4 in (6.1–8.6 cm); tail 1.2–1.7 in (3.0–4.3 cm); weight 0.2–0.4 oz (6–12 g). Grayish brown to reddish brown with a silverly luster on its back. The flanks and underside are brownish. The tail is brown on the dorsal surface and gray on the underside with long protruding hairs. The ears are visible. It has 28 completely white teeth (including three upper unicuspids). This species is considered to be more advanced than many other Crocidurinae genera.

DISTRIBUTION
Occurs in Morocco, Algeria, Tunisia, southern and western Europe including an Atlantic island off of France, and on the Mediterranean islands of Ibiza, Sardinia, and Pantelleria.

HABITAT
Lives in or near farmyards, yards, fields, and at the edges of cities. Close to the Mediterranean, these small shrews are found in brushlands, in the underbrush of cork and in olive groves and the overgrowth of vineyards.

BEHAVIOR
Generally aggressive and voracious. When provoked, they frequently crouch on the ground with head raised and emit a squeak. This species is more tolerant of conspecifics. During the cold of winter, individuals may share nests. With onset of spring, mature females resume their territorial behavior.

FEEDING ECOLOGY AND DIET
Their diet consists of invertebrates and the bodies of freshly killed animals. In captivity, the species will consume all but the skin, tail, and parts of the limbs of a small mammal; the brain is always eaten first. They also feed on frogs, toads and lizards.

REPRODUCTIVE BIOLOGY
Gestation lasts 27–30 days. Litter size varies from three to eleven young. At birth, young weigh 0.02–0.03 oz (0.8–0.9 g). The young open their eyes at 13 days. They are hairless for the first week and fully haired at 16 days. After weaning at around 17–22 days, the young are practically adults, ready to become independent. Young females of this species can conceive at approximately 30 days of age. The expected life span for this species is 34–38 months.

Common white-tooth shrew mothers and young have been observed in caravan, a behavior where a young shrew offspring of six days or older grabs onto the back of its mother, and other young shrews from the litter form a line of shrews by latching onto each other. May form monogamous pairs for breeding.

CONSERVATION STATUS
Not threatened.

SIGNIFICANCE TO HUMANS
None known. ◆

Savi's pygmy shrew
Suncus etruscus

TAXONOMY
Suncus etruscus (Savi, 1822), Pisa, Italy.

OTHER COMMON NAMES
English: Pygmy white-toothed shrew; French: Pachyure etrusque, musaraigne etrusque; German: Etruskerspizmaus.

PHYSICAL CHARACTERISTICS
Head and body length 1.4–2.1 in (3.6–5.3 cm); tail 0.8–1.2 in (2.1–3 cm). One of smallest mammals in the world; smallest in Europe. Grayish brown dorsal side, blackish brown or ash-gray ventral side. Ears clearly visible. Thirty completely white teeth. Individual long hairs on the tail.

DISTRIBUTION
Mediterranean to India and Sri Lanka. Includes southern Europe, northern Africa, Arabian Peninsula and Asia Minor to Iraq, Turkmenistan, Afghanistan, Pakistan, Nepal, Bhutan, Myanmar, Thailand, and Yunnan (China).

HABITAT
Lives in old vineyards, olive groves, brushwood yards, cultivated land with low rock walls, dumps, rock piles.

BEHAVIOR
Considered solitary and intolerant of each other. Their repertoire of vocalizations, made up of chirps and buzzes are regarded as aggressive signals. However, in the laboratory, pairs and young of this species have been observed living peacefully together during breeding season.

Crocidura russula
Feroculus feroculus
Diplomesodon pulchellum

■ Suncus etruscus
■ Solisorex pearsoni
■ Paracrocidura schoutedeni

■ Sylvisorex megalura
■ Ruwenzorisorex suncoides

FEEDING ECOLOGY AND DIET
Nothing known.

REPRODUCTIVE BIOLOGY
In laboratory studies the gestation for *Suncus etruscus* has been recorded as 27.5 days. Litter sizes of 4–6 were observed in Pakistan while smaller litters of 2–5 offspring have been recorded in captivity. Presumably monogamous when breeding.

CONSERVATION STATUS
Not threatened.

SIGNIFICANCE TO HUMANS
None known. ◆

Forest musk shrew
Sylvisorex megalura

TAXONOMY
Sylvisorex megalura (Jentink, 1888), Junk River, Schieffe-linsville, Liberia.

OTHER COMMON NAMES
English: Climbing shrew.

PHYSICAL CHARACTERISTICS
Head and body length 1.8–4 in (4.5–10 cm); tail 1.5–3.5 in (4–9 cm); weight 3–0.42 oz (12 g). Soft, velvety fur usually longer than that of *Suncus*. Lack long tail hairs. Upperparts are usually slate gray and underparts are slightly paler.

DISTRIBUTION
Tropical forest zone of Africa, from upper Guinea to Ethiopia, and south to Mozambique and Zimbabwe.

HABITAT
Inhabit deep forests. Individuals have been observed climbing in trees and in grasslands well away from trees.

BEHAVIOR
Solitary animal that has been found active during the day and at night.

FEEDING ECOLOGY AND DIET
Nothing is known.

REPRODUCTIVE BIOLOGY
Nothing is known.

CONSERVATION STATUS
Not threatened.

SIGNIFICANCE TO HUMANS
None known. ◆

Piebald shrew
Diplomesodon pulchellum

TAXONOMY
Diplomesodon pulchellum (Lichtenstein, 1823), eastern bank of Uraal River, Kazakhstan.

OTHER COMMON NAMES
None known.

PHYSICAL CHARACTERISTICS
Head and body length 3.5–5.0 in (9.0–12.5 cm); tail 3.5–4.5 in (8.9–11.0 cm). The piebald shrew has a striking color pattern; fur is grayish with an elongated patch of white fur at the nape of the neck. Underparts, feet and tail are white. All hairs are gray at the base. It has fringes of long, supple hair on the soles and toes of the forepaws and hind paw. These fringes increase

the support surface of the paws of this sand-dwelling species and facilitate its movement on the sand. The ears are comparatively large.

DISTRIBUTION
Inhabits desert environments within Turkmenistan, Uzbekistan, and southern Kazakhstan (between Lake Balkhash and the Volga River).

HABITAT
Inhabits the sandy soil of semideserts.

BEHAVIOR
The piebald shrew is active throughout the year, primarily at night, but sometimes during the day. It is able to dig through sand quickly.

FEEDING ECOLOGY AND DIET
Primarily consumes lizards and insects. Observed killing lizards by biting the head in captivity. The shrew consumed all parts of the lizard, including the skull and skeleton, but left the feet and tail. When eating insects, only the chitin is not consumed.

REPRODUCTIVE BIOLOGY
Nothing is known, but probably monogamous when breeding.

CONSERVATION STATUS
Not threatened.

SIGNIFICANCE TO HUMANS
None known. ◆

Forest shrew
Myosorex varius

TAXONOMY
Myosorex varius (Smuts, 1832), Cape of Good Hope, South Africa.

OTHER COMMON NAMES
English: Mouse shrew; French: Musaraigne des bois africaine; German: Sudafrikanische Waldspitzmaus.

PHYSICAL CHARACTERISTICS
Head and body length 2.8–3.8 in (7.2–9.5 cm), tail 1.2–2 in (3.1–5 cm), weight 0.3–0.7 oz (7–19 g). Dorsal side is brown; ventral side, somewhat lighter, tail relatively short; eyes small; ears reduced. Distinguished by its first upper incisor, which is enlarged and has a hook at the end. This tooth also has a cusp projecting ventrally at the base.

DISTRIBUTION
South Africa, from northwest Cape Province to eastern Transvaal; Lesotho and Orange Free State.

HABITAT
Moist areas, usually in forests, scrub, or other dense vegetation along the banks of open and forested streams.

BEHAVIOR
Diurnal and nocturnal, active throughout the year. Members of this genus are primarily solitary except during the breeding season. However, males and females have been observed eating

and sleeping together after a period of avoiding each other, aggressive squeaks and/or brief fighting.

FEEDING ECOLOGY AND DIET
Various invertebrates including insects, snails, spiders, centipedes. Small birds and mammals when available.

REPRODUCTIVE BIOLOGY
Usually monogamous. Litter size is two to five. Weight at birth is 0.03 oz (1 g). Wean young at 20–22 days. Life span averages 12 months.

CONSERVATION STATUS
Not threatened.

SIGNIFICANCE TO HUMANS
None known. ◆

Armored shrew
Scutisorex somereni

TAXONOMY
Scutisorex somereni (Thomas, 1910), Kyetume, Uganda.

OTHER COMMON NAMES
English: Hero shrew; French: Musaraigne cuirassee, musaraigne armee; German: Schildspitzmaus.

PHYSICAL CHARACTERISTICS
Head and body length 4.8–6in (12–15 cm); tail 2.7–3.8 in (6.8–9.5 cm); weight 1.1–3.2 oz (30–90 g). Long thick gray fur with thin, chamois-colored streaks. Distinguished by a unique spine having almost double the number of vertebrae of other shrews. The vertebrae have lateral interlocking spines in addition to dorsal and ventral spines. This spine structure may pro-

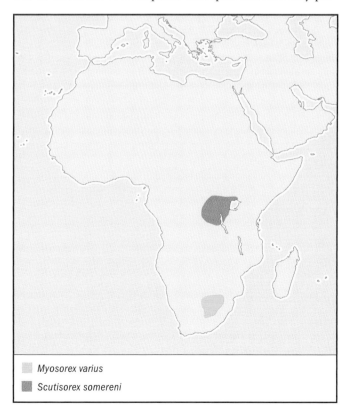

☐ *Myosorex varius*
■ *Scutisorex somereni*

vide considerable strength to the back, without limiting dorsal, ventral and lateral flexibility. Other characteristics include a naked tail, 30 completely white teeth, and four teats.

DISTRIBUTION
Tropical rainforests of the Zaire Basin and adjacent mountains in Uganda, Rwanda, and Burundi.

HABITAT
Edges of gallery forests; in fallen leaves and in swampy regions.

BEHAVIOR
The armored shrew is active both night and day.

FEEDING ECOLOGY AND DIET
Feeds on small mammals when available as well as insects and some plant matter.

REPRODUCTIVE BIOLOGY
Probably monogamous. Small litters with one to three off-spring.

CONSERVATION STATUS
Not threatened.

SIGNIFICANCE TO HUMANS
None known. ◆

Pearson's long-clawed shrew
Solisorex pearsoni

TAXONOMY
Solisorex pearsoni Thomas, 1924, Central Province, Sri Lanka.

OTHER COMMON NAMES
French: Pachyure aux longues griffes; German: Pearsons Langkrallenspitzmaus.

PHYSICAL CHARACTERISTICS
Head and body length 5–5.4 in (12.5–13.4 cm); tail 2.4–2.6 in (5.9–6.6 cm). The fur is soft, dark gray along the back with a lighter lustrous underside. Its small ears are fully furred and hidden in fur. Teeth are large and heavy and the anterior incisors are well developed. Claws on the forefeet are very long. The slender tail is closely haired and lacks the scattered long hairs. Twenty-eight completely white teeth.

DISTRIBUTION
Restricted to the central highlands of Sri Lanka, from an altitudinal range of 3,609–6,070 ft (1,100–1,850 m).

HABITAT
This species is reported to inhabit virgin forest though some individuals have been trapped in long grass.

BEHAVIOR
Nothing is known.

FEEDING ECOLOGY AND DIET
Feeds on invertebrates and vertebrates.

REPRODUCTIVE BIOLOGY
Nothing is known.

CONSERVATION STATUS
Endangered. *S. personi* is restricted to a small area of declining habitat that is being altered by human encroachment.

SIGNIFICANCE TO HUMANS
None known. ◆

Schouteden's shrew
Paracrocidura schoutedeni

TAXONOMY
Paracrocidura schoutedeni Heim de Balsac, 1959, Kasai, Democratic Republic of the Congo.

OTHER COMMON NAMES
French: Musaraigne a dents blanches du Congo, musaraigne a dents blanches de Schouteden; German: Schildspitzmaus.

PHYSICAL CHARACTERISTICS
Head and body length 2.6–3.8 in (6.5–9.5 cm); tail 1.4–1.7 in (3.4–4.4 cm); weight 0.3–0.7 oz (10–20 g). Short, fine, black fur, small ears covered with fine fur, short limbs, short claws. The tail has short hair with some long hairs. The teeth are completely white. *Paracrocidura* is identified as a separate genus on the basis of its distinct dental characteristics. It has 28 white teeth, of which three are maxillary unicuspids. An unusual feature is the straightness of the posterior borders of the fourth premolar and first and second molars, and the development of the first premolar. It also has aliform processes at the inner edge of the upper incisors and a bicuspid second lower premolar.

DISTRIBUTION
Occurs in the lowland primary forests in south Cameroon, Gabon, Democratic Republic of the Congo, and in Central Africa Republic.

HABITAT
Lowland rainforests.

BEHAVIOR
Nothing is known.

FEEDING ECOLOGY AND DIET
Nothing is known.

REPRODUCTIVE BIOLOGY
Litter size is small, one to three young.

CONSERVATION STATUS
Not threatened.

SIGNIFICANCE TO HUMANS
None known. ◆

Ruwenzori shrew
Ruwenzorisorex suncoides

TAXONOMY
Ruwenzorisorex suncoides (Osgood, 1936), Kalongi, Democratic Republic of the Congo.

OTHER COMMON NAMES
None known.

PHYSICAL CHARACTERISTICS
Head and body length 3.6–3.74 in (9.2–9.5 cm); tail 2.2–2.5 in (5.5–6.2 cm); weight 0.64 oz (18.2 g). Shiny black fur with paler sides and feet. Stocky body and short head. Hind feet are

long. The tail is almost naked and has no long hairs. The external ears are small and round. Primarily distinguished by cranial and dental features. The skull is relatively flat and rectangular in profile. The first lower molar is massive.

DISTRIBUTION
Endemic to the Albertine Rift, occurring from the Ruwenzori Mountains (Uganda), eastern Democratic Republic of the Congo, Burundi and Rwanda.

HABITAT
Damp, mossy mountain forests. All specimens have been found at elevations of 2,625–6,890 ft (800–2,100 m).

BEHAVIOR
Comparatively social as indicated by the presence of multiple shrews nesting together.

FEEDING ECOLOGY AND DIET
Nothing known.

REPRODUCTIVE BIOLOGY
There is some indication that Ruwenzori shrews breed twice a year. Probably monogamous when breeding

CONSERVATION STATUS
Vulnerable.

SIGNIFICANCE TO HUMANS
None known. ◆

Kelaart's long-clawed shrew
Feroculus feroculus

TAXONOMY
Feroculus feroculus (Kelaart, 1850), central mountains, Sri Lanka, at altitude of 6,000 ft (2,000 m). Initially, this species

was identified as a water shrew; more recently identified as a separate species (semi-fossorial).

OTHER COMMON NAMES
French: Pachyure aux longues griffes; German: Kelaarts Langkrallenspitzmaus.

PHYSICAL CHARACTERISTICS
Head and body length 4.2–4.7 in (10.6–11.8 cm); tail 2.2–2.9 in (5.6–7.3 cm); 1.2–1.3 oz (35–37 g). Known from only a few specimens. The back is slate black, underside is lighter, dark tail with a few whitish hairs at the tip and some individual long hairs, and the forefeet are off-white. It has 30 completely white teeth.

DISTRIBUTION
Endemic to central mountains of Sri Lanka.

HABITAT
Observed among weeds and other dense undergrowth in wet ravines. Primarily along hillsides of the central mountains of Sri Lanka. It appears to be confined to valleys and slopes of the central mountain range of Sri Lanka at elevations between 5,550 and 6,450 ft (1,850–2,150 m).

BEHAVIOR
Very little is known about the behavior of this species.

FEEDING ECOLOGY AND DIET
Feeds on invertebrates, vertebrates and plant material.

REPRODUCTIVE BIOLOGY
Nothing is known.

CONSERVATION STATUS
Endangered. Decline of this species may be due to encroachment of humans on its limited habitat.

SIGNIFICANCE TO HUMANS
None known. ◆

Common name / Scientific name/ Other common names	Physical characteristics	Habitat and behavior	Distribution	Diet	Conservation status
Poll's shrew *Congosorex polli*	Buffy, fawn, or shades of brown, gray, or black. Coat appears speckled due to light hairs interspersed with dark hairs. May have gray feet. Head and body length 2.4–4.3 in (6–11 cm), tail length 0.9–2.6 in (2.4–6.7 cm).	Moist areas, especially forests or scrub and the dense vegetation lining the banks of mountain streams. Generally solitary, except during breeding season.	Known only from type locality in southern Democratic Republic of the Congo (Zaire).	Mainly insects, but may also eat small birds and mammals.	Critically Endangered
Gray shrew *Crocidura attenuata* English: Indochinese shrew	Light brownish gray on upperparts, pale gray on underparts. Underside is faintly tinged with brown. Tail is dark above and light below. Backs of feet are thinly covered with short, pale hairs.	Variety of habitats, from seed-ling rice fields to cut-down forest-farmlands of weeds and grass. Solitary, except during breeding season.	Assam, India; Nepal; Bhutan; Burma; Thailand; Vietnam; Hainan, China; Taiwan; Peninsular Malaysia; Sumatra; Java; Christmas Islands (Indian Ocean); and Batan Islands, Philippines.	Mainly insects, but may also eat small birds and mammals.	Not threatened
Canary shrew *Crocidura canariensis* Spanish: Musarana canaria	Uniform chocolate brown. Head and body length 1.6–4.3 in (4–11 cm). Tail is long and covered with long, white and short, bristly hairs. Foreclaws are not enlarged.	Malpais (barren lava fields), as the species has adapted to the hot and dry conditions of the plains. Very shy animals, little known of reproductive cycles. Very low litter sizes.	Canary Islands, Spain.	Mainly insects.	Vulnerable
Dsinezumi shrew *Crocidura dsinezumi*	During winter pelage is pale gray to brown, summer, pelage is usually dark brown. Under parts are lighter. Head and body length 2.6–3.1 in (6.5–8 cm), tail length usually less than 70% of head and body length.	Along river banks and in the foothills with dense vegetation. Generally solitary, except during breeding season.	Japan; Quelpart Islands, Korea; and possibly Taiwan.	Insects and spiders.	Not threatened
Southeast Asian shrew *Crocidura fuliginosa*	Dark gray to blackish with a dull, silvery gloss. Underparts are lighter, tail is thin with a few, faint, white hairs. Ears are naked and prominent, eyes are small, feet covered with a few, short white hairs.	Various habitats from montane to lowland forest, cultivated areas, and even caves. Solitary, except during breeding season.	Northern India, Myanmar, adjacent China, Malaysian Peninsula and adjacent islands; perhaps also Borneo, Sumatra and Java; exact distribution unknown.	Mainly insects.	Not threatened
Horsfield's shrew *Crocidura horsfieldii*	Upperparts are neutral gray with bottom portions of hairs brown. Underside is dark gray. Tail is paler above than beneath.	Intermediate montane from 4,000 to 6,960 ft (1,220–2,120 m) in areas of fairly heavy cover. Solitary, except during breeding season.	Sri Lanka; northern Thailand to Vietnam; Nepal; Mysore and Ladak, India; Yunnan, Fukien, and Hainan Islands, China; Taiwan; and Ryukyu Islands, Japan.	Mainly insects.	Not threatened
Lesser white-toothed shrew *Crocidura suaveolens* English: Lesser shrew	Reddish gray above, slightly lighter under-side. Ears are short-haired and prominent, tail is covered with fine, long, white hairs. Tail length 0.9–1.7 in (2.4–4.4 cm), head and body length 2–3 in (5–7.5 cm).	Temperate woodlands and steppe. Generally solitary, not highly territorial, breeding season extends from March to September.	Entire Palearctic from Spain to Korea; Atlantic islands (Scilly, Jersey, Sark, Ushant, Yeu); many Mediterranean islands including Corsica, Crete, Cyprus, and Menorca; and Tsushima and Ullong Do between Korea and Japan.	Mainly insects.	Not threatened
Schaller's mouse shrew *Myosorex schalleri*	Buffy, fawn, or shades of brown, gray, or black. Underparts of hairs are brown, giving speckled effect. Head and body length 2.4–4.3 in (6–11 cm), tail length 6–2.6 in (2.4–6.7 cm).	Moist areas, especially forests or scrub and the dense vegetati on lining the banks of mountain streams. Generally solitary, except during breeding season.	Eastern Democratic Republic of the Congo (Zaire), Itombwe Mountains, "Nzombe (Mwenga)".	Mainly insects, but will also eat small birds and mammals.	Critically Endangered
Greater shrew *Paracrocidura maxima*	Short, large snout, short and visible ears, short limbs and claws. Body covered in thin, short, ashy black or dark brown hair, tail is black. Head and body length 2.6–3.8 in (6.5–9.6 cm), tail length 1.3–1.8 in (3.3–4.6 cm).	Rainforests at elevations of about 660–7,710 ft (200–2,350 m). Terrestrial.	Democratic Republic of the Congo (Zaire), Rwanda, and Uganda.	Mainly insects.	Not threatened

[continued]

Common name / Scientific name/ Other common names	Physical characteristics	Habitat and behavior	Distribution	Diet	Conservation status
Grauer's shrew *Paracrocidura graueri*	Short, large snout, short and visible ears, short limbs and claws. Body covered in thin, short, ashy black or dark brown hair, tail is black. Head and body length 2.6–3.8 in (6.5–9.6 cm), tail length 1.3–1.8 in (3.3–4.6 cm).	Rainforests at elevations of about 660–7,710 ft (200–2,350 m). Terrestrial.	Itombwe Mountains, Democratic Republic of the Congo (Zaire).	Mainly insects.	Critically Endangered
Aberdare shrew *Surdisorex norae*	Pelage is deep lustrous brown with pale frosting, under parts are paler and buffier. Tail is dark brown above and below. Fur is long, dense, and coarse. Head and body length 2.3–4.3 in (6–11 cm), tail length 0.9–2.6 in (2.4–6.7 cm).	Moss and heather cloud forests of the Kenyan mountains, at altitudes of 9,190–10,830 ft (2,800–3,300 m). Active both day and night. Stay mainly in burrow systems.	Aberdare Range, Kenya.	Mainly insects, but will also eat small mammals and birds.	Vulnerable
Mt. Kenya shrew *Surdisorex polulus*	Pelage is deep lustrous brown with pale frosting, under parts are paler and buffier. Tail is dark brown above and below. Fur is long, dense, and coarse. Head and body length 2.3–4.3 in (6–11 cm), tail length 0.9–2.6 in (2.4–6.7 cm).	Moss and heather cloud forests of the Kenyan mountains, at altitudes of 9,190–10,830 ft (2,800–3,300 m). Active both day and night. Stay mainly in burrow systems.	Mount Kenya, Kenya.	Mainly insects, but will also eat small mammals and birds.	Vulnerable
Volcano shrew *Sylvisorex vulcanorum*	Upperparts generally slaty gray, underparts are paler. Fur is soft and velvety. Head and body length 1.8–3.9 in (4.5–10 cm), tail length 1.6–2.6 in (4–6.5 cm).	Deep forest, as well as in grassy areas well away from trees. Arboreal, terrestrial, both diurnal and nocturnal. Solitary.	High altitude rainforest of eastern Democratic Republic of the Congo (Zaire), Uganda, Rwanda, and Burundi	Mostly insects taken through foraging.	Not threatened
Sunda shrew *Crocidura monticola*	Body uniform dull gray-brown. Tail paler with sparse, long, pale hair at the base, usually extending over 0.3 in (I cm) along the tail. Head and body length 1.6–7.1 in (4–18 cm), tail length 1.6–4.3 in (4–11 cm).	Damp and dry forests, grassland, cultivated areas, and occasionally human settlements and buildings. Aggressive. Litter size ranges from one to 10 offspring.	Borneo, Java, and Peninsular Malaysia.	Invertebrates and freshly killed animals.	Not threatened
Negros shrew *Crocidura negrina*	Small animal with long, slender tail. Pelage is dark, blackish, tinged with gray on upper parts. Underparts are dark brown with white spots.	This species has only been recorded from Cuernos de Negros mountain, Negros Island, the Philippines at an altitude range of 1,640–4,760 ft (500–1,450 m). Aggressive. Litter size ranges from one to 10 offspring. Solitary, except during breeding season.	Primary forest at 1,640–4,760 ft (500–1,450 m) on southern Negros Island, Philippines.	Mainly insects, invertebrates, and freshly killed animals.	Critically Endangered

Resources

Books

Corbet, G. B., and J. E. Hill. *The Mammals of the Indomalayan Region: A Systematic Review.* Oxford: Oxford University Press, 1992.

IUCN. *Eurasian Insectivores and Tree Shrews—Status Survey and Conservation Action Plan.* Gland, Switzerland: IUCN, 1995.

Merritt, J. F., G. L. Kirkland, and R. K. Roberts, eds. *Advances in The Biology of Shrews.* Pittsburgh: Carnegie Museum of Natural History, Special Publication 18, 1994.

Mitchell-Jones, A. J., et al. *The Atlas of European Mammals.* London: Poyser Natural History—Academic Press, 1999.

Wilson, D. E., and S. Ruff, eds. *The Smithsonian Book of North American Mammals.* Washington, DC, and London: Smithsonian Institution Press, 1999.

Wójcik, J. M., and M. Wolsan, eds. *Evolution of Shrews.* Bialowieza: Mammal Research Institute, Polish Academy of Sciences, 1998.

Organizations

IUCN Species Survival Commission, Insectivore Specialist Group, Dr. Werner Haberl, Chair. Hamburgerstrasse 11, Vienna, A-1050 Austria. Phone: +4315861094. Fax: +4315861094. E-mail: shrewbib@sorex.vienna.at Web site: <http://members.vienna.at/shrew/itses.html>

Corliss Karasov

Moles, shrew moles, and desmans

(Talpidae)

Class Mammalia

Order Insectivora

Family Talpidae

Thumbnail description
Small, often long- and narrow-snouted mammals, many with large forelegs, and small or hidden eyes suited to a fossorial lifestyle

Size
Average adult total lengths (including tail) range from about 2.4–17.0 in (6–43 cm), with tail lengths of 0.6–8.3 in (1.5–21.5 cm) and weights of about 0.4–7.8 oz (12–220 g)

Number of genera, species
17 genera; 42 species

Habitat
Depending on the species, they may prefer fossorial, terrestrial, or aquatic habitats

Conservation status
Critically Endangered: 2 species; Endangered: 5 species; Vulnerable: 4 species

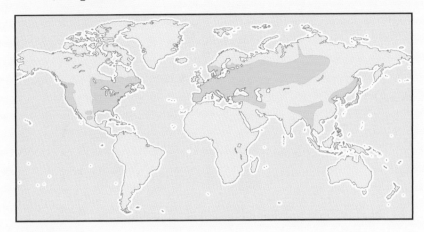

Distribution
North America and Eurasia

Evolution and systematics

The family Talpidae includes the moles, shrew moles, and desmans. Some taxonomists consider the desmans (*Desmana* and *Galemys* spp.) different enough from the other talpids to deserve a separate family status, but this chapter discusses them with the talpids.

The Talpidae is usually divided into three subfamilies—Uropsilinae, Desmaninae, and Talpinae—as it is here. Two other subfamilies, Scalopinae and Condylurinae, are occasionally separated out of the talpins, which is the largest subfamily.

The talpins contain 14 genera. Among that number are several genera, including *Mogera*, *Parascaptor*, and *Scaptochirus*, which have recently been split from the large *Talpa* genus.

The subfamily Uropsilinae has one genus. Taxonomists now regard its formerly lone taxon as four separate species. The subfamily Desmaninae has two monotypic genera.

Evolutionarily, the talpids are believed to have originated in Europe, and spread from there throughout Eurasia and into North America.

Physical characteristics

The typical talpid is a small, tube-shaped mammal with short, silky fur, and a narrow muzzle. The fossorial (burrowing) forms, which make up more than half the species in this family, have large, clawed hands specialized for digging, small or unseen eyes

suited to their dark habitat, and fur that lies flat regardless of whether it is pointing backward or forward on the body. The aquatic and the terrestrial, surface-dwelling species lack the exaggerated forefeet, and some aquatic taxa have webbed or enlarged hind feet that propel them through the water.

Talpids also have distinctive short necks and limbs. The enlarged upper arm bone, or humerus, articulates with the col-

A star-nosed mole (*Condylura cristata*) emerging from its hole. (Photo by Rod Plank/Photo Researchers, Inc. Reproduced by perission.)

A European mole (*Talpa europaea*) storing earthworms. (Illustration by Patricia Ferrer)

larbone, or clavicle, and the forefeet are turned outward rather than down. This combination of features permits a strong, sideways sweeping action that is efficient for digging or swimming.

Conspecific males and females are commonly similar in appearance, although the average male is often a bit larger.

Distribution

This Northern Hemisphere family occurs in North America, Europe, and Asia. In the New World, it ranges throughout the United States, and reaches into southern Canada and northern Mexico. In the Old World, talpids live in temperate climates from the Mediterranean Sea to Japan and north well into Russia.

Some species exist over a wide area. The European mole (*Talpa europaea*), for example, is spread throughout Europe and into Russia. More geographically limited species include the greater Japanese shrew mole (*Urotrichus talpoides*), which occurs only in Japan, and the Gansu mole (*Scaponulus oweni*), which lives in a small area within central China.

Habitat

The talpins are chiefly fossorial, the uropsilins prefer an above-ground lifestyle, and the desmanins are semi-aquatic.

The fossorial talpins exist in forests and/or fields with some opting for wet soils close to water, and a few, like the star-nosed mole (*Condylura cristata*), frequently leaving their tunnels for a swim. Besides the uropsilans, some species of the talpins, like the American shrew mole (*Neurotrichus gibbsii*) are mainly surface dwellers. These uropsilan and talpin moles that live above ground commonly shun open spaces, instead scooting beneath leaf litter or under a log, but many are known to climb into shrubs and trees. Semi-aquatic species generally favor freshwater, although a few species, such as the Russian desman (*Desmana moschata*), will sometimes venture into brackish water.

Behavior

As a whole, moles are best known for their tunnels, even though some talpids are not fossorial species. Evidence of tunnelers' activities is often visible as "mole runs" that zig-zag across an otherwise level lawn or forest trail. The runs are actually the roofs of the tunnels. These shallow tunnels are usually feeding runs, which the mole uses to seek out subterranean earthworms or other invertebrates. Moles can make the shallow tunnels rather quickly, with the eastern mole (*Scalopus aquaticus*) tunneling at a rate of up to 15 ft (4.6 m) an hour. Their activities are also frequently evident as molehills. While they pack some dirt to make the walls of the tunnel, fossorial

An eastern mole (*Scalopus aquaticus*) burrowing. (Photo by E. R. Degginger. Bruce Coleman, Inc. Reproduced by permission.)

moles typically push the leftover dirt from deep-tunnel excavations to the surface, where it forms small molehills. Molehills are usually only 6–12 in (15–30 cm) in diameter and 3–6 in (8–15 cm) tall. Deeper tunnels provide living quarters, breeding sites, and, in winter, protection from cold weather. Moles will also retreat to deeper tunnels during periods of summer drought.

Semi-aquatic species may also utilize burrows for mating or as winter shelter, but spend the bulk of their active hours in the water rather than underground. Some, like the Pyrenean desman (*Galemys pyrenaicus*), prefer fast-flowing streams and rivers, while others like the Russian desman, favor slow-moving streams and lakes. Surface-dwelling species are likely to find shelter under a log or leaf litter instead of a subterranean tunnel.

Active day and night, most moles are solitary animals, although some will share the same foraging grounds. In this case, they usually continue to maintain their distance from one another, often by covering the same area but at different times of the day. The Russian desman appears to be more social than other moles, and will not only share foraging tunnels but, in at least one case, its den.

Feeding ecology and diet

Talpids make good use of their Eimer's organs, which are sensory receptors on their snouts, to identify and perhaps to locate food items. The organs, which contain nerve cells, respond to touch and may also pick up seismic vibrations. The latter would help a fossorial mole, in particular, to hunt prey items in the dark tunnels. Some scientists suggest that talpid moles, including *Condylura* species, may also be able to feel vibrations through their forepaws. Despite these interesting adaptations possibly used in hunting, burrowing talpids seem to find most of their food by simply bumping into it while moving through their tunnels.

A star-nosed mole (*Condylura cristata*) foraging for worms. (Photo by Dwight Kuhn. Bruce Coleman, Inc. Reproduced by permission.)

A common Eurasian mole (*Talpa europaea*) heading out of its tunnel. (Photo by J-C Carton. Bruce Coleman, Inc. Reproduced by permission.)

A star-nosed mole (*Condylura cristata*) drinking. (Photo by Dwight Kuhn. Bruce Coleman, Inc. Reproduced by permission.)

A hairy-tailed mole (*Parascalops breweri*) showing front feet well suited for burrowing. (Photo by Animals Animals ©Zig Leszczynski. Reproduced by permission.)

Conservation status

Nearly a quarter of the species in this family are threatened. According to the IUCN, the list includes 10 of the approximately 42 talpid species. The Russian desman, Japanese mountain mole (*Euroscaptor mizura*), and Pyrenean desman are Vulnerable. The sado mole (*Mogera tokudae*), Echigo Plain mole (*Mogera etigo*), Ryukyu mole (*Nesoscaptor uchidai*), Yunnan shrew-mole (*Uropsilus investigator*), and Chinese shrew-mole (*Uropsilus soricipes*) are Endangered. The small-toothed mole (*Euroscaptor parvidens*) and Persian mole (*Talpa streeti*) are Critically Endangered.

Significance to humans

Burrowing moles are perhaps best known to humans for their tunnels, which are the bane of gardeners and farmers, as well as homeowners who desire a perfect lawn. Few realize that mole activities help turn over and aerate the soil, or that they feed on a considerable quantity of harmful insects, particularly beetle larvae and slugs.

Swimming species, on the other hand, do appear to engage in active hunting, and are able to catch even small fish. Several fossorial, terrestrial, and semi-aquatic talpids also eat vegetative matter, but the primary diet item is invertebrates.

Predators mainly are larger mammals, including domestic cats and dogs that will unearth fossorial moles. Talpids' strong musky odor, however, often repels attackers.

Reproductive biology

In general, moles in the family Talpidae mate from late winter to late spring, with a single litter born from mid- to early summer. The schedule within a species can be moved up by a month or so among populations in warmer climates. Some moles, such as the greater Japanese shrew mole (*Urotrichus talpoides*), have a second litter in the summer or early fall. Gestation typically lasts four to seven weeks, with the young weaned three to four weeks after birth. Litter size averages three or four young, but can range from just one to seven or more. The young attain sexual maturity within their first year. Life span averages three to four years. Mating system varies among species.

An American shrew mole (*Neurotrichus gibbsii*) foraging. (Photo by R. Wayne Van Devender. Reproduced by permission.)

1. Star-nosed mole (*Condylura cristata*); 2. Hairy-tailed mole (*Parascalops breweri*); 3. European mole (*Talpa europaea*); 4. Eastern mole (*Scalopus aquaticus*); 5. American shrew mole (*Neurotrichus gibbsii*); 6. Russian desman (*Desmana moschata*). (Illustration by Brian Cressman)

Species accounts

American shrew mole
Neurotrichus gibbsii

SUBFAMILY
Talpinae

TAXONOMY
Neurotrichus gibbsii (Baird, 1857), Naches Pass, Washington, United States. Three subspecies.

OTHER COMMON NAMES
English: Gibb's shrew-mole, least shrew-mole; German: Amerikanischer Spitzmull, Amerikanischer Spitzmausmaulwurf; Spanish: Topo musaraña americano.

PHYSICAL CHARACTERISTICS
The body length is 3.5–5.2 in (8.9–13.2 cm), and the tail is 1.0–1.6 in (2.5–4.0 cm) long. Average adult weight ranges from 0.32–0.39 oz (9–11 g). A shiny, black to dark gray mole with a thick, hairy tail and only faintly widened hands. Males and females are similar.

DISTRIBUTION
Along the North American coast from southwestern British Columbia in Canada to central California in the United States.

HABITAT
Occurs in forests and shrubby areas, usually near a water source.

BEHAVIOR
Although this mole does make burrows for resting, it also spends considerable amounts of time above ground, often scurrying through leaf litter or climbing into shrubs. It is also able to swim. Unlike more fossorial moles, the only evidence of its burrowing are the small entrances. It packs the dirt into the walls rather than pushing it out to form molehills.

This species is unusual in that it will travel in groups, instead of taking on the solitary lifestyle of most moles.

FEEDING ECOLOGY AND DIET
Active day and night, this mole searches for food along the substrate in leaf litter and above ground in shrubs. It eats earthworms, insects, other invertebrates, and occasionally fungus or vegetation.

REPRODUCTIVE BIOLOGY
Polygamous. Little is known about its reproductive biology, but it appears to produce several litters of one to four young each year.

CONSERVATION STATUS
Not threatened.

SIGNIFICANCE TO HUMANS
None known. ◆

Eastern mole
Scalopus aquaticus

SUBFAMILY
Talpinae

TAXONOMY
Scalopus aquaticus (Linnaeus, 1758), Philadelphia, Pennsylvania, United States. Sixteen subspecies.

OTHER COMMON NAMES
French: Taupe à queue glabre; German: Ostamerikanischer Maulwurf; Spanish: Topo de agua.

PHYSICAL CHARACTERISTICS
Adults range from 5.9–7.9 in (15.0–20.0 cm) in total length, and 0.8–1.5 in (2.0–3.8 cm) in tail length. Adults generally weigh 3.2–5.0 oz (90–143 g). On average, males are slightly larger than females. An often glistening, stocky mole with brownish gray, sometimes black fur. It has a short, hairless tail, and large, wide feet sporting long claws.

DISTRIBUTION
Eastern half of the United States, except the far northern reaches. Extends south to extreme northeastern Mexico.

HABITAT
Although it has webbed feet and its species name is *aquaticus*, this mole shuns aquatic habitats, opting instead for the moist, sandy or loamy soils of forests and fields, as well as lawns, and other cultivated areas.

BEHAVIOR
This is a solitary, fossorial mole, making shallow tunnels for foraging and deeper tunnels up to 2.5 ft (0.8 m) underground for winter denning. The shallow tunnels are evident as trails, or "mole runs," of loose dirt across the forest floor or a lawn. This mole also makes small molehills, which are piles of dirt pushed out of the tunnels and onto the surface.

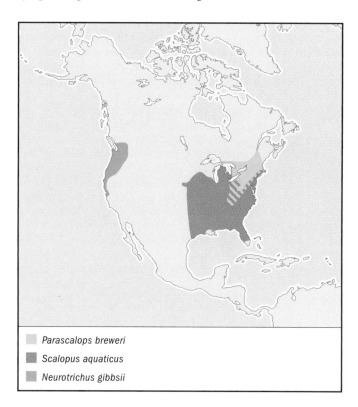

Parascalops breweri

Scalopus aquaticus

Neurotrichus gibbsii

During breeding season, moles make a large room, and line the floor with grass and leaves for a nest.

FEEDING ECOLOGY AND DIET
This species is an omnivore. Its diet includes mostly insect larvae and earthworms, but it also eats other invertebrates, including slugs and centipedes, as well as roots and seeds. Predators include hawks and owls during the rare occasions when the mole is on the surface, or digging mammals, such as foxes, and domestic cats and dogs.

REPRODUCTIVE BIOLOGY
Promiscuous. Mating occurs in early spring, with a litter of three to five altricial young born 40–45 days later. Females have only one litter per year. The young are weaned at four to five weeks and become sexually mature by the following spring.

CONSERVATION STATUS
Not threatened.

SIGNIFICANCE TO HUMANS
Moles are a pest species to many gardeners who lose crops to their foraging, and to homeowners who find their impeccably groomed lawns irregularly patterned by mole trails. ◆

European mole
Talpa europaea

SUBFAMILY
Talpinae

TAXONOMY
Talpa europaea Linnaeus, 1758, Engelholm, Sweden. Four subspecies.

OTHER COMMON NAMES
English: Mole, common mole; French: Taupe, taupe d'Europe; German: Europäischer Maulwurf; Spanish: Topo europeo.

PHYSICAL CHARACTERISTICS
The total length is 4.7–5.5 in (12–14 cm), with the tail 0.8–1.6 in (2–4 cm) long. Average adult weight ranges from 2.1–4.2 oz (60–120 g). Gray mole with a long snout, vertically oriented fur, and shovel-like hands. The back is dark gray, sometimes black, and the belly is lighter gray. On average, males are slightly larger and darker than females.

DISTRIBUTION
Temperate Europe to western Russia.

HABITAT
Ranges from forests to fields, not as common in farm fields, and seldom in sand dune areas.

BEHAVIOR
Fossorial moles that often use common, existing tunnels from previous generations. These common tunnels service many moles. Individuals do, however, construct shallow foraging tunnels that are used only by that one mole. Like many other fossorial talpids, this species builds a nest of leaves during breeding season in an underground chamber within the tunnel complex.

FEEDING ECOLOGY AND DIET
This carnivorous mole primarily eats earthworms, which it identifies mostly by touch. On occasion, it will also kill and eat small snakes, lizards, rodents, and birds. This species is known to maim earthworms so that they are unable to dig their own escape burrows, and store the captives alive for later consumption.

☐ *Talpa europaea*
■ *Condylura cristata*
▨ *Desmana moschata*

REPRODUCTIVE BIOLOGY

Probably promiscuous. Mating occurs in the spring, with a litter of three to four altricial young born 28 days later. Females typically have one litter per year. The young are weaned at four to five weeks and become sexually mature at six months.

CONSERVATION STATUS

Not threatened.

SIGNIFICANCE TO HUMANS

Gardeners, farmers, and homeowners typically regard this species as a pest because of its tunneling lifestyle. Its tunneling also aerates the soil, but this benefit is rarely acknowledged. ◆

Hairy-tailed mole
Parascalops breweri

SUBFAMILY

Talpinae

TAXONOMY

Parascalops breweri (Bachman, 1842), eastern North America.

OTHER COMMON NAMES

English: Brewer's mole; French: Taupe à queue velue; German: Bürstenmaulwurf, Haarschwanzmaulwurf.

PHYSICAL CHARACTERISTICS

The total length is 5.5–6.7 in (14.0–17.0 cm), with the tail 0.9–1.4 in (2.3–3.6 cm) long. Average adult weight ranges from 1.4–2.3 oz (40–65 g). Males are generally larger than females. Dark brown, sometimes black mole with a hairy tail. It has the wide, clawed hands and tiny eyes typical of burrowing talpids.

DISTRIBUTION

Eastern United States and southeastern Canada stretching from northwestern South Carolina and northern Georgia at its southern extreme to southern Ontario and Quebec in the north.

HABITAT

Prefers sandy or loamy forests, fields, lawns, and other cultivated areas.

BEHAVIOR

This fossorial mole spends most of its time in its shallow, foraging tunnels or in deeper tunnels that provide winter protection. It also employs the deeper tunnels as breeding sites, where it constructs a grassy or leafy nest. The shallow tunnels are narrow and just deep enough to at most only minimally disturb the surface. Its mole hills, however, are evident. Hairy-tailed moles are generally solitary, but not territorial, and conspecifics as well as other small mammals may also traverse its tunnels.

FEEDING ECOLOGY AND DIET

Day and night, this mole forages for food, which includes earthworms, beetles, and other invertebrates, including slugs and centipedes, and an occasional plant root. Predators are the same as those listed for the eastern mole, but also include short-tailed shrews (*Blarina brevicauda*), which sometimes prey on newborns.

REPRODUCTIVE BIOLOGY

Probably promiscuous. Mating occurs in early spring, with a litter of three–six altricial young born about 30–40 days later. Females typically have one litter per year. The young are weaned at about a month and become sexually mature at 10 months.

CONSERVATION STATUS

Not threatened.

SIGNIFICANCE TO HUMANS

This species' tunneling behavior can damage lawns and gardens. ◆

Russian desman
Desmana moschata

SUBFAMILY

Desmaninae

TAXONOMY

Desmana moschata (Linnaeus, 1758), "Habitat in Russiae aquosis."

OTHER COMMON NAMES

French: Desman de Russie; German: Russischer Desman, Bisamrüßler, Wychochol; Spanish: Desmàn almizclado.

PHYSICAL CHARACTERISTICS

The total length is 7.1–17.0 in (18–43 cm), with the tail 6.7–8.3 in (17–21 cm) long. Average adult weight ranges betwen 3.5–7.8 oz (100–220 g). Unlike other genera, the desmans have long guard hairs interspersed in their otherwise short fur coat, and webbed hind feet that are larger than their forefeet. Their pelage varies from a rusty brown dorsally to light gray below.

DISTRIBUTION

Major river basins of Russia, Belarus, eastern Ukraine, and Kazakhstan.

HABITAT

This semi-aquatic species lives in and near freshwater areas, including lakes, ponds, slow rivers, and marshes. It occasionally enters in brackish waters.

BEHAVIOR

This is a mainly aquatic species, although it uses shoreline burrows for both shelter and breeding. When water levels are high, it will move its nest to a drier site, sometimes into nearby trees or shrubs. It spends much of the year in shallow ponds, marshes, and streams, then travels to a deeper lake to spend the winter. Several members of this species may share a tunnel system, and even the same den.

FEEDING ECOLOGY AND DIET

A mainly nocturnal species, the Russian desman's diet includes insects, crustaceans, mollusks, amphibians, and small fish.

REPRODUCTIVE BIOLOGY

Little is known about the reproductive biology of this species, but mating is believed to occur in both spring and fall, with a litter of typically three to five young. Probably promiscuous.

CONSERVATION STATUS

Listed as Vulnerable by the IUCN. The major threats to this species include habitat loss, competition with introduced species (such as muskrats), and water pollution. Preservation efforts include nature reserves and refuges, as well as reintroduction attempts.

SIGNIFICANCE TO HUMANS

This species was once hunted for pelts, but that activity is now banned. ◆

Star-nosed mole
Condylura cristata

SUBFAMILY
Talpinae

TAXONOMY
Condylura cristata (Linnaeus, 1758), Pennsylvania, United States. Two subspecies.

OTHER COMMON NAMES
French: Condylure étoilé; German: Sternmull; Spanish: Topo de nariz estrellada.

PHYSICAL CHARACTERISTICS
Adults range from 6.1–8.1 in (15.5–20.5 cm) in total length, and 2.2–3.5 in (5.5–8.8 cm) in tail length. Adults weigh 1.1–3.0 oz (30–85 g). Males and females are similar. Brownish black, silky-furred mole with wide, shovel-like hands, distinctive tentacles surrounding the nostrils. The 22 fleshy tentacles are short and pink.

DISTRIBUTION
A North American species, this mole ranges from Labrador and Nova Scotia west to Manitoba, through the Great Lakes region of the United States and to South Carolina. A few spotty populations extend to coastal Georgia.

HABITAT
These semi-aquatic and fossorial moles live in damp- or soggy-soiled meadows or forests near water sources, such as marshes, swamps, streams, and lakes. Occasionally they will reside beneath lawns that are near water.

BEHAVIOR
The most noticeable feature of star-nosed moles is their tentacles, which are constantly in motion as the animal moves through its habitat. Once suspected of being capable to detect electric fields, the tentacles are now believed to have a tactile function, and help guide the animal through its tunnels or identify prey.

Evidence of star-nosed moles comes in the form of molehills, which are piles of dirt pushed out of the tunnels by the moles. Tunnels are generally shallow in the summer, and deeper in the winter. Some tunnels open into the water, where the mole will swim throughout the year, especially in colder months when terrestrial prey is scarce. During the winter, the star-nosed mole may also retreat into the deeper burrows to escape the cold, or emerge on land to burrow through the snow. This species commonly overwinters in small colonies, occasionally in male-female pairs.

FEEDING ECOLOGY AND DIET
Star-nosed moles are active foragers day and night, either finding earthworms, insect larvae, and other invertebrates in their tunnels, or swimming to hunt aquatic invertebrates, or an occasional small fish or crustacean. This mole also forages above ground.

Predators include birds of prey, snakes, fish, skunks, cats, and other mammals.

REPRODUCTIVE BIOLOGY
Mating occurs from late winter (in southern populations) to early summer and produces a single litter. May form monogamous pairs. In a dry nest made of leaves, grass, and other vegetation, the female gives birth to two to seven altricial young following a gestation of about 45 days. The young are independent at three to four weeks and become sexually mature at 10 months.

CONSERVATION STATUS
Not listed by the IUCN, but its habitat has decreased as wetlands have been drained for human uses.

SIGNIFICANCE TO HUMANS
This species typically lives in areas unsuitable for lawns, gardens, or farms, so is not usually known as a pest. ◆

Common name / Scientific name/ Other common names	Physical characteristics	Habitat and behavior	Distribution	Diet	Conservation status
Pyrenean desman *Galemys pyrenaicus* English: Iberian desman; French: Le desman des Pyrénées; German: Pyrenäen-Desman, Almizclero; Spanish: Desmán	Shiny, grayish brown mole, lighter on the belly, with a very long snout, tufted tail, and hands that are smaller than the feet. Body length 4.1–.3 in (10.5–13.5 cm); tail length 4.9–6.1 in (12.5–15.5 cm). Weight averages 1.6–2.8 oz (45–80 g).	Semiaquatic, mainly nocturnal animal that uses it webbed feet to swim among swift mountain streams, and occasionally slower moving bodies of water located in altitudes of 200–3,940 ft (60–1,200 m).	Southwestern Europe.	Invertebrates, including insect larvae and crustaceans.	Vulnerable
Large Japanese mole *Mogera robusta*	A brownish gray mole with lighter ventral pelage and yellowish feet. Body length 5.5–7.9 in (14.0–20.0 cm); tail length 0.8 in (2.0 cm).	Active day and night in fertile areas, including cultivated fields, this species makes shallow foraging tunnels, as well as deep, sheltering tunnels.	Japan and Korea north to southeastern Siberia.	Earthworms, insects, and other invertebrates.	Not listed by IUCN
Broad-footed mole *Scapanus latimanus* French: Taupe à larges pieds; German: Kalifornische Maulwurf, Kalifornischer Breitfußmaulwurf	Light gray to black moles with broad hands and a thick tail. Body length 3.4–4.4 in (8.6–11.1 cm); tail length 0.8–2.2 in (2.1–5.5 cm). Weight averages 1.4–1.8 oz (40–50 g). Males are typically slightly larger than females.	Extensive burrower, preferring the moist ground of lush forests or water associated areas.	North America from southern Oregon mostly along the coast to northern Baja California.	Invertebrates.	Not listed by IUCN

[continued]

Common name / Scientific name/ Other common names	Physical characteristics	Habitat and behavior	Distribution	Diet	Conservation status
Gansu mole *Scaponulus oweni* German: Kansu-Maulwurf, Ansumaulwurf	Long-snouted mole with shiny, gray fur tipped in brown, and fairly broad hands. Body length 3.9–4.3 in (9.8–10.8 cm); tail length 1.4–1.5 in (3.5–3.8 cm).	Live in coniferous forests. Behavior is unknown.	Central China.	Unknown.	Not listed by IUCN
Short-faced mole *Scaptochirus moschatus*	Grayish brown, short-tailed mole with a stubby muzzle. Body length about 5.5 in (14.0 cm); tail length 0.4–0.6 in (1.0–1.6 cm).	The behavior of this species is little known. It occurs in arid, sandy areas.	Northeastern China.	Uncertain, perhaps beetle larvae.	Not listed by IUCN
Long-tailed mole *Scaptonyx fuscicaudatus* German: Langschwanzmaulwurf	Mole with dark gray fur tipped in brown and hands that are only mildly broadened. Body length 2.4–3.5 in (6.0–9.0 cm); tail length 0.8–1.2 in (2.0–3.0 cm).	Lives in high altitude forests between 7,050 and 14,760 ft (2,150–4,500 m). Behavior is unknown.	Northern Myanmar to Sichuan and Yunnan, China.	Unknown.	Not listed by IUCN
Chinese shrew-mole *Uropsilus soricipes* English: Asiatic shrew-mole; German: Spitzmausmaulwurf	Long-tailed, shrew-like animal with a long snout and visible ears, but without the digging front limbs typical of many moles. Body length 2.5–3.5 in (6.3–8.8 cm); tail length 2.0–3.1 in (5.0–7.8 cm).	Lives in high altitude forests between 4,920 and 8,860 ft (1,500–2,700 m), probably spending much of its time beneath logs, leaf litter, or other debris. Its behavior is little known.	Central Sichuan, China.	Uncertain, but likely invertebrates.	Endangered
Greater Japanese shrew-mole *Urotrichus talpoides* German: Japanischer Spitzmull	Shiny, dark brown to black, shrew-like mole with mildly broadened hands and a hairy, often thick tail. Body length 2.5–4.0 in (6.4–10.2 cm); tail length 0.9–1.6 in (2.4–4.1 cm). Weight averages 0.5–0.7 oz (14–20 g).	Lives in forests and fields, and spends its time either in shallow burrows or above ground, where it ventures into shrubs and trees.	Japan.	Invertebrates, including worms, insects, and spiders.	Not listed by IUCN

Resources

Books

Gorman, M. L., and R. D. Stone. *The Natural History of Moles.* Ithaca, NY: Comstock Publishing Associates, 1990.

Kurta, A. *Mammals of the Great Lakes Region.* Ann Arbor: Universtiy of Michigan Press, 1995.

Nowak, R. *Walker's Mammals of the World.* Baltimore: Johns Hopkins University Press, 1999.

Wilson, D., and S. Ruff, eds. *The Smithsonian Book of North American Mammals.* Washington, DC: Smithsonian Institution Press, 1999.

Yates, T. L., and D. W. Moore. "Speciation and Evolution in the Family Talpidae (Mammalia: Insectivora)." In *Evolution of Subterranean Mammals at the Organismal and Molecular Levels.* New York: Alan Liss, 1990.

Periodicals

Catania, K. C. "A Comparison of the Eimer's Organs of Three North American Moles: The Hairy-tailed Mole (*Parascalops breweri*), the Star-nosed Mole (*Condylura cristata*), and the Eastern Mole (*Scalopus aquaticus*)." *Journal of Comparative Neurology* 354 (1995): 150–160.

Hebert, P.D.N., ed. "Star-nosed Mole, *Condylura cristata.*" *Canada's Aquatic Environments* University of Guelph, <http://www.aquatic.uoguelph.ca/mammals/freshwater/accounts/mole.htm>

Mason, Matthew J., and P. M. Narins. "Seismic Signal Use by Fossorial Mammals." *American Zoologist* 41, no. 5 (November 2001): 1171–84.

Yokohata, Y. "Biology of the Shrew Mole and Moles in Hiwa, Hiroshima Prefecture, Japan." *Recent Advances in the Biology of Japanese Insectivora, Proceedings of the Symposium on the Biology of Insectivores in Japan and on the Wildlife Conservation.* Hiba Society of Natural History and Hiwa Museum for Natural History. <http://yokohata.edu.toyama-u.ac.jp/Hiwarev.html>.

Organizations

IUCN Species Survival Commission, Insectivore Specialist Group, Dr. Werner Haberl, Chair. Hamburgerstrasse 11, Vienna, A-1050 Austria. E-mail: shrewbib@sorex.vienna.at Web site: <http://members.vienna.at/shrew/itses.html>

Other

IUCN 2002. *2002 IUCN Red List of Threatened Species.* <http://www.redlist.org>.

Mole Tunnel. <http://www.moletunnel.net>

Talpa europaea L. Mammalia, Insectivora, Talpidae. HYPP. <http://www.inra.fr/Internet/Produits/HYPPZ/RAVAGEUR/6taleur.htm>.

"Talpidae." *Discover Life.* <http://www.discoverlife.org/nh/tx/Vertebrata/Mammalia/Talpidae/>

Leslie Ann Mertz, PhD

Scandentia
Tree shrews
(Tupaiidae)

Class Mammalia
Order Scandentia
Family Tupaiidae
Number of families 1

Thumbnail description
Small, squirrel-like mammals with relatively dense fur and a prominent tail (bushy in Tupaiinae; naked and scaly with a terminal tuft of hairs in Ptilocercinae); ears large and membranous; colors range from gray to dark brown dorsally and white to yellowish brown ventrally; species in the subfamily Tupaiinae are diurnal, whereas the pen-tailed tree shrew (Ptilocercinae) is nocturnal

Size
Relatively small body size, ranging from the pen-tailed tree shrew (head/body length 5 in [13 cm]; tail length 4.5 in [11 cm]; body mass 1.5 oz [43 g]) to the Philippine tree shrew (head/body length 9.5 in [24 cm]; tail length 7 in [18 cm]; body mass 12 oz [340 g])

Number of genera, species
5 genera; 19 species

Habitat
Primarily inhabitants of evergreen tropical rainforests

Conservation status
Endangered: 2 species; Vulnerable: 4 species; Lower Risk/Near Threatened: 1 species

Distribution
Southern and Southeast Asia

Evolution and systematics

Tree shrews have attracted considerable interest because of the possibility that they might be related to primates. The family Tupaiidae was originally included within the order Insectivora. Then, in Simpson's seminal classification of 1945, tree shrews were formally transferred to the order Primates. This was particularly because of Le Gros Clark's reports of a series of morphological similarities in the skull and brain and because of Carlsson's account of similarities in the musculature. However, the postulated link between tree shrews and primates was increasingly questioned from 1965 onwards and, following Butler (1972), it is now customary to allocate tree shrews to their own order Scandentia. This change in interpretation was partly due to recognition of the fundamental principle that reliable reconstruction of phylogenetic relationships depends on identification of novel, derived characters to the exclusion of retained primitive features. In fact, the undoubted morphological similarities between tree shrews and primates are arguably attributable to retention of many primitive features augmented by a number of convergent adaptations for arboreal life. For

An albino common tree shrew (*Tupaia glis*). (Photo by © Frank W. Lane; Frank Lane Picture Agency/Corbis. Reproduced by permission.)

instance, tree shrews resemble primates in possessing a cecum in the digestive tract, whereas this feature is lacking from typical insectivores. However, it is highly likely that ancestral placental mammals already possessed a cecum, so retention of this feature does not indicate a specific link between tree shrews and primates. For certain other features, independent development of similar characters may have occurred. For example, both tree shrews and primates have a bony strut (postorbital bar) along the outer margin of the eye socket. This is not a primitive feature, as it was undoubtedly lacking from ancestral placental mammals, but it has been developed independently in several mammalian groups, including many hoofed mammals (ungulates), some carnivores, and hyraxes. The likelihood of convergent similarity between tree shrews and primates is increased by the observation that arboreal tree shrew species show closer similarity to primates than do terrestrial tree shrew species. All tree shrews lack all of the clearly defining features of primates that are identifiable in the skull, brain, and reproductive system. Moreover, given the general reliance on dental similarities in reconstructing mammalian evolution, it is surprising that any resemblance between the molar teeth of tree shrews and those of primates has never been proposed. Superficially, tree shrews do resemble strepsirrhine primates (lemurs and lorises) in possessing a tooth comb in the lower jaw. However, in tree shrews the comb is formed by six incisors, whereas in strepsirrhine primates it is formed by two canines and four incisors, so convergent evolution is again the most likely explanation. Analyses of sequences for both nuclear and mitochondrial DNA have consistently failed to indicate any link

A female common tree shrew (*Tupaia glis*) watching for predators. (Photo by Rod Williams. Bruce Coleman, Inc. Reproduced by permission.)

The pygmy tree shrew (*Tupaia minor*) is a diurnal mammal. (Photo by Ann and Rob Simpson. Reproduced by permission.)

between tree shrews and primates. Instead, there have been several indications that tree shrews may be related to Lagomorpha (rabbits, hares, and pikas).

When tree shrews were still included in the order Insectivora, they were united with the elephant shrews (Macroscelidea) in the suborder Menotyphla, partially because of shared possession of a caecum, while all other insectivores lacking a cecum were placed in the suborder Lipotyphla. Subsequently, it was suggested that the superorder Archonta should be established for Menotyphla, Chiroptera (bats), Dermoptera (colugos), and Primates together. Now that the tree shrews have generally been excluded from the order Primates, there have been several attempts to resurrect the superorder Archonta (while discarding the elephant shrews from this assemblage). However, this alternative attempt to link tree shrews to primates along with certain other mammals is subject to the same problems as inclusion of tree shrews in the order Primates. The characters supposedly linking primates, tree shrews, colugos, and bats can be attributed to a combination of retained primitive features and convergent arboreal adaptations. DNA sequences provide no convincing evidence for any link between bats, primates, or tree

A common tree shrew (*Tupaia glis*) pair grooming one another. (Photo by R. Williams. Bruce Coleman, Inc. Reproduced by permission.)

shrews, although there is some indication that colugos may be related to primates (but not tree shrews or bats).

The fossil record for tree shrews is very sparse, and many proposed fossil relatives have now been excluded from the group. It was originally believed that the Oligocene *Anagale* from North America was directly allied to tree shrews, but detailed examination instead suggested a link to lagomorphs (rabbits and their allies). However, some fragmentary material (partial skulls, isolated teeth, and possibly a ribcage) from Miocene deposits of India (*Palaeotupaia*), Pakistan (unnamed genus), and China (*Prodendrogale*) has been convincingly attributed to tree shrews. Furthermore, a possible early fossil relative of tree shrews (*Eodendrogale*) has also been reported from Eocene deposits of China.

Physical characteristics

Tree shrews are relatively small mammals that generally resemble squirrels in appearance, habitat, and behavior. All except the pen-tailed tree shrew have a squirrel-like bushy tail. Indeed, the Malay word *tupai* is used indiscriminately to refer to tree shrews and squirrels. As in squirrels, the eyes are moderate in size and oriented mainly laterally (except in the pen-tailed tree shrew, where they are rotated forward to some extent). In the skull, there is a bony strut (postorbital bar) on the outer margin of the eye socket. The dentition contains a total of 38 teeth with a dental formula of I2/3 C1/1 P3/3 M3/3. In the lower jaw, the crowns of the six incisors (three on each side) are an-

gled forward to form a dental comb that is used in feeding and in grooming the fur. All digits of the hand and foot bear sharp claws. Tree shrews are quadrupedal, scansorial mammals that range from essentially arboreal to essentially terrestrial in habits. Arboreal species are small with short snouts and relatively long tails while terrestrial species are large with long snouts and relatively short tails. Semi-terrestrial species (the majority) are intermediate in these features. Sexual dimorphism in body size is virtually absent in all tree shrew species.

Distribution

Range extends from northwestern India eastwards to Mindanao in the Philippines and from southern China southwards to Java, including most of the islands in the Malayan Archipelago.

Habitat

Generally inhabitants of evergreen tropical rainforests.

Behavior

Tree shrews of the subfamily Tupaiinae are diurnal, whereas the pen-tailed tree shrew is nocturnal. All tree shrews use nests of some kind. As far as is known, tree shrews live in monogamous pairs that occupy a common territory and show aggression towards unfamiliar conspecifics. Scent marking is carried out both with droplets of urine and with secretions of specialized skin glands located on the chest and belly. Like squirrels, tree shrews commonly squat on their hind quarters and hold food items in their hands while eating.

A terrestrial tree shrew (*Tupaia tana*) eating an insect. (Photo by R. William. Bruce Coleman, Inc. Reproduced by permission.)

A terrestrial tree shrew (*Tupaia tana*) on the forest floor. (Photo by Frans Lanting/Minden Pictures. Reproduced by permission.)

Feeding ecology and diet

The basic diet of all tree shrew species includes a mixture of fruit and arthropods, and in most species, a large proportion of the activity period is spent in fruiting trees or searching for fallen fruit on the ground. Processing of fruit is unusual because tree shrews often spit out fibers rather than swallowing them. Transit times through the digestive tract are also remarkably short (just a few hours), resembling the extreme condition found in fruit-eating bats. All tree shrews consume a variety of arthropod species, but there are marked interspecific differences in prey selection and other kinds of prey may be taken.

Reproductive biology

Tree shrews live in monogamous pairs. The reproduction of tree shrews is highly unusual because, at least in members of the Tupaiinae studied to date, the offspring are born in a separate nest and suckled only once every 48 hours. This "absentee" system of maternal care was first discovered with *Tupaia belangeri* in captivity and subsequently reported for *Tupaia minor* and *Tupaia tana* as well. Field observations have confirmed the separate nest and 48-hour suckling rhythm for *Tupaia tana*. It is not yet known whether the same pattern of maternal behavior occurs in *Anathana*, *Dendrogale*, *Urogale*, and/or *Ptilocercus*. Tree shrews give birth to small litters of poorly developed (altricial) offspring that are naked at birth, with their eyes and ears sealed with membranes. The only other mammals known to give birth to altricial offspring in a separate nest and suckle them at long intervals are rabbits, which have 24-hour suckling intervals. According to species, the typical litter size of tree shrews is one to three offspring and females correspondingly have one to three pairs of teats

(mammae). Gestation lasts between 43 and 56 days, with the average value differing by a few days between species. Tree shrews also have an unusual form of moderately invasive (endotheliochorial) placentation with twin placental discs attaching to special pads in the uterine wall. Development during the one-month nest phase and thereafter is rapid. In captivity, tree shrews reach sexual maturity at the age of four months, although under natural conditions breeding my be delayed until they are one year old.

Conservation status

Most species are relatively common, but some are subject to varying degrees of threat. Two species are Endangered (*Tupaia longipes* and *Tupaia nicobarica*), four are Vulnerable (*Dendrogale melanura*, *Tupaia chrysogaster*, *Tupaia palawanensis* and *Urogale everetti*) and one is Lower Risk/Near Threatened (*Anathana ellioti*).

Significance to humans

Tree shrews seem to be of no real significance to local human populations and it has in fact been reported that they are unpalatable.

A pygmy tree shrew (*Tupaia minor*) on a tree branch. (Photo by © Barry Slaven/Visuals Unlimited, Inc. Reproduced by permission.)

1. Pen-tailed tree shrew (*Ptilocercus lowii*); 2. Common tree shrew (*Tupaia glis*); 3. Terrestrial tree shrew (*Tupaia tana*); 4. Bornean smooth-tailed tree shrew (*Dendrogale melanura*); 5. Indian tree shrew (*Anathana ellioti*); 6. Philippine tree shrew (*Urogale everetti*). (Illustration by Jacqueline Mahannah)

Species accounts

Indian tree shrew
Anathana ellioti

SUBFAMILY
Tupaiinae

TAXONOMY
Anathana ellioti (Waterhouse, 1850), Andhra Pradesh, India.
Three subspecies recognized.

OTHER COMMON NAMES
French: Toupaie d'Elliot; German: Indisches Spithzhörnchen.

PHYSICAL CHARACTERISTICS
Head and body length 7.5 in (19 cm); tail length 7.5 in (19 cm); body mass 5.5 oz (160 g). Fur reddish brown to gray-brown dorsally and buff or white ventrally; oblique pale buff or white shoulder stripe present. Eyes surrounded by pale markings. Ears noticeably large and hairy. Tail equal in length to head and body combined. Relatively short snout. Canine teeth weakly developed.

DISTRIBUTION
Three main areas in peninsular India south of the Ganges River.

HABITAT
Evergreen tropical rainforests and thorny jungles.

BEHAVIOR
Diurnal and semiterrestrial. Use nests, sometimes in holes among rocks.

FEEDING ECOLOGY AND DIET
Forage mainly on the ground, eating a basic diet of fruit and insects and also digging for worms.

REPRODUCTIVE BIOLOGY
Females have three pairs of teats. No other specific information on reproduction is available, as this species has rarely been kept in captivity. Probably monogamous.

CONSERVATION STATUS
Lower Risk/Near Threatened.

SIGNIFICANCE TO HUMANS
None known. ◆

Bornean smooth-tailed tree shrew
Dendrogale melanura

SUBFAMILY
Tupaiinae

TAXONOMY
Dendrogale melanura (Thomas, 1892), Sarawak, Malaysia.

OTHER COMMON NAMES
German: Bergspitzhörnchen; Spanish: Tupaya de Borneo.

PHYSICAL CHARACTERISTICS
Head and body length 5 in (13 cm); tail length 4.5 in (11 cm); body mass 1.5 oz (43 g). Small-bodied. Dark brown fur dorsally and pale buff ventrally. Short snout; ear flaps large. Prominent orange-brown rings around eyes. Weakly marked facial streaks present on either side of the face, extending from the snout to the ear. No shoulder stripes present. Claws notably sharp. Tail covered with fine smooth hair and darkening towards the tip.

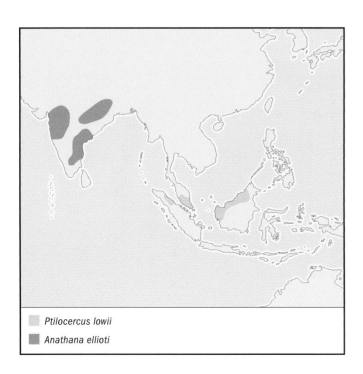

Ptilocercus lowii
Anathana ellioti

Tupaia glis
Dendrogale melanura

DISTRIBUTION
Northern Borneo.

HABITAT
Evergreen tropical rainforests.

BEHAVIOR
Diurnal and predominantly arboreal.

FEEDING ECOLOGY AND DIET
Little studied in the wild. Seems to feed predominantly on insects.

REPRODUCTIVE BIOLOGY
Monogamous. Females have one pair of teats. No other specific information on reproduction available.

CONSERVATION STATUS
Listed as Vulnerable.

SIGNIFICANCE TO HUMANS
None known. ◆

Common tree shrew
Tupaia glis

SUBFAMILY
Tupaiinae

TAXONOMY
Tupaia glis (Diard, 1820), Penang Island, Malaysia. Several subspecies recognized.

OTHER COMMON NAMES
French: Toupaie; German: Spitzhörnchen.

PHYSICAL CHARACTERISTICS
Head and body length 7.5 in (19.5 cm); tail length 6.5 in (16.5 cm). Body mass 5 oz (142 g). Fur dark brown and sometimes almost black dorsally and orange-rufous ventrally. No conspicuous markings on head, but an oblique pale shoulder stripe is usually recognizable. Ears relatively small. Tail very dark and somewhat shorter than the combined length of the head and body.

DISTRIBUTION
Range extends south of the Isthmus of Kra in Thailand, down the Malayan Peninsula and into Sumatra, Java, and surrounding islands.

HABITAT
Evergreen tropical rainforests.

BEHAVIOR
Diurnal and semi-terrestrial. Uses nests that are typically in tree hollows. Monogamous pairs live in defended territories.

FEEDING ECOLOGY AND DIET
Feeds primarily on the ground. Diet consists mainly of fruit and arthropods (including ants), but leaves have also been found in stomach contents.

REPRODUCTIVE BIOLOGY
Monogamous. Females have two pairs of teats. Gestation period approximately 46 days.

CONSERVATION STATUS
Relatively common and not immediately threatened.

SIGNIFICANCE TO HUMANS
None known. ◆

Terrestrial tree shrew
Tupaia tana

SUBFAMILY
Tupaiinae

TAXONOMY
Tupaia tana Raffles, 1821, Sumatra, Indonesia.

OTHER COMMON NAMES
French: Toupaie terrestre.

PHYSICAL CHARACTERISTICS
Head and body length 8.5 in (22 cm); tail length 7 in (18 cm). Body mass 7 oz (198 g). Large-bodied. Fur dark rufous brown dorsally and orange-red or rusty red ventrally. Well-marked, pale yellowish stripe present on each shoulder and a conspicuous dark brown to black midline streak along the back. Anteriorly, this dorsal stripe is fainter and highlighted by pale areas on either side. Tail bushy and distinctly shorter than the combined length of head and body. Snout markedly elongated; canine teeth well developed. Claws robust and elongated.

DISTRIBUTION
Borneo, Sumatra, and several nearby islands.

HABITAT
Evergreen tropical rainforests.

BEHAVIOR
Diurnal and essentially terrestrial.

FEEDING ECOLOGY AND DIET
Foraging typically takes place on the ground, including nosing through leaf litter and digging beneath it. The diet primarily includes fallen fruit and a large proportion of arthropods from a wide range of groups, including beetles, ants, spiders, orthopterans (cockroaches and crickets), centipedes, and millipedes. Also feeds regularly on earthworms.

REPRODUCTIVE BIOLOGY
Presumably monogamous. Females have two pairs of teats.

CONSERVATION STATUS
Relatively common and not immediately threatened.

SIGNIFICANCE TO HUMANS
None known. ◆

Philippine tree shrew
Urogale everetti

SUBFAMILY
Tupaiinae

TAXONOMY
Urogale everetti (Thomas, 1892), Mindanao, Philippines.

OTHER COMMON NAMES
French: Toupaie des Philippines; German: Philippinenspitzhörnchen.

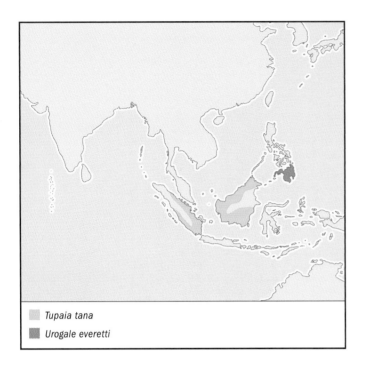

Tupaia tana
Urogale everetti

PHYSICAL CHARACTERISTICS
Head and body length 9.5 in (24 cm); tail length 7 in (18 cm). Body mass 9 oz (255 g). Largest-bodied, stocky tree shrew species. Fur dark brown dorsally and yellowish or rufous ventrally; pale shoulder stripe present. Tail markedly shorter than length of head and body combined and covered in dense rufous fur. Snout conspicuously elongated. Dentition distinctive in that the second pair of upper incisors is markedly enlarged to produce functional canines, while the third pair of lower incisors is reduced. Forelimb unique among tree shrews in showing adaptations for scratch-digging.

DISTRIBUTION
Mindanao Island and nearby islands (Dinigat and Siargao).

HABITAT
Predominantly found in the brush zone and the thick vegetation along river beds.

BEHAVIOR
Diurnal and fully terrestrial.

FEEDING ECOLOGY AND DIET
Little is known about the natural diet, which is said to be omnivorous. Probably feeds predominantly on insects and fruits, like other tree shrew species, but the canine-like incisors suggest a greater emphasis on predation, perhaps including vertebrates.

REPRODUCTIVE BIOLOGY
Probably forms monogamous pairs. Females have two pairs of teats. Gestation period reportedly about 55 days.

CONSERVATION STATUS
Listed as Vulnerable because of progressive population decline.

SIGNIFICANCE TO HUMANS
None known. ◆

Pen-tailed tree shrew
Ptilocercus lowii

SUBFAMILY
Ptilocercinae

TAXONOMY
Ptilocercus lowii Gray, 1848, Sarawak, Malaysia.

OTHER COMMON NAMES
French: Ptilocerque; German: Federschwanzspitzhörnchen.

PHYSICAL CHARACTERISTICS
Head and body length 5 in (13 cm); tail length 4.5 in (11 cm). Body mass 1.5 oz (43 g). Small-bodied. Fur dark gray dorsally and pale gray or buff ventrally. Dark facial stripe extends from the snout to behind the eye on each side. No shoulder stripe present. Short snout; upper incisors enlarged. Eyes more forward-facing than in other tree shrews but not enlarged. In association with their nocturnal habits, pen-tailed tree shrews have a reflecting layer (tapetum) behind the retina, producing a silvery eyeshine that is unique among mammals. Ear flaps large, membranous and mobile. Tail approximately equal in length to combined head and body length and covered with scales, except for the tip, which bears two rows of long, stiff white hairs on either side. In the foot, the first toe (hallux) is divergent but not opposable.

DISTRIBUTION
Disjunct distribution on the southern end of the Malayan peninsula, northwestern Borneo, northeastern Sumatra, and a few neighboring islands.

HABITAT
Evergreen tropical rainforests.

BEHAVIOR
Nocturnal and essentially arboreal. Reportedly rest in tree holes lines with dried leaves during daytime.

FEEDING ECOLOGY AND DIET
Food is sought exclusively in trees. Typically forage on the surfaces of tree trunks, branches, and lianas rather than in foliage. At least some fruits are included in the diet, but there may be a heavy concentration on arthropods.

REPRODUCTIVE BIOLOGY
Pen-tailed tree shrews have rarely been maintained in captivity and little is known about their reproduction. Females have two pairs of teats, indicating that the typical litter size is two offspring. Usually forms monogamous pairs.

CONSERVATION STATUS
Rarely encountered and little studied in the wild, but not currently recognized as endangered. Because of the disjunct distribution, the overall geographical range is quite large, but this species may be locally quite rare.

SIGNIFICANCE TO HUMANS
None known.

Common name / Scientific name/ Other common names	Physical characteristics	Habitat and behavior	Distribution	Diet	Conservation status
Northern smooth-tailed tree shrew *Dendrogale murina*	Small-bodied species. Fur light brown dorsally and pale buff ventrally. Black streak, highlighted by pale fur above and below, extends at eye level from the snout to the ear on either side of the face. Tail is thin and darkens distally. Claws small and blunt. Females have one pair of teats.	Inhabits evergreen tropical rainforests. Predominantly arboreal.	Eastern Thailand, southern Vietnam, and Cambodia.	Little studied in the wild. Probably feeds predominantly on arthropods.	Not listed by IUCN
Belanger's tree shrew *Tupaia belangeri* French: Toupaie de Belanger; German: Belangerspitzhörnchen	Medium-sized species. Fur varies from olivaceous to dark brown dorsally and from creamy white to orange-buff ventrally. Females have three pairs of teats.	Inhabits evergreen tropical rainforests. Semiterrestrial.	Northeastern India, Myanmar, China, and Thailand.	Feeds primarily on the ground. Diet consists mainly of fruit and arthropods.	Not listed by IUCN
Golden-bellied tree shrew *Tupaia dorsalis* French: Toupaie à raies; German: Streifenspitzhörnchen	Relatively large-bodied species, with an elongated snout. Fur dull brown dorsally and pale buff ventrally. Pale shoulder stripe present. Females have two pairs of teats.	Inhabits evergreen tropical rainforests. Terrestrial.	Northwestern Borneo.	Feeds on the ground and on surfaces of logs, rooting in and beneath leaf litter with the snout. Eats both fallen fruit and arthropods, especially ants, cockroaches, spiders, centipedes, and millipedes. Also eats earthworms.	Not listed by IUCN
Slender tree shrew *Tupaia gracilis* French: Toupaie grêle German: Schlankspitzhörnchen	Small-bodied species. Fur olivaceous dorsally and off-white ventrally. Females have two pairs of teats.	Inhabits evergreen tropical rainforests. Semi-arboreal.	Northern Borneo and neighboring islands.	Forages in the forest understory, on shrub foliage, and on the ground for fruits and arthropods, concentrating on caterpillars, crickets, and ants.	Not listed by IUCN
Javanese tree shrew *Tupaia javanica* French: Toupaie de Java; German: Javaspitzhörnchen	Small-bodied species. Fur olivaceous dorsally and off-white ventrally. Females have two pairs of teats.	Inhabits evergreen tropical rainforests. Essentially arboreal.	Sumatra, Java, Bali, and Nias.	Forages for fruits and arthropods in trees, on lianas, and in foliage. Concentrates on crickets, spiders, beetles, and caterpillars.	Not listed by IUCN
Bornean tree shrew *Tupaia longipes* English: Long-footed tree shrew; French: Toupaie à pieds longs; German: Langfuss-Spitzhörnchen; Spanish: Tupaya de pies largos	Medium-sized species. Fur dark brown and sometimes almost black dorsally and orange-rufous ventrally. An oblique pale shoulder stripe is usually recognizable.	Inhabits evergreen tropical rainforests. Semiterrestrial.	Northwestern and southern Borneo.	Feeds primarily on the ground and on log surfaces. Diet consists mainly of fruit and arthropods, particularly ants and termites.	Endangered
Pygmy tree shrew *Tupaia minor* English: Lesser tree shrew; French: Toupaie nain; German: Zwergspitzhörnchen	Small-bodied species. Fur olivaceous dorsally and off-white ventrally. Females have two pairs of teats.	Inhabits evergreen tropical rainforests. Essentially arboreal.	Borneo, Sumatra, southern Malay Peninsula, and neighboring islands.	Forages for fruits and arthropods in trees, on lianas and in foliage. Concentrates on crickets, spiders, beetles, and caterpillars.	Not listed by IUCN
Montane tree shrew *Tupaia montana* French: Toupaie des montagnes	Medium-sized species. Fur varies from olivaceous to dark brown dorsally and from creamy white to orange-buff below. Females have two pairs of teats.	Inhabits evergreen tropical rainforests. Semiterrestrial.	Mountains of northern Borneo.	Feeds primarily on the ground, foraging beneath logs and in leaf litter. Diet consists mainly of fruit and arthropods, notably ants, beetles, crickets, spiders, centipedes, and millipedes.	Not listed by IUCN
Nicobar tree shrew *Tupaia nicobarica* French: Toupaie des îles Nicobar; German: Nicobarspitzhörnchen; Spanish: Tupaya de Nicobar	Medium-sized species. Fur varies from olivaceous to dark brown dorsally and from creamy white to orange-buff below. Females have one pair of teats.	Inhabits evergreen tropical rainforests. Semiterrestrial.	Nicobar Islands.	Feeds primarily on the ground. Diet consists mainly of fruit and arthropods.	Endangered

[continued]

Common name / Scientific name/ Other common names	Physical characteristics	Habitat and behavior	Distribution	Diet	Conservation status
Palawan tree shrew *Tupaia palawanensis* French: Toupaie de l'île Palawan; German: Palawanspitzhörnchen; Spanish: Tupaya de Palawan	Medium-sized species. Fur varies from olivaceous to dark brown dorsally and from creamy white to orange-buff below. Females have two pairs of teats.	Inhabits evergreen tropical rainforests. Semiterrestrial.	Palawan, Busuanga, Cuyo, and Culion Islands in the Philippines.	Feeds primarily on the ground. Diet consists mainly of fruit and arthropods.	Vulnerable
Painted tree shrew *Tupaia picta* English: Ornate tree shrew	Medium-sized species. Fur varies from olivaceous to dark brown dorsally and from creamy white to orange-buff below. Conspicuous dark stripe runs along the back. Females have two pairs of teats.	Inhabits evergreen tropical rainforests. Semiterrestrial.	Lowlands of northern Borneo.	Feeds primarily on the ground. Diet consists mainly of fruit and arthropods.	Not listed by IUCN
Rufous-tailed tree shrew *Tupaia splendidula*	Medium-sized species. Fur varies from olivaceous to dark brown dorsally and from creamy white to orange-buff below. Females have two pairs of teats.	Inhabits evergreen tropical rainforests. Semiterrestrial.	Southwestern Borneo and northeastern Sumatra.	Feeds primarily on the ground. Diet consists mainly of fruit and arthropods.	Not listed by IUCN

Resources

Books

Butler, Percy M. "The Problem of Insectivore Classification." In *Studies in Vertebrate Evolution*, edited by Kenneth A. Joysey and Timothy S. Kemp, 253–265. Edinburgh: Oliver and Boyd, 1972.

Emmons, Louise H. *Tupai: A Field Study of Bornean Treeshrews.* Berkeley: University of California Press, 2000.

Le Gros Clark, Wilfred E. *The Antecedents of Man.* Edinburgh: Edinburgh University Press, 1971.

Luckett, W. Patrick. *Comparative Biology and Evolutionary Relationships of Tree Shrews.* New York: Plenum Press, 1980.

Martin, Robert D. *Primate Origins and Evolution: A Phylogenetic Reconstruction.* Princeton: Princeton University Press, 1990.

Wilson, Don E. "Order Scandentia." In *Mammal Species of the World: A Taxonomic and Geographic Reference*, edited by Don E. Wilson and DeeAnn M. Reeder. Washington, DC: Smithsonian Institution Press, 1993, 131–133.

Periodicals

Campbell, C., and G. Boyd. "On the Phyletic Relationships of the Tree Shrews." *Mammal Review* 4 (1974): 125–143.

D'Souza, Frances, and Robert D. Martin. "Maternal Behaviour and the Effects of Stress in Tree Shrews." *Nature* (London) 251 (1974): 309–311.

Emmons, Louise H. "Frugivory in Tree Shrews (*Tupaia*)." *American Naturalist* 138 (1991): 642–649.

Emmons, Louise H., and Alim Biun. "Malaysian Treeshrews: Maternal Behavior of a Wild Tree Shrew, *Tupaia tana*, in Sabah." *National Geographic Research* 7 (1991): 70–81.

Hill, John P. "On the Placentation of Tupaia." *Journal of Zoology* (London) 146 (1965): 278–304.

Kawamichi, Takeo, and Mieko Kawamichi. "Social System and Independence of Offspring of Tree Shrews." *Primates* 23 (1982): 189–205.

Lim, Boo L. "Note on the Food Habits of *Ptilocercus lowii* Gray (Pentail Tree-shrew) and *Echinosorex gymnurus* Raffles (Moonrat) in Malaya with Remarks on "Ecological Labelling" by Parasite Patterns." *Journal of Zoology* (London) 152 (1967): 375–379.

Martin, Robert D. "Reproduction and Ontogeny in Tree-shrews (*Tupaia belangeri*) with Reference to Their General Behaviour and Taxonomic Relationships." *Zeitschrift fur Tierzuchtung und Zuchtungsbiologie* 25 (1968): 409–532.

———. "Treeshrews: Unique Reproductive Mechanism of Systematic Importance." *Science* 152 (1966): 1402–1404.

Sargis, Eric J. "Functional Morphology of the Forelimb of Tupaiids (Mammalia, Scandentia) and Its Phylogenetic Implications." *Journal of Morphology* 253 (2002): 10–42.

———. "Functional Morphology of the Hindlimb of Tupaiids (Mammalia, Scandentia) and Its Phylogenetic Implications." *Journal of Morphology* 254 (2002): 149–185.

———. "The Postcranial Morphology of *Ptilocercus lowii* (Scandentia, Tupaiidae): An Analysis of Primatomorphan and Volitantian Characters." *Journal of Mammalian Evolution* 9 (2002): 137–160.

Schmitz, Jürgen, Martina Ohme, and Hans Zischler. "The Complete Mitochondrial Genome of *Tupaia belangeri* and the Phylogenetic affiliation of Scandentia to Other Eutherian Orders." *Molecular Biology and Evolution* 17 (2000): 1334–1343.

Simpson, George G. "The Principles of Classification and a Classification of Mammals." *Bulletin of the American Museum of Natural History* 85 (1945): 1–350.

Robert D. Martin, PhD

Dermoptera
Colugos
(Cynocephalidae)

Class Mammalia
Order Dermoptera
Family Cynocephalidae
Number of families 1

Thumbnail description
Specialized herbivorous gliding mammals with a furred membrane extending around almost the entire margin of the body

Size
Cat-sized mammals with an overall length of approximately 30 in (75 cm) and a body weight of about 3 lb (1.35 kg)

Number of genera, species
1 genus; 2 species

Habitat
Evergreen tropical rainforest

Conservation status
Vulnerable: 1 species

Distribution
Southeast Asia

Evolution and systematics

Colugos (also misleadingly labeled flying lemurs, although they are not lemurs and do not fly) exhibit a mosaic of features, some of which show individual similarities to bats, insectivores, or prosimian primates, while others are unique. Their most striking characteristics are adaptations for gliding. Because of their many unique features, the two colugo species have long been classified in an independent order of placental mammals (Dermoptera, literally meaning "skin-wings").

Although colugos are now generally regarded as an unusual and isolated mammalian lineage, they have been variously linked to insectivores (Insectivora), bats (Chiroptera), and/or primates (Primates). Some affinity to bats has often been suspected because of the potential link between gliding and actual flight. Perhaps the most influential suggestion has been that colugos, bats, tree shrews, and primates (with or without elephant shrews) should be allocated to a superorder labeled Archonta. However, the apparent morphological affinities between mammals allocated to the Archonta may be largely or exclusively attributable to retention of primitive adaptations for arboreal life that were present in ances-

The Malayan colugo (*Cynocephalus variegatus*) has large eyes and small ears. (Photo by C. J. Phillips/Mammal Images Library of the American Society of Mammalogists.)

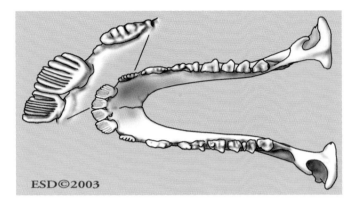

Colugos have comb-like lower incisors. (Illustration by Emily Damstra)

tral placental mammals. The unusual molar morphology of colugos, their small, very primitive brains, and their highly unusual ear region all indicate that there is no real link to primates. Molecular evidence concerning the affinities of colugos has been equivocal, reflecting the fact that it is difficult to resolve the position of an isolated, species-poor lineage that diverged at a very early stage. Indeed, evidence from complete mitochondrial DNA sequences seems to indicate that colugos are actually closer to higher primates than the latter are to prosimians. However, other molecular evidence clearly conflicts with this highly unlikely interpretation. While an early link between colugos and primates in the mammalian tree cannot be ruled out, there is still no strong evidence to support it.

As might be expected from the existence of only two modern species, the fossil record of colugos is very poorly documented. It was long accepted that the Plagiomenidae, a

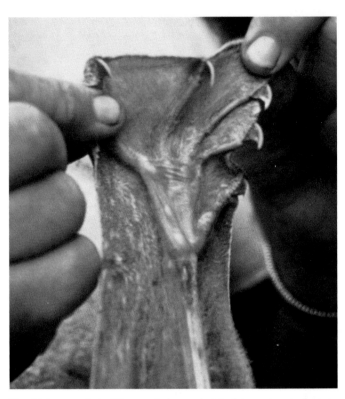

The Malayan colugo (*Cynocephalus variegatus*) has strong claws to help it cling to tree trunks. (Photo by C. J. Phillips/Mammal Images Library of the American Society of Mammalogists.)

poorly documented group of North American fossil mammals from mid-Paleocene to early Eocene deposits (approximately 50–60 million years old), are related to the colugos. However recent re-examinations of the Plagiomenidae indicate that supposed dental similarities to colugos are superficial and that there is no real affinity. A more recent alternative suggestion is that North American Paleocene forms in the family Mixodectidae are related to colugos, but this possibility requires further examination. In fact, a more likely prospect is provided by dental fossils discovered in Eocene deposits of Thailand, attributed to the genus *Dermotherium major* and identified more convincingly as related to colugos. There have been suggestions that some members of the Plesiadiformes (a largely Paleocene group of mammals from Europe and North America usually allocated to the order Primates) show affinities to colugos. Indirect evidence was cited to support the interpretation that some plesiadiforms (notably members of the family Paromomyidae) were adapted for gliding. This led to a variety of suggested relationships between colugos, Plesidapiformes, and primates of modern aspect. However, the supposed evidence for gliding adaptations in plesiadiforms has now been largely discredited and there is little else to suggest a link to colugos.

Physical characteristics

Colugos are medium-sized arboreal mammals with very soft, dense fur. The pelage shows considerable variation in color, although the dorsal fur is typically brown in males and grayish brown in females. The color of the ventral fur varies

A colugo's gliding membrane fully stretched out. (Illustration by Emily Damstra)

A Malayan colugo (*Cynocephalus variegatus*) with baby, in Malaysia. (Photo by Peter Ward. Bruce Coleman, Inc. Reproduced by permission.)

The gastrointestinal tract is very specialized as a reflection of the strictly herbivorous diet. Surprisingly, the stomach is relatively small and narrow, and there is an abbreviated small intestine, which is shorter than the colon. However, the caecum is greatly enlarged, providing a chamber for symbiotic bacteria that assist in digesting plant material. The dental formula is 2.1.2.3/3.1.3.3, giving a total of 34 teeth, and there are a number of dental peculiarities. All canine teeth and the posterior incisors in the upper jaw are double-rooted, an unusual feature otherwise found only in a few insectivores. In mammals generally, these teeth usually have only a single root. The first two incisors on either side of the lower jaw, which are tilted forwards (procumbent), are unique among mammals in that in each one the crown is notched and comblike, with up to 20 tines on each tooth. (There is a superficial resemblance here to the toothcomb found in lemurs and lorises among primates and in tree shrews, but in those mammals individual teeth form single tines of the "comb.") The function of the comblike lower incisors of colugos is unknown, although it has been suggested that they may be used in grooming the fur. It is highly likely that these special teeth are also used during feeding in some way. The molar teeth are relatively primitive, showing a simple three-cusped pattern in the upper jaw.

from yellow through bright orange to brownish red. The eyes are relatively large, reflecting adaptation for active movement at night, while the ears are small and almost naked. The snout is relatively long, giving the head a doglike appearance and accounting for the genus name *Cynocephalus* (literally "dog's head"). The skull is broad and relatively flat. The auditory bullae, which are also quite flat, are highly unusual among mammals both because of their bony composition and because the eardrum (tympanic membrane) is almost horizontal, a very primitive condition. The brain is also unusually small relative to body size and morphologically very primitive.

The most obvious gliding adaptation is a special membrane (patagium) extending around almost the entire margin of the body. All four limbs are relatively long and slender and the neck is long and mobile. During gliding, three dorsally furred membranes become stretched out on either side of the body: an anterior membrane (propatagium) between the side of the neck and the forelimb, an extensive lateral membrane (plagiopatagium) between the forelimb and hindlimb, and a large posterior membrane (uropatagium) between the hindlimb and the tail. Other gliding mammals (certain rodents and marsupials) lack a uropatagium and have a long tail, whereas the tail is quite short in colugos. Furthermore, the patagium in colugos extends between the fingers and toes, such that the term "mitten-gliders" has been used. In contrast to bats, the colugo's membranes are not used for actual flight but merely for gliding, so there is always some loss of height during aerial movement between trees. Some degree of steering is achieved by altering the positions of the limbs and tail. The fingers and toes bear prominent, strongly curving claws that are used to cling to trees.

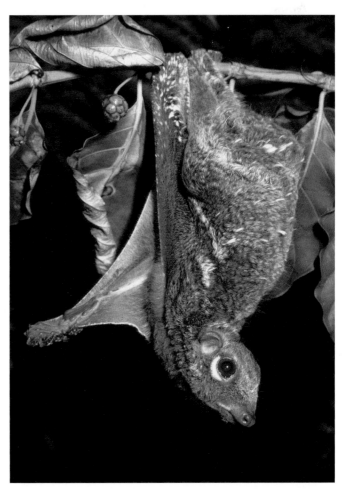

The Malayan colugo (*Cynocephalus variegatus*) has a thin membrane that it uses to glide from tree to tree. (Photo by C. & D. Frith/ANTPhoto.com.au. Reproduced by permission.)

A Malayan colugo (*Cynocephalus variegatus*) in Tawau Hills Park, Sabah, Malaysia. (Photo by © Fletcher & Baylis/Photo Researchers, Inc. Reproduced by permission.)

Distribution

Colugos are found throughout much of Southeast Asia, including southern Myanmar, Thailand, Laos, Cambodia, Vietnam, Malay Peninsula, Sumatra, Java, Borneo, and parts of the Philippines (southern Mindanao, Basilan, Samar, Leyte, Bohol).

Habitat

Primary evergreen rainforest.

Behavior

Colugos are nocturnal and apparently solitary in habits when active, although sometimes small nesting groups are found. During the day, they sleep in tree hollows or clinging to tree trunks or the undersides of large branches, using their well-developed claws. At night, they can move along the undersides of branches quite rapidly and scramble up tree trunks with a series of leaps, thus reaching the necessary height to glide to another tree. Single glides of up to 230 ft (70 m) are quite common, and a maxiumum of 450 ft (135 m) has been reported.

Feeding ecology and diet

Colugos are strictly herbivorous, eating leaves, buds, flowers, some fruit, and pods.

Reproductive biology

Colugos have rarely been kept in captivity, so few details of their reproduction are known. Birth of a single infant is typical, although twins can occur, and there is a single pair of teats (mammae) located near the armpits. In contrast to other medium-sized mammals with single births, the gestation period is short (about 60 days) and the newborn infant (neonate) is small (1.2 oz or 35 g) and very poorly developed, almost like a marsupial offspring. Up until weaning (at about six months of age), the infant is carried on the mother's belly, enclosed within her patagium. The mother may carry her infant with her even when gliding, although some reports indicate that the infant may also be "parked." The offspring grows slowly and does not reach maturity until about two to three years of age.

Conservation status

The Philippine colugo (*Cynocephalus volans*) is listed as Vulnerable by the IUCN. Although colugos are sometimes hunted for their meat and soft fur, they are not severely threatened by humans and are, in fact, protected in certain areas by local taboos. The greatest threat to these unusual animals comes from rampant deforestation in the areas they inhabit.

Significance to humans

Apart from occasional hunting, colugos seem to be of little significance to humans.

A Malayan colugo (*Cynocephalus variegatus*) displaying defensive behavior. (Photo by C. & D. Frith/ANTPhoto.com.au. Reproduced by permission.)

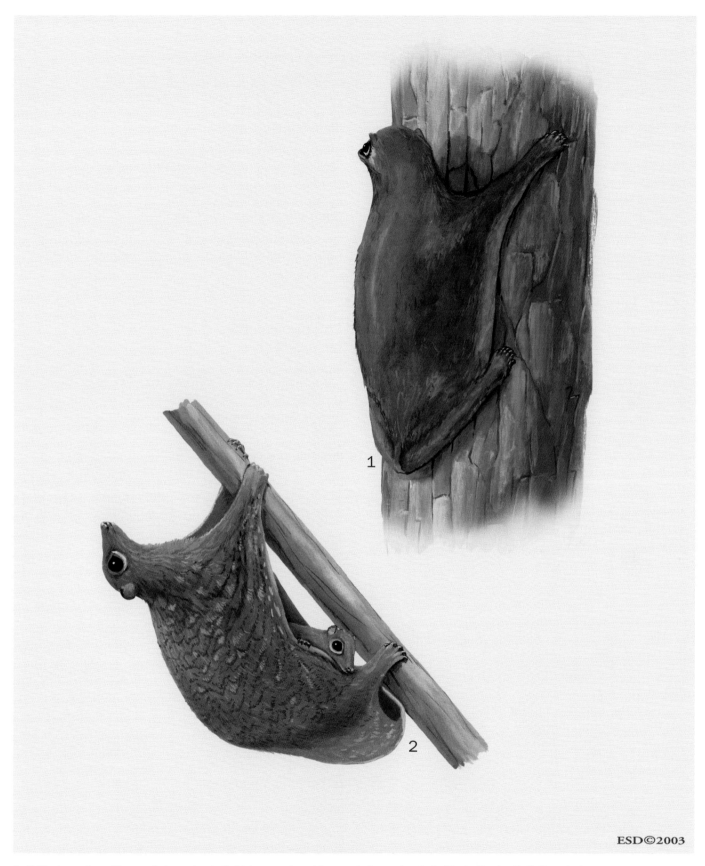

1. Philippine colugo (*Cynocephalus volans*); 2. Malayan colugo (*Cynocephalus variegatus*). (Illustration by Emily Damstra)

Species accounts

Malayan colugo
Cynocephalus variegatus

TAXONOMY
Galeopithecus variegatus (Audebert, 1799), Java, Indonesia. The Malayan colugo has sometimes been placed in the separate genus *Galeopterus*, but the degree of difference from the Philippine colugo is not really sufficient to justify separation at this level.

OTHER COMMON NAMES
English: Malayan flying lemur.

PHYSICAL CHARACTERISTICS
Head and body length 15 in (38 cm); tail length 10 in (25 cm); body weight 3.3 lb (1.5 kg). Width of spread patagium 28 in (70 cm). Dorsal fur typically brown/reddish brown in males and grayish brown in females, with conspicuous white spots, particularly on the patagium, in both sexes. Ventral fur orange-yellow to orange.

DISTRIBUTION
Thailand, Indochina, peninsular Malaya, Sumatra, Java, Borneo, and some adjacent islands.

HABITAT
Primary evergreen rainforest.

BEHAVIOR
Nocturnal and apparently solitary when active.

FEEDING ECOLOGY AND DIET
Strictly herbivorous, eating leaves, buds, flowers, some fruit, and pods.

REPRODUCTIVE BIOLOGY
Mating system is unknown. Single births and a single pair of teats typical. Gestation period is approximately 60 days. Infant small and poorly developed at birth. Weaning at about 6 months.

CONSERVATION STATUS
Not threatened.

SIGNIFICANCE TO HUMANS
Apart from occasional hunting, colugos seem to be of little significance to humans. ◆

Philippine colugo
Cynocephalus volans

TAXONOMY
Lemur volans (Linnaeus, 1758), Luzon, Philippines. There has been little discussion surrounding this species.

OTHER COMMON NAMES
English: Philippine flying lemur.

PHYSICAL CHARACTERISTICS
Head and body length 14 in (35 cm); tail length 10 in (25 cm); body weight 2.8 lb (1.25 kg). The Philippine colugo is somewhat smaller and darker than the Malayan species and lacks conspicuous white spots. Dorsal fur typically brown in males and grayish in females. Ventral fur orange to brownish red.

DISTRIBUTION
Philippine Islands (Mindanao, Basilan, Samar, Leyte, Bohol).

HABITAT
Primary evergreen rainforest.

BEHAVIOR
Nocturnal and apparently solitary when active.

FEEDING ECOLOGY AND DIET
Strictly herbivorous, eating leaves, buds, flowers, some fruit, and pods.

REPRODUCTIVE BIOLOGY
Mating system is unknown. Single births and a single pair of teats typical. Gestation period is approximately 60 days. Infant small and poorly developed at birth. Weaning at about 6 months.

CONSERVATION STATUS
Listed as Vulnerable by the IUCN, primarily because of deforestation.

SIGNIFICANCE TO HUMANS
Apart from occasional hunting, colugos seem to be of little significance to humans. ◆

Cynocephalus volans
Cynocephalus variegatus

Resources

Books

Beard, K. Christopher. "Phylogenetic Systematics of the Primatomorpha, with Special Reference to Dermoptera." In *Mammal Phylogeny*. Vol. 2, *Placentals*, edited by Frederick S. Szalay, Michael J. Novacek, and Malcolm C. McKenna. New York: Springer-Verlag, 1993.

Hayssen, Virginia D., Ari Van Tienhoven, and Ans Van Tienhoven. *Asdell's Patterns of Mammalian Reproduction: A Compendium of Species-Specific Data*. Ithaca, NY: Comstock Publishing Associates, Cornell University Press, 1993.

Lekagul, Boonsong, and Jeffrey A. McNeely. *Mammals of Thailand*. Bangkok: Sahakarnbhat Co., 1977.

Martin, Robert D. *Primate Origins and Evolution: A Phylogenetic Reconstruction*. Princeton: Princeton University Press, 1990.

Medway, Lord. *The Wild Mammals of Malaya and Singapore*. 2nd ed. London: Oxford University Press, 1978.

Nowak, Ronald M., and John L. Paradiso. *Walker's Mammals of the World*. 4th ed. Baltimore: Johns Hopkins University Press, 1983

Stucky, R. K., and Malcolm C. McKenna. "Mammalia." In *The Fossil Record II*, edited by Michael J. Benton, 739–771. London: Chapman & Hall, 1993.

Periodicals

Aimi, M., and H. Inagaki. "Grooved Lower Incisors in Flying Lemurs." *Journal of Mammalogy* 69 (1988): 138–140.

Arnason, Ulfur, et al. "Mammalian Mitogenomic Relationships and the Root of the Eutherian Tree." *Proceedings of the National Academy of Sciences U.S.A.* 99 (2002): 8151–8156.

Beard, K. "Gliding Behaviour and Palaeoecology of the Alleged Primate Family Paromomyidae (Mammalia, Dermoptera)." *Nature, London* 345 (1990): 340–341.

Ducrocq, Stéphane, et al. "First Fossil Flying Lemur: A Dermopteran from the Late Eocene of Thailand." *Palaeontology* 35 (1992): 373–380.

Hamrick, Mark W., Burt A. Rosenman, and Jason A. Brush. "Phalangeal Morphology of the Paromomyidae (Primates, Plesiadapiformes): The Evidence for Gliding Behavior Reconsidered." *American Journal of Physical Anthropology* 109 (1999): 397–413.

Hunt, Robert M., and William W. Korth. "The Auditory Region of Dermoptera: Morphology and Function Relative to Other Living Mammals." *Journal of Morphology* 164 (1980): 167–211.

Kay, Richard F., Richard W. Thorington, and Peter Houde. "Eocene Plesiadapiform shows Affinities with Flying Lemurs, not Primates." *Nature, London* 345 (1990): 342–344.

Krause, David W. "Were Paromomyids Gliders? Maybe, Maybe Not." *Journal of Human Evolution* 21 (1991): 177–188.

Lim, Boo Liat "Observations on the Food Habits and Ecological Habitat of the Malaysian Flying Lemur." *International Zoo Yearbook* 7 (1967): 196–197.

MacPhee, Russell D. E., Matt Cartmill, and Kenneth D. Rose. "Craniodental Morphology and Relationships of the supposed Eocene Dermopteran *Plagiomene* (Mammalia)." *Journal of Vertebrate Paleontology* 9 (1989): 329–349.

Martin, Robert D. "Some Relatives take a Dive." *Nature, London* 345 (1990): 291–292.

———. "Primate Origins: Plugging the Gaps." *Nature, London* 363 (1993): 223–234.

Murphy, William J., et al. "Resolution of the Early Placental Mammal Radiation using Bayesian Phylogenetics." *Science* 294 (December 14, 2001): 2348–2351.

Pirlot, P., and T. Kamiya. "Relative Size of Brain and Brain Components in Three Gliding Placentals (Dermoptera; Rodentia)." *Canadian Journal of Zoology* 60 (1982): 565–572.

Rose, Kenneth D., and Elwyn L. Simons. "Dental Function in the Plagiomenidae: Origin and Relationships of the Mammalian order Dermoptera." *Contributions of the Museum of Palentology of the University of Michigan* 24 (1977): 221–236.

Runestad, Jacqueline A., and Christopher B. Ruff. "Structural Adaptations for Gliding in Mammals with Implications for Locomotor Behavior in Paromomyids." *American Journal of Physical Anthropology* 98 (1995): 101–119.

Schmitz, Jürgen, Martina Ohme, Bambang Suryobroto, and Hans Zischler. "The Colugo (*Cynocephalus variegatus*, Dermoptera): The Primates' Gliding Sister?" *Molecular Biology and Evolution* 19 (2002): 2308–2312.

Stafford, B. J., and Frederick S. Szalay. "Craniodental Functional Morphology and Taxonomy of Dermopterans." *Journal of Mammalogy* 81 (2000): 360–385.

Szalay, Frederick S., and S. G. Lucas. "The postcranial Morphology of Paleocene *Chriacus* and *Mixodectes* and the Phylogenetic Relationships of Archontan Mammals." *Bulletin of the New Mexico Museum of Natural History and Science* 7 (1996): 1–47.

Wharton, Charles H. "Notes on the Life History of the Flying Lemur." *Journal of Mammalogy* 31 (1950): 269–273.

Wible, John R. "Cranial Circulation and Relationships of the Colugo *Cynocephalus* (Dermoptera, Mammalia)." *American Museum Novitates* 3072 (1993): 1–27.

Robert D. Martin, PhD

Chiroptera
(Bats)

Class Mammalia

Order Chiroptera

Number of families 19 living families (18 recognized by some researchers)

Number of genera, species 192 genera; 1,057 species

Photo: An Indian fruit bat (*Cynopterus sphinx*) nursing her young. (Photo by Harald Schütz. Reproduced by permission.)

Introduction

Bats are nocturnal, coming out at night. They are the only mammals capable of true (meaning, flapping) flight, because the other so-called flying mammals (for example, squirrels, lemurs, and sugar gliders) glide, they do not fly. Today, the diversity of bats is astonishing, with more than 1,000 species making them second only to rodents as the most diverse group of mammals. Although some bats have remarkable faces and behavior, wings are the most conspicuous features of the flying bats. Upon landing, bats immediately fold their wings so they appear to shrink in size. Small size is another distinctive feature of bats. While a few species weigh more than 3.5 oz (100 g) as adults, most weigh less than 1.7 oz (less than 50 g), and the majority less than 0.9 oz (less than 25 g). The smallest bats in the world (hog-nosed bats, *Craseonycteris thonglongyai*, family Craseonycteridae from Thailand and Myanmar) weigh 0.07 oz (2 g) as adults. The largest, the Indian flying foxes (*Pteropus giganteus*) from India, Pakistan, and Southeast Asia, weigh 52.9 oz (1,500 g).

Chiroptera means hand (*cheiro*) and wing (*ptera*). Bat wings are folds of skin supported by elongated arm, hand, and finger bones, and attached to the sides of the body. In most cases, bats' thumbs are relatively free of the wing membranes and bear claws, which are absent from the fingers. Flying foxes and their allies (family Pteropodidae) usually have claws on their second fingers, but some species (dawn bats, genus *Eonycteris*, and naked-backed fruit bats, genus *Dobsonia*) lack these claws. Second fingers with claws occur in some fossil bats.

Measurements made with Doppler radar indicate that bats fly about 6.5–49.2 ft (2–15 m) per second. Bats use nine pairs of muscles to power flight. Muscles that power the downstroke are located in the chest, and those responsible for the upstroke are located in the back. Although some bats have quite muscular forearms, their wing bones tend to be lightly muscled. In most bats, the folds of skin (wings) enclose some connective tissue, nerves, and blood vessels. Some free-tailed bats (Molossidae) also have sheets of muscle in the wing membranes. In mechanics and aerodynamics, the flight of bats is very similar to that of birds. The passage of air over the airfoil section of the wings generates lift. Movements of the wing tips generate propulsive thrust.

Bats show considerable variation in wing shape and flying abilities. In many species, the wings are relatively broad, providing good lift. Shorter-winged species tend to be more maneuverable than longer-winged ones. Many species of flower-visiting bats and some species that take animal prey from the ground can hover. While some species of bats can take off from the ground, those with longer, narrower wings cannot.

Although both can fly, bats differ from birds. While living birds such as ostriches, emus, and penguins, and fossils like elephant birds have lost the ability to fly, there are no living or fossil flightless bats. Bats have teeth, while living birds do not. Bats give birth to live young, while birds lay eggs. Since teeth are heavy, having them at the front end of a flying animal could create aerodynamic problems. Additionally, laying eggs could be more efficient and less costly for a flying animal than bearing live young. The diversity of birds (more than 8,000 species) suggests that their approach is more successful than that of bats (1,000 species).

In birds, the wings do not involve the hind limbs as they do in bats. The hind limbs of bats tend to be spindly and poorly muscled, and many bats are not mobile on the ground. Most birds are much more mobile on the ground, and their robust hind legs reflect this reality. Mobility in the air, on the

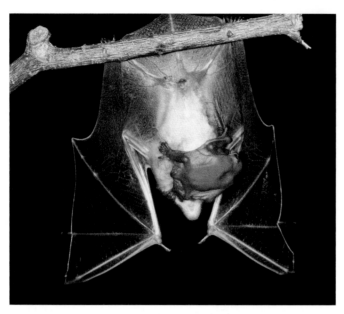

A pallid bat (*Antrozous pallidus*) mother and young. (Photo by James Hanken. Bruce Coleman, Inc. Reproduced by permission.)

ground, and beyond (as for penguins) may partly account for differences in diversity between bats and birds.

In addition to the obvious (i.e., bird wings are made of feathers, while bat wings are folds of skin), there are other differences between these animals. In bats, the division of power generation across nine pairs of muscles means that they are thin in profile through the chest. In birds, two pairs of muscles power flight: one the down-stroke, the other the up-stroke. Both sets of muscles are located on birds' chests and the upstroke muscles operate by a pulley system. A thin profile through the chest allows bats to squeeze into narrow cracks and crevices that serve as places to spend the day (day roosts). While a bird typically has a prominent keel on its breastbone (sternum), this feature is absent in most bats and never as well developed as it is in birds. Birds have wishbones (furcula), bats do not.

In most bats, the thumbs are free of the wings, appearing there as claws. In some bats, notably thumbless ones (family Furipteridae) and some ghost bats (genus *Diclidurus*, family Emballonuridae), the thumbs are greatly reduced in size and may lack claws. At the other extreme are the very long thumbs of vampire bats (*Desmodus rotundus*). Vampire bats' thumbs act like "throwing sticks," giving the bats extra leverage for taking off from the ground.

Bats' wing membranes usually join along the side of the body, but in two groups they meet in mid-back. The naked-backed bats belong to species in two families, Old World fruit bats (genus *Dobsonia*, family Pteropodidae) and naked-backed moustached bats (two species in the genus *Pteronotus*, family Mormoopidae). The function of the naked back remains unknown.

Wing membranes may attach to the bats' hind legs, extending as far down as the fifth toe of the hind feet. In other species, they attach higher up, at the ankle or knee. Carnivorous bats typically have an interfemoral membrane (between the hind legs) that encloses all or part of the tail. On the tail side, the calcar, a cartilaginous structure protruding from the ankle, supports the back edge of the interfemoral membrane. In flight, the interfemoral membrane acts as a rudder and also reduces oscillations of the body through each wing-beat cycle. But interfemoral membranes are not essential to bats: some plant-visiting and blood-feeding species have very narrow interfemoral membranes or lack them completely.

Most bats have tails. But while tails of some (e.g., mouse-tailed bats, Rhinopomatidae, and some Pteropodidae) are long and slender, others (e.g., free-tailed bats, Molossidae) are short and thick. Tails may or may not extend to the end of the interfemoral membrane. Some bats lack obvious tails.

The diversity of bats is reflected by the variety of trophic (ecological) roles they fill in ecosystems. While most species are mainly insectivorous, others eat plant products (fruit, leaves, seeds, nectar, or pollen). Other bats eat animals such as fish, frogs, birds, and even other bats. The most infamous of bats are the blood-feeding vampires, arguably the most remarkable of mammals. The ecological diversity of bats is reflected in their anatomy and behavior. The cheek teeth (molars and premolars) of carnivorous bats are quite different to those of frugivorous (fruit-eating) bats. Nectar- and pollen-feeding bats have teeth more specialized for crushing food, while the vampires have scalpel-sharp razors.

The facial features of bats reflect their remarkable diversity. Bats also are diverse in their selection of roosts, places they spend the day, and by the social systems that develop in these places.

Evolution and systematics

In spite of their small and delicate skeletons, bats have a long fossil record, something that was not obvious to biologists even 50 years ago. By the Eocene, there were species in

A vampire bat (*Desmodus rotundus*). (Photo by M. W. Larson. Bruce Coleman, Inc. Reproduced by permission.)

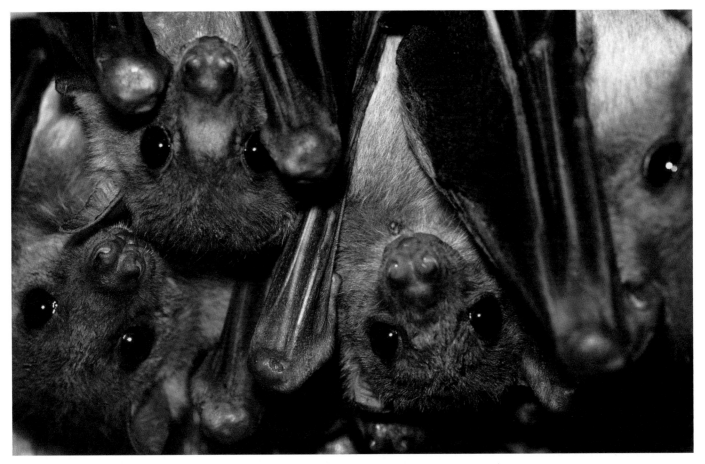

The faces and forearms of Egyptian rousettes (*Rousettus aegyptiacus*). (Photo by © Clive Druett/Papilio/Corbis. Reproduced by permission.)

at least 10 families of bats, four now extinct. Eocene bats are known from the United States, Germany, and Australia, as well as Pakistan.

It is assumed that bats evolved from nocturnal, arboreal, insectivorous animals that lived in forests. The combination of their small size, delicate skeletons, and the forest conditions make the ancestors of bats unlikely candidates for fossilization. There are no fossils of animals that are part bat, part something else, but it is speculated that a shrew-like animal would be a good candidate as a remote ancestor of bats.

The fact that bats appear fully formed and diverse in the Eocene means that it is not known when they first appeared. The very first bats could have shared the skies with the last of the pterosaurs, overlapping in time with the last dinosaurs. Furthermore, although the fossil record suggests a high level of variety, there are relatively few fossil bats and, for most living families, there is no fossil record.

Living species of bats are classified in two suborders: the Megachiroptera (flying foxes and their relatives, family Pteropodidae), which are the fruit- and flower-visiting bats of the Old World tropics, and the Microchiroptera, which include all of the other bats. The two groups are easy to distinguish. The Pteropodidae have dog-like faces, simple ears, and most have claws on their second fingers. Microchiroptera (18 families) lack claws on their second fingers, do not look like dogs, and their ears (and related structures) are more complex. The teeth of megachiropterans tend to be more specialized than those of microchiropterans (excepting vampire bats). Microchiropteran specializations for flight, particularly in the shoulder girdles, are more complex than those of megachiropterans. While the flying foxes and their relatives use flight to get from one place to another, many microchiropteran species feed on the wing and require higher levels of agility and maneuverability.

From the late nineteenth century, some biologists have questioned the closeness of the relationship between Megachiroptera and Microchiroptera. Two theories appeared. The monophyletic theory holds that the two suborders of Chiroptera are more closely related to one another than either is to any other group of mammals. The diphyletic theory proposes that the Megachiroptera are more closely related to some other group of mammals than they are to Microchiroptera. The neural details of how their eyes connect to their brains indicated that while megachiropterans were like primates, the microchiropterans were like all other mammals. Additional evidence about morphology and genetics has been presented. By the year 2000, the monophyletic theory was more accepted than the diphyletic one. New information could reopen and extend the debate.

Fruit bats sleeping in Kampala, Uganda. (Photo by Jen & Des Bartlett. Bruce Coleman, Inc. Reproduced by permission.)

Physical characteristics

In general anatomy and physiology, bats are like other mammals. Wings make bats distinct and flight imposes some physiological constraints that make them different in degree from other mammals. For example, the hearts of bats are larger than those of other mammals of comparable size. This difference reflects the higher capacity of the circulatory system associated with flight.

Bats are warm-blooded and, like other animals in this category (birds and mammals), expend energy to maintain body temperatures higher than ambient (surroundings). Bats face a challenge with respect to heat and energy, reflecting the realities of flight and small size. Flying bats shed excess heat generated by contractions of flight muscles, while roosting bats often conserve heat. In flying bats, wings act like radiators because direct connections between small arteries and small veins facilitate shedding heat. To conserve heat, bats roost in places that are at least warm if not hot, saving them the costs of thermoregulation, for example, shivering to keep warm. In temperate areas in summer, bats often use the warmest available roost sites, for example, in attics or tree hollows exposed to direct sunlight for much of the day, or heated rock crevices around hot springs.

Heterothermy, the ability of a warm-blooded animal to let its body temperature follow the ambient across some range, is another way that some bats conserve energy. Many species of plain-nosed (Vespertilionidae), horseshoe (Rhinolophidae), and free-tailed (Molossidae) bats are heterothermic, while other bats are not (e.g., slit-faced bats, Nycteridae; Old World leaf-nosed bats, Hipposideridae; and false vampire bats, Megadermatidae). The ability to hibernate (body tempera-

tures measuring below 50°F [10°C] to freezing for a prolonged period) is extreme heterothermy. Heterothermy excuses bats from paying the costs of maintaining their body temperatures above ambient. Heterothermy allows bats (and other animals) to wait out periods of inclement weather (short-term torpor) or to hibernate (long-term torpor).

Distribution

Absent only from very remote oceanic islands, the high Arctic, and the Antarctic, bats are extremely widespread. For the most part, heterothermic species are the bats of temperate regions.

Some species of bats migrate hundreds of miles (kilometers) to avoid inclement seasons, but there is detailed knowledge in only a few cases. Schreiber's long-fingered bats (*Miniopterus schreibersi*) in Australia, noctules (*Nyctalus noctula*) in Europe, or Brazilian free-tailed bats (*Tadarida brasiliensis*) in parts of the New World are species whose seasonal movements have been documented by band recoveries. Seasonal appearances and disappearances of straw-colored fruit bats (*Eidolon helvum*) at different locations in Africa suggest migrations, and the same patterns have been used to support the proposal that red bats (*Lasiurus borealis*), hoary bats (*Lasiurus cinereus*), and silver-haired bats (*Lasionycteris noctivagans*) migrate in North America. In many other species, the seasonality of captures suggest migrations, but there are not documented movements of individuals (e.g., wrinkle-faced bats, *Centurio senex*).

The migrations of Brazilian free-tailed bats, like those of many insectivorous birds, allow individuals to remain active year-round because they follow their food supply (insects). Migrations of other bats (e.g., little brown bats [*Myotis lu-*

The spectral bat (*Vampyrum spectrum*) bares its teeth, which it uses to capture and eat small animals. (Photo by © Gary Braasch/Corbis. Reproduced by permission.)

cifugus], Indiana bat [*Myotis sodalis*], gray myotis [*Myotis gris-escens*], and Daubenton's bats [*Myotis daubentonii*]) are from summer habitats to hibernation sites. Underground sites (caves or abandoned mines) often are used for hibernation because they provide consistent above-freezing temperatures, often combined with high relative humidity. There are differences in hibernation site requirements (temperature, relative humidity) between species. In temperate areas, other species of bats (e.g., big brown bats, *Eptesicus fuscus*) use hibernation sites that are close to their summer roosts. Species like noctules hibernate in hollow trees.

Habitat

Bats can be found in virtually every habitat available from rainforests to deserts, montane forests to seasides. Bats have two basic habitat requirements: roosts, or places to spend the day or hibernate, and places to feed. The actual selection of roosting or foraging habitat depends on the species and the time of year.

Roosts of bats can be divided into four broad categories: hollows, crevices, foliage, and "other." Hollows, situations where the roosting bat hangs free by its hind feet, may be inside trees, rocks (caves, mines), buildings, or even birds' nests. An unexpected discovery was of round-eared bats (genus *Tonatia*, family Phyllostomidae) roosting in hollows in the bases of arboreal termite nests. Crevices and cracks, situations where the bats' venter is against one surface (and its back may be close to the other), occur in rocks, trees (under bark or in wood), and buildings. Bats roosting in foliage may hang from branches of trees, or among or under leaves. The "other" category includes bats roosting in unfurled leaves or in tents.

In the New World, three species of disk-winged bats (genus *Thyroptera*) roost in unfurled leaves of plants like heliconia or bananas. These roosts are available for about a day, because as the leaves grow they unfurl, obliging the bats to move regularly to new leaves. Suction disks on the wrists and ankles allow the bats to move about on the smooth leaf surfaces. In Africa, rufous mouse-eared bats (*Myotis bocagei*) and African banana bats (*Pipistrellus nanus*) roost in unfurled banana leaves. These bats lack suction disks.

In the Neotropics, India, and Southeast Asia, a number of species of bats roost in tents. Tents, usually leaves modified by biting, shelter bats from direct sunlight and rain. One or two species of short-nosed fruit bats (genus *Cynopterus*, family Pteropodidae), one species of plain-nosed bat (a yellow bat, genus *Scotophilus*), and about 18 species of New World leaf-nosed bats (family Phyllostomidae) use tent roosts. In the New World tropics, individual tents can last for several months and be occupied by a succession of bats. There is relatively little information about how bats build tents. In India, male lesser short-nosed fruit bats (*Cynopterus brachyotis*) build tents that are occupied by the tent-builder and groups of females and their dependent young.

Flattened skulls are examples of morphological specializations associated with roosting in narrow spaces or in roosts with small entrances. Good examples occur among free-tailed bats (genera *Platymops*, *Sauromys*, and *Neoplatymops*) that roost

A vampire bat (*Desmodus rotundus*) feeds on a sow in Venezuela. (Photo by Animals Animals ©S. Dalton, OSF. Reproduced by permission.)

under rocks. Bamboo bats (plain-nosed bats, genus *Tylonycteris*) roost inside the hollows of bamboo stems. They enter these roosts through small holes made by bruchid beetles. These bats have very flattened skulls. Bats roosting on rough surfaces (e.g., under stones) may have wart-like projections on the forearms (e.g., *Platymops*, *Neoplatymops*). Species roosting on very smooth surfaces can have suction disks that are best developed in disk-winged bats, either from the New World (family Thyropteridae) or in Madagascar (family Myzopodidae).

Roosting bats readily exploit artificial structures as roosts. People may find bats roosting in their homes. Around buildings, bats can be found roosting in attics or eaves, behind shutters, even among the folds of rolled up patio umbrellas. Bats also roost in the expansion cracks of bridges, a famous example being the Congress Avenue Bridge in downtown Austin, Texas, United States. Bats frequently roost in abandoned mines and in active underground water conduit systems. Today, it is common for people to erect bat houses or bat boxes to provide additional roosting opportunities. Some bat houses have resident bats, while others do not. It remains to be determined if bat populations are limited in size by the availability of roosts.

A Seba's short-tailed leaf-nosed bat (*Carollia perspicillata*) in flight. (Photo by Animals Animals ©Joe McDonald. Reproduced by permission.)

Behavior

Among mammals, flight is a behavior unique to bats. Two other behaviors, echolocating and hanging upside down, are associated with bats, but not characteristic of them. Not all bats echolocate and not all bats use echolocation the same way. Furthermore, the echolocation calls of many species of bats are ultrasonic, which, by definition, is beyond the range of human hearing. But many species of bats echolocate with sounds readily audible to people.

Bats also use vocalizations in many social situations, and their social calls are often quite audible to people. Male fruit bats in Africa (African epauletted bats [genus *Epomops*]; epauletted fruit bats [genus *Epomophorus*]; hammer-headed fruit bats [*Hypsignathus monstrosus*]) call to attract females. Pipistrelles (*Pipistrellus pipistrellus*) use harsh-sounding calls to discourage others from feeding in small patches of insects. Greater spear-nosed bats (*Phyllostomus hastatus*) have group-specific screech calls that keep group members together. Researchers are only just beginning to explore the many ways in which bats use vocalizations.

When roosting, many bats hang upside down. This distinctive behavior makes take-off very easy: the bat just lets go, drops, and starts to fly. Hanging upside down may reflect specialization of the forelimbs as wings and attachments of wing membranes to the hind legs. Many other species of bats roost horizontally, their abdomens against the substrate. These bats, notably some species of free-tailed bats, are quite mobile on the ground, readily running or crawling. Compared to bats that probably never operate on the ground (e.g., some horseshoe bats), free-tailed bats have robust hind limbs, par-

ticularly thighbones (femora). Still other bats, for example, Spix's disk-winged bats (*Thyroptera tricolor*, family Thyropteridae), roost head-up.

Roosting upside down could present problems for hygiene. To relieve themselves (urinate or defecate), bats hanging upside down turn head up to minimize the chances of soiling themselves. Bats are very clean animals, spending time every day grooming. Grooming bats use their tongues, teeth, and toe claws. The claws are used to comb fur and teeth to remove groomed materials from the claws. Bats regularly lick their wings, reaching both sides by turning the wings inside out. Grooming bats also lick their fur. Bats roosting in groups may groom one another. Grooming also maintains coat condition in bats. Combing with the claws of the hind feet also may dislodge ectoparasites such as bat flies (families Streblidae, Nycteribiidae, order Diptera), fleas (order Siphonaptera), bedbugs (order Hemiptera), or lice (order Mallophaga). It is not clear if combing or licking affects mites, which can also be common ectoparasites of bats.

The level of behavioral interactions among roosting bats depends on the situation and the setting. Although a hibernaculum may harbor tens of thousands of individuals, the atmosphere is very quiet as the bats are asleep. This contrasts sharply with the levels of activity in other situations where many bats roost together. These roosts resound with vocalizations as neighbors jostle one another or vie for position. Some bats roost in large aggregations, for example, the several million Brazilian free-tailed bats roosting in Frio Cave in Texas. Many other bats roost alone or in small groups. Red and hoary bats roost alone or in small groups (a female and her dependent young), while Spix's disk-winged bats may roost alone or in groups of up to 10. It is not known where many of the 1,000 or so species of bats roost and there is no information about the social settings in which they operate.

Large aggregations of bats often attract predators because the small size of each bat is more than balanced by their large numbers. In many parts of the world, it is common to see birds of prey hunting among groups of bats emerging from (or returning to) roosts. The aerial chases are obvious, but on the ground other predators also exploit the rich patches of prey presented by large numbers of bats. The list includes mammals such as raccoons, skunks, foxes, civets, house cats, and even a variety of snakes. Although many predators take bats when the opportunity presents itself, few specialize on bats. Bat hawks (*Macheirhamphus alcinus*) are exceptions. These birds occur in Africa and Southeast Asia and feed heavily on bats. The same may be true of some bat-eating bats, but these are little-studied and details are lacking about the role bats play in their diets.

In many tropical species of bats, reproduction is an organizing influence in roosts. Typically, groups include a single adult male with a group of females and their young. In species forming larger colonies, for example, gray-headed flying foxes (*Pteropus poliocephalus*), males may be in contact with females only during the period of mating. In tropical species, groups of bachelor males, individuals not holding a territory suitable for breeding, are common (e.g., greater spear-nosed bats). In many temperate species, there is general isolation of males and females in summer, with the latter forming large nursery colonies, sites where they bear and raise young.

Feeding ecology and diet

The combination of small size and high metabolic rates means that bats consume enormous quantities of food. The heart of a flying little brown bat beats about 1,200 times a minute, reflecting the rate at which it burns energy. The same bat, having landed, has a heartbeat rate of less than 300 per minute. During seasons when they are active, little brown bats (like other bats) eat about half their weight in food every night. Nightly, fruit-eating bats may handle three times their weight in food. Lactating females have higher energy demands and nightly eat more than their weight in food.

Consumption of large amounts of food often means that bats eat a variety of prey species, whether insects or fruit. While some insectivorous bats may eat more soft (e.g., moths, flies) than hard (beetles, bugs) prey, there is little evidence of specialization by prey species. Insectivorous bats should not be thought of as consumers of mosquitoes. However, some smaller species, for example, Bodenheimer's pipistrelle (*Pipistrellus bodenheimeri*) from the Middle East, are known to regularly eat mosquitoes.

Bats normally do not eat food they find distasteful. The list of such prey includes at least arrowhead frogs and insects (like tiger moths) protected by unpleasant chemicals. While insectivorous and frugivorous bats learn to avoid food that has made them sick, this taste aversion does not occur in vampire bats. Whatever their food, bats have ways of avoiding ingesting indigestible materials. From insects, they typically bite off and drop wings and legs, from bats and birds, the wings, or cellulose fibers from leaves and fruit. Insects pass quickly through the digestive tracts of bats, 20–30 minutes for little brown bats. Weight reduction translates into lower costs of flight. There is no evidence that bat populations are limited in size by the availability of food.

Insect-eating is a recurring lifestyle in bats, from tropical South America to Alaska, northern Scandinavia to tropical Africa, Malaysia to Tasmania. While most species of insectivorous bats hunt flying insects and use echolocation to detect, track, and assess their targets, others (gleaners) take prey from surfaces such as foliage or the ground. Some bats such as New Zealand lesser short-tailed bats (*Mystacina tuberculata*) show great mobility on the ground where they search for food. Their menu includes animals as well as nectar and pollen.

Gleaning bats eat more than flying insects, consuming a wider range of prey, including walking insects and arthropods that do not fly. Larger gleaning bats take a wider range of prey by size than smaller ones. Gleaners such as pallid bats (*Antrozous pallidus*) from western North America eat large scorpions and centipedes, Jerusalem crickets, and even small pocket mice (genus *Perognathus*). Larger gleaning bats like the Australian false vampire bat (*Macroderma gigas*), spectral bat (*Vampyrum spectrum*), and large slit-faced bats (*Nycteris grandis*) also eat birds and even other bats.

Aerial-feeding bats tend to take smaller prey than gleaning bats. Among bats, it is rare to find aerial-feeding species

A Wahlberg's epauletted fruit bat (*Epomophorus wahlbergi*) feeding from a baobab flower (*Adansonia rubrostipa*). (Photo by Merlin D. Tuttle/Bat Conservation International/Photo Researchers, Inc. Reproduced by permission.)

taking vertebrate prey. The only known exception is the greater noctule (*Nyctalus lasiopterus*) from southern Europe. For at least part of the year, this 1.7 oz (50 g) bat preys on migrating birds.

Many species of animal-eating bats hunt along the water's surface. Called "trawlers," these bats (often in the family Vespertilionidae, the plain-nosed bats) have enlarged hind feet with which they gaff small fish. Mexican fishing bats (*Myotis vivesi*) from Baja California, and Rickett's big-footed bat (*Myotis rickettii*) from southern China are two examples. The best-known fishing bat is the greater bulldog bat (*Noctilio leporinus*, family Noctilionidae) from Central America, South America, and the West Indies. Other trawling bats do not have such large hind feet, but still take the occasional fish and even mosquito larvae. This list includes Daubenton's bat, pond bats (*Myotis dasycneme*), long-fingered bats (*Myotis capaccinii*), and large-footed myotis (*Myotis adversus*). Fish eating has been documented in two other bats, large slit-faced bats from Africa

A spotted bat (*Euderma maculatum*) echolocating in Utah, USA. (Photo by B. G. Thomson/Photo Researchers, Inc. Reproduced by permission.)

and pollen. In the Neotropics, the plant-visiting bats belong to the family Phyllostomidae. In the Old World tropics, the bats are in the family Pteropodidae. The two families show remarkable convergence in structure and behavior.

When eating fruit or leaves, bats typically chew their food thoroughly, all the while using their tongues to rub mashed food against prominent ridges on the roofs of their mouths (palates). During this process, the bats suck vigorously, removing the digestible parts of fruit and leaves before spitting out pellets of indigestible fibers.

Flower-visiting bats obtain sugars from nectar and proteins from pollen. Some Neotropical bat flowers have ultrasonic reflectors that guide nectar-feeding bats to nectar and pollen. By drinking their own urine, nectar-feeding bats create acidic conditions in their stomachs, ideal for digesting pollen. Some flower bats also obtain protein from insects.

The most infamous of bats are the blood-feeding vampires. There are three species: vampire bats, hairy-legged vampires (*Diphylla ecaudata*), and white-winged vampires (*Diaemus youngii*), all in the family Phyllostomidae. These bats only eat blood they obtain by making shallow bites on a prey's skin. The bites do not penetrate large blood vessels such as arteries or veins. Vampire bats use razor-sharp upper incisor teeth to remove a 0.2-in (5-mm) diameter divot of skin, creating a wound that bleeds readily. The bats enhance bleeding by the actions of their tongues and saliva. The saliva of vampire bats contains chemicals that inhibit the body's defenses against bleeding, including anticoagulants, anti-agglutinants, and chemicals that inhibit local vasoconstriction. Each vampire species appears to be a "one-stop shopper," getting each blood meal from one prey. The bats ingest about 2 tablespoonfuls (25 ml) of blood. Blood represents less than 10% of the mass of a bird or mammal, meaning that only victims larger than 4.4 lb (2 kg) are suitable hosts for vampire bats. Within two minutes of beginning to feed, vampire bats start to urinate. The urine consists mainly of the plasma from the current blood meal; therefore, it is very dilute. This is the bat's way of ridding itself of indigestible material.

Reproductive biology

In their life histories, bats are long-lived with low reproductive output. In the wild, individually marked bats (little brown bats and greater horseshoe bats [*Rhinolophus ferrumequinum*]) have survived more than 30 years, and females have the capacity to produce one young per year. In Britain, greater horseshoe females appear to have young only every second or third year. Furthermore, 70% of these bats born in any year do not survive their first winter. The litter size in bats is typically one, though a few species bear twins at least some of the time, and another few, notably red bats, may have litters of even three or four.

During birth, female bats turn heads-up to allow gravity to assist with the birth process. The ligaments holding the two halves of the pelvic girdle together are capable of great flexibility to allow birth. Young are born back-end first. Newborn bats are huge compared to their mothers: single young

and greater false vampire bats (*Megaderma lyra*) from India and Southeast Asia. These bats do not have enlarged hind feet and are thought to catch their fish directly in their mouths— but to date, they have never been observed fishing.

Other bats eat frogs. Most is known about the fringe-lipped bat (*Trachops cirrhosus*, family Phyllostomidae) of the New World tropics. This bat listens for the songs male frogs use when courting females, and uses them to find prey. Fringe-lipped bats grab singing frogs directly in their mouths and eat all of them, starting from the head. In south-central Africa, large slit-faced bats prey heavily on frogs, but there is no indication of their using the frogs' songs to find their prey. The same is true of heart-nosed bats (*Cardioderma cor*) that occur further north in Africa, or of greater false vampire bats in India. Large slit-faced bats also eat frogs from the head down, invariably leaving one leg from the ankle, and the toes of the other foot.

Throughout the tropics, some species of bats get food from plants. Included on the menu are fruits, seeds, leaves, nectar,

White bats (*Ectophylla alba*) rest under the sloping roof of a heliconia leaf that they shaped in the tropical rainforest of La Selva Biological Station, Costa Rica. (Photo by Gregory G. Dimijian/Photo Researchers, Inc. Reproduced by permission.)

are 25–30% of their mother's postpartum mass. Young consume their own weight in milk every day and grow quickly. In some bats, for example, little brown bats (0.3 oz [8 g]) from North America, young reach adult size (forearm length) by about age 18 days. By then, their milk teeth have been replaced by adult dentition, they have started to fly, and insects first appear in their diets. Big brown bats (0.5 oz [15 g]) take about 28 days to reach this stage, and young vampire bats continue to nurse until they are six months old.

Young bats have huge appetites for milk, which is expensive to produce. Female bats roosting in nurseries with hundreds or even thousands of others use a combination of spatial memory, voice, and odor to recognize their own young. This ensures that her young receives enough milk and maximizes its chances of survival. The challenge of recognizing her young depends upon the female's situation. A red bat roosting only with her own young, has a different task than the Daubenton's bat roosting with tens of other Daubenton's bats. Female Brazilian free-tailed bats with nurseries numbering in the millions have a huge challenge in this regard—one they regularly meet and overcome.

Gestation periods in bats range from 60–100 days, and in most bats, fertilization follows copulation. Most species of bats are monestrus, with females having one reproductive event per year. Some tropical species are diestrus, have two reproductive events per year, and females in a few species (e.g., lesser-crested mastiff bats, *Chaerephon pumila*) may bear up to five young per year (one per estrous cycle).

Some species of bats extend the time between mating (typically polygynous) and birth. Sometimes fertilization follows copulation, but development or implantation of the fertilized egg is delayed, extending the gestation period. This occurs in some New World leaf-nosed bats (e.g., California leaf-nosed bats, *Macrotus californicus*) and plain-nosed bats (some populations of Schreiber's long-fingered bats). The other approach, known from plain-nosed bats and horseshoe bats, is to delay fertilization. In this case, females store sperm in the uterus after copulation. Storage can last from less than 20 days in some tropical species, to almost 200 days for north temperate forms. Delayed fertilization does not extend the gestation period. Extension of the time between mating and birth ensures that young are born at the most productive (in terms of food) time of the year.

Conservation

There are three categories of threats to the survival of bats: the general threat of habitat destruction, specific threats to habitats or habitat features important to bats, and threats to bats themselves.

General loss of habitat is the most pressing threat to the survival of most species of bats. Habitat loss typically reflects human population density either directly (urban sprawl) or indirectly (harvesting of resources that generates habitat destruction or disruption). Most species of bats occur in tropical areas, often those with rapidly increasing human populations. Extensive harvesting of rainforests occurs in many parts of the world, no doubt affecting the survival of bats. Detailed information is lacking about the distribution of many species of bats, and there is no accurate information about the sizes of their populations. Nor is it known which habitats or habitat features are vital to bats. This level of ignorance means that specific data cannot be provided about the impact of habitat loss on most of the world's bat species.

Some habitat disruption, specifically forestry or other operations that remove roosts used by bats, can imperil their survival. In other cases, programs to close caves or old mines in the interest of human safety can deprive bats of vital roost sites. In still other cases, habitat connections such as hedgerows are vital for bats like lesser horseshoe bats (*Rhinolophus hipposideros*) so that agricultural and other land-use practices can threaten the survival of some bats. In the United States, species of bats listed as Endangered experienced population declines as the result of disturbance in their cave roosts. For gray myotis, disturbances were to both nursery and hibernating colonies. For Indiana bats, disturbances were to hibernating animals. Bats are known to survive hibernation by going long periods without arousing. In little brown bats, each arousal from hibernation costs the energy that would support 60 days of hibernation. Survival of bat species that regularly hibernate depends on protecting them in their hibernacula.

Public perception can jeopardize bats. The association of bats with blood-feeding and diseases like rabies can make them the objects of persecution. Continued access to roosts for bats using buildings depends on human attitudes. When bats are perceived as dangerous, human occupants of their building roosts are more likely to take steps to evict them. If bats have moved into buildings in the wake of loss of natural roosts, then eviction may be tantamount to a death sentence. The level of protection accorded bats varies considerably. In the United Kingdom and much of Europe, bats enjoy considerable protection. In the United States and Canada, protection is not nearly as effective. In these countries, bats have more often been associated with rabies, coloring their status with respect to protection, people, and public health. In too many countries, bats have little practical or effective protection.

Their small size means that bats are rarely hunted by people, although in some parts of the world, bats are regular components of people's diets. On some South Pacific islands, hunting pressure has driven some species of bats to extinction. The situation in Guam, for example, demonstrates how the use of bats as festive food affected neighboring populations. After bat populations in Guam had been hunted to very low levels or to extinction, bats were then imported from as far away as the Philippines and New Guinea.

Significance to humans

For the most part, bats interact little with people although many species exploit human structures as roosts or feed in rich patches of food people create. But bat-people interactions are not entirely benign. Bats are commonly associated with two diseases that can afflict humans, histoplasmosis and rabies.

Histoplasmosis, a fungus disease of the lungs, can be contracted when people inhale the spores of the fungus *Histoplasma capsulatum*. In warmer parts of the world, these spores are often associated with bat droppings. Their occurrence in bird droppings, including those of pigeons and chickens, is much more widespread. Although histoplasmosis typically gives flu-like symptoms, it can cause severe illness and even death. By wearing a mask that filters out particles larger than 10 microns (0.0004 in), people working in areas where they could encounter the spores of *H. capsulatum* can avoid exposure.

Rabies, a disease of the nervous system, is caused by a lyssavirus. Normally associated with mammals, rabies is usually fatal. Rabies virus tends to accumulate in the saliva of infected animals. Transmission usually occurs by biting when saliva with virus enters a wound. Today, rabies is an uncommon disease in the developed world. Elsewhere, rabies is usually associated with dogs and some other Carnivora and annually accounts for 30,000–70,000 human deaths.

Using molecular techniques, strains of rabies occurring in bats can be distinguished from those in other mammals. Human deaths from bat strains of rabies have been reported in the New World (27 cases in the United States and Canada between 1980 and 2000) and in Europe. In Australia, at least one human death was caused by another lyssavirus reported from bats. Biting appears to be the main route of infection, and strains of bat rabies have been found in other mammals. People bitten by bats should obtain post-exposure rabies vaccinations as soon as possible after the incident.

Images of bats abound in some human cultures, from depictions in Chinese art, on military emblems, and on coats-of-arms. Bats may be positive or negative symbols, but often people do not know what they represent. It is obvious that at least some people have a long fascination with bats.

In China, the "wu fu" (five bats) is commonly portrayed on dishes and robes. In this case, the bats are arranged in a circle facing inward and they depict the five blessings: good health, long life, wealth, love of virtue, and a peaceful death. Chinese bats are often shown in red, the color of joy, and they may carry other positive symbols such as blossoms or fruit. Bats carrying swastikas are jarring images for those unfamiliar with the underlying symbolism. In some Chinese dialects, the word for swastika sounds the same as the word for 10,000. The bat image symbolizes a blessing, but the swastika image it carries turns it into 10,000 blessings.

The Maya god of the underworld, Zotz has the head of a vampire bat on a human body. Zotz usually carries a bleeding heart. Other Mayan portrayals of bats reflect knowledge of different species, from leaf-nosed bats (family Phyllostomidae) to ghost-faced bats (genus *Mormoops*, family Mormoopidae). The significance of these other bats to the Maya remains unclear, but the Maya unmistakably associated vampire bats with blood and the underworld. Further south in the area that today is northern Colombia and Venezuela, the Taironan people associated vampire bats with human fertility. A woman who had "been bitten by the bat" had started to menstruate. The connection here was to the fertility of women. In some areas of New Guinea, long penises make bats symbols of male fertility.

The connection between bats and blood is strong and recurring, but things are not what they seem in the area of vampires, bats, and blood. In the late nineteenth century, when he was writing *Dracula*, Bram Stoker wrote bats into the book, perhaps because blood-feeding bats were in the news. European explorers and naturalists had long been intrigued by blood-feeding bats and called them vampires. Indeed, many bats that eat fruit or animals are called vampyressa, vampyrops, or vampyrum, reflecting this fascination. In Africa, India, Southeast Asia, and Australia, there are false vampire bats.

In human folklore, vampires are people who come back from the dead to feed on the blood of living people. Folklore about vampires is widespread in parts of the world where it typically has nothing to do with bats. Vampire bats occur only in the New World tropics (parts of Central and South America). For Europeans, the name vampire goes from human folklore to the bat, not the other way around. Vampire bats, the blood-feeders, do not occur in Transylvania, Africa, India, or Australia.

Modern military units with bats on their emblems are often those associated with electronic warfare. The parallel is with bats and echolocation (or biosonar). At least one British unit has a tiger moth on its emblem, reflecting the defensive behavior of these moths: some tiger moths use acoustic signals to thwart the attacks of bats. The most famous bat in the world, the Bacardi bat, also has a military connection. This bat comes from Spain where it was associated with a victory of the Spanish over the Moors. On the eve of the battle, the bat that flew into the tent of James I of Aragon proved to be a good omen. The bat was then placed on the city of Valencia's coat of arms. This story trail ends up with a bat symbol on a bottle of rum.

But most of the 1,000 or so species of bats have little to do with people and vice versa. A few fruit-eating bats impact economically as pests of commercial crops, and vampire bats may be responsible for spreading rabies among livestock. On balance, other bats pollinate plants that are ecologically (and sometimes economically) important, while still others disperse seeds and play a vital role in reforestation. Insect-eating bats consume vast quantities of insects every year, including some agricultural pests.

Although bats are occasionally harvested as human food, and may be important economically as pollinators or agents of reforestation, they are rarely exploited economically. One important exception is the harvesting of bat guano, an activity that may disturb bats. In many parts of the world, there is a long tradition of harvesting bat guano for fertilizer. Today in Canada, some garden stores sell bat guano from the Philippines. In the past, bat guano has been a source of saltpeter for gunpowder. During the War of 1812, American forces depended upon bat guano for some of their gunpowder. Later, during the Civil War in the United States, Confederate forces were likewise partly dependent upon bats.

Resources

Books

Allen, G. M. *Bats.* Cambridge: Harvard University Press, 1939

Altringham, J. D. *Bats: Biology and Behaviour.* London: Oxford University Press, 1996.

Barber, P. *Vampires, Burial, and Death: Folklore and Reality.* New Haven: Yale University Press, 1998.

Bates, P. J. J., and D. L. Harrison. *Bats of the Indian Subcontinent,* Sevenoaks, England: Harrison Zoological Museum, 1997.

Bonaccorso, F. J. *Bats of Papua New Guinea.* Washington: Conservation International, 1998.

Brosset, A. *La Biologie des Cchiroptères.* Paris: Masson et Cie, 1966.

Fenton, M. B. *Communication in the Chiroptera.* Bloomington: Indiana University Press, 1985.

———. *Bats: Revised Edition.* New York: Facts On File Inc., 2001.

Findley, J. S. *Bats: A Community Perspective.* Cambridge: Cambridge University Press, 1993.

Fleming, T. H., and A. Valiente-Banuet, eds. *Columnar Cacti and Their Mutualists.* Tucson: University of Arizona Press, 2002.

Greenhall, A. M., and U. Schmidt, eds. *The Natural History of Vampire Bats.* Boca Raton: CRC Press, 1988.

Griffin, D. R. *Listening in the Dark.* New Haven: Yale University Press, 1958.

Hill, J. E., and J. D. Smith. *Bats: A Natural History.* London: British Museum of Natural History, 1984.

Hutson, A. M., S. P. Mickelburgh, and P. A. Racey. *Microchiropteran Bats—Global Status Survey and Conservation Action Plan.* Gland, Switzerland: IUCN, 2001.

Jackson, A. C., and W. H. Wunner, eds. *Rabies.* New York: Academic Press, 2002.

Kunz, T. H., ed. *Ecology of Bats.* New York: Plenum Press, 1982.

Kunz, T. H., and M. B. Fenton, eds. *Bat Ecology.* Chicago: University of Chicago Press, 2003.

Marshall, A. G. *The Ecology of Ectoparasitic Insects.* London: Academic Press, 1981.

Resources

Neuweiler, G. *Biology of Bats*. Oxford: Oxford University Press, 2000.

Norberg, U. M. *Vertebrate Flight, Mechanics, Physiology, Morphology, Ecology and Evolution*. Berlin: Springer-Verlag, 1989.

Nowak, R. M. *Walker's Mammals of the World*, 6th edition. Baltimore: Johns Hopkins Press, 1999.

Popper, A. N., and R. R. Fay, eds. *Hearing by Bats*. New York: Springer-Verlag, 1995.

Ransome, R. D. *The Natural History of Hibernating Bats*. London: Christopher Helm, 1990.

Reid, F. A. *A Field Guide to the Mammals of Central America and Southeast Mexico*. New York: Oxford University Press, 1997.

Roeder, K. D. *Nerve Cells and Insect Behavior, Revised Edition*. Cambridge: Harvard University Press, 1967.

Schober, W., and E. Grimmberger. *The Bats of Europe and North America*. Neptune City, FL: TFH Publications Inc., 1997.

Taylor, P. J. *Bats of Southern Africa*. Pietermaritzberg, South Africa: University of Natal Press, 2000.

Tupinier, D. *La Chauve-souris et l'Homme*. Paris: Editions L'Harmattan, 1989.

Tuttle, M. D. *America's Neighborhood Bats*. Austin: University of Texas Press, 1988.

Wimsatt, W. A., ed. *Biology of Bats*, Volume 1. New York: Academic Press, 1970.

———, ed. *Biology of Bats*, Volume 2. New York: Academic Press, 1970.

———, ed. *Biology of Bats*, Volume 3. New York: Academic Press, 1977.

Periodicals

Simmons, N. B., and J. H. Geisler. "Phylogenetic Relationships of *Icaronycteris*, *Archaeonycteris*, *Hassianycteris*, and *Palaeochiropteryx* with Comments on the Evolution of Echolocation and Foraging Strategies in Microchiroptera." *Bulletin of the American Museum of Natural History* 235 (1998): 1–182.

Melville Brockett Fenton, PhD

Old World fruit bats I
(Pteropus)

Class Mammalia

Order Chiroptera

Suborder Megachiroptera

Family Pteropodidae

Thumbnail description
The largest of all bats and best known of the pteropodids, with a dog-like facial appearance and very large, forward-facing eyes, hence the common name "flying foxes"; coloration ranges from light to dark brown and some have very distinctly colored mantles

Size
Head and body length 6.7–16 in (17–41 cm); forearm 3.3–9 in (8.5–23 cm); weight 0.4–3.5 lb (0.2–1.6 kg); wingspan 2–6 ft (0.6–1.8 m)

Number of genra, species
1 genus, 60 species

Habitat
Subtropical and tropical forests, caves, and swamps

Conservation status
Extinct: 5 species; Critically Endangered: 7 species; Endangered: 3 species; Vulnerable: 16 species; Lower Risk/Near Threatened: 3 species; Data Deficient: 2 species

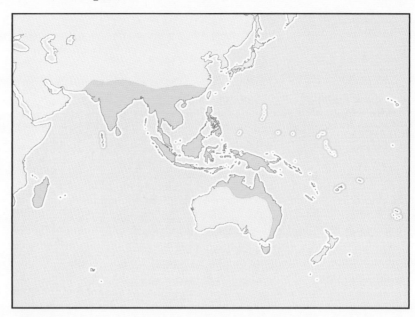

Distribution
Islands of the Pacific and Indian Oceans, and from Pakistan across Southeast Asia to Australasia

Evolution and systematics

The family Pteropodidae is divided into two subfamilies, Macroglossinae and Pteropodinae, the latter of which includes the genus *Pteropus*, also known as the flying foxes. *Pteropus* is further divided into 17 species groups and about 60 total species, depending on authority. Fossil records of this and other chiropteran groups are scarce due to the delicate nature of the skeletal structure of bats, but pteropid fossils have been found in Europe from the middle Oligocene and Miocene periods, in Africa from the Miocene period, and in Madagascar and the East Indies from the Pleistocene period. *Pteropus* is assumed to have arisen in the Australo-Pacific during the early Miocene. Their closest relatives are those of the genus *Acerodon*, a similar group in both size and morphology.

Physical characteristics

Pteropus species are the largest of all bats, weighing up to 3.5 lb (1.6 kg) and with a wingspan of up to 6 ft (1.8 m) Fur is dense and coloration is grayish brown or black. These mammals are characterized by a yellow or grayish yellow contrasting mantle (covering portions of the head, neck, and upper shoulders). Variations among species do occur, such as in the spectacled flying fox (*P. conspicillatus*), which has a light ring around its eyes. The external ears are small and do not have a tragus, and the tail is absent. The second finger is in-

dependent of the third and a claw is present on the thumb. Eyes are very large, forward-facing, and highly adapted to both nocturnal and day vision—this allows flying foxes to easily recognize light colors, which assists in locating food sources.

Distribution

These mammals primarily inhabit islands of the Indian and Pacific Oceans, from Madagascar north to the Maldives and Sri Lanka, across Indonesia, and into the middle Pacific on the island groups of Caroline, Tonga, and Samoa, as far east as the Cook Islands. On the mainland, *Pteropus* species are found from Pakistan in the west across India and Southeast Asia to Australia.

Habitat

Flying foxes inhabit tropical coastal areas such as mangrove forests, primary and secondary growth rainforest, lowland dry forest, swamps, and occasionally caves. Most species roost high above the canopy in a ridge of emergent trees.

Behavior

Flying foxes are most active in the evening and at night. They roost in trees by day, and many of the larger species do

An Indian flying fox (*Pteropus giganteus*) in flight. (Photo by Stephen Dalton/Photo Researchers, Inc. Reproduced by permission.)

A black flying fox (*Pteropus alecto*) roosts in tree. (Photo by Animals Animals ©Austin J. Stevens. Reproduced by permission.)

so in extremely large groups called "camps," which range in size from a few dozen individuals up to 250,000. When resting in the daylight hours, they hang from branches by one or both feet with wings wrapped around their bodies, though there is still sizable activity among the camp as the bats move from one spot to another. At dusk, when the time to forage arrives, *Pteropus* species will flap their wings until their bodies are parallel with the ground—only then do they release the branch and initiate flight.

Migration among flying foxes depends primarily on the seasonal availability of food sources. They do not migrate over particularly long distances, but instead travel between winter and summer roosts when fruits or blossoms are ready for the season. Mainland species will travel about 30 mi (50 km) to reach a new feeding site, and island groups may relocate to neighboring islands or to an accessible mainland area. Colonies will often use the same roosting sites year after year.

Flying fox vocalizations are in the range of 4–6 kHz. Vocalizations play an important role in feeding, mating, territorial disputes, and interaction with infants. In the case of the gray-headed flying fox (*P. poliocephalus*), at least 30 different kinds of calls have been documented.

Feeding ecology and diet

Pteropus species primarily consume fruit, nectar, and pollen. They are able to locate food using highly developed senses of vision and smell (like most fruit bats, members of this genus do not orient themselves using echolocation). Flying foxes employ optimal foraging (seeking the greatest ratio of benefit versus the amount of time and energy spent) as well as searching and handling techniques when going out to feed.

Once food is acquired, the bat will take it to a nearby roost and eat while hovering, or hang from a branch while using one foot to hold the fruit. Juice is the chief source of food for these mammals; it is consumed by compressing bits of pulp against the rigid palate of the mouth, swallowing the juice, and spitting out the pulp and seeds. If the pulp is soft, however, it may be occasionally eaten as well. They drink while traveling to or from a feeding location, skimming the surface of the water during flight. Some species drink seawater in an effort to acquire minerals that are unavailable in other food sources.

Owing to the pursuit of nectar, flying foxes have developed a working relationship with several plants within their habitat. Flower pollination and dispersal of seeds is of course beneficial to these organisms, and certain fruits and trees are specialized to attract fruit bats—some are lightly colored (sea almond trees, *Terminalia catappa*) while others have a strong odor (mangos, *Mangfiera indiaca*). Trees such as the durian (*Durio zibethines*) bloom only at night, easily able to attract the eye of a passing pteropid at peak feeding time.

Reproductive biology

During the summer, when fruit and blossoms are mature and in good supply, flying foxes organize camps. Mating

A little red flying fox (*Pteropus scapulatus*) roosts alone in Atherton, Queensland, Australia. (Photo by B. G. Thomson/Photo Researchers, Inc. Reproduced by permission.)

takes place at this time and small groups, or harems, form. Males soon become very territorial over the females and the roosts, marking their areas using a scent gland located on the throat.

Females are seasonal breeders and usually produce one young per year; they begin to breed at two years. During mating season, flying foxes will mate more than once per day and over the course of several days. Ovulation takes place from February to April, and births occur from September to November. Lactating lasts about six weeks, and most of the female's time during the remainder of the year is spent caring for the young.

The gestation period of the flying fox is six months. During that time, the sexes begin to segregate and pregnant females form a colony; each female then helps to care for the others by mutual grooming. Birth occurs during the day—when it is imminent, the female hangs by her thumbs and feet and licks her genital area until the pup's head begins to emerge—this can last up to several hours. After birth, the pup moves itself into a suckling position and attaches itself to a

nipple. The mother will fly with her young for about two to three weeks.

The pup has light fur, the eyes are closed, and the ear flaps are down. The mother keeps her wings wrapped around the pup for warmth. After approximately three weeks, it becomes too heavy to carry with her and is left with the other young. Upon the mother's return, she is able to recognize her offspring by its unique vocalizations. In about a month, the young become better coordinated and begin to explore, and by January and February, they begin to form small groups near their mothers. Once they become able to care for themselves, the mother will again begin to accept the advances of a male.

Conservation Status

The IUCN Red List ranks seven species as Critically Endangered, three species as Endangered, 16 species as Vulnerable, three species as Lower Risk/Near Threatened, two species as Data Deficient, and five species as Extinct.

A spectacled flying fox (*Pteropus conspicillatus*) roosting on tree branch in Queensland, Australia. (Photo by Animals Animals ©Steven David Miller. Reproduced by permission.)

Significance to humans

All over their range, flying foxes have been considered a delicacy for centuries, and this is especially so for the Chamorro people of Guam and the nearby Northern Marianas Islands. Traditionally, the animal (fur, wings, and innards included) is boiled in coconut milk and eaten in its entirety, usually during ceremonial or otherwise special occasions. Samoan islanders use branches bound to the end of long poles to snag the animal and pull it to the ground, while aboriginal Australians have also been known to use specialized methods to hunt these mammals for food.

Flying foxes have occasionally been considered beneficial for medical use. In the 1970s, Indian flying foxes (*P. giganteus*) in Pakistan were harvested for their fat, which was thought to be a cure for rheumatism. Still, by far the most notable contribution is that of pollination and seed dispersal. As a whole, *Pteropus* plays an integral role in the survival of 300 species of plants across its range, about half of which are regularly used by humans for nourishment, materials, and medicine.

A Madagascar flying fox (*Pteropus rufus*) hanging from a branch in Berenty, Madagascar. (Photo by Nigel J. Dennis/Photo Researchers, Inc. Reproduced by permission.)

The most serious threat to flying fox populations is probably deforestation. The removal of primary forest not only limits habitat in the most basic sense, but it also encourages additional loss—the logging processes used in these areas tend to inhibit growth of new canopy, and the elimination of large sections of forest leaves the remaining habitat even more vulnerable to the tropical storms that frequently strike island environments. The mass conversion of mangrove swamps into shrimp farms has also had a devastating effect on certain species, most notably the Pohnpei flying fox (*P. molossinus*).

Some *Pteropus* species are also losing ground due to illegal wildlife trade, human use for food, and extermination—flying foxes are often considered by orchard growers to be destructive agricultural pests. Populations can sometimes be disturbed by predation as well. An example is the brown tree snake (*Boiga irregularis*), which was introduced on Guam in the 1940s and has had a significant impact on bat populations since. Other enemies include predatory birds, such as owls and falcons.

A spectacled flying fox (*Pteropus conspicillatus*) hangs from a small branch. (Photo by David Hosking/Photo Researchers, Inc. Reproduced by permission.)

1. Island flying fox (*Pteropus hypomelanus*); 2. Tongan flying fox (*Pteropus tonganus*); 3. Livingstone's fruit bat (*Pteropus livingstonii*); 4. Madagascar flying fox (*Pteropus rufus*); 5. Marianas fruit bat (*Pteropus mariannus*); 6. Blyth's flying fox (*Pteropus melanotus*). (Illustration by Marguette Dongvillo)

1. Indian flying fox (*Pteropus giganteus*); 2. Black flying fox (*Pteropus alecto*); 3. Spectacled flying fox (*Pteropus conspicillatus*); 4. Rodricensis flying fox (*Pteropus rodricensis*); 5. Little red flying fox (*Pteropus scapulatus*); 6. Big-eared flying fox (*Pteropus macrotis*). (Illustration by Marguette Dongvillo)

Species accounts

Island flying fox
Pteropus hypomelanus

SUBFAMILY
Pteropodinae

TAXONOMY
Pteropus hypomelanus Temminck, 1853, Moluccas Islands, Indonesia.

OTHER COMMON NAMES
English: Variable flying fox, small flying fox, Condoro Island flying fox.

PHYSICAL CHARACTERISTICS
Head and body length 8–9 in (20–22 cm), forearm 5.5–6 in (14–15 cm), wingspan 3.5–4 ft (1–1.2 m). Color patterns vary. The fur pattern on the face can be black with the crown light to dark brown. Mantle can be light brown, russet brown, or red-tinged over a brown base. It is lighter in the dorsum, mixed with gray and black hairs. The stomach is a buff color, medium brown on the lower shoulders, and dark brown to black on the flanks and ventrum.

DISTRIBUTION
Small islands of the coast of Southeast Asia, Indonesia, New Guinea, and the Solomons.

HABITAT
Small to medium offshore islands, lowland and disturbed forests. It roosts in tall broadleaf trees, coconut trees (*Cocos nucifera*), and orchards.

BEHAVIOR
This species will forage over great distances during the night. It flies approximately 20–30 mi (32–48 km) over land. During this time it will fly low or skim through the troughs of waves to lower wind resistance. It roosts singly, in small family groups, or in large colonies of up to several thousand. If it is particularly hot, these bats may use a wing to fan themselves while panting. In overcast weather, the wings are wrapped around their body and face.

FEEDING ECOLOGY AND DIET
Diet includes fruits, flowers and foliage. This includes figs (*Ficus carica*), sea almond, kapok (*Ceiba pentandra*), chico (*Poaloria sapota*), eucalyptus flowers (*Eucalyptus globulus*), bananas (*Musa* spp.), and paw paws (*Asimina triloba*).

REPRODUCTIVE BIOLOGY
Males and females are sexually mature at 1.5 years. Gestation is 180–210 days and births take place from April to May. Polygynous.

CONSERVATION STATUS
Not threatened.

SIGNIFICANCE TO HUMANS
Hunted for food. Performs valuable pollination and seed dispersal of plants. ◆

Madagascar flying fox
Pteropus rufus

SUBFAMILY
Pteropodinae

TAXONOMY
Pteropus rufus Tiedemann, 1808, Madagascar.

OTHER COMMON NAMES
English: Malagasy flying fox.

PHYSICAL CHARACTERISTICS
Head and body length 9–10 in (23–25 cm), weight 1.2–1.5 lb (533–700 g). Fur coloration on top of the head is yellow or orange. Nose, dorsum, and ventrum are brown. Ears and nose are long and pointed.

DISTRIBUTION
Madagascar.

HABITAT
Coastal forests, low-lying central plains, and small off-shore islands.

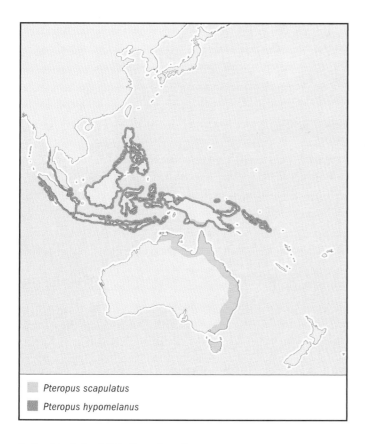

Pteropus scapulatus
Pteropus hypomelanus

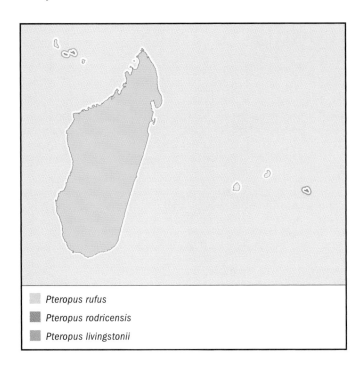

Pteropus rufus
Pteropus rodricensis
Pteropus livingstonii

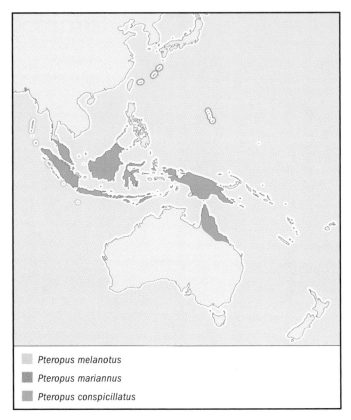

Pteropus melanotus
Pteropus mariannus
Pteropus conspicillatus

BEHAVIOR
Nocturnal, colonial, and very gregarious, with screeching vocalizations. Males tend to their territories quite vigorously and are very aggressive.

FEEDING ECOLOGY AND DIET
Various fruits, blossoms, and leaves. Tamarind (*Tamarindus indica*) pods are a favorite.

REPRODUCTIVE BIOLOGY
Monogamous. Birth weight is approximately 1.5 oz (43 g). Pups are weaned by the eleventh week and volant by three to four months.

CONSERVATION STATUS
Not threatened.

SIGNIFICANCE TO HUMANS
Hunted for food. Performs valuable pollination and seed dispersal of plants. ◆

Blyth's flying fox
Pteropus melanotus

SUBFAMILY
Pteropodinae

TAXONOMY
Pteropus melanotus Blyth, 1863, Nicobar Islands, India.

OTHER COMMON NAMES
English: Black-eared flying fox.

PHYSICAL CHARACTERISTICS
Forearm 6–7 in (15–18 cm). Fur coloration on the head is brown to black. The mantle can be golden tawny, reddish, buff, or light rufous. The ventral surface colors range from

pale and light or dark brown to blackish brown. Ears are large with broad, round tips.

DISTRIBUTION
Andaman Islands, Nicobar Islands, Enggano, Nias Islands off western Sumatra, and Christmas Island.

HABITAT
Roosts in large colonies in mangrove forests, usually near a body of water.

BEHAVIOR
Nothing is known.

FEEDING ECOLOGY AND DIET
Mangoes, papayas (*Carica papaya*), guava (*Psidium guajava*).

REPRODUCTIVE BIOLOGY
Probably polygynous. Females mature rapidly, and are able to breed at at six months. Males reach maturity at 18 months. There is a single breeding season with most births taking place in February.

CONSERVATION STATUS
Not threatened.

SIGNIFICANCE TO HUMANS
Hunted for food. Performs valuable pollination and seed dispersal of plants. ◆

Livingstone's fruit bat

Pteropus livingstonii

SUBFAMILY
Pteropodinae

TAXONOMY
Pteropus livingstonii Gray, 1866, Comoro Islands.

OTHER COMMON NAMES
English: Livingstone's flying fox, Comoro black flying fox.

PHYSICAL CHARACTERISTICS
Head and body length 12 in (30 cm), weight 1.1–1.8 lb
(500–800 g), wingspan 5 ft (1.5 m). Patches of golden fur on
chest.

DISTRIBUTION
Johanna and Moheli Islands in the Comoros.

HABITAT
Dense mountain forest with steep-sided valleys.

BEHAVIOR
There is little information on the behavior of these bats in the
wild. The known social structure has small groups roosting together and forming harems. These bats fly with a slow wing
beat and glide on thermals, using them to help extend their
soaring time.

FEEDING ECOLOGY AND DIET
Blossoms and fruits such as figs are the major food sources.

REPRODUCTIVE BIOLOGY
Polygynous. Females have an annual birth season from July
through October.

CONSERVATION STATUS
Critically Endangered.

SIGNIFICANCE TO HUMANS
Hunted for food. Performs valuable pollination and seed dispersal of plants. ◆

Marianas fruit bat

Pteropus mariannus

SUBFAMILY
Pteropodinae

TAXONOMY
Pteropus mariannus Desmaret, 1822, Mariana Islands, Guam.

OTHER COMMON NAMES
English: Micronesian flying fox, Marianas flying fox, Marianna
flying fox.

PHYSICAL CHARACTERISTICS
Dorsum and wings are brown to blackish with silver hairs. The
mantle and sides are yellow to bright gold.

DISTRIBUTION
Okinawa, Ryukyu Islands, Guam, Mariana Islands.

HABITAT
Subtropical and tropical areas, primarily riparian. Marianas bats
roost in lowland swamp forest and trees such as the Banyon
(*Ficus virens*) and she-oak (*Casuarina* spp.) that rise above the
canopy line.

BEHAVIOR
This species is primarily sedentary and nonmigratory. The
roosting size of the colonies are frequently 60–800 individuals,
smaller groups of 10–12, smaller bachelor groups of 10–15,
and solitary individuals. Marianas bats are not as vocal as other
flying foxes.

FEEDING ECOLOGY AND DIET
These bats are primarily frugivorous but do eat flowers and
leaves. The primary foraging area is the agroforest and pan-
damas savanna. They feed on 53 species of fruit, 23 species of
flowers, and one species of leafy plant.

REPRODUCTIVE BIOLOGY
Polygynous. Breeding colonies are made up of harem groups.
One male to several females.

CONSERVATION STATUS
Endangered. Habitat loss due to timber removal, invasive non-
native species, natural disasters (cyclones and tornadoes), hunting for food, and the wildlife trade market.

SIGNIFICANCE TO HUMANS
Hunted for food. Performs valuable pollination and seed dispersal of plants. ◆

Tongan flying fox

Pteropus tonganus

SUBFAMILY
Pteropodinae

☐ *Pteropus tonganus*
■ *Pteropus giganteus*

TAXONOMY
Pteropus tonganus Quay-Gaimard, 1830, Tonga Islands.

OTHER COMMON NAMES
English: White-necked fruit bat, insular fruit bat, Tongan fruit bat, Pacific flying fox.

PHYSICAL CHARACTERISTICS
Forearm 5–6 in (13–15 cm), wingspan up to 3 ft (0.9 m), weight 10.5–21 oz (300–600 g). The fur is black or brown with numerous white hairs on its head, dorsum, and ventrum. Mantle is red and yellow-brown with a strip on the dorsum between the wings.

DISTRIBUTION
Guam.

HABITAT
The preferred habitat is riparian. The Tongan flying fox uses coconut palm trees, broadleaf trees, and forest remnants. There are both day and night roosts.

BEHAVIOR
Roosts singly in small groups or large communal groups, hanging from branches in the shade. The size and structure of the roost sizes appear to be organized by reproductive status. Bachelor males, clusters of females defended by a male, groups of females, and young. Compared to some of the other Pteropus species, it is a relatively quiet bat. Foraging begins about and hour before dark with short flights between various trees.

FEEDING ECOLOGY AND DIET
Frugivorous. Primarily feeds on fruit, flowers, nectar, and sap.

REPRODUCTIVE BIOLOGY
Polygynous. Births can occur year-round but is most common from June to August. Gestation lasts for five months. Single or twin births can be expected. Young are weaned at three months but will stay with the mother until they reach adult size.

CONSERVATION STATUS
Not threatened.

SIGNIFICANCE TO HUMANS
Hunted for food. Performs valuable pollination and seed dispersal of plants. ◆

Indian flying fox
Pteropus giganteus

SUBFAMILY
Pteropodinae

TAXONOMY
Pteropus giganteus (Brünnich, 1782), Bengal, India.

OTHER COMMON NAMES
English: Giant Indian fruit bat.

PHYSICAL CHARACTERISTICS
Head and body length 8–12 in (20–30 cm), forearm 6–7 in (15–18 cm), wingspan 4 ft (1.2 m), weight 3.3 lb (1.5 kg).

DISTRIBUTION
Pakistan, Nepal, India, and the Maldives through to Myanmar. A single specimen was recorded in Tsinghai, China.

HABITAT
The preferred habitat is well established trees and swamp areas near large bodies of water. Some of the tree species favored for roosting sites are semal (*Bombax malabaricus*), peepal (*Ficus bengalensis*), and mango.

BEHAVIOR
A colonial species that lives in large diurnal roosts. These bats form a hierarchy within the male population, defined by strength and size. Within the hierarchy, each male has a rank and his own roosting spot. Roosting takes place out in open. Colony size varies seasonally, shrinking in number during the summer and increasing during the rainy season. Colonies of these species tend to favor roosting areas near towns or villages.

FEEDING ECOLOGY AND DIET
These bats will leave the roosting site soon after sunset and return about 45 minutes after sunrise. They may break into small feeding groups after leaving and when entering a feeding area. Before feeding they will visit a lake or river to drink and they skim it while they are flying.

Diet includes fruits, flowers, and leaves during the non-fruiting season. Examples are flower buds from the silk cotton tree (*Gossampinus malabarocus*), tulip tree (*Spathodea campunulata*), guava, and the soft leaves and twigs of the tamarind trees.

REPRODUCTIVE BIOLOGY
Polygynous. Mating takes place from July to October. Within that span, copulation will take place three times. Gestation is 140–150 days. When they are ready to give birth, females will congregate in the upper branches of the roosts. The pup is able to fly by the eleventh week and is weaned within two to six months.

CONSERVATION STATUS
Not threatened.

SIGNIFICANCE TO HUMANS
Hunted for food. Performs valuable pollination and seed dispersal of plants. ◆

Black flying fox
Pteropus alecto

SUBFAMILY
Pteropodinae

TAXONOMY
Pteropus alecto Temminck, 1837, Sulawesi, Indonesia.

OTHER COMMON NAMES
English: Pygmy fruit bat, gray fruit bat.

PHYSICAL CHARACTERISTICS
Head and body length (19–28 cm), forearm 6–7.5 in (15–19 cm), wingspan up to 3.3 ft (1 m) weight 1.1–2.2 lb (500–1,000 g). The fur on the head is black, the mantle ranges from chocolate brown to reddish brown, and white hairs frequently appear over the body, including the underside.

DISTRIBUTION
Sulawesi, Salayer Island, Baweam and Kangean islands to the Java Sea, Lombok, Sumba and Savu islands, southern New Guinea, northern and eastern Australia.

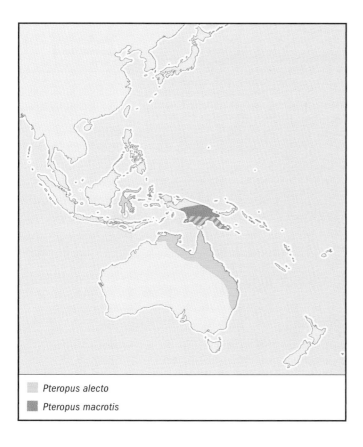

Pteropus alecto

Pteropus macrotis

HABITAT
Found in tropical and subtropical forest and woodlands.

BEHAVIOR
Males will establish territory and will groom themselves on a daily basis. Groups will return to the same site to roost year after year. On occasion, they will share their roosts with gray-headed flying foxes (*P. poliocephalus*).

FEEDING ECOLOGY AND DIET
These groups will travel up 30 mi (50 km) to forage. Nectar, fruit, and tree blossoms are the principal diet. They do not eat citrus fruits.

REPRODUCTIVE BIOLOGY
Polygynous. Camps congregate from early to late summer. Birthing season is October to November in southern Queensland and January to February in the Northern Territory. This difference relates to the availability of food resources. The young are carried by the mothers until approximately fours weeks and then are left at the roost site. Juvenile or immature bats do not leave but remain together and form winter camps.

CONSERVATION STATUS
Not threatened.

SIGNIFICANCE TO HUMANS
Hunted for food. Performs valuable pollination and seed dispersal of plants. ◆

Rodricensis flying fox
Pteropus rodricensis

SUBFAMILY
Pteropodinae

TAXONOMY
Pteropus rodricensis Dobson, 1878, Rodrigues, Mascarene Islands.

OTHER COMMON NAMES
English: Rodrigues flying fox.

PHYSICAL CHARACTERISTICS
Wingspan 3 ft (0.9 m), weight 0.7 lb (300 g). The fur is dark brown and it covers most of the body. Head, neck, and shoulders are a golden color.

DISTRIBUTION
Island of Rodriguez in the Mauritius.

HABITAT
Dense forests and dry woodlands.

BEHAVIOR
Colonial in nature. Camps are very gregarious and vocal. They are not skilled fliers.

FEEDING ECOLOGY AND DIET
Various fruits, such as tamarind pods and mangos.

REPRODUCTIVE BIOLOGY
Males will pick out territories and form harems during breeding season. During this time, the males will be physically aggressive with each other by biting. Young can fly by three to four months and remain with their mothers for up to a year.

CONSERVATION STATUS
Critically Endangered.

SIGNIFICANCE TO HUMANS
Hunted for food. Performs valuable pollination and seed dispersal of plants. ◆

Spectacled flying fox
Pteropus conspicillatus

SUBFAMILY
Pteropodinae

TAXONOMY
Pteropus conspicillatus Gould, 1850, Fitzroy Island, Australia.

OTHER COMMON NAMES
English: Spectacled fruit bat.

PHYSICAL CHARACTERISTICS
Head and body length 8.7–9.8 in (22–25 cm), forearm 6.3–7 in (16–18 cm), weight 0.9–1.3 lb (400–600 g). A large black flying fox with pale yellow or straw-colored fur around its eyes. The mantle is pale yellow and goes across the back, neck, and shoulders. Some specimens have been found to have pale yellow fur on the face and top of the head.

DISTRIBUTION
New Guinea, Irian Jaya, Indonsia, Louisiade Archipelego, D'Entrecasteaus and Trobriand Islands, and northeastern Queensland.

HABITAT
Spectacled flying foxes are forest dwellers. They prefer to roost in the upper canopies of rainforest. Most roost trees are she-oaks, but these bats have also been known to roost in the upper branches of broadleaf trees for protection from hunters. They have also been observed stripping the upper leaves from trees. This behavior allows better visual observation of the surrounding area.

BEHAVIOR
Spectacled flying foxes roost together in mixed colonies with different species, such as the variable flying fox (*Pteropus hympomelanus*). The size of the camp will range from the hundreds to thousands. They are very vocal over roosting sites, food, mating, and group or individual territories. Camps can be found quite easily due to the level of noise they produce.

FEEDING ECOLOGY AND DIET
Spectacled flying foxes feed on several different species of fruits, blossoms, and foliage, as well as the occasional insect. The foraging distance for these bats is about 30 mi (50 km). As with many of the flying foxes, this species drinks seawater on the way to feeding sites.

REPRODUCTIVE BIOLOGY
Polygynous. Breeding takes place between March and May. Females give birth to a single offspring between October and December.

CONSERVATION STATUS
Not threatened.

SIGNIFICANCE TO HUMANS
Hunted for food. Performs valuable pollination and seed dispersal of plants. ◆

Big-eared flying fox
Pteropus macrotis

SUBFAMILY
Pteropodinae

TAXONOMY
Pteropus macrotis Peters, 1867, Indonesia.

OTHER COMMON NAMES
English: Black-bearded flying fox.

PHYSICAL CHARACTERISTICS
Forearm 5–6 in (13–15 cm), weight 0.7–0.9 lb (300–400 g). The fur on the head is medium to dark brown, throat is brown, dorsum and rump are dark brown, ventrum is black or reddish brown, and the mantle is reddish to yellow.

DISTRIBUTION
New Guinea, Boigu Islands of Australia.

HABITAT
Lowland forest; inland forests.

BEHAVIOR
Known to forage in drier areas than other *Pteropus* species.

FEEDING ECOLOGY AND DIET
Large numbers of bats will fly to the mainland of New Guinea to feed. They eat coconut palm flowers and a variety of fruit and blossoms in dry monsoon scrub. It will invade plantations to feed on coconut and sago palm (*Cycas revoluta*) blossoms.

REPRODUCTIVE BIOLOGY
Nothing is known, but probably polygynous.

CONSERVATION STATUS
Not threatened.

SIGNIFICANCE TO HUMANS
Hunted for food. Performs valuable pollination and seed dispersal of plants. ◆

Little red flying fox
Pteropus scapulatus

SUBFAMILY
Pteropodinae

TAXONOMY
Pteropus scapulatus Peters, 1862, Cape York, Australia.

OTHER COMMON NAMES
None known.

PHYSICAL CHARACTERISTICS
Head and body length 5–8 in (13–20 cm), forearm 4.3–5.5 in (11–14 cm).

DISTRIBUTION
Extreme southern New Guinea, northern and eastern Australia, rarely on Tasmania, one record from New Zealand.

HABITAT
Broad range of habitat, from tropical to semiarid and monsoon forests, temperate eucalypt forests, and paperback swamps.

BEHAVIOR
Roost size can range into the thousands in early summer. They are a very nomadic species and normally do not stay in a camp for long periods of time. Roosts are located near a body of water.

FEEDING ECOLOGY AND DIET
Their primary source of food is the blossoms of plants, including trees and shrubs. Other foods that may be eaten are insects, sap, and fruit.

REPRODUCTIVE BIOLOGY
Mating occurs from November to January. Harems are then formed, two to five females with a male. Gestation is five months and birth takes places between April and May. Young are carried by the mother for the first month and then in roost while she forages and returns through the night. The young are volant at two months but females continue to care for them while they develop adult skills.

CONSERVATION STATUS
Not threatened.

SIGNIFICANCE TO HUMANS
Hunted for food. Performs valuable pollination and seed dispersal of plants. ◆

Common name / Scientific name/ Other common names	Physical characteristics	Habitat and behavior	Distribution	Diet	Conservation status
Admiralty flying fox *Pteropus admiralitatum*	Coloration is grayish brown or black. Area between the shoulders is often yellow or grayish yellow. Head and body length 6.7–16 in (17–40.6 cm), forearm length 3.3–9 in (8.5–22.8 cm), and wingspan 24–66.9 in (61–170 cm).	Forests and swamps, often on small islands near coasts. Roost in colonies in trees.	Solomon Islands; Admiralty Islands, New Britain, and Tabar Islands, Bismarck Archipelago.	Fruits and their juices.	Not threatened
Ambon flying fox *Pteropus argentatus* Spanish: Zorro volador argénteo	Coloration is grayish brown or black. Area between the shoulders is often yellow or grayish yellow. Head and body length 6.7–16 in (17–40.6 cm), forearm length 3.3–9 in (8.5–22.8 cm), and wingspan 24–66.9 in (61–170 cm).	Forests and swamps, often on small islands near coasts. Roost in colonies in trees.	Perhaps Amboina Island.	Fruits and their juices.	Data Deficient
Ryukyu flying fox *Pteropus dasymallus* Spanish: Zorro volador de Ryu-kyu	Dorsal fur varies in coloration. Head is usually brown, neck is cinnamon. Lack a tragus and noseleaf. Head and body length 8.7 in (22.1 cm), forearm length 5.3 in (13.4 cm), weight 1.4–1.6 oz (40–45 g).	Trees in tropical rain and deciduous forests during the day, as this is where they roost. Some live in small groups, while others form large colonies.	Taiwan; Ryukyu Islands, Daito Islands, and extreme southern Kyushu, Japan.	Fruits, plants, and possibly flowers.	Endangered
Dusky flying fox *Pteropus brunneus* Spanish: Zorro volador de las islas Percy	Extinct and known from a single specimen. Possibly: coloration grayish brown or black, with area between the shoulders yellow or grayish yellow. Head and body length 6.7–16 in (17–40.6 cm), forearm length 3.3–9 in (8.5–22.8 cm), and wingspan 24–66.9 in (61–170 cm).	Found on Percy Island. Only one type specimen found.	Percy Island, Australia.	Unknown.	Extinct
North Moluccan flying fox *Pteropus caniceps*	Coloration is grayish brown or black. Area between the shoulders is often yellow or grayish yellow. Head and body length 6.7–16 in (17–40.6 cm), forearm length 3.3–9 in (8.5–22.8 cm), and wingspan 24–66.9 in (61–170 cm).	Forests and swamps, often on small islands near coasts. Roost in colonies in trees.	Halmahera Islands, Sulawesi, and Sula Islands, Indonesia. The Sulawesi record is dubius, and a Sangihe Island record is erroneus.	Fruits.	Not threatened
Lyle's flying fox *Pteropus lylei* Spanish: Zorro volador de Luzón	Coloration of back is seal brown, underparts are blackish seal brown, mantle is reddish buff, top of head is same as mantle. Forearm length 6 in (15.2 cm), wingspan up to 3 in (7.6 cm),	Lives in mangrove swamps. Forms large colonies.	Thailand and Vietnam.	Mainly fruits.	Not threatened
Sanborn's flying fox *Pteropus mahaganus* Spanish: Zorro volador de Bougainville	Small with short, brown fur, long pointed ears, and whorls of thick, light-colored fur on shoulders. Head and body length 6.7–16 in (17–40.6 cm), forearm length 3.3–9 in (8.5–22.8 cm), wingspan 24–66.9 in (61–170 cm), weight 43–44.1 oz (1,220–1,250 g).	Forests and swamps, often on small islands near coasts. Strong flier.	Bougainville and Ysabel Islands, Solomon Islands.	Mainly fruits.	Vulnerable
Greater Mascarene flying fox *Pteropus niger* Spanish: Zorro volador negro de Mauricio	Small with short, brown fur, long pointed ears, and whorls of thick, light-colored fur on shoulders. Head and body length 6.7–16 in (17–40.6 cm), forearm length 3.3–9 in (8.5–22.8 cm), and wingspan 24–66.9 in (61–170 cm).	Roosts mainly in primary forests. Nothing known of reproductive and behavioral patterns.	Mascarene Islands of Reunion Island, Mauritius Island, and a subfossil on Rodrigues Island.	Feeds on native and cultivated fruits in Mauritius: kapok, mango, and lychee.	Vulnerable
Banks flying fox *Pteropus fundatus* German: Banks-Flughund; Spanish: Zorro volador de las islas Banks	Coloration is grayish brown or black. Area between the shoulders is often yellow or grayish yellow. Head and body length 6.7–16 in (17–40.6 cm), forearm length 3.3–9 in (8.5–22.8 cm), and wingspan 24–66.9 in (61–170 cm).	Forests and swamps, often on small islands near coasts. Roost in colonies in trees.	Banks Islands, north Vanuatu.	Fruits.	Vulnerable
Little golden-mantled flying fox *Pteropus pumilus* Spanish: Zorro volador de Taylor	Coloration is brown with reddish tufts on chest and belly. Wingspan 30 in (76.2 cm), weight 7 oz (198 g).	Primary and well-developed secondary lowland forests from sea level to about 3,610 ft (1,100 m), rarely to 4,100 ft (1,250 m), rarely outside of forests. Live in small groups or individually, but will aggregate in small numbers to feed.	Philippines.	Mainly fruits.	Vulnerable

[continued]

Common name / Scientific name/ Other common names	Physical characteristics	Habitat and behavior	Distribution	Diet	Conservation status
Samoan flying fox *Pteropus samoensis* French: Roussette des Îles Samoa	Coloration of body and wings is dark brown with variations from blond to gray on head, neck, and shoulders. Forearm length 5.1–5.9 in (13–15 cm), wingspan 33.9 in (86 cm), weight 14.1–17.6 oz (400–500 g).	Primary forests along ridge tops, usually roosting in trees. Form monogamous couples, one offspring produced per year. Most active in morning and late afternoon.	Fiji Islands, Samoan Islands.	Various types of fruits, flowers, and leaves.	Vulnerable
Gilliard's flying fox *Pteropus gilliardi* Spanish: Zorro volador de Gilliard	Coloration is grayish brown or black. Area between the shoulders is often yellow or grayish yellow. Head and body length 6.7–16 in (17–40.6 cm), forearm length 3.3–9 in (8.5–22.8 cm), and wingspan 24–66.9 in (61–170 cm).	Forests and swamps, often on small islands near coasts. Roost in colonies in trees.	New Britain Island, Bismarck Archipelago.	Fruits.	Vulnerable
Mearns's flying fox *Pteropus mearnsi* Spanish: Zorro volador de Mearns	Coloration is grayish brown or black. Area between the shoulders is often yellow or grayish yellow. Head and body length 6.7–16 in (17–40.6 cm), forearm length 3.3–9 in (8.5–22.8 cm), and wingspan 24–66.9 in (61–170 cm).	Frests and swamps, often on small islands near coasts. Roost in colonies in trees.	Mindanao and Basilan Islands, Philippines.	Fruits.	Data Deficient
Guam flying fox *Pteropus tokudae* Spanish: Zorro volador de Tokuda	Abdomen and wings are dark brown with a few whitish hairs. Mantle and sides of neck are brown to light gold. Top of head is grayish to yellowish brown with prominent ears; throat and chin are dark brown. Head and body length 5.5–5.9 in (14–5.1 cm), wingspan 25.6–27.9 in (65–70.9 cm), weight 5.4 oz (152 g).	Last specimen found in mature limestone forest. Nothing known of reproductive and behavioral patterns.	Known only from Guam.	Fruits and flowers from evergreen shrubs.	Extinct

Resources

Books

Altringham, John D. *Bats, Biology, and Behavior*. New York: Oxford University Press, 1996.

Bonnacorso, Frank J. *Bats of Papua New Guinea*. Chicago: University of Chicago Press, 1998.

Buchmann, Stephen L., and Gary Paul Nabhan. *The Forgotten Pollinators*. Washington, DC: Island Press, 1997.

Crichton., Elizabeth G., and Phillip H. Krutzsch, eds. *Reproductive Biology of Bats*. New York: Academic Press, 2000.

Hall, Leslie, and Greg Richards. *Flying Foxes, Fruit and Blossom Bats of Australia*. Malabar, FL: Krieger Publishing Company, 2000.

Kunz, Thomas, and Paul Racey, eds. *Bat Biology and Conservation*. Washington, DC: Smithsonian Institution Press, 1998.

Mickleburgh, Simon, Anthony M. Hutson, and Paul Racey. *Old World Fruit Bats: An Action Plan for Their Conservation*. Gland, Switzerland: IUCN, 1992.

Neuweiler, Gerhard. *The Biology of Bats*. New York: Oxford University Press, 2000.

Nowak, Ronald, ed. *Walker's Mammals of the World*, 6th ed. Baltimore and London: The John Hopkins Unversity Press, 1999.

Taylor, Peter John. *Bats of Southern Africa*. Pietermaritzburg: University of Natal Press, 2000.

Periodicals

Banack, Sandra Anne. "Diet Selection and Resource Use by Flying Foxes (Genus *Pteropus*)." *Ecology* 79 (1998): 1949–1967.

Fujita, Marty. "Flying Foxes and Economics." *Bats* 6, no. 1 (1998): 49.

Rainey, William E. "The Flying Foxes: Becoming a Rare Commodity. *Bats* 8, no. 1 (1990): 69.

Other

"Family Pteropodidae." *University of Michigan Animal Diversity Web*. [23 June 2003]. <http://animaldiversity.ummz.umich.edu>.

Martin, Len. "The Effects of Culling the Flying Foxes, *Pteropus conspicillatus* in Northern Queensland, and *Pteropus poliocephalus* in Victoria, NSW and Southeast Queensland." [23 June 2003]. <http://www.austrop.org.au/ghff/home.htm>.

Thatcher, Oliver. "Destruction of Fruit Bat Habitat." [23 June 2003]. <http://www.biology.leeds.ac.uk/staff/dawa/bats/Fruitbats/deforest.html>.

Kate Kretschmann
Robin L. Hayes

Old World fruit bats II
(*All other genera*)

Class Mammalia
Order Chiroptera
Suborder Megachiroptera
Family Pteropodidae

Thumbnail description
Small to the largest bats, dog-like or lemur-like faces, relatively small external ear with no tragus, no echolocation with one exception, visually oriented with color vision, claws on first and second digits, tail is short, vestigial or absent; includes fruit bats, flying foxes, blossom bats, rousette bats, and tube-nosed bats

Size
Head and body length, 2–15.7 in (5–40 cm); weight, 0.4–42.3 oz (12–1,200 g)

Number of genera, species
38–41 genera, approximately 106 species (excluding *Pteropus*)

Habitat
Forest, woodland, montane forests, savanna, scrub, swampy forests, and mangrove

Conservation status
Extinct: 3 species; Critically Endangered: 6 species; Endangered: 3 species; Vulnerable: 20 species; Lower Risk/Near Threatened: 15 species; Data Deficient: 3 species

Distribution
Tropical and subtropical regions of the Old World (i.e., the Paleotropics) ranging from Africa east to Australia and the Caroline and Cook Islands

Evolution and systematics

The evolution of bats is not well known. Because bat bones are fragile, their fossil record is very incomplete. The oldest fossil bats date back to 50 million years ago (mya) and were found at sites in Wyoming and Germany. These specimens are fully evolved bats. The oldest megachiropteran fossils date to around 35–25 mya in the Oligocene of Italy.

In the late 1980s Jack Pettigrew of the University of Queensland presented evidence that the megachiropterans were a convergence with the Microchiroptera (insectivorous bats). He pointed out that fruit bats and primates shared derived traits, in particular the neural pathway from the retina of the eye to the tectal (roof of the midbrain) portion of the brain. He concluded that the order Chiroptera was polyphyletic and that the megachiropterans were actually flying primates. This proposal stimulated discussion for several years. By 1994 Simmons presented molecular evidence from over 30 studies that supported the chiropteran monophyly. The two suborders, Microchiroptera and Megachiroptera, are much more closely related to each other than to primates. Nevertheless, the primates are close relatives, along with the colugos and tree shrews. These four mammal groups are of-

ten grouped together in the grandorder Archonta. It is interesting to note that Linnaeus, the founder of taxonomy, included all four of the archontan taxa in the order Primates.

Physical characteristics

Body size of fruit bats, flying foxes, rousettes, tube-nosed bats, and blossom bats ranges from small (around 0.4 oz/12 g) to large (42.3 oz/1,200 g). They have dog-like faces with long snouts, which is the reason that some species are called flying foxes. Fruit bats differ in a number of features from the small insectivorous bats of the order Microchiroptera. Fruit bats are primarily visually oriented and have large forward facing eyes that give them depth perception. The retina contains cones for color vision. These bats have good daytime and nocturnal vision, but they are inactive in complete darkness. Except for the rousette bat, fruit bats do not echolocate. Consequently, most have a simple and relatively small external ear without a tragus. The sense of smell is also well developed in fruit bats. All fruit bats have claws on the first digit (thumb) and, unlike microchiropterans, most have claws on the second digits of the wing skeleton. Wings tend to be broad. These bats do not fly as fast as microchiropterans nor

Madagascan fruit bats (*Eidolon dupreanum*) flying at dusk. (Photo by John Vissor. Bruce Coleman, Inc. Reproduced by permission.)

species, such as the hammer-headed fruit bat, these ridges are highly developed.

The most common pelage (fur coat) color is dark brown, but this is highly variable. The ventrum (the belly side) is often a lighter color, such as off white or yellow.

Sexual dimorphism is often present. The most common difference between males and females is body size, but males may also differ in pelage patterns, especially on the head. Other male characteristics are hair tufts on the shoulders (epaulettes) and large pharyngeal sacs in the thoracic region.

Distribution

The fruit bats are confined to the Old World, ranging from Africa to Southeast Asia to Australasia and the islands of the western Pacific.

Habitat

Most fruit bats live in humid forests of the tropical and subtropical regions of the Old World (Paleotropics).

do they perform the aerobatics of their relatives. They do travel long distances and have good hovering ability; some are even able to fly backwards. The wing is mostly devoid of fur. The wing membrane of one genus extends to the vertebral column creating a naked back. The tail is short, vestigial, or absent altogether. The uropatagium, a segment of membrane between the legs that helps provide lift, is not present in those species with a reduced or absent tail. The legs are splayed to the side like a reptile rather than underneath the body like other mammals. The hind paws are completely clawed. They are used predominantly for hanging upside down and, in conjunction with the clawed thumb, are used for climbing in trees.

The number of teeth is variable and ranges from *Rousettus* with a dental formula of (I2/2 C1/1 P3/3 M2/3) × 2 = 34 teeth, to *Nyctimene* and *Paranyctimene* having a dental formula of (I1/0 C1/1 P3/3 M1/2) × 2 = 24 teeth. The teeth of other species are intermediate in number: 32, 30, or 28 teeth in total. The incisors are small, canines are always present, and the cheek teeth (premolars and molars) tend to be flat and wide, suitable for crushing soft fruit. Tongues are sometimes long and mobile, especially in the nectarivorous species. In fruit eating species the tongue is used to crush food against the transverse ridges of the palate (roof of the mouth). In some

The Philippine tube-nosed fruit bat's (*Nyctimene rabori*) nostrils are typically about 0.6 in (6 mm) long. (Photo by © Jay Dickman/Corbis. Reproduced by permission.)

A long-tailed fruit bat (*Notopteris macdonaldi*) resting on moss in Fiji. (Photo by Pavel German. Reproduced by permission.)

A Queensland tube-nosed fruit bat (*Nyctimene robinsoni*) baby hangs on a tree branch in northern Australia. (Photo by B. G. Thomson/Photo Researchers, Inc. Reproduced by permission.)

Behavior

For the most part fruit bats are gregarious animals that form roosting colonies (often called camps) during the day of from 10 to over one million animals. Most roost in trees, but some occupy caves or human-built structures. Many species have grooming sessions just before they take flight. Some of the smaller species and a few of the nectarivorous species roost alone. In many cases when fruit bats take flight from the roost they forage alone.

Fruit bats are crepuscular, meaning they are most active at dawn and dusk. They are occasionally active during the day when moving from one location to another in a tree for thermoregulation.

Feeding ecology and diet

Fruit bats, as the name implies, are mainly frugivorous. However, some genera consume mainly nectar and pollen. Unlike the microchiropterans, there are no insectivorous fruit bats. Fruit bats obtain a piece of fruit and fly to a feeding tree.

A naked-backed fruit bat (*Dobsonia magna*) hanging at night in vegetation. (Photo by B. G. Thomson/Photo Researchers, Inc. Reproduced by permission.)

A Gambian epauletted fruit bat (*Epomophorus gambianus*) mother and pup. (Photo by Merlin D. Tuttle/Photo Researchers, Inc. Reproduced by permission.)

A southern blossom bat (*Syconycteris australis*). (Photo by © Hans & Judy Beste/Lochman Transparencies. Reproduced by permission.)

A male lesser short-nosed fruit bat (*Cynopterus brachyotis*). (Photo by © Bryan Rogers/Visuals Unlimited, Inc. Reproduced by permission.)

There they hang upside down by one foot, press the fruit to their chest with the other foot, and bite off pieces. They ingest the fruit, crush it against the palate with the tongue, consume the juice, and spit out the pulp and seeds. Because of this feeding behavior they are important seed dispersers. Nectarivorous bats, such as the blossom bats, have a long narrow snout and a long protrusible tongue that enables them to reach nectar in flowers. Some are pollinators and have a brushlike tip on the tongue for collecting pollen. During feeding some of the pollen sticks to their noses and the fur on the snout. When they visit the next flower they leave some pollen from the first flower.

Reproductive biology

Most fruit bats have two breeding seasons a year, although the females of most species only conceive in one of the seasons. They tend to be polygamous with males attempting to mate with as many females as possible (although many do not mate at all) and females mating with two or more males. There is at least one species that is believed to be monogamous. Delayed

A Queensland tube-nosed fruit bat (*Nyctimene robinsoni*) in a native guava tree with fruit. (Photo by © Hans & Judy Beste/Lochman Transparencies. Reproduced by permission.)

implantation occurs in some species. The ovum is fertilized, but the zygote goes into stasis rather than implanting in the uterine wall. When conditions are right the zygote is implanted and begins development. Births are usually synchronized to correspond with the period of greatest food availability. Single births are the norm. Gestation is between four and six months.

Conservation status

Old World fruit bats are threatened as a group. Three recent species have already become Extinct. Six are Critically Endangered, and three are Endangered. Twenty species are Vulnerable. Fifteen species have Near Threatened status. Thus, of the approximately 102 species of megachiropterans outside of the genus *Pteropus*, 46% are on the World Con-

servation Union (IUCN) Red List of species of concern. The main threats to Old World fruit bats include habitat destruction, hunting and sale of bush meat, and extermination because they are perceived as agricultural pests.

Significance to humans

Old World fruit bats are often the major pollinators of a particular region. Removal of these bats can actually harm agriculture. At the same time fruit bats are viewed as pests that destroy crops. In at least one case this has been demonstrated not to be true; the bats were being blamed for damage caused mainly by monkeys. More study needs to be done on this issue. Old World fruit bats are exploited commercially as meat in some parts of their range.

1. Wahlberg's epauletted fruit bat (*Epomophorus wahlbergi*); 2. Indian fruit bat (*Cynopterus sphinx*); 3. Egyptian rousette (*Rousettus aegyptiacus*); 4. Golden-crowned flying fox (*Acerodon jubatus*); 5. Dwarf epauletted fruit bat (*Micropteropus pusillus*); 6. Hammer-headed fruit bat (*Hypsignathus monstrosus*); 7. Singing fruit bat (*Epomops franqueti*); 8. Straw colored fruit bat (*Eidolon helvum*). (Illustration by Brian Cressman)

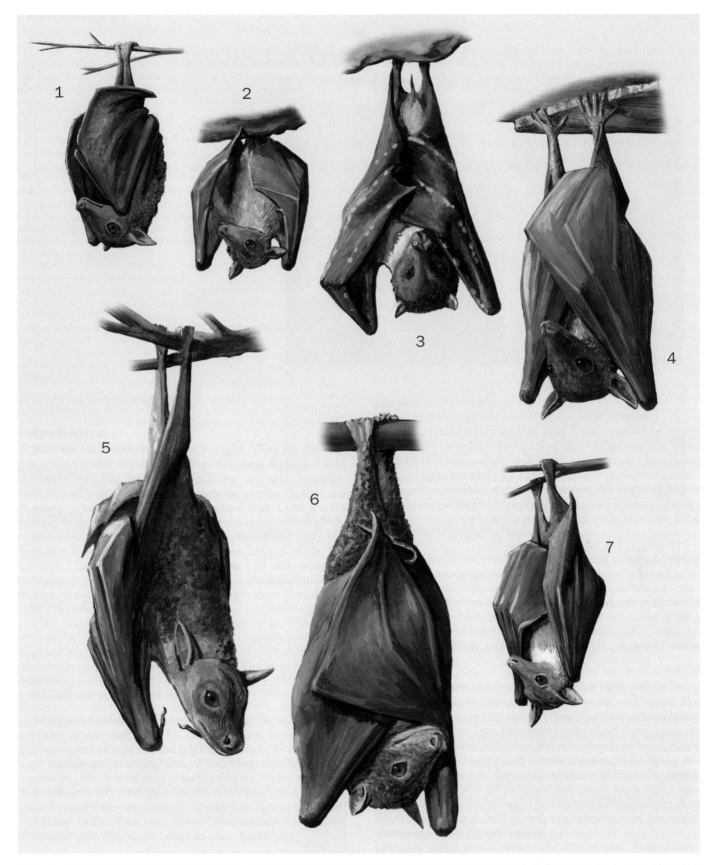

1. Greater long-tongued fruit bat (*Macroglossus sobrinus*); 2. Southern blossom bat (*Syconycteris australis*); 3. Queensland tube-nosed bat (*Nyctimene robinsoni*); 4. Dawn fruit bat (*Eonycteris spelaea*); 5. Dyak fruit bat (*Dyacopterus spadiceus*); 6. Harpy fruit bat (*Harpyionycteris whiteheadi*); 7. African long-tongued fruit bat (*Megaloglossus woermanni*). (Illustration by Brian Cressman)

Species accounts

Egyptian rousette
Rousettus aegyptiacus

SUBFAMILY
Pteropodinae

TAXONOMY
Pteropus aegyptiacus (Geoffroy, 1810), Giza, Egypt.

OTHER COMMON NAMES
English: Egyptian fruit bat, Arabian rousette, Cape rousette, West African rousette; German: Agyptischer Flughund.

PHYSICAL CHARACTERISTICS
Head and body length, 4.5–5 in (11.4–12.7 cm); forearm length 3.3–3.5 in (8.4–8.9 cm); wingspan, 23.6 in (60 cm); tail length, 0.4–0.9 in (1–2.2 cm); weight, 2.8–6 oz (80–170 g). Pelage is brownish gray. Ventrum is a lighter gray.

DISTRIBUTION
Southern, western, and eastern Africa, Egypt to Turkey, Cyprus, Arabian peninsula east to Pakistan.

HABITAT
Egyptian rousettes prefer slightly humid dark roosts. Most are found in caves, where they occupy the walls and ceilings close to the opening. Other roosting places are trees, rock crevices, human-built structures, including ancient ruins (the type speci-

men was collected from the Great Pyramid), wells, and underground irrigation tunnels. They also have been observed in savannas.

BEHAVIOR
Gregarious, size of camp varies. In Pakistan roosting colonies may be small, approximately 20–40 individuals. In South Africa large camps were observed having from 7,000 to 9,000 bats. Unlike the other megachiropterans, this fruit bat employs echolocation. Unlike microchiropterans, which echolocate by ultrasonic signals produced in the larynx and emitted through nostrils or the mouth, rousettes make an audible clicking sound with their tongues.

FEEDING ECOLOGY AND DIET
Food preference is very ripe fruit. Farmers often blame these bats for destruction of their crops, but they will not consume green fruit growing on a farm.

REPRODUCTIVE BIOLOGY
Varies depending on environment. Most populations appear to have two breeding seasons. Gestation is about six months and one pup a year the norm. In some locations breeding seasons appear to correspond to the end of a rainy season. Thought to be polygamous.

CONSERVATION STATUS
Not threatened.

SIGNIFICANCE TO HUMANS
Regarded as a pest over much of its range. In particular it is falsely perceived as a major threat to crops. Egyptian rousettes serve as pollinators. ◆

Straw-colored fruit bat
Eidolon helvum

SUBFAMILY
Pteropodinae

TAXONOMY
Vespertilo vampyrus helvus (Kerr, 1792), Senegal.

OTHER COMMON NAMES
English: Yellow fruit bat.

PHYSICAL CHARACTERISTICS
Head and body length, 5.5–7.9 in (14–20 cm); forearm length, 4.3–5.1 in (11–13 cm); tail length, 0.2–0.8 in (0.4–2 cm); weight, 8.1–12.3 oz (230–350 g). Pelage is brown, yellow, or reddish gray, yellowish ventrum, and a wide yellowish collar.

DISTRIBUTION
Africa south of the Sahara, north into Ethiopia and Egypt, southwestern Arabia.

HABITAT
Tropical forests, but also found in urban areas where human activity does not seem to disturb it. Savannas. Found up to 6,562 ft (2,000 m) altitude. Prefers tall trees for roosts.

Rousettus aegyptiacus

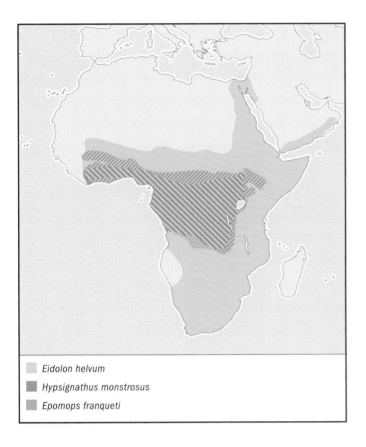

Eidolon helvum

Hypsignathus monstrosus

Epomops franqueti

BEHAVIOR
Gregarious. There have been reports that mixed sex colonies may actually number from 100,000 to one million. Eisentraut reported 10,000, which is probably the more usual number. Camps break up from around June to September. Migrate seasonally in small groups, but return to same roosting sites year after year.

FEEDING ECOLOGY AND DIET
A variety of ripe fruit including domestic crops such as mango, bananas, and papaya. Appears to travel long distances when foraging.

REPRODUCTIVE BIOLOGY
Polygamous. Appears to have a single breeding season, but delayed implantation may occur during the dry season. Consequently, females all give birth at the same time during the rainy season. Actual gestation believed to last four months and a single pup is the norm.

CONSERVATION STATUS
Not threatened.

SIGNIFICANCE TO HUMANS
Perceived as a threat to agriculture. Hunted as bushmeat. In some parts of Africa, local populations believe eating the meat of these bats increases women's fertility. A vector for rabies. ◆

Golden-crowned flying fox
Acerodon jubatus

SUBFAMILY
Pteropodinae

TAXONOMY
Pteropus jubatus (Eschscholtz, 1831), Manila, Philippines.

OTHER COMMON NAMES
English: Golden-capped fruit bat; Spanish: Zorro volador filipino.

PHYSICAL CHARACTERISTICS
Head and body length, 7–11.4 in (17.8–29 cm); forearm length, 4.9–7.9 in (12.5–20 cm); wingspan, 4.9–5.6 ft (1.5–1.7 m); no tail; weight 1–2.6 lb (450–1,200 g). Pelage is variable, but ranges from brown to black. The crown of the head is comprised of golden-yellow fur as indicated by the common name.

DISTRIBUTION
Philippines.

HABITAT
Primary and secondary forest, montane forest, swamps, and mangroves. Has been observed at 3,609 ft (1,100 m) altitude.

BEHAVIOR
Gregarious. In the past camps were reported to contain up to 150,000 animals, but recent reports estimate around 5,000 individuals in roosting colonies. Observed association with large flying foxes (*Pteropus vampyrus*).

FEEDING ECOLOGY AND DIET
Mainly figs, occasional cultivated fruits. Appear to travel long distances when foraging. Small foraging units consist of from two to six animals.

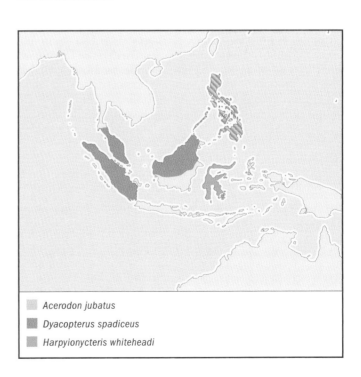

Acerodon jubatus

Dyacopterus spadiceus

Harpyionycteris whiteheadi

REPRODUCTIVE BIOLOGY
Polygamous. Appear to have two breeding seasons, but females only become pregnant during one of them. Single births are the norm. Females reach sexual maturity in two years.

CONSERVATION STATUS
Endangered.

SIGNIFICANCE TO HUMANS
Subsistence hunting, commercially exploited for meat. ◆

Hammer-headed fruit bat
Hypsignathus monstrosus

SUBFAMILY
Pteropodinae

TAXONOMY
Hypsignathus monstrosus Allen, 1861, Gabon.

OTHER COMMON NAMES
English: Horse-faced bat; German: Hammerkopf.

PHYSICAL CHARACTERISTICS
Head and body length, 7.9 in (20 cm); forearm length, 4.6–5.4 in (11.8–13.7 cm); wingspan, 35.4 in (90 cm). These bats exhibit the greatest sexual dimorphism of any member of the order Chiroptera. Males are twice the size of females. They weigh 14.8 oz (420 g), while females weigh 8.3 oz (234 g). This is the largest African fruit bat. Pelage is grayish brown with a lighter ventrum. Females have a fox-like face and resemble *Epomophorus*. Males have a very different appearance, probably as a result of a form of sexual selection called "female choice." The skull is elongated presenting a hammer-like appearance. The muzzle is inflated and widened at the front and covered by large cheek pouches, which serve as a resonating chamber. There is an enlarged larynx that fills much of the thoracic cavity and enables them to make a booming frog-like croaking. The large extensible lips form a "megaphone."

DISTRIBUTION
Western and central Africa.

HABITAT
Found primarily in tropical rainforest. Also located in swamps, mangroves, and gallery forest.

BEHAVIOR
The most significant behavior was described by Bradbury in 1977. Hammer-headed bats are one of the few vertebrates that use arenas or leks for mating. Males aggregate twice a year in an area and fight for territories. The leks consist of from 25 to 130 males. When the females arrive the males display by making honking calls and flapping their wings. The females choose the males they wish to mate with. Seventy-nine percent of the matings are performed by only 6% of the males in any one breeding season.

FEEDING ECOLOGY AND DIET
Males eat figs, females and juveniles consume softer fruits. It has been suggested that males consume the more nutritious figs because they need the additional energy for displays.

REPRODUCTIVE BIOLOGY
Polygamous. Two breeding seasons, June–August and December–January. One offspring is the norm. Females mature within six months of birth, males within 18 months. See behavior above.

CONSERVATION STATUS
Not threatened.

SIGNIFICANCE TO HUMANS
Consumed as bushmeat. ◆

Singing fruit bat
Epomops franqueti

SUBFAMILY
Pteropodinae

TAXONOMY
Epomophorus franqueti (Tomes, 1860), Gabon.

OTHER COMMON NAMES
English: Franquet's epauletted bat.

PHYSICAL CHARACTERISTICS
Head and body length, 5.3–7.1 in, (13.5–18 cm); forearm length, 2.8–3.9 in (7–10 cm); no tail; weight, 2–4 oz (56–115 g). Pelage is pale brown to cream, the sides are dark brown with large white stomach spots. White or yellow shoulder epaulets are hidden in a pouch until skin muscles evert the tufts for display. Males have two large pharyngeal sacs and an enlarged larynx that enables them to make a high-pitched sound.

DISTRIBUTION
Western through central Africa.

HABITAT
Tropical rainforests, dry lowland forests, and woodland-savanna mosaics.

BEHAVIOR
Solitary or in small groups of two or three. Males "sing" by making a high pitched sound that has the quality of a musical note. When netted, both males and females emit defensive squeaking noises.

FEEDING ECOLOGY AND DIET
Fruit. They obtain nourishment by placing their lips around fruit and sucking out juice and the soft parts of fruit.

REPRODUCTIVE BIOLOGY
Polygamous. There appears to be two breeding seasons. Offspring are born at the beginning of each of the two rainy seasons. Gestation is five to six months. A single pup is the norm. Females are sexually mature at six months and can conceive at one year, males become sexually mature at 11 months.

CONSERVATION STATUS
Not threatened.

SIGNIFICANCE TO HUMANS
Considered to be a great agricultural pest. ◆

Wahlberg's epauletted fruit bat
Epomophorus wahlbergi

SUBFAMILY
Pteropodinae

TAXONOMY
Pteropus wahlbergi (Sundevall, 1846), vicinity of Durban, Natal, South Africa.

OTHER COMMON NAMES
None known.

PHYSICAL CHARACTERISTICS
Sexually dimorphic with males slightly larger than females. Head and body length, 5.5–9.8 in (14–25 cm); forearm length, males 2.8–3.7 in (7.2–9.5 cm), females 2.7–3.5 in (6.8–8.8 cm); wingspan, males 201–236 in (510–600 cm), females 179.5–212.5 in (456–540 cm); weight, males 2.1–4.4 oz (60–124 g), females 1.9–4.4 oz (54–125 g). Pelage is dark brown, in females the ventrum is lighter and the neck is white. Males have white tufts at the base of the pinna (the fleshy external ear), white shoulder epaulettes, and a small pharyngeal sac. The shoulder epaulets are hidden in a pouch until skin muscles evert the tufts for display. Both sexes have expandable pendulous lips.

DISTRIBUTION
Eastern and southeastern Africa.

HABITAT
Edges of forest, woodlands, and savanna.

BEHAVIOR
Roosts in camps of 3–100 individuals. About 30 minutes before flight time there is intensive grooming. Little interaction is observed among individuals outside the roosting time. Males hang on branches, evert their tufts, and make calls after leaving the day roost. This appears to have a territorial function to separate males. During the breeding season it probably serves to attract females.

FEEDING ECOLOGY AND DIET
Consume figs, guava, bananas, and nectar. Stomach contents of several individuals contained beetles suggesting some insectivory. These bats fly considerable distances from roosts to feeding areas.

REPRODUCTIVE BIOLOGY
Polygamous. Two breeding seasons a year. Gestation lasts 5–6 months. Single births are the norm.

CONSERVATION STATUS
Not threatened.

SIGNIFICANCE TO HUMANS
Believed to cause some crop damage. Considered a pest because their calls keep people awake at night and their guano droppings fall on human-built structures. ◆

Dwarf epauletted fruit bat
Micropteropus pusillus

SUBFAMILY
Pteropodinae

TAXONOMY
Epomorphorus pusillus (Peters, 1867), Yoruba, Nigeria.

OTHER COMMON NAMES
English: Peters's dwarf epauletted fruit bat.

PHYSICAL CHARACTERISTICS
Head and body length, 2.6–3.7 in (6.7–9.5 cm); forearm length, 1.8–2.2 (4.6–5.6 cm); tail length variable, lacking in some individuals to 0.2 in (0.4 cm) in others; weight 0.8–1.2 oz (24–35 g). Males larger than females. Pelage light brown with a lighter ventrum. Males have white tufts at the base of the pinna and white shoulder epaulettes. The shoulder epaulets are hidden in a pouch until skin muscles evert the tufts for display. Similar in morphology to *Epomorphorus*, but smaller in body size and a shorter muzzle.

DISTRIBUTION
Western, southwestern, and central Africa.

HABITAT
Savanna woodlands and forest edge.

BEHAVIOR
Usually solitary or with one other conspecific. Sometimes found in groups up to 10. Males display with erect epaulets and croaking calls. Nomadic and does not return to regular sleeping or feeding trees. Cryptic species that is not easily disturbed or observed.

FEEDING ECOLOGY AND DIET
Consumes small fruits, nectar, and pollen. They obtain nourishment by placing their lips around fruit and sucking.

■ *Epomophorus wahlbergi*
■ *Micropteropus pusillus*
■ *Megaloglossus woermanni*

REPRODUCTIVE BIOLOGY
Polygamous. Two breeding seasons. Single young is the norm. Mating behavior is unknown, but it is presumed that males make displays.

CONSERVATION STATUS
Not threatened.

SIGNIFICANCE TO HUMANS
None known. ◆

Indian fruit bat
Cynopterus sphinx

SUBFAMILY
Pteropodinae

TAXONOMY
Vespertilio sphinx (Vahl, 1797), Tranquebar, Madras, India.

OTHER COMMON NAMES
English: Short-nosed fruit bat, dog-faced fruit bat.

PHYSICAL CHARACTERISTICS
Head and body length, 2.8–5.1 in (7–13 cm); forearm length, 2.5–3.1 in (6.4–7.9 cm); wingspan, 12–18.9 in (30.5–48 cm); tail length, 2.4–5.9 in (6–15 cm); weight, 0.9–3.5 oz (25–100 g). Pelage color ranges from rusty brown to olive. These bats have pronounced tubular nostrils (although they are not part of the tube-nosed fruit bat subfamily).

DISTRIBUTION
Indian subcontinent, Southeast Asia, Islands of Southeast Asia to Borneo and Sulawesi.

HABITAT
Forests, farms, and city parks.

BEHAVIOR
Variable. May be found singly; small roosting groups of three or four individuals are common, but camps of up to 25 have

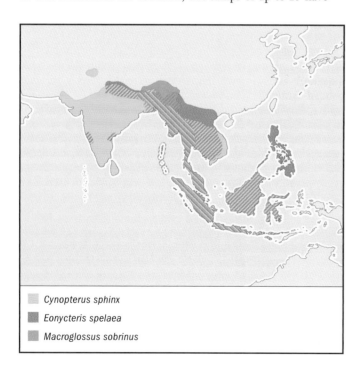

Cynopterus sphinx
Eonycteris spelaea
Macroglossus sobrinus

been observed. Older males roost alone, younger males roost with females. In some locations one-male units (or harems) exists during breeding season. Males construct elaborate shelters called "stem tents" over a period of one to two months. One male attracts 2–20 females. The stem tent is used for roosting during the day and as a nursery for the pups. After the pups are weaned the males and females separate into unisex groups.

FEEDING ECOLOGY AND DIET
Ripe fruit such as figs, soursop, and mango, flowers, nectar, and pollen.

REPRODUCTIVE BIOLOGY
Polygamous. Two breeding seasons. Females produce one offspring per year. Gestation lasts about four months. Females reach sexual maturity at about five months, males at around 15 months.

CONSERVATION STATUS
Not threatened.

SIGNIFICANCE TO HUMANS
Some Asian populations use these bats for medicinal purposes. ◆

Dyak fruit bat
Dyacopterus spadiceus

SUBFAMILY
Pteropodinae

TAXONOMY
Cynopteropus spadiceus (Thomas, 1890), Baram, Sarawak, Malaysia.

OTHER COMMON NAMES
None known.

PHYSICAL CHARACTERISTICS
Head and body length, 3.9-5.9 in (10–15 cm); forearm length, 3–3.5 in (7.6–9 cm); tail length, 0.5–0.7 in (1.3–1.8 cm); weight, 2.5–3.5 oz (70–100 g). Pelage brown back and flanks, off-white ventrum, light yellow shoulders.

DISTRIBUTION
Islands of Southeast Asia including the southern Philippines.

HABITAT
Lowland primary forest and montane forest. Has been observed in open areas.

BEHAVIOR
Little known. May be monogamous.

FEEDING ECOLOGY AND DIET
Little known, but skull and dental morphology suggests fruit. One group was observed consuming figs.

REPRODUCTIVE BIOLOGY
Little known. Some captured males had functional mammary glands and were lactating. This suggests that males may be involved in care of offspring. Thought to be polygamous.

CONSERVATION STATUS
Lower Risk/Near Threatened.

SIGNIFICANCE TO HUMANS
None known. ◆

Harpy fruit bat

Harpyionycteris whiteheadi

SUBFAMILY
Harpyionycterinae

TAXONOMY
Harpyionycteris whiteheadi Thomas, 1896, Mindoro Island, Philippines.

OTHER COMMON NAMES
German: Whitehead-Spitzzahn-Flughund.

PHYSICAL CHARACTERISTICS
Head and body length, 5.5–6 in (14–15.3 cm); forearm length, 3.2–3.6 in (8.2–9.2 cm); no tail; weight, 2.9–5 oz (83–142 g). Pelage is chocolate to dark brown, ventrum is lighter. Differs from all other fruit bats in its dental structure. The molars have 5–6 cusps, the lower canines have three cusps, and the incisors are directed forward.

DISTRIBUTION
Philippines, Sulawesi.

HABITAT
Undisturbed primary rainforest and lower montane forest up to 5,906 ft (1,800 m).

BEHAVIOR
Nothing known.

FEEDING ECOLOGY AND DIET
Fruit. Observed consuming pandan fruit and figs. Forage in canopy.

REPRODUCTIVE BIOLOGY
Polygamous. Two breeding seasons. Females give birth to one pup twice a year. Gestation 4–5 months. Females appear to become sexually mature within the first year of life.

CONSERVATION STATUS
Not threatened.

SIGNIFICANCE TO HUMANS
None known. ◆

Dawn fruit bat

Eonycteris spelaea

SUBFAMILY
Macroglossinae

TAXONOMY
Macroglossus spelaeus (Dobson, 1873), Moulmein, Tenasserim, Myanmar.

OTHER COMMON NAMES
English: Cave fruit bat, common nectar-feeding fruit bat, Dobson's long-tongued dawn bat.

PHYSICAL CHARACTERISTICS
Head and body length, 3.3–4.9 in (8.5–12.5 cm); forearm length, 2.6–3.1 in (6.6–7.8 cm); tail length, 0.5–1.3 in (1.2–3.3 cm). Sexual dimorphism in body weight, males weighing between 1.9–2.9 oz (55–82 g) and females weighing 1.2–2.8 oz

(35–78 g). Brown pelage with gray mottled ventrum. Males have a frill of longer hairs on the sides of their necks. Has long protrusible tongue with brush at the end.

DISTRIBUTION
Southeast Asia from Myanmar through Indonesia to the Philippines and Sulawesi.

HABITAT
Primary forest and cultivated land.

BEHAVIOR
Gregarious with roosting colonies in caves that number from a dozen to over 10,000, which are, in turn, divided into sexually segregated clusters. Roost in caves. Form associations with Leschenault's rousette (*Rousettus leschenaulti*). Some, perhaps most, of the individuals observed in roosting camps may be of this species.

FEEDING ECOLOGY AND DIET
Nectar and pollen from night flowering plants.

REPRODUCTIVE BIOLOGY
Females appear to come into estrus twice a year, but it is not synchronized with other females or with any particular season. Gestation in the dawn fruit bat has been reported to be 3–4 months; however, other reports suggest gestation is slightly longer than six months. Females become sexually mature by six months, males by one year of age. Thought to be polygamous.

CONSERVATION STATUS
Not threatened.

SIGNIFICANCE TO HUMANS
Pollinator of many important commercial plant species. Hunted for meat. ◆

African long-tongued fruit bat

Megaloglossus woermanni

SUBFAMILY
Macroglossinae

TAXONOMY
Megaloglossus woermanni Pagenstecher, 1885, Sibange farm, Gabon.

OTHER COMMON NAMES
English: Woermann's bat.

PHYSICAL CHARACTERISTICS
Head and body length, 2.8–3 in (7–7.5 cm); forearm length, 1.5–2 in (3.7–5 cm); wingspan, 9.8 in (25 cm); no tail; weight, 0.3–0.6 oz (8.4–15.6 g). Pelage is dark brown with cream ventrum; a dark dorsal crown stripe runs longitudinally to the nape of the neck.

DISTRIBUTION
Along the tropical belt from Guinea to Uganda in Africa.

HABITAT
Primary and secondary lowland forest.

BEHAVIOR
Poorly known. Appears to be solitary, but will congregate in unisex or mixed-sex clusters at flowering trees.

FEEDING ECOLOGY AND DIET
Pollen and nectar.

REPRODUCTIVE BIOLOGY
Poorly known. May not have a distinct breeding season or may have two breeding seasons. Thought to be polygamous.

CONSERVATION STATUS
Not threatened.

SIGNIFICANCE TO HUMANS
None known. ◆

Greater long-tongued fruit bat
Macroglossus sobrinus

SUBFAMILY
Macroglossinae

TAXONOMY
Macroglossus minimus sobrinus Andersen, 1911, Gunong Igari, Perak, Malaysia.

OTHER COMMON NAMES
English: Hill long-tongued fruit bat, greater long-tongued nectar bat.

PHYSICAL CHARACTERISTICS
Head and body length, 2.4–3.3 in (6–8.5 cm); forearm length, 1.9–2 in (4.7–5.2 cm); wingspan, 9.8 in (25 cm); no tail; weight, 0.7–0.8 oz (18.5–23 g). Pelage reddish brown, buffy brown ventrum. Elongated muzzle, long tongue with brushlike structures at the end.

DISTRIBUTION
Northeast India to the islands of Southeast Asia to Java and Bali.

HABITAT
Lowland and montane forest up to 6,562 ft (2,000 m), villages with bamboo, cultivated fruit orchards.

BEHAVIOR
Solitary, but can sometimes be found roosting in groups up to 10. Sometimes found rolled up in leaves.

FEEDING ECOLOGY AND DIET
Pollen, nectar, and some soft fruit.

REPRODUCTIVE BIOLOGY
Polygamous. Believed to breed throughout the year. Single young, but it has been reported that females appear to produce two pups per year.

CONSERVATION STATUS
Not threatened.

SIGNIFICANCE TO HUMANS
Pollinator; may pollinate cultivated plants. ◆

Southern blossom bat
Syconycteris australis

SUBFAMILY
Macroglossinae

TAXONOMY
Macroglossus minimus var. *australis* (Peters, 1867), Rockhampton, Queensland, Australia.

OTHER COMMON NAMES
English: Queensland blossom bat, common blossom bat.

PHYSICAL CHARACTERISTICS
Head and body length, 2–3 in (5–7.5 cm); forearm length, 1.5–1.7 in (3.8–4.3 cm); very rudimentary tail; weight, 0.4–0.9 oz (11.5–25 g). Pelage is reddish brown with lighter ventrum. Elongated snout with long protrusible tongue tipped with a brush.

DISTRIBUTION
Moluccas through New Guinea to Bismarck Archipelago to eastern Australia.

HABITAT
Rainforest.

BEHAVIOR
Solitary.

FEEDING ECOLOGY AND DIET
Pollen and nectar. May occasionally consume fruit.

REPRODUCTIVE BIOLOGY
Poorly known, but it has been suggested that they breed throughout the year. One pup is the norm. Believed to be polygamous.

CONSERVATION STATUS
Not threatened.

SIGNIFICANCE TO HUMANS
Pollinato; may pollinate cultivated plants. ◆

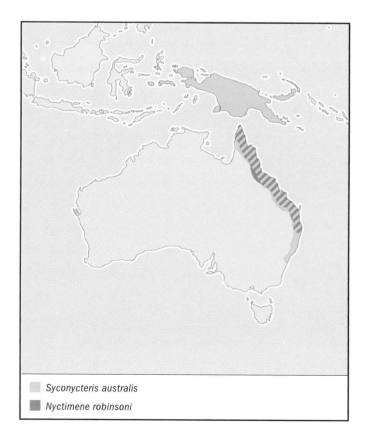

☐ *Syconycteris australis*
■ *Nyctimene robinsoni*

Queensland tube-nosed bat
Nyctimene robinsoni

SUBFAMILY
Nyctimeninae

TAXONOMY
Nyctimene robinsoni var. *australis* Thomas, 1904, Cooktown, Queensland, Australia.

OTHER COMMON NAMES
English: Eastern tube-nosed fruit bat.

PHYSICAL CHARACTERISTICS
Head and body length, 3–5.1 in (7.5–13 cm); forearm length, 2.4–2.8 in (6–7 cm); tail length, 0.8–1 in (2–2.5 cm); weight, 1.1–1.8 oz (30–50 g). Pelage light brown with a dark dorsal stripe, lighter ventrum, brown wings with yellowish blotches and spots, believed to function as camouflage. Distinctive feature is tubed nostrils which protrude for up to 1 in (2.5 cm), the function of which is not known for certain.

DISTRIBUTION
Tropical and subtropical eastern Australia.

HABITAT
Tropical rainforest, subtropical rainforest remnants.

BEHAVIOR
Solitary.

FEEDING ECOLOGY AND DIET
Eats fruit.

REPRODUCTIVE BIOLOGY
Polygamous with one breeding season. One pup the norm. Gestation 4.5–5 months. Females sexually mature at seven months.

CONSERVATION STATUS
Not threatened.

SIGNIFICANCE TO HUMANS
None known. ◆

Common name / Scientific name/ Other common names	Physical characteristics	Habitat and behavior	Distribution	Diet	Conservation status
Small-toothed fruit bat *Neopteryx frosti*	Tawny or brownish pelage, medial stripe from tip of nose to between the eyes, off-white stripes along muzzle. The wing membrane extends to the vertebral column creating the impression of a naked or bare back. No claw on second digit. Palate is narrowed and canines are reduced in size. Head and body length 4.1 in (10.5 cm), forearm length 4.1–4.4 in (10.5–11.1 cm), no tail, weight 6.6 oz (190 g).	Lowland tropical rainforest. Behavior poorly known.	Northern and western Sulawesi.	Unknown. Morphology of rostrum and teeth unique among the fruit bats, making an extrapolation of diet difficult.	Vulnerable
Striped-faced fruit bat *Styloctenium wallace*	Pelage is gray with reddish brown ventrum. White lateral stripes along the sides of the muzzle. Head and body length 5.9–7.1 in (15–18 cm), forearm length 3.5–4.1 in (9–10.3 cm), no tail, weight 6.0–7.6 oz (175–220 g).	Tropical rainforest. Natural history poorly known.	Sulawesi.	Fruit.	Lower Risk/Near Threatened
Naked-backed fruit bat *Dobsonia magna* English: Greater naked-backed bat	Pelage is brown to grayish black. The wing membrane extends to the vertebral column creating the impression of a naked or bare back. Head and body length 4.0–9.5 in (10.2–24.2 cm), forearm length 5.3–6.1 in (13.5–15.5 cm), short tail, weight 12.1–17.2 oz (350–500 g).	Gregarious, although has been observed roosting alone. Occupies forested areas. Roosts near the mouth of caves, old mines, and abandoned structures. In camps, these bats vocalize continuously. However, they forage alone and silently. The wing morphology makes this bat an excellent and highly maneuverable flyer that can even fly backwards. Its flying ability allows it to take fruits below the canopy level not available to *Pteropus* species in the area.	Northern Queensland, Australia to New Guinea to Moluccas.	Wide variety of fruits.	Not threatened
Bulmer's fruit bat *Aproteles bulmerae*	Very similar to *Dobsonia* except in size. Brown pelage, lighter on the ventral side. The wing membrane extends to the vertebral column creating the impression of a naked or bare back. Head and body length 9.5 in (24.2 cm), forearm length 6.5 in (16.6 cm), tail length 1.3 in (3.2 cm), weight 20.7 oz (600 g).	Occupies caves in montane forest. Behavior unknown.	Central highlands of New Guinea.	Fruit.	Critically Endangered; threatened by human hunting
Little flying cow *Nanonycteris veldkampi* English: Veldkamp's dwarf fruit bat, Veldkamp's dwarf epauletted bat	Reddish brown pelage with lighter ventrum. Males possess pouches that evert to display epaulettes. Head and body length 2.1–3.0 in (5.4–7.5 cm), forearm length 1.7–2.13 in (4.3–5.4 cm), rudimentary tail present, weight 0.7–1.1 oz (19–33 g).	Forest and savannas. Migrates from forest environments to savannas during the rainy season. Roosts alone or in small groups.	Guinea to Central African Republic.	Nectar.	Not threatened
Short-palate fruit bat *Casinycteris argynnis*	Pelage mainly light brown, but muzzle and wings are orange. There is a white medial eye patch between the eyes and lateral white patches behind the eyes. White tufts are present at the base of the pinnae. The palate is shortened and the dentition reduced in size. Head and body length 3.5–3.7 in (9.0–9.5 cm), forearm length 2.0–2.5 in (5.0–6.3 cm), rudimentary tail, weight 0.9–1.0 oz (26–30 g).	Tropical rainforest. Behavior largely unknown. Single individuals have been observed roosting.	Cameroon to north-eastern Democratic Republic of the Congo (Zaire).	Presumably fruit.	Lower Risk/Near Threatened
Fischer's pygmy fruit bat *Haplonycteris fischeri*	Cinnamon pelage with lighter ventrum. Head and body length 2.7–3.1 in (6.8–8.0 cm), forearm length 1.7–2.1 in (4.4–5.3 cm), no tail, weight 0.6–0.7 oz (16–21 g).	Primary and secondary forest. Behavior poorly known.	Philippines.	Fruit.	Vulnerable; threatened by habitat destruction

[continued]

Common name / Scientific name/ Other common names	Physical characteristics	Habitat and behavior	Distribution	Diet	Conservation status
Little collared fruit bat *Myoncyteris torquata* English: Ringed fruit bat	Physically very similar to rousette bats except they have a pronounced shortening of the facial structure. Pelage is various shades of brown, a lighter ventrum, and a light collar around the neck. Head and body length 3.5–6.5 in (9–16.5 cm) forearm length 2.2–2.75 in (5.5–7.0 cm), tail length 0.2–0.5 in (0.4–1.3 cm), weight 0.9–1.9 oz (27–54 g).	Tropical rainforest, woodland-savanna mosaic. Poorly known, but appears to be solitary.	Western and central Africa.	Fruit.	Not threatened
Zenker's fruit bat *Scotonycteris zenkeri*	Pelage ranges from light to dark brown with a lighter ventrum. A white spot appears above the nose in front of the eyes, and two additional spots at the upper lateral border of the eyes. Head and body length 2.6–3.1 in (6.5–8.0 cm), forearm length 2–2.2 in (5–5.6 cm), tail is rudimentary, weight 0.6–0.9 oz (18–27 g).	Primary rainforest. Solitary.	Western to west central Africa.	Small fruits.	Not threatened
Ratanaworabhan's fruit bat *Megaerops niphanae*	Pelage is grayish brown, lighter brown on shoulders, with gray ventrum. Head and body length 2.8–3.9 in (7–10 cm), forearm length 2.0–2.5 in (5.2–6.3 cm), wingspan 16.7 in (42.4 cm), no tail, weight 0.6–1.3 oz (18–38 g).	Lowland and montane forest up to 5,900 ft (1,800 m) in altitude. Behavior is unknown.	Northeastern India, Thailand, and Vietnam.	Unknown, but presumably fruit.	Not listed by IUCN
Salim Ali's fruit bat *Latidens salimalii*	Pelage dark brown to black, grizzled light fur on shoulders, between the eyes, and cheeks. Head and body length, 4.0–4.3 in (10.2–10.9 cm), forearm length 2.6–2.7 in (6.6–6.9 cm), no tail.	Broadleaf montane forest. Behavior unknown.	Southern India.	Unknown, but dentition suggests hard fruits and seeds.	Critically Endangered; may be fewer than 50 individuals
Blanford's fruit bat *Sphaerias blanfordi*	Pelage grayish brown. Head and body length 2.5–3.2 in (6.4–8.0 cm), forearm length 2.0–2.4 in (5.2–6.0 cm), no tail.	Montane forest between 2,620 and 8,860 ft (800–2,700 m). Behavior unknown.	Northeastern India, southern Tibet, north-western Thailand, and southwestern China.	Unknown.	Not listed by IUCN
Queensland tube-nosed bat *Nyctimene robinsoni* English: Eastern tube-nosed fruit bat	Pelage light brown with a dark dorsal stripe, lighter ventrum, brown wings with yellowish blotches and spots, believed to function as camouflage. Distinctive feature is tubed nostrils which protrude for up to 1 in (3 cm), the function of which is not known for certain. Head and body length 3.0–5.1 in (7.5–13.0 cm), forearm length, 2.4–2.8 in (6.0–7.0 cm), tail length 0.8–1.0 in (2.0–2.5 cm), weight 1.0–1.7 oz (30–50 g).	Tropical rainforest, sub-tropical rainforest remnants. Solitary.	Tropical and subtropical eastern Australia.	Fruit.	Not threatened

Resources

Books

Bates, P. J. J., and D. L. Harrison. *Bats of the Indian Subcontinent.* Sevenoaks, UK: Harrison Zoological Museum Publication, 1997.

Corbet, G. B., and J. E. Hill. *A World List of Mammalian Species.* 3rd ed. New York: Oxford University Press, 1991.

Eisentraut, M. "The Old World Fruit Bats." In *Grzimek's Animal Life Encyclopedia,* Vol. 11, edited by B. Grzimek. New York: Van Nostrand Reinhold Company, 1972.

Macdonald, D., ed. *The Encyclopedia of Mammals.* New York: Facts on File Publications, 1984.

Mickleburgh, S. P., A. M. Hutson, and P. A. Racey. *Old World Fruit Bats: An Action Plan for their Conservation.* Gland, Switzerland: International Union for the Conservation of Nature, 1992.

Nowak, R. M. *Walker's Mammals of the World.* 6th ed. Baltimore: Johns Hopkins University Press, 1999.

Qumsiyeh, M. B. *The Bats of Egypt.* Lubbock, TX: Texas University Press, 1985.

Periodicals

Acharya, L. "*Epomophorus wahlbergi.*" *Mammalian Species* 394 (1992): 1–4.

Resources

Balasingh, J., S. Suthakar-Isaac, and R. Subbaraj. "Tent Roosting by the Frugivorous Bat *Cynopterus sphinx* in Southern India." *Current Science* 65 (1993): 418.

Bradbury, J. W. "Lek Mating Behavior in the Hammer-headed Bat." *Zeitschrift fur Tierpsychologie* 45 (1977): 225–255.

Owen-Ashley, N. T., and D. E. Wilson. "*Micropteropus pusillus.*" *Mammalian Species* 577 (1998): 1–5.

Simmons, N. B. "The Case for Chiropteran Monophyly." *American Museum Novitiates* 3103 (1994): 1–54.

Tuttle, M. D. "Fruit Bats Exonerated." *Bats* 1 (1984): 1–2.

Other

International Union for the Conservation of Nature. "Red List of Threatened Species." 2000 [June 11, 2003]. <http://www.redlist.org>.

Marcus Young Owl, PhD

Mouse-tailed bats

(Rhinopomatidae)

Class Mammalia
Order Chiroptera
Suborder Microchiroptera
Family Rhinopomatidae

Thumbnail description
The mouse-tailed bats are the only microbats with a tail nearly as long as the head and body, which extends well beyond the end of the interfemoral membrane; the large ears are connected by a band of skin across the forehead, and extend beyond the nostrils when the ears are projected forward

Size
Small to medium-sized bats with forearms ranging from 1.8 to 3 in (4.5–7.5 cm) in length and weighing 0.2–1.6 oz (6–45 g)

Number of genera, species
1 genus; 4 species

Habitat
Mouse-tailed bats are mainly found in arid habitats but also in moist, forested regions

Conservation status
Vulnerable: 1 species; Lower Risk/Least Concern: 2 species

Distribution
North Africa to Asia

Evolution and systematics

Included in the superfamily Rhinopomatoidea with the hog-nosed bats (Craseonycteridae) family, mouse-tailed bats are relatively unspecialized compared to most other groups of microchiropterans. They are not known as fossils and no sub-families have been described.

Physical characteristics

These small to medium-sized bats have long slender tails making them among the most distinctive of microchiropterans. The valvular nostrils are without equivalent among bats, but the small flap of a noseleaf is somewhat matched by the condition in some plain-nosed bats (Vespertilionidae).

Distribution

Mouse-tailed bats occur across the Sahara from western Africa through the middle east to India, Myanmar, Thailand, and Sumatra.

Habitat

Mouse-tailed bats are mainly associated with arid habitats, although they also range into moist, forested regions. As expected from animals of arid regions, mouse-tailed bats have kidneys specialized in the production of highly concentrated

urine. The urine of mouse-tailed bats is four to five times more concentrated than that of humans. These bats roost in hollows, typically in caves, abandoned mines, pyramids and wells, but also in buildings. Mouse-tailed bats emerge from their roosts around dark and forage well above (16–32 ft, or 5–10 m) the ground or vegetation. Foraging individuals can be observed over forested areas as well as over arid habitats mainly devoid of vegetation.

Behavior

Rhinopomatids often form colonies consisting of hundreds of individuals. Colonies echo with a mixture of social and echolocation calls, and individuals may or may not roost in contact with conspecifics. At some times of the year these bats lay down large fat deposits. These deposits may equal the bats' normal body mass and appear to support them through times when prey is scarce or absent (sometimes winter). Heavier individuals with body weight exceeding 1.4 oz (40 g) are those with large fat accumulations and males typically are heavier than females. In some parts of their range, some species migrate between summer and winter roosts.

Feeding ecology and diet

These aerial-feeding bats use echolocation to detect, track and assess the flying insects that they eat. The bats' echolo-

A side view of Hardwicke's lesser mouse-tailed bat (*Rhinopoma hardwickei*). (Photo by © Merlin D. Tuttle, Bat Conservation International. Reproduced by permission.)

cation calls consist of several harmonics, with most energy in the third. Most of the sound energy is in the ultrasonic range, with species-specific variations in the actual frequencies dominating the calls. Foraging animals do not appear to be territorial, perhaps reflecting the dispersed nature of their insect prey. Their diet includes insects ranging from flying ants and termites to beetles, bugs and moths. When insect food is not easily available, or during cold seasons, they undergo periods of torpor. During winter, fat that was accumulated in the fall, is metabolized and the bat is able to survive for several weeks without water.

Reproductive biology

These bats are thought to be polygamous. Females bear a single young each year after a gestation period of about 120 days. Young are fully grown in about six weeks when they are weaned. Females attain sexual maturity in their second year.

Reproduction periods differ depending on the geographic habitat of the bat and are dependent on the seasonal movement due to torpor. Young mouse-tailed bats are born in June and July in Egypt, females in Egypt and Sudan are pregnant in late March and lactate during August.

Conservation status

Two species, the greater mouse-tailed bat (*Rhinopoma microphyllum*) and small mouse-tailed bat (*Rhinopoma muscatellum*), are categorized by IUCN as Lower Risk/Least Concern. One, MacInnes's mouse-tailed bat (*Rhinopoma macinnesi*), is identified as Vulnerable because of the decline in available habitat for its small and restricted population.

Significance to humans

They eat many insects that humans consider pests.

A male Hardwicke's lesser mouse-tailed bat (*Rhinopoma hardwickei*) perched on rock. (Photo by © Merlin D. Tuttle, Bat Conservation International. Reproduced by permission.)

Species accounts

Hardwicke's lesser mouse-tailed bat
Rhinopoma hardwickei

TAXONOMY
Rhinopoma hardwickei Gray, 1831, India. Two subspecies are recognized.

OTHER COMMON NAMES
None known.

PHYSICAL CHARACTERISTICS
Smaller than some other mouse-tailed species and larger than others. Forearms ranging in length from 2.0 to 2.5 in (5.2 to 6.4 cm); weighing 0.4–0.5 oz (11–14 g).

Rhinopoma hardwickei

DISTRIBUTION
Lesser mouse-tailed bats are found in North Africa (Morocco and Senegal to Egypt, Somalia, and Kenya), the middle east (Israel, Syria, Jordan, Lebanon, Saudi Arabia, and Yemen), Afghanistan, India, and Pakistan, as well as on Socotra Island.

HABITAT
A species of more arid areas, lesser mouse-tailed bats are also found in habitats ranging from deserts to dry woodland.

BEHAVIOR
Lesser mouse-tailed bats roost in hollows (caves, mines, pyramids, wells, buildings) and their colonies often number in the thousands of individuals. Roosting in groups numbering from one to 10 also are common. These bats retire to sheltered roosts in extremely hot weather. During these periods they

Rhinopoma hardwickei

may be relatively inactive, depending upon accumulated fat deposits for energy and water.

FEEDING ECOLOGY AND DIET
These bats eat flying insects (moths, beetles, and neuropterans) which they detect and track using echolocation. When several individuals forage together, different ones use echolocation calls dominated by different frequencies. This jamming avoidance behavior may minimize interference between bats.

REPRODUCTIVE BIOLOGY
Female lesser mouse-tailed bats bear a single young annually after a gestation period of 90–100 days. Maximum lifespan ranges between 1 to 2 years. Thought to be polygamous.

CONSERVATION STATUS
Rated by IUCN as Lower Risk/Least Concern.

SIGNIFICANCE TO HUMANS
Lesser mouse-tailed bats eat insects, many of which are considered pests by humans.

Resources

Books

Bates, P. J. J., and D. L. Harrison. *Bats of the Indian Subcontinent.* Sevenoaks, UK: Harrison Zoological Museum, 1997.

Hutson, A. M., S. P. Mickelburgh, and P. A. Racey. *Global Status Survey and Conservation Action Plan, Microchiropteran*

Bats. Gland, Switzerland: IUCN SSC Chiroptera Specialist Group, 2001.

Kingdon, J. *Mammals of East Africa: an Atlas of Evolution.* Vol. 2b. New York: Academic Press, 1974.

Nowak, R. M. *Walker's Mammals of the World.* Vol. 1. Baltimore: Johns Hopkins University Press, 1999.

Periodicals

Schlitter, D. A., and M. B. Qumsiyeh. "*Rhinopoma microphyllum.*" *Mammalian Species* 542 (1996): 1–5.

Habersetzer, J. "Adaptive Echolocation Sounds in the Bat *Rhinopoma hardwickei.*" *Journal of Comparative Physiology* A144 (1981): 559–566.

Simmons, J. A., S. Kick, and B. D. Lawrence. "Echolocation and Hearing in the Mouse-tailed Bat, *Rhinopoma hardwickei*: Acoustic Evolution in the Echolocation of Bats." *Journal of Comparative Physiology* A154 (1984): 347–356.

Melville Brockett Fenton, PhD

Sac-winged bats, sheath-tailed bats, and ghost bats

(Emballonuridae)

Class Mammalia
Order Chiroptera
Suborder Microchiroptera
Family Emballonuridae

Thumbnail description
Small, insect-feeding bats with mostly brown or gray fur and relatively large eyes; many emballonurid bats roost at almost vertical substrates with the folded forearms supporting the body

Size
Head and body length: 1.4–6.3 in (36–160 mm); forearm: 1.4–3.9 in (36–100 mm); weight: 0.1–3.5 oz (3–100 g)

Number of genera, species
12 genera, 47 species

Habitat
Daytime roosts in well-lit portions of hollow trees, buttress cavities, rock shelters, or caves. Foraging habitats in evergreen forests, semi-deciduous forests, or savannas

Conservation status
Critically Endangered: 2 species; Endangered: 2 species; Vulnerable: 10 species; Lower Risk/Near Threatened: 7 species

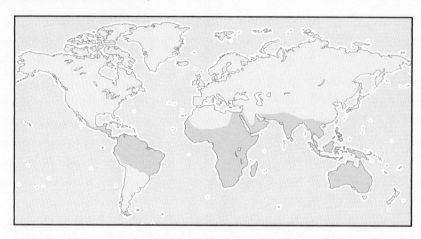

Distribution
In tropical and subtropical regions from Mexico to Argentina in the New World. In the Old World in Africa, Madagascar, the Indian subcontinent, and Southeast Asia to Australasia

Evolution and systematics

Emballonurid bats are first recorded in Europe from the middle Eocene to early Miocene, in Africa from early Miocene to Recent, in South and Central America from Pleistocene to Recent, and in other areas of their geographical range from Recent. Two subfamilies are distinguished: Taphozoinae and Emballonurinae. Some authors consider the genera *Diclidurus* and *Cyttarops* as a separate subfamily Diclidurinae.

Physical characteristics

Emballonurid bats are small bats with relatively large eyes. The ears have a tragus and, in some species, the ears are connected. Males of some New World species (sac-winged bats) have a sac-like organ in the frontal wing membrane that contains a strong smelling liquid. Females have only rudiments of this organ and it is not known if females use it in a behavioral context. Position and size of these sacs vary among species. The ghost bats, genus *Diclidurus*, have a sac-like organ in the wing-tail membrane. These bats are also an exception within the whole family as they have white fur. Within the genus *Taphozous*, some members possess a gland at the chest. In general, emballonurid bats have grayish to brown fur. The tail emerges free through the tail membrane and projects above its dorsal surface.

Distribution

The family Emballonuridae occurs both in the New World and in the Old World. There are 18 species in the New World with a distribution range from southern Mexico to northern Argentina. The 29 species of the Old World inhabit Africa, the Indian subcontinent, and Southeast Asia to Australasia. Most species are restricted to the lowland regions. The species diversity of emballonurid bats increases towards the equator.

Habitat

Emballonurid bats inhabit humid rainforests, seasonal semi-deciduous forests, and savannas. Most species roost in well-lit places like entries to caves and temples, at the outside of buildings, or in hollow trees and buttress cavities of large trees.

Greater sac-winged bats (*Saccopteryx bilineata*) roosting over water, Barra del Colorado National Wildlife Refuge, Costa Rica. (Photo by Janis Burger. Bruce Coleman, Inc. Reproduced by permission.)

The brown-bearded sheath-tail bat (*Taphozous achates*) is considered Vunerable. (Photo by Pavel German. Reproduced by permission.)

specialized holding sacs in the front wing membrane in which the fragrances are stored. The scents emanating from wing sacs of emballonurid bats smell differently in different species. Histological studies proved that holding sacs of male *S. bilineata* and other emballonurid bats do not contain any glandular tissues. Male *S. bilineata* actively fill fragrances into the wing sacs each afternoon during a stereotyped behavioral sequence. During the first phase, which has been interpreted as cleaning, males swallow their urine and lick the holding

Behavior

Colonies of some emballonurid species are easily found because these bats emit social calls audible to humans. The roosting posture of emballonurid bats is characteristic as they support their body from the surface by the thumbs of the folded wings.

The social behavior of the greater sac-winged bat can be observed easily from some distance. The mating system has been described as harem-polygynous, but recent studies using molecular genetic techniques proved that harem males father only 30% of the offspring within their harems. However, harem holders have on average a higher reproductive success than non-harem males and, therefore, males benefit from defending a group of females. The courtship of male *S. bilineata* includes visual, acoustic, and olfactory displays. Besides a large number of social calls that have been described for this species, males emit songs that can last for up to an hour and that include many different syllables. Songs of male *S. bilineata* are partly in the audible range of humans. In addition to singing, males also perform hovering flights during which fragrances are fanned towards roosting females. Males possess

The greater dog-faced bat (*Peropteryx kappleri*). (Photo by © Merlin D. Tuttle, Bat Conservation International. Reproduced by permission.)

sacs intensively. After approximately 10–20 minutes of cleaning, males rest for a few minutes and then switch to a second phase. During the second phase, males press their chin onto the penis and deposit a small droplet at their chin. Afterwards, males smear this droplet into one of the holding sacs with a sideward movement of the head. This secretion probably originates from the preputial glands. Additional droplets of the gular gland are also added to the holding sacs. When transferring genital and gular secretions to the holding sacs, males alternate between the two sacs, thus refilling almost equal amounts of fragrances into both wing sacs. The second phase of perfume-blending can last up to 20–30 minutes. It is unknown if other emballonurid species with wing sacs in the male sex show a similar behavior.

Feeding ecology and diet

Emballonurid bats are aerial insectivorous bats that can be easily observed hunting for insects in a slow butterfly-like flight. Larger emballonurid species, like the genus *Taphozous*, have a more pronounced, powerful flight. Emballonurids are among the first bats to start foraging in the evening. During periods of bad weather, some species may even start foraging in the afternoon. Occasionally, some species also glean insects from leaves. The diet of neotropical emballonurids consists mostly of small insects, primarily beetles and flies. In *S. bilineata*, colony members forage in an area where only they have access to. When insect abundance is declining, the whole colony moves to a different foraging site. Within foraging areas, single individuals forage in beats of 32.8–98.4 ft (10–30 m). The proboscis bat, *Rhynchonycteris naso*, hunts insects above or close to water surfaces. Ghost bats, genus *Diclidurus*, forage above the

A Hildegarde's tomb bat (*Taphozous hildegardeae*) in Tanzania. (Photo by David Hosking/FLPA–Images of Nature. Reproduced by permission.)

canopy. The Old World members are similar in their diet to the New World members. The tomb bats, genus *Taphozous*, forage more in open spaces.

A small Asian sheath-tailed bat (*Emballonura alecto*) on cave wall. (Photo by Pavel German. Reproduced by permission.)

Instead of a throat sac, the black-bearded tomb bat (*Taphozous melanopogon*) has pores that open into the throat. (Photo by Pavel German. Reproduced by permission.)

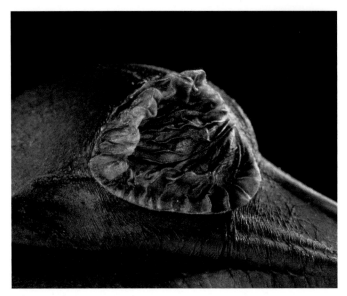

The wing pouch of an adult male greater sac-winged bat (*Saccopteryx bilineata*). (Photo by © Merlin D. Tuttle, Bat Conservation International. Reproduced by permission.)

An adult male greater sac-winged bat (*Saccopteryx bilineata*). (Photo by © Merlin D. Tuttle, Bat Conservation International. Reproduced by permission.)

Reproductive biology

Bats of the family Emballonuridae usually give birth to a single offspring per year. Exceptions to this rule may be small species like the proboscis bat that reproduce twice each year. Most emballonurid bats show a seasonal pattern of reproduction with females giving birth to their offspring at the beginning of the rainy season. Sperm storage or delayed embryonic development occurs in some Old World members within the family Emballonuridae. The mating system varies by species.

Emballonurid bats exhibit a variety of different mating systems. Similar to other mammalian groups, polygynous mating patterns are most common. However, exceptions are for example the monogamous mating system of *Cormura brevirostris* and possibly also of the greater dog-faced bat, *Peropteryx kappleri*, and some members of the genus *Taphozous*. The proboscis bat is considered to have a promiscuous mating pattern. Among emballonurid bats, the greater sac-winged bat, *Saccopteryx bilineata*, is the best studied species.

Conservation status

The IUCN Chiroptera Specialist group lists two species as Critically Endangered (*Coleura seychellensis*, which is endemic to the Seychelles, and *Taphozous troughtoni*, of which only six specimen have ever been collected), two species as Endangered (*Balantiopteryx infusca* is threatened by habitat destruction and *Emballonura semicaudata* by development and cyclone damage), and 10 species as Vulnerable. In tropical regions, forest specialists seem to face an uncertain future if habitat destruction and forest fragmentation continues. Opportunistic species like *Rhynchonycteris naso*, *Saccopteryx bilineata*, or some species within the genus *Taphozous* may even live around buildings when humans do not distrub them.

Significance to humans

None known.

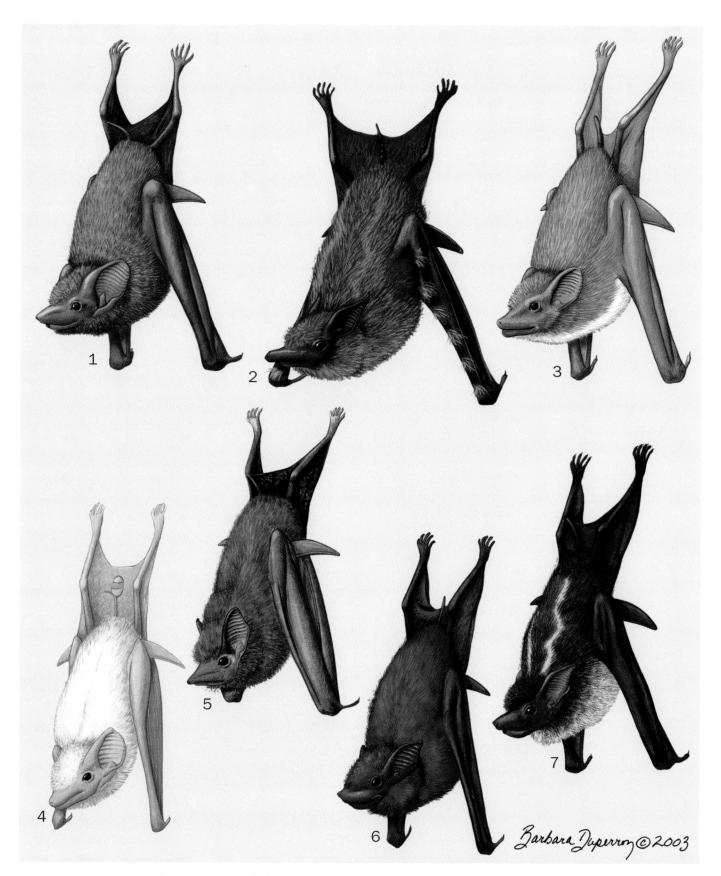

1. Gray sac-winged bat (*Balantiopteryx plicata*); 2. Proboscis bat (*Rhynchonycteris naso*); 3. Mauritian tomb bat (*Taphozous mauritianus*); 4. Northern ghost bat (*Diclidurus albus*); 5. Greater dog-faced bat (*Peropteryx kappleri*); 6. Lesser sheath-tailed bat (*Emballonura monticola*); 7. Greater sac-winged bat (*Saccopteryx bilineata*). (Illustration by Barbara Duperron)

Species accounts

Greater sac-winged bat

Saccopteryx bilineata

SUBFAMILY
Emballonurinae

TAXONOMY
Saccopteryx bilineata (Temminck, 1838), Suriname.

OTHER COMMON NAMES
English: Greater white-lined bat, greater two-lined bat; German: Große Taschenflügelfledermaus, Große Sackflügelfledermaus; Spanish: Murciélago de listas.

PHYSICAL CHARACTERISTICS
Head and body length 1.8–2.2 in (47–56 mm); forearm 1.7–1.9 in (44–48 mm); weight 0.2–0.35 oz (6–10 g).Dorsal fur dark brown with two distinct white lines and dark gray ventral fur; dark wings and tail membrane; males with wing sacs in the front wing membrane; females slightly larger than males. *Saccopteryx* bats turn the next to the last joint on their third fingers up rather than down when resting.

DISTRIBUTION
From south Mexico to southeast Brazil. From lowlands to up to 1,640 ft (500 m) in elevation.

HABITAT
Inhabit lowland evergreen or semi-deciduous forests and roost in well-lit portions of hollow tress, buttress cavities, and occa-sionally at buildings. Colonies are divided into smaller territories in which one to nine females roost. Larger colonies can count up to 60 individuals. Each colony defends its own foraging territory, and colonies shift seasonally between different foraging habitats.

BEHAVIOR
During a hovering display, males fan scents from the wing membrane towards roosting females. In addition, males emit songs to attract and retain females into their harem.

FEEDING ECOLOGY AND DIET
Feed exclusively on insects. Usually individuals emerge from the colony before sunset and hunt below the canopy along forest edges or small gaps. After dark they move higher, sometimes close to or even above the canopy. Subsequent echolocation calls alternate between 44 and 47 kHz.

REPRODUCTIVE BIOLOGY
In Costa Rica and Panama, the mating season is restricted to a few weeks in December and January. Females give birth to a single offspring at the onset of the rainy season in July or August. In the Ecuadorian Amazon, females give birth to their offspring in December to January. Believed to be polygynous.

CONSERVATION STATUS
Not threatened.

SIGNIFICANCE TO HUMANS
None known. ◆

Diclidurus albus

Saccopteryx bilineata

Gray sac-winged bat

Balantiopteryx plicata

SUBFAMILY
Emballonurinae

TAXONOMY
Balantiopteryx plicata Peters, 1867, Puntarenas, Costa Rica.

OTHER COMMON NAMES
German: Graue Sackflügelfledermaus.

PHYSICAL CHARACTERISTICS
Head and body 1.8–2.0 in (47–53 mm); forearm 1.5–1.7 in (40–45 mm); weight 0.17–0.21 oz (5–6 g). Dorsal fur pale or smoky gray, ventral fur slightly lighter gray.

DISTRIBUTION
Northwest Mexico to northwest Costa Rica along the Pacific slope. Additionally in some areas in east Mexico and northern Colombia.

HABITAT
Commonly found in deciduous forests and in areas with dry thorn scrubs.

BEHAVIOR
Daytime roosts are found in caves, rock crevices, or tree hollows. Colony members perform a swarming flight around the

Rhynchonycteris naso
Balantiopteryx plicata

roost during which low frequency sounds are noticeable. These sounds are probably not produced through vocalization.

FEEDING ECOLOGY AND DIET
Feed on insects that they hunt usually above the canopy.

REPRODUCTIVE BIOLOGY
A single pup is born at the beginning of the rainy season. Believed to be polygynous.

CONSERVATION STATUS
Not threatened.

SIGNIFICANCE TO HUMANS
None known. ◆

Proboscis bat
Rhynchonycteris naso

SUBFAMILY
Emballonurinae

TAXONOMY
Rhynchonycteris naso (Wied-Neuwied, 1820), Bahia, Brazil.

OTHER COMMON NAMES
English: Long-nosed bat, sharp-nosed bat; German: Nasenfledermaus; Spanish: Murciélago de trompa.

PHYSICAL CHARACTERISTICS
Head and body length 1.4–1.9 in (36–48 mm); forearm 1.4–1.6 in (36–40 mm); weight 0.10–0.21 oz (3–6 g). Very small bat with two wavy, faint dorsal stripes. Upperparts with grizzled, whitish fur. Nose projecting beyond mouth. Tufts of grayish hair on forearms and wing membrane. Wing and tail membrane brown.

DISTRIBUTION
From Veracruz, Mexico, to east Brazil, north Bolivia, and Peru. From lowlands to elevations up to 1,640 ft (500 m).

HABITAT
Lives close to rivers, lakes, or mangroves. Daytime roosts are found under logs, tree trunks, or branches hanging over water.

BEHAVIOR
Colonies count up to 50 bats. Individuals align in vertical rows and maintain a minimum distance of approximately 3.9 in (10 cm) to the neighbor. If one approaches, the whole colony may suddenly fly off simultaneously and land in a similar place at some distance. The mating system is probably promiscuous as no pattern of association between individuals was detected in a study in Costa Rica.

FEEDING ECOLOGY AND DIET
Feed on small flies and other insects that they catch in a slow flight over water. Echolocation calls vary between 90 and 95 kHz.

REPRODUCTIVE BIOLOGY
Polygynous. In some areas, females give birth to a single pup early in the wet season. Males have no seasonal spermatogenic pattern.

CONSERVATION STATUS
Not threatened.

SIGNIFICANCE TO HUMANS
None known. ◆

Greater dog-faced bat
Peropteryx kappleri

TAXONOMY
Peropteryx kappleri Peters, 1867, Suriname.

OTHER COMMON NAMES
English: Greater doglike bat.

PHYSICAL CHARACTERISTICS
Head and body length 2.5–2.9 in (63–75 mm); forearm 1.8–1.9 in (45–50 mm); weight 0.24–0.46 oz (7–13 g). Dorsal fur dark or reddish brown, head with distinct tuft on crown, ears black, wing and tail membrane black; males possess small sacs in the front wing membrane.

DISTRIBUTION
From Veracruz, Mexico, to east Brazil and Peru; from lowlands to up to 4,920 ft (1,500 m) in elevation.

HABITAT
Inhabits lowland evergreen forests; daytime roosts in small caves and under fallen logs within close distance to the ground.

BEHAVIOR
Males sit on top of females. This roosting behavior implies a form of mate-guarding; thus, a monogamous mating system is most likely for this species. However, has also been found to roost in larger groups and not in close body contact.

FEEDING ECOLOGY AND DIET
Forage in long beats of 66–98 ft (20–30 m) in the subcanopy. The diet consists of insects. Echolocation calls vary between 30 and 35 kHz.

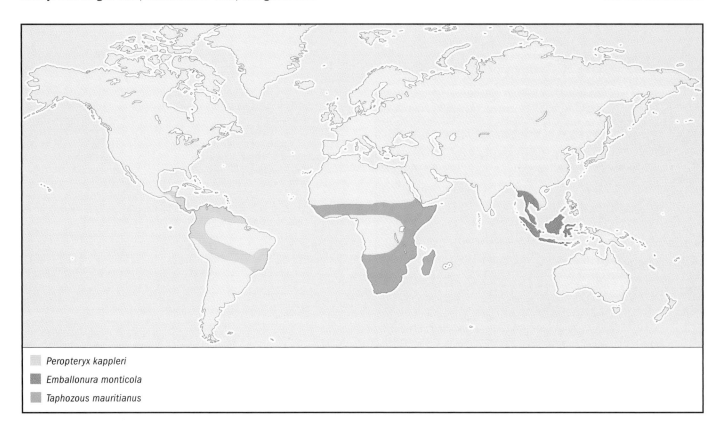

Peropteryx kappleri

Emballonura monticola

Taphozous mauritianus

REPRODUCTIVE BIOLOGY
Monogamous. A single pup is born at the beginning of the rainy season.

CONSERVATION STATUS
Not threatened.

SIGNIFICANCE TO HUMANS
None known. ◆

Northern ghost bat
Diclidurus albus

SUBFAMILY
Emballonurinae

TAXONOMY
Diclidurus albus Wied-Neuwied, 1820, Bahia, Brazil.

OTHER COMMON NAMES
English: White bat, jumbie bat; German: Geisterfledermaus, Weisse fledermaus; Spanish: Murciélago blanco, murciélago albino.

PHYSICAL CHARACTERISTICS
Head and body length 2.7–3.2 in (68–82 mm); forearm 2.4–3.6 in (63–93 mm); weight 0.6–0.8 oz (17–24 g). Large white bat with yellowish ears; grayish hair bases may be visible; ears short and rounded; tail membrane with single brown sac close to the tip of the tail.

DISTRIBUTION
From Nayarit, Mexico, to east Brazil and north Peru; from lowland to montane areas up to 4,920 ft (1,500 m).

HABITAT
Daytime roosts consist of small groups or single individuals under palm fronds.

BEHAVIOR
The social behavior of this bat is virtually unknown.

FEEDING ECOLOGY AND DIET
Forage in the open space above the canopy. Occasionally, these bats are attracted insects swarming around street lamps. Echolocation calls vary between 22 and 25 kHz.

REPRODUCTIVE BIOLOGY
Nothing is known. The reproductive biology of this species has not yet been studied.

CONSERVATION STATUS
Not threatened.

SIGNIFICANCE TO HUMANS
None known. ◆

Lesser sheath-tailed bat
Emballonura monticola

SUBFAMILY
Emballonurinae

TAXONOMY
Emballonura monticola Temminck, 1838, Java, Indonesia.

OTHER COMMON NAMES
None known.

PHYSICAL CHARACTERISTICS

Head and body length 1.6–1.8 in (40–45 mm); forearm 1.7–1.8 in (43–45 mm); weight 0.14–0.17 oz (4–5 g). Fur dark brown to reddish brown, wing and tail membrane black; large eyes.

DISTRIBUTION

Malay Peninsula, including south Myanmar and Thailand, Sumatra, Java, Banka, Billiton, Engano, Babi, Anamba, Batu, Nias, Mentawi, Borneo, Sulawesi, and Karimata.

HABITAT

Inhabits forests; daytime roosts may be found at exposed sites like hollow logs, caves, or rock shelters.

BEHAVIOR

Sizes of daytime roosts range from two to 20 individuals. However, larger colonies with up to 100–150 bats have also been found in caves.

FEEDING ECOLOGY AND DIET

Hunts small insects in dense forests. This species has also been observed to forage during the day in dense shade.

REPRODUCTIVE BIOLOGY

Observations indicate two birth periods per year in Malay Peninsula in February–March and October–November; females give birth to a single young during each birth period. Believed to be polygynous.

CONSERVATION STATUS

Not threatened.

SIGNIFICANCE TO HUMANS

None known. ◆

Mauritian tomb bat
Taphozous mauritianus

SUBFAMILY

Taphozoinae

TAXONOMY

Taphozous mauritianus Geoffroy, 1818, Mauritius.

OTHER COMMON NAMES

German: Mauritius Grabfledermaus.

PHYSICAL CHARACTERISTICS

Head and body length 2.9–3.6 in (75–93 mm); forearm 2.3–2.5 in (58–65 mm); weight 0.7–1.3 oz (20–36 g). Dorsal fur brownish gray speckled with white; ventral side almost pure white. Both sexes may have a throat pocket. In Nigeria and Mozambique, this pocket is present only in males, whereas it can be found in both sexes in Sudan (although more pronounced in the male sex). In West Africa, males have a functioning sac and females have a vestigial pouch.

DISTRIBUTION

Sub-Saharan Africa, Madagascar, Mauritius, Assumption Island, Aldabra Island, and Reunion.

HABITAT

Daytime roosts in tree trunks, rock faces, or the external walls of buildings. Predominantly open savanna, sometimes semi-deciduous forests.

BEHAVIOR

Initially discovered in ancient Egyptian tombs by the scientists who accompanied Napoleon on his campaign. The name tomb bat is basically unjustified, because these animals are not generally associated with tombs. Daytime roosts are found in rather open spaces that are occasionally penetrated by daylight, such as underneath roofs or on walls. In colonies, individuals roost as pairs in close association. Individuals move quickly sideways or upwards when disturbed.

FEEDING ECOLOGY AND DIET

A fast aerial hunter of insects, mostly moths. A long-range feeding strategy has been observed.

REPRODUCTIVE BIOLOGY

Monestrous in some areas of its distribution range and polyestrous in others. Most likely polygynous.

CONSERVATION STATUS

Not threatened.

SIGNIFICANCE TO HUMANS

None known. ◆

Common name / Scientific name/ Other common names	Physical characteristics	Habitat and behavior	Distribution	Diet	Conservation status
Lesser sac-winged bat *Saccopteryx leptura* English: Lesser white-lined bat; German: Kleine Taschenflügelfledermaus, Kleine Sackflügelfledermaus; Spanish: Murciélago de la Delgada, murciélago de listas y de cola corta	Dorsal fur chocolate-brown with two distinct yellowish or whitish lines, underparts brown, brown wings and tail membrane. Females slightly larger than males. Head and body length 1.5–2 in (3.8–5.1 cm), forearm 1.4–1.7 in (3.7–4.4 cm), weight 0.1–0.2 oz (3–6 g).	Lowland deciduous and evergreen forests. Roosts below large branches or at tree trunks in groups of one to nine animals. Each group defends a foraging territory against neighbors. Echolocation calls at about 55 kHz.	From Chiapas and Tabasco (Mexico) and Belize to Brazil, including Peru, Guianas, Magarita Islands, Trinidad, and Tobago.	Small insects.	Vulnerable
Lesser dog-faced bat *Peropteryx macrotis* English: Doglike bat, lesser doglike bat, Peters's sac-winged bat; Spanish: Murciélago orejon	Dorsal fur reddish brown or dark brown, ventral fur grayish brown. Face naked with a long fringe of hair on forehead. Wing membranes blackish. Head and body length 1.6–2.1 in (4.2–5.5 cm), forearm 1.5–1.7 in (3.8–4.45 cm), weight 0.14–0.25 oz (4–7 g).	Colonies usually consist of 10 to 20 individuals, but sometimes as many as 80 bats may be found. Daytime roosts in caves, shallow crevices, Maya buildings, and churches. Often near water. Lowlands to 2,300 ft (700 m).	Oaxaca, Guerrero, and Yucatán (Mexico) to Peru, Paraguay, and south and east Brazil, Tobago, Magarita, Aruba, Trinidad, Grenada (Lesser Antilles).	Insects.	Not listed by IUCN

[continued]

Common name / Scientific name / Other common names	Physical characteristics	Habitat and behavior	Distribution	Diet	Conservation status
Chestnut sac-winged bat *Cormura brevirostris* English: Wagner's sac-winged bat; German: Kastanienfarbige Sackflügelfledermaus; Spanish: Murciélago chato	Upperparts chestnut red-brown, or dark, black brown, wing membrane black, lower face naked. Large wing sac (extending almost from edge of propatagium to near elbow). Females slightly larger than males. Head and body length 1.8–2.3 in (4.6–5.8 cm), forearm 1.8–1.9 in (4.6–4.8 cm), weight 0.14–0.39 oz (7–11 g).	Roosts in small groups in large hollow rotting logs, under fallen trees, or in tree hollows. Individuals, probably the males, often roost on top of other bats (probably the females).	Nicaragua to Peru, Amazonian Brazil, and Guianas.	Insects.	Not listed by IUCN
Shaggy-haired bat *Centronycteris centralis* English: Thomas's bat; Spanish: Murciélago de Thomas	Body yellow or gray brown, wing membranes black, facial skin pink. No wing sacs. Head and body length 1.9–2.3 in (4.9–5.9 cm), forearm 1.7–1.8 in (4.3–4.6 cm), weight 0.18–0.21 oz (5–6 g).	Roosts in tree holes and on trunks of trees. Found in evergreen and semideciduous forest and secondary growth.	South Veracruz in Mexico, to Peru, Brazil, and Guianas.	Insects.	Not listed by IUCN
Seychelles sheath-tailed bat *Coleura seychellensis*	Fur is reddish to brown, ventral fur lighter. No wing sacs. Head and body length 2.2–2.6 in (5.5–6.5 cm), forearm 1.7–2.2 in (4.5–5.6 cm), weight 0.35–0.38 oz (10–11 g).	During daytime, roost in cliffs facing the sea or in houses. Daytime roosts in caves counted up to several hundreds of individuals. Colonies are divided in smaller subgroups of 20 individuals, probably harems.	Endemic to Seychelles and Mahe Islands.	Insects.	Critically Endangered; fewer than 50 individuals
Egyptian tomb bat *Taphozous perforatus* German: Ägyptische Grabfledermaus	Dark or brown colored fur, ventral fur lighter than the dorsal fur. No gular sac. Head and body length 2.8–3.3 in (7.1–8.5 cm), forearm 2.3–2.6 in (6.0–6.7 cm), weight 0.77–0.88 oz (22–25 g).	Semi-arid habitats, abundant along the Nile River. Daytime roosts in narrow cracks of rocks or human building. In large colonies, individuals may roost also in more open areas in close association to each other.	Northwest India, Arabia, Eastern and Central Africa, Sudan, and Botswana.	Moths and beetles.	Not listed by IUCN
Black-bearded tomb bat *Taphozous melanopogon* German: Schwarzbartige Grabfledermaus	White hair with distinct pale brown to reddish tips. Males with elongated blackish hairs on the underside of chin and throat. During the mating season secretions run over the beard. Head and body length 2.7–3.1 in (7.0–8.0 cm), forearm 2.1–2.5 in (5.5–6.5 cm), weight 0.71–0.88 oz (20–25 g).	Hilly areas near water. Daytime roosts in caves or vertical faults in cliffs with up to 4,000 individuals. Males defend a small territory with a female and young in it. Within roosts, individuals occupy territories.	Indian subcontinent, Myanmar, Yunnan, Laos, Thailand, Indochina, Malaysia, Indonesia, and Borneo.	Insects, possibly also small fruits.	Not listed by IUCN
Giant pouched bat *Saccolaimus peli*	Large with broad flat head and shoulders, large eyes, and relatively small ears; short greasy fur; both sexes carry a gular sac.	Forests. Daytime roosts in hollow trees; individuals maintain distance from one another in roost.	From Liberia to Congo, Uganda, and western Kenya.	Moths and beetles.	Lower Risk/Near Threatened
Pacific sheath-tailed bat *Emballonura semicaudata*	Small bat, forearm 1.7–1.9 in (4.3–4.7 cm), weight 0.18–0.21 oz (5–6 g).	Daytime roosts in caves, overhanging cliffs, or lava tubes	Once widespread on Polynesian and Micronesian Islands. Recent records only from Samoa, Fiji, Mariana, Palau, and a small number of other islands.	Small insects.	Endangered

Resources

Books

Eisenberg, J. F. *Mammals of the Neotropics 1. The Northern Neotropics.* Chicago and London: The University of Chicago Press, 1989.

Eisenberg, J. F., and K. H. Redford. *Mammals of the Neotropics 3. The Central Neotropics.* Chicago and London: The University of Chicago Press, 1999.

Emmons, L. H. *Neotropical Rainforest Mammals—A Field Guide.* Chicago and London: The University of Chicago Press, 1990.

Hutson, A. M., S. P. Mickleburgh, and P. A. Racey. *Global Status Survey and Conservation Plan—Microchiropteran Bats.* IUCN/SSC Chiroptera Specialist Group. Gland, Switzerland, and Cambridge, UK: IUCN Publications, 2001.

Resources

Kingdon, J. *East African Mammals*. Chicago and London: The University of Chicago Press, 1984.

Reid, F. A. *A Field Guide to the Mammals of Central America and Southeast Mexico*. New York and Oxford: Oxford University Press, 1997.

Nowak, R. M. *Walker's Mammals of the World*. 5th ed. Baltimore and London: The Johns Hopkins University Press, 1991.

Voigt C. C., G. Heckel, and O. von Helversen. "Conflicts and Strategies in the Mating System of the Sac-winged Bat." In *Functional and Evolutionary Ecology of Bats*, edited by G. F. McCracken, A. Zubaid, and T. H. Kunz. Oxford: Oxford University Press, 2003.

Periodicals

Bradbury, J. W., and L. Emmons. "Social Organization of some Trinidad bats. I. Emballonuridae." *Zeitschrift für Tierpsychologie* 36 (1974): 137.

Bradbury, J. W., and S. L. Vehrencamp. "Social Organization and Foraging in Emballonurid Bats. I. Field Studies." *Behavioural Ecology and Sociobiology* 1 (1976): 337.

———. "Social Organization and Foraging in Emballonurid Bats. III. Mating Systems." *Behavioural Ecology and Sociobiology* 2 (1977): 1.

Heckel, G., C. C. Voigt, F. Mayer, and O. von Helversen. "Extra-harem Paternity in the White-lined Bat *Saccopteryx bilineata*." *Behaviour* 136 (1999): 1173.

Voigt, C. C. "Individual Variation of Perfume-blending in Male Sac-winged Bats." *Animal Behaviour* 63 (2002): 907.

Voigt, C. C., and O. von Helversen. "Storage and Display of Odor by Male *Saccopteryx bilineata* (Chiroptera; Emballonuridae)." *Behavioral Ecology and Sociobiology* 47 (1999): 29.

Christian C. Voigt, PhD

Kitti's hog-nosed bats

(Craseonycteridae)

Class Mammalia
Order Chiroptera
Suborder Microchiroptera
Family Craseonycteridae

Thumbnail description
World's smallest bat, thickened snout with two crescent shaped nostrils, eyes are minute, ears large with well developed tragus, no external tail, pelage is light buffy brown above, paler below

Size
Head and body 1.3 in, 34 mm; no tail; forearm 0.9 in, 24–25 mm; weight 2.0–2.6 grams

Number of genera, species
1 genus; 1 species

Habitat
Caves in limestone outcrops usually located near rivers, bamboo, deciduous and evergreen forest, paddy and cassava fields, orchards

Conservation status
Endangered

Distribution
Thailand and Myanmar (Burma)

Evolution and systematics

The evolutionary history of Kitti's hog-nosed bat is still unclear and there is no fossil record. It shares morphological characters with two Old World bat families belonging to the superfamily Emballonuroidea (which includes the Rhinopomatidae and Emballonuridae). It has the same dental formula, general skeletal design and skull morphology as the mouse-tailed bats (Rhinopomatidae) but specific traits of the skeleton and arrangement of the premaxillae are closer to the sheath-tailed bats (Emballonuridae). Unlike both families, it lacks a tail and has a more inflated braincase and relatively larger incisors. However, recent molecular evidence suggests that the hog-nosed bat should be included in the superfamily Rhinolophoidea (including the Nycteridae, Megadermatidae, Rhinolophidae, and Hipposideridae), suggesting that it is more closely related to elements of the leaf-nosed bats (Hipposideridae) and particularly the trident roundleaf bat *Aselliscus stoliczkanus*.

The taxonomy for this species is *Craseonycteris thonglongyai* Hill, 1974, Kanchanaburi, Thailand.

Physical characteristics

Kitti's hog-nosed bat is so small that it is considered to be the smallest mammal in the world, and for this reason, it is also known as the bumblebee bat. It weighs between 0.7 oz and 0.9 oz (2.0 and 2.6 g). The tail is absent, although there are two caudal vertebrae. There is a large interfemoral membrane but no calcar. The snout is thickened and there are two clearly defined, crescent-shaped, hog-like nostrils. The eyes are minute and largely concealed by hair. The ears are large, with well-defined but rounded tips; they are not connected to one another and each has a well-developed tragus. In males, there is a large glandular swelling on the lower part of the throat; in females it is less developed or absent. The wings are relatively long and broad, their structure similar to those of bats in the superfamily Rhinolophoidea (slit-faced bats, false vampires, horseshoe, and leaf-nosed bats). The hairs on the back are light buffy brown. They are slightly paler on the belly. In the skull, the braincase is inflated, with a prominent sagittal crest and enlarged tympanic bullae. The premaxillae are a unique character of the family. They are not fused with the maxillae but form a separate ring-like structure. There are 28 teeth. There is one

Kitti's hog-nosed bat (*Craseonycteris thonglongyai*). (Illustration by Jonathan Higgins)

pair of upper and two pairs of lower incisors, one pair of upper and lower canines; one pair of upper and two pairs of lower premolars and three pairs of upper and lower premolars. When hunting, Kitti's hog-nosed bat uses 3.5 ms long multiharmonic constant frequency (CF) search signals with a prominent second harmonic at 73 kHz repeated at around 22 Hz. This can be used to acoustically indentify the bats.

Distribution

The global distribution of Kitti's hog-nosed bat is currently thought to be restricted to two small areas of South-East Asia. In Thailand, it has been recorded in 21 caves, most of which are located in Sai Yok National Park, Kanchanaburi Province with the remainder in adjacent areas on the Kwae Noi (River Kwai). In March 2001, a second population was found in a cave in Mon State, Myanmar (Burma). This represents a range extension of about 155 mi (250 km).

Habitat

Kitti's hog-nosed bat roosts in limestone caves, preferably those with many chambers and domed roofs and located near rivers. In Thailand, the land adjacent to the caves was formerly dry deciduous hardwood forest, with some dry evergreen forest and giant bamboo, but much of the forest is now cleared and given over to agriculture, particularly cassava and kapok plantations and to a lesser extent bananas, corn, and orchards of mango, jackfruit, and lemon. In Myanmar, the hog-nosed bat was found in a cave in a large limestone outcrop in an agricultural area of rice paddy and toddy palms.

Behavior

The bats hide in small holes or in crevices formed by stalactites in caves. Each bat maintains a certain distance from other individuals. In Thailand, the number of individuals in one roosting site varies between 1 and 500, with an average colony size of 100. There is some seasonal movement between caves. They have two brief activity periods, one in the morning and one in the evening. They circle at the cave entrance about one minute before leaving. About 10 minutes after sun-

set, they leave the cave and circle over the entrance. About one minute later they separate into small groups, each group using a flyway to a feeding area. The flyways are often no more than 16.4 ft (5 m) wide and the foraging area is usually within 820 ft (250 m) of the cave. They feed for about 30 minutes in the evening. Then they return to the cave where they remain until about 40 minutes before sunrise, when they again feed for about 18 minutes. There are some minor variations in activity patterns dependent on season, particularly related to cool temperatures and heavy rain, which discourage activity.

Feeding ecology and diet

Some have suggested that the hog-nosed bat is a foliage gleaner, taking insects directly from leaves and twigs, or even that it hunts near the ground. However, according to other observations, it is thought to catch insects on the wing as it flies around trees and bamboo. This view is supported by an analysis of its feeding buzzes, which are 3.5 ms-long and rather intense and not suited to a gleaning mode of feeding strategy. The stomach contents of one adult male included small beetles and other small insects.

Reproductive biology

In Thailand, a female with a single young was collected in May and an apparently pregnant female was caught in April. The time of birth would therefore appear to coincide with the onset of the summer rainy season. This species is thought to be polygamous.

The Kitti's hog-nosed bat (*Craseonycteris thonglongyai*) is about the same size as a bumblebee. (Photo by © Merlin D. Tuttle, Bat Conservation International. Reproduced by permission.)

Conservation status

Endangered, since it is considered to have a small distribution, a continuing decline in habitat quality and a continuing decline of a small, single population. The population in Thailand has been estimated at about 2,000 individuals. Its status has not been reassessed since the discovery of a second population in Myanmar.

Significance to humans

In Thailand, many visitors come to the Sai Yok National Park, which was declared a protected area in 1980 specifically to help conserve Kitti's hog-nosed bat. In the past this has led to some disturbance of the bat's roosting sites.

Resources

Books

Hutson, A. M., S. P. Mickleburgh, and P. A. Racey. *Microchiropteran Bats: Global Status Survey and Conservation Action Plan.* IUCN/SSC Chiroptera Specialist Group. Gland, Switzerland and Cambridge, UK: IUCN, 2001.

Periodicals

Bates, P. J. J., T. Nwe, K. M. Swe, and S. S. Hla Bu. "Further New Records of Bats from Myanmar (Burma), including *Craseonycteris thonglongyai* Hill 1974 (Chiroptera: Craseonycteridae)." *Acta Chiropterologica* 3, 1 (2001): 33–41.

Duangkhae, S. "Ecology and Behavior of Kitti's Hog-nosed Bat (*Craseonycteris thonglongyai*) in Western Thailand." *Natural History Bulletin of the Siam Society* 38 (1990): 135–161.

Hill, J. E. "A New Family, Genus and Species of Bat (Mammalia: Chiroptera) from Thailand." *Bulletin of the British Museum (Natural History) Zoology* 27 (1974): 303–336.

Hulva, P. and I. Horacek. "*Craseonycteris thonglongyai* (Chiroptera: Craseonycteridae) is a Rhinolophoid: Molecular Evidence from Cytochrome b." *Acta Chiropterologica* 4, 2 (2002): 107–120.

Surlykke, A., L. A. Miller, B. Mohl, B. B. Andersen, J. Christensen-Dalsgaard, and M. B. Jorgensen. "Echolocation in Two Very Small Bats from Thailand: *Craseonycteris thonglongyai* and *Myotis siligorensis*." *Behavioral Ecology* 33 (1993): 1–12.

Paul J. J. Bates, PhD

Slit-faced bats

(Nycteridae)

Class Mammalia
Order Chiroptera
Suborder Microchiroptera
Family Nycteridae

Thumbnail description
Small- to medium-sized bats with large ears and a tail ending in a diagnostic T-shaped cartilage; face split by a distinctive and diagnostic groove

Size
Forearms range 1.2–2.5 in (3.2–6.4 cm); body mass 0.2–1.2 oz (6–36 g)

Number of genera, species
1 genus; 14 species

Habitat
Occur from rainforest to savanna as well as in arid habitats

Conservation status
Vulnerable: 2 species; Lower Risk/Near Threatened: 3 species; Data Deficient: 1 species

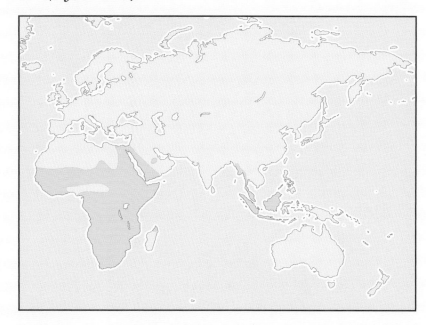

Distribution
Africa, Madagascar, Southeast Asia, and East Indies

Evolution and systematics

Although slit-faced bats are included in the superfamily Rhinolophoidea, the closeness of their relationship to the other families in the group (false vampires, the Megadermatidae, horseshoe bats, the Rhinolophidae, and Old World leaf-nosed bats, the Hipposideridae) has been questioned. Apart from recent material, there is no fossil record of slit-faced bats. No subfamilies are recognized.

Physical characteristics

Slit-faced bats are small to medium in size, and have broad wings and large ears. Their fur is long and fine and ranges in color from gray to red. The T-shaped tail cartilage, large ears, and slit-faces make them distinctive.

Distribution

Most species of slit-faced bats occur in Africa, one ranging from the north (Israel and adjacent countries) to the south (the Cape). Two other species occur in Southeast Asia, from Myanmar, Thailand, and Malaysia to Sarawak, Sumatra, Java, Borneo, and Bali. One species has been reported from Madagascar.

Habitat

Most species of slit-faced bats are found in rainforest in Africa or in Southeast Asia, but other species occur in drier areas, from savanna woodlands to desert.

Behavior

Slit-faced bats roost in hollows by day. The hollows include caves and mines, those in trees, as well as others associated with buildings or other artificial structures. Roosting slit-faced bats are usually not in physical contact with one another. They produce low-intensity echolocation calls, which

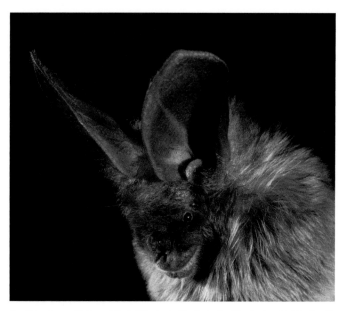

An Egyptian slit-faced bat (*Nycteris thebaica*). (Photo by © Merlin D. Tuttle, Bat Conservation International. Reproduced by permission.)

A slit-faced bat (*Nycteris* sp.) roosting in the Samburu National Reserve, Kenya. (Photo by Ann and Rob Simpson. Reproduced by permission.)

A large-eared slit-faced bat (*Nycteris macrotis*) in flight with young. (Photo by Merlin D. Tuttle, Bat Conservation International. Reproduced by permission.)

A slit-faced bat's (*Nycteris* sp.) arm joints with wing membrane. (Photo by Bruce Davidson/Naturepl.com Reproduced by permission.)

A female Egyptian split-faced bat (*Nycteris thebaica*) holding her young. (Photo by Manuel Ruedi. Reproduced by permission.)

A large-eared slit-faced bat (*Nycteris macrotis*) nursing her young. (Photo by Merlin D. Tuttle, Bat Conservation International. Reproduced by permission.)

An Egyptian slit-faced bat (*Nycteris thebaica*) feeding on an insect. (Photo by Brock Fenton. Reproduced by permission.)

they may not depend upon to find their prey, relying instead on sound cues such as the songs or footfalls of prey. Slit-faced bats also take flying prey. Accumulations of discarded pieces of prey under feeding roosts provide biologists with a picture of the diets of slit-faced bats. Unlike other species of bat, slit-faced bats are warm-blooded and cannot enter torpor, a state of total inactivity.

Feeding ecology and diet

Smaller species of slit-faced bats feed almost entirely on arthropods, typically insects such as moths, beetles, and crickets, but also spiders, centipedes, and scorpions. Larger species of nycterids also eat small vertebrates such as fish, frogs, birds, and other bats. Slit-faced bats often hunt from a perch, dropping to the ground to grab passing prey, or snatching it from the foliage or branches or trunks of trees.

Reproductive biology

A single young is born each year, typically at the beginning of the rainy season. The slit-faced bats are most likely promiscuous.

Conservation status

The intermediate slit-faced bat (*Nycteris intermedia*), Wood's slit-faced bat (*Nycteris woodi*), and *N. aurita* are considered by the IUCN to be Lower Risk/Near Threatened. The Javan slit-faced bat (*Nycteris javanica*) and Ja slit-faced bat (*Nycteris major*) are considered Vulnerable, the former because of declining range, the latter because of restricted range. Too little is known about the Madagascar slit-faced bat (*Nycteris madagascariensis*) to assess its conservation status, and it is listed as Data Deficient.

Significance to humans

None known.

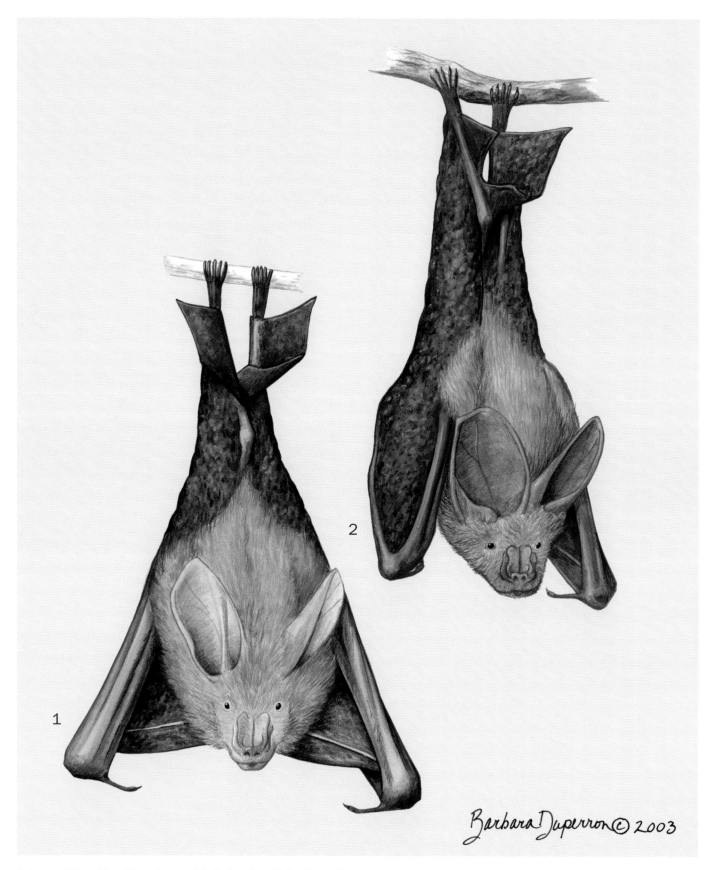

1. Large slit-faced bat (*Nycteris grandis*); 2. Egyptian slit-faced bat (*Nycteris thebaica*). (Illustration by Barbara Duperron)

Species accounts

Large slit-faced bat
Nycteris grandis

TAXONOMY
Nycteris grandis Peters, 1865, Guinea.

OTHER COMMON NAMES
None known.

PHYSICAL CHARACTERISTICS
The largest of slit-faced bats: forearms 2.2–2.6 in (5.7–6.6 cm); weight 0.8–1.2 oz (23–36 g). Long, fine fur is gray to red.

DISTRIBUTION
Occur from Sierra Leone in the west to Lake Victoria in the east. There are two isolated populations, one on the coast in Tanzania, the other from Zambia south into Zimbabwe along the Zambezi River.

HABITAT
Most often found in rainforest, but outlying populations occur in areas of savanna woodlands.

BEHAVIOR
Roost in hollows in trees, caves, and mines, or those in artificial structures such as buildings and unused military bunkers. Roosting individuals, other than females and their dependent young, are not in physical contact with one another. Along the Zambezi River in Zimbabwe, they use the same day roosts year after year, including hollows in acacia trees as well as those in buildings and in military bunkers; the roosts provide shelter

from the extreme heat of the day. Along the Zambezi, they use feeding roosts, sites that offer protection from above. Typically, feeding roosts are under thatched roofs, on porches, and in rooms that they enter through open doors or windows. Some day roosts also serve as feeding roosts.

FEEDING ECOLOGY AND DIET
May hang from perches and wait for passing prey, or fly in search of food. In either case, they depend upon the sounds of prey to locate their targets and usually take prey from surfaces. Along the Zambezi River, they feed heavily on vertebrates, usually frogs (representing seven species), but occasionally on birds and fish. They also eat other species of bats, including Egyptian slit-faced bats. Large slit-faced bats also eat large arthropods, including sun spiders, moths, and beetles. Prey is killed with a bite to the head, and inedible parts dropped below feeding roosts. When they eat frogs, they usually discard one foot bitten off at the ankle, the other leg bitten off at the knee.

REPRODUCTIVE BIOLOGY
Females bear a single young each year. Probably promiscuous.

CONSERVATION STATUS
Not listed by the IUCN.

SIGNIFICANCE TO HUMANS
None known. ◆

Egyptian slit-faced bat
Nycteris thebaica

TAXONOMY
Nycteris thebaica E. Geoffroy, 1818, Egypt. Seven subspecies are recognized.

OTHER COMMON NAMES
None known.

PHYSICAL CHARACTERISTICS
A medium-sized bat; forearm ranging 1.6–2 in (4.2–5.1 cm); weight 0.2–0.4 oz (7–12 g). Long, fine fur is gray to red. Large ears.

DISTRIBUTION
Widespread in savanna woodlands of sub-Saharan Africa from Sierra Leone in the west through East Africa and north to the middle east on either side of the Red Sea; occur south in Africa to the Cape.

HABITAT
Occur in savanna woodlands and more arid areas.

BEHAVIOR
Roost in hollows, whether in trees, buildings, caves, or abandoned mines. Within roosts, daytime temperatures can be around 86°F (30°C) when outside temperatures are well over 104°F (40°C). Roosting individuals are not usually in physical contact with one another, so the size of a roosting group depends upon the available space. In some cave roosts, they roost within 3.2–6.5 ft (1–2 m) of active bee hives.

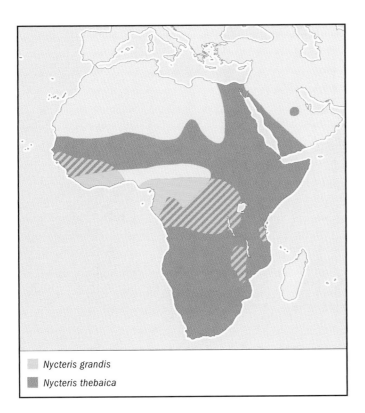

☐ *Nycteris grandis*
■ *Nycteris thebaica*

FEEDING ECOLOGY AND DIET

Take prey from surfaces (the ground or vegetation) as well as flying prey. The usual diet is arthropods, from sun spiders to scorpions, and insects such as orthopterans, moths, and beetles. In Zimbabwe, Egyptian slit-faced bats ate 1.1 in (3 cm) beetles in less than two minutes. From South Africa, there is one record of an Egyptian slit-faced bat taking a lizard. These bats appear to use sounds generated by prey to detect and assess their targets. The role of echolocation in hunting remains unclear. They produce bird-like chirps when foraging at night, but the function of these calls remains unknown.

REPRODUCTIVE BIOLOGY

Most likely promiscuous. Females produce a single young annually after a gestation period of 150 days. The birth period is typically timed to coincide with the start of the rains. Females take small young with them when they leave the roost to forage, but leave them in another secure place while actually hunting.

CONSERVATION STATUS

Not listed by the IUCN.

SIGNIFICANCE TO HUMANS

None known. ◆

Common name / Scientific name/ Other common names	Physical characteristics	Habitat and behavior	Distribution	Diet	Conservation status
Bate's slit-faced bat *Nycteris arge*	Light brown above, ventral side is lighter brown or grayish white. Medium-sized bat with large ears and well-developed calcar. Muzzle has deep median furrow. Head and body length 3.8–4.3 in (9.7–11 cm), tail length 1.8–2.2 in (4.7–5.8 cm), forearm length 1.6–1.9 in (4.3–4.7 cm).	Woodland savanna or dry country. Solitary.	Sierra Leone to southern and eastern Democratic Republic of the Congo; western Kenya; south-western Sudan; north-eastern Angola; and Bioko.	Variety of arthropods, including moths, butterflies, other insects, spiders, and sun spiders.	Not threatened
Gambian slit-faced bat *Nycteris gambiensis*	Pelage varies from rich brown or russet to pale brown or grayish. Muzzle has longitudinally divided nostrils, which are long, deep pit in forehead. T-shaped tip at end of tail. Head and body length 1.5–3.7 in (4–9.3 cm), tail length 1.7–3 in (4.3–7.5 cm), forearm length 1.2–2.4 in (3.2–6 cm).	Woodland savanna or dry country. Very gregarious.	Senegal, Gambia, Guinea, Sierra Leone, Ghana, Togo, Benin, and Burkina Faso.	Variety of arthropods, including moths, butterflies, other insects, spiders, and sun spiders.	Not threatened
Hairy slit-faced bat *Nycteris hispida*	Pelage varies from rich brown or russet to pale brown or grayish. Muzzle has longitudinally divided nostrils, which are long, deep pit in forehead. T-shaped tip at end of tail. Head and body length 1.5–3.7 in (4–9.3 cm), tail length 1.7–3 in (4.3–7.5 cm), forearm length 1.2–2.4 in (3.2–6 cm).	Woodland savanna or dry country. Group size may reach up to 20 individuals. Females begin new gestation while nursing the first offspring. May roost in hollow trees, dense foliage, rocky outcrops, caves, buildings, ruins, culverts, and abandoned wells.	Senegal to Somalia and south to Angola and South Africa; Zanzibar; and Bioko.	Variety of arthropods, including moths, butterflies, other insects, spiders, and sun spiders.	Not threatened
Intermediate slit-faced bat *Nycteris intermedia* French: Nyctère moyen	Pelage varies from rich brown or russet to pale brown or grayish. Muzzle has longitudinally divided nostrils, which are long, deep pit in forehead. T-shaped tip at end of tail. Head and body length 1.5–3.7 in (4–9.3 cm), tail length 1.7–3 in (4.3–7.5 cm), forearm length 1.2–2.4 in (3.2–6 cm).	Woodland savanna or dry country. May roost in hollow trees, dense foliage, rocky outcrops, caves, buildings, ruins, culverts, and abandoned wells.	Liberia to western Tanzania and south to Angola.	Variety of arthropods, including moths, butterflies, other insects, spiders, and sun spiders.	Lower Risk/Near Threatened
Javan slit-faced bat *Nycteris javanica* French: Nyctère de Java	Varies from rich brown or russet to pale brown or grayish. Muzzle has longitudinally divided nostrils, which are long, deep pit in forehead. T-shaped tip at end of tail. Head and body length 1.5–3.7 in (4–9.3 cm), tail length 1.7–3 in (4.3–7.5 cm), forearm length 1.2–2.4 in (3.2–6 cm).	Dense forests. Breeds through-out year and females mate again shortly after giving birth.	Java, Bali, and Kangean Islands, Indonesia.	Variety of arthropods, including moths, butterflies, other insects, spiders, and sun spiders.	Vulnerable
Large-eared slit-faced bat *Nycteris macrotis* German: Großohr-Schlitznase	Varies from rich brown or russet to pale brown or grayish. Exhibits considerable orange color variation in Democratic Republic of the Congo (Zaire). Muzzle has longitudinally divided nostrils, which are long, deep pit in forehead. T-shaped tip at end of tail. Head and body length 1.5–3.7 in (4–9.3 cm), tail length 1.7–3 in (4.3–7.5 cm), forearm length 1.2–2.4 in (3.2–6 cm).	Woodland savanna or dry country. Females nurse young for 45–60 days. Generally shelters in caves in pairs or alone.	Senegal to Ethiopia, south to Zimbabwe, Malawi and Mozambique; Zanzibar; and Madagascar.	Variety of arthropods, including moths, butterflies, other insects, spiders, and sun spiders.	Not threatened

[continued]

Common name / Scientific name/ Other common names	Physical characteristics	Habitat and behavior	Distribution	Diet	Conservation status
Ja slit-faced bat *Nycteris major*	Pelage varies from rich brown or russet to pale brown or grayish. Muzzle has longitudinally divided nostrils, which are long, deep pit in forehead. T-shaped tip at end of tail. Head and body length 1.5–3.7 in (4–9.3 cm), tail length 1.7–3 in (4.3–7.5 cm), forearm length 1.2–2.4 in (3.2–6 cm).	Woodland savanna or dry country. Little known of reproductive and behavioral patterns.	Liberia to Zambia.	Variety of arthropods, including moths, butterflies, other insects, spiders, and sun spiders.	Vulnerable

Resources

Books

Garbutt, N. *Mammals of Madagascar*. New Haven: Yale University Press, 1999.

Hutson, A. M., S. P. Mickelburgh, and P. A. Racey. *Global Status Survey and Conservation Action Plan, Microchiropteran Bats*. Gland, Switerzland: IUCN SSC Chiroptera Specialist Group, 2001.

Kingdon, J. *Mammals of East Africa: An Atlas of Evolution*, Volume 2b. New York: Academic Press, 1974.

Nowak, R. M. *Walker's Mammals of the World*, Volume 1. Baltimore: Johns Hopkins University Press, 1999.

Payne, J., C. M. Francis, and K. Phillips. *A Field Guide to the Mammals of Borneo*. Kuala Lumpur: The Sabah Society with World Wildlife Fund Malaysia, 1985.

Taylor, P. J. *Bats of Southern Africa*. Pietermaritzburg, South Africa: University of Natal Press, 2000.

Periodicals

Aldridge, H. D. J. N., M. Obrist, H. G. Merriam, and M. B. Fenton. "Roosting, Vocalizations and Foraging by the African Bat, *Nycteris thebaica*." *Journal of Mammalogy* 71 (1990): 242–246.

Fenton, M. B., C. L. Gaudet, and M. L. Leonard. "Feeding Behaviour of the Bats *Nycteris grandis* and *Nycteris thebaica* (Nycteridae) in Captivity." *Journal of Zoology* 200 (1983): 347–354.

Fenton, M. B., C. M. Swanepoel, R. M. Brigham, J. Cebek, and M. B. C. Hickey. "Foraging Behaviour and Prey Selection by Large Slit-faced Bats (*Nycteris grandis*; Chiroptera: Nycteridae)." *Biotropica* 22 (1990): 2–8.

Gray, P. A., M. B. Fenton, and V. Van Cakenberghe. "*Nycteris thebaica*." *Mammalian Species* 612 (1999): 1–8.

Hickey, M. B. C., and J. M. Dunlop. "*Nycteris grandis*." *Mammalian Species* 632 (2000).

Melville Brockett Fenton, PhD

False vampire bats

(Megadermatidae)

Class Mammalia
Order Chiroptera
Suborder Microchiroptera
Family Megadermatidae

Thumbnail description
The bifurcate tragus and the absence of upper incisor teeth make false vampire bats distinctive

Size
Medium-sized to large bats with forearms that range from 1.9–4.6 in (5–11.8 cm); 0.7–5.9 oz (20–170 g)

Number of genera, species
4 genera; 5 species

Habitat
From forest to open woodland and desert

Conservation status
Vulnerable: 1 species; Lower Risk/Near Threatened: 1 species

Distribution
Africa, Asia, through the East Indies into Australia

Evolution and systematics

False vampire bats belong to the superfamily Rhinolophoidea. They first appeared as fossils in Eocene deposits, but the fossil record is not extensive. No subfamilies are recognized.

Physical characteristics

Medium-sized to large bats with broad wings. False vampire bats have large ears, large eyes, and distinctive noseleaves. Their long and silky fur tends to be gray in color. The bifurcate tragus and lack of upper incisors are diagnostic for the family.

Distribution

Megadermatids occur widely in Africa and from eastern Afghanistan through India and Sri Lanka, and on through Myanmar, southwestern China, Malaysia, and the East Indies (Java, Borneo, the Philippines, Bali, Sulawesi, Sangihe Islands, Togian Islands, Lombok, the Moluccas). They are also found in the Andaman Islands, as well as in parts of Australia.

Habitat

False vampire bats live in open, arid habitats in Australia, and in savanna woodlands or the more arid areas of Africa and India.

A greater false vampire bat (*Megaderma lyra*) diving for a mouse. (Photo by Stephen Dalton/Photo Researchers, Inc. Reproduced by permission.)

A heart-nosed bat (*Cardioderma cor*) swooping down on its prey. (Photo by Merlin D. Tuttle/Bat Conservation International/Photo Researchers, Inc. Reproduced by permission.)

A lesser false vampire bat (*Megaderma spasma*) in flight with a colored moth. (Photo by Merlin D. Tuttle/Bat Conservation International/Photo Researchers, Inc. Reproduced by permission.)

A heart-nosed bat (*Cardioderma cor*) roosting. (Photo by Harald Schütz. Reproduced by permission.)

A group of Australian false vampire bats (*Macroderma gigas*), also known as ghost bats, hanging in a cave. (Photo by © Eric and David Hosking/Corbis. Reproduced by permission.)

African yellow-winged bat (*Lavia frons*) roosting on a small tree branch. (Photo by Michael Fogden. Bruce Coleman, Inc. Reproduced by permission.)

Three Australian false vampire bats (*Macroderma gigas*) roosting in a cave. (Photo by Tom McHugh/Photo Researchers, Inc. Reproduced by permission.)

A close-up of a greater false vampire bat (*Megaderma lyra*). (Photo by Brock Fenton. Reproduced by permission.)

Behavior

False vampire bats roost in a variety of situations, from caves and mines, to tree hollows and buildings. The African yellow-winged bat (*Lavia frons*) roosts by hanging from tree branches. Unlike some other species of bat, false vampire bats are warm-blooded and cannot enter torpor, a state of total inactivity.

Feeding ecology and diet

Broad-winged bats with low-wing loading, false vampire bats are capable of flying and maneuvering in areas of vegetation. They may take prey from surfaces (the ground or vegetation) or in flight. False vampire bats eat animals ranging from arthropods to vertebrates, including bats, birds, fish, frogs, and lizards, with larger species taking larger prey. False vampire bats use at least three approaches to finding prey. Often they identify prey-generated sounds such as the footfalls of large arthropods or mice. They may also use vision and, at least in experimental situations, greater false vampire bats used echolocation to find frogs. Prior to foraging in the evening, false vampire bats as well as heart-nosed bats (*Car-*

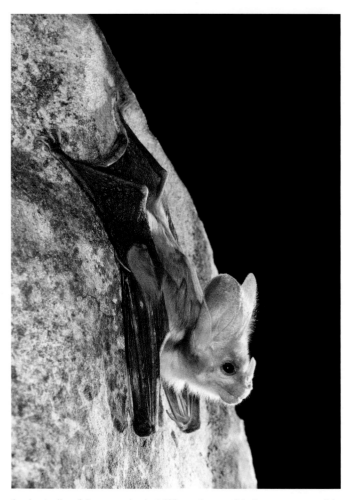

A greater false vampire bat (*Megaderma lyra*) roosting. (Photo by Harald Schütz. Reproduced by permission.)

An Australian false vampire bat (*Macroderma gigas*) on a cave wall in Kakadu National Park, Northern Territory, Australia. (Photo by B. G. Thomson/Photo Researchers, Inc. Reproduced by permission.)

dioderma cor) sing from perches in their hunting areas. Their songs appear to be territorial advertisements.

Reproductive biology

Female false vampire bats produce a single young each year. The gestation period is about 90 days and young are born at the beginning of the rainy season. Mating system varies by species.

Conservation status

The IUCN lists the heart-nosed bat as being Lower Risk/Near Threatened. The Australian false vampire bat (*Macroderma gigas*) is listed as Vulnerable.

Significance to humans

None known.

1. Australian false vampire bat (*Macroderma gigas*); 2. Yellow-winged bat (*Lavia frons*). (Illustration by Barbara Duperron)

Species accounts

Yellow-winged bat
Lavia frons

TAXONOMY
Megaderma frons (Geoffroy, 1810), Senegal.

OTHER COMMON NAMES
French: Chauve-souris orangée.

PHYSICAL CHARACTERISTICS
These are medium-sized bats, with forearm lengths ranging from 2.0–2.5 in (5.3–6.4 cm), and weighing 0.88–1.23 oz (25–35 g). The combination of yellowish wings and ears and a long, blunt nose-leaf distinguish yellow-winged bats from the only other African megadermatid, *Cardioderma cor*. The bat's large ears, bifurcate tragi, and broad wings are typical for the family.

DISTRIBUTION
Widespread in sub-Saharan Africa, occurring from Gambia and Senegal through west Africa to Sudan, Ethiopia, and south to Malawi and Zambia. Also through Nigeria, Cameroon, Central African Republic, Congo, Gabon, and northern Angola.

HABITAT
Inhabit savanna woodland.

BEHAVIOR
Roost by day 16.4–32.8 ft (5–10 m) above the ground, hanging from small branches in acacia trees. Roosting bats are alert and difficult to approach. Typically, one pair (adult male and adult female) occupies a territory and both members of the pair often move between roosts, apparently to follow shade. They use social calls that are audible to people, and also higher frequency calls in echolocation.

FEEDING ECOLOGY AND DIET
Insectivorous, they hunt well above the ground, taking flying insects as well as walking insects and other arthropods from surfaces such as foliage, the trunks, and limbs of trees. Other prey include termites, beetles, moths, katydids, and flies.

REPRODUCTIVE BIOLOGY
Appear to be monogamous, with pairs (one adult male, one adult female) remaining together at least through one reproductive season. At the end of the dry season after three-months gestation, females bear a single young annually. At first, foraging females carry their young, but later leave them alone when hunting. As they approach adult size, young bats hang by their hind feet from their mothers' shoulders and flap their wings, in preparation for their own flights.

CONSERVATION STATUS
Not threatened.

SIGNIFICANCE TO HUMANS
None known. ◆

Australian false vampire bat
Macroderma gigas

TAXONOMY
Megaderma gigas Dobson, 1880, Queensland, Australia.

OTHER COMMON NAMES
English: Ghost bat.

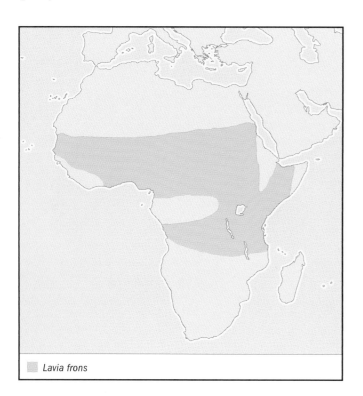

Lavia frons

Macroderma gigas

PHYSICAL CHARACTERISTICS

One of the largest bats, they have forearms 3.7–4.6 in (9.6–11.8 cm) long, and weigh 2.6–5.0 oz (74–144 g). They have huge ears, large eyes, and prominent nose-leaves, with fur that varies from light brown to almost white (hence the name "ghost bat."

DISTRIBUTION

Today their populations occur in north Queensland, along the north central coast, and in the northwest.

HABITAT

Occur in habitats from arid hillsides, grasslands, and monsoon forest, to savanna woodland and other forests.

BEHAVIOR

Roost in hollows, usually in caves or abandoned mines, sometimes forming colonies of more than 1,000 individuals. Very vocal in their roosts, producing both lower frequency signals audible to humans, and higher frequency echolocation calls.

FEEDING ECOLOGY AND DIET

Consume animals ranging in size from insects such as cockroaches to vertebrates from frogs and geckoes to birds and other species of bats. When hunting, these bats sometimes hang in wait from a branch and attack passing prey, taking it either from the ground or in flight. Their hunting perches can be recognized by distinctive claw marks left on small tree branches.

REPRODUCTIVE BIOLOGY

Mating occurs in May and each female bears a single young in July. Mothers stay with their young for some time, including flying together to forage after the young is large enough. May be promiscuous.

CONSERVATION STATUS

Vulnerable.

SIGNIFICANCE TO HUMANS

Australian false vampire bats have great spiritual significance to Australian aborigines. ◆

Resources

Books

Churchill, S. *Australian Bats.* Sydney: Reed New Holland, 1998.

Hutson, A. M., S. P. Mickelburgh, and P. A. Racey. *Global Status Survey and Conservation Action Plan, Microchiropteran Bats.* Gland, Switzerland: IUCN SSC Chiroptera Specialist Group, 2001.

Kingdon, J. *Mammals of East Africa: An Atlas of Evolution,* Volume 2b. New York: Academic Press, 1974.

Nowak, R. M. *Walker's Mammals of the World,* Vol. 1. Baltimore: Johns Hopkins University Press, 1999.

Periodicals

Csada, R. "*Cardioderma cor.*" *Mammalian Species* 519 (1998): 1–4.

Hudson, W. S., and D. E. Wilson. "*Macroderma gigas.*" *Mammalian Species* 260 (1986): 1–4.

Vonhof, M. J., and M. C. Kalcounis. "*Lavia frons.*" *Mammalian Species* 614 (1999): 1–4.

Melville Brockett Fenton, PhD

Horseshoe bats
(*Rhinolophidae*)

Class Mammalia
Order Chiroptera
Suborder Microchiroptera
Family Rhinolophidae

Thumbnail description
Small to medium-sized, insectivorous bats with complex, horseshoe-shaped noseleaf, large, pointed ears, broad, rounded wings, and short tail enclosed within the flight membrane

Size
Head and body length 1.4–4.3 in (3.5–11 cm); tail 0.6–2.2 in (1.5–5.6 cm); forearm 1.2–3 in (3–7.5 cm); weight 0.15–13.8 oz (4–35 g)

Number of genera, species
1 genus; up to 70 species

Habitat
Open, scrub, savanna, woodland, or forested habitats with roosting sites (caves, rock outcrops, tunnels, mines, or buildings) at low to high altitudes in tropical and temperate zones

Conservation status
Critically Endangered: 1 species; Endangered: 2 species; Vulnerable: 8 species; Lower Risk/Near Threatened: 20 species; Data Deficient: 7 species

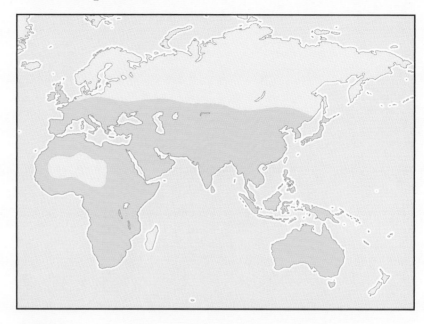

Distribution
Throughout tropical and temperate regions of the Old World, from western Europe and Africa through Asia to Japan, Australia, and western Pacific islands

Evolution and systematics

The horseshoe bats (Rhinolophidae) are included here in the superfamily Rhinolophoidea with the Old World leaf-nosed bats (Hipposideridae). However, many authorities believe there is strong evidence that the two groups are sister taxa and recognize them as subfamilies within the Rhinolophidae.

Up to 70 species are recognized within the single genus *Rhinolophus.* The most recent classification (Nowak 1999) lists 62 species separated into 12 groups.

Fossil Rhinolophidae are known from the early to middle Eocene and early Oligocene of Europe, the Miocene of Australia and the Miocene and late Pliocene deposits of Africa. In 1992, Bogdanowicz and Owen suggested that the family originated in Southeast Asia, that the *megaphyllus* group includes the most primitive species and that the Palearctic and African species are generally the most advanced.

Physical characteristics

Horseshoe bats are characterized by a complex noseleaf expansion of the skin surrounding the nostrils and consisting of three parts. The horseshoe-shaped lower part covers the up-

per lip, surrounds the nostrils, and has a central notch in the lower edge. Above the nostrils is the lancet, a pointed, erect structure attached by its base. The horseshoe and the lancet are flattened from front to back. The sella is located between the horseshoe and the lancet, is flattened from side to side and is connected at its base by means of folds and ridges. The form of the noseleaf is often diagnostic in species identification. Horseshoe bats generally fly with the mouth closed and emit ultrasonic sounds through the nostrils. The sounds are channeled by the noseleaf structure to achieve a maximum intensity at a point of focus ahead of the bat, and the noseleaf also shields the ears from the direct reception of the impulses. In some species, such as Hildebrandt's horseshoe bat (*R. hildebrandti*), the noseleaf has a distinctly arranged pattern of sensory hairs.

The large ears are widely separated, usually pointed, and lack a tragus, but the antitragal lobe is much enlarged and folds across the open base of the ear. The ears are capable of independent movement. The eyes are quite small, and the field of vision seems to be partly obstructed by the noseleaf, so sight is probably of little importance.

The fur of horseshoe bats is long, loose, and soft. The most common colors are gray-brown to rufous-brown, but the color varies from black or dark brown to bright orange-red

Blyth's horseshoe bat (*Rhinolophus lepidus*) roosting in Bhutan. (Photo by Harald Schütz. Reproduced by permission.)

that are noticeable by their coloring. These males also have well-developed nipples, which do not produce milk, but perhaps secrete odor-producing substances.

Young horseshoe bats shed milk teeth before birth. The teeth of adults exhibit the normal cuspidate pattern found in insectivorous bats. The dental formula is: (I1/2, C1/1, PM2/3, M3/3) × 2 = 32. The upper incisors are mounted in a projection of the palatine bone, well forward of the canines. The lower incisors are trifid. The first upper and second lower premolars are small, usually displaced externally, and may be missing. The most characteristic feature of the skull is the dome just above the nasal aperture.

Distribution

Horseshoe bats occur throughout the temperate and tropical zones of the Old World, from the British Isles across southern and central Europe, through Arabia and southern Asia, east to China and Japan, and throughout the Indian subcontinent, Southeast Asia, Indonesia, and the Philippines to Australasia and western Pacific islands. They also occur throughout Africa, except in the most arid regions, but not in Madagascar. The most wide-ranging species is the greater horseshoe bat (*R. ferrumequinum*), which occurs through the entire southern Palaearctic region from Great Britain across Europe, North Africa, and southern Asia east to China and Japan. In contrast, some island species have very restricted ranges. For example,

or cream-yellow. The underparts are usually more pale than the upperparts. The woolly (*R. luctus*) and lesser woolly (*R. beddomei*) horseshoe bats have a very long, woolly pelage, unusual in the genus.

The wings are short, broad, and rounded, the second finger consisting of the metacarpal only, without phalanges. The third, fourth and fifth fingers each have two phalanges, which fold under the wing when the bat is at rest. The vertebrae of the tail end at the posterior fringe of the interfemoral membrane, which is supported on either side by curved calcanea arising from the ankles. The tail and interfemoral membrane fold upwards when the bat is resting. The hind limbs are poorly developed and these bats cannot walk quadrupedally. All the toes have three bones, except the first, which has two; the Hipposideridae, in contrast, have only two bones in each toe.

Females have two mammary glands in the pectoral region and two "false nipples" on the lower abdomen, just anterior to the genital orifice, to which the young cling while they are carried around by their mother during flight. In two African species, the small Lander's horseshoe bat and the larger *R. alcyone*, the males have glandular hair tufts in the armpit area

A grouping of three eastern horseshoe bats (*Rhinolophus megaphyllus*) roosting in a large tree hollow in Raluma Range, Queensland, Australia. (Photo by B. G. Thomson/Photo Researchers, Inc. Reproduced by permission.)

A large group of greater horseshoe bats (*Rhinolophus ferrumequinum*). (Photo by Jose Luis G. Grande/Photo Researchers, Inc. Reproduced by permission.)

the Andaman horseshoe bat (*R. cognatus*) is confined to three known areas on the Andaman Islands in the Indian Ocean, while *R. monoceros* occurs only on Taiwan and *R. imaizumii* only on Iriomote Island in the southern Ryukyu Islands, south of Japan. A few continental species also have very small ranges. The mitred horseshoe bat (*R. mitratus*) is known from only one specimen collected in Bihar, eastern India, while *R. paradoxolophus* is known by only single specimens from northern Vietnam and eastern Thailand, and in Africa the Cape horseshoe bat (*R. capensis*) is endemic to South Africa and *R. maclaudi* is confined to Guinea in West Africa.

Habitat

The Rhinolophidae occur throughout the temperate and tropical zones of the Old World, being found in a great variety of habitats at both high and low altitudes. They are found in forest, woodland, savanna, scrub, open areas, and sometimes even in deserts, but it seems that the availability of suitable shelters for daytime roosting, nurseries, and hibernation is often a more important factor governing habitat suitability than is the type of vegetation occurring in the occupied area. The variety of sites used for such shelters is extensive, and includes caves, rock outcrops and crevices, overhangs, mines,

tunnels, buildings (disused and occupied), cellars, culverts, hollow trees, and dense foliage. Hildebrandt's horseshoe bat of eastern Africa even uses disused aardvark (*Orycteropus afer*) and warthog (*Phacochoerus aethiopicus*) holes for daytime roosts, but also roosts in hollow trees, especially the baobab (*Adansonia digitata*), and buildings.

Forest-inhabiting species include the rufous horseshoe bat (*R. rouxii*), whose diurnal roosts tend to be humid and include caves, tunnels, hollow trees, wells, temples, old houses, and barns. The little-known little Nepalese horseshoe bat (*R. subbadius*) is recorded from bamboo clumps in dense jungle. Although Blyth's horseshoe bat (*R. lepidus*) is normally associated with forested country, it is also recorded from a desert biome in India; its diurnal roosts include subterranean silos. The woolly horseshoe bat occurs in dense forests on precipitous mountains in the Kathmandu Valley, where it roosts in caves. The lesser woolly horseshoe bat is also restricted to forest, where it roosts in hollow trees, small caves, and under ledges; it also uses dungeons, old houses, barracks, and tunnels. The trefoil horseshoe bat (*R. trifoliatus*) of southern, southeastern Asia and Malaysia lives in dense evergreen jungle and roosts in thick foliage. Lander's horseshoe bat (*R. landeri*) of Africa is predominantly a forest species but in the south of its range it inhabits savanna woodland with riverine vegetation and

An agitated large-eared horseshoe bat (*Rhinolophus philippinensis*) hanging from a branch at Peach Creek, Cape York, northern Australia. (Photo by B. G. Thomson/Photo Researchers, Inc. Reproduced by permission.)

well-watered areas. The presence of substantial shelters (caves, mines, boulders, hollow trees) is probably a critical habitat requirement.

The African endemic Ruppell's horseshoe bat (*R. fumigatus*) is widespread in African open savanna woodland but is absent from desert and semidesert areas. In contrast, Geoffroy's horseshoe bat (*R. clivosus*), which occurs in Africa and Asia, inhabits savanna woodland but also occurs in deserts. In southern Africa its absence from semidesert parts of Botswana may be due to the lack of suitable roosting sites such as caves, rock crevices, and mines. The bushveld horseshoe bat (*R. simulator*) of African savanna woodland is dependent on the availability of caves and mine shafts for shelter.

Some horseshoe bat species occur in association with man. The greater horseshoe bat uses caves but has adapted to larger buildings for nurseries, especially in the northern parts of its European range, while in southern Asia it roosts in temples, outhouses, and ruins. Also in Europe, the nurseries of the lesser horseshoe bat are predominantly in warm caves on southern regions, but mostly in roofs of buildings in the north. This species hibernates in caves, mines, and cellars.

Behavior

Roosting habits within the family are diverse. Most species are predominantly cave dwellers, but some have adapted to human-made structures such as mines and buildings (see Habitat section). Historically, in Europe the lesser horseshoe bat roosted all year round in caves, but it has changed its behavior markedly and has adopted buildings as summer roosts in many areas. Most species are gregarious, roosting in small to large colonies, but the woolly horseshoe bat normally lives in pairs. Some species, such as the greater, rufous big-eared (*R. macrotis*), intermediate (*R. affinis*) and Blyth's, may roost together. Blyth's and intermediate may roost with bats from other genera (*Hipposideros*, *Rhinopoma*, *Taphozous*). In some species the sexes live together all year, whereas in others, such as the greater horseshoe bat, the females form single-sex maternity colonies. Eastern horseshoe bat (*R. megaphyllus*) colonies are usually small (fewer than 20 individuals), but in the nonbreeding season may reach 2,000 individuals, while in South Africa, Geoffroy's horseshoe bat occurs in roosts of up to 10,000 individuals.

When resting, the bats hang freely from the ceiling of the cave or other shelter, some species with individuals well separated and others with individuals close together. They do not close their wings alongside their body as do most bats, but wrap them wholly or partially around the body. The small bare patch on the back at the base of the tail is covered by the upturned tail and membrane, so that the bat is enclosed in its flight membranes.

Those species that live in temperate zones hibernate during the cold season, although they interrupt their hibernation from time to time and may change their hibernating places occasionally. Species such as the eastern horseshoe bat, which occupy both tropical and temperate zones, remain active throughout the year in the tropical parts of their range. Most hibernating species leave their daytime roosts at the end of the summer and seek alternative winter hibernation sites (hibernacula) in more sheltered situations. Some species commonly move over distances of 3.1–6.2 mi (5–10 km) between summer and winter sites. Greater horseshoe bats often start hibernating near the entrance of caves, then move to sites deeper in the caves as the temperature drops and winter advances. Most species hibernate colonially but some (such as the eastern horseshoe bat) hibernate singly. Hibernacula are often humid, usually have an ambient temperature of around 42.7–50°F (6–10°C) (measured in Europe), and have restricted airflow. Close access to sheltered winter foraging areas is important, as is freedom from disturbance. Daily torpor is also common in horseshoe bat species in response to cooling and/or reduced food availability. In summer, torpor is slight and bats can fly off quickly.

Feeding ecology and diet

Horseshoe bats are aerial foragers, catching flying prey on the wing, and some are also gleaners, snatching stationary prey from branches, foliage, stones, or the ground. Many are clutter foragers, using their broad, short wings and high echolocation frequencies to fly slowly and maneuver through dense vegetation. Many species also "flycatch" or "perch-hunt" by hanging from a perch and making rapid sallies to snatch prey detected flying past. Some species can hover. Horseshoe bats usually hunt within 16.5–19.5 ft (5–6 m) of

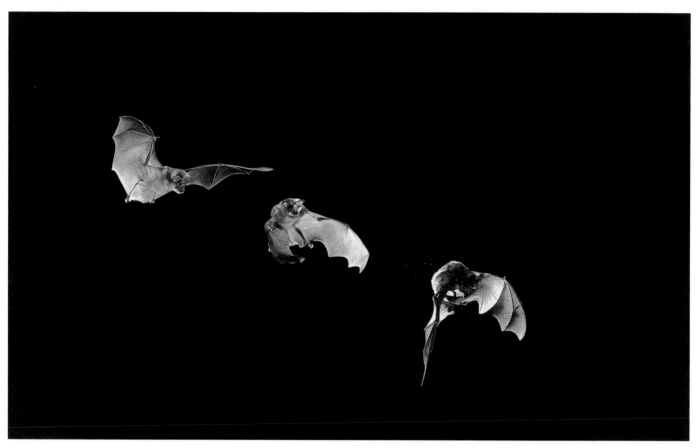

A greater horseshoe bat (*Rhinolophus ferrumequinum*) catching a moth in flight. (Photo by Stephen Dalton/Photo Researchers, Inc. Reproduced by permission.)

the ground, and will also feed on the ground. Prey can be caught in the wing membrane and may be stored for brief intervals in cheek pouches. The relatively short tail and small tail membrane are not large enough to form a pouch for holding insects. When a large insect is caught in flight, it may be tucked into the wing membrane under the arm while the bat manipulates it with its mouth.

Horseshoe bats begin foraging later in the evening than most bats and often return to the roost, or to a feeding perch, to eat captured prey. Such perches can often be located by the piles of insect fragments that collect on the ground beneath them. In general *Rhinolophus* species are solitary hunters, while *Hipposideros* species forage in small groups. Like many other bats, horseshoe bats generally have regular feeding territories or hunting areas, and greater horseshoe bats may hunt regularly around the periphery of the same trees, low bushes, and buildings. Blyth's horseshoe bats have small, well-defined foraging territories near their roosts. They explore the foliage of trees, making frequent stops to pick insects off leaves. Some species are attracted to insects flying around lights and may enter buildings in search of prey. Hildebrandt's horseshoe bat hunts on the wing during the early part of the night, after which it flycatches throughout the night. The rufous horseshoe bat adopts a similar feeding strategy. The greater horseshoe bat changes its foraging habitat seasonally, hunting in woodland in the spring and over

pasture in late summer. This behavior is linked to prey availability: the ambient temperature in woodlands is higher than that over open ground, leading to higher insect numbers in spring, while *Aphodius* dung beetles, a favored prey, increase in pasture during the summer as cattle dung accumulates.

Echolocation pulses are emitted through the nostrils with the mouth closed. The echolocation calls are of high duty cycle (56–60%), are of constant frequency with a short frequency drop at the end, and are often exceptionally long (20 to over 100 ms) in duration. To a certain extent, frequency can be used to identify a species, since individual species emit signals over a characteristic limited frequency band. Signal intensity (loudness) 4 in (10 cm) in front of the bat has been recorded at 27.0 N/m^2 in the greater horseshoe bat and 2.0 N/m^2 in the Mediterranean horseshoe bat *R. euryale*. The former species can differentiate between two identical targets, one placed as little as 0.47–0.51 in (12–13 mm) in front of the other, while the latter can detect a 0.12 in (3 mm) wire at a range of 4.6 ft (1.4 m) and a 0.002 in (0.05 mm) wire at 8 in (20 cm).

Prey items include Lepidoptera (almost entirely moths, including microlepidoptera), Coleoptera (including cockchafers and dung beetles), Hemiptera, Diptera (including mosquitoes and craneflies), Orthoptera (grasshoppers), Hymenoptera (wasps), Isoptera (termites), and spiders. The greater horseshoe bat drinks during low-level flight or while hovering.

A greater horseshoe bat (*Rhinolophus ferrumequinum*) chasing a moth in flight. (Photo by Animals Animals ©S. Dalton, OSF. Reproduced by permission.)

Reproductive biology

Courtship and mating behavior are poorly known, but lesser horseshoe bats chase each other as a preliminary to mating, the male then hanging himself behind and over the female for the very brief copulation. Polygyny may be common, a male mating with a harem of females. In some species, dominant males guard favored mating caves, and in Britain individual greater horseshoe bats are known to guard the same spot at cave entrances for a large number of years, and to be visited by a number of females during the mating period. The existence of strictly monogamous family units in any species has not been proved, but the woolly horseshoe bat apparently lives in pairs—whether these pairs are permanent is not clear.

In some species, including those that hibernate, mating occurs during the autumn, and may continue during the winter, but ovulation is delayed so that fertilization does not occur until the spring. In other species mating and fertilization occur in the spring, and in tropical regions births usually take place during the warm, wet months. In South Africa, Geoffroy's horseshoe bat copulates in May, at the end of the summer, and sperm are stored in the female's oviducts and uterine horns during winter hibernation. Ovulation and fertilization take place in spring (August) and parturition in December. In the cape horseshoe bat, sperm storage devolves upon the male. Spermatogenesis takes place in October–May and sperm are released to the cauda epididymis in April–May. At this time, females are in estrus, but copulation and ovulation are delayed until August–September, after the winter hibernation. Male eastern horseshoe bats of Australia undergo spermatogenesis in February, produce mature sperm in March, and store the sperm until mating, ovulation and fertilization take place in June.

Gestation periods vary from 7 weeks to 3–4.5 months, reaching over 5 months in the rufous horseshoe bat in India. A single young is normally produced, but in India the greater horseshoe bat apparently regularly gives birth to twins. Birth weight is around 0.07–0.2 oz (2–6 g) and lactation continues for 1–3 months. Young are independent at 6–8 weeks. After weaning, the young are usually abandoned by the mothers, but may stay at the breeding site for some time afterwards. Females of some species become sexually mature in their first year, males often not until the second year. In some species, sexual maturity is delayed: male greater horseshoe bats mature at 3 years (but often do not acquire a territory for another 1–2 years) and females at 3–4 years. One breeding season appears to be the norm, but in Malaysia the intermediate horseshoe bat apparently breeds twice a year, with pregnant females collected in April–May and again in October. Longevity in the genus is typically 4–7 years, but there are several records of much longer lifespans. The record is that of a greater horseshoe bat that was captured in France in 1982, 29 years after initial banding, and was then seen again in 1983. In this species, most mortality occurs in the first winter, or when females first breed.

Conservation status

Many horseshoe bat species are poorly known and their conservation status is therefore difficult to assess. However, all the species are vulnerable to loss of foraging habitat through destruction or modification, while they are also (and more significantly) threatened by the loss of roost sites, nurseries, and hibernacula as a result of destruction, disturbance, or vandalism. The widespread use of insecticides is also a threat to some species. The IUCN Red List contains 38 species, 57% of the total species in the family, including 7 that are Data Deficient and 20 that are Lower Risk/Near Threatened. The only Critically Endangered species is the recently described *R. convexus* of Malaysia. The two Endangered species both have a very restricted island distribution, the Kai horseshoe bat (*R. keyensis*) being confined to the Kai Islands (Indonesia), and *R. imaizumii* being restricted to one island in the Ryukyu group south of Japan. The Vulnerable Andaman horseshoe bat also has a restricted island distribution and a small, possibly declining population.

Although the survival of the more widespread species may not be of serious concern globally, regionally the situation may well be more serious. For example, the greater horseshoe bat is globally classed as Lower Risk/Near Threatened but is

A Hildebrandt's horseshoe bat (*Rhinolophus hildebrandti*) catching prey in midflight. (Photo by Merlin D. Tuttle/Bat Conservation International/Photo Researchers, Inc. Reproduced by permission.)

regarded as endangered in Europe, where it is threatened with extinction in northern parts of its range. It is declining rapidly as a result of disturbance to its roosts in caves and buildings, vandalism, habitat modifications leading to the loss of large insect prey, and the increasing use of insecticides. Similar problems have beset the lesser horseshoe bat, which is already extinct in northern central Europe and northern Britain, and is threatened with extinction in Germany. This species, with Blasius's (*R. blasii*), Mediterranean (*R. euryale*), and Mehely's (*R. mehelyi*) horseshoe bats, have also been listed as endangered in Europe. A slow recovery of the Mediterranean horseshoe bat has been noted since the most dangerous pesticides were banned in the 1980s, while in Britain the decline of the greater horseshoe bat has been halted by protecting maternity roosts and hibernacula.

The 18 horseshoe bat species that occur in the Philippines are poorly studied there but are known to have been impacted by the widespread disturbance of caves. Other species roost in large hollow trees, especially in lowland dipterocarp forest, and have been severely affected by logging that destroys both the roosting trees and the foraging habitat.

Significance to humans

The folklore, superstitions, and legends that surround bats in general also apply to the Rhinolophidae, but without any specific reference to any species in this family. Horseshoe bats have no significance to humans in terms of public health, or nuisance factors, and relatively few species roost extensively in human-made structures such as occupied buildings.

A close-up of Dobson's horseshoe bat (*Rhinolophus yunanensis*) on a cave wall in Southeast Asia. (Photo by Merlin D. Tuttle/Photo Researchers, Inc. Reproduced by permission.)

1. Lesser horseshoe bat (*Rhinolophus hipposideros*); 2. Blasius's horseshoe bat (*Rhinolophus blasii*); 3. Mediterranean horseshoe bat (*Rhinolophus euryale*); 4. Dent's horseshoe bat (*Rhinolophus denti*). (Illustration by Emily Damstra)

1. Greater horseshoe bat (*Rhinolophus ferrumequinum*); 2. Cape horseshoe bat (*Rhinolophus capensis*); 3. Rufous horseshoe bat (*Rhinolophus rouxii*); 4. Lesser woolly horseshoe bat (*Rhinolophus beddomei*); 5. Eastern horseshoe bat (*Rhinolophus megaphyllus*). (Illustration by Emily Damstra)

Species accounts

Eastern horseshoe bat
Rhinolophus megaphyllus

TAXONOMY
Rhinolophus megaphyllus Gray, 1834, New South Wales, Australia.

OTHER COMMON NAMES
English: Smaller horseshoe bat.

PHYSICAL CHARACTERISTICS
Head and body length 1.7–2.1 in (4.4–5.3 cm); tail 0.9–1.1 in (2.2–2.8 cm); forearm 1.8–1.9 in (4.45–4.9 cm); weight 0.2–0.4 oz (7–10.5 g). Fur grayish brown, lighter on belly; wings pinkish gray. Rufous form occurs in Queensland, Australia.

DISTRIBUTION
Thailand, Malaysia, Moluccas, New Guinea, Lesser Sunda Islands, and east coast of Australia.

HABITAT
Tropical and temperate rainforest, deciduous vine forest, sclerophyll forest, open woodland, coastal scrub, and grassland.

BEHAVIOR
Roosts in caves, mines, rock outcrops, abandoned buildings, and road culverts. Colonies usually small (fewer than 20 individuals), but in nonbreeding season may reach 2,000 individuals. In temperate regions, disperse in winter to roost singly in torpid state; in tropics, active all year.

FEEDING ECOLOGY AND DIET
Eats mainly moths, also bugs, beetles, flies, and wasps. Flight slow and fluttery; can hover, and maneuver through dense foliage. Also flycatch from a perch, and glean spiders from ground; often use feeding perch.

REPRODUCTIVE BIOLOGY
Females congregate in maternity colonies of 15–2,000 individuals in spring and summer, in warm, humid caves. Sperm produced February–March; copulation, ovulation and fertilization late June; gestation 4–4.5 months; single young born in November. Young nursed for about 8 weeks; fully grown at 5–6 weeks. Species thought to be polygynous.

CONSERVATION STATUS
Not globally threatened; widespread and apparently locally common.

SIGNIFICANCE TO HUMANS
None known. ◆

Rufous horseshoe bat
Rhinolophus rouxii

TAXONOMY
Rhinolophus rouxii Temminck, 1835, Pondicherry and Calcutta, India.

OTHER COMMON NAMES
English: Rufous leaf bat.

PHYSICAL CHARACTERISTICS
Head and body length 1.7–2.6 in (4.2–6.6 cm); tail 0.8–1.3 in (2.1–3.3 cm); forearm 1.7–2.1 in (4.4–5.3 cm); weight 0.5–0.6 oz (14–16.5 g). Fur soft and silky, color variable, from orange to russet brown or buffy brown.

DISTRIBUTION
Nepal, India, and Sri Lanka to southeastern China and Vietnâm.

HABITAT
Forested regions in higher rainfall areas.

BEHAVIOR
Roosts in caves, tunnels, hollow trees, wells, and buildings. Colonies vary from a few to several hundred individuals. Sexes live separately for at least part of the year (probably when young present), males living alone or in small groups, females in large colonies. Hibernates in colder parts of range.

FEEDING ECOLOGY AND DIET
Begins feeding after sunset, catching insects on the wing for 30–60 minutes. Flight low, often through bushes. Then rests for 60–120 minutes before foraging throughout night, "flycatching" from a perch. Eats primarily grasshoppers, moths, beetles, termites, mosquitoes, and other Diptera.

REPRODUCTIVE BIOLOGY
Polygynous. Copulation occurs in December; implantation delayed; gestation 150–160 days; single young born April–June (possibly also September in Sri Lanka).

■ *Rhinolophus rouxii*
■ *Rhinolophus megaphyllus*

CONSERVATION STATUS
Not globally threatened. Widespread; common in Indian subcontinent.

SIGNIFICANCE TO HUMANS
None known. ◆

Lesser woolly horseshoe bat
Rhinolophus beddomei

TAXONOMY
Rhinolophus beddomei Andersen, 1905, Madras, India.

OTHER COMMON NAMES
None known.

PHYSICAL CHARACTERISTICS
Head and body length 2.6–3 in (6.5–7.5 cm); tail 1.5–1.9 in (3.9–4.8 cm); forearm 2.2–2.5 in (5.5–6.4 cm); weight 0.6–0.7 oz (18–19 g). Fur long, woolly and dark, usually black with paler hair tips.

DISTRIBUTION
Southern and western peninsular India, and Sri Lanka.

HABITAT
Restricted to forested areas.

BEHAVIOR
Normally roosts singly, in pairs or threes, in hollow trees, small caves, overhanging ledges, buildings, or tunnels; hangs by one foot, with wings wrapped round body.

FEEDING ECOLOGY AND DIET
Emerges in late evening; flies low to seek prey among bushes, along forest edges or tracks, and in forest glades. Eats beetles, termites, and other flying insects.

☐ *Rhinolophus beddomei*
■ *Rhinolophus ferrumequinum*

REPRODUCTIVE BIOLOGY
May be monogamous. Pregnant female collected Sri Lanka in January; female with young seen India in May; no other information.

CONSERVATION STATUS
Lower Risk/Near Threatened. Low density populations and its dependence on forests suggest it is very vulnerable to habitat destruction.

SIGNIFICANCE TO HUMANS
None known. ◆

Greater horseshoe bat
Rhinolophus ferrumequinum

TAXONOMY
Vespertilio ferrum-equinum (Schreber, 1774), France.

OTHER COMMON NAMES
French: Grand rhinolophe fer à cheval; German: Große hufeisennase; Spanish: Murciélago grande de herradura.

PHYSICAL CHARACTERISTICS
Head and body length 2.2–3.1 in (5.6–7.9 cm); tail 1.2–1.7 in (3–4.2 cm); forearm 2–2.4 in (5.1–6.2 cm); wingspan 13.8–15.6 in (35–40 cm); weight 0.5–1.2 oz (13–34 g). Upperparts gray-brown, tinged reddish; underparts gray-white to yellow-white; wings light gray-brown; juvenile grayer on upperparts.

DISTRIBUTION
Throughout southern Palearctic, from Great Britain across central and southern Europe, North Africa, Arabia, and southern Asia east to southern China and Japan.

HABITAT
Associated with a mixture of pasture, scrub, and woodland. Lives in caves and mine tunnels; has also adapted to larger buildings.

BEHAVIOR
Colonial; both sexes occur together in diurnal roosts; females form separate colonies at parturition; often moves to more sheltered roosts for winter. Hangs from toes, wraps wings round body. Hibernates September/October–April, in caves and tunnels, at 44.6–50°F (7–10°C); body temperature 46.4°F (8°C) in hibernation, 104°F (40°C) when active. Has rather deep, chirping or scolding calls.

FEEDING ECOLOGY AND DIET
Flies at dusk; flight slow and fluttering, with short glides; usually flies low. Takes small to large insects, especially cockchafers, dung beetles, grasshoppers, and moths. Sometimes hangs from perch, and pursues passing prey; can take food from ground. Forages up to 9 mi (15 km) from roost, often only within 1.2–1.9 mi (2–3 km) of roost.

REPRODUCTIVE BIOLOGY
Females mature at 3–4 years, males at 2–3 years. Mates autumn to spring; ovum implanted in spring; gestation about 9 weeks, single young born June–July (Europe). Young fly at 3 weeks, independent at 7–8 weeks. Lifespan up to 30 years. Species thought to be polygynous.

CONSERVATION STATUS
Lower Risk/Near Threatened, but declining rapidly, especially in Europe, through disturbance of roosts, vandalism, loss of

prey through habitat changes, and use of insecticides. In Great Britain, population has crashed, with only 1–2% (about 4,000 individuals) surviving in 1995.

SIGNIFICANCE TO HUMANS
None known. ◆

Cape horseshoe bat
Rhinolophus capensis

TAXONOMY
Rhinolophus capensis Lichtenstein, 1823, Cape of Good Hope, South Africa.

OTHER COMMON NAMES
None known.

PHYSICAL CHARACTERISTICS
Head and body length 2.3–2.4 in (5.8–6.2 cm); tail 0.9–1.3 in (2.4–3.2 cm); forearm 1.9–2.1 in (4.8–5.2 cm); weight not recorded. Upperparts dark brown, hairs cream basally; underparts light fawn-gray; wings dark brown.

DISTRIBUTION
Coastal belt of western and southern South Africa.

HABITAT
Coastal habitats associated with caves.

BEHAVIOR
Gregarious; roosts in caves and disused mines, hanging in clusters. Hibernates in winter. Migrations of 6.2 mi (10 km) noted.

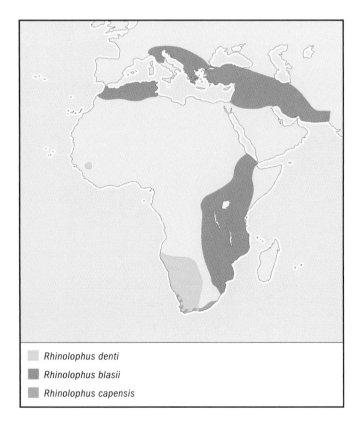

Rhinolophus denti
Rhinolophus blasii
Rhinolophus capensis

FEEDING ECOLOGY AND DIET
Eats mainly beetles; feeds on the wing, flying slowly in dense vegetation; also perch-hunts; may also glean.

REPRODUCTIVE BIOLOGY
Spermatogenesis occurs October–May; male stores sperm. Copulation and ovulation in spring (August–September); gestation 3–4 months; single young born November–December. Young cling to mothers during day. Probably polygynous.

CONSERVATION STATUS
Locally common to abundant, with roosts of thousands recorded, but considered Vulnerable because of its restricted distribution and relatively few suitable underground roosts.

SIGNIFICANCE TO HUMANS
None known. ◆

Dent's horseshoe bat
Rhinolophus denti

TAXONOMY
Rhinolophus denti Thomas, 1904, Kuruman, Cape Province, South Africa.

OTHER COMMON NAMES
None known.

PHYSICAL CHARACTERISTICS
Head and body length 1.7–2.2 in (4.3–5.7 cm); tail 0.75–0.9 in (1.9–2.3 cm); forearm 1.6–1.7 in (4–4.4 cm); mean weight 0.2 oz (6.2 g). Upperparts pale gray, pale brown, or pale cream; underparts off-white; wings pale brown, edged in white.

DISTRIBUTION
Namibia south to northwestern Botswana and west-central South Africa; isolated population in Guinea (West Africa).

HABITAT
Arid habitats with caves or rock outcrops.

BEHAVIOR
Gregarious; roosts in clusters in caves, crevices, and rocky caverns, and also under thatched roofs.

FEEDING ECOLOGY AND DIET
Insectivorous; feeding habits not described.

REPRODUCTIVE BIOLOGY
Species is probably polygynous; nothing else is known.

CONSERVATION STATUS
Not threatened, but poorly known.

SIGNIFICANCE TO HUMANS
None known. ◆

Blasius's horseshoe bat
Rhinolophus blasii

TAXONOMY
Rhinolophus blasii Peters, 1866, Italy.

OTHER COMMON NAMES
English: Peak-saddle horseshoe bat; French: Rhinolophe de Blasius; German: Blasius hufeisennase.

PHYSICAL CHARACTERISTICS
Head and body length 1.7–2.5 in (4.4–6.4 cm); tail 0.8–1.4 in (2–3.5 cm); forearm 1.6–2 in (4.1–5 cm); wingspan 10.2–12.2 in (26–31 cm); weight 0.3–0.6 oz (7.5–16 g). Upperparts gray-brown; underparts almost white; wings broad, brown.

DISTRIBUTION
Italy to Palestine, Arabia, Iran, Afghanistan, and Pakistan (one specimen); Morocco to Tunisia; Eritrea and Ethiopia; eastern Democratic Republic of the Congo and Zambia south to eastern South Africa.

HABITAT
Savanna woodland, open and scrub habitats. Roosts in caves and mines.

BEHAVIOR
Roosts in small to large groups; hangs freely. Presumed sedentary.

FEEDING ECOLOGY AND DIET
Insectivorous; feeding not described.

REPRODUCTIVE BIOLOGY
In Europe, cave nurseries occupied by up to 200 females. Single young; probably polygynous.

CONSERVATION STATUS
Lower Risk/Near Threatened. Poorly known and possibly overlooked, but threatened by disturbance and destruction in caves.

SIGNIFICANCE TO HUMANS
None known. ◆

Rhinolophus euryale

Rhinolophus hipposideros

bernates in caves and mine tunnels, temperature around 50°F (10°C). Has deep chirping, squeaking, or scolding calls. Usually sedentary.

FEEDING ECOLOGY AND DIET
Leaves roost at late dusk; hunts low over ground on warm hillsides and also in tree cover and scrub. Flight slow, fluttering; can hover. Eats moths and other insects; often uses feeding sites.

REPRODUCTIVE BIOLOGY
Poorly known. Nursery roosts may contain 50–100 females, with males also present. Young fly from mid-July; females also pregnant at the same time. Thought to be polygynous.

CONSERVATION STATUS
Vulnerable. Has declined in northern parts of range, particularly in France and Czechoslovakia, partly due to disturbance in caves.

SIGNIFICANCE TO HUMANS
None known. ◆

Mediterranean horseshoe bat
Rhinolophus euryale

TAXONOMY
Rhinolophus euryale Blasius, 1853, Milan, Italy.

OTHER COMMON NAMES
French: Rhinolophe euryale; German: Mittelmeerhufeisennase.

PHYSICAL CHARACTERISTICS
Head and body length 1.5–2.3 in (3.7–5.8 cm); tail 0.9–1.3 in (2.2–3.3 cm); forearm 1.7–2 in (4.2–5.1 cm); wingspan 11.4–12.8 in (29–32.5 cm); weight 0.3–0.6 oz (8–18 g). Upperparts gray-brown, tinged reddish or lilac; underparts gray-white to yellowish white; wings light gray; juveniles gray.

DISTRIBUTION
Mediterranean region of Europe and North Africa; Balkan peninsula; east to Iran and Turkmenistan.

HABITAT
Well-wooded country close to water, with caves.

BEHAVIOR
Colonial; hang free, often with bodies in contact, embracing each other with wing membranes and licking each other's faces and heads. Often roosts with other horseshoe bat species. Hi-

---

Lesser horseshoe bat
Rhinolophus hipposideros

TAXONOMY
Vespertilio hipposideros (Bechstein, 1800), France.

OTHER COMMON NAMES
French: Petit rhinolophe fer à cheval; German: Kleine Hufeisennase.

PHYSICAL CHARACTERISTICS
Head and body length 1.5–1.8 in (3.7–4.7 cm); tail 0.9–1.3 in (2.3–3.3 cm); forearm 1.4–1.7 in (3.5–4.3 cm); wingspan 7.6–10 in (19–25.5 cm); weight 0.1–0.4 oz (4–10 g). Small and delicate. Upperparts smoky brownish; underparts gray to gray-white; wings light gray-brown; juvenile dark gray.

DISTRIBUTION
British Isles to Arabian peninsula and central Asia (east to Kashmir); northern Africa from Morocco to Sudan, Ethiopia and Eritrea.

HABITAT
Pasture, woodland edge, forest, and wetlands. Lives in caves, mines, and buildings.

BEHAVIOR
Colonial; summer roosts often in buildings (Europe); colonies show some daytime activity. Hibernates in underground roosts (e.g., caves, tunnels) September/October–April, at 42.8–48.2°F (6–9°C) (Europe). Sedentary; moves 3.1–6.2 mi (5–10 km) between summer and winter roosts. Has chirping or scolding calls.

FEEDING ECOLOGY AND DIET
Flight skillful, fairly fast, with almost whirring wings; flies low. Catches insects, especially moths, mosquitoes, craneflies, and beetles; also spiders. Feeds mainly in woodlands; sometimes takes prey from stones and branches. Often eats prey at night perch.

REPRODUCTIVE BIOLOGY
Matures in first year. Mates in autumn, sometimes in winter; moves to nurseries from April; males also occur in nurseries. Single young born June–July (Europe). Eyes open at around 10 days, independent at 6–7 weeks. Average lifespan 4 years; 21 years recorded. Thought to be polygynous.

CONSERVATION STATUS
Lower Risk/Near Threatened, with European populations generally in decline, sometimes seriously; locally extinct in some northern areas. Threatened by disturbance and destruction of roosts, and use of insecticides; climatic changes may have affected northern populations.

SIGNIFICANCE TO HUMANS
None known. ◆

Resources

Books

Bates, P. J., and D. L. Harrison. *Bats of the Indian Subcontinent*. Sevenoaks, UK: Harrison Zoological Museum, 1997.

Churchill, S. *Australian Bats*. Sydney: Reed New Holland, 1998.

Corbett, G. B., and J. E. Hill. *The Mammals of the Indomalaysian Region*. Oxford: Oxford University Press, 1992.

Hill, J. E., and J. D. Smith. *Bats: A Natural History*. London: British Museum (Natural History), 1984.

Koopman, K. F. "Order Chiroptera." In *Mammal Species of the World*, edited by D. E. Wilson and D. M. Reeder. Washington, DC: Smithsonian Institution Press, 1983.

Kunz, T. H., and P. A. Racey, eds. *Bat Biology and Conservation*. Washington, DC: Smithsonian Institution Press, 1998.

Nowak, Ronald M. *Walker's Mammals of the World*. 6th ed. Baltimore: The Johns Hopkins University Press, 1999.

Schober, W., and E. Grimmberger. *Bats of Britain and Europe*. London: Hamlyn, 1993.

Skinner, R., and R. H. N. Smithers. *The Mammals of the Southern African Subregion*. 2nd ed. Pretoria, South Africa: University of Pretoria, 1998.

Periodicals

Bogdanowicz, W., and R. D. Owen. "Phylogenetic Analyses of the Bat Family Rhinolophidae." *Zeitschrift für zoologische Systematik und Evolutionsforschung* 30 (1992): 142–160.

Organizations

Australian Museum. 6 College Street, Sydney, New South Wales 2010 Australia. Phone: (2) 9320 6000. Web site: <http://www.amonline.net.au>

Bat Conservation International. P.O. Box 162603, Austin, TX 78716 USA. Phone: (512) 327-9721. Fax: (512) 327-9724. E-mail: batinfo@batcon.org Web site: <http://www.batcon.org>

Eurobats. Martin-Luther-King Str. 8, Bonn, D-53175 Germany. E-mail: eurobats@uno.de Web site: <http://www.eurobats.org/>

IUCN Species Survival Commission, Chiroptera Specialist Group. Web site: <http://www.iucn.org>

Barry Taylor, PhD

Old World leaf-nosed bats
(Hipposideridae)

Class Mammalia
Order Chiroptera
Suborder Microchiroptera
Family Hipposideridae

Thumbnail description
This large family is characterized by elaborate modifications of the nose and muzzle, forming leaf-like projections that are thought to help focus echolocation signals emitted through the nose

Size
Range in size from small to very large, with head and body lengths of 1.1–4.3 in (28–110 mm), and forearms a similar length

Number of genera, species
9 genera; 66 species

Habitat
Occupy a variety of habitats, from arid to tropical areas throughout much of the Old World

Conservation status
Critically Endangered: 2 species; Endangered: 1 species; Vulnerable: 15 species; Lower Risk/Near Threatened: 23 species; Data Deficient: 7 species

Distribution
Old World tropics and subtropics

Evolution and systematics

Similar in diversity, size, and characteristics to horseshoe bats (family Rhinolophidae), Old World leaf-nosed bats are sometimes considered a subfamily (Hipposiderinae, but sometimes referred to as the subfamily Rhinonycterinae) of Rhinolophidae. Recent analyses of large molecular and morphological datasets place Hipposideridae in a large group along with the bat families Craseonycteridae, Rhinopomatidae, Nycteridae, Megadermatidae, and Rhinolophidae, all of which are exclusively Old World in distribution.

A recent study employing primarily nuclear sequence data suggests that the group, including Hipposideridae (with the exception of Nycteridae), is more closely related to flying foxes (Megachiroptera) than to other microbats (Microchiroptera). This study also provides some morphological support for their assertions. If this finding is supported by additional evidence and analyses, it would mean that either the general characteristics associated with microbats (echolocation and various morphological features) have evolved twice independently or that they have been lost in the megachiropteran lineage.

A systematic analysis, based on morphological features, of relationships among bats in the family Hipposideridae suggested that the genus *Hipposideros* is actually composed of

three distinct groups, which may deserve generic status. In general, relationships among hipposiderid bats are poorly understood and additional research is needed to fully understand the pattern of evolutionary history in this group. These bats are known from the Eocene to Oligocene in the fossil record.

Physical characteristics

Bats in this family vary greatly in size, from small to very large. Head and body lengths are 1.1–4.3 in (2.8–11 cm) and forearms are a similar length. One of the largest insectivorous bat species is a hipposiderid, Commerson's leaf-nosed bat (*Hipposideros commersoni*). They are characterized by their elaborate, leaf-like nose leaves, which are composed of an anterior, horseshoe-shaped portion and a posterior portion that is often lobed. Lateral leaflets are also present in many species. These elaborate facial appendages seem to be related to their use of nasal echolocation, where nose leaves act to focus and modify emitted echolocation signals. The ears of these bats vary in size but always lack a tragus, the anterior ear appendage found in many microbats. Members of the families Hipposideridae and Rhinolophidae share a unique feature of the premaxillary bones, whereby premaxillae on either side of the skull are not fused to each other or to the maxillae. Fur coloration is generally shades of brown and red. Tail length ranges from zero (nonexistant) to 2.4 in (6 cm).

A diadem roundleaf bat (*Hipposideros diadema*) hanging in a rainforest. (Photo by B. G. Thomson. Reproduced by permission.)

A Commerson's leaf-nosed bat (*Hipposideros commersoni*) roosting. (Photo by Harald Schütz. Reproduced by permission.)

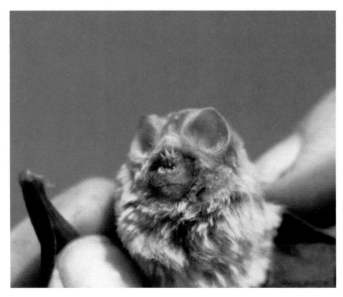

Percival's trident-nosed bat (*Cloeotis percivali*) has three pointed protrusions that extend from its nose. (Photo by Brock Fenton. Reproduced by permission.)

Their teeth are much like those of other insectivorous bats, the dental formula is I1/2 C1/1 P1–2/2–3 M3/3.

Distribution

Hipposiderids are found mainly in tropical and subtropical regions of the Old World, including Africa, Asia, and Australia.

Habitat

These bats are found in a variety of habitats, from deserts to tropical rainforests. Most are found in moist, lowland areas.

Behavior

The majority of species in the family Hipposideridae are only poorly known. They are a diverse group and exhibit a variety of life history strategies, social structures, and behaviors. Some northern populations of this family hibernate, such as some of the genus *Hipposideros*, while others are active year-round, and

A roundleaf horseshoe bat (*Hipposideros larvatus*). (Photo by Gerry Ellis/Minden Pictures. Reproduced by permission.)

A Sundevall's roundleaf bat (*Hipposideros caffer*) roosting in east Africa. (Photo by David Hosking/FLPA–Images of Nature. Reproduced by permission.)

The orange form of a dusky leaf-nosed bat (*Hipposideros ater*) flying out of an old mine. (Photo by B. G. Thomson. Reproduced by permission.)

A northern leaf-nosed bat (*Hipposideros stenotis*) roosting in a cave. (Photo by © Jiri Lochman/Lochman Transparencies. Reproduced by permission.)

A northern leaf-nosed bat (*Hipposideros stenotis*). (Photo by B. G. Thomson. Reproduced by permission.)

A Commerson's leaf-nosed bat (*Hipposideros commersoni*). (Photo by © Merlin D. Tuttle, Bat Conservation International. Reproduced by permission.)

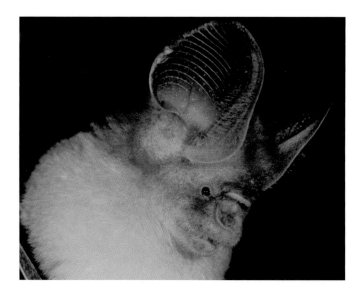

Hipposideros bicolor fly lower than most bats to feed on ground-dwelling insects. (Photo by Brock Fenton. Reproduced by permission.)

A dusky leaf-nosed bat (*Hipposideros ater*) using its claws to hang upside down. (Photo by B. G. Thomson. Reproduced by permission.)

only one species is thought to migrate. Most roost in groups varying in size from small (as few as 12) to very large (5,000) congregations, though some are solitary. Roosting often occurs in caves and tunnels, but some species also roost in hollow trees, human structures, and the burrows of animals. The fulvus roundleaf bat (*Hipposideros fulvus*) roosts in African porcupine (*Hystrix*) burrows.

Many hipposiderid species have a small sac that sits behind the nose leaf. The sac secretes a waxy substance and is mainly found in males, suggesting the possibility that it is used in social or reproductive interactions for attracting mates or for male competiton.

Feeding ecology and diet

Hipposiderids are generally insectivorous. Little information is available on the specific diets of most species, although most seem to capture insects in flight. Many species return to a roost to eat captured prey.

These bats seem to fly with their mouths closed, emitting ultrasonic pulses through the nose. The many and various modifications of the nose leaf are presumably adaptations to specific modes and frequencies of nasal echolocation. Indeed, some evidence supports a proportional relationship between call frequency and nose leaf width in several species of hipposiderids. This relationship was independent of both body size and generic affiliation. Where studied, echolocation calls tend to be constant frequency and frequency modulated, with frequencies potentially varying with sex and age of individuals.

Reproductive biology

In general, very little is known about reproduction in these bats, but they may be polygynous. Females typically give birth to a single young each year, with breeding occurring seasonally. Hipposiderids mate during the fall and females store the sperm internally during the winter, becoming fertilized and giving birth the following year. Age of weaning, age at first flight, and age at sexual maturity may be correlated with latitude, with tropical species taking longer than temperate species to achieve all of the above landmarks. In *Hipposideros*, eight tropical species reached sexual maturity at 16–24 months, one subtropical species at less than 12 months, and two temperate species at 6–8 months. The subtropical species (*H. terasensis*) is weaned at seven weeks old, while tropical species are weaned at 8–20 weeks old.

Conservation status

The majority of bats in this large family are poorly studied, so potential conservation concerns are unknown. Some species are common, others are rare and restricted to islands, making them especially vulnerable to threats such as habitat and roost destruction. The IUCN lists two species as Critically Endangered, one as Endangered, 15 as Vulnerable, 23 as Lower Risk/Near Threatened, and 7 as Data Deficient.

Significance to humans

These primarily insectivorous bats help to control insect pest populations throughout their ranges. Their dung may also be used locally as a fertilizer.

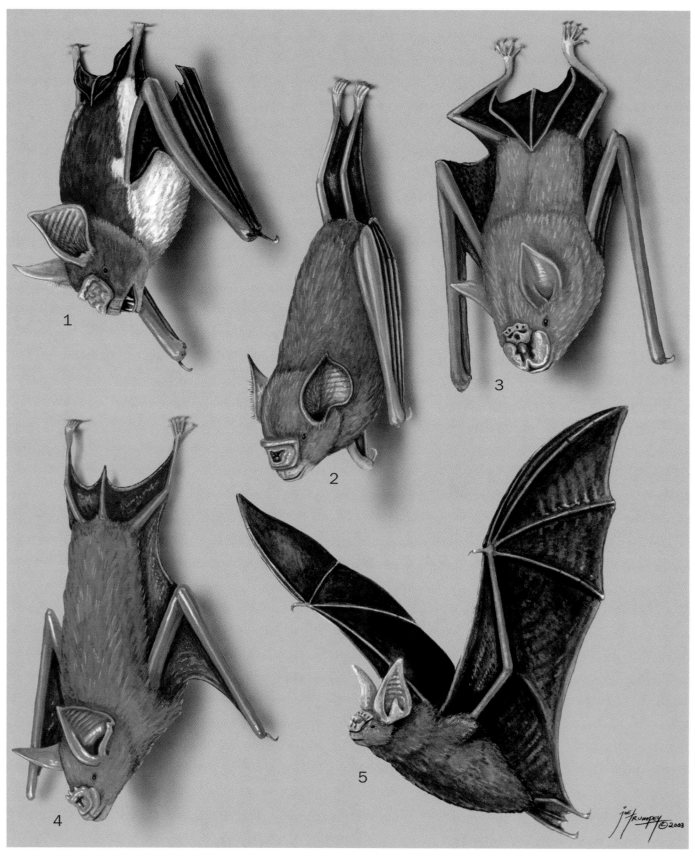

1. Diadem roundleaf bat (*Hipposideros diadema*); 2. Noack's roundleaf bat (*Hipposideros ruber*); 3. Golden horseshoe bat (*Rhinonicteris aurantia*); 4. Trident leaf-nosed bat (*Asellia tridens*); 5. Fulvus roundleaf bat (*Hipposideros fulvus*). (Illustration by Joseph E. Trumpey)

Species accounts

Trident leaf-nosed bat
Asellia tridens

TAXONOMY
Asellia tridens (Geoffroy, 1813), Egypt.

OTHER COMMON NAMES
None known.

PHYSICAL CHARACTERISTICS
Medium-sized bats. Total body length 1.8–4.2 in (46–62 mm); forearm length 1.7–2.0 in (45–52 mm); tail length 0.7–2.0 in (18–27 mm); weight 0.2–0.35 oz (6–10 g). Fur coloration varies considerably, ranging from pale yellow to buffy gray and orange-brown, membranes are slightly darker, and the large ears and face are pale. The nose leaf is distinctive, having a large leaf behind the nostrils, with three toothed projections.

DISTRIBUTION
Found throughout Africa north of the Sahara desert, the Arabian Peninsula, and into Pakistan.

HABITAT
One of the most arid-adapted and common of bats in the areas in which they occur. They have been observed roosting in caves, wells, irrigation culverts, and other man-made structures. They are thought to travel between summer roosts and hibernation sites in caves and tombs. Foraging seems to occur in palm groves and other vegetated areas, the bats sometimes flying many miles (kilometers) across barren terrain to suitable feeding sites.

BEHAVIOR
A gregarious species, roosts of several hundred individuals have been routinely observed, with one roost numbering as many as 5,000. They fly in small groups when exiting roosts and fly close to the ground when traveling to and from roosts.

FEEDING ECOLOGY AND DIET
Eat primarily coleopterans (beetles) and hymenopterans (bees, ants, and wasps) that are taken during flight. They have been recorded foraging mainly in cluttered environments, implying that they are agile in flight and may be able to take prey from the ground and other surfaces.

REPRODUCTIVE BIOLOGY
Available evidence suggests that females are pregnant in the spring and give birth to a single young in early summer. Gestation time is estimated at 9–10 weeks, and the young nurse for 40 days after birth, after which they become independent. The time of mating is unknown. Most likely polygynous.

CONSERVATION STATUS
Common throughout their range, though disturbance to roosting sites could severely impact local populations.

SIGNIFICANCE TO HUMANS
Control insect pest populations throughout their range and are likely to benefit agricultural interests. ◆

Diadem roundleaf bat
Hipposideros diadema

TAXONOMY
Hipposideros diadema (Geoffroy, 1813), Indonesia.

OTHER COMMON NAMES
English: Large Malayan leaf-nosed bat.

PHYSICAL CHARACTERISTICS
Medium-sized bat. Weight 1.1–1.7 oz (34–50 g). The fur is dark brown to black and a white spot occurs at the juncture of the forearm and shoulder. The ears are large and the nose leaf is fairly simple, with an anterior horseshoe portion and a posterior, uninterrupted, ridge-like leaflet.

DISTRIBUTION
Found throughout Southeast Asia from Myanmar and Vietnam through Thailand, western Malaysia, and Indonesia to New Guinea, the Bismarck Archipelago, Solomon Islands, Philippines, Nicobar islands, and northeast and north-central Australia.

HABITAT
Occur primarily in moist, tropical forests.

BEHAVIOR
Little is known of behavior in this species. They are gregarious, roosting in groups in caves.

FEEDING ECOLOGY AND DIET
No detailed study of diet has been conducted. Insectivorous, likely to hunt from perches, capturing prey when it is detected by using echolocation.

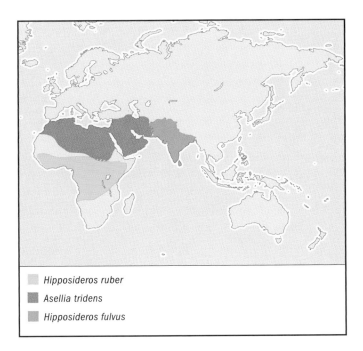

Hipposideros ruber
Asellia tridens
Hipposideros fulvus

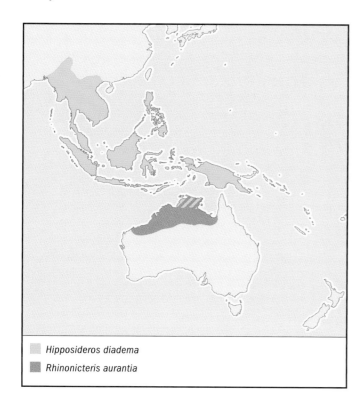

Hipposideros diadema

Rhinonicteris aurantia

REPRODUCTIVE BIOLOGY
Females congregate in March and April, giving birth to a single young during that time. Young probably are weaned and become able to fly between two and three months of age. Nothing is known of their mating behavior; but they are most likely polygynous.

CONSERVATION STATUS
Not threatened. They are widespread and common throughout their range.

SIGNIFICANCE TO HUMANS
Consume insect pests. ◆

Fulvus roundleaf bat
Hipposideros fulvus

TAXONOMY
Hipposideros fulvus Gray, 1838, India.

OTHER COMMON NAMES
English: Fulvous leaf-nosed bat.

PHYSICAL CHARACTERISTICS
Medium-sized bats, with simple nose leaves.

DISTRIBUTION
South Asia, from Pakistan and Vietnam south to Sri Lanka.

HABITAT
Tropical, lowland forested habitats.

BEHAVIOR
Found roosting in the burrows of porcupines (*Hystrix*). Both smaller (10–20) and larger (50–100) colonies have been observed in caves.

FEEDING ECOLOGY AND DIET
Insectivorous, though detailed studies of their diet have not been conducted.

REPRODUCTIVE BIOLOGY
In India, they mate in mid November and give birth to single young in late April and early May. The gestation period is 150–160 days. The female carries her young for 20–22 days and sexual maturity is reached at 18–19 months old. A banded female was captured at an age of at least 12 years and was pregnant at the time of capture. This species is probably polygynous.

CONSERVATION STATUS
Has not been evaluated by the IUCN. They are common throughout their range, though population trends are unknown.

SIGNIFICANCE TO HUMANS
Consume insect pests. ◆

Noack's roundleaf bat
Hipposideros ruber

TAXONOMY
Hipposideros ruber (Noack, 1893), Tanzania.

OTHER COMMON NAMES
English: Noack's African leaf-nosed bat.

PHYSICAL CHARACTERISTICS
These bats have a simple nose leaf; the anterior and posterior leaflets are simple in outline, without indentations. Their fur is long, silky, and light brown, and the ears are large and round.

DISTRIBUTION
Central Africa, from Senegal to Ethiopia and south to Angola, Zambia, Malawi, and Mozambique, including Bioko and São Tomé and Principe.

HABITAT
Tropical lowland areas.

BEHAVIOR
Use a constant frequency echolocation call that differs between the sexes. Some evidence suggests that frequencies used correspond with body condition indices, making it possible for females to discern which mates are in the best condition. Frequency was also found to change with humidity. Little is known about roosting and social behaviors.

FEEDING ECOLOGY AND DIET
Feed on insects that are captured from hunting perches when detected by echolocation.

REPRODUCTIVE BIOLOGY
Reproduction dynamics are unknown in this species; but they are most likely polygynous.

CONSERVATION STATUS
Has not been evaluated by the IUCN. They are common throughout their range, although population trends are unknown.

SIGNIFICANCE TO HUMANS
Controls insect pests. ◆

Golden horseshoe bat

Rhinonicteris aurantia

TAXONOMY

Rhinonicteris aurantia (Gray, 1845), Northern Territory, Australia.

OTHER COMMON NAMES

Orange leaf-nosed bat.

PHYSICAL CHARACTERISTICS

Head and body length 1.7–2.0 in (45–53 mm); forearm length 1.8–1.9 in (47–50 mm); tail length 0.9–1.1 in (24–28 mm; weight 0.28–0.35 oz (8–10 g). The fur is long, silky, and typically bright orange above and paler below, though specimens vary from dark brown to almost white. Characterized by their unique nose leaf; the anterior portion is a divided horseshoe and the posterior portion is deeply indented, making it appear a bit like lace.

DISTRIBUTION

Across northern Australia, including north Western Australia, Northern Territory, and northwestern Queensland.

HABITAT

Found in a variety of habitats, from open grassland to dense forests. They roost in hot, humid mines and caves.

BEHAVIOR

Do not hibernate and seem unable to become torpid. They become lethargic at temperatures below 68°F (20°C). They emerge up to 1.5 hours after sunset and forage up to 9.8 ft (3 m) above vegetation. Their flight is described as somewhat faster than most hipposiderids. They congregate in roosts of several hundred individuals, with up to 5,000 being observed in a single roost.

FEEDING ECOLOGY AND DIET

Eat primarily moths.

REPRODUCTIVE BIOLOGY

Breeding occurs during the wet season, from October–April. Probably polygynous.

CONSERVATION STATUS

Considered Vulnerable by the IUCN, populations seem to be in decline. Several important roosts have suffered from human disturbance. Cave gating has led to modest increases in the populations of previously declining roosts.

SIGNIFICANCE TO HUMANS

Controls insect pests. ◆

Common name / Scientific name/ Other common names	Physical characteristics	Habitat and behavior	Distribution	Diet	Conservation status
Persian trident bat *Triaenops persicus*	Coat color highly variable, ranging from brown to gray to reds, with some individuals being almost white. The anterior portion of the nose leaf is similar to *Rhinonicteris*, which is thought to be closely related. The posterior portion of the nose leaf, however, is tridentate. Head and body length 1.4–2.4 in (3.5–6.2 cm), tail length 0.8–1.3 in (2.0–3.4 cm), and forearm length 1.8–2.2 in (4.5–5.5 cm), adults weigh 0.3–0.5 oz (8–15 g).	These bats roost in smaller groups, often associated with larger groups of other species. They emerge early in the evening, before dusk, to forage and typically fly close to the ground. They have been observed roosting in caves and underground tunnels. Births have been recorded in January, but may occur at other times as well.	Africa, including Somalia, Ethiopia, Eritrea, Kenya, Tanzania Uganda, Angola, Zanzibar, Mozambique, the Congo Republic, and Egypt. The Arabian Peninsula, including Oman, Yemen, and Iran.	Insects.	Not listed by IUCN
Flower-faced bat *Anthops ornatus*	This species is known only from a few specimens. The fur is gray-buff, long, and silky. The tail is much reduced, otherwise this species is similar to *Hipposideros* species. The nose leaf is elaborate, with a tri-lobed posterior portion. Head and body length about 2 in (5 cm), forearm length 1.9–2.0 in (4.8–5.1 cm).	Nothing known.	Solomon Islands.	Presumably insects, though no information is available.	Vulnerable
East Asian tailless leaf–nosed bat *Coelops frithi*	Fur is variable, from brown to black. Ears are short and rounded and the nose leaf lacks lateral leaflets. Head and body length 1.1–2.0 in (2.8–5.0 cm) and forearm length 1.3–1.9 in (3.3–4.7 cm), weight 0.25–0.32 oz (7–9 g).	These bats have been recorded roosting in small numbers in caves, human structures, and hollow trees. They may hibernate in caves. Young have been recorded in March, though births may occur at other times of the year as well.	South and Southeast Asia, from India to southern China and Vietnam, and into Malaysia, Taiwan, Java, and Bali.	Presumably insects, though no information is available.	Not listed by IUCN

[continued]

Common name / Scientific name/ Other common names	Physical characteristics	Habitat and behavior	Distribution	Diet	Conservation status
Temminck's trident bat *Aselliscus tricuspidatus*	Fur is bright brown above and buffy below. There are lateral nose leaflets on either side of the anterior portion of the nose leaf and the posterior portion is divided into three, bluntly toothed, portions. The tail extends beyond the tail membrane. Smaller bats, head and body length 1.5–1.8 in (3.8–4.5 cm), forearm length 1.3–1.8 in (3.5–4.5 cm), tail length 0.8–1.6 in (2.0–4.0 cm). Weights have been recorded at 0.1–0.14 oz (3.5–4 g).	These bats roost in caves. They hang singly and evenly spaced from other bats, but in distinct groups of usually 40 to 50 bats. Females were recorded both pregnant with single young and lactating in May and June.	Pacific islands, including the Moluccas, New Guinea, Bismarck Archipelago, Solomon Islands, Vanuatu, and adjacent islands.	Presumably insects, though no information is available.	Not listed by IUCN
Percival's trident bat *Cloeotis percivali*	Two subspecies are recognized based on fur coloration, which varies from sooty to grayish brown and dark to buffy brown, with lighter tips. The tail is well-developed and the nose leaf has three, distinctly pointed processes arising from the posterior nose leaf. The ears are small and round. Head and body length 1.3–2.0 in (3.3–5.0 cm), forearm length 1.2–1.4 in (3.0–3.6 cm), tail length 0.9–1.3 in (2.2–3.3 cm). These small bats weigh 0.13–0.21 oz (3.8–5.9 g).	These bats roost in large numbers in narrow-mouthed caves. Individuals roost in tight clusters. Pregnant females have been recorded in October.	Sub-Saharan Africa, including Kenya, Tanzania, Zaire, Mozambique, Zambia, Zimbabwe, Botswana, Swaziland, and the Transvaal of South Africa.	In some areas these bats have been reported to eat almost exclusively moths.	Lower Risk/Near Threatened
Ashy roundleaf bat *Hipposideros cineraceus*	These bats weigh 0.25–0.28 oz (7–8 g), making them smaller members of the genus *Hipposideros*	A gregarious species, they can be found roosting in large numbers. Reproduction is seasonal, with mating in October, births occurring in April, and the young capable of flight by June.	South and Southeast Asia, from Pakistan and northern India to Vietnam, Borneo, and adjacent small islands. May occur in the Philippines.	Little information available, presumably insectivorous.	Not listed by IUCN
Commerson's leaf-nosed bat *Hipposideros commersoni* English: Giant leaf-nosed bat	This is one of the largest insectivorous bats worldwide, weighing from 2.5 to 3.5 oz (74–100 g).	These bats roost in large numbers of several hundred bats. Pregnant females have been recorded in August and October.	Africa, from Gambia and Ethiopia south to South Africa, and Madagascar.	These bats are sit-and-wait predators, hanging from a perch and flying to intercept large prey (up to 2.6 in [60 mm] long) that are detected via echolocation. Prey are detected at distances up to 65 ft (20 m) and are taken back to the perch for consumption. They have been observed eating beetle larvae inside wild figs and incidentally take fig pulp as well.	Not listed by IUCN
Ridley's roundleaf bat *Hipposideros ridleyi* English: Singapore roundleaf horseshoe bat; Ridley's leaf-nosed bat	No specific information available.	Little information is available. Survives only in small patches of heavily logged lowland peat forest. Females give birth in April.	Western Malaysia and Borneo.	Unknown, presumably insects.	Vulnerable

Resources

Books

Bogdanowicz, Wieslaw, and Robert D. Owen. "In the Minotaur's Labryinth: Phylogeny of the Bat Family Hipposideridae." In *Bat Biology and Conservation*, edited by Thomas H. Kunz and Paul A. Racey. Washington, DC: Smithsonian Institution Press, 1998.

Hill, John E., and James D. Smith. *Bats, A Natural History*. London: British Museum of Natural History, 1984.

Koopman, Karl F. "Chiroptera: Systematics." In *Handbook of Zoology*. Volume VIII, *Mammalia*, edited by J. Niethammer, H. Schliemann, and D. Starck. Berlin: Walter de Gruyter, 1994.

McKenna, Malcolm C., and Susan K. Bell, eds. *Classification of Mammals Above the Species Level*. New York: Columbia University Press, 1997.

Nowak, Ronald M. *Walker's Bats of the World*. Baltimore: Johns Hopkins University Press, 1994.

Resources

Simmons, Nancy B., and Jonathan H. Geisler. "Phylogenetic Relationships of *Icaronycteris*, *Archaeonycteris*, *Hassianycteris*, and *Palaeochiropteryx* to Extant Bat Lineages, with Comments on the Evolution of Echolocation and Foraging Strategies in Microchiroptera." *Bulletin of the American Museum of Natural History, Number 235.* New York: American Museum of Natural History, 1998.

Periodicals

Brosset, A. "Recherches sur la Biologie des Chiropteres Troglophiles dans le Nord-est du Gabon." *Biologie du Gabon* 5 (1969): 93–116.

Cheng, H. C., and L. L. Lee. "Postnatal Growth, Age Estimation, and Sexual Maturity in the Formosan Leaf-nosed Bat (*Hipposideros terasensis*)." *Journal of Mammalogy* 83 (2002): 785–793.

Churchill, S. H. "Reproductive Ecology of the Orange Horseshoe Bat, *Rhinonycteris aurantius* (Chiroptera: Hipposideridae), a Tropical Cave-dweller." *Wildlife Research* 22 (1995): 687–698.

Habersetzer, J., and G. Marimuthu. "Ontogeny of Sounds in the Echolocating Bat *Hipposideros speoris.*" *Journal of Comparative Physiology, A. Sensory, Neural, and Behavioral Physiology* 158 (1986): 247–257.

Hill, J. E. "A Revision of the Genus *Hipposideros.*" *Bulletin of the British Museum of Natural History, Zoology* 11 (1963): 1–129.

Jones, G., M. Morton, P. M. Hughes, and R. M. Budden. "Echolocation, Flight Morphology, and Foraging Strategies of Some West African Hipposiderid Species." *Journal of the Zoological Society of London* 230 (1993): 385–400.

Robinson, Mark F. "A Relationship between Echolocation Calls and Nose-leaf Widths in Bats of the Genera *Rhinolophus* and *Hipposideros.*" *Journal of the Zoological Society of London* 239 (1996): 389–393.

Teeling, Emma C., Ole Madsen, Ronald A. Van Den Bussche, Wilfried W. de Jong, and Michael J. Stanhope. "Microbat Paraphyly and the Convergent Evolution of a Key Innovation in Old World Rhinolophoid Microbats." *Proceedings of the National Academy of Sciences* 99 (2002): 1431–1436.

Organizations

Gulf of Guinea Conservation Group. CP289, Sao Tome, Sao Tome e Principe. Phone: (230) 225428. E-mail: info@ggcg.st Web site: <http://www.ggcg.st/>

Tanya Dewey

American leaf-nosed bats
(Phyllostomidae)

Class Mammalia
Order Chiroptera
Suborder Microchiroptera
Family Phyllostomidae

Thumbnail description
Small to relatively large bats, most of which have a fleshy, triangular nose-leaf projecting above the nostrils

Size
Head and body length 1.6–5.3 in (40–135 mm); tail 0–2.2 in (0–55 mm); forearm 1.2–4.1 in (31–105 mm); weight 0.2–6.8 oz (5–190 g)

Number of genera, species
49 genera; 151 species

Habitat
Mainly forests, but also deserts

Conservation status
Extinct: 1 species; Endangered: 4 species; Vulnerable: 25 species; Lower Risk/Near Threatened: 36 species; Data Deficient: 5 species

Distribution
Subtropical and tropical America from southern Arizona and the West Indies to northern Argentina

Evolution and systematics

Members of this family are among the most common mammals in American tropical forests. Rainforests in Central and South America contain 31–49 species of phyllostomids; tropical forests receiving less rainfall contain 20–30 species; and dry regions contain two to three species. In addition to being one of the most taxonomically diverse families of bats, the Phyllostomidae is the most ecologically diverse group of bats. Food habits range from insects, flowers, and fruit to blood and other vertebrates. Sizes of social groups range from monogamous pairs to colonies containing several hundred thousand individuals.

Along with the two small American families, Noctilionidae (one genus, two species) and Mormoopidae (two genera, eight species), the Phyllostomidae is classified in superfamily Noctilionoidea of suborder Microchiroptera. Food habits of noctilionids (also known as bulldog bats) include insects and fish, while mormoopids (also known as moustached or leaf-chinned bats) are insectivorous. These two families are unknown in the fossil record other than from Pleistocene deposits. The earliest known fossil, phyllostomid (*Notonycteris*), comes from

the Miocene of Colombia (about 20 million years ago). It is likely that each of these families evolved in South America. All three families occur in the West Indies as well as in mainland subtropical and tropical America.

Reflecting its substantial dietary diversity, the Phyllostomidae is currently classified into eight subfamilies containing a total of 49 genera and 151 species. Subfamilies and their diversity include: Phyllostominae (11 genera, 37 species), Lonchophyllinae (3, 9), Brachyphyllinae (1, 2), Phyllonycterinae (2, 4), Glossophaginae (10, 23), Carolliinae (2, 7), Stenodermatinae (17, 66), and Desmodontinae (3, 3). Phyllostomines include insectivores and carnivores; lonchophyllines, phyllonycterines (a West Indian group), and glossophagines are nectarivores; brachyphyllines (a West Indian group), carolliinines, and stenodermatines are frugivores; and desmodontines are blood-feeding vampires.

Physical characteristics

Sizes and facial characteristics of American leaf-nosed bats differ substantially among the subfamilies and reflect their di-

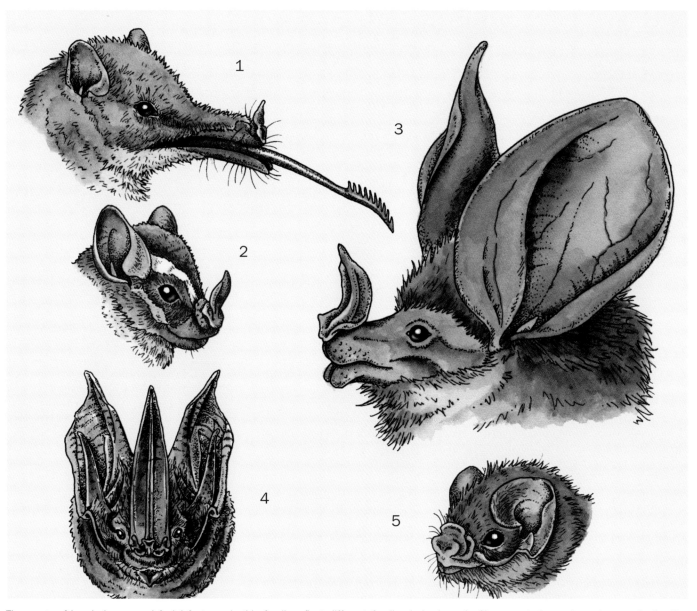

The range of head shapes and facial features in this family reflect different feeding behaviors. 1. *Choeronycteris mexicana* is a nectar-feeding bat with a long snout and bristled tongue; 2. *Platyrrhinus helleri* is a fruit-eating bat with a short snout; 3. *Chrotopterus auritus* is carnivorous, and 4. *Lonchorhina aurita* is an insect gleaner; both have large ears and a nose leaf; 5. *Diphylla ecaudata* is a blood drinker with a short snout and stubby nose leaf. (Illustration by Gillian Harris)

verse feeding adaptations. Phyllostomines include the largest members of the family (3.2–6.7 oz; 90–190 g) and generally have the longest nose leaves. The nose leaf of the sword-nosed bat (*Lonchorhina aurita*), for example, is as long as its long, pointed ears. Large members of this subfamily have robust canines and molars for killing and chewing vertebrate prey. At the other end of the size spectrum are flower-visiting bats (Lonchophyllinae and Glossophaginae), which weigh 0.2–0.8 oz (7–25 g). These bats have elongated muzzles, small nose-leaves, long tongues, and dentition that are reduced in size and number. Fruit-eating bats (Stenodermatinae) weigh 0.2–2.8 oz (5–80 g), have medium-sized nose leaves, and generally have flattened faces with dagger-like canines and broad cheek teeth for grabbing and crushing fruit. Vampire bats

(Desmodontinae) weigh 0.2–1.7 oz (20–50 g) and have much-reduced nose leaves and a reduced number of morphologically specialized teeth. The two upper incisors are sharp and chisel-like for making incisions in the skin of mammals or birds.

The fur color of phyllostomid bats is generally brown or gray, but one species (*Ectophylla alba*) is white. Many genera of stenodermatines, which often roost in foliage by day, have white facial stripes, and a few have a white mid-dorsal stripe.

The nose-leaves that give this family its common name are also found in several families of Old World Microchiroptera (e.g., Nycteridae, Rhinolophidae, Hipposideridae, and Megadermatidae). All of these bats emit echolocation sounds

A sword-nosed bat (*Lonchorhina aurita*) hangs from a tree branch. (Photo by Merlin D. Tuttle/Bat Conservation International/Photo Researchers, Inc. Reproduced by permission.)

through their nostrils rather than through their mouths. It is thought that these structures serve as an acoustic lens that focuses the outgoing sound into a narrow beam.

Distribution

Phyllostomid bats are currently distributed from the southwestern United States and the West Indies south to northern Argentina and central Chile. Two subfamilies (Phyllonycterinae and Brachyphyllinae) are restricted to the West Indies, which generally lack members of the Phyllostominae and Carolliinae. Lonchophyllinae and Desmodontinae are absent from the West Indies, although the common vampire, *Desmodus rotundus* (Desmodontinae), is known as a fossil from Cuba. In the late Pleistocene, vampire bats had a much broader distribution and occurred across the southern United States. Highest diversities of phyllostomid bats, in terms of number of subfamilies and species, occur in the lowland rainforests of northwestern South America and adjacent Central America. Phyllostomid diversity declines with altitude, latitude, and increasing aridity.

Habitat

Most phyllostomid bats are forest-dwellers. They live in a wide range of forest habitats, including tropical rainforests, tropical dry forests, and subtropical cloud forests. A few species, including two species of *Leptonycteris* and *Choeronycteris mexicana* (Glossophaginae) in Mexico and *Platalina genovensium* (Lonchophyllinae) in Peru, inhabit deserts or very dry tropical forests where they are important pollinators of columnar cacti and century plants (agaves).

Daytime roost structures of American leaf-nosed bats are quite diverse. Most species live in caves and/or hollow trees, but alternate roosts include mines, culverts, hollow logs, under tree roots, undercut river banks, houses, abandoned termite nests, and tree foliage. Suitable roosts are extremely important in the ecology of these bats and can sometimes limit their abundance and distribution. The relatively recent local extinction of many species of mormoopid and phyllostomid bats in the West Indies, for example, was likely caused by the inundation of extensive cave systems as a result of post-Pleistocene increases in sea level.

In addition to providing protection from inclement weather and predators, cave roosts provide stable microclimates. Most cave-dwelling phyllostomids live in roosts that are at or slightly below outside ambient temperatures. These temperatures are often below the thermoneutral zones of the bats, which forces them to expend energy to maintain a constant, high body temperature (98.6–100.4°F; 37–38°C). Certain glossophagine bats, however, including *Leptonycteris curasoae* and *Monophyllus redmani*, live in hot, humid caves where temperatures reach 91°F (33°C) within their thermoneutral zones. These caves are hot because they trap the body heat of tens of thousands of phyllostomid and mormoopid bats. In addition to reducing individual daily energy costs and rates of evaporative water loss,

A Pallas's long-tongued bat (*Glossophaga soricina*) pollinating a banana flower. (Photo by Merlin D. Tuttle/Bat Conservation International/Photo Researchers, Inc. Reproduced by permission.)

A velvety fruit-eating bat (*Artibeus hartii*) eats a banana in Costa Rica. (Photo by Animals Animals ©Richard La Val. Reproduced by permission.)

this roosting strategy increases the developmental rates of embryos and lactating babies.

Most members of the Stenodermatinae usually roost either solitarily or in small groups in foliage and, hence, are constantly exposed to (shaded) ambient temperatures. In addition, several species (e.g., *Uroderma bilobatum*, *Artibeus watsoni*, and *Ectophylla alba*) are known to construct "tents" by clipping the leaves of banana-like herbs, philodendrons, and palms to form shelters from rain and predators. Species of *Rhinophylla* (Carolliinae) also use tents as day roosts. Tent roosts are short-lived, which forces groups of bats to constantly change roost sites.

Behavior

Most phyllostomid bats are colonial rather than solitary roosters, but roost size varies tremendously within and between species. Modal roost sizes within most species are relatively small and range from a few individuals to a few thousand individuals. A few species (the hot cave bats), in contrast, are highly colonial and roost in groups of tens of thousands to several hundred thousand individuals. Except for the flower-visiting, hot-cave bats, there does not appear to be a strong correlation between feeding habits and average roosting group size in this family. Vampires, fruit-eaters, many flower visitors, and certain carnivore/omnivores (e.g., *Phyllostomus hastatus*) sometimes live in roosts containing thousands of individuals. Small roost groups occur in pure carnivores, insectivorous phyllostomines, and a variety of flower-visiting and fruit-eating species (especially the tent-making stenodermatines).

Seasonal sexual segregation commonly occurs in many bats, including phyllostomids. Segregation usually involves the formation of maternity roosts by females. This behavior is most strongly developed in migratory glossophagines such as *Leptonycteris curasoae*, in which females sometimes form maternity roosts containing tens of thousands of adults. Single-sex maternity roosts also occur in non-migratory glossophagines and members of other subfamilies.

Like other Microchiroptera, phyllostomids use ultrasonic sounds for foraging and communication. Unlike other microbats, however, most phyllostomids produce very low intensity sounds that contain only about one-thousandth the sound energy as similar-sized vespertilionid bats. Additional characteristics of their echolocation sounds include multiple harmonics, frequency modulation, and short duration. These sounds provide short-range (3.2–6.5 ft; 1–2 m) information about potential insect or other prey items in areas of high vegetation clutter. In addition to echolocation information, many phyllostomids use other sensory modes (e.g., vision such as the ground-feeding insectivore, *Macrotus californicus*, and many flower-visiting bats, or olfaction such as many fruit-eating bats) to find food. Certain phyllostomines, for example, locate prey using prey-generated sounds (e.g., singing katydids, *Tonatia sylvicola*, and singing male frogs, *Trachops cirrhosus*). Careful experiments have shown that plant-visiting phyllostomids use a combination of vision, olfaction, and echolocation to locate and gain access to their food.

Echolocation and other kinds of vocalizations are also used for communication in these bats. Babies communicate with their mothers using "double-note" calls (rapidly repeated se-

A Jamaican fruit bat (*Artibeus jamaicensis*) at a balsa flower. (Photo by Merlin D. Tuttle/Bat Conservation International/Photo Researchers, Inc. Reproduced by permission.)

A group of fringe-lipped bats (*Trachops cirrhosus*) huddles for warmth on Barro Colorado Island, Panama. (Photo by Merlin D. Tuttle/Bat Conservation International/Photo Researchers, Inc. Reproduced by permission.)

though most species are sedentary and do not migrate among habitats during the year, a few species are known to undergo seasonal migrations. Relatively short-distance (<62 mi; 100 km) altitudinal movements are known to occur in three Mexican or Central American species—two frugivores, *Carollia perspicillata* and *Sturnira lilium*, and the nectar bat, *Leptonycteris curasoae*. Longer distance latitudinal migrations occur in three arid-zone nectar bats: *Leptonycteris curasoae*, *L. nivalis*, and *Choeronycteris mexicana*. Movements of over 620 mi (1,000 km) occur in females of *L. curasoae* as they move between spring maternity roosts in the Sonoran Desert and their late fall mating sites in west central Mexico. The lonchophylline nectar bat (*Platalina genovensium*) is also thought to undergo substantial migrations in the Andes of Peru.

Feeding ecology and diet

American leaf-nosed bats exhibit a wide variety of feeding habits and much of their morphological and behavioral diversity reflects adaptations for exploiting different kinds of food. The ancestral feeding mode in this family is undoubtedly insectivory, but blood-feeding was an early offshoot and nectarivory and frugivory are widespread in the family.

Foraging styles range from species that have small home ranges (many insectivorous or carnivorous phyllostomines and many frugivorous carolliinines and stenodermatines) to species with very large foraging ranges (the arid-zone glossophagine, *Leptonycteris curasoae*). Most phyllostomids commute only 0.6–1.2 mi (1–2 km) from their day roosts to feed, but individuals of *L. curasoae* sometimes commute up to 18.6 mi (30 km). Solitary foraging predominates in the family, but

ries of long and short notes); these calls serve to reunite mother and baby in the roost after a female returns from foraging. Female harem-mates in *Phyllostomus hastatus* communicate while foraging with loud screeching calls. Certain stenodermatine bats (e.g., *Artibeus jamaicensis* and *Uroderma bilobatum*) produce intense warning calls that attract conspecifics or other species when they are captured in Japanese mist nets or while being handled.

Territorial behavior away from day roosts appears to be uncommon in phyllostomids bats. Male home ranges of the tent-roosting frugivore (*Rhinophylla pumilio*) do not overlap and are thought to be territories. The nectar-feeding bat (*Glossophaga soricina*) defends flowering stalks of agaves, at least in a suburban tropical setting, but no other evidence exists to suggest that flower-visiting or fruit-eating phyllostomids defend their food plants. Instead, it is common to see many individuals of several species feeding together in fruiting or flowering trees. The nectar bat (*Leptonycteris curasoae*) sometimes feeds in large numbers in isolated patches of flowering columnar cactus plants. Small groups of two to four bats of this species sometimes forage together at cactus and agave blossoms.

Because they live in tropical and subtropical habitats, phyllostomid bats do not hibernate and are active year-round. Al-

A California leaf-nosed bat (*Macrotus californicus*) showing its large ears. (Photo by Merlin D. Tuttle/Bat Conservation International/Photo Researchers, Inc. Reproduced by permission.)

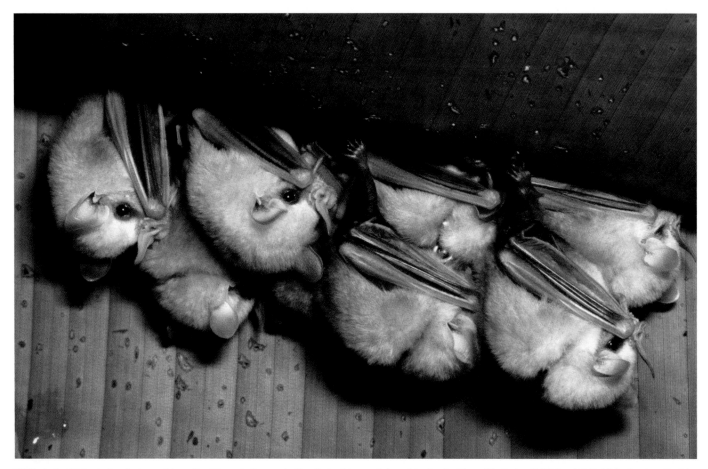

White bats (*Ectophylla alba*) roost in a "tent" made from a heliconia leaf in a rainforest in Costa Rica. (Photo by © Michael & Patricia Fogden/Corbis. Reproduced by permission.)

group foraging also occurs. Mother-young pairs forage together in the common vampire (*Desmodus rotundus*); female harem-mates forage together in *Phyllostomus hastatus*; and adults (and possibly young) of *L. curasoae* and *Phyllostomus discolor* form small foraging groups when visiting flowers.

Fruit-eating phyllostomids typically harvest one fruit at a time and carry it to a night roost located 65.6–656 ft (20–200 m) from the fruiting plant to eat. As a result, substantial piles of fruit remains and seeds can accumulate under these roosts. When eating fruit, phyllostomids use two different feeding methods. Glossophagines, carolliinines, and species of *Sturnira* are rapid feeders; they consume the pulp and seeds of a fruit in one to three minutes and then harvest another fruit. They tend to feed on succulent fruit produced by early successional plants (e.g., *Cecropia* species and *Muntingia calabura*) and understory shrubs (e.g., species of *Piper*, *Solanum*, and *Vismia*). In contrast, stenodermatines are slow feeders and carefully chew fruits into pellets while swallowing fruit juice and a small amount of pulp and seeds. They specialize on fig fruits, which contain high amounts of fiber, and other fruits produced by forest canopy trees.

Food sharing is known to occur in the carnivorous bat (*Vampyrum spectrum*); it also occurs in the common vampire.

Adult female vampires share blood meals with their young and with unrelated adult females in their roost. Blood-sharing among adults is thought to prevent individuals that have been unsuccessful in finding food on a particular night from starving. This form of food-sharing is one of the few examples of reciprocal altruism known to occur in non-primate mammals.

The feeding and foraging behavior of plant-visiting phyllostomids bats has considerable ecological and economic importance. Flower-visiting species, for example, are known to pollinate nearly 1,000 species of Neotropical plants, including trees, shrubs, vines, bromeliads, and arid-zone succulents such as columnar cacti and agaves. Economically important trees that are pollinated by these bats include balsa, kapok, and calabash, as well as agaves. Likewise, fruit-eating species disperse the seeds of hundreds of species of trees and shrubs. Important tree families containing bat-dispersed species include Anacardiaceae (cashew), Saptoaceae (chicle), Moraceae (figs), and Arecaceae (palms). Important understory plant families that are bat-dispersed include Piperaceae (pepper) and Solanaceae (nightshade). Bats such as *Carollia perspicillata*, *Sturnira lilium*, and *Artibeus jamaicensis* are very important for dispersing seeds to disturbed habitats where secondary plant succession and forest regeneration can begin.

A southern long-nosed bat (*Leptonycteris curasoae*) feeds on cactus fruit pulp. (Photo by Merlin D. Tuttle/Photo Researchers, Inc. Reproduced by permission.)

Reproductive biology

In part, roosting group size reflects the mating systems of American leaf-nosed bats. At one extreme is the carnivore *Vampyrum spectrum*, which roosts in small family groups consisting of a pair of adults and up to three of their recent offspring. This species has a monogamous mating system, the only known example so far in this family, and adults share the prey they capture with each other and their offspring. It would not be surprising to learn that other carnivorous species (e.g., *Chrotopterus auritus*) are also monogamous. At the other extreme are colonially roosting species with harem polygynous mating systems, the most common mating system in this family. In these systems, single males aggressively defend groups of females against the intrusions of other males. Only a fraction (about 20% in *Carollia perspicillata*) of adult males possesses a harem at any given time; the majority of adult males are bachelors. Harems either form seasonally or remain stable in their female composition year-round. The most stable harems occur in the greater spear-nosed bat (*Phyllostomus hastatus*), in which groups of unrelated females may spend their entire adult lives together. More ephemeral groups of females occur in two common fruit-eating species, *Carollia perspicillata* and *Artibeus jamaicensis*. Genetic studies indicate that harem males father most, but not all, of the babies born in their harem.

Another variation on the theme of polygyny in phyllostomids is the multiple-male/multiple-female group. An example of this mating system occurs in the common vampire, *Desmodus rotundus*. In this species, stable groups of eight to 12 adult females live together in roosts with several adult males, which form a dominance hierarchy regarding mating rights. When roosting in hollow trees, males fight with each other for access to the top of the roost, where most matings take place. In the phyllostomine (*Macrotus californicus*), seasonal aggregations of adult males and females form in which males defend preferred roosting sites against other males. They attract females for mating with wing flapping and vocalization displays.

Like all bats, American leaf-nosed bats are low-fecundity animals. Females almost always give birth to a single young once or twice a year. Insectivorous or carnivorous species and vampires tend to be monestrous and undergo a single pregnancy each year. Most plant-visiting species, in contrast, are polyestrous and undergo two pregnancies a year. In Central America, one birth occurs in the late dry season (March–April) and another occurs in the middle of the wet

The vampire bat (*Desmodus rotundus*) feeds exclusively on the blood of invertebrates. (Photo by © Michael & Patricia Fogden/Corbis. Reproduced by permission.)

season (July–August) in these species. Gestation periods are relatively long and last about four months in carolliinines and many stenodermatines. It lasts about seven months in the common vampire. In at least two species, *Macrotus californicus* and *Artibeus jamaicensis*, gestation is prolonged as a result of delayed embryonic development.

The timing of reproduction in the glossophagine (*Leptonycteris curasoae*) shows interesting intraspecific variation. Its general pattern is monestry, but the timing of births varies geographically. In northern South America and northern Mexico, births occur in May after a five-to-six-month gestation period. In southern Mexico, births occur in December after a similar gestation period. May births coincide with the flowering seasons of columnar cacti in the Sonoran Desert of northern Mexico and in the arid regions of northern South America. December births in southern Mexico coincide with the flowering seasons of tropical dry forest trees and shrubs. In Mexico, most of the mating activity that produces these babies occurs in caves located in south central Mexico, far from the maternity sites. After mating, females migrate north (in the spring) or south (in the fall) to form maternity colonies. Genetic studies indicate that populations belonging to these reproductive demes currently undergo substantial gene flow. Different reproductive schedules have not resulted in genetic isolation in this species in Mexico.

With a few notable exceptions, parental care is not extensive in this family. Most young phyllostomids are weaned about six weeks after birth, and few young have any further contact with their mothers (or fathers). Extended parent-offspring contact occurs in at least two species, the spectral (*Vampyrum spectrum*) and the common vampire (*Desmodus rotundus*). In the former species, young bats remain with their parents long after weaning and are provisioned with vertebrate prey while they are learning to hunt for them-

selves. Young vampires remain with their mothers for up to a year after birth. Their mothers sometimes feed them regurgitated blood after they reach three months of age, and they continue to forage with their mothers until they are one year old.

Conservation status

In terms of relative abundance, American leaf-nosed bats range from rare to very common. Rarity or commonness in these bats is mostly associated with their food habits or trophic position. As expected, carnivores such as *Vampyrum spectrum* are rare. Insectivorous phyllostomines in general are also far less common in any habitat than frugivorous carolliinines or stenodermatines. Flower visitors tend to be substantially less common than fruit-eaters. The local abundance of common vampires varies tremendously and is correlated with the availability of domestic animals. In primary forest, vampires are uncommon, but they can be very common in disturbed habitats whenever they have regular access to livestock.

In 2001, the IUCN/SSC Chiroptera Specialist Group listed four species of phyllostomids as Endangered and 25 species as Vulnerable. The endangered species included *Phyllonycteris aphylla* on Jamaica, *Chiroderma improvisum* on Guadeloupe and Montserrat, and *Sturnira thomasi* on Guadeloupe, as well as *Leptonycteris nivalis* in Mexico. The vulnerable species included seven phyllostomines, five lonchophyllines, four glossophagines, and nine stenodermatines. Only two of these species, *Ariteus flavescens* and *Stenoderma rufum*, occur on islands.

Phyllostomid bats, and Latin American bats in general, suffer from the "vampire problem," meaning most Latin Americans consider all bats to be *vampiros* that should be destroyed. Millions of cave-dwelling bats in Mexico alone have been killed in recent decades as a result of misguided vampire-control programs. Until enlightened vampire-control methods become widespread in Latin America, all colonially roosting bats are vulnerable to local destruction. Bat Conservation International, located in Austin, Texas, United States, has made a major effort to disseminate information about the selective control of vampires in areas where they pose an economic threat in Latin America.

Significance to humans

Because it sometimes transmits rabies to livestock, the common vampire is the most notorious phyllostomid bat. It has been estimated that 100,000 or more cattle die annually, at a cost of $40 million, as a result of vampire-transmitted rabies in Latin America. On the other hand, other phyllostomid bats can have an important positive impact on humans through their pollination and seed dispersal activities. Many of the market fruits in both the New and Old World tropics are bat-dispersed species. Sisal and tequila plants originally relied on phyllostomids bats for their pollination. Forest regeneration resulting from seed dispersal and enhanced fruit and seed set resulting from pollination are two of the beneficial ecosystem services provided by phyllostomid bats.

1. Southern long-nosed bat (*Leptonycteris curasoae*); 2. Tent-making bat (*Uroderma bilobatum*); 3. Jamaican fruit-eating bat (*Artibeus jamaicensis*); 4. White bat (*Ectophylla alba*); 5. Little yellow-shouldered bat (*Sturnira lilium*); 6. Dwarf little fruit bat (*Rhinophylla pumilio*); 7. Wrinkle-faced bat (*Centurio senex*); 8. Long-snouted bat (*Platalina genovensium*); 9. Seba's short-tailed bat (*Carollia perspicillata*). (Illustration by Gillian Harris)

1. Spectral bat (*Vampyrum spectrum*); 2. Pallas's long-tongued bat (*Glossophaga soricina*); 3. Vampire bat (*Desmodus rotundus*); 4. Geoffroy's tailless bat (*Anoura geoffroyi*); 5. Antillean fruit-eating bat (*Brachyphylla cavernarum*); 6. Fringe-lipped bat (*Trachops cirrhosus*); 7. Buffy flower bat (*Erophylla sezekorni*); 8. California leaf-nosed bat (*Macrotus californicus*); 9. Greater spear-nosed bat (*Phyllostomus hastatus*). (Illustration by Gillian Harris)

Species accounts

California leaf-nosed bat
Macrotus californicus

SUBFAMILY
Phyllostominae

TAXONOMY
Macrotus californicus Baird, 1858, California, United States.

OTHER COMMON NAMES
None known.

PHYSICAL CHARACTERISTICS
Head and body length 2.1–2.5 in (53–64 mm); tail 1.4–1.6 in (35–41 mm); forearm 1.9–2.1 in (48–54 mm); weight 0.4–0.7 oz (12–20 g); upper body is brown or gray, lower body is brown or buff with a silvery wash.

DISTRIBUTION
Southwestern United States and northwestern Mexico, including Baja California.

HABITAT
Arid subtropical lowlands; roosts in caves, mines, and abandoned buildings.

BEHAVIOR
A relatively sedentary, nonmigratory species that is active year-round at its northern distributional limit, where it roosts in geothermally heated mines in the winter; forms roosts of a few hundred to 1,000 or more individuals; occurs in many small, scattered colonies throughout its range.

FEEDING ECOLOGY AND DIET
Feeds on insects, including relatively large orthopterans and moths, which it captures in flight and by gleaning from vegetation and the ground; begins to forage well after sunset.

REPRODUCTIVE BIOLOGY
This species is monestrous and probably polygynous. Males and females form colonies in which mating takes place in September and October; males attract females with wing-flapping and vocal displays. Embryonic development is slow during the winter and faster in spring; gestation period is eight months. Females form maternity colonies, and babies are born in June.

CONSERVATION STATUS
Considered Vulnerable in the United States by the IUCN because of its small roost sizes.

SIGNIFICANCE TO HUMANS
None known. ◆

Fringe-lipped bat
Trachops cirrhosus

SUBFAMILY
Phyllostominae

TAXONOMY
Trachops cirrhosus (Spix, 1823), Pernambuco, Brazil.

OTHER COMMON NAMES
None known.

PHYSICAL CHARACTERISTICS
Head and body length 2.6–3.5 in (65–88 mm); tail 0.4–0.8 in (10–20 mm); forearm 2.2–2.6 in (57–65 mm); weight 0.9–1.3 oz (24–36 g); upper body is dark reddish brown or cinnamon brown, lower body is dull brown washed with gray.

DISTRIBUTION
Southern Mexico to Bolivia and southern Brazil, and Trinidad.

HABITAT
Humid tropical lowlands; roosts in caves, hollow trees, culverts, and abandoned buildings.

BEHAVIOR
Relatively sedentary and nonmigratory. Roosts in small, mixed-sex colonies.

FEEDING ECOLOGY AND DIET
Diet includes insects, frogs, and lizards. Flies low to the ground in foraging. Preys on singing male frogs that it captures on the wing; can discriminate between palatable and unpalatable frogs on the basis of vocalizations.

Trachops cirrhosus
Macrotus californicus

REPRODUCTIVE BIOLOGY
Monestrous and polygynous, with babies being born in May or June. Mating system is unknown.

CONSERVATION STATUS
Not threatened. Widespread and locally common to uncommon; vulnerable to cave disturbance and habitat fragmentation.

SIGNIFICANCE TO HUMANS
None known. ◆

Greater spear-nosed bat
Phyllostomus hastatus

SUBFAMILY
Phyllostominae

TAXONOMY
Phyllostomus hastatus (Pallas, 1767), Suriname.

OTHER COMMON NAMES
None known.

PHYSICAL CHARACTERISTICS
Head and body length 4.1–4.9 in (103–124 mm); tail 0.4–1.1 in (10–29 mm); forearm 3.1–3.7 in (80–93 mm); weight 2.8–3.9 oz (78–110 g); upper body is dark brown or reddish brown, lower body is somewhat paler.

DISTRIBUTION
Honduras to Paraguay and southeastern Brazil, and Trinidad.

Phyllostomus hastatus
Brachyphylla cavernarum

HABITAT
Lowland tropical forests; roosts in caves, hollow trees, culverts, and abandoned buildings.

BEHAVIOR
Relatively sedentary and nonmigratory; roosts in colonies of dozens to thousands of individuals. Females communicate with harem-mates by giving audible screeching calls.

FEEDING ECOLOGY AND DIET
Omnivorous and eats insects (beetles), small vertebrates (reptiles, mammals), nectar, and pollen (especially *Ochroma*), and fruit (especially *Cecropia*). Individuals forage within 6.2 mi (10 km) of their roosts; female harem-mates may forage in groups but not with their harem males. Different groups of females within a roost forage in different areas.

REPRODUCTIVE BIOLOGY
Monestrous and polygynous, with mating and birth being quite synchronous within a roost. Births occur in April or May. Mating system is harem-polygynous with single harem males defending groups of up to about 20 females. Female harem composition is very stable with most individuals remaining together their entire lives. Recently weaned females form new social groups and do not join groups of older females. Harem male turnover rates are relatively high.

CONSERVATION STATUS
Not threatened. Widespread and locally common to uncommon; vulnerable to cave disturbance and habitat fragmentation.

SIGNIFICANCE TO HUMANS
Important pollinator and seed-disperser of certain tropical trees. ◆

Spectral bat
Vampyrum spectrum

SUBFAMILY
Phyllostominae

TAXONOMY
Vampyrum spectrum (Linnaeus, 1758), Suriname.

OTHER COMMON NAMES
None known.

PHYSICAL CHARACTERISTICS
Head and body length 5.3–6.0 in (135–152 mm); forearm 3.9–4.3 in (98–110 mm); weight 5.2–6.8 oz (145–190 g); upper body is reddish brown, lower body is slightly paler.

DISTRIBUTION
Southern Mexico to Peru and central Brazil, and Trinidad.

HABITAT
Lowland tropical forests; roosts in hollow trees.

BEHAVIOR
Sedentary and nonmigratory; roosts in single family groups.

FEEDING ECOLOGY AND DIET
Carnivorous species that feeds on a wide range of vertebrates, including lizards, birds, and mammals (rodents, other bats). Adults feed within several hundred feet (meters) of their roost

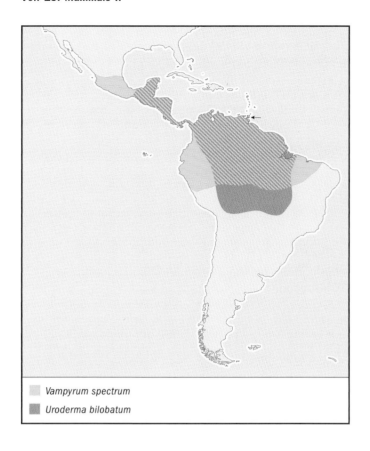

☐ *Vampyrum spectrum*
■ *Uroderma bilobatum*

☐ *Desmodus rotundus*
■ *Erophylla sezekorni*

and bring prey back to the day roost to share with roost-mates. In one roost studied in detail, bats fed mostly on birds ranging from 0.7–5.3 oz (20–150 g) in weight; major prey included one species of parakeet and the groove-billed ani, two species that sleep in groups at night.

REPRODUCTIVE BIOLOGY
Monestrous and monogamous; young bats remain in their natal roost long after they are weaned.

CONSERVATION STATUS
Not threatened, though uncommon throughout its range. Roost trees and foraging habitat threatened by forest destruction.

SIGNIFICANCE TO HUMANS
None known. ◆

Vampire bat
Desmodus rotundus

SUBFAMILY
Desmodontinae

TAXONOMY
Desmodus rotundus (E. Geoffroy, 1810), Asuncion, Paraguay.

OTHER COMMON NAMES
None known.

PHYSICAL CHARACTERISTICS
Head and body length 2.7–3.7 in (68–93 mm); forearm 2.1–2.6 in (53–65 mm); weight 0.7–1.5 oz (20–43 g); upper body is

dark gray-brown, lower body is paler and sometimes with a buffy wash. Nose leaf is reduced in size.

DISTRIBUTION
Northern Mexico to central Chile, Argentina, and Uruguay, and Trinidad.

HABITAT
A wide variety of tropical and subtropical habitats but most common where livestock densities are high; not common in primary forest. Roosts in caves, hollow trees, mines, and abandoned buildings.

BEHAVIOR
A sedentary and nonmigratory bat that lives in roosts containing a few dozen up to about 2,000 individuals. Highly social animals that often groom each other, a behavior that is uncommon in most other bats.

FEEDING ECOLOGY AND DIET
Morphologically, behaviorally, and physiologically specialized for feeding on vertebrate blood (mostly mammals but also birds). Flies only during the darkest part of the night and avoids flying when the moon is bright. Has long, narrow wings for flying quickly from roost to feeding areas up to 12.4 mi (20 km) away; often flies near the ground. Once it has located a sleeping mammal, it often approaches it on the ground. Its elongated thumbs serve as front feet and make this species one of the most agile bats on the ground. Makes a bite with its sharp upper incisors on the ears, neck, anus, or ankles of its victim. Laps blood from free-flowing wound and ingests about 0.7 fl oz (20 ml) of liquid. Its stomach and kidneys are adapted for quickly removing excess water (ballast) from its blood meal. Vampires will not survive if they miss more than one night of

feeding. Adult females sometimes share a blood meal with other adult roost-mates that have not fed successfully.

REPRODUCTIVE BIOLOGY
Monestrous, and single babies are born after a gestation period of about seven months. The mating system involves polygyny, but males do not defend harems. Instead, they compete for locations in roosts (e.g., the highest points inside hollow trees), which give them access to the largest number of females. Females live in stable groups of up to about 20 individuals. When they disperse, young females sometimes join mixed-age groups of females; young males disperse from their natal roosts. Young bats remain with their mothers for some time after they are weaned and sometimes share a blood meal with them. They also feed from the same wounds with their mothers.

CONSERVATION STATUS
Not threatened. Can be too common in areas of high livestock density. In this situation, other species of bats can suffer for two reasons: vampires are aggressive and can supplant other bats from roosts and unselective vampire-control programs often involve the destruction of bat roosts; many kinds of bats die when this happens. Effective methods for controlling vampires involve applying anticoagulants directly on the fur of netted bats or injecting anticoagulants into livestock. Other vampires will ingest the anticoagulant whenever they groom each other or when they receive a blood meal. Selective control of vampires is possible because of its high degree of sociality.

SIGNIFICANCE TO HUMANS
These bats cause substantial (tens of millions of dollars annually) economic damage to livestock by transmitting rabies. Humans sometimes die from vampire-transmitted rabies. ◆

Antillean fruit-eating bat
Brachyphylla cavernarum

SUBFAMILY
Brachyphyllinae

TAXONOMY
Brachyphylla cavernarum Gray, 1934, St. Vincent.

OTHER COMMON NAMES
None known.

PHYSICAL CHARACTERISTICS
Head and body length 2.6–4.6 in (65–118 mm); tail vestigial; forearm 2–2.7 in (51–69 mm); weight 1.4–1.8 oz (40–50 g); upper body is ivory-yellow with black-tipped hairs, lower body is brown. Nose-leaf reduced in size.

DISTRIBUTION
Puerto Rico, Virgin Islands, and the Lesser Antilles south to St. Vincent and Barbados.

HABITAT
Tropical forest. Roosts primarily in caves but also in buildings.

BEHAVIOR
A colonial bat that lives in colonies of a few thousand individuals. Bats are strong and aggressive among themselves and other species.

FEEDING ECOLOGY AND DIET
Diet is generally broad and includes insects as well as fruit, nectar, and pollen. Like most phyllostomids, probably takes fruit to a night roost some distance from the fruiting plant to eat.

REPRODUCTIVE BIOLOGY
Monestrous; babies born in late May or early June. Nothing is known about mating system, but most likely is polygynous.

CONSERVATION STATUS
Not threatened. Common throughout its range but vulnerable to habitat and roost destruction.

SIGNIFICANCE TO HUMANS
None known. ◆

Buffy flower bat
Erophylla sezekorni

SUBFAMILY
Phyllonycterinae

TAXONOMY
Erophylla sezekorni (Gundlach, 1860), Cuba.

OTHER COMMON NAMES
None known.

PHYSICAL CHARACTERISTICS
Head and body length 2.6–3.0 in (65–75 mm); tail 0.5–0.7 in (12–17 mm); forearm 1.8–2.2 in (45–55 mm); weight 0.6 oz (16–18 g); upper body is yellowish brown or buffy, lower body is slightly paler. Nose leaf larger than that of *Brachyphylla*, but still small.

DISTRIBUTION
Bahamas, Greater Antilles, and Grand Cayman.

HABITAT
Subtropical and tropical forests, including pine woodlands. Roosts in caves.

BEHAVIOR
Likely to be a relatively mobile bat that sometimes appears briefly on the coast of south Florida near Miami. In the northern Bahamas, females may move between islands seasonally to form maternity roosts. Roosts contain a few hundred to a few thousand individuals.

FEEDING ECOLOGY AND DIET
Probably omnivorous, although its somewhat elongated muzzle suggests that it regularly visits flowers. Feces sometimes contain insects and seeds (*Piper* on Puerto Rico; *Ficus, Solanum,* and *Tetrazygia* on Grand Bahama).

REPRODUCTIVE BIOLOGY
Monestrous; babies born in late May to early June. Mating system likely involves harem-polygyny. Single males seen with groups of females in solution holes in cave ceilings. Females appear to form maternity colonies to rear their young. One such colony on Grand Bahama contained adults and young that were well spaced out from each other on a cave wall.

CONSERVATION STATUS
Though not threatened, habitat loss is a long-term threat. Has managed to avoid extinction on islands that have lost other species of bats owing to sea level rise and flooding of hot caves.

SIGNIFICANCE TO HUMANS
None known. ◆

Pallas's long-tongued bat
Glossophaga soricina

SUBFAMILY
Glossophaginae

TAXONOMY
Glossophaga soricina (Pallas, 1766), Suriname.

OTHER COMMON NAMES
None known.

PHYSICAL CHARACTERISTICS
Head and body length 1.8–2.3 in (45–59 mm); tail 0.2–0.4 in (5–10 mm); forearm 1.3–1.5 in (33–38 mm); weight 0.3–0.4 oz (7–12 g); upper body is dark brown to reddish brown, lower body is slightly paler.

DISTRIBUTION
Northern Mexico to Paraguay and northern Argentina, Trinidad, Grenada, and Jamaica.

HABITAT
Widespread and common in many tropical lowland habitats; more common in dry forests than in wet forests where it is replaced by *Glossophaga commissarisi*. Roosts in wide variety of sites, including caves, hollow trees, mines, culverts, and abandoned houses.

■ *Ectophylla alba*
■ *Glossophaga soricina*

BEHAVIOR
A colonial species that lives in colonies of a few hundred to a few thousand individuals. Sedentary and nonmigratory.

FEEDING ECOLOGY AND DIET
Despite having an elongated snout and a long tongue, this bat is an omnivore. Its primary food is nectar and pollen from many species of trees, shrubs, and epiphytes, but it switches to eating fruit (of early successional trees and shrubs) when flower availability is low and eats many insects. Individuals forage solitarily, probably relatively close to their day roosts. Flies for about four hours and for about 28 mi (45 km) to feed each night.

REPRODUCTIVE BIOLOGY
This is a polyestrous, seasonal breeder. Females undergo two pregnancies a year with births occurring in December–February and April–June in Costa Rica. The mating system has not been described but undoubtedly involves polygyny, though perhaps not harem polygyny. Females form maternity colonies to produce their young.

CONSERVATION STATUS
Given its generally high abundance and tolerance of disturbed habitats, this species is not currently threatened with extinction.

SIGNIFICANCE TO HUMANS
It is an important pollinator and seed disperser of many tropical plants. ◆

Geoffroy's tailless bat
Anoura geoffroyi

SUBFAMILY
Glossophaginae

TAXONOMY
Anoura geoffroyi Gray, 1838, Rio de Janeiro, Brazil.

OTHER COMMON NAMES
None known.

PHYSICAL CHARACTERISTICS
Head and body length 2.3–2.9 in (58–73 mm); forearm 1.6–1.8 in (40–45 mm); weight 0.5–0.6 oz (13–18 g); upper body dull brown with a silvery wash, lower body is gray-brown.

DISTRIBUTION
Mexico to southeastern Brazil and northwestern Argentina, and Trinidad.

HABITAT
Relatively common in tropical forests, especially at mid-montane elevations; occurs from 820–8,200 ft (250–2,500 m) above sea level. Roosts in caves and tunnels.

BEHAVIOR
Forms relatively small colonies of 100 individuals or fewer. Likely to undergo seasonal altitudinal migrations.

FEEDING ECOLOGY AND DIET
Although it frequently visits flowers of trees, shrubs, and epiphytes, it routinely eats many insects and occasionally eats fruit. Often visits more than one species of flower during a foraging bout. Efficiently digests pollen and prefers sucrose nectar

Anoura geoffroyi

Artibeus jamaicensis

PHYSICAL CHARACTERISTICS
Head and body length 2.6–3.4 in (67–86 mm); forearm 2.1–2.2 in (53–57 mm); weight 0.7–1.0 oz (20–27 g); upper body reddish brown, lower body cinnamon. Moderately elongated snout and small nose leaf.

DISTRIBUTION
Southern Arizona and New Mexico to El Salvador, Aruba, Bonaire, Curaçao, and adjacent Colombia and Venezuela.

HABITAT
Relatively common in semiarid and arid habitats in Mexico and the southwestern United States; more common in lowlands than in montane drylands where it is replaced by its sister species, *L. nivalis*. Roosts in caves and mines; is a hot cave bat.

BEHAVIOR
A very colonial bat forming colonies containing tens to hundreds of thousands of individuals. Not highly social (no allogrooming or food-sharing) in day roosts. Many populations are highly migratory; latitudinal and altitudinal migrations are known.

FEEDING ECOLOGY AND DIET
Highly specialized flower visitor that subsists on nectar and pollen during the maternity season; also eats fruit but rarely insects. Feeds heavily on flowers of columnar cacti and paniculate agaves, when available. Also visits flowers of tropical trees and shrubs. Wide-ranging forager that spends about five hours in flight and flies about 62 mi (100 km) each night; sometimes commutes 12.4–18.6 mi (20–30 km) from day roosts to feeding areas. Although it leaves its day roost at sunset, it visits flowers mostly between midnight and 2 A.M. Sometimes forages in small groups of two to four when visiting cactus and agave

to equicaloric solutions of fructose and glucose in lab trials. A related species (*A. caudifer*) flies for about five hours and for about 43 mi (70 km) while foraging each night.

REPRODUCTIVE BIOLOGY
Monestrous; babies are born late in the year, during or just before the dry season, on Trinidad. Laboratory studies show that the seasonal male testicular cycle follows an endogenous rhythm that is not regulated by photoperiod. Mating system unknown but likely involves some form of polygyny since bats roost in small clusters in their roosts. Females form maternity roosts separate from males prior to giving birth. Young grow rapidly and are weaned by five to six weeks of age.

CONSERVATION STATUS
Vulnerable to roost disturbance and habitat destruction, but its numbers are currently high in many locations and it is not considered threatened.

SIGNIFICANCE TO HUMANS
Pollinates tropical plants. ◆

Southern long-nosed bat
Leptonycteris curasoae

SUBFAMILY
Glossophaginae

TAXONOMY
Leptonycteris curasoae Miller, 1900, Curaçao.

OTHER COMMON NAMES
None known.

Leptonycteris curasoae

Platalina genovensium

flowers. Juvenile bats likely forage with their mothers after they are weaned.

REPRODUCTIVE BIOLOGY
Monestrous, but timing of mating and births varies geographically. In Mexico, births occur in mid-May in the Sonoran Desert and in December in tropical dry forest in Chiapas. Both sexes gather seasonally at specific mating caves. Mating system unknown but likely involves promiscuous mating in both males and females. Gestation is relatively long and lasts five to six months; females migrate while pregnant.

CONSERVATION STATUS
Listed as Vulnerable by the IUCN; declared federally Endangered in the United States in 1988; received official protection in Mexico in 1994. Current numbers and genetic evidence, however, indicate that its population sizes are large. Because of its highly colonial roosting behavior, it is very vulnerable to roost disturbance. Protection is also needed for the habitat and food plants (columnar cacti, agaves) that support long-distance migrations.

SIGNIFICANCE TO HUMANS
Important pollinators of columnar cacti and paniculate agaves, two keystone plants of arid habitats in Mexico, southwestern United States, and northern South America. ◆

Long-snouted bat
Platalina genovensium

SUBFAMILY
Lonchophyllinae

TAXONOMY
Platalina genovensium Thomas, 1928, Peru.

OTHER COMMON NAMES
None known.

PHYSICAL CHARACTERISTICS
Head and body length 2.8 in (70–72 mm); tail 0.3 in (8–9 mm); forearm 1.7–1.9 in (44–47 mm); weight 0.6–0.8 oz (16–23 g); upper and lower body color pale brown. Very long snout and small nose-leaf.

DISTRIBUTION
Peru.

HABITAT
Arid regions from sea level to about 8,200 ft (2,500 m) in the Andes; mostly a montane species. Roosts in caves and mines.

BEHAVIOR
Roosts in small groups of 20 individuals or less. Nowhere is it common, and it is poorly represented in museum collections. Undoubtedly undergoes latitudinal and altitudinal migrations but these have not yet been documented.

FEEDING ECOLOGY AND DIET
Carbon stable isotope data suggest that this bat is a specialized feeder at flowers of columnar cacti. They abandon roosts whenever cactus flower supplies in the area decrease.

REPRODUCTIVE BIOLOGY
Probably monestrous, but this topic is unstudied, as is the form of its mating system. Most likely polygynous. Pregnant females have been captured in September.

CONSERVATION STATUS
Listed as Vulnerable by the IUCN, this species clearly warrants concern because of its low numbers and mobile lifestyle.

SIGNIFICANCE TO HUMANS
None known. ◆

Seba's short-tailed bat
Carollia perspicillata

SUBFAMILY
Carolliinae

TAXONOMY
Carollia perspicillata (Linnaeus, 1758), Suriname.

OTHER COMMON NAMES
None known.

PHYSICAL CHARACTERISTICS
Head and body length 1.9–2.8 in (48–70 mm); tail 0.3–0.6 in (8–16 mm); forearm 1.6–1.8 in (41–45 mm); weight 0.5–0.9 oz (15–25 g); upper body dark brown with silvery wash, lower body lighter brown.

DISTRIBUTION
Southern Mexico to Paraguay and southern Brazil, Trinidad and Tobago, and Grenada.

☐ *Sturnira lilium*
◼ *Carollia perspicillata*

HABITAT
Tropical forests of all kinds, mostly in the lowlands. Roosts in many sites, including caves, hollow trees, mines, culverts, and abandoned houses.

BEHAVIOR
One of the most common bats in Latin America. Roosts contain dozens to a few thousand individuals. A relatively sedentary bat, but in western Costa Rica, females make seasonal altitudinal migrations.

FEEDING ECOLOGY AND DIET
Mostly a fruit-eater but also visits flowers opportunistically and eats insects. Individuals forage within about 1.2 mi (2 km) of their day roost. Harvests one fruit at a time and eats it in a sheltered night roost. Feeds selectively on fruits of understory shrubs (especially *Piper* and *Vismia*) and early successional trees.

REPRODUCTIVE BIOLOGY
Polyestrous; females give birth to a single young twice a year (March–April and July–August in Central America); gestation period is about four months. Degree of synchrony of births is relatively low within populations. Mating system involves harem-polygyny with single males defending groups of up to about 20 females. Most adult males in a roost are bachelors. Sexes usually do not segregate during the maternity period.

CONSERVATION STATUS
Not threatened. Thrives in disturbed habitats but is vulnerable to roost destruction.

SIGNIFICANCE TO HUMANS
Important seed disperser that helps to promote tropical forest regeneration resulting from natural or human disturbances. ◆

▢ *Centurio senex*

▨ *Rhinophylla pumilio*

Dwarf little fruit bat
Rhinophylla pumilio

SUBFAMILY
Carolliinae

TAXONOMY
Rhinophylla pumilio Peters, 1865, Bahia, Brazil.

OTHER COMMON NAMES
None known.

PHYSICAL CHARACTERISTICS
Head and body length 1.7–1.9 in (43–48 mm); forearm 1.1–1.5 in (29–37mm); weight 0.3–0.4 oz (8–10 g); upper body grayish brown, lower body paler brown.

DISTRIBUTION
Colombia, Ecuador, Peru, and Bolivia to the Guianas and eastern Brazil.

HABITAT
Lowland tropical forests. Roosts in abandoned tents made by stenodermatine bats. Much less common than species of *Carollia* when they occur together.

BEHAVIOR
A sedentary bat that lives in small groups (fewer than 10) in forest understory tents formed from the leaves of a variety of plants (*Heliconia*, *Philodendron*, and palms). Group-foraging ranges are small (24.7–37 acres; 10–15 ha). Males may defend these areas from the intrusions of other males.

FEEDING ECOLOGY AND DIET
Mostly frugivorous but, unlike species of *Carollia*, does not feed on fruits of *Piper* shrubs; instead eats fruits of epiphytes (*Macgravia* and *Philodendron*).

REPRODUCTIVE BIOLOGY
Probably polyestrous, but not yet studied in detail. Mating system likely involves harem-polygyny with a single adult male and two to three females and their young living in stable social units in tents. These groups remain together when they change tent locations.

CONSERVATION STATUS
Not threatened, though, because of their specialized roosting behavior, they are vulnerable to habitat destruction.

SIGNIFICANCE TO HUMANS
None known. ◆

Little yellow-shouldered bat

Sturnira lilium

SUBFAMILY
Stenodermatinae

TAXONOMY
Sturnira lilium (E. Geoffroy, 1810), Asuncion, Paraguay.

OTHER COMMON NAMES
None known.

PHYSICAL CHARACTERISTICS
Head and body length 2.1–2.6 in (54–65 mm); forearm 1.5–1.7 in (37–42 mm); weight 0.5–0.6 oz (13–18 g); upper body buff orange with a brown wash, lower body paler; males have conspicuous tufts of stiff yellow or red hairs on the shoulders.

DISTRIBUTION
Northwestern and northeastern Mexico to northern Argentina, southern Lesser Antilles, and Jamaica.

HABITAT
Most common in lowland forests, especially disturbed sites.

BEHAVIOR
A relatively sedentary bat that roosts in small, inconspicuous colonies in hollow trees, clusters of vines, and palm fronds. Can be a very common bat, especially in habitats dominated by *Piper* and *Solanum* shrubs and treelets. Undergoes seasonal altitudinal migrations in southwestern Mexico and probably elsewhere.

FEEDING ECOLOGY AND DIET
Mostly frugivorous but also occasionally visits flowers. Morphologically adapted for feeding on relatively hard fruits. Selectively feeds on fruits of Solanum and Piper shrubs and treelets.

REPRODUCTIVE BIOLOGY
Polyestrous with females giving birth to a single baby twice a year. In Costa Rica, most births occur in May and December. Mating system of this species is unknown, but, to judge from other fruit-eating phyllostomids, likely involves some form of polygyny.

CONSERVATION STATUS
Not currently threatened, but habitat destruction and loss of hollow roost trees are two major threats.

SIGNIFICANCE TO HUMANS
Has an important role in forest regeneration through its seed dispersal activities. ◆

Tent-making bat

Uroderma bilobatum

SUBFAMILY
Stenodermatinae

TAXONOMY
Uroderma bilobatum Peters, 1866, São Paulo, Brazil.

OTHER COMMON NAMES
None known.

PHYSICAL CHARACTERISTICS
Head and body length 2.3–2.7 in (59–69 mm); forearm 1.6–1.7 in (40–44 mm); weight 0.5–0.7 oz (13–20 g); upper and lower body pale gray, two bright white stripes on top of head, a single mid-dorsal white stripe.

DISTRIBUTION
Southern Mexico to Bolivia and southeastern Brazil; Trinidad.

HABITAT
Wide variety of lowland tropical forests containing palms that it uses for tent roosts.

BEHAVIOR
Roosts solitarily or in small groups in leaf tents that it fashions from palm or banana-like leaves in the forest understory. Often uses palmate-leafed palms for tent sites. Occupancy of a particular tent can last as long as two months.

FEEDING ECOLOGY AND DIET
Mostly frugivorous and, like many stenodermatines, feeds heavily on fig fruits. On Barro Colorado Island, Panama, feeds mostly on small figs. Details of its foraging behavior are unknown, but this species sometimes feeds in swarms at fruiting fig trees.

REPRODUCTIVE BIOLOGY
Seasonally polyestrous with females producing a single baby twice a year. Birth peaks occur in February and July in Panama; degree of birth synchrony is high. Mating system is unstudied but undoubtedly involves harem-polygyny. Single adult males and a few females and their young occupy one tent.

CONSERVATION STATUS
Not currently threatened, though because of its specialized roosting requirements, it is vulnerable to deforestation.

SIGNIFICANCE TO HUMANS
It is a disperser of fig seeds. ◆

White bat

Ectophylla alba

SUBFAMILY
Stenodermatinae

TAXONOMY
Ectophylla alba H. Allen, 1892, Honduras.

OTHER COMMON NAMES
None known.

PHYSICAL CHARACTERISTICS
Head and body length 1.6–1.9 in (40–47 mm); forearm 0.9–1.2 in (23–31 mm); weight 0.1–0.3 oz (4–7 g); entire body white, ears and nose leaf edged in yellow.

DISTRIBUTION
Eastern Honduras to western Panama.

HABITAT
Moist or wet lowland tropical forests. Roosts in understory in tents.

BEHAVIOR
Constructs tents about 6.5 ft (2 m) above the ground using banana-like leaves in disturbed or wet areas. Males and females use the same tent.

FEEDING ECOLOGY AND DIET
Foraging behavior unstudied. Feeds on figs and other fruit.

REPRODUCTIVE BIOLOGY
Details unstudied, but may be monestrous. Young born in April in Costa Rica. Mating system likely involves harem polygyny.

CONSERVATION STATUS
Listed as Lower Risk/Near Threatened. Vulnerable to habitat destruction.

SIGNIFICANCE TO HUMANS
None known. ◆

Jamaican fruit-eating bat
Artibeus jamaicensis

SUBFAMILY
Stenodermatinae

TAXONOMY
Artibeus jamaicensis Leach, 1821, Jamaica.

OTHER COMMON NAMES
None known.

PHYSICAL CHARACTERISTICS
Head and body length 2.8–3.3 in (70–85 mm); forearm 2.2–2.6 in (55–67 mm); weight 1.0–1.8 oz (29–51 g); upper body gray or gray-brown, lower body paler with hairs frosted with silver. Faint facial stripes on head.

DISTRIBUTION
Central Mexico to Paraguay and central Brazil, Trinidad and Tobago, Greater and Lesser Antilles, and southern Bahamas.

HABITAT
Lives in wide variety of habitats, mostly in the lowlands.

BEHAVIOR
One of the most common bats in Latin America. Often roosts in foliage, less often in caves.

FEEDING ECOLOGY AND DIET
Mostly frugivorous, but also occasionally visits flowers of canopy trees and eats protein-rich leaves. Diet is known to contain many kinds of fruits but concentrates on fruits produced by the Moraceae (the fig family). In some locations, most of diet comes from figs. Eats primarily large figs in central Panama, where bats commute 0.6–1.2 mi (1–2 km) from their day roosts to fruiting fig trees. Full moon suppresses foraging activity.

REPRODUCTIVE BIOLOGY
Seasonally polyestrous with females producing a single baby twice a year. Birth peaks occur in March and July in Panama; births are highly synchronous. Babies born in March are con-

ceived in August; delayed development produces an eight-month gestation period. Mating system involves harem polygyny with adult males defending groups of two to 18 females. Some large harems contain a dominant and a subordinate male, which are likely to be related. Stability of the female composition of harems is relatively low. It is unlikely that young females recruit into their natal harems.

CONSERVATION STATUS
Not threatened. A very adaptable bat that can tolerate deforestation better than most species.

SIGNIFICANCE TO HUMANS
An extremely important disperser of the seeds of forest trees. ◆

Wrinkle-faced bat
Centurio senex

SUBFAMILY
Stenodermatinae

TAXONOMY
Centurio senex Gray, 1842, Chinandega, Nicaragua.

OTHER COMMON NAMES
None known.

PHYSICAL CHARACTERISTICS
Head and body length 2.1–2.7 in (54–68 mm); forearm 1.6–1.8 in (41–45 mm); weight 0.5–0.9 oz (13–26 g); upper body yellowish brown with a white spot on shoulder, lower body paler; face grotesquely wrinkled with no nose leaf; flap of skin under chin hangs over face when bat is at rest.

DISTRIBUTION
Northern coastal Mexico south to Venezuela and Trinidad and Tobago.

HABITAT
Lowland tropical forests, including second growth, up to about 4,595 ft (1,400 m). Roosts in foliage and vine tangles.

BEHAVIOR
An uncommon bat that roosts in small groups of less than a dozen individuals. Bats are inconspicuous in their roosts. They apparently change roost sites regularly; new roosts are often located just several hundred feet (meters) from old roosts.

FEEDING ECOLOGY AND DIET
Thought to eat soft fruit and use its facial wrinkles to direct fruit juices into its mouth.

REPRODUCTIVE BIOLOGY
Details unknown but records of pregnant or lactating females (February–August in Mexico and Central America) suggest seasonal polyestry. Mating system is likely to involve harem polygyny.

CONSERVATION STATUS
Not currently threatened, but vulnerable to habitat destruction.

SIGNIFICANCE TO HUMANS
None known. ◆

Common name / Scientific name	Physical characteristics	Habitat and behavior	Distribution	Diet	Conservation status
Common big-eared bat *Micronycteris microtis*	Reddish to gray-brown dorsally and paler brown underneath. Head and body length 1.4–2.0 in (3.5–5.1 cm); tail 0.3–0.6 (0.8–1.5 cm); forearm 1.3–1.5 (3.2–3.7 cm); weight 0.1–0.3 oz (4–9 g).	Lowland deciduous and evergreen forests. Roosts in small groups in caves, mines, hollow trees, logs, and abandoned buildings.	Northern coastal Mexico south through Central America to Colombia, French Guiana, and Brazil.	Primarily insects (of at least 13 orders) but also fruit.	Not threatened
Pale spear-nosed bat *Phyllostomus discolor*	Reddish to gray-brown dorsally and paler brown underneath. Head and body length 2.6–3.8 in (6.6–9.7 cm); tail 0.5–0.9 (1.2–2.3 cm); forearm 2.7–3.3 (6.9–8.3 cm); weight 1.8–2.3 oz (51–65 g).	Lowland deciduous and evergreen forest. Roosts in hollow trees or caves in colonies of several hundred individuals.	Southern Mexico to southern Brazil and Paraguay, also Trinidad.	Nectar, pollen, fruit, and insects; more nectarivorous than the greater spear-nosed bat, *P. hastatus*.	Not threatened
White-throated round-eared bat *Tonatia sylvicola*	Gray or gray-brown dorsally and paler gray underneath. Head and body length 1.8–1.9 in (4.6–4.9 cm); tail 0.4–0.9 (1.0–2.2 cm); forearm 2.0–2.2 (5.0–5.6 cm); weight 0.9–1.4 oz (25–39 g).	Lowland primary forests. Roosts in small groups in arboreal termite nests and sometimes in caves.	Honduras to Bolivia, northern Argentina, and eastern Brazil.	Primarily insectivorous (especially large orthopterans) but also small vertebrates and fruit.	Not threatened
Woolly false vampire bat *Chrotopterus auritus*	Dark gray or gray-brown dorsally and silvery gray underneath. Fur long and woolly. Head and body length 3.7–4.4 in (9.3–11.3 cm); tail 0.2–0.6 (0.6–1.5 cm); forearm 3.0–3.3 (7.7–8.3 cm); weight 2.2–3.3 oz (61–92 g).	Lowland evergreen forests. Roosts in small family groups in caves or hollow trees.	Southern Mexico to southern Brazil and northern Argentina.	Small vertebrates, including rodents, birds, frogs, and reptiles.	Not threatened
Hairy-legged vampire bat *Diphylla ecaudata*	Gray-brown dorsally and gray underneath. Head and body length 2.7–3.2 in (6.9–8.2 cm); tail 0; forearm 1.9–2.2 in (4.9–5.6 cm); weight 0.6–1.2 oz (18–33 g).	Lowland deciduous and evergreen forests. Roosts in small colonies in caves and mines.	Southern Texas, United States, and eastern Mexico to Venezuela, Peru, and eastern Brazil.	The blood of birds (rarely mammals).	Lower Risk/Near Threatened
Orange nectar bat *Lonchophylla robusta*	Orange dorsally and buffy underneath. Head and body length 2.2–3.0 in (5.6–7.5 cm); tail 0.2–0.4 (0.6–1.1 cm); forearm 1.6–1.8 (4.0–4.5 cm); weight 0.5–0.7 oz (14–19 g).	Lowland evergreen forest. Roosts in small colonies in caves or mines.	Nicaragua to Venezuela and Ecuador.	Nectar, pollen, and insects.	Not threatened
Mexican hog-nosed bat *Choeronycteris mexicana*	Gray-brown dorsally and paler gray underneath. Head and body length 2.7–3.7 in (6.8–9.3 cm); tail 0.2–0.5 in (0.6–1.2 cm); forearm 1.7–1.9 in (4.3–4.9 cm); weight 0.5–0.7 oz (14–19 g).	Lowland to mid-montane elevations in desert and deciduous and pine-oak forest. Roosts in small colonies in caves or mines. Females are migratory.	Southwestern United States and along Pacific coast of Mexico to southern Honduras.	Nectar and pollen of flowers of columnar cacti, agaves, and tropical trees; occasionally fruit.	Lower Risk/Near Threatened
Southern long-tongued bat *Glossophaga longirostris*	Light brown dorsally and paler underneath. Head and body length 2.4–2.8 in (6.0–7.0 cm); tail 0.1–0.4 in (0.3–0.8 cm); forearm 1.4–1.5 in (3.5–3.9 cm); weight 0.4–0.5 oz (11–15 g).	Lowland to mid-montane deserts and riparian and deciduous tropical forests. Roosts in small colonies in caves, mines, and hollow trees.	Northeastern Colombia and northern Venezuela to Guyana.	Nectar, pollen, fruit, and insects.	Not threatened
Cuban flower bat *Phyllonycteris poeyi*	Grayish white dorsally and ventrally. Head and body length 2.5–3.3 in (6.4–8.3 cm); tail 0.3–0.5 in (0.7–1.2 cm); forearm 1.7–2.0 in (4.4–5.0 cm); weight 0.6–1.0 oz (16–28 g).	Disturbed and primary forests. Roosts in "hot" caves in large numbers.	Cuba.	Nectar, pollen, fruit, and insects.	Lower Risk/Near Threatened
Chestnut short-tailed bat *Carollia castanea*	Reddish brown dorsally and paler underneath. Head and body length 1.9–2.4 in (4.8–6.0 cm); tail 0.3–0.6 in (0.7–1.4 cm); forearm 1.3–1.5 in (3.4–3.8 cm); weight 0.4–0.5 oz (12–14 g).	Lowland secondary and primary evergreen forests. Roosts in small colonies in caves, mines, hollow trees, and under tree roots.	Honduras to Venezuela, Bolivia, and western Brazil.	Mostly fruit (especially those of *Piper* shrubs) and insects.	Not threatened
Highland yellow-shouldered bat *Sturnira ludovici*	Gray-brown or orange dorsally and paler gray underneath. Male shoulder patches orange to dark red. Head and body length 2.6–2.8 in (6.6–;7.0 cm); tail 0; forearm 1.6–1.8 in (4.1–4.5 cm); weight 0.6–0.8 oz (17–23 g).	Moist to wet montane forests. Probably roosts in hollow trees in small groups.	Mexican highlands to Guyana and west of the Andes to Ecuador.	Fruit mostly *Solanum* and *Piper* and insects.	Not threatened

[continued]

Common name / Scientific name	Physical characteristics	Habitat and behavior	Distribution	Diet	Conservation status
Great fruit-eating bat *Artibeus lituratus*	Brown or tan dorsally and gray-brown underneath. Head and body length 3.4–4.0 in (8.7–10.1 cm); tail 0; forearm 2.7–3.1 in (6.9–7.8 cm); weight 1.9–2.6 oz (53–73 g).	Lowland deciduous and evergreen forests. Roosts in small groups in foliage, hollow trees, and sometimes in larger groups in caves and mines.	Mexico to southern Brazil and northern Argentina; Trinidad and Tobago, southern Lesser Antilles.	Fruit (especially figs) of canopy trees; sometimes pollen and nectar.	Not threatened
Hairy big-eyed bat *Chiroderma villosum*	Entirely gray or gray-brown. Head and body length 2.4–3.1 in (6.2–7.9 cm); tail 0; forearm 1.7–1.9 in (4.2–4.7 cm); weight 0.5–1.0 oz (15–28 g).	Lowland deciduous and evergreen forests. Roosts in foliage in small groups.	South central Mexico to southern Brazil, Bolivia, and Peru; Trinidad and Tobago.	Fruit (especially figs).	Not threatened

Resources

Books

Altringham, J. D. *Bats: Biology and Behaviour.* Oxford: Oxford University Press, 1996.

Dobat, K. *Bluten and Fledermause.* Frankfurt am Main: Dr. Waldemar Kramer, 1985.

Fleming, T. H. *The Short-tailed Fruit Bat.* Chicago: University of Chicago Press, 1988.

Fleming, T. H., and J. Nassar. "Population Biology of the Lesser Long-nosed Bat (*Leptonycteris curasoae*) in Mexico and Northern South America." In *Columnar Cacti and Their Mutualists,* edited by T. H. Fleming and A. Valiente-Banuet. Tucson: University of Arizona Press, 2002.

Greenhall, A. M., and U. Schmidt, eds. *Natural History of Vampire Bats.* Boca Raton: CRC Press, 1988.

Heideman, P. D. "Environmental Regulation of Reproduction." In *Reproductive Biology of Bats,* edited by E. G. Crichton and P. H. Krutzsch. San Diego: Academic Press, 2000.

McCracken, G. F., and G. S. Wilkinson. "Bat Mating Systems." In *Reproductive Biology of Bats,* edited by E. G. Crichton and P. H. Krutzsch. San Diego: Academic Press, 2000.

Nowak, R. M. *Walker's Bats of the World.* Baltimore: Johns Hopkins University Press, 1994.

Wilkinson, G. S. "Information Transfer in Bats." In *Ecology, Evolution and Behaviour of Bats,* edited by P. A. Racey and S. M. Swift. Oxford: Clarendon Press, 1995.

Periodicals

Handley Jr., C. O., D. E. Wilson, and A. L. Gardner, eds. "Demography and Natural History of the Common Fruit Bat, *Artibeus jamaicensis,* on Barro Colorado Island, Panama." *Smithsonian Contributions to Zoology* 511 (1991).

Simmons, N. B., and R. S. Voss. "The Mammals of Paracou, French Guiana: A Neotropical Lowland Rainforest Fauna, Part 1. Bats." *Bulletin of the American Museum of Natural History* 237 (1998).

Wetterer, A. L., M. V. Rockman, and N. B. Simmons. "Phylogeny of Phyllostomid Bats (Mammalia: Chiroptera): Data From Diverse Morphological Systems, Sex Chromosomes, and Restriction Sites." *Bulletin of the American Museum of Natural History* 248 (2000).

Organizations

Bat Conservation International. P.O. Box 162603, Austin, TX 78716 USA. Phone: (512) 327-9721. Fax: (512) 327-9724. E-mail: batinfo@batcon.org Web site: <http://www.batcon .org>

Theodore H. Fleming, PhD

Moustached bats
(*Mormoopidae*)

Class Mammalia
Order Chiroptera
Suborder Microchiroptera
Family Mormoopidae

Thumbnail description
The lips of moustached bats are ornamented with flaps and folds of skin surrounded by bristle-like hairs giving the appearance of a moustache

Size
Moustached bat are small to medium in size with forearms ranging from 1.4 to 2.6 in (3.5–6.5 cm) in length and weighing 0.2–0.9 oz (6–26 g)

Number of genera, species
2 genera; 8 species

Habitat
Moustached bats are found in lowland Neotropical areas below 10,000 ft (3,000 meters) from rainforest to forest and in more open, arid areas

Conservation status
Vulnerable: 1 species; Lower Risk/Near Threatened: 2 species

Distribution
Central America and parts of South America

Evolution and systematics

Moustached bats belong to the Noctilionoidea superfamily with the bulldog bats (Noctilionidae) and American leaf-nosed bats (Phyllostomidae). Mormoopids are not known as other than recent fossils.

Physical characteristics

Generally small in size with body weight not exceeding 0.9 oz (26 g), moustached bats have flaps and folds of skin around the mouth as well as moustache-like hairs making them distinctive. There are three distinct forms of moustached bats. Ghost-faced bats, species of the genus *Mormoops*, have very well developed flaps of skin around the mouth. In the skulls of ghost-faced bats, the braincase protrudes over the rostrum. In other moustached bats, the flaps of skin are not so well developed and the braincase does not protrude over the rostrum. Among these moustached bats, the wings either join at the side of the body (sooty moustached bat [*Pteronotus quadridens*], Wagner's moustached bat [*Pteronotus personatus*], Parnell's moustached bat [*Pteronotus parnellii*], MacLeay's moustached bats [*Pteronotus macleayii*]) or at mid-back (Davy's naked-backed

bat [*Pteronotus davyi*], big naked-backed bat [*Pteronotus gymnonotus*]). The significance of the expression "naked backs," a species in the genus *Dobsonia* (family Pteropodidae), remains unknown. The fur of moustached bats can be gray or bright orange in color.

Distribution

Moustached bats occur in the West Indies, Central America, and in South America ranging into Brazil and Peru.

Habitat

Moustached bats occur in lowland Neotropical areas from rainforest to forest and in more open, arid areas.

Behavior

Moustached bats typically roost together in large colonies, in hollows such as caves, mines, or tunnels, and probably also in hollow trees. Roosting individuals may or may not be in physical contact with one another. Aerial-feeding species that

A Parnell's moustached bat (*Pteronotus parnellii*) resting on a rock. (Photo by © Rick & Nora Bowers/Visuals Unlimited, Inc. Reproduced by permission.)

Parnell's moustached bat (*Pteronotus parnellii*) in La Selva, Costa Rica. (Photo by Animals Animals ©Richard La Val. Reproduced by permission.)

eat flying insects, moustached bats use echolocation to detect, track, and evaluate their targets. When echolocating, seven species of mormoopids separate pulse and echo in time. But Parnell's moustached bat separates pulse and echo in the frequency domain, making them more like the horseshoe (Rhinolophidae) and Old World leaf-nosed bats (Hipposideridae) of the Old World.

Feeding ecology and diet

Moustached bats, like all mormoopids, are exclusively insect eaters. They eat a wide range of flying insects, from flies to beetles and moths. In forested areas, these bats typically hunt along trails and roads sometimes flying very close to within 3.3 ft (1 m) the ground and vegetation.

Reproductive biology

Females bear a single young each year at the beginning of the rainy season. Gestation appears to last about 60 days, and in the West Indies copulations occur in January and February.

Conservation status

The IUCN considers two species (Antillean ghost-faced bat [*Mormoops blainvillii*] and sooty moustached bat [*Pteronotus quadridens*]) as Lower Risk/Near Threatened. MacLeay's moustached bat is considered Vulnerable because of its declining habitat.

The mouth of Wagner's moustached bat (*Pteronotus personatus*) is shaped like a funnel. (Photo by Brock Fenton. Reproduced by permission.)

A Davy's naked-backed bat (*Pteronotus davyi*). (Photo by © Rick & Nora Bowers/Visuals Unlimited, Inc. Reproduced by permission.)

The wings of Davy's naked-backed bat (*Pteronotus davyi*) attach mid-back and give the appearance of there being fur on the back. (Photo by Brock Fenton. Reproduced by permission.)

Parnell's moustached bat (*Pteronotus parnellii*) has no fur on its chin. (Photo by © Rick & Nora Bowers/Visuals Unlimited, Inc. Reproduced by permission.)

A close-up of a moustached bat (*Pteronotus* sp.). (Photo by © Michael & Patricia Fogden/Corbis. Reproduced by permission.)

Significance to humans

As all bats, mormoopids pollinate numerous plants, including economically important crops such as banana, mango, and avocado. In the tropics, bats are responsible for 70–95% of all seeds dispersed, thus playing an important role in forest regeneration.

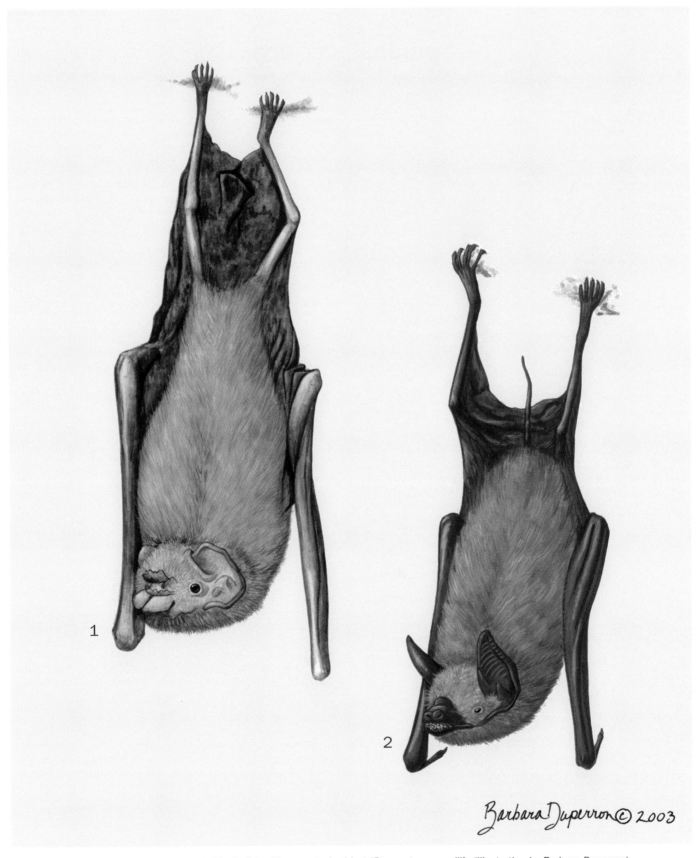

1. Ghost-faced bat (*Mormoops megalophylla*); 2. Parnell's moustached bat (*Pteronotus parnellii*). (Illustration by Barbara Duperron)

Species accounts

Parnell's moustached bat
Pteronotus parnellii

TAXONOMY
Phyllodia parnellii (Gray, 1843), Jamaica. Nine subspecies are currently recognized, four occurring in the West Indies.

OTHER COMMON NAMES
Spanish: Murcielago bigotudo.

PHYSICAL CHARACTERISTICS
Smaller, medium-sized bats with forearms ranging from 2.2 to 2.5 in (5.5–6.3 cm), and weighing 0.4–0.9 oz (12–26 g).

DISTRIBUTION
Found throughout the Greater Antilles, and in the mainland of Central America from southern Sonora and Tamaulipas, the south of Mexico to northern South America east of the Andes. They also occur in northern Colombia, Venezuela, the Guianas, Brazil, and Peru.

HABITAT
These bats occur in habitats ranging from arid to humid, tropical forest.

BEHAVIOR
Roost in hollows, usually in caves or abandoned mines. Colonies can consist of hundreds of individuals. Their presence in areas without caves indicates that they also roost in hollows in trees. When bats are active, they are extremely active, to an extent that they require to spend much time sleeping. They are also strongly heterothermic, meaning that their body temperature is highly variable. They can reach and maintain body temperatures appropriate to their physiological and behavioral needs, thus conserving energy. When they are actively feeding, metabolic activity and body temperature are high. When they are resting, both metabolism and temperature are low. This process of resting under lowered metabolic activity and body temperature is called *torpor*. And they may enter torpor for a few hours or several months. The intensity of torpor ranges from shallow to deep. Deep, long-term torpor occurring in winter is termed hibernation. Bats spend their periods of torpor in a roost, where they can hang protected from dangers.

FEEDING ECOLOGY AND DIET
These bats hunt flying insects that they detect by Doppler-shifted echoes, separating pulse and echo in frequency. Their echolocation calls are dominated by one frequency, usually around 60 kHz. The bats eat mainly beetles and moths. In forested areas, they often can be observed hunting along trails and roads. Their distinctive echolocation calls makes them easy to recognize with a bat detector tuned to about 60 kHz. In flight, the production of echolocation calls is synchronized to the wingbeat and partly driven by movements of the viscera against the diaphram. Contractions of muscles in the middle ear contribute to the bats' avoiding deafening themselves during production of echolocation signals. Acoustic information acquired during echolocation is represented in the cerebral cortex.

REPRODUCTIVE BIOLOGY
Females bear a single young annually after a gestation period of about 50 days. The young are naked and helpless at birth. The timing of reproduction varies across the species' range, with births usually peaking around the start of the rainy season. For example, if mating occurs in autumn, the sperm is typically stored by females throughout hibernation, sometimes up to seven months, in the uterus. Within a few days of leaving their winter shelter, females ovulate one egg, and sperm are released. Fertilization and implantation then take place shortly afterwards. Typically, females of a population form a maternity colony at a site different than the hibernation site where breeding occurred. Gestation usually lasts from 40 to 50 days and results in a single offspring, usually in the late spring. Birth is a rather uneasy process: hanging inverted, mothers grab the newborn as it emerges from the birth canal and the newborn in turn grabs the abdominal fur of the mother with its hind feet, pulling to facilitate its own birth. Infants usually begin nursing almost immediately after birth. Some reports are indicative of females helping others with the birth of young. Healthy species in the wild live from five to ten years. Probably polygynous.

CONSERVATION STATUS
Classified as Lower Risk/Least Concern by the IUCN.

SIGNIFICANCE TO HUMANS
These bats have occasionally been reported with rabies. Insectivorous bats such as mormoopids eat large quantities of insects, including those that are harmful to crops and humans.

Pteronotus parnellii
Mormoops megalophylla

For example, a 0.5 oz (13 g) individual could easily consume more than 1,000 insects, including mosquitoes, on an average night. ◆

Ghost-faced bat
Mormoops megalophylla

TAXONOMY
Mormoops megalophylla (Peters, 1864), Coahuila, Mexico. Four subspecies are currently recognized.

OTHER COMMON NAMES
Spanish: Murcielago de labios festoneados.

PHYSICAL CHARACTERISTICS
These bats have a peculiar upturned nosed and extensive flaps of skin around the mouth. Smaller medium-sized bats, ghost-faced bats have forearms 2.0–2.2 in (5.1–5.7 cm) long. They weigh from 0.4 to 0.7 oz (12–19 g). They are larger than Antillean ghost-faced bats, the other species in the genus.

DISTRIBUTION
Found from southwestern Texas and Arizona southwards through Baja California, Mexico, Honduras and El Salvador. There is a second population in South America, the northern parts of Colombia and Venezuela, the Dutch Antilles and Trinidad, and another along the Pacific coasts of Colombia, Ecuador and northern Peru.

HABITAT
Occur in humid through semi-arid and arid regions from tropical forests to riparian forests and arid coastal regions. Ghost-faced bats roost in hollows, typically in caves and abandoned mines.

BEHAVIOR
Strong, fast fliers that forage over water, land, and in forests. Males and females may roost in different parts of the same cave or mine. Roosting individuals usually not in physical contact with one another. Colonies of ghost-faced bats can number thousands of individuals which tend to emerge together in streams.

FEEDING ECOLOGY AND DIET
They eat flying insects, including moths, beetles and flies, reflecting on the availability of prey where they forage. Ghost-faced bats eat flying insects, taking a wide range of prey according to what is available where they forage.

REPRODUCTIVE BIOLOGY
Females bear a single young each year, with the timing of birth reflecting local rainy seasons. This species is most likely polygynous.

CONSERVATION STATUS
Not listed by the IUCN.

SIGNIFICANCE TO HUMANS
These bats are subject to cyclical outbreaks of rabies that cause mass mortality. ◆

Common name / Scientific name	Physical characteristics	Habitat and behavior	Distribution	Diet	Conservation status
Antillean ghost-faced bat *Mormoops blainvillii*	Light brown upperparts, buffy underparts. In dark phase, upperparts are dark brown, underparts are ochraceous tawny. Lower lip has fleshy peg-like projections. Tail is well developed. Head and body length 2–2.9 in (5–7.3 cm), tail length 0.7–1.2 in (1.8–3.1 cm), forearm length 1.8–2.4 in (4.5–6.1 cm), average adult weight 0.4–0.6 oz (12–18g).	Variety of habitats from desert scrub to tropical forest. Extremely swift flight, movement of wings makes humming sound. Dwells deeper in caves than any other Jamaican bat.	Greater Antilles and adjacent small islands.	Most likely consists mainly of insects.	Lower Risk/Near Threatened
Davy's naked-backed bat *Pteronotus davyi*	Delicate, naked bat. Wings joined on back. Coat color is usually coffee, sometimes orange. Head and body length 1.6–2.2 in (4.2–5.5 cm), tail length 0.7–1 in (1.8–2.5 cm), forearm length 1.7–1.9 (4.3–4.9 cm), weight 0.17–0.35 oz (5–10 g).	Dry territories of the province of Guanacaste, and the environs of Quepos to the northern Caribbean slope from sea level to 1,310 ft (400 m). Take refuge in caverns with high temperatures. Nocturnal.	Northwestern Peru and northern Venezuela to southern Baja California, southern Sonora, and Nuevo Leon, Mexico; Trinidad; and southern Lesser Antilles.	Mainly insects.	Not threatened
Big naked-backed bat *Pteronotus gymnonotus*	Naked back, wings united on mean line, hindquarters are naked with a very small coat. Coat color is orange coffee. Head and body length 2.1–2.7 in (5.5–6.9 cm), tail length 0.8–1.1 in (2.1–2.8 cm), forearm length 2–2.2 in (5–5.5 cm), weight 0.38–0.63 oz (11–18 g).	Low territories of the Caribbean and Pacific slopes, from sea level to 4,920 ft (1,500 m). Refuge is usually caverns. Gregarious and nocturnal.	Southern Veracruz, Mexico, to Peru, north-eastern Brazil, and Guyana.	Insects caught in the air.	Not threatened
Macleay's moustached bat *Pteronotus macleayii*	Skin folds are on chin and lower lip. Brownish in color. Hindquarters are naked with a small coat. Average weight 0.14–0.21 oz (4–6 g), wingspan 9.8–11 in (25–28 cm).	Large chambers and passageways far from the cave entrance. Roost in large colonies.	Habana and Guanabacoa, Cuba.	Mainly insects.	Vulnerable
Wagner's moustached bat *Pteronotus personatus*	Coat color is orange or coffee, both equally common. Head and body length 1.7–2.2 in (4.3–5.5 cm), tail length 0.6–0.8 in (1.5–2 cm), forearm length 1.6–1.9 in (4.2–4.8 cm).	Dry and humid low territories of the Pacific slope, from sea level to 1,310 ft (400 m). Roost in large colonies. Nocturnal.	Colombia, Peru, Brazil, and Suriname to south-ern Sonora and southern Tamaulipas, Mexico; and Trinidad.	Insects.	Not threatened
Sooty moustached bat *Pteronotus quadridens*	Coloration is variable, oftentimes light or dark brown, grayish brown, or ochraceous orange. Underparts are usually paler. Head and body length 1.5–3 in (4–7.7 cm), tail length 0.6–1.2 in (1.5–3 cm), forearm length 1.4–2.6 in (3.5–6.5 cm).	Roost in caves and tunnels, as well as houses. May also shelter in hollows of plants. Seek darker recesses when in caves. Generally hang singly rather than in compact masses.	Cuba, Jamaica, Hispaniola, and Puerto Rico.	Mainly insects.	Lower Risk/Near Threatened

Resources

Books

Hutson, A. M., S. P. Mickelburgh, and P. A. Racey. *Global Status Survey and Conservation Action Plan, Microchiropteran bats.* Gland, Switzerland: IUCN SSC Chiroptera Specialist Group, 2001.

Nowak, R. M. *Walker's Mammals of the World.* Vol. 1. Baltimore: Johns Hopkins University Press, 1999.

Reid, F. A. *A Field Guide to the Mammals of Central America and Southeast Mexico.* New York: Oxford University Press, 1997.

Periodicals

Adams, J. K. "*Pteronotus davyi.*" *Mammalian Species* 346 (1989): 1–5.

Herd, R. M. "*Pteronotus parnellii.*" *Mammalian Species* 209 (1983): 1–5.

Lancaster, W. C., and E. K. V. Kalko. "*Mormoops blainvilli.*" *Mammalian Species* 544 (1996): 1–5.

Rezsutek, M., and G. N. Cameron. "*Mormoops megalophylla.*" *Mammalian Species* 448 (1993): 1–5.

Rodriguez-Duran, A., and T. H. Kunz. "*Pteronotus quadridens.*" *Mammalian Species* 395 (1992): 1–4.

Melville Brockett Fenton, PhD

Bulldog bats

(Noctilionidae)

Class Mammalia

Order Chiroptera

Suborder Microchiroptera

Family Noctilionidae

Thumbnail description

Noctilionids get their common name from their distinctive full and pendant lips and cheek pouches, which makes them look much like a bulldog; their legs are long and their feet and claws are large; give off a pungent, fishy odor

Size

Large bats ranging 2.1–3.6 in (54–92 mm) in forearm length and weighing 0.6–7.8 oz (18–78 g)

Number of genera, species

1 genus; 2 species

Habitat

Tropical lowland areas near water

Conservation status

Overall not threatened, though regional populations may be at risk

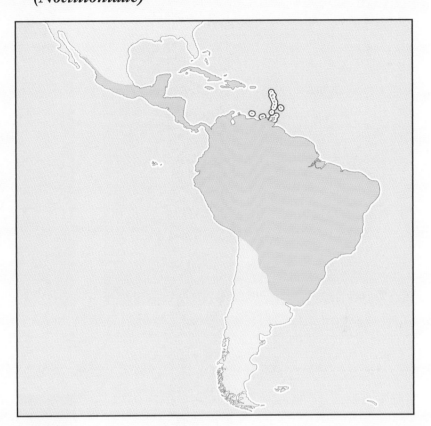

Distribution

Neotropical

Evolution and systematics

The systematic affinities of the family Noctilionidae were debated for many years. Recent analyses of large molecular and morphological datasets provide strong evidence that these bats are closely allied with the other exclusively neotropical bat families, Mormoopidae and Phyllostomidae, and with the endemic New Zealand family, Mystacinidae, that together form the superfamily Noctilionoidea.

Analyses of molecular variation among populations of the two noctilionid species and their close relatives, the mormoopids, suggest that the evolution of a fish-eating lifestyle in *Noctilio leporinus* is relatively recent, within three million years. These data also suggest the possibility that *Noctilio albiventris* is actually composed of two, distinct lineages that diverged from each other approximately one to three million years ago. Additional data are needed to evaluate fully this possibility.

Fossils of an extinct species of *Noctilio* are now known from the late Miocene in the Amazon Basin.

Physical characteristics

The muzzle and elongate feet and legs of bulldog bats are their most distinctive features. The nose and muzzle are simple, but a projecting nose-pad, enlarged and swollen lips, pronounced cheek pouches, and folds of skin on the lower lip give these bats a bulldog-like appearance. The ears are separate, long, and narrow, with a lobed tragus. The tail emerges dorsally from the tail membrane about one-third of its length from the body. The calcar, or heel extension, of these bats is long and bony. Both species have long legs with well-developed feet and claws, but the modification of the claws and feet into large, razor-sharp, and hydrodynamically efficient gaffing mechanisms in *Noctilio leporinus* is truly astounding.

Bulldog bats are fairly large, with short pelage ranging in color from bright orange to orange-brown and gray-brown. Fur color was previously thought to represent differences between sexes, with males being orange and females brown. It is now understood that color is quite variable within each species. A white line, which may or may not be distinct, runs

The greater bulldog bat (*Noctilio leporinus*) does not have a nose leaf. (Photo by © Merlin D. Tuttle, Bat Conservation International. Reproduced by permission.)

A greater bulldog bat (*Noctilio leporinus*) searches for fish under the water's surface. (Photo by Frans Lanting/Minden Pictures. Reproduced by permission.)

The lesser bulldog bat (*Noctilio albiventris*) feeds mostly on insects. (Photo by P. V. August/Mammal Images Library of the American Society of Mammologists.)

A greater bulldog bat (*Noctilio leporinus*) in low search flight. (Photo by Stephen Dalton/Photo Researchers, Inc. Reproduced by permission.)

A greater bulldog bat (*Noctilio leporinus*) fishing. (Photo by N.H.P.A./ANTPhoto.com.au. Reproduced by permission.)

down the center of the back from the shoulder region to the rump. Wing and tail membranes are brown.

Total body lengths range 2.2–5.1 in (57–132 mm), forearm lengths 2.1–3.6 in (54–92 mm), and weights range 0.6–7.8 oz (18–78 g). Males are significantly larger than females in both species.

Both species have well-developed teeth typical of most insectivorous bats. The dental formula is: I2/1 C1/1 P1/2 M3/3.

Distribution

Bulldog bats are found throughout much of the Neotropics, from southern Mexico to northern Argentina and southeastern Brazil, and in the Antilles.

Habitat

Bulldog bats are found in a variety of habitats, most commonly in moist, tropical lowland areas that are close to water. They have been recorded roosting in hollow trees and sea caves.

The lesser bulldog bat (*Noctilio albiventris*) stores food in its cheek pouches while flying. (Photo by Brock Fenton. Reproduced by permission.)

A greater bulldog bat, or fishing bat (*Noctilio leporinus*), cruises above the water and uses echolocation to locate fish near the surface of the water. It then scrapes the surface with its feet and grabs the fish. (Photo by Animals Animals ©Partridge, OSF. Reproduced by permission.)

A greater bulldog bat (*Noctilio leporinus*) in flight with a fish in its feet. (Photo by © Merlin D. Tuttle, Bat Conservation International. Reproduced by permission.)

Behavior

Information on behavior is limited to general observations on roosting and activity patterns. Few studies of social or mating behaviors have been conducted. The two species differ in activity patterns throughout the night, but are similar in their choice of roosts (hollow trees, caves, and human-made structures), reproductive biology, and general foraging behavior, though *N. leporinus* preys mainly on fish, and *N. albiventris* primarily on insects.

Feeding ecology and diet

Noctilio leporinus is best known for its remarkable set of adaptations to piscivory. Only one other bat species worldwide is known to prey on fish: *Myotis vivesi* (Vespertilionidae). *N. albiventris* is primarily insectivorous, and *N. leporinus* also includes a large proportion of insects in the diet. Both species use their large cheek pouches to store partially chewed food, possibly allowing them to continue foraging without having to stop to consume their prey fully, as in most bats.

Bulldog bats use constant frequency and frequency-modulated echolocation to detect prey at the surface of the water and in flight. The rear feet are dragged through the water and used to gaff prey at the surface. The stiff, elongate calcars are used to elevate the tail membrane above the surface of the water, minimizing drag. Prey are sometimes transferred to the tail membrane and then to the mouth during flight.

Reproductive biology

Field studies suggest that females of both species give birth to a single young each year, with the majority of mating in November and December and births occurring in April through June. However, pregnancy and births have been recorded outside of these seasons in different parts of their ranges and a secondary reproductive peak has been recorded for *Noctilio leporinus* in Cuba. Young are able to fly at about one month of age. Males and females seem to attend the young during their first month of life, suggesting extensive biparental care. A single male typically lives with a harem of females in a polygynous organization. Mating behaviors are unknown, but a warty out-pocket of skin on the scrotum in males, and scent-producing glands associated with it, may be involved.

Conservation status

Populations seem to be stable, although bulldog bats will be negatively affected by water contamination and habitat destruction. *Noctilio leporinus* has been proposed as an indicator species for water quality because of the likelihood that contaminants will accumulate in their tissues through predation on aquatic insects and fish.

Significance to humans

Bulldog bats eat insects that may be considered pests by humans.

The bulldog bat (*Noctilio* sp.) secretes an oil that has a musky odor. (Photo by B. G. Thomson. Reproduced by permission.)

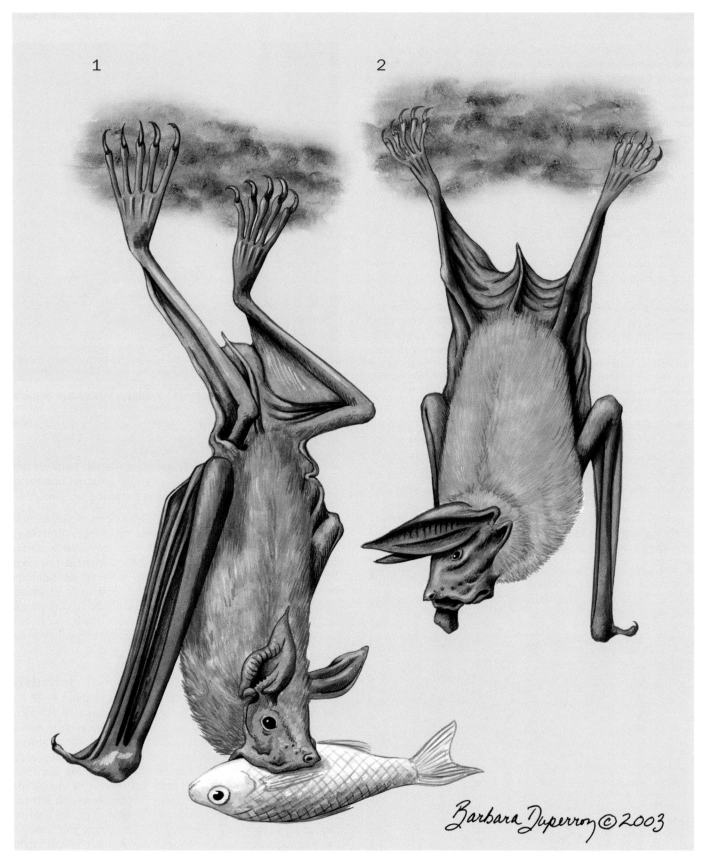

1. Greater bulldog bat (*Noctilio leporinus*); 2. Lesser bulldog bat (*Noctilio albiventris*). (Illustration by Barbara Duperron)

Species accounts

Greater bulldog bat
Noctilio leporinus

TAXONOMY
Noctilio leporinus (Linnaeus, 1758), Suriname. Three subspecies are recognized.

OTHER COMMON NAMES
English: Fishing bat, bulldog bat.

PHYSICAL CHARACTERISTICS
Head and body length 4.6–5 in (119–127 mm); forearm length longer than 2.9 in (75 mm); hindfoot length >0.9 in (>25 mm); combined foot and tibia length: >1.9 in (>50 mm); wingspan 19.6 in (500 mm); tail length 0.9–1.1 in (25–28 mm); ear length 1.10–1.16 in (28–29.5 mm); weighing more than 1.7 oz (50 g). Vary in size geographically, with larger individuals in the northern and southern parts of the range, and smaller individuals in the Amazon Basin. Males are larger than females; they can be distinguished from their smaller congeners, lesser bulldog bats, by their larger size and more highly modified feet.

DISTRIBUTION
Found from Sinaloa and Veracruz in Mexico south to northern Argentina. They are also found throughout many of the Antillean islands.

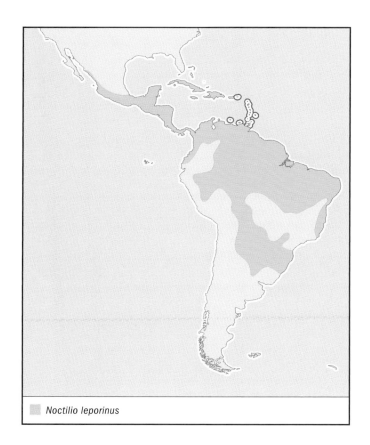

Noctilio leporinus

HABITAT
Restricted to moist, lowland, and coastal areas within their range, including major river basins, coastal embayments, and lakes.

BEHAVIOR
Colonies of up to several hundred individuals have been found roosting in hollow trees and sea caves. They have been found roosting in association with several other species of bats. They are active throughout the night; groups of five to 15 animals emerge to forage together. Time of roost emergence may be influenced by temperature.

FEEDING ECOLOGY AND DIET
Famous for their fish-eating habits. Aquatic and flying insects also comprise a large portion of the diet and they have been known to eat shrimp and crabs. They have several notable adaptations to their piscivorous lifestyle, including their greatly enlarged feet and claws, elongated calcar, robust and sharp teeth, and specializations of the stomach that allow them to ingest very large prey, including fish up to 4.7 in (120 mm) long. The bats fly low over water bodies while foraging and emit echolocation pulses, which allow them to detect tiny surface disturbances. They then use their long, sharp claws to drag through the water and gaff their prey.

REPRODUCTIVE BIOLOGY
Information on reproduction is mainly available from Central American and Antillean sites and breeding season may vary regionally. Mating seems to coincide with wet seasons, when insects and fish are most abundant; it typically begins in November and December, with births occurring from April through June. A second mating season has been suggested in parts of the range, with mating occurring in the summer and births from October–December. Males and females may contribute to caring for offspring. Young begin to emerge from roosts and fly at about one month of age. A captive lived to 11 years and six months old. This species is probably polygynous.

CONSERVATION STATUS
Currently abundant throughout their range.

SIGNIFICANCE TO HUMANS
Control insect populations and may act as indicator species of water contaminant levels. ◆

Lesser bulldog bat
Noctilio albiventris

TAXONOMY
Noctilio albiventris Desmarest, 1818, Bahia, Brazil. Formerly known as *Noctilio labialis*, four subspecies are recognized.

OTHER COMMON NAMES
None known.

Noctilio albiventris

PHYSICAL CHARACTERISTICS
Head and body length 2.5–2.6 in (65–68 mm); forearm length 2.7 in (70 mm); foot length <0.7 in (<20 mm); combined foot and tibia length <1.5 in (<40 mm); tail length 0.5–0.6 in (13–16 mm); ear length 0.8–0.9 in (22–24 mm); wingspan 11.2–14.9 in

(285–380 mm). Males are significantly larger than females in body measurements and in the development of the sagittal crest. Distinguished from greater bulldog bats by their smaller size and less extravagantly modified feet.

DISTRIBUTION
Occurs from Honduras and western Guatemala in the north to Argentina, Paraguay, and coastal Brazil in the south.

HABITAT
Forage over rivers, streams, and wetlands in wet, tropical forests up to an altitude of 3,610 ft (1,100 m).

BEHAVIOR
Available evidence suggests that there are two activity peaks in a night, one soon after dusk and another after midnight. Foraging occurs in small groups over bodies of water. Roosts are occupied by small to large groups of bats; they can be found roosting with *Molossus* species.

FEEDING ECOLOGY AND DIET
Primarily insectivorous. Analyses of stomach contents and feces suggest that they eat a wide variety of insects, some of which are taken from the water. In some areas, individuals have been found with fish scales, pollen, and fruit seeds in their stomachs as well. They forage in groups of eight to 15.

REPRODUCTIVE BIOLOGY
Most likely polygynous. Females give birth to one young per year. Breeding season may vary regionally, but available evidence suggests that breeding occurs once yearly. Pregnant females have been recorded primarily from February through May. A captive infant was nursed for nearly three months.

CONSERVATION STATUS
Currently abundant throughout their range.

SIGNIFICANCE TO HUMANS
Control populations of potential insect pests. ◆

Resources

Books

Hill, John E., and James D. Smith. *Bats, A Natural History*. London: British Museum of Natural History, 1984.

Koopman, Karl F. "Chiroptera: Systematics." In *Handbook of Zoology, Volume VIII, Mammalia*, edited by J. Niethammer, H. Schliemann, and D. Starck. Berlin: Walter de Gruyter, 1994.

McKenna, Malcolm C., and Susan K. Bell, eds. *Classification of Mammals Above the Species Level*. New York: Columbia University Press, 1997.

Nowak, Ronald M. *Walker's Bats of the World*. Baltimore: Johns Hopkins University Press, 1994.

Simmons, Nancy B., and Jonathan H. Geisler. "Phylogenetic Relationships of *Icaronycteris, Archaeonycteris, Hassianycteris*, and *Palaeochiropteryx* to Extant Bat Lineages, with Comments on the Evolution of Echolocation and Foraging Strategies in Microchiroptera." *Bulletin of the American Museum of Natural History, Number 235*. New York: American Museum of Natural History, 1998.

Periodicals

Altenbach, J. Scott. "Prey Capture by the Fishing Bats *Noctilio leporinus* and *Myotis vivesi*." *Journal of Mammalogy* 70 (1989): 421–424.

Brooke, Anne P. "Diet of the Fishing Bat, *Noctilio leporinus* (Chiroptera: Noctilionidae)." *Journal of Mammalogy* 75 (1994): 212–218.

Czaplewski, Nicholas J. "Opossums (Didelphidae) and Bats (Noctilionidae and Molossidae) from the Late Miocene of the Amazon Basin." *Journal of Mammalogy* 77 (1996): 84–94.

Fenton, M. B., et al. "Activity Patterns and Roost Selection by *Noctilio albiventris* (Chiroptera: Noctilionidae) in Costa Rica." *Journal of Mammalogy* 74 (1993): 607–613.

Fish, Frank E., Brad R. Blood, and Brian D. Clark. "Hydrodynamics of the Feet of Fish-catching Bats: Influence of the Water Surface on Drag and Morphological Design." *The Journal of Experimental Zoology* 258 (1991): 164–173.

Resources

Hood, Craig, S., and Jay Pitocchelli. "*Noctilio albiventris*." *Mammalian Species* 197 (1983): 1–5.

Hood, Craig S., and J. Knox Jones, Jr. "*Noctilio leporinus*." *Mammalian Species* 216 (1984): 1–7.

Lewis-Oritt, Nicole, Ronald A. Van Den Bussche, and Robert J. Baker. "Molecular Evidence for Evolution of Piscivory in *Noctilio* (Chiroptera: Noctilionidae)." *Journal of Mammalogy* 82 (2001): 748–759.

Mendez, L., and S. T. Alvarez-Castañeda. "Comparative Analysis of Heavy Metals in Two Species of Ichthyophagous Bats *Myotis vivesi* and *Noctilio leporinus*." *Environmental Contamination and Toxicology* 65 (2000): 51–54.

Roverud, Roald C., and Mark A. Chappell. "Energetic and Thermoregulatory Aspects of Clustering Behavior in the Neotropical Bat *Noctilio albiventris*." *Physiological Zoology* 64 (1991): 1527–1541.

Van Den Bussche, Ronald A., and Steven R. Hoofer. "Further Evidence for Inclusion of the New Zealand Short-tailed Bat (*Mystacina tuberculata*) within Noctilionoidea." *Journal of Mammalogy* 81 (2000): 865–874.

Van Den Bussche, Ronald A., Steven R. Hoofer, and Nancy B. Simmons. "Phylogenetic Relationships of Mormoopid Bats Using Mitochondrial Gene Sequences and Morphology." *Journal of Mammalogy* 83 (2002): 40–48.

Tanya Dewey

New Zealand short-tailed bats

(Mystacinidae)

Class Mammalia
Order Chiroptera
Suborder Microchiroptera
Family Mystacinidae

Thumbnail description
New Zealand short-tailed bats are medium-sized, robust bats; they have evolved a suite of unusual characteristics, including a unique wing-folding mechanism, additional spurs below each claw, and robust legs that permit efficient movement along the ground and tree trunks; only bats demonstrated to be omnivorous

Size
Forearm length 1.5–1.8 in (40–48 mm); total length 2.3–3.5 in (60–90 mm); wingspan 11–12.2 in (280–310 mm); ear lengths 0.6–0.7 in (17.5–18.6 mm); weight 0.4–1.2 oz (11–35 g)

Number of genera, species
1 genus; 2 species

Habitat
Native forests and scrub habitats of New Zealand

Conservation status
Extinct: 1 species; Vulnerable: 1 species

Distribution
New Zealand

Evolution and systematics

Mystacinidae has been aligned with several other bat superfamilies, including Vespertilionoidea, Emballonuroidea, and Noctilionoidea. Traditional classifications have primarily aligned Mystacinidae with the vespertilionid family Molossidae. Assessing relationships between Mystacinidae and other bat families with morphology has proven difficult since so many mystacinid morphological features are unique. Recent analyses of molecular evidence, including DNA hybridization, immunology, and mitochondrial DNA sequences, along with systematic analyses of morphological traits, have provided strong evidence that Mystacinidae is a member of the Noctilionoidea, including the bat families Noctilionidae, Mormoopidae, and Phyllostomidae. This finding lends support to the hypothesis that the ancestral lineage leading to Mystacinidae originated in the New World tropics.

Physical characteristics

New Zealand short-tailed bats are robustly built, with stocky bodies and short, stout legs. The feet are broad, short, and are positioned under the body during movement on the ground and in trees, unlike other bats. The feet have grooved soles and needle-sharp claws, which are further modified with denticles, or talons, at the base of each claw and the thumb. These modifications of the legs, feet, and claws make New Zealand short-tailed bats capable of running on the ground, climbing, and burrowing.

The dark brown fur is short, thick, and velvety, with white-tipped hairs that give it a frosted appearance. The head tapers to a slender, truncated muzzle with prominent, oblong nostrils and a set of stiff whiskers. The ears and tragi are long, slender, and pointed. The wing membranes are unique among bats, being thickened and leathery along the body, forearm, and lower leg. The remainder of the wing membrane can be folded under this thickened portion, protecting it during climbing and running. The tail emerges from the dorsal surface of the tail membrane.

The teeth are cuspidate, with chisel-like incisors. The number of teeth is reduced through loss of incisors and premolars to a total of 28 (I1/1 C1/1 P2/2 M3/3). The tongue is extensible and covered with small projections (papillae), and there is a gap between the front incisors.

A lesser New Zealand short-tailed bat (*Mystacina tuberculata*) on wood. (Photo by Geoff Moon/FLPA–Images of Nature. Reproduced by permission.)

Distribution

New Zealand short-tailed bats are found only in New Zealand. Both species are thought to have once occurred throughout the islands of New Zealand. *Mystacina tuberculata* is now restricted to a portion of its former range. *Mystacina robusta* was restricted to a single locality in recent history and is now presumed extinct, having not been observed since 1965.

Habitat

Both species of *Mystacina* occupy the moist forests of New Zealand and muttonbird scrub (*Olearia* sp.) habitats found on certain islands. Roosts are found solely within native, broadleaf forests, although foraging may occur in scrub habitats and along coastlines. Roosts are typically in trees but may also occur in caves, burrows, and houses.

Behavior

Mystacinid bats are unusual for their ability to run, climb, and burrow. They have been described as having rodent-like agility and are frequently seen scurrying along tree branches and the ground. Their robust and modified limbs and feet are clearly adaptations for a more terrestrial lifestyle than is found

in most bats, except perhaps vampire bats. They burrow under leaf litter in search of food and excavate their own tunnels and roosts in rotten logs.

They are active primarily at night, emerging from roosts several hours after dusk. Flight is typically slow and low, with bats flying 6.5–9.8 ft (2–3 m) off the ground. Echolocation calls are frequency modulated sweeps with maximum energy between 60 and 65 kHz. They become inactive during cold weather but do not hibernate, as they emerge to forage in winter when the weather is warm.

Feeding ecology and diet

New Zealand short-tailed bats have an unusually broad diet: they are essentially omnivorous, which is unusual among bats. Foods eaten include arthropods taken in flight and from surfaces, nectar, pollen, and fruit. They have also been observed eating nestling and adult birds and chewing meat and fat from muttonbirds killed by humans.

Reproductive biology

These bats reproduce once yearly, giving birth to one young. Births may be synchronized. The season of mating and births seems to vary with latitude. The mating system of these bats is not known.

A lesser New Zealand short-tailed bat (*Mystacina tuberculata*) in flight. (Photo by Nic Bishop. Reproduced by permission.)

Conservation status

New Zealand short-tailed bat populations have been negatively affected by introduced species such as rats, stoats, and cats, which prey on them in roosts and while foraging. They are particularly vulnerable because they forage on the ground and may occupy roosts accessible to predators. The introduction of rats to islands occupied by greater New Zealand short-tailed bats is likely to have resulted in their extinction. Destruction of native forests also threatens these bats, as they rely on native trees for roost sites and food.

Significance to humans

If *Mystacina robusta* is, indeed, extinct, then *Mystacina tuberculata* is the only surviving member of a unique bat family endemic to New Zealand. These bats represent a unique evolutionary history and are part of the rich cultural and natural heritage of New Zealand. Their nectar-feeding and insect-eating habits also make them valuable pollinators of native trees and predators of insect pests.

1. Lesser New Zealand short-tailed bat (*Mystacina tuberculata*); 2. Greater New Zealand short-tailed bat (*Mystacina robusta*). (Illustration by Barbara Duperron)

Species accounts

Lesser New Zealand short-tailed bat
Mystacina tuberculata

TAXONOMY

Mystacina tuberculata Gray, 1843, New Zealand. Three subspecies are recognized.

OTHER COMMON NAMES

English: Lesser short-tailed bat, northern short-tailed bat, New Zealand long-eared bat.

PHYSICAL CHARACTERISTICS

The smallest of the New Zealand short-tailed bats. Total length is 2.3–2.6 in (60–68 mm); forearm length 1.5–1.7 in (40–45 mm); wingspan 11–11.4 in (280–290 mm); weight 0.38–0.52 oz (11–15 g) (up to 0.65 oz [18.5 g] in pregnant females). There is considerable variation in size among the three subspecies, with body size increasing toward the south. Fur is dark brown, short, thick, and velvety with white-tipped hairs.

DISTRIBUTION

Based on current distribution and the locations of subfossils, the former distribution of *M. tuberculata* included all of the New Zealand islands. Populations of the three subspecies are now found only in portions of their former range. Kauri forest short-tailed bats (*M. t. aupourica*) are found in Omahuta kauri forest, Northland, and Little Barrier Island. Volcanic plateau short-tailed bats (*M. t. rhyacobia*) are found in the podocarp-hardwood and *Nothofagus* forests of the volcanic plateau. Southern short-tailed bats (*M. t. tuberculata*) are currently found only in Tararua Forest Park on North Island, in Northwest Nelson Forest Park on South Island, and on Codfish Island. (Specific distribution map not available.)

HABITAT

Found primarily in the native forests of New Zealand. They are sometimes observed flying along coastlines and foraging in the scrub habitats (*Olearia* sp.) in which petrel and muttonbird breeding colonies occur. Roosts have been found in kauri (*Agathis australis*), rimu (*Dacrydium cupressinum*), totara (*Podocarpus totara* and *P. hallii*), southern rata (*Metrosideros umbellata*), kamahi (*Weinmannia racemosa*), and beech (*Nothofagus* sp.) trees and in granitic and pumice caves, as well as houses.

BEHAVIOR

Roost in small groups, emerging one to two hours after dark to forage in forested and scrub habitats.

During the breeding season, males and females establish separate colonies. Individual males travel, after dark, to small, hollow trees where they utter a repetitive, high-intensity call, which has been likened to a song. Radio-tagged females were recorded visiting calling males each evening before beginning their foraging flights. This provocative observation suggests a lek-mating system in Mystacina, but requires further investigation.

FEEDING ECOLOGY AND DIET

Eat both flying and non-flying arthropods, nectar, pollen, fruit, and other plant material. They have also been observed eating nestling and adult birds and chewing meat and fat from harvested seabirds.

REPRODUCTIVE BIOLOGY

One young is born each year. Mating has been observed in autumn, but may occur throughout winter and spring. Births occur in the austral summer (December and January) in northern populations and later (April–May) in southern populations. Young develop quickly, being able to fly at 4–6 weeks and reaching adult size at 8–12 weeks. Mating system not known, but thought to be a lek-mating system.

CONSERVATION STATUS

Restricted to a portion of their former range, leaving them in isolated populations. They are vulnerable to predation by introduced species such as stoats, cats, and rats, and to the destruction of their forest habitat. They are currently listed as Vulnerable by the IUCN and in the New Zealand Red Data Book, but may be moved into the Endangered category.

SIGNIFICANCE TO HUMANS

Likely pollinators of native trees and may control arthropod pest populations. ◆

Greater New Zealand short-tailed bat
Mystacina robusta

TAXONOMY

Mystacina robusta Dwyer, 1962, New Zealand.

OTHER COMMON NAMES

English: Greater short-tailed bat, southern short-tailed bat, Stewart Island short-tailed bat.

PHYSICAL CHARACTERISTICS

The larger of the two mystacinid species. Total length 3.5 in (90 mm); forearm length 1.7–1.8 in (45–48 mm); wingspan 11.4–12.2 in (290–310 mm). Size may have decreased from north to south. Appearance is similar to *M. tuberculata*.

DISTRIBUTION

Subfossil finds indicate these bats once occurred throughout the New Zealand islands. However, they have not been collected from the three main islands since the beginning of European colonization in 1840. Until 1965, living greater short-tailed bats were known only from two, rat-free islands off Stewart Island. (Specific distribution map not available.)

HABITAT

The islands on which these bats were found in recent times are primarily composed of a scrub habitat, with a central area of broadleaf forest. Known roost sites were in granitic caves along the coastline.

BEHAVIOR

May have been even more terrestrial than their smaller congener. They were known to occupy seabird burrows, dig in the soil, and were remarkably agile on the ground. They emerged from roosts one to two hours after dark and tended to fly low to the ground. They did not appear to hibernate, being observed flying throughout the austral winter.

FEEDING ECOLOGY AND DIET
Probably shared the broad diet of their living relative. Stomach analysis of two specimens yielded pollen and fern spores and they were observed eating the meat and fat of harvested seabirds.

REPRODUCTIVE BIOLOGY
The few existing observations suggest these bats had one young yearly, but the timing of mating and births is unknown.

CONSERVATION STATUS
Have not been observed since 1965 and are presumed Extinct.

SIGNIFICANCE TO HUMANS.
Were likely pollinators of native trees and predators of insect pests. ◆

Resources

Books
Daniels, M. J. "Greater Short-tailed Bat." In *The Handbook of New Zealand Mammals*, edited by Carolyn M. King. Auckland: Oxford University Press, 1995.

Daniels, M. J. "Lesser Short-tailed Bat." In *The Handbook of New Zealand Mammals*, edited by Carolyn M. King. Auckland: Oxford University Press, 1995.

Hill, John E., and James D. Smith. *Bats, A Natural History.* London: British Museum of Natural History, 1984.

Koopman, Karl F. "Chiroptera: Systematics." In *Handbook of Zoology, Volume VIII, Mammalia*, edited by J. Niethammer, H. Schliemann, and D. Starck. Berlin: Walter de Gruyter, 1994.

McKenna, Malcolm C., and Susan K. Bell, eds. *Classification of Mammals Above the Species Level.* New York: Columbia University Press, 1997.

Nowak, Ronald M. *Walker's Bats of the World.* Baltimore: Johns Hopkins University Press, 1994.

Simmons, Nancy B., and Jonathan H. Geisler. "Phylogenetic Relationships of *Icaronycteris, Archaeonycteris, Hassianycteris,* and *Palaeochiropteryx* to Extant Bat Lineages, with Comments on the Evolution of Echolocation and Foraging Strategies in Microchiroptera." *Bulletin of the American Museum of Natural History*, Number 235. New York: American Museum of Natural History, 1998.

Periodicals
Arkins, A. M., A. P. Winnington, S. Anderson, and M. N. Clout. "Diet and Nectarivorous Foraging Behaviour of the Short-tailed Bat (*Mystacina tuberculata*)." *Journal of Zoology* 247 (1999): 183–187.

Hill, J. C., and M. J. Daniel. "Systematics of the New Zealand Short-tailed Bat *Mystacina* Gray, 1843 (Chiroptera: Mystacinidae)." *Bulletin of the British Museum of Natural History, Zoology* 48 (1985): 279–300.

Mayer, G. C., J. A. W. Kirsch, J. M. Hutcheon, F. J. Lapointe, and J. Gingras. "On the Valid Name of the Lesser New Zealand Short-tailed Bat (Mammalia: Chiroptera)." *Proceedings of the Biological Society of Washington* 112 (1999): 470–490.

Parson, Stuart. "Search-phase Echolocation Calls of the New Zealand Lesser Short-tailed Bat (*Mystacina tuberculata*) and Long-tailed Bat (*Chalinolobus tuberculatus*)." *Canadian Journal of Zoology* 75 (1997): 1487–1494.

Van Den Bussche, Ronald A., and Steven R. Hoofer. "Further Evidence for Inclusion of the New Zealand Short-tailed Bat (*Mystacina tuberculata*) within Noctilionoidea." *Journal of Mammalogy* 81 (2000): 865–874.

Van Den Bussche, Ronald A., Steven R. Hoofer, and Nancy B. Simmons. "Phylogenetic Relationships of Mormoopid Bats Using Mitochondrial Gene Sequences and Morphology." *Journal of Mammalogy* 83 (2002): 40–48.

Webb, P. I., J. A. Sedgeley, and C. F. J. O'Donnell. "Wing Shape in New Zealand Lesser Short-tailed Bats (*Mystacina tuberculata*)." *Journal of Zoology* 246 (1998): 462–465.

Worthy, T. H., M. J. Daniel, and J. E. Hill. "An Analysis of Skeletal Size Variation in *Mystacina robusta* Dwyer, 1962 (Chiroptera: Mystacinidae)." *The Royal Society of New Zealand* 23 (1996).

Tanya Dewey

▲
Funnel-eared bats
(Natalidae)

Class Mammalia
Order Chiroptera
Suborder Microchiroptera
Family Natalidae

Thumbnail description
Small slim-bodied bats with soft long fur; legs proportionally very long; large pale ears with distinctive papillae; facial skin pale; eyes tiny; tail longer than head and body

Size
Head and body length: 2 in (5 cm); tail 1.7–2.4 in. (4.3–6.0 cm); forearm length 1.4–1.8 in (3.5–4.5 cm); weight 1.4–2.5 oz. (4–7 g)

Number of genera, species
1 genus; 5 species

Habitat
Dry and semi-deciduous forest, secondary growth; occasionally in primary forest

Conservation status
Vulnerable: 1 species; Lower Risk/Near Threatened: 1 species

Distribution
Central and South America, Caribbean Islands

Evolution and systematics

Known by fossils and subfossils back to the Pleistocene, this family speciated in the Antilles, which have three of today's five species. One species has a very restricted distribution, being known only from two small islands in the Bahamas. This island-by-island speciation pattern parallels that of Old World fruit bats in Southeast Asia. In diet, flight pattern, and body form, this genus shows remarkable parallels with the unrelated Old World bat genus *Kerivoula*, the painted bats of Africa and Southeast Asia. Recent analysis of DNA has supported the conclusions of anatomists that the families Natalidae, Furipteridae, and Thyropteridae (along with the Old World Myzopodidae) are a very closely related quartet. Two of the species were formerly placed in a separate genus, *Chilonatalus*, while a third was considered to belong to *Nyctiellus*. The two are now generally considered to be subgenera of *Natalus*. The generic name means "related to one's birth," and refers to the small size of the adults, which look like young bats even when fully grown.

Physical characteristics

Adult male natalids have a gland-like structure in the center of the forehead. Though characteristic of the family, the precise function of this free-floating disc is uncertain. The thumb is short and nearly completely enveloped in a skin of the wing (in the closely related family Furipteridae, it is completely enclosed). Possibly as a means of providing extra flexibility in flight or perhaps to avoid wing damage in the cluttered under-story in which natalids usually fly, the third joint of the third finger remains cartilaginous, even in adults. Exceptionally long, the legs can be longer than the head and body combined. The ears are broader than high, and shaped like three-quarters of a funnel. Seeming to dominate the face, they probably serve to focus the very slight sounds of moth flight to the hunting bat. Like *Kerivoula*, the cone-shaped ears of natalids have small papillae on the inner surface. These may improve auditory sensitivity in some as-yet unknown way. All natalids lack a true nose leaf. However, at the tip of the snout there is a hairy protuberance that resembles a nose leaf.

A funnel-eared bat (*Natalus stramineus*) roosts in a cave. (Photo by © Merlin D. Tuttle, Bat Conservation International. Reproduced by permission.)

The ear of the funnel-eared bat (*Natalus stramineus*) has glandular papillae on its surface. (Photo by Heather A. York. Reproduced by permission.)

spaced colonies of up to 300. Northern populations of *N. stramineus* may migrate in winter. In between foraging bouts, natalids may roost under overhanging rock ledges outside caves.

As befits very small bats, natalids emit very high-pitched calls, up to 170 kHz.

Distribution

One species is widely spread, occurring from northern Mexico to eastern Brazil and certain Caribbean islands. Another occurs in northern South America and adjacent offshore islands. The other three are restricted to islands or island groups in the Caribbean.

Habitat

Dry and seasonally deciduous forests, rarely above 984 ft (300 m), though there are verified records up to 7,874 ft (2,400 m).

Behavior

All natalids roost deep in caves where it is hot and humid and the climate changes little. Individuals hang in widely

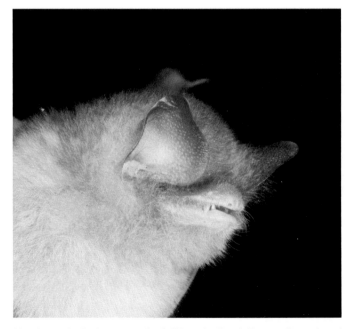

Natalus major lacks a nose leaf. (Photo by Brock Fenton. Reproduced by permission.)

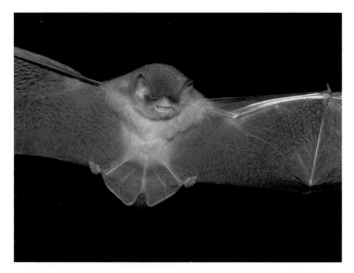

Funnel-eared bats (*Natalus stramineus*) have long slender wings. (Photo by Heather A. York. Reproduced by permission.)

The small-footed funnel-eared bat (*Natalus micropus*) is the smallest bat in the Americas. (Photo by Brock Fenton. Reproduced by permission.)

Feeding ecology and diet

Natalids generally leave their roost in groups 30 minutes after sunset. The greatest feeding activity occurs in forest two hours after sunset. Natalids are very agile fliers, able to fly in and out of dense under-story vegetation in search of insects. Their flight is fast and fluttery. Frequency of their echolocation calls exceeds 85 kHz, providing a very detailed "image" of the world. Because of this, natalids are rarely caught in mist nets.

Reproductive biology

Single offspring are often born late in the dry season. At this time, females establish separate maternity colonies. Species of this family are most likely polygynous.

Conservation status

Populations may be limited by dependence on deep caves as roosts. IUCN classifies *N. tumidifrons* as Vulnerable and *N. lepidus* as Lower Risk/Near Threatened.

Significance to humans

Humans have severely impacted populations of *Natalus*. Fossils show that today's very patchy distribution reflects past extinctions of formerly more-widespread populations. Cave-inhabiting humans may have been especially damaging, but many populations were probably affected by post-ice age climatic change and sea level rises that flooded caves and altered the environment within those that remained.

A close-up of a funnel-eared bat (*Natalus stramineus*). (Photo by © Merlin D. Tuttle, Bat Conservation International. Reproduced by permission.)

1. Bahamian funnel-eared bat (*Natalus tumidifrons*); 2. Gervais' funnel-eared bat or butterfly bat (*Natalus lepidus*); 3. Small-footed funnel-eared bat (*Natalus micropus*); 4. Funnel-eared bat (*Natalus stramineus*); 5. White-bellied funnel-eared bat (*Natalus tumidirostris*). (Illustration by Michelle Meneghini)

Species accounts

Funnel-eared bat
Natalus stramineus

TAXONOMY
Natalus stramineus Gray, 1838, type locality unknown, probably Antigua, Lesser Antilles. Seven subspecies are currently recognized.

OTHER COMMON NAMES
French: Vespertilion à couleur de paille.

PHYSICAL CHARACTERISTICS
Over most of its continental range, it occurs in two color phases: a light phase with a buffy back and a darker phase of reddish brown. The belly is lighter, but correspondingly tinted in each phase. The adaptive significance of this chromatic polymorphism is currently unclear. However, the populations show great fidelity to particular cave systems as roost sites. Since these may be quite isolated, this has resulted in some inter-population difference in coloration and average measurements. It is currently unclear if this reflects random drift or microevolution to precise local conditions. There is a black "moustache" of stiffer hairs above the upper lip and a white one below the lower lip. The natalid organ is bell-shaped and covers the entire muzzle. The specific name means "made of straw" and refers to the color of the body in the first-described subspecies, *N. s. stramineus*.

DISTRIBUTION
Northeastern Mexico (Baja California) to eastern Brazil, Cuba, Hispaniola, Jamaica, Lesser Antilles, and Tres Marias islands off western Mexico. Also known from fossil remains in cave deposits on the islands of Andros, Grand Caicos, and New Providence in the Bahamas, and Isle of Pines, near Cuba, and Grand Cayman Island south of Cuba. The New Providence deposit is some 8,000 years old; those on Andros are less than 4,500 years old.

HABITAT
Dry and seasonally deciduous forests, and gallery forests. Occasionally entering moister forest types.

BEHAVIOR
At higher altitudes, some populations may go into seasonal torpor.

FEEDING ECOLOGY AND DIET
Nothing is known.

REPRODUCTIVE BIOLOGY
Females migrate to special maternity roosts during the breeding season. Breeding occurs in the late dry season. Embryonic development is slow, with gestation lasting 10 months. Babies are proportionately large, weighing up to 0.07 oz (2.1 g) at birth, or more than 50% of the mother's weight. Thought to be polygynous.

CONSERVATION STATUS
Not threatened. The Cuban subspecies, *primus*, is known only from skeletal remains from Cueva de los Indios, Daiquiri, and is considered Extinct.

SIGNIFICANCE TO HUMANS
None known. ◆

Natalus lepidus
Natalus stramineus
Natalus tumidifrons

White-bellied funnel-eared bat
Natalus tumidirostris

TAXONOMY
Natalus tumidirostris Miller, 1900, Hatto, Curaçao.

OTHER COMMON NAMES
English: Trinidadian funnel-eared bat; French: Vespertilion à couleur de paille; German: Trichterohr.

PHYSICAL CHARACTERISTICS
Weight is 0.1 oz (3.3 g). Dorsally, a rusty brown; belly white; snout and lips pink; ears black outside, while pink with black rims inside.

DISTRIBUTION
Colombia, Venezuela, Suriname, Trinidad, and Curaçao. Subfossil remains have been found in caves on Andros, Cat, Great Exuma, and New Providence Islands in the Bahamas.

HABITAT
Roosts deep in caves.

BEHAVIOR
Roosts in caves, but a roost also recorded in a hollow rubber tree. Colonies reported from 100 to several thousand.

Natalus micropus

Natalus tumidirostris

re-designated as a subgenus of *Natalus*. Later, along with three other taxa, *N. tumidifrons* was considered to be a subspecies of *N. micropus* (*N. m. tumidifrons*). While it was recently re-elevated to species status, three other former species are now considered to be subspecies of *N. micropus*. It is the only bat endemic to the Bahamas. Fossils bones of *N. tumidifrons*, 8,000–12,000 years old from caves on Andros and New Providence, are indistinguishable from those of living animals.

OTHER COMMON NAMES
None known.

PHYSICAL CHARACTERISTICS
Very similar in color and appearance to *N. micropus*, but has a larger forearm and body size.

DISTRIBUTION
Great Abaco and Watling (San Salvador) islands in the northern Bahamas. Formerly occurred on Andros and other islands in the southern Bahamas, as part of an extensive 16-species mammalian fauna (15 of which were bats). Though some human impacts from cave use may have occurred, the most significant impacts on ancient populations of this bat appear to have come from post-glacial changes in climate and sea level, which flooded many caves and disrupted the thermal ecology of most others. Most bat species that became extinct at this time were, like *N. tumidifrons*, species that prefer to roost deep in caves where the climate is hot, humid, and stable. *N. lepidus* appears in the fossil record in some islands of the Bahamas at the same time *N. tumidifrons* disappears, leading to suggestions that competition may have hastened its disappearance. *N. tumidifrons* now only occurs on islands from which *N. lepidus* is absent.

HABITAT
Occurs only in the Bahaman dry forest. The forest on Great Abaco is lusher and taller than that on Watling Island, which is low, scrubby, and with an understory of cacti.

BEHAVIOR
Nothing is known.

FEEDING ECOLOGY AND DIET
Nothing is known.

REPRODUCTIVE BIOLOGY
Mating system is not known, but most likely polygynous.

CONSERVATION STATUS
Vulnerable. It is one of three mammals endemic to the Bahamas. Of the other two, the Bahaman Raccoon (*Procyon maynardi*) is Endangered and the Bahaman Hutia (*Geocapromys ingrahami*) is Vulnerable.

SIGNIFICANCE TO HUMANS
None known. ◆

FEEDING ECOLOGY AND DIET
Has the slow fluttery flight of all natalids, reported to make a soft regular vocalization when hunting, audible to children, and described as sounding like a sewing machine. It uses the tail membrane to catch insects.

REPRODUCTIVE BIOLOGY
On Curaçao, young are born in October, at the start of the rainy season. This species is most likely polygynous.

CONSERVATION STATUS
Not threatened.

SIGNIFICANCE TO HUMANS
Species of the bacterial genus *Borellia* have been isolated from *N. tumidirostris*. *B. recurrentis* causes relapsing fever in humans; it is transmitted by the bite of insects and ticks from wild reservoirs such as bats and mice. The effects of the *Borellia* isolated from *N. tumidirostris* (if any) are not known. Many bats, including *N. tumidirostris*, have organisms associated with their guano, which cause disease in humans. These include several fungi, including *Blastomyces dermatitidis*, which causes blastomycosis (Gilchrist's disease, an infection of the skin, lungs, and lymph nodes). A species of yeast-like fungus, *Candida chiropterorum* has been found in the organs of several bat species, including *N. tumidirostris*. ◆

Bahamian funnel-eared bat
Natalus tumidifrons

TAXONOMY
Natalus tumidifrons (Miller, 1903), Watling Island, Bahamas. This species has had an involved taxonomic history. It was originally described as a member of the genus *Chilonatalus*. This was

Small-footed funnel-eared bat
Natalus micropus

TAXONOMY
Natalus micropus Dobson, 1880, Kingston, Jamaica. Sometimes placed in the subgenus *Chilonatalus*.

OTHER COMMON NAMES
English: Cuban funnel-eared bat.

PHYSICAL CHARACTERISTICS

With head and body length of 1.5 in (4 cm) and a weight of 0.7–1.0 oz (2–3 g), this is the smallest member of the family; it is also the smallest New World bat. Dorsal fur is pale yellow at the base and reddish or chestnut-brown at the tips. Belly hair is a uniform yellowish brown. The sexes are the same size. The lower lip is reflected outward and possesses a fleshy projection, giving the appearance that there are two lower lips. Characteristically, the natalid organ is rounded and located near the base of the muzzle. A small nub of flesh on the top of the nose resembles a rudimentary nose leaf.

DISTRIBUTION

Cuba, Isle of Pines, Jamaica, Hispaniola, and islands off east coast of Nicaragua. Subfossils are also known from several cave sites in Jamaica and on Cuba. These suggest there has been little chance in the appearance of this species since the Pleistocene. The population from Old Providence Island off the Nicaraguan coast is often given subspecific status as *N. m. brevimanus*. Other subspecies are *N. m. macer* from Cuba, and *N. m. micropus* from Jamaica. All of these subspecies have, at various times, been given species rank.

HABITAT

Prefers deep moist caves, where they roost in loose colonies of up to several hundred under low ledges. Some populations may enter summer torpor (estivation). One colony inhabits the St. Clair cave in Jamaica, an area of rugged relief surrounded by rainforest. It shares the cave with eight other bat species. When exiting, they fly within 3.2 ft (1 m) of the cave floor.

BEHAVIOR

Nothing is known.

FEEDING ECOLOGY AND DIET

Little known. Small, lightweight, and agile flyers, these bats are rarely caught in mist nets.

REPRODUCTIVE BIOLOGY

Mating system is not known, but most likely polygynous.

CONSERVATION STATUS

Not threatened.

SIGNIFICANCE TO HUMANS

Biologists consider the genus to be of great interest in puzzling out how species evolve on islands and how biological communities have developed in the Caribbean. ◆

Gervais' funnel-eared bat
Natalus lepidus

TAXONOMY

Natalus lepidus (Gervais, 1837), Cuba. Because of a relatively flattened braincase, a small natalid organ (situated in the middle of the muzzle), and an entire (not grooved) lip, this species is placed in the subgenus *Nyctiellus*.

OTHER COMMON NAMES

English: Butterfly bat; Spanish: Murcielago mariposa.

PHYSICAL CHARACTERISTICS

Forearm length is 1.0–1.3 in (2.7–3.4 cm). Fur yellowish with a buffy wash. Distinguished from *N. micropus* by the absence of the lip-like ridge on the chin and the lack of prominent hair-covered nodules on the snout.

DISTRIBUTION

Cuba, Isle of Pines, plus Cat, Eleuthera, and Great Exuma Islands in the Bahamas. Subfossils in cave deposits show it also formerly occured on Bahamian island of Andros, but that it arrived within the last few thousand years and was not among the old Pleistocene fauna of the Bahamas.

HABITAT

Cuban dry forest.

BEHAVIOR

An obligate cave dweller, this bat requires a warm humid cave environment and prefers caves not occupied by other bat species.

FEEDING ECOLOGY AND DIET

Nothing is known.

REPRODUCTIVE BIOLOGY

Mating system is not known, but most likely polygynous.

CONSERVATION STATUS

Lower Risk/Near Threatened.

SIGNIFICANCE TO HUMANS

Proudly promoted as "the world's smallest bat," a large *N. lepidus* colony in the Escambray range in south-central Cuba may be threatened by future tourism developments. Cuban dry forest (an umbrella term for a variety of plant communities) formerly covered 50% of the region; it is now 10% or less on most islands. ◆

Resources

Books

Eisenberg, J. F., and K. H. Redford. *Mammals of the Neotropics. Volume 3, The Central Tropics: Ecuador, Peru, Bolivia, Brazil.* Chicago: University of Chicago Press, 1999.

Reid, F. A. *A Field Guide to the Mammals of Central America and Southeast Mexico.* Oxford: Oxford University Press, 1997.

Woods, C. A., and F. E. Sergile, eds. *Biogeography of the West Indies: Patterns and Perspectives.* Boca Raton: CRC Press, 2001.

Periodicals

Buden, D. W. "A Guide to the Identification of the Bats of the Bahamas." *Caribbean Journal of Science* 23 (1987): 362–367.

Organizations

Bat Conservation International. P.O. Box 162603, Austin, TX 78716 USA. Phone: (512) 327-9721. Fax: (512) 327-9724. E-mail: batinfo@batcon.org Web site: <http://www.batcon.org>

Adrian A. Barnett, PhD

Smoky bats

(Furipteridae)

Class Mammalia

Order Chiroptera

Suborder Microchiroptera

Family Furipteridae

Thumbnail description

Tiny coarse-furred bats with short, broad, dish-like ears, reduced eyes, a short upturned snout, a vestigial thumb, and functionless wing claw

Size

Head and body length 1.4–2.6 in (3.5–5.8 cm); tail 0.9–1.4; in (2.4–3.6 cm); forearm 1.8–1.6 in (3.0–4.0 cm); 0.1 oz (3 g). Females are lightly larger than males

Number of genera, species

2 genera; 2 species

Habitat

Amorphochilus is known from isolated populations in dryland coastal forests; *Furipterus* prefers moist lowland tropical rainforests below 492 ft (150 m)

Conservation status

Vulnerable: 1 species

Distribution

Central and South America

Evolution and systematics

There are no known fossils of this bat family. Belonging to the superfamily Vespertilionoidea, they are probably most closely related to Central and South American disk-winged bats (Thyropteridae), funnel-eared bats (Natalidae), and the New World sucker-footed bat (Thyropteridae).

Physical characteristics

Among the smallest Neotropical bats, furipterids have a delicate appearance. The broad wings are long for the body, an adaptation for a fluttering flight. This is aided by a well-developed uropatigium, which is stiffened by a long tail that does not reach the tail membrane's trailing edge. The translucent uropatigium bears transverse lines. The skull is distorted into a helmet-like shape to accommodate the enlarged ears. Dish-shaped, the ears enclose the eyes and extend almost to the lower jawline. The pig-like snout is short and upturned at the tip. There is no nose leaf. The tiny eyes are nearly hidden by fur. A thumb is present, but is so small as to be invisible. It is enclosed in the wing membrane and only a small functionless claw protrudes. The legs and feet are short and weak, but the claws are powerful. The two genera are best

told apart by the presence of pronounced wart-like outgrowths around the mouth and lips in *Amorphochilus*. In the field, confusion is unlikely since the ranges of the two genera do not overlap and the physical characteristics of the family are unmistakable. The family name means "winged furies," and comes from hideous avenging deities of Greek mythology. Why this small inoffensive bat should excite such contempt is unknown.

Distribution

Amorphochilus occurs west of the Andes, and is known from a number of scattered and isolated sites from central coastal Ecuador south to northern Chile. *Furipterus* has a much broader and continuous distribution, occurring from Costa Rica through lowland Brazil and Peru. It is also found in Trinidad.

Habitat

Of the two species, *Amorphochilus* appears to have the broader habitat tolerances. It has been recorded in a number of different vegetation types, ranging from primary forest to semidesert brush and cultivated land. *Furipterus* either prefers or requires primary lowland moist forest.

Behavior

The great aerial agility of furipterids means they easily avoid mist nets. With spotty distributions and a small number of museum specimens, these bats have a reputation for being rare and difficult to study. This means little is known about them. Most information comes from studies of their roosts. Both *Amorphochilus* and *Furipterus* roost colonially. Up to 300 animals have been found roosting together.

Feeding ecology and diet

Furipterids mostly hunt small moths within the forest undergrowth. Such hunting requires both agility and low flight speeds. This is achieved with wings that are both broad (a low stalling speed) and proportionately long (smaller turning circle). Both are assisted by a well-developed uropatigium that allows for a low stalling speed and slow flight.

Reproductive biology

Little is known. Like other bats, female bears a single young one. Unusual among bats, furipterid nipples are positioned abdominally. As a result, the young position themselves head-up on the head-down roosting mother.

Conservation status

The IUCN considers *Amorphochilus* to be Vulnerable. *Furipterus* is not currently thought to be in danger.

Significance to humans

None known, beyond the usual insect-removal services provided by any small insectivorous chiropteran. No known legends or religious significance.

The smoky bat (*Furipterus horrens*) has a reduced thumb that is included in the wing membrane. (Photo by Maarten J. Vonhof. Reproduced by permission.)

1. Smoky bat (*Furipterus horrens*); 2. Schnabeli's thumbless bat (*Amorphochilus schnablii*). (Illustration by John Megahan)

Species accounts

Smoky bat
Furipterus horrens

TAXONOMY

Furia horrens (F. Cuvier, 1828), Mana River, French Guiana. In Latin, *horrens* means "bristle," and, indeed, its muzzle is very bristly.

OTHER COMMON NAMES

English: Thumbless bat, lesser thumbless bat.

PHYSICAL CHARACTERISTICS

The smaller of the two species in this family, it is one of the smallest bats in the Neotropics. Females are significantly larger than males by 10–15%. Body fur is dense, with the fur on the head especially long and thick, enough to almost conceal the mouth. Color of the back varies from fine slate blue to brownish gray, with a paler belly. The ears are dark and stiff, and the snout is black.

DISTRIBUTION

Costa Rica to southern Brazil, including Venezuela and Colombia, but not west of the Andes. Also on Trinidad, but no other Caribbean islands.

HABITAT

Most commonly collected in humid lowland rainforest, often near streams. However, some have been netted in village clearings.

BEHAVIOR

Roosts of have been found in caves, hollow trees, and in or beneath fallen rotting logs, all inside forest. Though poorly known and infrequently encountered, studies in French Guiana suggest they can be quite commonly found if roost sites are searched for. It has a wide distribution and is probably more retiring than rare.

FEEDING ECOLOGY AND DIET

Has been observed foraging over the forest floor at heights of 3.2–16.4 ft (1–5 m). The recorded diet consisted solely of small moths. Individuals leave the roost only when darkness is complete.

REPRODUCTIVE BIOLOGY

Roosting aggregations are mostly mixed sex, but all-male groups have been found, suggesting that females may use special sites for raising young. Mating system is not known.

CONSERVATION STATUS

Not threatened.

SIGNIFICANCE TO HUMANS

None known. ◆

Schnabeli's thumbless bat
Amorphochilus schnablii

TAXONOMY

Amorphochilus schnablii Peters, 1877, Tumbes, Tumbes State, Peru.

OTHER COMMON NAMES

English: Western thumbless bat, greater thumbless bat.

PHYSICAL CHARACTERISTICS

Distinguished by the presence of warty outgrowths around the mouth. Fur of the back has a brown wash. It is slightly larger than *Furipterus*, and has ears that are brownish gray rather than black.

DISTRIBUTION

From Puna Island in the Gulf of Guayaquil, Ecuador, south through Peru to northern Chile. Probably temperature limited in southern part of range, and limited in the north by increasing moisture and rainfall.

HABITAT

Mostly in dry coastal forests, but occasionally inland if suitable habitat is available. Within its range area, it appears to prefer the lower seasonally deciduous rain-fed forests and avoids those forests on hilltops kept nearly perpetually moist by coastal fog. In southern Ecuador and northern Peru, it occurs in coastal oases at the mouths of rivers and in associated gallery forests.

BEHAVIOR

As befits its flexible nature, it has been found roosting in many different places, including disused buildings, irrigation tunnels,

Furipterus horrens

Amorphochilus schnablii

road culverts, and even a wine store. In unspoiled habitat, it has been found roosting in rock fissures.

FEEDING ECOLOGY AND DIET
There is little information. Analysis of five individuals from a colony in desert scrub at the mouth of Ecuador's Rio Javita found each stomach to contain the remains of small moths. Elsewhere, individuals have been collected whose stomachs contained both moths and small flies.

REPRODUCTIVE BIOLOGY
Timing of birth is very restricted and appears to be timed to coincide with the onset of the local rainy season. This allows energy-stressed lactating females to forage when food is most abundant. Mating system is not known.

CONSERVATION STATUS
Because it inhabits seasonally dry forests, it may be adapted to surviving in the hot, dry, more open habitats generally created by human action. However, it requires some type of forest cover and the IUCN expects populations to decline by 20% over the next decade due to human disturbance. It is listed as Vulnerable.

SIGNIFICANCE TO HUMANS
None known. ◆

Resources

Books
Eisenberg, J. F., and K. H. Redford. *Mammals of the Neotropics.* Vol. 3, *The Central Tropics: Ecuador, Peru, Bolivia, Brazil.* Chicago: University of Chicago Press, 1999.

Reid, F. A. *A Fieldguide to the Mammals of Central America and Southeast Mexico.* Oxford: Oxford University Press, 1997.

Periodicals
Ibanez, C. "Notes on *Amorphochilus schnabelii* Peters (Chiroptera, Furipteridae)." *Mammalia* 49 (1985): 584–87.

Simmons, W., and R. S. Voss. "The Mammals of Paracou, French Guiana: A Neotropical Lowland Rainforest Fauna.

Part 1. Bats." *Bulletin of the American Museum of Natural History* 237 (1998): 1–297.

Uieda, W., I. Sazima, and A. S. Filho. "Aspectos da Biologia do Morcego *Furipterus horrens* (Mammalia, Chiroptera, Furipteridae)." *Revista Brasiliera do Biologia* 40 (1980): 59–66.

Organizations
Bat Conservation International. P.O. Box 162603, Austin, TX 78716 USA. Phone: (512) 327-9721. Fax: (512) 327-9724. E-mail: batinfo@batcon.org Web site: <http://www.batcon .org>

Adrian A. Barnett, PhD

Disk-winged bats
(*Thyropteridae*)

Class Mammalia
Order Chiroptera
Suborder Microchiroptera
Family Thyropteridae

Thumbnail description
Tiny dark-backed bats with moist fleshy "suction-cups" at the base of the thumb and heel; the feet are tiny, and have the toes fused together; the tail membrane is large and obvious and the face long and delicately pointed

Size
Head and body length 1.2–2.3 in (3.0–5.7 cm); tail 0.9–1.3 in (2.4–3.3 cm); forearm 1.2–1.4 in (3.1–3.5 cm); weight 0.10–0.17 oz (3–5 g)

Number of genera, species
1 genus; 3 species

Habitat
Lowland rainforest, often by water

Conservation status
Vulnerable: 1 species

Distribution
Central America and northern South America

Evolution and systematics

There are no known fossil thyropterids. In a remarkable example of convergent evolution, bats have evolved foot suckers on several occasions. They may be found in Madagascar's endemic bat family, the Myzopodidae, and some vespertilionid bats (e.g., *Eudiscopus denticulatus* from Southeast Asia; the club-footed bats, *Tylonycteris*; and the thick-thumbed pipistrelles, *Glishropus*). None of these foot suckers, however, are as well developed as in *Thyroptera*. A larger third species of *Thyroptera*, *T. laveli*, is recognized by some authorities. Much darker furred than the other two species, it is known only from the type series of four individuals, collected near Loreto in the Peruvian Amazon. Apparently highly restricted in distribution, this little-known species is considered Vulnerable by the IUCN. Unlike other thyropterids, it may roost in canopy vegetation.

Physical characteristics

These small delicate bats have long fluffy fur and a long triangular muzzle, like the tip of an ice cream cone. There is no nose leaf, though there is a small wart-like projection between and above the nostrils. The ears are large and funnel-like. The front edge of the ears reaches forward, though not sufficiently to conceal the eyes as in the related families Natalidae and Furipteridae. The outer edge of the ear attaches to the head near the mouth. The eyes are small. The tail membrane is longer than the legs and comes to a point; the tail is longer than the membrane and pokes out beyond it. The disks are stalked and those on the thumbs are larger than those on the feet. Loosely translated, the family name means "disk foot" from the Greek *thureos*, a door-shaped shield, and "wing" from the Greek *pteron*.

Distribution

From Chiapas, Mexico, south through Central America across northern South America to the edge of Amazonian Brazil and Peru, and in Atlantic coastal forest, stopping at the *Auricaria* zone. Though widely spread in Central and South America, the genus does not occur west of the Andes.

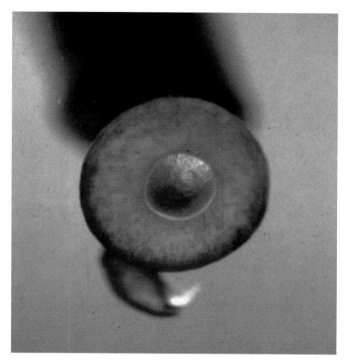

The disk-winged bat (*Thyroptera tricolor*) clings to the underside of leaves with the suction cups at the ends of its limbs. (Photo by Brock Fenton. Reproduced by permission.)

Habitat

Evergreen forest and tall secondary growth. Absent from areas with a prolonged or pronounced dry season. On Costa Rica's Osa Peninsula, *Thyroptera* occur at densities of up to four colonies per 2.5 acres (1 ha).

Behavior

Disk-winged bats use their suction cups to cling to near-vertical smooth surfaces of the rolled-up leaves inside which they roost. While moving around a roost, neither the feet nor claws touch the smooth leaf surfaces, with the cups alone being responsible for grip. Situated on short stalks, the cups generate sufficient suction to allow captive *Thyroptera* to move easily and without slippage across a clean pane of glass. The suction is not generated passively; modified sweat glands in the disks produce a sticky secretion and there is also a tendon leading from a cartilaginous plate in the disk to muscles outside it; this helps keep the shape appropriate. A bat will also lick its disks to aid adhesion. Using this combination of suction and wet adhesion, a single disk can support the bat's entire weight. Suction requires constant muscular expenditure to keep the cup in the right shape. This could be energetically demanding over a night's roosting. So, the wet adhesion may be an energy-saving device. Experiments have shown that, in specializing to roosting on smooth surfaces, *Thyroptera* has lost the ability to roost on rough surfaces (rock, bark) as most bats can and do. The roosting sites protect the tiny bats from the rain and from predators. In captivity, disk-winged bats with no suitable substrate will hang head-up by their thumb-claws, rather than try to suspend themselves from their tiny weak feet. Within the order Chiroptera, the suction cups of *Thy-*

roptera are the most specialized organs of their kind. Some African bats (*Myotis bocagei* and *Glishropus nanus*) also roost in rolled-up banana leaves. The latter has thickened pads at the wrist. These are reported to allow adhesion to the leaf surface. *M. bocagei* uses its fine sharp claws for attachment to the leaf.

Feeding ecology and diet

Disk-winged bats are insectivorous. They emit feeding echolocation calls that are low in intensity with broadband, multi-harmonic, and narrowband components.

Reproductive biology

A female moves to a maternity roost in a hollow log to give birth and wean her offspring. This species is most likely polygynous.

Conservation status

Thyroptera laveli is listed as Vulnerable by the IUCN; the other two species are not threatened.

Significance to humans

None known, beyond natural insect control.

The disk-winged bat (*Thyroptera tricolor*) roosts with its head upright. (Photo by Brock Fenton. Reproduced by permission.)

1. Peters's disk-winged bat (*Thyroptera discifera*); 2. Spix's disk-winged bat (*Thyroptera tricolor*). (Illustration by Marguette Dongvillo)

Species accounts

Spix's disk-winged bat
Thyroptera tricolor

TAXONOMY
Thyroptera tricolor Spix, 1823, Amazon River, Brazil. Three subspecies are recognized.

OTHER COMMON NAMES
German: Haftscheiben-Fledermaus; Spanish: Murcielago de ventosas, murcielago tricolor con mamantones.

PHYSICAL CHARACTERISTICS
Back (and sometimes throat) dark brown to reddish brown. Belly white or yellowish; flanks are frequently an intermediate color; ears blackish. Calcar has one cartilaginous bump. Tail long, extending (0.19–0.31 in [5–8 mm]) beyond the uropatagium. Females are slightly larger than males.

DISTRIBUTION
From Veracruz, Mexico, to southeast Brazil. Apparently absent from El Salvador and Nicaragua.

HABITAT
Individuals captured in a number of different habitats within rainforest, including primary forest, swamp, and man-made clearings. Not recorded above 4,265 ft (1,300 m), and usually below 2,625 ft (800 m).

BEHAVIOR
All known roosts have been in foliage. In French Guiana, most roosts were found in unrolled new leaves of *Heliconia* plants or, to a lesser extent, of *Phenakospermum*. A smaller number of roosts were found in old, dead, scrolled leaves of *Phenakospermum*. Elsewhere, also recorded in rolled-up arrowroot (*Calathea*, Marantaceae) leaves. Most roosts are near water and none are out of direct sunlight. Preferred leaves form vertical tubes 1.9–3.9 in (50–100 mm) in diameter and do not touch any other vegetation (so reducing the danger of predation by snakes). Such roosts are ephemeral, because the leaves generally unroll within 24 hours. Consequently, the bats must find a new roost once every few days. A stable, socially cohesive group will sequentially occupy all favorable roosts in an area as they become available and defend their patch against other groups. Roosts are generally occupied by one to nine individuals; it is rare for there to be more than one adult male in a roost. Within the rolled-up leaf, individuals roost with the head pointing upward toward the opening, making for a swifter escape if danger threatens. When roosting together, individuals are aligned one above the other. A group's home range may average some 32,290 ft² (3,000 m²). Groups are clearly social.

FEEDING ECOLOGY AND DIET
Slow, fluttering, agile flight and a tendency to fly low indicate a diet of insects caught close to the ground. Small beetles and flies may be important diet components. Each individual consumes 0.03 oz (1 g) of insects a night (one-quarter of its body weight).

REPRODUCTIVE BIOLOGY
Polygynous. Births probably occur during the peak of the rainy season. Gestation is about two months. Once born, young cannot fly for a month and the female flies and forages with them

☐ *Thryoptera tricolor*
■ *Thyroptera discifera*

attached. By the end of the month, the single baby's weight may be half that of the mother. After the young learn to fly, they may fly with the mother for a few days until fully weaned. Breeding occurs twice a year.

CONSERVATION STATUS
Not threatened.

SIGNIFICANCE TO HUMANS
None; roosting occurs in banana plants but does not endanger any economic interests. ◆

Peters's disk-winged bat
Thyroptera discifera

TAXONOMY
Thyroptera discifera (Lichtenstein and Peter, 1855), Puerto Cabello, Carabobo, Venezuela. Two subspecies are generally recognized. *Abdita* means "hidden" and refers to the fact that the subspecies was described 84 years after the specimens were collected.

OTHER COMMON NAMES
Spanish: Murcielago con mamantones de Peter.

PHYSICAL CHARACTERISTICS
Dorsal fur brown or reddish, belly gray-brown or yellowish. Tail membrane (uropatigium) hairy for first half of its length;

tail extends to 0.15 in (4 mm) beyond trailing edge of uropatagium. The calcar (the ankle spur that stiffens the uropatagium's trailing edge) is also characteristic, having one anterior nodule. The ears are yellowish.

DISTRIBUTION
Central America (few records), and northern South America (southeastern Brazil and northern Peru).

HABITAT
Evergreen forest and banana plantations.

BEHAVIOR
Roosts underneath leaves, rather than within rolled-up ones. Uses bananas, heliconias, and palms.

FEEDING ECOLOGY AND DIET
Nothing known.

REPRODUCTIVE BIOLOGY
Little known, but most likely polygynous. Disks and claws are not well developed in newborns. Consequently, females have broad strap-like nipples to which the young cling with their teeth. One youngster was observed to cling to its mother for 20 minutes without using its claws.

CONSERVATION STATUS
Not threatened.

SIGNIFICANCE TO HUMANS
None known. ◆

Resources

Books

Eisenberg, J. F., and K. H. Redford. *Mammals of the Neotropics. Vol. 3, The Central Tropics: Ecuador, Peru, Bolivia, Brazil.* Chicago: University of Chicago Press, 1999.

Reid, F. A. *A Fieldguide to the Mammals of Central America and Southeast Mexico.* Oxford: Oxford University Press, 1997.

Periodicals

Findlay, J. S., and D. E. Wilson. "Observations on the Neotropical Disk-winged Bat, *Thyroptera tricolor* Spix." *Mammalia* 55 (1974): 562–571.

Pine, R. H. "A New Species of *Thyroptera* Spix (Mammalia: Chiroptera: Thyropteridae) from the Amazon Basin of Northeastern Peru." *Mammalia* 57 (1993): 213–225.

Riskin, D. K., and M. B. Fenton. "Sticking Ability in the Spix's Disk-winged Bat *Thryoptera tricolor* (Microchiroptera: Thyropteridae)." *Canadian Journal of Zoology* 79 (2001): 2261–2267.

Organizations

Bat Conservation International. P.O. Box 162603, Austin, TX 78716 USA. Phone: (512) 327-9721. Fax: (512) 327-9724. E-mail: batinfo@batcon.org Web site: <http://www.batcon .org>

Adrian A. Barnett, PhD

Old World sucker-footed bats

(Myzopodidae)

Class Mammalia
Order Chiroptera
Suborder Microchiroptera
Family Myzopodidae

Thumbnail description
A small bat, characterized by horseshoe shaped
sucker-like pads on the thumb and soles of the
feet; the ears are separate and large, each with
a mushroom-like tragus, comprising a kidney-
shaped fleshy expansion surmounting a short
stalk; upper lip extends significantly beyond the
lower; the pelage is moderately dense and mid-
brown to golden brown, with some russet tinges

Size
Head and body length 2.3 in (57 mm); tail
length 1.9 in (48 mm); forearm 1.9–2.0 in
(47–50 mm); weight 0.3 oz (8 g)

Number of genera, species
1 genus; 1 species

Habitat
Primary and secondary rainforest of eastern
Madagascar, also agricultural land and urban
areas

Conservation status
Vulnerable

Distribution
Endemic to Madagascar

Evolution and systematics

Although restricted to Madagascar today, *Myzopoda* has
been found as a fossil in the early Pleistocene deposits of
Olduvai 1 in Tanzania on mainland Africa. The Old World
sucker-footed bat (*Myzopoda aurita*) was described in 1878 and
was originally included in the family of Old World evening
bats, the Vespertilionidae. Subsequently, in 1904, it was transferred into a separate family, the Myzopodidae. It resembles
the New World sucker-footed bats, Thryopteridae, of Central and South America, and the humerus and shoulder joint
have a similar structure to this latter family and to the New
World funnel-eared bats, the Natalidae. It has been included
in the superfamily Nataloidea, which in addition to the Myzopodidae, includes the Natalidae, Furipteridae, and Thryopteridae. Recent molecular studies have challenged the
monophyletic origin of the Nataloidea, suggesting that the
Myzopodidae is not a sister taxon to the other families but
evolved at a much earlier date as a distinct lineage that cannot satisfactorily be grouped with any other bat family.

The taxonomy for this species is *Myzopoda aurita* Milne-
Edwards and A. Grandidier, 1878, Madagascar.

Physical characteristics

A small bat, characterized by the presence of conspicuous
horseshoe shaped sucker-like pads on the thumb and soles of
the feet. The ears are separate and large, some 1.18–1.38 in
(30–35 mm) in length. Each ear has a tragus, which is fused
along its anterior edge to the pinna; the meatus is partly closed
by this conspicuous mushroom-shaped process, which comprises a kidney-shaped fleshy expansion surmounting a short
stalk. The upper lip extends significantly beyond the lower.
The pelage is moderately dense and mid-brown to golden
brown, with some russet tinges. For this reason, the bat is also
known as the golden bat. The ears are very large, and the
thumb is quite small and has a vestigial claw. The toes of the
feet have only two phalanges and are syndactylous (joined together) for much of their length. The tail projects noticeably
beyond the free edge of the interfemoral membrane. The skull
is short, broad, and rounded. The tympanic bullae and cochleae
are rather large but not peculiar in structure. The hamular
processes are unusually long. There are 38 teeth, including
two pairs of upper and three pairs of lower incisors, one pair
of upper and lower canines, three pairs of upper and lower
premolars and three pairs of upper and lower molars.

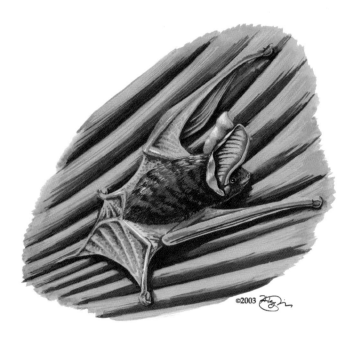

Old World sucker-footed bat (*Myzopoda aurita*). (Illustration by Jonathan Higgins)

Distribution

It is now restricted to Madagascar, where it is primarily found on the east coast, in the former rainforest region. In addition, there has been one record from the west of the country: Mahajanga. There are fewer than 20 locality records. These stretch the full length of the eastern sea board of Madagascar.

Habitat

Its preferred habitat is thought to be primary and secondary rainforest of eastern Madagascar. It is also known to frequent paddies, marshlands, vanilla plantations, streams, and even urban areas.

Behavior

Very little is known of its behavior. One individual was mist-netted over a small stream, another in a vanilla plantation, and one in a sparsely forested area over a path close to a stream. In August 1992, a single individual was caught in a mist net set at 6:55 P.M., about one hour after sunset. It has also been recorded, based on acoustic identification, flying over urban areas. A single specimen was collected from the unrolled leaf of a traveler's palm, *Ravenala madagascariensis* (Musaceae). Like the New World disk-winged bats, the Thryopteridae, it probably roosts in a variety of palm species and other similar types of vegetation (for example Araceae). It roosts with its head uppermost and uses its stiff projecting tail as a prop. The suction pads present on the thumbs and feet allow it to cling to smooth vertical surfaces, such as the leaves of *Ravenala*. The pads contain rows of glands that secrete directly onto the surface of the suction pads. This increases the adhesion and allows the whole body weight to be supported. Its flight is highly maneuverable.

Feeding ecology and diet

The species possesses a complex echolocation system and produces remarkably long calls, which are presumably suited to its feeding habits. It emits complex frequency modulated (FM) echolocation calls, which are composed of two to four distinct elements of increased amplitude and long call durations. Due to different numbers of elements and the presence of up to four harmonics, the call structure is highly variable. The second harmonic is the strongest. It decreases from 42 to 24 kHz with a shallow FM sweep (about 0.6–0.9 kHz/ms) during the first three elements and a steep FM sweep (about 2 kHz/ms) during the fourth element. Analysis of fecal pellets of a single individual showed that its diet included moths (Microlepidoptera). It is not known whether this is a specialization or an abundance of this type of prey at that particular moment. It has been observed to spend relatively long periods of time, probably feeding, over freshly dug and planted paddy fields and also within forest clearings in primary rainforest. However, no feeding buzzes were recorded.

Reproductive biology

Nothing is known of its breeding biology, although a female collected on 13 August 1992 had well developed nipples.

Conservation status

Vulnerable. It is generally considered to be rare, based on the very few specimens that have been collected. However, recent studies based on ultrasonic evidence have found it to be widespread in the Antongil Bay area of the Masoala Peninsula in northeast Madagascar. Here it inhabits areas not only close to forest but also urban areas. This suggests that further studies elsewhere in Madagascar using bat detectors may find larger populations than previously thought. In general, loss of forest is thought to be a probable threat but the species is still poorly known and it is therefore difficult to determine accurately its conservation requirements. If the species does roost primarily in the traveler's palm, then the bat may be more common than currently thought as this plant is widespread in primary and secondary forest. Bats, including this species, are not protected in Madagascar.

Significance to humans

This species has little significance to humans. Like many bat species it probably helps to reduce agricultural pests by feeding on potentially harmful insects.

Resources

Books

Garbutt, N. *Mammals of Madagascar.* New Haven, CT: Yale University Press, 1999.

Hutson, A. M., S. P. Mickleburgh, and P. A. Racey. *Microchiropteran Bats: Global Status Survey and Conservation Action Plan.* IUCN/SSC Chiroptera Specialist Group. Gland, Switzerland and Cambridge, UK: IUCN, 2001.

Periodicals

Gopfert, M. C., and L. T. Wasserthal. "Notes on Echolocation Calls, Food and Roosting Behaviour of the Old World Sucker-footed Bat *Myzopoda aurita* (Chiroptera, Myzopodidae)." *Zeitschrift für Säugetierkunde* 60, no. 1 (1994): 1–8.

Schliemann, H., and B. Maas. "*Myzopoda aurita.*" *Mammalian Species* 116 (1978): 1–2.

Van Den Bussche, R. A., and S. R. Hoofer. "Evaluating Monophyly of Nataloidea (Chiroptera) with Mitochondrial DNA Sequences." *Journal of Mammalogy* 82, no. 2 (2001): 320–327.

Paul J. J. Bates, PhD

Free-tailed bats and mastiff bats

(Molossidae)

Class Mammalia
Order Chiroptera
Suborder Microchiroptera
Family Molossidae

Thumbnail description
Free-tailed and mastiff bats, with a thick tail extending well beyond the tail membrane and long, narrow wings adapted for rapid flight in open airspace; a broad snout, often with wrinkles, which projects well over the lower lip; ears usually somewhat flattened and stiff, lying low over the head, and tilted forward; body covered with short, dense fur

Size
Small- to large-sized with forearms ranging 1.1–3.4 in (2.7–8.5 cm) in length and weighing 0.2–3.8 oz (5–167 g)

Number of genera, species
12 genera; 90 species

Habitat
Tropical, subtropical, and warmer temperate regions; arid as well as moist, forested habitats

Conservation status
Critically Endangered: 3 species; Endangered: 1 species; Vulnerable: 15 species; Lower Risk/Near Threatened: 21 species; Data Deficient: 4 species

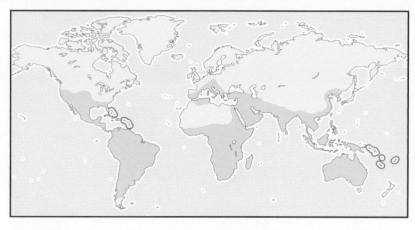

Distribution
Throughout the warmer parts of the world, on all continents except Antarctica, on the Malay Archipelago and southern Pacific islands east to Fiji, and throughout much of the Caribbean

Evolution and systematics

Fossils of the family date from the late Eocene in Europe, late Oligocene or early Miocene in South America, Miocene in Africa, and Pleistocene in Asia, Australia, North America, and the East and West Indies. Morphological and molecular data place free-tailed and mastiff bats in the superfamily Vepertilionoidea allied with the vespertilionid bats (Vespertilionidae) and the funnel-eared bats (Natalidae). Other authors place them in the superfamily Molossoidea.

Physical characteristics

Free-tailed and mastiff bats (molossids) are small- to large-sized bats, distinctly characterized by a thick tail that protrudes well beyond the tail membrane. The tail is not so long as in the mouse-tailed bats (Rhinopomatidae). Most species have a broad face, with a wrinkled snout and lips, and the snout projects well over the lower lip. Their eyes are small. Their umbrella-like ears vary in size, but typically the ears lie low over the head, are stiff, rounded, and project forward, often with a lateral orientation. The ears are often connected over the forehead. These bats have long, narrow wings, and the wing and tail membranes are tough and leathery. The hind legs are short

and strong and the feet are broad and fringed with long bristles. Some species are well endowed with glands on their chins, throats, and chest regions, and they often have a distinctive odor. Most free-tailed and mastiff bats have short brown fur, but in some, the fur is gray or black. Males of some species of the genus *Chaerephon* develop head crests of erectile hair during the mating season, while others have tufts of hair associated with chest glands. Members of the genus *Cheiromeles* (one or two species) are almost completely hairless. These "naked" bats also have wing pouches, or flaps of skin along the sides of the body, into which they place their folded wings. A few crevice-roosting species in the genera *Molossops* and *Mormopterus* have distinctly flattened skulls.

Distribution

Except for bats of the family Vespertilionidae, molossids have the widest distribution of any family of bats. They are found throughout the warmer parts of the world, including southern Europe, much of Africa, southern Asia, Malaysia, Australia, the Australasian region east to Fiji, in central and southern North America, Central America, the Caribbean Islands, and all except the southern-most portion of South America.

Greater house bat (*Molossus ater*) face, seen in Ecuador. (Photo by S. C. Bisserot. Bruce Coleman, Inc. Reproduced by permission.)

Habitat

They occur in a wide range of habitats and are common in natural, rural, and urban areas. They reach their greatest abundance in arid and semi-arid habitats. Natural roosting sites include caves, rock crevices, tree cavities, bark, rotting logs, foliage, and holes in the ground (*Cheiromeles*). These bats also commonly roost in human-made structures, including

Close-up of a Brazilian free-tailed bat (*Tadarida brasiliensis*) foot, clinging onto tree bark. (Photo by R. Wayne Van Devender. Reproduced by permission.)

Beccari's mastiff bat (*Mormopterus beccarii*) is found in Northern Territory and Queensland, Australia. (Photo by Frithfoto. Bruce Coleman, Inc. Reproduced by permission.)

buildings, mines, tunnels, culverts, under bridges, and in bat houses. They are commonly found under corrugated steel roofs, roof tiles, or in attics of tropical houses, and they tolerate high temperatures that can exceed 130°F (55°C).

Behavior

Molossids tend to be active throughout the year. Populations of Brazilian free-tailed bats (*Tadarida brasiliensis*) in

A Brazilian free-tailed bat (*Tadarida brasiliensis*) rests on a tree in Cochise County, Arizona, USA. (Photo by John Hoffman. Bruce Coleman, Inc. Reproduced by permission.)

A Brazilian free-tailed bat (*Tadarida brasiliensis*) (Photo by Animals Animals ©C. W. Schwartz. Reproduced by permission.)

temperate, seasonal habitats are known to engage in long-distance, annual migrations that exceed 800 mi (1,300 km). However, other populations of the same species are known to remain in place or to engage only in short-distance seasonal movements and to utilize torpor to survive cold temperatures during relatively mild winters. Most molossids are colonial, with colony sizes typically reported as a few tens to a few hundreds of individuals. There are some reports of solitary bats, and numerous accounts of colonies into the thousands of bats. Brazilian free-tailed bats in the southwestern United States and northern Mexico form cave colonies of tens of millions of bats, which are the largest known aggregations of mammals. The behavior of this large family of bats is characterized by diversity and plasticity.

Feeding ecology and diet

These strong-flying bats typically pursue insects in open, uncluttered airspace above the canopy and they can fly to high altitudes. Studies in Africa, Australia, and North America document foraging by molossids at altitudes of several hundred

feet (meters) above ground level. Radar shows that Brazilian free-tailed bats fly to altitudes of up to 2 mi (3.2 km) over central Texas, and research has confirmed that large numbers of these bats are actively feeding on insects at altitudes of at least 4,000 ft (1,219 m) above the ground. Molossids detect and pursue insects using relatively low-frequency echolocation calls (typically <30kHz, but <10kHz in some species) that travel long distances in open airspace. Recent studies suggest remarkable diversity in their echolocation calls.

Molossids are known to forage in groups and to exploit patches of insects such as emerging swarms of termites, winged ants, and large migratory populations of moths. They also forage around streetlights that attract concentrations of insects. These bats prey on a great variety of insects. Recent studies of Brazilian free-tailed bats document that their insect prey consists of at least 12 orders and 35 families of insects. Moths (Lepidoptera) and beetles (Coleoptera) provide the bulk of their prey, but all evidence indicates that Brazilian free-tails, in particular, and molossids, in general, are highly opportunistic feeders that exploit a diverse diet of insects.

A naked bat (*Cheiromeles torquatus*) in Deer Cave, Gunung Mulu National Park, Sarawak, Malaysia. (Photo by Simon D. Pollard/Photo Researchers, Inc. Reproduced by permission.)

Brazilian free-tailed bats (*Tadarida brasiliensis*) near Bracken Cave, Texas, USA. (Photo by © W. Perry Conway/ Corbis. Reproduced by permission.)

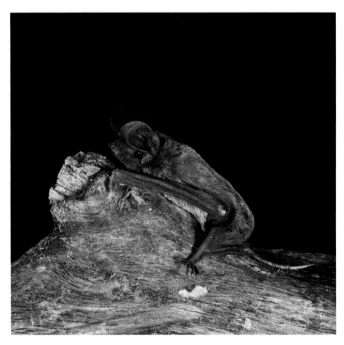

A European free-tailed bat (*Tadarida teniotis*) on an old log. (Photo by Eric & David Hosking/FLPA–Images of Nature. Reproduced by permission.)

Reproductive biology

Little is known regarding mating behavior of most molossids. *Chaerephon pumila* is reported to roost and mate in stable harem groups of about 20 females attended by a single male. Evidence suggests that *Tadarida brasiliensis* mates promiscuously during a brief period in spring when males and females assemble at specific sites. Many reports of the use by molossids of low-frequency vocal communication, the abundance of scent glands, and the existence of obvious structures for social displays such as head crests all suggest that molossids engage in a diversity of social interactions and mating systems that are, as yet, unstudied.

Females of most species appear to give birth to a single young annually. However, some species are reported to be polyestrus, giving birth twice (*Molossus ater* and *M. molossus*) or three times (*Chaerephon pumila*) annually, in parts of their geographic ranges. *Cheiromeles* is additionally unique among mollosids in giving birth to twins during a single annual reproductive period. Where known, gestation is usually two to three months in length, and the period from birth to weaning typically lasts five to six weeks. Studies on milk composition and reproductive energetics in Brazilian free-tailed bats demonstrate an extremely high-fat content in the milk of fe-

A white-striped free-tailed bat (*Tadarida australis*) on a tree branch near Alice Springs, Northern Territory, Australia. (Photo by B. G. Thomson/Photo Researchers, Inc. Reproduced by permission.)

or suspected loss and degradation of habitat, and/or because populations are small and restricted in their area of occupancy. Thus, the loss of habitat, coupled with restricted distributions, contributes the most concerns for the conservation of these bats. Although it still forms the largest aggregations of mammals in existence, *Tadarida brasiliensis* is the only bat listed on Appendix I (Endangered) of the Bonn Convention of Migratory Animals. This listing results from the extreme vulnerability of these bats to so many individuals being aggregated at only a limited number of sites, and the fact that many formerly huge cave colonies in the United States and Mexico either no longer exist or have suffered severe reductions in the populations. Habitat destruction, disturbance, vandalism, and poisoning from pesticides are the major risks to these bats.

Significance to humans

Where they are abundant, molossid bats can provide important service to humans by consuming huge numbers of insects that are agricultural pests. The 100 million Brazilian free-tailed bats that occupy Texas each summer consume an estimated 1,000 tons of insects each night, with many of these

males, allowing for the rapid growth of their young. During the period of peak lactation, it is estimated a lactating female Brazilian free-tailed bat has energy demands of 106 kj/day to meet her needs and those of her growing young, requiring that the female consume approximately 70% of her body weight in insects each night.

During pregnancy and lactation, females typically roost in maternity colonies, separated from adult males. But, even in the largest maternity colonies that contain tens of millions of individuals, females relocate and selectively nurse their own young. The mating system is not known for all species, but most are thought to be polygynous.

Conservation status

Extremely fragmented or limited distributions, coupled with near-term concern for the integrity of critical habitat, are responsible for the listing of *Chaerephon gallagheri*, *Mops niangarae* (both African), and *Otomops wroughtoni* (Asian) as Critically Endangered, and for the listing of *Mormopterus phrudus* (South American) as Endangered. All 15 species listed as Vulnerable are considered to be at risk because of projected

Free-tailed bat (*Tadarida fulminans*) adults deposit juveniles before leaving to forage. (Photo by Harald Schütz. Reproduced by permission.)

insects known to be adult cotton bollworms, fall armyworms, and other moths that are major crop pests. The guano of molossid bats that live in large colonies is harvested commercially by local farmers as a rich source of nitrogen for fertilizer.

As with all mammals, bats can contract and transmit rabies virus. The rabies virus associated with Brazilian free-tailed bats has been implicated in the deaths of approximately 12 people in North and South America over the last three decades. Other human health concerns involve *Histoplasma capsulatum*, a fungus that commonly grows in bat (and bird) guano that can infect humans and cause histoplasmosis, typically of the human respiratory system via inhalation. The habits of molossids of roosting in houses and other buildings may result in human contacts and their risks of exposure to rabies or histoplasmosis.

1. Giant mastiff bat (*Otomops martiensseni*); 2. Greater house bat (*Molossus ater*); 3. Brazilian free-tailed bat (*Tadarida brasiliensis*); 4. White-striped free-tailed bat (*Tadarida australis*); 5. Lesser crested mastiff bat (*Chaerephon pumila*); 6. Naked bat (*Cheiromeles torquatus*). (Illustration by Brian Cressman)

Species accounts

Lesser-crested mastiff bat

Chaerephon pumila

TAXONOMY

Chaerephon pumila (Cretzschmar, 1830), Massawa, Eritrea.

OTHER COMMON NAMES

English: Crested free-tailed bat.

PHYSICAL CHARACTERISTICS

Forearms ranging in length 2.5–2.9 in (6.2–7.2 cm); weighing 1.0–1.3 oz (31–39 g). It has long ears and very long narrow wings. Males develop a head crest of hair during the mating season.

DISTRIBUTION

Throughout much of sub-Saharan Africa, from Senegal to Yemen and south to South Africa. Also in Madagascar.

HABITAT

Present from sea level to over 6,560 ft (2,000 m), from semi-arid to humid montane forest, and in urban habitats.

BEHAVIOR

Roosts in caves, tree hollows, and buildings. Most known colonies consist of up to a few tens of individuals, but colonies of hundreds have been reported from lava tubes in Kenya. It is reported to mate in year-round harems of three to 21 females attended by a single adult male, with young females recruited into their natal groups.

FEEDING ECOLOGY AND DIET

Forage high above the canopy with very rapid flight and using low-frequency echolocation calls. Known to eat moths, beetles, and grasshoppers.

REPRODUCTIVE BIOLOGY

Polygynous. Reproductive schedule may vary geographically; in Kenya, they mate in August with a single young born in December–January.

CONSERVATION STATUS

Listed by the IUCN as Vulnerable due to documented and projected population declines and disturbance and destruction of known roost sites.

SIGNIFICANCE TO HUMANS

Guano is mined for fertilizer from formerly large cave roost sites in Kenya. May consume insects that are agricultural pests. ◆

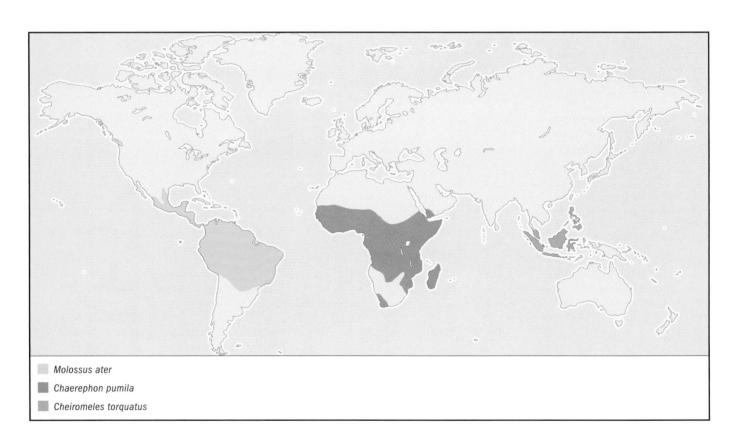

Molossus ater

Chaerephon pumila

Cheiromeles torquatus

Naked bat
Cheiromeles torquatus

TAXONOMY
Cheiromeles torquatus Horsfield, 1824, Penang, Malaysia. May be synonymous with the only other species in the genus, *C. parvidens*.

OTHER COMMON NAMES
English: Hairless bat, naked bulldog bat.

PHYSICAL CHARACTERISTICS
The largest molossid and the most distinctive with forearms ranging in length 2.8–3.6 in (7–9 cm); weighing 3.2–5.7 oz (96–170 g). Naked bats are almost completely devoid of hair with dark gray to black skin. Deep flaps of skin on the sides of the body from the upper wing bones to lower leg bones form a wing pouch into which the bats push their folded wings using the hind feet. The first toe of each foot is opposable and has a flattened nail rather than a claw.

DISTRIBUTION
Peninsular Malaysia, Borneo, Java, Sumatra, the Philippines, and associated smaller islands.

HABITAT
Tropical forest.

BEHAVIOR
Roosts in caves, rock crevices, tree cavities, and holes in the ground. Formerly found in colonies of up to 200,000 in large caves, and up to 1,000 individuals in tree cavities. They are capable of crawling swiftly on the ground.

FEEDING ECOLOGY AND DIET
A strong, fast-flying bat that forages above the canopy and in open areas on a variety of insects, particularly winged ants and termites.

REPRODUCTIVE BIOLOGY
Typically produces twins. Mating system is not known, but most likely polygynous.

CONSERVATION STATUS
Several large cave populations have declined substantially due to human disturbance, collection for food by local human populations, and persecution because these bats were mistakenly considered a pest of crops. Rated by IUCN as Lower Risk/Near Threatened.

SIGNIFICANCE TO HUMANS
Eaten by local human populations. Have a unique symbiotic association with a distinct suborder of earwigs (Insecta, Dermaptera, Arexiniina) which resulted in Niah Cave in Sarawak being designated as an earwig sanctuary. ◆

Greater house bat
Molossus ater

TAXONOMY
Molossus ater Geoffroy, 1805, Cayenne, French Guiana.

OTHER COMMON NAMES
English: Velvety free-tailed bat, Guianan mastiff bat.

PHYSICAL CHARACTERISTICS
Forearms ranging in length 1.9–2.1 in (4.8–5.2 cm); weighing 0.5–1.3 oz (14–40 g). It has short, rounded ears that are connected across the forehead. Range in color from reddish brown to black, often with two color phases.

DISTRIBUTION
Northern Mexico to Peru, northern Argentina, Brazil, and the Guianas, also on Trinidad.

HABITAT
Moist to dry areas in forested and open habitats, as well as in rural and urban habitats.

BEHAVIOR
Roosts in tree hollows, foliage, and caves and very commonly in buildings, where it can be found under galvanized roofing or in attics where temperatures can exceed 130°F (55°C). Colonies range in size from a few bats to several hundred individuals.

FEEDING ECOLOGY AND DIET
Characterized by fast erratic flight in pursuit of insects from near the ground to high in open air. This bat fills its large cheek pouches with insects and returns to the roost to consume its food.

REPRODUCTIVE BIOLOGY
Produces a single young, however, reported to produce two young per year in different seasons in several parts of its range. Males and females segregate and may roost separately even when roosting at the same site. Mating system is not known, but most likely polygynous.

CONSERVATION STATUS
Rated by IUCN as Lower Risk/Least Concern.

SIGNIFICANCE TO HUMANS
Can be a nuisance due to guano accumulations and odor resulting from its habit of roosting in buildings. May consume insects that are agricultural pests. ◆

Giant mastiff bat
Otomops martiensseni

TAXONOMY
Otomops martiensseni (Matschie, 1897), Magrotto Plantation, Tanga, Tanzania.

OTHER COMMON NAMES
English: Big-eared free-tailed bat.

PHYSICAL CHARACTERISTICS
Forearms ranging in length 2.5–2.9 in (6.2–7.2 cm); weighing 1.0–1.3 oz (31–39 g). A good-sized bat with long ears and very long narrow wings.

DISTRIBUTION
Widely distributed in eastern Africa, including Central African Republic, Djibouti, Ethiopia, Democratic Republic of the Congo, Kenya, Tanzania, Angola, Zimbabwe, Malawi, and Natal. Also reported from Madagascar, Ghana, and Yemen.

HABITAT
Present from sea level to over 6,560 ft (2,000 m), from semi-arid to humid montane forest, and in urban habitats.

Tadarida australis

Tadarida brasiliensis

Otomops martiensseni

BEHAVIOR

Roosts in caves, tree hollows, and buildings. Most known colonies consist of up to a few tens of individuals, but colonies of hundreds have been reported from lava tubes in Kenya. In buildings in South Africa small groups of females and young roost with single adult males suggesting a harem mating system.

FEEDING ECOLOGY AND DIET

Forage high above the canopy with very rapid flight and using low-frequency echolocation calls. Known to eat moths, beetles, and grasshoppers.

REPRODUCTIVE BIOLOGY

Reproductive schedule may vary geographically; in Kenya, they mate in August with a single young born in December–January. Thought to be polygynous.

CONSERVATION STATUS

Listed by the IUCN as Vulnerable due to documented and projected population declines and disturbance and destruction of known roost sites.

SIGNIFICANCE TO HUMANS

Guano is mined for fertilizer from formerly large cave roost sites in Kenya. May consume insects that are agricultural pests. ◆

White-striped free-tailed bat

Tadarida australis

TAXONOMY

Tadarida australis (Gray, 1839), New South Wales, Australia.

OTHER COMMON NAMES

None known.

PHYSICAL CHARACTERISTICS

Forearms ranging in length 2.4–2.5 in (5.9–6.3 cm); weighing 1.1–1.5 oz (33–44 g). This bat has dark fur and two white ventral to lateral stripes.

DISTRIBUTION

Endemic to Australia. Found throughout southern Australia, approximately south of the tropic of Capricorn, but not in Tasmania.

HABITAT

Found in forest, scrub, grassland, and urban habitats.

BEHAVIOR

Roost in tree cavities alone or in groups of up to a few tens of individuals. Strong hind legs and retractable tail membrane allow them to scurry on the ground with surprising agility.

FEEDING ECOLOGY AND DIET

Forages above the canopy mostly on moths, true bugs, and grasshoppers. Many of its prey are agricultural pests.

REPRODUCTIVE BIOLOGY

Mate in August, with the birth of a single young in December–January, young wean in May. Mating system is not known.

CONSERVATION STATUS

Listed by the IUCN as Lower Risk/Near Threatened.

SIGNIFICANCE TO HUMANS

Consumes insects that are agricultural pests. ◆

Brazilian free-tailed bat

Tadarida brasiliensis

TAXONOMY

Tadarida brasiliensis (Geoffroy, 1824), Parana, Brazil.

OTHER COMMON NAMES

English: Guano bat, Mexican free-tailed bat.

PHYSICAL CHARACTERISTICS

Forearms ranging in length 1.4–1.8 in (3.6–4.6 cm); weighing 0.3–0.5 oz (10–15 g).

DISTRIBUTION

One of the most widespread bats in the Western Hemisphere, found throughout the southern half of the United States, Mexico, Central America, South America to southern Chile and Argentina, and much of the Lesser and Greater Antilles.

HABITAT

Brazilian free-tailed bats are most abundant in arid and semiarid habitats, but are common in urban areas, and present in moist forest and scrub habitats.

BEHAVIOR

Maternity colonies have been estimated at up to 30 million bats. They form the largest, densest aggregations of mammals known to exist, and provide spectacular nightly emergences and flights to high altitudes in search of insect prey.

FEEDING ECOLOGY AND DIET

Their diverse diet includes at least 12 orders and 35 families of insects, many of which are agricultural pests.

REPRODUCTIVE BIOLOGY

Single young born annually in May–July, wean at six weeks of age. Lifespan to 10 years. Mating system thought to be promiscuous.

CONSERVATION STATUS

Listed by the IUCN as Lower Risk/Near Threatened. Listed on Appendix 1 of the Bonn Convention for the Conservation of Migratory Animals due to documented loss and decline of numerous colonies and the vulnerability of such large aggregations.

SIGNIFICANCE TO HUMANS

Consumes enormous numbers of agricultural pests. Guano provides valuable fertilizer. These bats are implicated in several human rabies infections. ◆

Common name / Scientific name/ Other common names	Physical characteristics	Habitat and behavior	Distribution	Diet	Conservation status
Wrinkle-lipped free-tailed bat *Chaerephon plicata*	Reddish brown to almost black coloration. Pelage is dense and soft, face covered with black bristles, underparts slightly paler. Head and body length 1.8–4.8 in (4.5–12.1 cm), tail length 0.8–2.4 in (2–6 cm), forearm length 1.1–2.6 in (2.7–6.6 cm), weight 0.6–1.1 oz (17–31 g).	Found in caves in groups of 200,000 or more individuals. Range from sea level to 660 ft (200 m) within forested habitat. Roosts in hollow trees, crevices, caves, or roofs. A single offspring is produced at a time.	India and Sri Lanka to south China and Vietnam, southeast to Philippines, Borneo, and Lesser Sunda Islands; Hainan, China; and Cocas Keeling Islands, Indian Ocean.	Mostly small moths and beetles.	Not threatened
Northern mastiff bat *Chaerephon jobensis*	Medium brown and white upperparts and slightly grayer underparts. Heavily wrinkled upper lip overhangs lower jaw. Head and body length 3.1–3.5 in (8–9 cm), tail length 1.4–1.8 in (3.5–4.5 cm), forearm length 1.8–2 in (4.6–5.2 cm), weight 0.7–1.1 oz (20–30 g).	Found in open forests, savanna, and agricultural areas, sometimes in mountains from sea level to 4,590 ft (1,400 m). Colonies up to 350 individuals.	New Guinea, north and central Australia, Solomon Islands, Vanuatu, Fiji, and perhaps Bismarck Archipelago.	Moths, grasshoppers, termites, and other insects that are caught in fast, direct aerial flight.	Not threatened
Lesser naked bat *Cheiromeles parvidens*	Nearly devoid of hair. Dark brown, thick, elastic skin, great development of conspicuous glandular throat sac, and wing pouches. Head and body length 1.8–4.5 in (4.5–11.5 cm), tail length 2–2.8 in (5–7.1 cm), forearm length 2.8–3.4 in (7–8.6 cm).	Found in agricultural areas from sea level to 660 ft (200 m). Roosts in colonies of about 20,000 individuals in hollow trees, rock crevices, and holes in the earth. Normally two offspring produced per year.	Sulawesi and the Philippines; Mindanao, Misamis Oriental, and South Cotabato, and Negros.	Consists of termites, or other insects caught in open air, such as grasshoppers and moths.	Lower Risk/Near Threatened
Western bonneted bat *Eumops perotis* English: Western mastiff bat	Mostly consistent coloration of dark gray to brownish gray. Very narrow wings. Largest bat found in United States. Head and body length up to 6.9 in (17.5 cm), forearm length 2.8–3.2 in (7.2–8.2 cm), weight 2.3–2.6 oz (64–74 g)	Found in many habitats from rainforest to arid scrub. Roosts in small groups in tree holes, cliffs, and human dwellings. Forages at great heights. Single offspring born in a year.	California and Texas, United States, to Zacatecas and Hidalgo, Mexico; Colombia to northern Argentina and eastern Brazil; and Cuba.	Species feeds extensively on Hymenoptera. Specializes on butterflies and moths.	Not threatened
Wagner's bonneted bat *Eumops glaucinus*	Upperparts are cinnamon brown to black, underparts slightly paler brown. Large ears rounded or angular. Head and body length 1.6–5.1 in (4–13 cm), tail length 1.4–3.1 in (3.5–8 cm), forearm length 1.5–3.3 in (3.7–8.3 cm).	Found in moist habitats and multistratal tropical evergreen forest. Group sizes can range from 10–20 to 70. Adult males and females do not segregate.	Jalisco, Mexico, to Peru, northern Argentina, and Brazil; Jamaica; Cuba; and Florida, United States.	Consists of small insects, mainly members of the order Hymenoptera, caught near ground level to tree top height.	Not threatened

[continued]

Common name / Scientific name/ Other common names	Physical characteristics	Habitat and behavior	Distribution	Diet	Conservation status
Greenhalli's dog-faced bat *Molossops greenhalli*	Upperparts are yellowish brown to black, underparts gray. Broad face, widely separated eyes, no development of wrinkles on lips. Head and body length 1.6–3.7 in (4–9.5 cm), tail length 0.6–1.5 in (1.4–3.7 cm), forearm length 1.1–2 in (2.8–5.1 cm).	Found mostly at low elevations. Roosts in hollow branches of large trees in colonies of 50–75 individuals. Males and females remain together throughout year.	Nayarit, Mexico, to Ecuador and north-eastern Brazil; and Trinidad.	Consists mainly of insects, mainly moths.	Not threatened
Pallas's mastiff bat *Molossus molossus*	Gray brown to dark brown coloration. Many have two color phases: bi-colored or one, consistent color. Head and body length 2–3.7 in (5–9.5 cm), tail length 0.8–2.8 in (2–7 cm), forearm length 1.4–1.6 in (3.6–4.1 cm), weight 0.4–0.5 oz (12–15 g).	Found in both moist and dry areas within a variety of forested and open habitats. Roosts in buildings, hollow trees, logs, and holes in rocks or trees. Generally two offspring produced per year.	Sinaloa and Coahuila, Mexico, to Peru, northern Argentina, Uruguay, Brazil, and Guianas; Greater and Lesser Antilles; Margarita Island, Venezuela; Curacao and Bonaire, Netherlands Antilles; and Trinidad and Tobago.	Feeds on insects, mainly moths, beetles, and flying ants.	Not threatened
Miller's mastiff bat *Molossus pretiosus*	Dorsal pelage is dark brown, almost black. Largest species of *Molossus*. Head and body length 1.9–4.5 in (4.8–11.5 cm), tail length 1.5–2.1 in (3.8–5.4 cm), forearm length 1.8–1.9 in (4.6–4.9 cm).	Found in open areas, such as grassland savannas, dry woodlands, and cactus and thorn scrub. Polyestrus. Form small colonies under palm leaves, in hollow trees, and under roofs.	Guerrero, Oaxaca, Mexico; and Nicaragua to Colombia, Venezuela, and Guyana.	Consists of hard items, such as beetles, and other aerial insects, mainly moths.	Not threatened
Spurelli's free-tailed bat *Mops spurelli*	Upperparts vary from reddish brown to almost black, underparts are paler. Head and body length 2–4.8 in (5.2–12.1 cm), tail length 1.3–2.2 in (3.4–5.6 cm), forearm length 1.1–2.6 in (2.9–6.6 cm), weight 0.2–2.3 oz (7–64 g).	Includes forest, woodland, savanna, and dry brushland.	Liberia; Ivory Coast; Ghana; Togo; Benin; Rio Muni and Bioko, Equatorial Guinea; and Democratic Republic of the Congo (Zaire).	Hard-bodied insects, such as beetles.	Not threatened
Midas free-tailed bat *Mops midas*	Upperparts reddish to black, underparts paler. Ears are joined over top of head by band of skin, very wrinkled lips. Head and body length 2–4.8 in (5.2–12.1 cm), tail length 1.3–2.2 in (3.4–5.6 cm), forearm length 1.1–2.6 in (2.9–6.6 cm), weight 0.2–2.3 oz (7–64 g).	Prefers open woodland with scattered, tall trees and open spaces. Emerge after sunset. High, fast flight while foraging. Female may have two offspring in one year.	Senegal to Saudi Arabia, south to Botswana and Transvaal, South Africa; and Madagascar.	Hard-bodied insects, such as beetles.	Not threatened
Natal free-tailed bat *Mormopterus acetabulosus*	Upperparts are dark brown, grayish brown, or charcoal. Underparts are paler. Head and body length 1.7–2.6 in (4.3–6.5 cm), tail length 1.1–1.6 in (2.7–4 cm), forearm length 1.1–1.6 in (2.9–4.1 cm), weight 0.2–0.7 oz (6–19 g).	Found in tropical forests, woodlands, open areas, and cities, roost mainly in roofs and tree hollows. Roost in colonies of fewer than 10 to several hundred individuals. A single offspring is produced each year.	Reunion and Mauritius, Mascarene Islands; Madagascar, South Africa; and Ethiopia.	Mainly insects above tree canopy, water holes, or creeks. Sometimes prey on the ground.	Vulnerable; threatened by fragmented population and declining habitat
Beccari's mastiff bat *Mormopterus beccarii*	Upperparts dark red brown or charcoal, underparts are paler. Wings are long, narrow, and tapered. Head and body length 1.7–2.6 in (4.3–6.5 cm), tail length 1.1–1.6 in (2.7–4 cm), forearm length 1.1–1.6 in (2.9–4.1 cm), weight 0.2–0.7 oz (6–19 g).	Found within habitat range from sea level to 980 ft (300 m). Occurs in sclerophyll woodland within openings of tropical forest. Roosts in colonies of up to 50 individuals.	Molucca Islands, New Guinea, adjacent small islands, and northern Australia.	Consists of moths, beetles, dipterans, orthopterans, and homopterans. Forages mostly aerial insects over water.	Not threatened
Bini free-tailed bat *Myopterus whitleyi*	Upperparts dark brown, underparts light reddish yellow to white. Ears are shorter than head, muzzle projects beyond jaws, end of nose is separate from upper lip. Head and body length 2.2–2.6 in (5.6–6.6 cm), tail length 1–1.3 in (2.5–3.3 cm), forearm length 1–3.3 in (2.5–3.3 cm).	Found only in rainforest zone. Solitary. Flies in forest at night to hunt.	Ghana, Nigeria, Cameroon, Democratic Republic of the Congo (Zaire), and Uganda.	Consists of small, soft-bodied prey.	Not threatened

[continued]

Common name / Scientific name/ Other common names	Physical characteristics	Habitat and behavior	Distribution	Diet	Conservation status
Pocketed free-tailed bat Nyctinomops femorosaccus	Upperparts brownish to grayish brown, whitish basal area. Thicker, smaller ears close to crown. Head and body length 3.9–4.3 in (10–11 cm), tail length 1.3–1.7 in (3.4–4.4 cm), weight 0.4–0.6 oz (11.5–18 g).	Colonial, roosts mainly in crevices of cliffs, slopes, and rocky outcrops. Squeaky chatter much of the time in day roosts. Females characteristically give birth to one offspring annually.	Guerrero, Mexico, to New Mexico, Arizona, and California, United States, and Baja California, Mexico.	Consists mostly of large moths (only bodies are eaten), and smaller prey species, including flying ants and leafhoppers (entire insect is eaten).	Not threatened
Wroughton's free-tailed bat Otomops wroughtoni	Upperparts are reddish brown, pale brown, or dark brown with grayish or whitish area on the back of the neck and upper back. Head and body length 2.4–3.9 in (6–10 cm), tail length 1.2–2 in (3–5 cm), and forearm length 1.9–2.8 in (4.9–7 cm).	Roosts in caves, hollow trees, and human-made structures. Usually solitary or associate in male groups. Breeding season occurs near end of autumn.	Southern India.	Insects, as far as known.	Critically Endangered; single habitat location is declining, being degraded
Big-eared mastiff bat Otomops papuensis	Dorsal fur is red brown, paler at base. Pale brown upperparts, light brown underparts.	Endemic to Papua New Guinea, found from sea level to 980 ft (300 m). Usually solitary or associate in small groups.	Southeastern New Guinea, in two localities: Mai-u River and Vailala River.	Aerial insects found above forest canopy.	Vulnerable; very little known distribution
Big crested mastiff bat Promops centralis	Upperparts are drab brown to glossy black, underparts slightly paler. Largest of Promops species. Short, broad skull, short and rounded ears, throat sacs present. Head and body length 2.4–3.5 in (6–9 cm), forearm length over 2 in (5 cm).	Colonies of up to six individuals found roosting under palm leaves. Not as gregarious as other molossid bats.	Jalisco and Yucatán, Mexico, to Peru, northern Argentina, and Suriname; and Trinidad.	Insects, as far as known.	Not threatened
European free-tailed bat Tadarida teniotis	Coloration from reddish brown to almost black, heavy crest of long straight hairs on back of membrane uniting ears. "Bulldog" face, large ears, and very long, narrow wings. Head and body length 1.8–4.8 in (4.5–12.1 cm), tail length 0.8–2.4 in (2–6 cm), forearm length 1.1–2.6 in (2.7–6.6 cm), weight 0.4–0.5 oz (10–15 g).	Found in mountainous forests or in fissures in the sides of cliffs, natural rock formations, or roofs of caves. Can fly high and travel long distances.	France, Portugal and Morocco to Japan, southern China, and Taiwan; Madeira, Portugal, and Canary Islands, Spain.	Small insects, mainly moths.	Not threatened
Egyptian free-tailed bat Tadarida aegyptiaca English: Egyptian tomb bat; German: Faltlippen Fledermäuse; Spanish: Murciélagos guaneros	Coloration varies from reddish brown to almost black. Ears are separated. Head and body length 2.4–3.9 in (6.5–10 cm), tail length 1.2–2.3 in (3–5.9 cm), forearm length 1.8–2.6 in (4.5–6.6 cm), weight 0.5–1.4 oz (14–39 g).	Found in forest or open country, generally roosts in trees and buildings.	South Africa to Nigeria, Algeria, and Egypt to Yemen and Oman, east to India and Sri Lanka.	Small insects, mainly moths.	Not threatened

Resources

Books

Altringham, John D. *Bats: Biology and Behaviour.* Oxford: Oxford University Press, 1996.

Barbour, R. W., and W. H. Davis. *Bats of America.* Lexington: University of Kentucky Press, 1969.

Churchill, Sue. *Australian Bats.* Sydney: New Holland Publishers, 1998.

Crichton, E. G., and P. H. Krutzsch, eds. *Reproductive Biology of Bats.* London: Academic Press, 2000.

Hill, J. E., and J. D. Smith. *Bats: A Natural History.* Austin: University of Texas Press, 1984.

Hutson, A. M., et al. *Microchiropteran Bats: Global Status Survey and Action Plan.* Cambridge, UK: International Union for the Conservation of Nature and Natural Resources, 2001.

Neuweiler, Gerhard. *The Biology of Bats.* Oxford: Oxford University Press, 1998.

Nowack, R. M. *Walker's Bats of the World.* Baltimore: Johns Hopkins University Press, 1991

Wilson, Don E., and DeeAnn M. Reeder, eds. *Mammal Species of the World: A Taxonomic and Geographic Reference.* 2nd edition. Washington, DC: Smithsonian Institution Press, 1993.

Periodicals

Jones, Kate E., et al. "A Phylogenetic Supertree of Bats (Mammalia: Chiroptera)." *Biological Reviews* 77 (2002): 223–259.

Organizations

Bat Conservation International. P.O. Box 162603, Austin, TX 78716 USA. Phone: (512) 327-9721. Fax: (512) 327-9724.

Resources

E-mail: batinfo@batcon.org Web site: <http://www.batcon.org>

The Bat Conservation Trust. 15 Cloisters House, 8 Battersea Park Rd., London, SW8 4BG UK. Phone: 020 7627 2629.

Fax: 020 7627 2628. E-mail: enquiries@bats.org.uk Web site: <http://www.bats.org.uk/aboutbct.htm>

IUCN Species Survival Commission, Chiroptera Specialist Group. Web site: <http://www.iucn.org>

Gary F. McCracken, PhD

Vespertilionid bats I
(Vespertilioninae)

Class Mammalia
Order Chiroptera
Suborder Microchiroptera
Family Vespertilionidae

Thumbnail description
Small- to medium-sized bats with well-developed tails that in most species are completely covered by tail membranes

Size
Range about 1.4–5.5 in (3.5–14 cm) in body length, and about 0.09–1.59 oz (2.5–45 g) in weight

Number of genera, species
30 genera; about 267 species

Habitat
Forests and open fields, both moist and dry regions, lowlands and mountains up to the tree line

Conservation status
Extinct: 2 species; Critically Endangered: 6 species; Endangered: 17 species; Vulnerable: 46 species; Lower Risk: 59 species

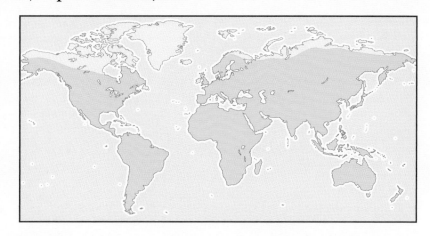

Distribution
Old and New World from the tropics into temperate zones

Evolution and systematics

The subfamily of vespertilionines is by far the largest of the five subfamilies within the family Vespertilionidae. With 30 genera and about 267 species, it makes up more than 90% of all the species within Vespertilionidae, which itself is the largest bat family.

In the subfamily Vespertilioninae, the major genera include:

- *Myotis*, 84 species

- *Pipistrellus*, 50 species

- *Eptesicus*, 32 species

- *Chalinolobus*, 15 species

- *Scotophilus*, 10 species

Eighteen of the remaining genera each have fewer than five species, and half have only one representative species. The two-species genus *Antrozous*, or the pallid bats, is notable in that some systematists place it in its own family, Antrozoidae, or in its own subfamily, Antrozoinae, within the vespertilionids.

Phylogenetic studies indicate that the family Vespertilionidae is likely most closely related to the families Mystacinidae, or the New Zealand short-tailed bats, and Molossidae, or the free-tailed and mastiff bats.

Physical characteristics

Vespertilionines are wide-ranging. The bamboo bat (*Tylonycteris pachypus*) and African banana bat (*Pipistrellus nanus*) are not only the smallest in the subfamily at about 0.1 oz (3 g) in weight, but also two of the tiniest bats in the world. At the opposite end of the spectrum is the subfamily's largest member: the large mouse-eared bat (*Myotis myotis*) with its nearly 14-in (35-cm) wingspan.

Although a large number of vespertilionines are brown, a few are frosted with silver, yellow, or reddish tips. The hoary bat (*Lasiurus cinereus*), for example, has a luxuriant pelage highlighted in white. Fur in other species within this subfamily may be golden like that of the yellow bat (*Rhogeessa anaeus*) of Belize, ginger and white like that of Welwitch's hairy bat (*M. welwitschii*) of Africa, or vivid orange like that of *Myotis formosus* of Southeast Asia.

Vespertilionines have some features in common. All lack the fleshy nose ornaments common to so many other bat families. In fact, the family Vespertilionidae is often called "the plain-faced bats." Most of the vespertilionines also have small eyes, although a few, such as the pallid bat (*Antrozous pallidus*), have noticeably larger eyes. Ear size can be small or large, with many species having tiny, rounded ears smaller than their head while others have enormous ears that extend almost the length of their bodies. The long-eared bat

Dew may collect on hibernating bats. Shown here is a hibernating little brown bat (*Myotis lucifugus*). (Illustration by Emily Damstra)

(*Plecotus auritus*) has extraordinarily long ears, but it can tuck them so far under the wing (done during the daily sleep and in hibernation) that only the pointed ear cover (tragus) can be seen.

Other distinctive features of vespertilionines—indeed all species within the family Vespertilionidae—are a well developed tragus that reaches up from the base of the ear, and a nearly naked patagium, or flight membrane, that covers the relatively long tail. Vespertilionine tails are commonly half as long as the body.

Distribution

Vespertilionines spread almost around the globe. The pipistrelles (*Pipistrellus*) include more than three dozen species in North America, Europe, Africa, Madagascar, Asia, and Australia. The noctule bats (*Nyctalus*) are distributed from the Azores through Europe and Asia to the Philippines, and the big brown bats (genus *Eptesicus*) are distributed nearly worldwide. The northern bat (*E. nilssoni*) has an extraordinary resistance to cold; its distribution extends into the Arctic. The mouse-eared bats (*Myotis* spp.) are especially noted for their wide distribution, and members of this genus are found almost everywhere bats exist.

Habitat

Given their nearly global distribution, it is not surprising that the habitat preferences of the vespertilionines vary greatly. While a large number of the temperate members inhabit caves during winter, their summer haunts can range from caves to woodlands, and riparian areas to deserts. A few vespertilionines stay away from disturbed areas, but many, like the serotine bat (*Eptesicus serotinus*), will make use of cracks and crevices in human-made structures for roosts. Many species utilize tree hollows and loose bark for daytime roosts in the summer, with various southern species even taking advantage of these spots for hibernation. A few more tropical species make use of vegetation for their roosts. The African banana bat (*Pipistrellus nanus*), for example, invariably spends the day in young banana leaves that are still rolled up.

Behavior

Temperate vespertilionines hibernate—sometimes alone, sometimes in the hundreds. Commonly, individuals awaken periodically during the winter. If the outside temperature is warm enough, they will travel outside of the hibernaculum,

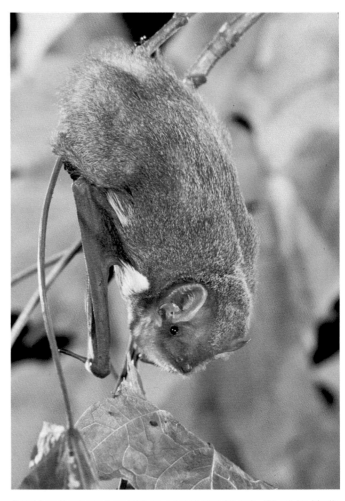

A red bat (*Lasiurus borealis*) on a small tree branch. (Photo by Merlin D. Tuttle/Bat Conservation International/Photo Researchers, Inc. Reproduced by permission.)

A little brown bat (*Myotis lucifugus*) in flight. (Photo by © Joe McDonald/Corbis. Reproduced by permission.)

which may be a cave, attic, or tunnel, and look for food. Following hibernation, male bats in this subfamily will typically spend the summer alone, while females will group together in maternity colonies to bear and raise their young. A maternity colony of pipistrelles (*Pipistrellus pipistrellus*), for example, may include more than 100 females. Unlike most vespertilionines, pipistrelle nursing colonies may include some males. After the young are independent, the bats abandon these sites.

Mouse-eared bats generally return to the same summer and winter roosts, which may be as much as 125 mi (200 km) away from each other. Another change of roosts is occasionally made in summer or in winter. Roost changes are also typical of other vespertilionine species.

Most species begin mating in the fall, but very little is known about courtship behaviors. The little brown bat (*Myotis lucifugus*), which is one of the most well-known bats in the New World, engages in no courtship. Mating simply involves the male grasping the female by the nape of her neck during copulation, performed upside down. The two separate after copulation, often to find additional partners.

In the fall, many temperate vespertilionines disappear from northern habitats, assumedly to migrate south, although some species' seasonal movements are little known. The hoary bat is one species whose migration pattern is generally understood,

although specifics are still lacking. These bats begin migrating in late summer to early fall, with often-large, mixed-sex groups traveling to the Gulf states and Mexico to spend the winter. However, the destination of some individual populations is still in question. For instance, populations of the red bat (*Lasiurus borealis*) of North America that spend their summers in the upper Great Lakes may not fly as far south as more

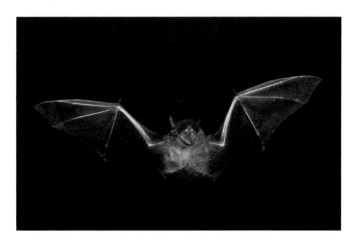

A big brown bat (*Eptesicus fuscus*) foraging in flight. (Photo by Animals Animals ©Fred Whitehead. Reproduced by permission.)

A pallid bat (*Antrozous pallidus*) in flight with prey. (Photo by Merlin D. Tuttle/Photo Researchers, Inc. Reproduced by permission.)

southern populations, perhaps migrating only as far as the Ohio River valley for the colder months instead of the Gulf states or Mexico. Part of the mystery surrounding many bat migratory movements stems from their hibernacula, which are often hidden, remote, and undiscovered. Return migrations for many species, including the hoary bat, begin when gravid females head north. The males follow shortly thereafter.

Vocalizations are used for communication and carry a variety of information. Acoustic studies on hoary bats show that they use mainly multiharmonic signals with considerable intra- and inter-individual variability in five signal variables (call duration, call interval, highest, lowest frequency, and frequency with maximum energy) to recognize each other and communicate with one another. Echolocation behavior is influenced by the presence of conspecifics. When bats hunt together, call duration decreases and call interval increases. While hunting, the pallid bat flies slowly, and close to the ground, with rhythmic dips and rises. Instead of echolocating, the desert-adapted pallid bat relies on sounds made by its prey to locate and capture a meal, often crickets or scorpions.

Some species (pallid bats) are known to produce a musky skunk-like odor from glands on the muzzle. There have been no experimental studies to determine the function of this odor—it may be a defense mechanism for repelling predators.

Feeding ecology and diet

The bats in this group are almost exclusively insectivorous. Armed with echolocation capability, these skilled fly-

ers are typically able to catch moths, beetles, flying ants, and other insects on the wing, but many will also glean leafhoppers, spiders, or other arthropods off of foliage. A few have expanded their diet to include fish or other vertebrates. Remains of food under feeding sites of the pallid bat (*Antrozous pallidus*) have included bugs, crickets, and locusts, and occasionally scorpions, lizards, and geckos. The Mexican fishing bat (*Pizonyx visesi*) employs its long hindfoot claws to spear fish. Daubenton's bat (*Myotis daubentonii*) also occasionally eats fish.

Vespertilionines commonly rest during the day in roosts, and forage at night. Some species, like the western barbastelle (*Barbastella barbastellus*) become active early in the evening and hunt for a few hours. Noctules leave their sleeping site early, and may begin the hunting flight in the afternoon, particularly in the fall. The spotted bat (*Euderma maculatum*) is active through the night. All typically rest between foraging flights to aid digestion.

Predation on these bats is generally rare. Predators may include owls who catch and kill a bat while it is foraging for insects, or mammals, like raccoons or skunks, and snakes that may find a daytime roost.

Reproductive biology

This subfamily of bats generally begins mating in the fall. Females store sperm in the reproductive tract after mating and during hibernation and then ovulate in the spring. In some species,

A harlequin bat (*Scotomanes ornatus*) feeds on an insect. (Photo by Harald Schütz. Reproduced by permission.)

such as the big brown bat (*Eptesicus fuscus*), mating also occurs in the spring. Gestation averages about two months, with altricial young typically born in late spring or early summer.Litter size is commonly one or two pups, although some species produce more. During birth, most females turn right side up and catch the infant in the tail membrane. Hoary bats and red bats have four teats, and occasionally raise four pups. On average, females lactate for one to two months. During the same period, the young learn to fly from their mothers just three weeks after birth on average and begin to forage on their own. The young may become independent during the first year, or spend the winter with the family unit. They are often left in roosts with hundreds of other bats. Females of many species become sexually mature the first year, while males typically mature the following year. Compared to mammals of a similar size, bats can live a very long time. They average 10 years or so but records of bats living 15 years and longer under natural conditions are not uncommon and there is one record of a little brown bat surviving for at least 32 years in the wild.

Conservation status

Many vespertilionines are experiencing population declines, particularly due to habitat destruction. A large number of these bats roost in dying or dead trees, as well as abandoned buildings, which are often targeted for removal.

A noctule bat (*Nyctalus noctula*) rests on a tree stump. (Photo by Nils Reinhard/OKAPIA/Photo Researchers, Inc. Reproduced by permission.)

A brown long-eared bat (*Plecotus auritus*) in flight. (Photo by Stephen Dalton/Photo Researchers, Inc. Reproduced by permission.)

Legislative, organizational and grassroots efforts are now under way to protect populations. The Indiana bat (*Myotis sodalis*), for example, is listed as Endangered by the IUCN. The U. S. Fish and Wildlife Service lists the causes of the decline as habitat disturbance, destruction and degradation, and pesticide use. Also listed as endangered under the U. S. Endangered Species Act, the species is protected from hunting or harassment. In addition, ongoing programs are managing bat habitats to ensure that species have sufficient roosting and hibernation sites, and to help educate the public on the bat's plight.

Concerned individuals have also become involved in the decline of vespertilionines and other species, and made various artificial roosts, like bat houses, more popular. According to Paul A. Racey of the U. K.'s Bat Conservation Trust, "Instead of trying to get rid of bats in their attics or other spaces, many home owners frequently ask bat workers how they can attract bats to their house. To accommodate bats, special roof-

ing tiles and clay bricks that facilitate the entry of bats to roof spaces are now marketed in the U. K."

Significance to humans

Bats have major cultural importance, especially in Asia. But perhaps the greatest significance of bats to humans stems from their insectivorous diet. A single vespertilionine can eat thousands of insects a night, and many of these arthropods are seen as pest species. For instance, bats help control populations of mosquitoes, flies, moths, beetles, and other insects. While decreases in fly or mosquito populations are readily seen as a benefit for all humans, unchecked populations of beetles, moths, and other pest species can cause significant damage to agricultural crops and forests.

In addition, many vespertilionine bats are very susceptible to pesticides and other chemicals, making them good bioindicators. These bats carry rabies and other zoonotic diseases of concern to humans.

A mouse-eared bat (*Myotis myotis*) catches a cricket on a stump. (Photo by Stephen Dalton/Photo Researchers, Inc. Reproduced by permission.)

A silver-haired bat (*Lasionycteris noctivagans*) rests on a tree trunk. (Photo by Alvin E. Staffan/Photo Researchers, Inc. Reproduced by permission.)

Himalayan pipistrelle (*Pipistrellus babu*) communal roosting spot. (Photo by Harald Schütz. Reproduced by permission.)

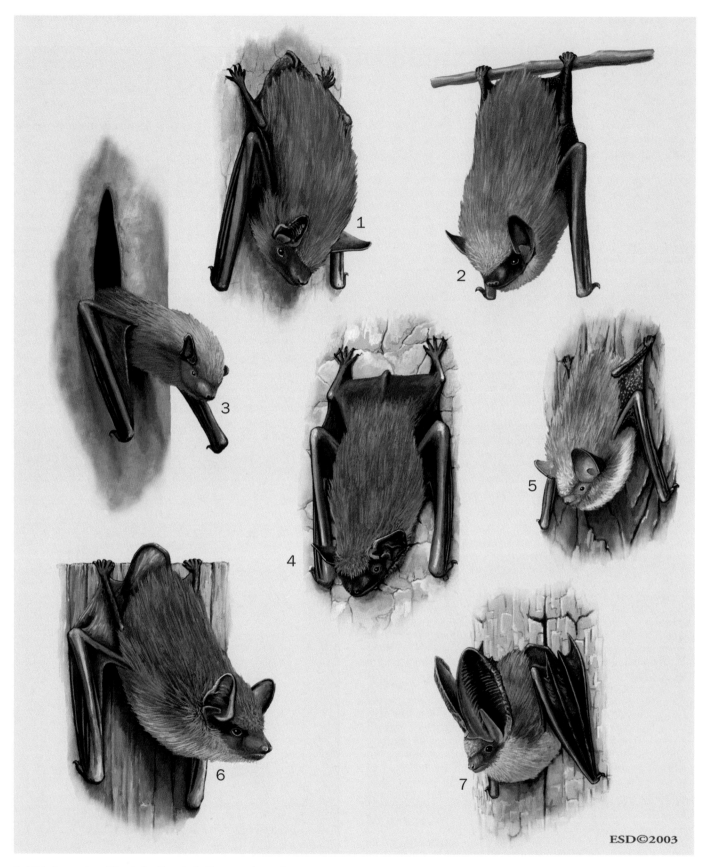

1. Little brown bat (*Myotis lucifugus*); 2. Eastern pipistrelle (*Pipistrellus subflavus*); 3. Bamboo bat (*Tylonycteris pachypus*); 4. Noctule (*Nyctalus noctula*); 5. Daubenton's bat (*Myotis daubentonii*); 6. Evening bat (*Nycticeius humeralis*); 7. Brown long-eared bat (*Plecotus auritus*). (Illustration by Emily Damstra)

1. Big brown bat (*Eptesicus fuscus*); 2. Pallid bat (*Antrozous pallidus*); 3. Allen's big-eared bat (*Idionycteris phyllotis*); 4. Spotted bat (*Euderma maculatum*); 5. Hoary bat (*Lasiurus cinereus*); 6. Silver-haired bat (*Lasionycteris noctivagans*); 7. Western barbastelle (*Barbastella barbastellus*). (Illustration by Emily Damstra)

Species accounts

Pallid bat
Antrozous pallidus

SUBFAMILY
Vespertilioninae

TAXONOMY
Antrozous pallidus (La Conte, 1856), El Paso County, Texas, United States. Six subspecies.

OTHER COMMON NAMES
French: Chauve-souris blonde; Spanish: Murcielago pálido.

PHYSICAL CHARACTERISTICS
Body length ranges from 3.6–5.5 in (9.2–14 cm), wingspan about 14 in (35.5 cm), and forearm length from about 1.8–2.4 in (4.5–6 cm). Adults weigh from 0.46 to 1.02 oz (13–29 g). A large-eared, yellowish bat with big eyes compared to other vespertilionids. The dental formula is (I1/2 C1/1 P1/2 M3/3) × 2 = 28.

DISTRIBUTION
North America from central Mexico through the western to west-central continental United States and into extreme southwest Canada. Also in western Cuba.

HABITAT
Found in deserts, and other drier regions, they roost in small openings in cliffs, trees, and buildings.

BEHAVIOR
In addition to the normal echolocation calls used in hunting, pallid bats also have intraspecies communication calls. These "directive" calls bring mothers and their young together, and also facilitate the gathering of adults in a population to a new roost. Pallid bats roost in caves, rock crevices, mines, hollow trees, and buildings and they have been reported to hibernate in some areas.

FEEDING ECOLOGY AND DIET
An insectivorous bat that can live in an arid habitat in part by obtaining necessary water from its insect prey and employing water-conservation behaviors, such as folding its wings. Besides insects, they also eat scorpions and centipedes, as well as flower nectar and pollen.

REPRODUCTIVE BIOLOGY
Promiscuous. Mating starts in late October. Following delayed fertilization, gestation lasts about 60 days, and litters are born in late spring to early summer. Litter size is typically one or two altricial young per female. The young begin to fly at about one to one-and-a-half months. Young females attain sexual maturity during the first year, and males a year later.

CONSERVATION STATUS
Not threatened.

SIGNIFICANCE TO HUMANS
Aid cross-fertilization of plants. ◆

Antrozous pallidus

Idionycteris phyllotis

Western barbastelle
Barbastella barbastellus

SUBFAMILY
Vespertilioninae

TAXONOMY
Barbastella barbastellus (Schreber, 1774), Burgundy, France. Two subspecies.

OTHER COMMON NAMES
French: Barbastelle d'Europe, barbastelle commune; German: Mopsfledermaus; Spanish: Murcielago de bosque.

PHYSICAL CHARACTERISTICS
Body length ranges 1.8–2.4 in (4.5–6.0 cm), wingspan 9.4–11.8 in (24–30 cm), and forearm length about 1.2–1.8 in (3.1–4.5 cm). Adults weigh from 0.2 to 0.42 oz (6–12 g). A small-to-medium-sized bat with black, notched ears that are about as big as its head and are connected at the base across the forehead. The head and back are covered with light-tipped blackish fur. Ventral fur is a bit lighter in color. The dental formula is (I2/3 C1/1 P2/2 M3/3) × 2 = 34.

DISTRIBUTION
Found in scattered pockets throughout much of central and northern Europe.

HABITAT
Forests often near water, roosting in trees or human buildings, and hibernating in tree hollows and caves.

Pipistrellus subflavus

Barbastella barbastellus

BEHAVIOR
This bat is quite rare and its behavior is little known. Scattered reports, however, indicate that the western barbastelle hibernates from fall to spring, but frequently awakens to fly outside of its hibernaculum. During the summer, they roost alone or in small maternity colonies typically under bark, in slight hollows in trees, or in tree stumps, but they will also crawl into tight crevices in buildings. From one to about six individuals usually share a given roost, but during the breeding season barbastelles apparently congregate in fairly large numbers. Individuals are also known to migrate up to 60 mi (100 km).

FEEDING ECOLOGY AND DIET
Diet is mainly moths and other flying insects taken while on the wing, but also includes insects and other small arthropods plucked off plants. It appears to become active earlier than most vespertilionids, sometimes emerging from its roost before sunset.

REPRODUCTIVE BIOLOGY
Little known, but reported to mate in autumn, with one young born in early summer, and weaning occurring about a month-and-a-half later. Thought to be promiscuous.

CONSERVATION STATUS
Listed as Vulnerable by the IUCN.

SIGNIFICANCE TO HUMANS
Helps to control insect populations. ◆

Big brown bat
Eptesicus fuscus

SUBFAMILY
Vespertilioninae

TAXONOMY
Eptesicus fuscus (Beauvois, 1796), Philadelphia, Pennsylvania, United States. Ten subspecies.

OTHER COMMON NAMES
French: Chauve-souris brune; German: Große braune Fledermaus; Spanish: Murcielago ali-oscuro.

PHYSICAL CHARACTERISTICS
Body length ranges 3.5–5.3 in (9–13.5 cm), and forearm length about 1.6–2.2 in (4–5.5 cm). Adults weigh from 0.39 to 0.88 oz (11–25 g), with the heavier weights typical just before hibernation and the lighter weights immediately afterward. The dental formula of the big brown bat is (I2/3 C1/1 P1/2 M3/3) × 2 = 32. A small- to medium-sized bat with brown to reddish brown or tan dorsal pelage, a lighter-colored underside, and brownish black wings.

DISTRIBUTION
Throughout the continental United States, southern Canada, most of Mexico, the Caribbean, much of Central America, and northwest South America.

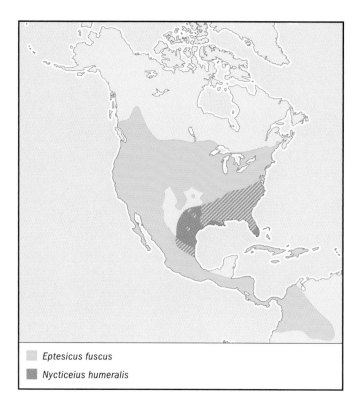

Eptesicus fuscus

Nycticeius humeralis

CONSERVATION STATUS
Not threatened.

SIGNIFICANCE TO HUMANS
Big brown bats consume large quantities of pest insects. ◆

Spotted bat
Euderma maculatum

SUBFAMILY
Vespertilioninae

TAXONOMY
Euderma maculatum (J. A. Allen, 1891), Santa Clara Valley, California, United States.

OTHER COMMON NAMES
French: L'Oreillard maculé; Spanish: Murcielago moteado.

PHYSICAL CHARACTERISTICS
Adults weigh about 0.53–0.71 oz (15–20 g), and their body length ranges 4.1–5.0 in (10.5–12.6 cm). A distinctive bat with three, large white spots on black dorsal fur. It has huge, pink ears that are about the same length as its 1.8–2.2 in (4.5–5.5 cm) forearms. Its underside is mostly white, and its face is brown and black.

DISTRIBUTION
North America from northwest Mexico, through the western United States and into British Columbia in Canada.

HABITAT
Ponderosa pine and other forests, typically near water and rocky cliffs.

HABITAT
Big brown bats spend much of their summer in the hollows of trees, and small hiding places beneath tree bark or leaves. They are also common in attics, barns, and in other human-built structures. They are typically found near water and/or woods or meadows, but also are quite common in desert habitats. During the winter, they usually hibernate in caves.

BEHAVIOR
Females will form maternity colonies, with three dozen or more females roosting together in an attic, tree hollow or other location. In the fall, big brown bats move to caves, where they soon enter hibernation. A hibernaculum may include only a handful of bats.

Their courtship and mating behaviors are unknown, although genetic differences between litter mates show that they may have different fathers.

FEEDING ECOLOGY AND DIET
Their diet includes scarab beetles, as well other flying and ground-dwelling insects. Big brown bats typically hunt on the wing, picking moths and other flying insects out of the air or snatching grasshoppers and beetles from the floor of an open forest. Each night, a big brown bat can eat several thousand insects, often equal to its body weight.

REPRODUCTIVE BIOLOGY
Mating commonly occurs in early spring, although some mate the previous fall or winter. Early mating females delay fertilization until the spring. Gestation lasts 50 to 60 days, and the young are born in late spring or early summer. Litter size is typically one or two altricial young per female. The young are weaned and begin to fly at about one to one-and-a-half months. Young females attain sexual maturity at four months, while males mature at four to 16 months. May be polygynous.

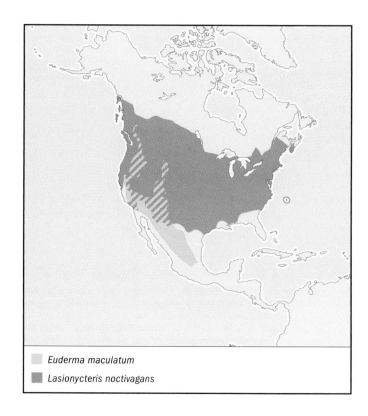

Euderma maculatum

Lasionycteris noctivagans

BEHAVIOR
Active through the night, spotted bats have echolocation calls that are audible to humans, including buzzes made while feeding. During the daytime, they rest in roosts set in small openings in steep cliffs sides.

FEEDING ECOLOGY AND DIET
Insectivores, these bats capture moths and other prey insects in fight, sometimes diving within a few feet of the ground after a low-flying arthropod.

REPRODUCTIVE BIOLOGY
Little is known of the reproductive biology of this species. The young are born in late spring to early summer. Litter size is typically one altricial young per female.

CONSERVATION STATUS
Not threatened.

SIGNIFICANCE TO HUMANS
Helps to control insect populations. ◆

Allen's big-eared bat
Idionycteris phyllotis

SUBFAMILY
Vespertilioninae

TAXONOMY
Idionycteris phyllotis (G. M. Allen, 1916), San Luis Potosí, Mexico. Two subspecies.

OTHER COMMON NAMES
English: Mexican big-eared bat, lappet-brown bat; Spanish: Murcielago de grandes orejas de Allen.

PHYSICAL CHARACTERISTICS
Average adult body length is 2–2.3 in (5–6 cm), weight is 0.28–0.56 oz (8–16 g), and forearm length is 1.7–1.9 in (4.2–4.9 cm). These long-eared bats are distinguished by the presence of lappets, which are fleshy, horizontal flaps lying over the forehead. They have dark-brown to black dorsal fur tipped with tan, reddish or yellowish brown. Like other big-eared bats, they can curl up their ears to lie them quite flat against their bodies. When unfurled, the ears are about two-thirds the length of the body.

DISTRIBUTION
Interior southwestern United States and Mexico.

HABITAT
Mountainous woods, particularly ponderosa and other pines, frequently near waterways, but also occasionally in arid scrubs and grasslands.

BEHAVIOR
Nocturnal bats that are highly skilled flyers, even capable of flying vertically. During the summer, males likely travel alone, while females will form maternity colonies of up to 150 individuals in tunnels, rock piles, and mine shafts. These bats disappear late in the fall, but little is known about possible migratory patterns or their winter whereabouts.

FEEDING ECOLOGY AND DIET
Excellent flyers, these bats feed on insects, especially in-flight moths, but also various beetles taken from foliage. While echolocating at constant frequency, they often make loud peeps and clicks that are audible to humans.

REPRODUCTIVE BIOLOGY
Litter size is generally one, and pups are born in early summer. Little else about their reproductive biology is known.

CONSERVATION STATUS
Not threatened.

SIGNIFICANCE TO HUMANS
Helps control agricultural pest insect populations. ◆

Silver-haired bat
Lasionycteris noctivagans

SUBFAMILY
Vespertilioninae

TAXONOMY
Lasionycteris noctivagans (Le Conte, 1831), eastern United States.

OTHER COMMON NAMES
French: Chauve-souris argentée; German: Silberhaar-Fledermaus; Spanish: Murcielago plateado.

PHYSICAL CHARACTERISTICS
Average adult body length is 3.5–4.7 in (9–12 cm), weight is 0.32–0.39 oz (9–11 g), and forearm length is 1.5–1.8 in (3.7–4.5 cm). Medium-sized brown to black bat with noticeable white fur tips, accounting for its "silver" appearance. The dental formula is (I2/3 C1/1 P2/3 M3/3) × 2 = 36.

DISTRIBUTION
North America from west-central Canada to Nova Scotia, through much of the continental United States except Florida, and into extreme north-central Mexico.

HABITAT
Found in forests near water.

BEHAVIOR
Normally solitary bats that will sometimes travel in twos or threes, and will form small groups during migration. Migrations occur in late fall and spring when they travel between their summer home in the northern United States and Canada, and their winter environs in southern areas. Those that migrate to warmer climates typically hibernate beneath loose bark, in crevices in trees or rocky cliffs, in buildings, or in piles of lumber. Those that spend their winters farther north may hibernate in caves. Silver-haired bats are known to form maternity colonies, which they make in small tree crevices.

FEEDING ECOLOGY AND DIET
This slow-flying, nocturnal bat typically searches for food over open areas, including ponds, and eats flies, moths, mosquitoes, and other mainly small arthropods. Predators include large owls and skunks.

REPRODUCTIVE BIOLOGY
Their reproductive biology is little known, but mating is believed to occur in autumn, and following delayed fertilization, litters of two naked young are born in early summer.

CONSERVATION STATUS
Not threatened.

SIGNIFICANCE TO HUMANS
Assist in controlling pest insect populations. ◆

Hoary bat
Lasiurus cinereus

SUBFAMILY
Vespertilioninae

TAXONOMY
Lasiurus cinereus (Beauvois, 1796), Philadelphia, Pennsylvania,
United States. Three subspecies.

OTHER COMMON NAMES
French: Chauve-souris cendrée; German: Weißgraue Fledermaus; Spanish: Murcielago blanquizco.

PHYSICAL CHARACTERISTICS
Body length ranges 5.1–5.9 in (13–15 cm), weight about
0.88–1.23 oz (25–35 g), and forearm length 1.8–2.2 in (4.6–5.5
cm). Large, dark brown to dark gray bats with white-tipped fur
and a wingspan that can reach 12 in (30.5 cm). Unusually, its
tail membrane and parts of its wings are lined with fur. The
dental formula is (I1/3 C1/1 P2/2 M3/3) × 2 = 32.

DISTRIBUTION
Throughout the continental United States, Mexico, and Cen-
tral America, north into south-central and southeastern
Canada, and as far south as Argentina and Chile in South

America. They are also found in Hawaii, where they are the
only native land mammal.

HABITAT
Wide-ranging from deserts to tundra to mountains, but most
common in deciduous and coniferous forests.

BEHAVIOR
Migratory bats that leave their summer haunts in late summer
to early fall, and return in the spring. They migrate in large
groups of sometimes 200 or more. Except for mother-and-
young families, these bats are solitary animals. While other
bats seek caves, tree hollows, or other cover during hibernation, hoary bats frequently hibernate in the open, often on the
side of a tree. They use their furred tail membrane as a blanket
to protect them from the winter weather.

FEEDING ECOLOGY AND DIET
This nocturnal bat's diet is mainly moths, supplemented by occasional flies, beetles, and other insects. The Hawaiian subspecies prefers termites. An agile and fast flier, it often hunts in
open areas. Predation on hoary bats is rare, but raptors and
snakes are known to pose risks.

REPRODUCTIVE BIOLOGY
Assumed to mate in flight in late summer or early fall, with
gestation delayed until winter to early spring. The young are
born in late spring to early summer. Litter size is typically two
altricial young per female, but may be as high as four. The
young begin to fly at about one to one-and-a-half months.
Young females attain sexual maturity at four months, while
males mature at four to 16 months. Mating system is not known.

CONSERVATION STATUS
Not threatened.

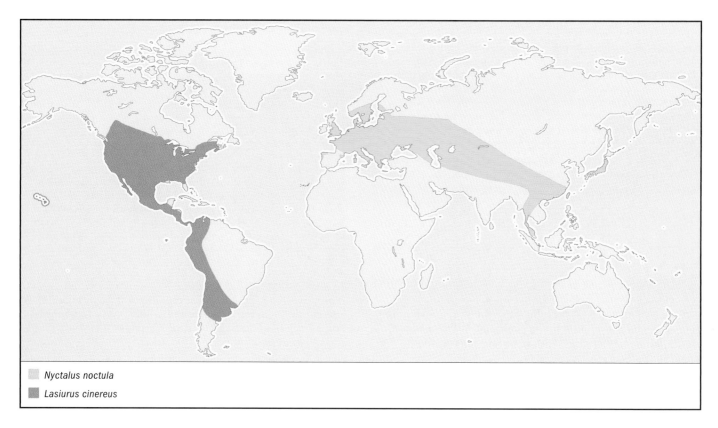

Nyctalus noctula
Lasiurus cinereus

SIGNIFICANCE TO HUMANS
Assist in controlling pest insect populations. ◆

Little brown bat
Myotis lucifugus

SUBFAMILY
Vespertilioninae

TAXONOMY
Myotis lucifugus (Le Conte, 1831), near Riceboro, Georgia, United States. Six subspecies.

OTHER COMMON NAMES
English: Little brown myotis; French: Petite Chauve-souris brune; German: Kleine braune Fledermaus; Spanish: Murcielago marrón Americano.

PHYSICAL CHARACTERISTICS
Wingspan varies from 7.9 to 10.6 in (20–27 cm), and forearms about 1.4–1.6 in (3.5–4 cm). With a body length of 3.1–3.7 in (8–9.5 cm), they are smaller than the big brown bat (*Eptesicus fuscus*), with which they are often confused. Adults weigh 0.21–0.49 oz (6–14 g), tending toward the higher end before entering hibernation and the lower end when awakening in the spring. A small bat that ranges in color from light to dark brown dorsally and light tan to whitish on its belly. They are similar in appearance to the Indiana bat (*M. sodalis*), but the little brown bat lacks the keel present on the calcar of the Indiana bat. The dental formula of the little brown bat is (I2/2 C1/1 P3/3 M3/3) × 2 = 38.

DISTRIBUTION
Found through much of North America, including southern and south-central Alaska, the southern two thirds of Canada, all but the extreme southeastern, south-central and southwestern United States, and north-central Mexico.

HABITAT
Little brown bats are found in a wide variety of habitats. They spend much of their summer in the hollows of trees, or in attics, barns, between wooden vents, and in other human-built structures. They are typically found near water and/or woods. During the winter, they hibernate in caves.

BEHAVIOR
Males are typically solitary during the summer. Females, however, will form maternity colonies with a dozen to more than 1,000 bats roosting together often in hot locations, such as attics, where temperatures can top 100°F (38°C). During the winter, males and females roost together in caves, sometimes migrating more than 150 mi (250 km) between their winter and summer roosts. A single hibernaculum draws bats from a wide area, often totaling 200,000 bats or more.

FEEDING ECOLOGY AND DIET
These are insectivorous bats that hunt in flight for their favorite prey items—moths, mayflies, and chironomid flies—by either snatching the insect with their jaws or scooping them into their wings and bringing them to their mouths. Adults may eat close to, and sometimes more than, their body weight in insects in a single night. Little brown bats are most active and do the bulk of their feeding early in the evening and at dawn.

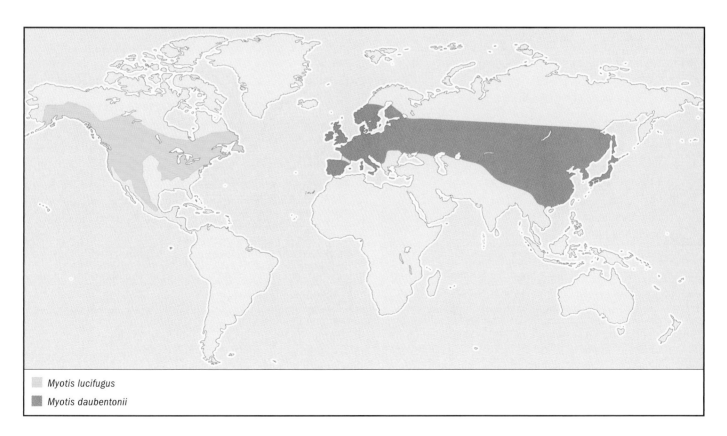

☐ *Myotis lucifugus*
■ *Myotis daubentonii*

Predators may include raptors that catch a bat in flight, or raccoons that reach a roost. However, predation on little brown bats is not prevalent.

REPRODUCTIVE BIOLOGY

A promiscuous species, mating commonly occurs in late summer and early fall as bats begin to move into the hibernaculum. Fertilization is delayed until the following spring. Gestation lasts 50 to 60 days, and the young are born in late spring to mid-summer. Litter size is typically one altricial young per female. After clinging to the mother for a day or two, the young bat hangs in the roost until it reaches about three to four weeks old, when it begins to fly and is weaned. The young reach full size at about two months old, and attain sexual maturity at one to two years. A long-lived bat, individuals in the wild have reached 33 years of age.

CONSERVATION STATUS

Not listed by the IUCN, however, some populations are experiencing declines. Numbers of the subspecies *M. l. occultus*, for instance, have dropped precipitously, particularly through habitat destruction.

SIGNIFICANCE TO HUMANS

These voracious insectivores help control pest insect populations, and also serve as bio-indicators. ◆

Noctule
Nyctalus noctula

SUBFAMILY

Vespertilioninae

TAXONOMY

Nyctalus noctula (Schreber, 1774), France. Seven subspecies.

OTHER COMMON NAMES

French: Noctule; German: Abendsegler; Spanish: Nóctulo común.

PHYSICAL CHARACTERISTICS

A medium-sized bat with yellowish to dark brown dorsal fur and slightly lighter-colored fur ventrally. Body length ranges from 2.6 to 3.2 in (6.5 to 8.2 cm), weight from about 0.53 to 1.23 oz (15 to 35 g), and forearm length from 1.9 to 2.3 in (4.7 to 5.8 cm).

DISTRIBUTION

Throughout Europe and much of temperate Asia, perhaps as far south as Singapore.

HABITAT

Forests and fields, often near water.

BEHAVIOR

They migrate—sometimes more than 400 mi (670 km) and possibly much farther—to winter hibernation sites in caves, tree hollows, and building crevices. Hundreds may hibernate together. They frequently awaken during the winter and leave the hibernaculum in search of food. After the spring migration, groups may temporarily roost together in buildings, frequently emitting screeching trills that are audible to humans. The groups disperse, and individuals separate to roost in small tree hollows and rock crevices. Males become territorial during breeding season, and release pheromones to attract females.

FEEDING ECOLOGY AND DIET

Fly over open areas, foraging for winged insects, including moths and beetles.

REPRODUCTIVE BIOLOGY

Mating commonly occurs both in early fall and in the spring; most likely polygynous. Delayed fertilization follows early mating, so that only one litter is produced in late spring to early summer. Gestation lasts 50–70 days. Litter size ranges from one to three altricial young per female, although three is rare. Weaned at six weeks of birth, the young reach full size at about two months old. Females may attain sexual maturity as early as three months of age, but most females and males do not become sexually active until the following year.

CONSERVATION STATUS

Not listed by the IUCN, however, habitat destruction appears to be reducing their numbers.

SIGNIFICANCE TO HUMANS

Assist in controlling pest insect populations. ◆

Evening bat
Nycticeius humeralis

SUBFAMILY

Vespertilioninae

TAXONOMY

Nycticeius humeralis (Rafinesque, 1818), Kentucky, United States.

OTHER COMMON NAMES

English: Twilight bat, black-shouldered bat; French: Chauve-souris vespérale.

PHYSICAL CHARACTERISTICS

Adults range from 3.5 to 4.3 in (9–11 cm) in body length, about 0.28–0.53 oz (8–15 g) in weight, and 1.3–1.5 in (3.3–3.9 cm) in forearm length. A medium-sized dark brown bat with short, black ears. It has a single pair of incisors, rather than two pair as are present in the similar appearing, although larger, big brown bat (*Eptesicus fuscus*). The dental formula is (I1/3 C1/1 P1/2 M3/3) $\times$ 2 = 30.

DISTRIBUTION

North America, including northeastern Mexico and about a third of the United States from the Gulf states to the Midwest and the mid-Atlantic coast states.

HABITAT

Forested areas.

BEHAVIOR

Evening bats roost in tree crevices and buildings in the summer, and become active in early evening. Solitary animals, but females and their pups may gather in large, multi-family nursery colonies. Little is known about this species' migratory behavior, but they do accumulate fat stores before disappearing in the fall, indicating that they may migrate south. Cave surveys have not found evening bats, suggesting that this species either uses something other than caves for its hibernacula, or does not hibernate.

FEEDING ECOLOGY AND DIET
Evening bats begin foraging soon after sunset, seeking beetles, moths, flying ants, and other insects.

REPRODUCTIVE BIOLOGY
Polygynous; mating occurs in the fall, with birth in early summer. Litter size is usually two naked young per female. The young begin to fly within a month of their birth, and are weaned at about two months old. Young males leave their mothers upon weaning, but females remain for at least a few more months.

CONSERVATION STATUS
Not threatened.

SIGNIFICANCE TO HUMANS
Assist in controlling pest insect populations, notably the chrysomelid beetle. Its larvae, called the corn rootworm, is well-known to farmers for a destructive potential. ◆

Eastern pipistrelle
Pipistrellus subflavus

SUBFAMILY
Vespertilioninae

TAXONOMY
Pipistrellus subflavus (F. Cuvier, 1832), Georgia, United States.

OTHER COMMON NAMES
French: Pipistrelle de l'est.

PHYSICAL CHARACTERISTICS
Adults range from 3 to 3.5 in (7.5–9 cm) in length, 0.18–0.28 oz (5–8 g) in weight, and 1.2–1.4 in (3.1–3.6 cm) in forearm length. Medium-sized, yellowish to reddish brown bat with an orangish forearm that is a start contrast to its dark wing. The individual dorsal hairs bear a characteristic yellowish patch in the middle. The dental formula is (I2/3 C1/1 P2/2 M3/3) × 2 = 34.

DISTRIBUTION
The eastern half of Mexico and the United States, except for much of Minnesota, Michigan, and southern Florida. Also extends into extreme southern Ontario and Quebec, and south to Honduras.

HABITAT
Wooded areas near water.

BEHAVIOR
The eastern pipistrelle remains quite active all year in southern climates, but enters a deep hibernation in northern areas, usually opting to spend its winter in the same cave or mine from year to year. It is one of the first bats to hibernate, and awakens infrequently during the cold months. During the summer, this bat becomes active early in the evening, sometimes at sunset.

FEEDING ECOLOGY AND DIET
Diet includes small, flying insects, which the bat catches in flight, often swiping them up with its wing or tail membrane and drawing them to its mouth.

REPRODUCTIVE BIOLOGY
Mating commonly occurs in early fall and spring, with delayed fertilization following the fall mating. Young are born in early summer. Gestation lasts at least 44 days. Litter size is typically two altricial young per female, although it can range from one to three. Weaned within a month of birth, the young begin flying before they reach one month old. Most likely polygynous.

CONSERVATION STATUS
Not threatened.

SIGNIFICANCE TO HUMANS
Assist in controlling pest insect populations. ◆

Brown long-eared bat
Plecotus auritus

SUBFAMILY
Vespertilioninae

TAXONOMY
Plecotus auritus (Linnaeus, 1758), Sweden.

OTHER COMMON NAMES
English: Whispering bat; French: Oreillard brun; German: Braunes Langohr; Spanish: Orejudo septentrional.

PHYSICAL CHARACTERISTICS
Adults range from 1.5 to 2 in (3.7–5 cm) in length, 0.18–0.42 oz (5–12 g) in weight, and 1.3–1.7 in (3.4–4.2 cm) in forearm length. Medium-sized, light-brown bat that rests with its long ears curled along its body or hidden beneath the wings. When outstretched, the ears are almost as long as the bat's body.

DISTRIBUTION
Throughout all but far southern Europe, east through temperate Asia to northern China and Nepal.

HABITAT
Open forests and park-like settings.

Plecotus auritus

Tylonycteris pachypus

BEHAVIOR
During the summer, they roost individually or in nursery colonies in tree crevices or in buildings. Males often join the nursery colonies. They leave the roost for nighttime feeding well after sunset. They typically enter hibernation in late fall, opting to spend the winter in small crevices in trees or man-made structures, although they will sometimes hibernate in caves.

FEEDING ECOLOGY AND DIET
A slow, but skillful flyer, this bat forages for insects in flight and by picking earwigs and spiders off of plants. Research has shown that this species uses taste and/or smell to select food items. Predators include ground mammals, such as house cats, that catch the bats while they are gleaning arthropods from vegetation.

REPRODUCTIVE BIOLOGY
Promiscuous; mating commonly occurs in fall and spring, with delayed fertilization in early breeding females. Young are born in early summer. Litter size is typically one altricial young per female. The young begin flying before they reach one month old and are weaned at about a month-and-a-half. Females become sexually mature their first year, and males the following spring.

CONSERVATION STATUS
Listed as Vulnerable by the IUCN.

SIGNIFICANCE TO HUMANS
A bio-indicator species that is particularly sensitive to pesticides. ◆

Bamboo bat
Tylonycteris pachypus

SUBFAMILY
Vespertilioninae

TAXONOMY
Tylonycteris pachypus (Temminck, 1840), Java, Indonesia.

OTHER COMMON NAMES
English: Club-footed bat, lesser flat-headed bat.

PHYSICAL CHARACTERISTICS
Adults range from 1.6 to 2.4 in (4–6 cm) in length, 0.09–0.21 oz (2.5–6 g) in weight, and 0.8–1.2 in (2–3 cm) in forearm length. One of the smallest bats, the bamboo bat is brown to reddish dorsally, slightly lighter ventrally, and has a flattened head.

DISTRIBUTION
Southeast Asia from India to southern China and into Indonesia.

HABITAT
Bamboo forests.

BEHAVIOR
Most notably, these bats are able to climb into and actually roost in internodal hollows of the large bamboo, *Gigantochloa scortechinii*. Their flattened skulls combined with small overall body size provide access to the tiny openings in the hollow stem joints, and pads on their thumbs and feet help them cling to the sides of the stems. Males may travel alone, with a few

other males, or in transient harems of one male, several females, and numerous young.

FEEDING ECOLOGY AND DIET
Termites appears to be a preferred food item.

REPRODUCTIVE BIOLOGY
Young are born in February to May. Gestation lasts about three months. Litter size is typically two altricial young per female. The young are weaned at about a month-and-a-half. Females and males become sexually mature their first year. Most likely polygynous.

CONSERVATION STATUS
Not threatened.

SIGNIFICANCE TO HUMANS
Assist in controlling termite populations. ◆

Daubenton's bat
Myotis daubentonii

SUBFAMILY
Vespertilioninae

TAXONOMY
Myotis daubentonii Kuhl, 1817, Hanau, Hessen, Germany.

OTHER COMMON NAMES
English: Water bat; French: Vespertilion de Daubenton; German: Wasserfledermaus; Spanish: Murcielago ribereño.

PHYSICAL CHARACTERISTICS
Adults range from 1.8 to 2.2 in (4.5–5.5 cm) in length, 0.25–0.42 oz (7–12 g) in weight, and 1.3–1.6 in (3.4–4.1 cm) in forearm length. Medium-sized brown bat with lighter ventral fur.

DISTRIBUTION
Eurasia.

HABITAT
Riparian areas, often near woods.

BEHAVIOR
In addition to mother-and-young maternal roosts, adult males and females may roost together during the summer. Roosts are usually underground tunnels, cellars, and caves. They hibernate in the winter in similar underground locations, but remain quite active, leaving the hibernaculum frequently.

FEEDING ECOLOGY AND DIET
Often feeds in flight, capturing insects over water. Also known to occasionally take a small fish.

REPRODUCTIVE BIOLOGY
Species most likely promiscuous. Mating commonly occurs in the fall and winter, with delayed fertilization. Young are born in early summer. Litter size is typically one. The young are weaned by two months of age.

CONSERVATION STATUS
Not threatened.

SIGNIFICANCE TO HUMANS
May help control pest insect populations. ◆

Common name / Scientific name/ Other common names	Physical characteristics	Habitat and behavior	Distribution	Diet	Conservation status
Butterfly bat *Chalinolobus variegatus* English: African bat	Yellow bat with brown face and variegated wings. Head and body length 2–2.4 in (5.0–6.0 cm), forearm length 1.6–1.8 in (4.1–4.5 cm), weight 0.4–0.5 oz (10–15 g).	Found in open woodlands. Roost among leaves or thatched building roofs, sometimes wrapping their wings around themselves. Maternal groups of a dozen or more females and young roost together.	Central to south central Africa.	Insects.	Not listed by IUCN
Serotine bat *Eptesicus serotinus* French: Grande sérotine; German: Breitflügelfledermaus; Spanish: Murcielago hortelano	Black face and ears, dark brown dorsal pelage. Ventral fur lighter. Head and body length 2.3–3.1 in (5.8–8.0 cm), forearm length 1.9–2.2 in (4.8–5.5 cm), weight 0.5–1.2 oz (15–35 g).	Found in woodlands, meadows and cultivated areas. Often roost in human-made structures in summer and probably the winter. Males are usually solitary in the summer, and females form small maternity colonies, averaging about one to three dozen bats. Become active before sunset, and forage into the night.	Europe, southern Asia into China, and northern Africa.	Moths, beetles, and other insects.	Not listed by IUCN
Long-fingered bat *Myotis capaccinii* English: Cave-dwelling bat; French: Vespertilion de Capaccini; German: Langfußfledermaus; Spanish: Murcielago patudo	Brownish gray with slightly pinkish brown face, ears, and forelimbs. Head and body length 1.9–2.1 in (4.7–5.3 cm), forearm length 1.5–1.7 in (3.9–4.4 cm), weight 0.2–0.5 oz (6–15 g).	Found in forests and shrubby areas. Roost in caves. Known to hibernate.	Northwest Africa, along the Mediterranean in Europe, and east to southeastern China.	Insects.	Vulnerable
Pond bat *Myotis dasycneme* French: Vespertilion des marais; German: Teichfledermaus; Spanish: Murcielago lagunero	Light brown bat with medium brown face, ears, and wing membranes. Head and body length 2.2–2.7 in (5.7–6.8 cm), forearm length 1.6–1.9 in (4.1–4.9 cm), weight 0.4–0.8 oz (11–23 g).	Found in forests near water. Often roosts in trees or in human-made structures in the summer, making feeding forays over still water at night. Maternity roosts can reach 600 bats, but a few dozen is more common. Hibernate in caves and other underground structures.	Northern and central Europe through Russia, the Ukraine and Kazakhstan, perhaps to Manchuria, China.	Insects.	Vulnerable
Greater mouse-eared bat *Myotis myotis* French: Grand murin; German: Großes Mausohr; Spanish: Murcielago ratonero grande	Brown bat, whitish below. Head and body length 2.4–3.6 in (6.3–9.1 cm), forearm length 2.2–2.6 in (5.5–6.6 cm), weight 0.2–0.3 oz (6–8 g).	Found primarily in forest areas. Slow but straight flyers that emerge at night to forage. Usually have one primary roosting site, and visit secondary sites occasionally. Females leave nursery colonies to visit male roosts for mating purposes.	Central and southern Europe, Ukraine, Israel, and Asia Minor.	Insects.	Lower Risk/Near Threatened
Whiskered bat *Myotis mystacinus* French: Vespertilion à moustaches; German: Kleine Bartfledermaus; Spanish: Murcielago bigotudo	Dark brown dorsal pelage sometimes tipped with gold. Ventral fur usually light to medium gray. Head and body length 1.4–2.0 in (3.5–5 cm), forearm length 1.2–1.6 in (3–4 cm), weight 0.2–0.3 oz (5–9 g).	Found in woodlands, meadows, and cultivated areas, often near water. Often roost in trees or in human-made structures in the summer, and caves in the winter. Roost in colonies ranging in size from three dozen bats to 200. Females in temperate populations bear young in early summer, but those in more tropical populations may have a litter at other times of the year. Nocturnal, emerging around sunset.	Temperate Europe and Asia.	Insects, especially moths, and spiders.	Not listed by IUCN

[continued]

Common name / Scientific name/ Other common names	Physical characteristics	Habitat and behavior	Distribution	Diet	Conservation status
Natterer's bat *Myotis nattereri* English: Red-armed bat; French: Vespertilion de Natterer; German: Fransenfledermaus; Spanish: Murcielago de Natterer	Light brown bat with a whitish underside, pinkish limbs, and bristles at the edge of the tail membrane. Head and body length 1.6–2.0 in (4.0–5.0 cm), forearm length 1.4–1.7 in (3.6–4.3 cm), weight 0.2–.4 oz (7–12 g).	Found in woods and cultivated areas, often near water. Often roost in trees or in human-made structures in the summer, and caves in the winter. Nocturnal, emerging after sunset.	All but northern Europe, east through south central Asia, and into northwest Africa.	Insects.	Not listed by IUCN
Greater noctule *Nyctalus lasiopterus* French: Grand noctule; German: Riesenabendsegler; Spanish: Nóctulo gigante	Reddish brown bat with dark brown to blackish face, ears, and wing membranes. Head and body length 2.0–2.8 in (5.0–7.0 cm), forearm length 1.5–1.9 in (3.8–4.7 cm), weight 0.4–0.7 oz (12–20 g).	Found in or near forested areas. Unusually, this nocturnal bat hunts birds during its migratory flights, probably overtaking their prey, and killing or disabling the birds in mid-air.	Most of Europe, temperate Asia, and parts of Southeast Asia.	Insects, but also birds during migration.	Lower Risk/Near Threatened
Leisler's bat *Nyctalus leisleri* English: Hairy-armed bat; French: Noctule de Leisler; German: Kleiner Abendsegler; Spanish: Nóctulo pequeño	Yellowish brown, heavily furred bat with fur extending onto wings and forearms. Head and body length 1.6–2.3 in (4.0–5.8 cm), forearm length 1.5–1.8 in (3.7–4.5 cm), weight 0.2–0.5 oz (5–14 g).	Found in woods and cultivated areas. This migratory species often roosts in trees or in human-made structures in the summer, and trees and caves in the winter. Nocturnal, usually emerging after sunset, but occasionally seen during the daytime in the spring.	Much of Europe, east through south central Asia, and into northwest Africa.	Insects.	Lower Risk/Near Threatened
Kuhl's pipistrelle bat *Pipistrellus kuhlii* French: Pipistrelle de Kuhl; German: Weißrandfledermaus; Spanish: Murcielago de borde claro	Yellowish brown bat with a white margin on the wing. Head and body length 1.5–2.0 in (3.8–5 cm), forearm length 1.1–1.3 in (2.7–3.2 cm), weight 0.2–0.3 oz (7–8 g).	Found in lowland forests. Roost in trees cavities, as well as human-made, wooden structures. Sympatric with *P. nanus*, but produces echolocation calls of a different frequency.	Northern and eastern Africa, west to the Canary Islands, through Europe to southwest Asia.	Insects, especially small beetles, moths, and flies.	Not listed by IUCN, though declining in Europe due to habitat destruction and exposure to toxins in treated lumber
Nathusius's pipistrelle bat *Pipistrellus nathusii* French: Pipistrelle de Nathusius; German: Rauhhautfledermaus; Spanish: Falso murcielago común	Yellowish brown bat with a darker brown face and ears. Head and body length 1.5–2.0 in (3.8–5.0 cm), forearm length 1.2–1.4 (3.1–3.6 cm), weight 0.2–0.5 oz (6–15 g).	Found primarily in woods. Often roost in trees or in human-made structures in the summer, and caves and small crevices in the winter.	From southern England through western Europe and to western Asia Minor.	Small flying insects.	Not listed by IUCN
Gray long-eared bat *Plecotus austriacus* French: Oreillard gris; German: Graues Langohr; Spanish: Orejudo meridional	Very similar to *P. auritus*, but more gray than brown. Head and body length 1.6–2.3 in (4.0–5.8 cm), forearm length 1.5–1.8 in (3.7–4.5 cm), weight 0.2–0.5 oz (5–14 g).	Found in lowland forests. Behavior is little known, but roosting sites have been found in human-made structures in the summer and winter, when they prefer underground locations.	Northern Africa through southern Europe and east to western China.	Moths and other insects.	Not listed by IUCN, though declining in parts of Europe due to habitat destruction and possibly exposure to toxins in treated lumber
House bat *Scotophilus kuhlii* English: Lesser yellow bat, yellow house bat	Noted for its two color phases: reddish brown and greenish brown. Head and body length 2.6–2.8 in (6.7–7.2 cm), forearm length 1.9–2.0 in (4.8–5.2 cm), weight 0.5–.8 oz (15–22 g).	Found in forests, but are adapted to people. Frequently roost in houses, but will also construct day roosts, or tents, of foliage. Begin foraging shortly after sunset. Maternity colonies commonly number about three dozen bats, but may exceed 200 individuals.	Pakistan east to Southeast Asia and East Indies islands.	Insects, especially beetles, termites, and moths.	Not listed by IUCN
Parti-colored bat *Vespertilio murinus* French: Sérotine bicolore; German: Zweifarbfledermaus; Spanish: Murcielago bicolor	Dark brown dorsal pelage tipped with white to give an overall tan appearance. Head and body length 2.0–2.5 in (5–6.6 cm), forearm length 1.6–1.8 in (4–4.6 cm), weight 0.4–0.5 oz (10–14 g).	Found in wooded areas and "steppes," these nocturnal bats often roost in large colonies of several thousand bats in summer and winter. Roosts are often human-made structures.	Through central Europe and central Russia, as far north as Siberia.	Insects.	Not listed by IUCN

Resources

Books

Altringham, J. *Bats: Biology and Behavior.* Oxford: Oxford University Press, 1996.

Fenton, M. *Bats.* New York: Checkmark Books, 2001.

Kunz, T., and P. Racey, eds. *Bat Biology and Conservation.* Washington, DC: Smithsonian Institution Press, 1998.

Kurta, A. *Mammals of the Great Lakes Region.* Ann Arbor: The University of Michigan Press, 1995.

Nowak, R. *Walker's Mammals of the World.* Baltimore: Johns Hopkins University Press, 1999.

Periodicals

Adams, R., and S. Pedersen, "Wings on Their Fingers." *Natural History* 103 (January 1994): 48–55

Arizona Game and Fish Department. "*Antrozous pallidus.*" Unpublished abstract compiled and edited by the Heritage Data Management System, Arizona Game and Fish Department. Phoenix: 1994.

Bradley, P. "Nevada's Night Fliers." *Natural History* 105 (February 1996): 72–6.

Milius, S. "Bat Bites Bird—In Migration Attacks." *Science News* 160, no. 6 (2001): 86.

Petit, E., L. Excoffier, and F. Mayer. "No Evidence of Bottleneck in the Postglacial Recolonization of Europe by the Noctule Bat (*Nyctalus noctula*)." *Evolution* 53, no. 4 (1999): 1247–1258.

Racey, P. "The Conservation of Bats in Europe." *Bats* 10, no. 4 (1992): 4–10.

Whitaker, John O., Jr. "Food of the Big Brown Bat *Eptesicus fuscus* from Maternity Colonies in Indiana and Illinois." *The American Midland Naturalist* 134 (1995): 346–360.

Organizations

Bat Conservation International. P.O. Box 162603, Austin, TX 78716 USA. Phone: (512) 327-9721. Fax: (512) 327-9724. E-mail: batinfo@batcon.org Web site: <http://www.batcon.org>

The Bat Conservation Trust. 15 Cloisters House, 8 Battersea Park Rd., London, SW8 4BG UK. Phone: 020 7627 2629. Fax: 020 7627 2628. E-mail: enquiries@bats.org.uk Web site: <http://www.bats.org.uk/aboutbct.htm>

The Organization for Bat Conservation at Cranbrook Institute of Science. 39221 Woodward Avenue, P.O. Box 801, Bloomfield Hills, MI 48303-0801 USA. Phone: (800) 276-7074. E-mail: obcbats@aol.com Web site: <http://batconservation.org/>

Leslie Ann Mertz, PhD

Vespertilionid bats II
(Other subfamilies)

Class Mammalia
Order Chiroptera
Suborder Microchiroptera
Family Vespertilionidae

Thumbnail description
Small- to medium-sized bats, frequently brown in color, but occasionally with strikingly colored fur

Size
Range about 1.2–3.0 in (3.1–7.5 cm) in body length, and about 0.07–0.7 oz (2–20 g) in weight

Number of genera, species
5 genera; at least 48 species

Habitat
Forested areas, frequently near rocky outcrops and caves

Conservation status
Critically Endangered: 1 species; Endangered: 3 species; Vulnerable: 6 species; Lower Risk: 14 species

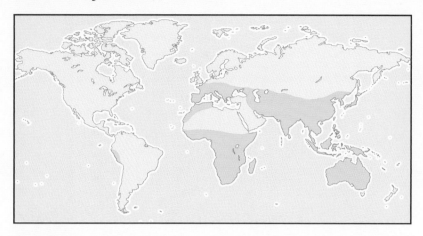

Distribution
Southern Europe, southern and eastern Africa, southern Asia, Indonesia, New Guinea, Australia, and Peru

Evolution and systematics

The family Vespertilionidae is commonly split into five subfamilies, including the enormous Vespertilioninae, which is treated in a separate chapter. The remaining four subfamilies are:

- Miniopterinae
- Murininae
- Tomopeatinae
- Kerivoulinae

The subfamily Kerivoulinae is the largest with one genus (*Kerivoula*) and 22 species, followed by the subfamily Murininae with two genera (*Harpiocephalus* and *Murina*) and 15 species. The subfamily Miniopterinae has one genus (*Miniopterus*) and 10 species, and the subfamily Tomopeatinae includes a single species, *Tomopeas ravus*. Although the subfamilies are relatively small, taxonomists are still struggling with their taxonomy. For example, some scientists consider the murinines a separate family, or split the *Kerivoula* genus and place some of the species in a separate *Phoniscus* genus. In addition, controversy exists over whether to place *T. ravus* in the family Molossidae or Vespertilionidae. Based on its morphology, which includes features of both the vespertilionids and molossids, many believe *T. ravus* may represent an ancestor to both families.

Physical characteristics

Since all four subfamilies are in the family Vespertilionidae, they share some features, including a well-developed tragus that reaches up from the base of the ear, a nearly naked patagium, or flight membrane, that covers a relatively long tail, and

A golden-tipped bat (*Kerivoula papuensis*) on a log. (Photo by Pavel German. Reproduced by permission.)

A little long-fingered bat (*Miniopterus australis*). (Photo by Pavel German. Reproduced by permission.)

A tube-nosed insectivorous bat (*Murina leucogaster*). (Photo by Brock Fenton. Reproduced by permission.)

the lack of the fleshy nose ornament common to so many other bat families. Unlike other vespertilionids, however, species in the subfamily Murininae have nostrils appearing at the end of tube-shaped nose. *Tomopeas* lacks the anterior basal earlobe present in the other subfamilies. Miniopterines have an extremely long third finger that they can fold, earning them the moniker bent-wing bats. Members of the subfamily Kerivoulinae all share a long, slender, pointed tragus.

Distribution

The species of the subfamily Kerivoulinae extend through central and southern Africa, and from India east to China, throughout southeast Asia, Indonesia, New Guinea and northeastern Australia. The subfamily Miniopterinae reaches into central and southern Africa, as well as Madagascar, and from southern Europe and Asia north into China, and also New Guinea, Indonesia, and Australia. Murinines are only

A golden-tipped bat (*Kerivoula papuensis*) resting on moss. (Photo by Pavel German. Reproduced by permission.)

Common bentwing bats (*Miniopterus schreibersi*) roosting in a cave. (Photo by Michael Pitts/Naturepl.com. Reproduced by permission.)

A common bentwing bat (*Miniopterus schreibersi*) in flight. (Photo by Brock Fenton. Reproduced by permission.)

slightly less widespread, reaching from southern Asia north into China, and through Indonesia, New Guinea, and northeast Australia. *Tomopeas* is found strictly in western Peru.

Habitat

Species in these four subfamilies range in their habitat choices from subtropical to temperate forests and fields, frequently selecting areas that are near caves or rocky outcrops commonly used for roosts. Members of the *Miniopterus* species, for example, use caves for both summer and winter roosts. Some *Murina* species, on the other hand, can roost among the leaves of trees and vines, and do not require underground sites. Some *Kerivoula* species also use bird nests for roosting sites.

Behavior

The more temperate bats in these four subfamilies, including many *Miniopterus* species, migrate 100 mi (160 km) or more in the winter and spend much of the season in hibernation, while those that live in more tropical climates may stay in one general area and remain active throughout the year. In the spring, when young are typically born, females of most species form maternity colonies. Males usually do not roost within these large groups. Overall, little is known about the behavior of most of these bats.

Feeding ecology and diet

These bats are nocturnal, typically becoming active shortly after sunset and feeding on various insects well into the night. Recent studies reveal that the diet of *Kerivoula papuensis* may primarily be spiders rather than insects. Analyses of fecal matter indicate that more than 90% of its diet consists of arachnids.

Common bentwing bats (*Miniopterus schreibersi*) are thought to be extremely gregarious. (Photo by Brock Fenton. Reproduced by permission.)

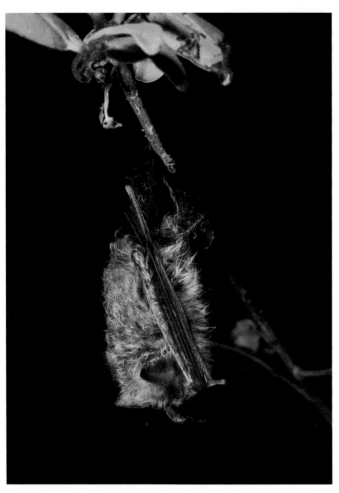

Kerivoula lanosa has long woolly hair. (Photo by Brock Fenton. Reproduced by permission.)

Reproductive biology

Little is known about the reproductive biology of many of these species. It is assumed, however, that at least the temperate species mate in the fall. Most vespertilionids in the temperate regions have delayed fertilization, with sperm being stored in the winter within the female reproductive tract. Females then ovulate and fertilization occurs in the spring. With Miniopterinae, following delayed implantation of the embryo until late winter or early spring, females give birth to litters averaging one or two pups in the late spring or early summer. The capture of gravid warmer-climate females suggests that these species are capable of producing offspring all year long.

Conservation status

Fully half of the species in these four subfamilies are listed as Lower Risk, Vulnerable, Endangered, or Critically Endangered by the IUCN. The population declines are often accompanied by reports of habitat destruction. The golden-tipped bat (*Kerivoula papuensis*) is an example of a bat that may be facing a threat from habitat loss. A major challenge to scientists now is to learn more about this and other species that have historically been difficult to find. In fact, the golden-tipped bat was considered extirpated from Australia until it was rediscovered in 1981. With new trapping and monitoring techniques, scientists are beginning to gather data about habitat and feeding requirements. This information is critical in developing effective conservation plans.

Significance to humans

Like most bats, the species within these four subfamilies are insectivorous, making them beneficial in controlling various beetles and other agricultural pest populations.

1. Painted bat (*Kerivoula picta*); 2. Common bentwing bat (*Miniopterus schreibersi*); 3. Brown tube-nosed bat (*Murina suilla*); 4. Harpy-headed bat (*Harpiocephalus harpia*). (Illustration by Brian Cressman)

Species accounts

Harpy-headed bat
Harpiocephalus harpia

SUBFAMILY
Murininae

TAXONOMY
Harpiocephalus harpia (Temminck, 1840), Mt. Gede, Java, Indonesia. Two subspecies.

OTHER COMMON NAMES
English: Harpy-winged bat, hairy-winged bat.

PHYSICAL CHARACTERISTICS
Average adult body length is 2.2–3.0 in (5.5–7.5 cm), and forearm length is 1.6–2.1 in (4.0–5.4 cm). Reddish orange, furry bat, somewhat mottled with gray, and a lighter underside. It has a tube-shaped nose.

DISTRIBUTION
India through Southeast Asia and Indonesia.

HABITAT
Hilly areas and lowland forest.

BEHAVIOR
Little known, but are believed to roost in foliage.

FEEDING ECOLOGY AND DIET
Little known, but have been reported to eat beetles.

REPRODUCTIVE BIOLOGY
Nothing is known.

CONSERVATION STATUS
Not threatened.

SIGNIFICANCE TO HUMANS
May help control populations of crop-damaging insects. ◆

Brown tube-nosed bat
Murina suilla

SUBFAMILY
Murininae

TAXONOMY
Murina suilla (Temminck, 1840), Java, Indonesia.

OTHER COMMON NAMES
English: Brown murine bat.

PHYSICAL CHARACTERISTICS
Average adult body length is 1.3–2.4 in (3.3–6.0 cm), weight is 0.11–0.18 oz (3–5 g), and forearm length is 1.0–1.8 in (2.6–4.5 cm). Dark grayish brown, furry bat with a tube-shaped nose. Its underside is whitish, and it has brown wing membranes.

DISTRIBUTION
Western Malaysia and Indonesia, including Borneo.

HABITAT
Hilly areas, often near cultivated and grassy fields.

BEHAVIOR
Forage flights take them close to the ground, often barely over the tops of fields. Often roost together in foliage.

FEEDING ECOLOGY AND DIET
Insectivorous.

REPRODUCTIVE BIOLOGY
Nothing is known.

CONSERVATION STATUS
Not threatened.

SIGNIFICANCE TO HUMANS
May help control insects that damage crops. ◆

Common bentwing bat
Miniopterus schreibersi

SUBFAMILY
Miniopterinae

TAXONOMY
Miniopterus schreibersi (Kuhl, 1817), Banat, Romania. Two subspecies.

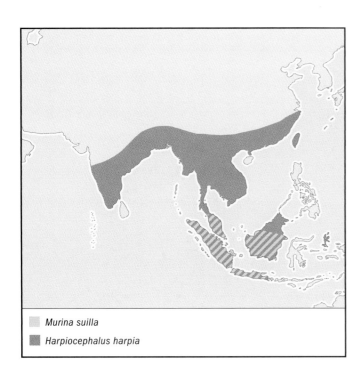

■ *Murina suilla*
■ *Harpiocephalus harpia*

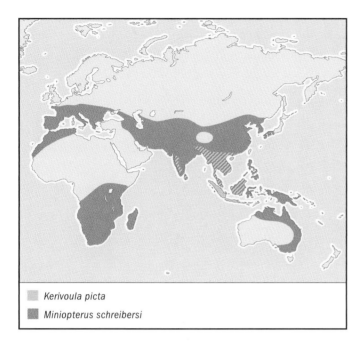

Kerivoula picta

Miniopterus schreibersi

REPRODUCTIVE BIOLOGY
Most likely polygynous. Mating commonly occurs in the fall, with delayed implantation until winter. Gestation lasts about 115–130 days after implantation. The young are born in early summer. In more tropical populations, implantation may occur almost immediately with young born up to four months earlier than in temperate areas. Litter size is typically one young per female. The young are weaned at about two months. They attain sexual maturity at 1 year old.

CONSERVATION STATUS
Listed as Lower Risk/Near Threatened by the IUCN. Population declines, including the disappearance of entire colonies, have been reported in western Europe.

SIGNIFICANCE TO HUMANS
Eat beetles and other insects that are potentially damaging to agricultural crops. ◆

Painted bat
Kerivoula picta

SUBFAMILY
Kerivoulinae

TAXONOMY
Kerivoula picta (Pallas, 1767), Ternate Island, Molucca Islands, Indonesia.

OTHER COMMON NAMES
None known.

PHYSICAL CHARACTERISTICS
Average adult body length is 1.6–1.7 in (4.0–4.3 cm), forearm length is 1.3–1.5 in (3.4–3.9 cm), and weight about 0.21–0.28 oz (6–8 g). A vivid orange to reddish, furry bat with black wings set off by orange fingers.

DISTRIBUTION
From India to southern China, south to Indonesia.

HABITAT
Forests, often roosting in tall grass or among foliage.

BEHAVIOR
These bats will roost in nearly any cavity, including the spaces between leaves and flowers, and in bird nests. They often roost alone or in small groups of up to a half dozen bats. Nocturnal, they become active after dark, and forage above low vegetation.

FEEDING ECOLOGY AND DIET
Insectivorous.

REPRODUCTIVE BIOLOGY
Nothing is known.

CONSERVATION STATUS
Not threatened.

SIGNIFICANCE TO HUMANS
May help control insects that damage crops.◆

OTHER COMMON NAMES
English: Schreiber's long-fingered bat, Schreiber's bent-winged bat, greater bent-winged bat, long-winged bat; French: Minoptère à longues ailes; German: Langflügelfledermaus; Spanish: Murcielago de cueva.

PHYSICAL CHARACTERISTICS
Average adult body length is 2.0–3.1 in (5.1–7.8 cm), weight is 0.28–0.56 oz (8–16 g), and forearm length is 1.7–2.0 in (4.2–5.0 cm). Gray to golden or reddish brown, thickly furred bat with a long tail; long, slender wings; and short ears.

DISTRIBUTION
Along the Mediterranean Sea in Europe and Africa, Africa south of the Sahara, southern Asia, New Guinea, and northern and eastern Australia.

HABITAT
Found in forests in the vicinity of caves or other rocky areas, usually near a subterranean water source, in summer and winter.

BEHAVIOR
Tens of thousands of common bentwing bats may roost together in the winter, and northernmost populations may hibernate. After migrating up to 200 mi (322 km) north in the spring, the males disperse, but females and their young continue to roost together in maternity colonies of many thousand bats. The body heat of so many bats helps to maintain a tolerable temperature in the cool summer caves. Sometimes, individuals or small groups may roost in a nearby secondary roost. Shortly after sunset, the bats emerge from their roosts to forage throughout the night.

FEEDING ECOLOGY AND DIET
The diet of these speedy, nocturnal bats includes small insects, especially beetles.

Common name / Scientific name/ Other common names	Physical characteristics	Habitat and behavior	Distribution	Diet	Conservation status
Peruvian crevice-dwellling bat *Tomopeas ravus* English: Crevice bat, blunt-eared bat	Light brown to darker gray dorsal pelage, whitish below, with black face, ears, and wing and tail membranes. Head and body length 2.9–3.3 in (7.3–8.5 cm), forearm length 1.2–1.4 in (3.1–3.5 cm), weight 0.07–0.12 oz (2–3.5 g).	Found in forests, often near rocky outcrops. Behavior little known.	Western Peru.	Insects.	Vulnerable
Papillose bat *Kerivoula papillosa*	Furry bat light-brown dorsally, paler below, with brown wing membranes. Head and body length 1.8–2.2 in (4.5–5.5 cm), forearm length 1.6–1.8 in (4–4.5 cm), weight 0.3–0.4 oz (9–10 g).	Found in forests, sometimes roosting in pairs in hollow bamboo stems. Behavior little known.	Eastern India, east through China, and south to Indonesia.	Insects.	Not listed by IUCN
Southeast Asian bent-winged bat *Miniopterus medius*	Dark brown to black bat, sometimes with reddish patches, and black wing membranes. Head and body length 2.0–2.2 in (5.0–5.6 cm), forearm length 1.6–1.7 in (4.0–4.4 cm), weight 0.3–0.4 oz (8–11 g).	Found in lowland forests, often roosting in large numbers in caves.	Thailand, Philippines, and Indonesia to New Guinea and southwest Pacific islands.	Insects.	Not listed by IUCN
Hutton's tube-nosed bat *Murina huttoni*	Tube-nosed bat with grayish brown dorsal pelage, whitish underside. Head and body length 1.9–2.0 in (4.7–5.0 cm), forearm length 1.1–1.4 in (2.9–3.5 cm), weight 0.2–0.3 oz (6–9 g).	Often found near fields and agricultural areas, where it flies low over vegetation during foraging.	A narrow band running through northern India to southeast China, and to Indochina and Malaysia.	Insects.	Lower Risk/Near Threatened

Resources

Books

Altringham, J. *Bats: Biology and Behavior.* Oxford: Oxford University Press, 1996.

Fenton, M. *Bats.* New York: Checkmark Books, 2001.

Medway, L. *The Wild Mammals of Malaya (Peninsular Malaysia) and Singapore.* 2nd ed. Oxford: Oxford University Press, 1978.

Nowak, R. *Walker's Mammals of the World.* Baltimore: Johns Hopkins University Press, 1999.

Periodicals

Bernard, R., F. Cotterill, and R. Fergusson. "On the occurrence of a short period of delayed implantation in Schreiber's long-fingered bat (*Miniopterus schreibersii*) from a tropical latitude in Zimbabwe." *Journal of Zoology: Proceedings of the Zoological Society of London* 238, no. 1 (1996): 13.

Hannah, D., and M. Schulz. "Relative abundance, diet and roost selection of the tube-nosed insect bat, *Murina florium,* on the Atherton Tablelands, Australia." *Wildlife Research* 25, no. 3 (1998): 261.

Schulz, M. "Diet and foraging behavior of the golden-tipped bat, *Kerivoula papuensis*: A spider specialist?" *Journal of Mammalogy* 81, no. 4 (2000): 948–957.

Organizations

Bat Conservation International. P.O. Box 162603, Austin, TX 78716 USA. Phone: (512) 327-9721. Fax: (512) 327-9724. E-mail: batinfo@batcon.org Web site: <http://www.batcon.org>

The Bat Conservation Trust. 15 Cloisters House, 8 Battersea Park Rd., London, SW8 4BG UK. Phone: 020 7627 2629. Fax: 020 7627 2628. E-mail: enquiries@bats.org.uk Web site: <http://www.bats.org.uk/aboutbct.htm>

The Organization for Bat Conservation at Cranbrook Institute of Science. 39221 Woodward Avenue, P.O. Box 801, Bloomfield Hills, MI 48303-0801 USA. Phone: (800) 276-7074. E-mail: obcbats@aol.com Web site: <http://batconservation.org/>

Leslie Ann Mertz, PhD

For further reading

Alcock, J. *Animal Behavior*. New York: Sinauer, 2001.

Alderton, D. *Rodents of the World*. New York: Facts on File, 1996.

Alterman, L., G. A. Doyle, and M. K. Izard, eds. *Creatures of the Dark: The Nocturnal Prosimians*. New York: Plenum Press, 1995.

Altringham, J. D. *Bats: Biology and Behaviour*. New York: Oxford University Press, 2001.

Anderson, D. F., and S. Eberhardt. *Understanding Flight*. New York: McGraw-Hill, 2001.

Anderson, S., and J. K. Jones Jr., eds. *Orders and Families of Recent Mammals of the World*. John Wiley & Sons, New York, 1984.

Apps, P. *Smithers' Mammals of Southern Africa*. Cape Town: Struik Publishers, 2000.

Attenborough, D. *The Life of Mammals*. London: BBC, 2002.

Au, W. W. L. *The Sonar of Dolphins*. New York: Springer-Verlag, 1993.

Austin, C. R., and R. V. Short, eds. *Reproduction in Mammals*. 4 vols. Cambridge: Cambridge University Press, 1972.

Avise, J. C. *Molecular Markers, Natural History and Evolution*. London: Chapman & Hall, 1994.

Barber, P. *Vampires, Burial, and Death: Folklore and Reality*. New Haven: Yale University Press, 1998.

Barnett, S. A. *The Story of Rats*. Crows Nest, Australia: Allen & Unwin, 2001.

Baskin, L., and K. Danell. *Ecology of Ungulates. A Handbook of Species in Eastern Europe, Northern and Central Asia*. Heidelberg: Springer-Verlag, 2003.

Bates, P. J. J., and D. L. Harrison. *Bats of the Indian Subcontinent*. Sevenoaks, U. K.: Harrison Zoological Museum, 1997.

Bekoff, M., C. Allen, and G. M. Burghardt, eds. *The Cognitive Animal*. Cambridge: MIT Press, 2002.

Bennett, N. C., and C. G. Faulkes. *African Mole-rats: Ecology and Eusociality*. Cambridge: Cambridge University Press, 2000.

Benton, M. J. *The Rise of the Mammals*. New York: Crescent Books, 1991.

Berta, A., and L. Sumich. *Marine Mammals: Evolutionary Biology*. San Diego: Academic Press, 1999.

Bonaccorso, F. J. *Bats of Papua New Guinea*. Washington, DC: Conservation International, 1998.

Bonnichsen, R, and K. L. Turnmire, eds. *Ice Age People of North America*. Corvallis: Oregon State University Press. 1999.

Bright, P. and P. Morris. *Dormice* London: The Mammal Society, 1992.

Broome, D., ed. *Coping with Challenge*. Berlin: Dahlem University Press, 2001.

Buchmann, S. L., and G. P. Nabhan. *The Forgotten Pollinators*. Washington, DC: Island Press, 1997.

Burnie, D., and D. E. Wilson, eds. *Animal*. Washington, DC: Smithsonian Institution, 2001.

Caro, T., ed. *Behavioral Ecology and Conservation Biology*. Oxford: Oxford University Press, 1998.

Carroll, R. L. *Vertebrate Paleontology and Evolution*. New York: W. H. Freeman and Co., 1998.

Cavalli-Sforza, L. L., P. Menozzi, and A. Piazza. *The History and Geography of Human Genes*. Princeton: Princeton University Press, 1994.

Chivers, R. E., and P. Lange. *The Digestive System in Mammals: Food, Form and Function*. New York: Cambridge University Press, 1994.

Clutton-Brock, J. *A Natural History of Domesticated Mammals*. 2nd ed. Cambridge: Cambridge University Press, 1999.

Conley, V. A. *The War Against the Beavers*. Minneapolis: University of Minnesota Press, 2003.

Cowlishaw, G., and R. Dunbar. *Primate Conservation Biology*. Chicago: University of Chicago Press, 2000.

Craighead, L. *Bears of the World* Blaine, WA: Voyager Press, 2000.

Crichton., E. G. and P. H. Krutzsch, eds. *Reproductive Biology of Bats.* New York: Academic Press, 2000.

Croft, D. B., and U. Gansloßer, eds. *Comparison of Marsupial and Placental Behavior.* Fürth, Germany: Filander, 1996.

Darwin, C. *The Autobiography of Charles Darwin 1809–1882 with original omissions restored.* Edited by Nora Barlow. London: Collins, 1958.

Darwin, C. *On The Origin of Species by Means of Natural Selection, or The Preservation of Favoured Races in the Struggle for Life.* London: John Murray, 1859.

Darwin, C. *The Zoology of the Voyage of HMS* Beagle *under the Command of Captain Robert FitzRoy RN During the Years 1832-1836.* London: Elder & Co., 1840.

Dawson, T. J. *Kangaroos: The Biology of the Largest Marsupials.* Kensington, Australia: University of New South Wales Press/Ithaca, 2002.

Duncan, P. *Horses and Grasses.* New York: Springer-Verlag Inc., 1991.

Easteal, S., C. Collett, and D. Betty. *The Mammalian Molecular Clock.* Austin, TX: R. G. Landes, 1995.

Eisenberg, J. F. *Mammals of the Neotropics.* Vol. 1, *The Northern Neotropics.* Chicago: University of Chicago Press, 1989.

Eisenberg, J. F., and K. H. Redford. *Mammals of the Neotropics.* Vol. 3, *The Central Neotropics.* Chicago: University of Chicago Press, 1999.

Ellis, R. *Aquagenesis.* New York: Viking, 2001.

Estes, R. D. *The Behavior Guide to African Mammals.* Berkeley: The University of California Press, 1991.

Estes, R. D. *The Safari Companion: A Guide to Watching African Mammals.* White River Junction, VT: Chelsea Green, 1999.

Evans, P. G. H., and J. A. Raga, eds. *Marine Mammals: Biology and Conservation.* New York: Kluwer Academic/Plenum, 2001.

Ewer, R. F. *The Carnivores.* Ithaca, NY: Comstock Publishing, 1998.

Feldhamer, G. A., L. C. Drickamer, A. H. Vessey, and J. F. Merritt. *Mammalogy. Adaptations, Diversity, and Ecology.* Boston: McGraw Hill, 1999.

Fenton, M. B. *Bats.* Rev. ed. New York: Facts On File Inc., 2001.

Findley, J. S. *Bats: A Community Perspective.* Cambridge: Cambridge University Press, 1993.

Flannery, T. F. *Mammals of New Guinea.* Ithaca: Cornell University Press, 1995.

Flannery, T. F. *Possums of the World: A Monograph of the Phalangeroidea.* Sydney: GEO Productions, 1994.

Fleagle, J. G. *Primate Adaptation and Evolution.* New York: Academic Press, 1999.

Frisancho, A. R. *Human Adaptation and Accommodation.* Ann Arbor: University of Michigan Press, 1993.

Garbutt, N. *Mammals of Madagascar.* New Haven: Yale University Press, 1999.

Geist, V. *Deer of the World: Their Evolution, Behavior, and Ecology.* Mechanicsburg, PA: Stackpole Books, 1998.

Geist, V. *Life Strategies, Human Evolution, Environmental Design.* New York: Springer Verlag, 1978.

Gillespie, J. H. *The Causes of Molecular Evolution.* Oxford: Oxford University Press, 1992.

Gittleman, J. L., ed. *Carnivore Behavior, Ecology and Evolution.* 2 vols. Chicago: University of Chicago Press, 1996.

Gittleman, J. L., S. M. Funk, D. Macdonald, and R. K. Wayne, eds. *Carnivore Conservation.* Cambridge: Cambridge University Press, 2001.

Givnish, T. I. and K. Sytsma. *Molecular Evolution and Adaptive Radiations.* Cambridge: Cambridge University Press, 1997.

Goldingay, R. L., and J. H. Scheibe, eds. *Biology of Gliding Mammals.* Fürth, Germany: Filander Verlag, 2000.

Goodman, S. M., and J. P. Benstead, eds. *The Natural History of Madagascar.* Chicago: The University of Chicago Press, 2003.

Gosling, L. M., and W. J. Sutherland, eds. *Behaviour and Conservation.* Cambridge: Cambridge University Press, 2000.

Gould, E., and G. McKay, eds. *Encyclopedia of Mammals.* 2nd ed. San Diego: Academic Press, 1998.

Groves, C. P. *Primate Taxonomy.* Washington, DC: Smithsonian Institute, 2001.

Guthrie, D. R. *Frozen Fauna of the Mammoth Steppe.* Chicago: University of Chicago Press. 1990.

Hall, L., and G. Richards. *Flying Foxes: Fruit and Blossom Bats of Australia.* Malabar, FL: Krieger Publishing Company, 2000.

Hancocks, D. *A Different Nature. The Paradoxical World of Zoos and Their Uncertain Future.* Berkeley: University of California Press, 2001.

Hartwig, W. C., ed. *The Primate Fossil Record.* New York: Cambridge University Press, 2002.

Hildebrand, M. *Analysis of Vertebrate Structure.* 4th ed. New York: John Wiley & Sons, 1994.

Hillis, D. M., and C. Moritz. *Molecular Systematics.* Sunderland, MA: Sinauer Associates, 1990.

Hoelzel, A. R., ed. *Marine Mammal Biology: An Evolutionary Approach.* Oxford: Blackwell Science, 2002.

Hunter, M. L., and A Sulzer. *Fundamentals of Conservation Biology.* Oxford, U. K.: Blackwell Science, Inc., 2001.

Jefferson, T. A., S. Leatherwood, and M. A. Webber, eds. *Marine Mammals of the World.* Heidelberg: Springer-Verlag, 1993.

Jensen, P., ed. *The Ethology of Domestic Animals: An Introductory Text.* Oxon, MD: CABI Publishing, 2002.

Jones, M. E., C. R. Dickman, and M. Archer. *Predators with Pouches: The Biology of Carnivorous Marsupials.* Melbourne: CSIRO Books, 2003.

Kardong, K. V. *Vertebrates: Comparative Anatomy, Function, Evolution.* Dubuque, IA: William C. Brown Publishers, 1995.

King, C. M. *The Handbook of New Zealand Mammals.* Auckland: Oxford University Press, 1990.

Kingdon, J. *The Kingdon Field Guide to African Mammals.* London: Academic Press, 1997.

Kingdon, J., D. Happold, and T. Butynski, eds. *The Mammals of Africa: A Comprehensive Synthesis.* London: Academic Press, 2003.

Kinzey, W. G., ed. *New World Primates: Ecology, Evolution, and Behavior.* New York: Aldine de Gruyter, 1997.

Kosco, M. *Mammalian Reproduction.* Eglin, PA: Allegheny Press, 2000.

Krebs, J. R., and N. B. Davies. *An Introduction to Behavioural Ecology.* 3rd ed. Oxford: Blackwell Scientific Publications, 1993.

Kunz, T. H., and M. B. Fenton, eds. *Bat Ecology.* Chicago: University of Chicago Press, 2003.

Lacey, E. A., J. L. Patton, and G. N. Cameron, eds. *Life Underground: The Biology of Subterranean Rodents.* Chicago: University of Chicago Press, 2000.

Lott, D. F. *American Bison: A Natural History.* Berkeley: University of California Press, 2002.

Macdonald, D. W. *European Mammals: Evolution and Behavior.* London: Collins, 1995.

Macdonald, D. W. *The New Encyclopedia of Mammals.* Oxford: Oxford University Press, 2001.

Macdonald, D. W. *The Velvet Claw: A Natural History of the Carnivores.* London: BBC Books, 1992.

Macdonald, D. W., and P. Barrett. *Mammals of Britain and Europe.* London: Collins, 1993.

Martin, R. E. *A Manual of Mammalogy: With Keys to Families of the World.* 3rd ed. Boston: McGraw-Hill, 2001.

Matsuzawa, T., ed. *Primate Origins of Human Cognition and Behavior.* Tokyo: Springer-Verlag, 2001.

Mayr, E. *What Evolution Is.* New York: Basic Books, 2001.

McCracken, G. F., A. Zubaid, and T. H. Kunz, eds. *Functional and Evolutionary Ecology of Bats.* Oxford: Oxford University Press, 2003.

McGrew, W. C., L. F. Marchant, and T. Nishida, eds. *Great Ape Societies.* Cambridge: Cambridge University Press, 1996.

Meffe, G. K., and C. R. Carroll. *Principles of Conservation Biology.* Sunderland, MA: Sinauer Associates, Inc., 1997.

Menkhorst, P. W. *A Field Guide to the Mammals of Australia.* Melbourne: Oxford University Press, 2001.

Mills, G., and M. Harvey. *African Predators.* Cape Town: Struik Publishers, 2001.

Mills, G., and L. Hes. *Complete Book of Southern African Mammals.* Cape Town: Struik, 1997.

Mitchell-Jones, A. J., et al. *The Atlas of European Mammals.* London: Poyser Natural History/Academic Press, 1999.

Neuweiler, G. *Biology of Bats.* Oxford: Oxford University Press, 2000.

Norton, B. G., et al. *Ethics on the Ark.* Washington, DC: Smithsonian Institution Press, 1995.

Nowak, R. M. *Walker's Bats of the World.* Baltimore: The Johns Hopkins University Press, 1994.

Nowak, R. M. *Walker's Mammals of the World.* 6th ed. Baltimore: Johns Hopkins University Press, 1999.

Nowak, R. M. *Walker's Primates of the World.* Baltimore: The Johns Hopkins University Press, 1999.

Payne, K. *Silent Thunder: The Hidden Voice of Elephants.* Phoenix: Wiedenfeld and Nicholson, 1999.

Pearce, J. D. *Animal Learning and Cognition.* New York: Lawrence Erlbaum, 1997.

Pereira, M. E., and L. A. Fairbanks, eds. *Juvenile Primates: Life History, Development, and Behavior.* New York: Oxford University Press, 1993.

Perrin, W. F., B. Würsig, and J. G. M. Thewissen. *Encyclopedia of Marine Mammals.* San Diego: Academic Press, 2002.

Popper, A. N., and R. R. Fay, eds. *Hearing by Bats.* New York: Springer-Verlag, 1995.

Pough, F. H., C. M. Janis, and J. B. Heiser. *Vertebrate Life.* 6th ed. Upper Saddle River, NJ: Prentice Hall, 2002.

Premack, D., and A. J. Premack. *Original Intelligence: The Architecture of the Human Mind.* New York: McGraw-Hill/Contemporary Books, 2002.

Price, E. O. *Animal Domestication and Behavior.* Cambridge, MA: CAB International, 2002.

Racey, P. A., and S. M. Swift, eds. *Ecology, Evolution and Behaviour of Bats.* Oxford: Clarendon Press, 1995.

Redford, K. H., and J. F. Eisenberg. *Mammals of the Neotropics.* Vol. 2, *The Southern Cone.* Chicago: University of Chicago Press, 1992.

Reeve, N. *Hedgehogs.* London: Poyser Natural History, 1994.

Reeves, R., B. Stewart, P. Clapham, and J. Powell. *Sea Mammals of the World.* London: A&C Black, 2002.

Reynolds, J. E. III, and D. K. Odell. *Manatees and Dugongs.* New York: Facts On File, 1991.

Reynolds, J. E. III, and S. A. Rommel, eds. *Biology of Marine Mammals.* Washington, DC: Smithsonian Institution Press, 1999.

Rice, D. W. *Marine Mammals of the World.* Lawrence, KS: Allen Press, 1998.

Ridgway, S. H., and R. Harrison, eds. *Handbook of Marine Mammals.* 6 vols. New York: Academic Press, 1985-1999.

Riedman, M. *The Pinnipeds.* Berkeley: University of California Press, 1990.

Rijksen, H., and E. Meijaard. *Our Vanishing Relative: The Status of Wild Orang-utans at the Close of the Twentieth Century.* Dordrecht: Kluwer Academic Publishers, 1999.

Robbins, C. T. *Wildlife Feeding and Nutrition.* San Diego: Academic Press, 1992.

Robbins, M. M., P. Sicotte, and K. J. Stewart, eds. *Mountain Gorillas: Three Decades of Research at Karisoke.* Cambridge: Cambridge University Press, 2001.

Roberts, W. A. *Principles of Animal Cognition.* New York: McGraw-Hill, 1998.

Schaller, G. B. *Wildlife of the Tibetan Steppe.* Chicago: University of Chicago Press, 1998.

Seebeck, J. H., P. R. Brown, R. L. Wallis, and C. M. Kemper, eds. *Bandicoots and Bilbies.* Chipping Norton, Australia: Surrey Beatty & Sons, 1990.

Shepherdson, D. J., J. D. Mellen, and M. Hutchins. *Second Nature: Environmental Enrichment for Captive Animals.* Washington, DC: Smithsonian Institution Press, 1998.

Sherman, P. W., J. U. M. Jarvis, and R. D. Alexander, eds. *The Biology of the Naked Mole-rat.* Princeton: Princeton University Press, 1991.

Shettleworth, S. J. *Cognition, Evolution, and Behavior.* Oxford: Oxford University Press, 1998.

Shoshani, J., ed. *Elephants.* London: Simon & Schuster, 1992.

Skinner, R., and R. H. N. Smithers. *The Mammals of the Southern African Subregion.* 2nd ed. Pretoria, South Africa: University of Pretoria, 1998.

Sowls, L. K. *The Peccaries.* College Station: Texas A&M Press, 1997.

Steele, M. A. and J. Koprowski. *North American Tree Squirrels.* Washington, DC: Smithsonian Institution Press, 2001.

Sunquist, M. and F. Sunquist. *Wild Cats of the World* Chicago: University of Chicago Press, 2002.

Sussman, R. W. *Primate Ecology and Social Structure.* 3 vols. Needham Heights, MA: Pearson Custom Publishing, 1999.

Szalay, F. S., M. J. Novacek, and M. C. McKenna, eds. *Mammalian Phylogeny.* New York: Springer-Verlag, 1992.

Thomas, J. A., C. A. Moss, and M. A. Vater, eds. *Echolocation in Bats and Dolphins.* Chicago: University of Chicago Press, 2003.

Thompson, H. V., and C. M. King, eds. *The European Rabbit: The History and Biology of a Successful Colonizer.* Oxford: Oxford University Press, 1994.

Tomasello, M., and J. Calli. *Primate Cognition.* Chicago: University of Chicago Press, 1997.

Twiss, J. R. Jr., and R. R. Reeves, eds. *Conservation and Management of Marine Mammals.* Washington, DC: Smithsonian Institution Press, 1999.

Van Soest, P. J. *Nutritional Ecology of the Ruminant.* 2nd ed. Ithaca, NY: Cornell University Press, 1994.

Vaughan, T., J. Ryan, and N. Czaplewski. *Mammalogy.* 4th ed. Philadelphia: Saunders College Publishing, 1999.

Vrba, E. S., G. H. Denton, T. C. Partridge, and L. H. Burckle, eds. *Paleoclimate and Evolution, with Emphasis on Human Origins.* New Haven: Yale University Press, 1995.

Vrba, E. S., and G. G. Schaller, eds. *Antelopes, Deer and Relatives: Fossil Record, Behavioral Ecology, Systematics and Conservation.* New Haven: Yale University Press, 2000.

Wallis, Janice, ed. *Primate Conservation: The Role of Zoological Parks.* New York: American Society of Primatologists, 1997.

Weibel, E. R., C. R. Taylor, and L. Bolis. *Principles of Animal Design.* New York: Cambridge University Press, 1998.

Wells, R. T., and P. A. Pridmore. *Wombats.* Sydney: Surrey Beatty & Sons, 1998.

Whitehead, G. K. *The Whitehead Encyclopedia of Deer.* Stillwater, MN: Voyager Press, 1993.

Wilson, D. E., and D. M. Reeder, eds. *Mammal Species of the World: a Taxonomic and Geographic Reference.* 2nd ed. Washington, DC: Smithsonian Institution Press, 1993.

Wilson, D. E., and S. Ruff, eds. *The Smithsonian Book of North American Mammals.* Washington, DC: Smithsonian Institution Press, 1999.

Wilson, E. O. *The Diversity of Life.* Cambridge: Harvard University Press, 1992.

Wójcik, J. M., and M. Wolsan, eds. *Evolution of Shrews.* Bialowieza, Poland: Mammal Research Institute, Polish Academy of Sciences, 1998.

Woodford, J. *The Secret Life of Wombats.* Melbourne: Text Publishing, 2001.

Wrangham, R. W., W. C. McGrew, F. B. M. de Waal, and P. G. Heltne, eds. *Chimpanzee Cultures.* Cambridge: Harvard University Press, 1994.

Wynne, C. D. L. *Animal Cognition.* Basingstoke, U. K.: Palgrave, 2001.

FOR FURTHER READING

Organizations

African Wildlife Foundation
1400 16th Street, NW, Suite 120
Washington, DC 20036 USA
Phone: (202) 939-3333
Fax: (202) 939-3332
E-mail: africanwildlife@awf.org
<http://www.awf.org/>

The American Society of Mammalogists
<http://www.mammalsociety.org/>

American Zoo and Aquarium Association
8403 Colesville Road, Suite 710
Silver Spring, MD 20910 USA.
Phone: (301) 562-0777
Fax: (301) 562-0888
<http://www.aza.org/>

Australian Conservation Foundation Inc.
340 Gore Street
Fitzroy, Victoria 3065 Australia
Phone: (3) 9416 1166
<http://www.acfonline.org.au>

The Australian Mammal Society
<http://www.australianmammals.org.au/>

Australian Regional Association of Zoological Parks and
Aquaria
PO Box 20
Mosman, NSW 2088
Australia
Phone: 61 (2) 9978-4797
Fax: 61 (2) 9978-4761
<http://www.arazpa.org>

Bat Conservation International
P.O. Box 162603
Austin, TX 78716 USA
Phone: (512) 327-9721
Fax: (512) 327-9724
<http://www.batcon.org/>

Center for Ecosystem Survival
699 Mississippi Street, Suite 106
San Francisco, 94107 USA
Phone: (415) 648-3392
Fax: (415) 648-3392

E-mail: info@savenature.org
<http://www.savenature.org/>

Conservation International
1919 M Street NW, Ste. 600
Washington, DC 20036
Phone: (202) 912-1000
<http://www.conservation.org>

The European Association for Aquatic Mammals
E-mail: info@eaam.org
<http://www.eaam.org/>

European Association of Zoos and Aquaria
PO Box 20164
1000 HD Amsterdam
The Netherlands
<http://www.eaza.net>

IUCN-The World Conservation Union
Rue Mauverney 28
Gland 1196 Switzerland
Phone: ++41(22) 999-0000
Fax: ++41(22) 999-0002
E-mail: mail@iucn.org
<http://www.iucn.org/>

The Mammal Society
2B, Inworth Street
London SW11 3EP United Kingdom
Phone: 020 7350 2200
Fax: 020 7350 2211
<http://www.abdn.ac.uk/mammal/>

Mammals Trust UK
15 Cloisters House
8 Battersea Park Road
London SW8 4BG United Kingdom
Phone: (+44) 020 7498 5262
Fax: (+44) 020 7498 4459
E-mail: enquiries@mtuk.org
<http://www.mtuk.org/>

The Marine Mammal Center
Marin Headlands
1065 Fort Cronkhite
Sausalito, CA 94965 USA
Phone: (415) 289-7325

Fax: (415) 289-7333
<http://www.marinemammalcenter.org/>

National Marine Mammal Laboratory
7600 Sand Point Way N.E. F/AKC3
Seattle, WA 98115-6349 USA
Phone: (206) 526-4045
Fax: (206) 526-6615
<http://nmml.afsc.noaa.gov/>

National Wildlife Federation
11100 Wildlife Center Drive
Reston, VA 20190-5362 USA
Phone: (703) 438-6000
<http://www.nwf.org/>

The Organization for Bat Conservation
39221 Woodward Avenue
Bloomfield Hills, MI 48303 USA
Phone: (248) 645-3232
E-mail: obcbats@aol.com
<http://www.batconservation.org/>

Scripps Institution of Oceanography
University of California-San Diego
9500 Gilman Drive
La Jolla, CA 92093 USA
<http://sio.ucsd.edu/gt;

Seal Conservation Society
7 Millin Bay Road
Tara, Portaferry
County Down BT22 1QD
United Kingdom

Phone: +44-(0)28-4272-8600
Fax: +44-(0)28-4272-8600
E-mail: info@pinnipeds.org
<http://www.pinnipeds.org>

The Society for Marine Mammalogy
<http://www.marinemammalogy.org/>

The Wildlife Conservation Society
2300 Southern Boulevard
Bronx, New York 10460
Phone: (718) 220-5100

Woods Hole Oceanographic Institution
Information Office
Co-op Building, MS #16
Woods Hole, MA 02543 USA
Phone: (508) 548-1400
Fax: (508) 457-2034
E-mail: information@whoi.edu
<http://www.whoi.edu/>

World Association of Zoos and Aquariums
PO Box 23
Liebefeld-Bern CH-3097
Switzerland
<http://www.waza.org>

World Wildlife Fund
1250 24th Street N.W.
Washington, DC 20037-1193 USA
Phone: (202) 293-4800
Fax: (202) 293-9211
<http://www.panda.org/>

ORGANIZATIONS

Dr. Fritz Dieterlen
Zoological Research Institute and A.
Koenig Museum
Bonn, Germany

Dr. Rolf Dircksen
Professor, Pedagogical Institute
Bielefeld, Germany

Josef Donner
Instructor of Biology
Katzelsdorf, Austria

Dr. Jean Dorst
Professor, National Museum of
Natural History
Paris, France

Dr. Gerti Dücker
Professor and Chief Curator,
Zoological Institute, University of
Münster
Münster, Germany

Dr. Michael Dzwillo
Zoological Institute and Museum,
University of Hamburg
Hamburg, Germany

Dr. Irenäus Eibl-Eibesfeldt
Professor and Director, Institute of
Human Ethology, Max Planck
Institute for Behavioral Physiology
Percha/Starnberg, Germany

Dr. Martin Eisentraut
Professor and Director, Zoological
Research Institute and A. Koenig
Museum
Bonn, Germany

Dr. Eberhard Ernst
Swiss Tropical Institute
Basel, Switzerland

R. D. Etchecopar
Director, National Museum of
Natural History
Paris, France

Dr. R. A. Falla
Director, Dominion Museum
Wellington, New Zealand

Dr. Hubert Fechter
Curator, Lower Animals, Zoological
Collection of the State of Bavaria
Munich, Germany

Dr. Walter Fiedler
Docent, University of Vienna, and
Director, Schönbrunn Zoo
Vienna, Austria

Wolfgang Fischer
Inspector of Animals, Animal Park
Berlin, Germany

Dr. C. A. Fleming
Geological Survey Department of
Scientific and Industrial Research
Lower Hutt, New Zealand

Dr. Hans Frädrich
Zoological Garden
Berlin, Germany

Dr. Hans-Albrecht Freye
Professor and Director, Biological
Institute of the Medical School
Halle a.d.S., Germany

Günther E. Freytag
Former Director, Reptile and
Amphibian Collection, Museum of
Cultural History in Magdeburg
Berlin, Germany

Dr. Herbert Friedmann
Director, Los Angeles County
Museum of Natural History
Los Angeles, California, U.S.A.

Dr. H. Friedrich
Professor, Overseas Museum
Bremen, Germany

Dr. Jan Frijlink
Zoological Laboratory, University of
Amsterdam
Amsterdam, The Netherlands

Dr. H. C. Karl Von Frisch
Professor Emeritus and former
Director, Zoological Institute,
University of Munich
Munich, Germany

Dr. H. J. Frith
C.S.I.R.O. Research Institute
Canberra, Australia

Dr. Ion E. Fuhn
Academy of the Roumanian Socialist
Republic, Trajan Savulescu Institute of
Biology
Bucharest, Romania

Dr. Carl Gans
Professor, Department of Biology,
State University of New York at
Buffalo
Buffalo, New York, U.S.A.

Dr. Rudolf Geigy
Professor and Director, Swiss Tropical
Institute
Basel, Switzerland

Dr. Jacques Gery
St. Genies, France

Dr. Wolfgang Gewalt
Director, Animal Park
Duisburg, Germany

Dr. H. C. Viktor Goerttler
Professor Emeritus, University of Jena
Jena, Germany

Dr. Friedrich Goethe
Director, Institute of Ornithology,
Heligoland Ornithological Station
Wilhelmshaven, Germany

Dr. Ulrich F. Gruber
Herpetological Section, Zoological
Research Institute and A. Koenig
Museum
Bonn, Germany

Dr. H. R. Haefelfinger
Museum of Natural History
Basel, Switzerland

Dr. Theodor Haltenorth
Director, Mammalogy, Zoological
Collection of the State of Bavaria
Munich, Germany

Barbara Harrisson
Sarawak Museum, Kuching, Borneo
Ithaca, New York, U.S.A.

Dr. Francois Haverschmidt
President, High Court (retired)
Paramaribo, Suriname

Dr. Heinz Heck
Director, Catskill Game Farm
Catskill, New York, U.S.A.

Dr. Lutz Heck
Professor (retired), and Director,
Zoological Garden, Berlin
Wiesbaden, Germany

Dr. H. C. Heini Hediger
Director, Zoological Garden
Zurich, Switzerland

Dr. Dietrich Heinemann
Director, Zoological Garden, Münster
Dörnigheim, Germany

Dr. Helmut Hemmer
Institute for Physiological Zoology,
University of Mainz
Mainz, Germany

Dr. W. G. Heptner
Professor, Zoological Museum,
University of Moscow
Moscow, Russia

Dr. Konrad Herter
Professor Emeritus and Director
(retired), Zoological Institute, Free
University of Berlin
Berlin, Germany

Dr. Hans Rudolf Heusser
Zoological Museum, University of
Zurich
Zurich, Switzerland

Dr. Emil Otto Höhn
Associate Professor of Physiology,
University of Alberta
Edmonton, Canada

Dr. W. Hohorst
Professor and Director, Parasitological
Institute, Farbwerke Hoechst A.G.
Frankfurt-Höchst, Germany

Dr. Folkhart Hückinghaus
Director, Senckenbergische Anatomy,
University of Frankfurt a.M.
Frankfurt a.M., Germany

Francois Hüe
National Museum of Natural History
Paris, France

Dr. K. Immelmann
Professor, Zoological Institute,
Technical University of Braunschweig
Braunschweig, Germany

Dr. Junichiro Itani
Kyoto University
Kyoto, Japan

Dr. Richard F. Johnston
Professor of Zoology, University of
Kansas
Lawrence, Kansas, U.S.A.

Otto Jost
Oberstudienrat, Freiherr-vom-Stein
Gymnasium
Fulda, Germany

Dr. Paul Kähsbauer
Curator, Fishes, Museum of Natural
History
Vienna, Austria

Dr. Ludwig Karbe
Zoological State Institute and
Museum
Hamburg, Germany

Dr. N. N. Kartaschew
Docent, Department of Biology,
Lomonossow State University
Moscow, Russia

Dr. Werner Kästle
Oberstudienrat, Gisela Gymnasium
Munich, Germany

Dr. Reinhard Kaufmann
Field Station of the Tropical Institute,
Justus Liebig University, Giessen,
Germany
Santa Marta, Colombia

Dr. Masao Kawai
Primate Research Institute, Kyoto
University
Kyoto, Japan

Dr. Ernst F. Kilian
Professor, Giessen University and
Catedratico Universidad Australia,
Valdivia-Chile
Giessen, Germany

Dr. Ragnar Kinzelbach
Institute for General Zoology,
University of Mainz
Mainz, Germany

Dr. Heinrich Kirchner
Landwirtschaftsrat (retired)
Bad Oldesloe, Germany

Dr. Rosl Kirchshofer
Zoological Garden, University of
Frankfurt a.M.
Frankfurt a.M., Germany

Dr. Wolfgang Klausewitz
Curator, Senckenberg Nature
Museum and Research Institute
Frankfurt a.M., Germany

Dr. Konrad Klemmer
Curator, Senckenberg Nature
Museum and Research Institute
Frankfurt a.M., Germany

Dr. Erich Klinghammer
Laboratory of Ethology, Purdue
University
Lafayette, Indiana, U.S.A.

Dr. Heinz-Georg Klös
Professor and Director, Zoological
Garden
Berlin, Germany

Ursula Klös
Zoological Garden
Berlin, Germany

Dr. Otto Koehler
Professor Emeritus, Zoological
Institute, University of Freiburg
Freiburg i. BR., Germany

Dr. Kurt Kolar
Institute of Ethology, Austrian
Academy of Sciences
Vienna, Austria

Dr. Claus König
State Ornithological Station of Baden-
Württemberg
Ludwigsburg, Germany

Dr. Adriaan Kortlandt
Zoological Laboratory, University of
Amsterdam
Amsterdam, The Netherlands

Dr. Helmut Kraft
Professor and Scientific Councillor,
Medical Animal Clinic, University of
Munich
Munich, Germany

Dr. Helmut Kramer
Zoological Research Institute and A.
Koenig Museum
Bonn, Germany

Dr. Franz Krapp
Zoological Institute, University of
Freiburg
Freiburg, Switzerland

Dr. Otto Kraus
Professor, University of Hamburg,
and Director, Zoological Institute and
Museum
Hamburg, Germany

Dr. Hans Krieg
Professor and First Director (retired),
Scientific Collections of the State of
Bavaria
Munich, Germany

Dr. Heinrich Kühl
Federal Research Institute for
Fisheries, Cuxhaven Laboratory
Cuxhaven, Germany

Dr. Oskar Kuhn
Professor, formerly University
Halle/Saale
Munich, Germany

Dr. Hans Kumerloeve
First Director (retired), State
Scientific Museum, Vienna
Munich, Germany

Dr. Nagamichi Kuroda
Yamashina Ornithological Institute,
Shibuya-Ku
Tokyo, Japan

Dr. Fred Kurt
Zoological Museum of Zurich
University, Smithsonian Elephant
Survey
Colombo, Ceylon

Dr. Werner Ladiges
Professor and Chief Curator,
Zoological Institute and Museum,
University of Hamburg
Hamburg, Germany

Leslie Laidlaw
Department of Animal Sciences,
Purdue University
Lafayette, Indiana, U.S.A.

Dr. Ernst M. Lang
Director, Zoological Garden
Basel, Switzerland

Dr. Alfredo Langguth
Department of Zoology, Faculty of
Humanities and Sciences, University
of the Republic
Montevideo, Uruguay

Leo Lehtonen
Science Writer
Helsinki, Finland

Bernd Leisler
Second Zoological Institute,
University of Vienna
Vienna, Austria

Dr. Kurt Lillelund
Professor and Director, Institute for
Hydrobiology and Fishery Sciences,
University of Hamburg
Hamburg, Germany

R. Liversidge
Alexander MacGregor Memorial
Museum
Kimberley, South Africa

Dr. Konrad Lorenz
Professor and Director, Max Planck
Institute for Behavioral Physiology
Seewiesen/Obb., Germany

Dr. Martin Lühmann
Federal Research Institute for the
Breeding of Small Animals
Celle, Germany

Dr. Johannes Lüttschwager
Oberstudienrat (retired)
Heidelberg, Germany

Dr. Wolfgang Makatsch
Bautzen, Germany

Dr. Hubert Markl
Professor and Director, Zoological
Institute, Technical University of
Darmstadt
Darmstadt, Germany

Basil J. Marlow , BSc (Hons)
Curator, Australian Museum
Sydney, Australia

Dr. Theodor Mebs
Instructor of Biology
Weissenhaus/Ostsee, Germany

Dr. Gerlof Fokko Mees
Curator of Birds, Rijks Museum of
Natural History
Leiden, The Netherlands

Hermann Meinken
Director, Fish Identification Institute,
V.D.A.
Bremen, Germany

Dr. Wilhelm Meise
Chief Curator, Zoological Institute
and Museum, University of Hamburg
Hamburg, Germany

Dr. Joachim Messtorff
Field Station of the Federal Fisheries
Research Institute
Bremerhaven, Germany

Dr. Marian Mlynarski
Professor, Polish Academy of
Sciences, Institute for Systematic and
Experimental Zoology
Cracow, Poland

Dr. Walburga Moeller
Nature Museum
Hamburg, Germany

Dr. H. C. Erna Mohr
Curator (retired), Zoological State
Institute and Museum
Hamburg, Germany

Dr. Karl-Heinz Moll
Waren/Müritz, Germany

Dr. Detlev Müller-Using
Professor, Institute for Game
Management, University of Göttingen
Hannoversch-Münden, Germany

Werner Münster
Instructor of Biology
Ebersbach, Germany

Dr. Joachim Münzing
Altona Museum
Hamburg, Germany

Dr. Wilbert Neugebauer
Wilhelma Zoo
Stuttgart-Bad Cannstatt, Germany

Dr. Ian Newton
Senior Scientific Officer, The Nature
Conservancy
Edinburgh, Scotland

Dr. Jürgen Nicolai
Max Planck Institute for Behavioral
Physiology
Seewiesen/Obb., Germany

Dr. Günther Niethammer
Professor, Zoological Research
Institute and A. Koenig Museum
Bonn, Germany

Dr. Bernhard Nievergelt
Zoological Museum, University of
Zurich
Zurich, Switzerland

Dr. C. C. Olrog
Institut Miguel Lillo San Miguel de
Tucumán
Tucumán, Argentina

Alwin Pedersen
Mammal Research and Arctic Explorer
Holte, Denmark

Dr. Dieter Stefan Peters
Nature Museum and Senckenberg
Research Institute
Frankfurt a.M., Germany

Dr. Nicolaus Peters
Scientific Councillor and Docent,
Institute of Hydrobiology and
Fisheries, University of Hamburg
Hamburg, Germany

Dr. Hans-Günter Petzold
Assistant Director, Zoological Garden
Berlin, Germany

Dr. Rudolf Piechocki
Docent, Zoological Institute,
University of Halle
Halle a.d.S., Germany

Dr. Ivo Poglayen-Neuwall
Director, Zoological Garden
Louisville, Kentucky, U.S.A.

Dr. Egon Popp
Zoological Collection of the State of
Bavaria
Munich, Germany

Dr. H. C. Adolf Portmann
Professor Emeritus, Zoological
Institute, University of Basel
Basel, Switzerland

Hans Psenner
Professor and Director, Alpine Zoo
Innsbruck, Austria

Dr. Heinz-Siburd Raethel
Oberveterinärrat
Berlin, Germany

Dr. Urs H. Rahm
Professor, Museum of Natural History
Basel, Switzerland

Dr. Werner Rathmayer
Biology Institute, University of
Konstanz
Konstanz, Germany

Walter Reinhard
Biologist
Baden-Baden, Germany

Dr. H. H. Reinsch
Federal Fisheries Research Institute
Bremerhaven, Germany

Dr. Bernhard Rensch
Professor Emeritus, Zoological
Institute, University of Münster
Münster, Germany

Dr. Vernon Reynolds
Docent, Department of Sociology,
University of Bristol
Bristol, England

Dr. Rupert Riedl
Professor, Department of Zoology,
University of North Carolina
Chapel Hill, North Carolina, U.S.A.

Dr. Peter Rietschel
Professor (retired), Zoological
Institute, University of Frankfurt a.M.
Frankfurt a.M., Germany

Dr. Siegfried Rietschel
Docent, University of Frankfurt;
Curator, Nature Museum and
Research Institute Senckenberg
Frankfurt a.M., Germany

Herbert Ringleben
Institute of Ornithology, Heligoland
Ornithological Station
Wilhelmshaven, Germany

Dr. K. Rohde
Institute for General Zoology, Ruhr
University
Bochum, Germany

Dr. Peter Röben
Academic Councillor, Zoological
Institute, Heidelberg University
Heidelberg, Germany

Dr. Anton E. M. De Roo
Royal Museum of Central Africa
Tervuren, South Africa

Dr. Hubert Saint Girons
Research Director, Center for
National Scientific Research
Brunoy (Essonne), France

Dr. Luitfried Von Salvini-Plawen
First Zoological Institute, University
of Vienna
Vienna, Austria

Dr. Kurt Sanft
Oberstudienrat, Diesterweg-
Gymnasium
Berlin, Germany

Dr. E. G. Franz Sauer
Professor, Zoological Research
Institute and A. Koenig Museum,
University of Bonn
Bonn, Germany

Dr. Eleonore M. Sauer
Zoological Research Institute and A.
Koenig Museum, University of Bonn
Bonn, Germany

Dr. Ernst Schäfer
Curator, State Museum of Lower
Saxony
Hannover, Germany

Dr. Friedrich Schaller
Professor and Chairman, First
Zoological Institute, University of
Vienna
Vienna, Austria

Dr. George B. Schaller
Serengeti Research Institute, Michael
Grzimek Laboratory
Seronera, Tanzania

Dr. Georg Scheer
Chief Curator and Director,
Zoological Institute, State Museum of
Hesse
Darmstadt, Germany

Dr. Christoph Scherpner
Zoological Garden
Frankfurt a.M., Germany

Dr. Herbert Schifter
Bird Collection, Museum of Natural
History
Vienna, Austria

Dr. Marco Schnitter
Zoological Museum, Zurich
University
Zurich, Switzerland

Dr. Kurt Schubert
Federal Fisheries Research Institute
Hamburg, Germany

Eugen Schuhmacher
Director, Animals Films, I.U.C.N.
Munich, Germany

Dr. Thomas Schultze-Westrum
Zoological Institute, University of
Munich
Munich, Germany

Dr. Ernst Schüt
Professor and Director (retired), State
Museum of Natural History
Stuttgart, Germany

Dr. Lester L. Short , Jr.
Associate Curator, American Museum
of Natural History
New York, New York, U.S.A.

Dr. Helmut Sick
National Museum
Rio de Janeiro, Brazil

Dr. Alexander F. Skutch
Professor of Ornithology, University
of Costa Rica
San Isidro del General, Costa Rica

Dr. Everhard J. Slijper
Professor, Zoological Laboratory,
University of Amsterdam
Amsterdam, The Netherlands

Bertram E. Smythies
Curator (retired), Division of Forestry
Management, Sarawak-Malaysia
Estepona, Spain

Dr. Kenneth E. Stager
Chief Curator, Los Angeles County
Museum of Natural History
Los Angeles, California, U.S.A.

Dr. H. C. Georg H. W. Stein
Professor, Curator of Mammals,
Institute of Zoology and Zoological
Museum, Humboldt University
Berlin, Germany

Dr. Joachim Steinbacher
Curator, Nature Museum and
Senckenberg Research Institute
Frankfurt a.M., Germany

Dr. Bernard Stonehouse
Canterbury University
Christchurch, New Zealand

Dr. Richard Zur Strassen
Curator, Nature Museum and
Senckenberg Research Institute
Frandfurt a.M., Germany

Dr. Adelheid Studer-Thiersch
Zoological Garden
Basel, Switzerland

Dr. Ernst Sutter
Museum of Natural History
Basel, Switzerland

Dr. Fritz Terofal
Director, Fish Collection, Zoological
Collection of the State of Bavaria
Munich, Germany

Dr. G. F. Van Tets
Wildlife Research
Canberra, Australia

Ellen Thaler-Kottek
Institute of Zoology, University of
Innsbruck
Innsbruck, Austria

Dr. Erich Thenius
Professor and Director, Institute of
Paleontolgy, University of Vienna
Vienna, Austria

Dr. Niko Tinbergen
Professor of Animal Behavior,
Department of Zoology, Oxford
University
Oxford, England

Alexander Tsurikov
Lecturer, University of Munich
Munich, Germany

Dr. Wolfgang Villwock
Zoological Institute and Museum,
University of Hamburg
Hamburg, Germany

Zdenek Vogel
Director, Suchdol Herpetological
Station
Prague, Czechoslovakia

Dieter Vogt
Schorndorf, Germany

Dr. Jiri Volf
Zoological Garden
Prague, Czechoslovakia

Otto Wadewitz
Leipzig, Germany

Dr. Helmut O. Wagner
Director (retired), Overseas Museum,
Bremen
Mexico City, Mexico

Dr. Fritz Walther
Professor, Texas A & M University
College Station, Texas, U.S.A.

John Warham
Zoology Department, Canterbury
University
Christchurch, New Zealand

Dr. Sherwood L. Washburn
University of California at Berkeley
Berkeley, California, U.S.A.

Eberhard Wawra
First Zoological Institute, University
of Vienna
Vienna, Austria

Dr. Ingrid Weigel
Zoological Collection of the State of
Bavaria
Munich, Germany

Dr. B. Weischer
Institute of Nematode Research,
Federal Biological Institute
Münster/Westfalen, Germany

Herbert Wendt
Author, Natural History
Baden-Baden, Germany

Dr. Heinz Wermuth
Chief Curator, State Nature Museum,
Stuttgart
Ludwigsburg, Germany

Dr. Wolfgang Von Westernhagen
Preetz/Holstein, Germany

Dr. Alexander Wetmore
United States National Museum,
Smithsonian Institution
Washington, D.C., U.S.A.

Dr. Dietrich E. Wilcke
Röttgen, Germany

Dr. Helmut Wilkens
Professor and Director, Institute of
Anatomy, School of Veterinary
Medicine
Hannover, Germany

Dr. Michael L. Wolfe
Utah, U.S.A.

Hans Edmund Wolters
Zoological Research Institute and A.
Koenig Museum
Bonn, Germany

Dr. Arnfrid Wünschmann
Research Associate, Zoological Garden
Berlin, Germany

Dr. Walter Wüst
Instructor, Wilhelms Gymnasium
Munich, Germany

Dr. Heinz Wundt
Zoological Collection of the State of
Bavaria
Munich, Germany

Dr. Claus-Dieter Zander
Zoological Institute and Museum,
University of Hamburg
Hamburg, Germany

Dr. Fritz Zumpt
Director, Entomology and
Parasitology, South African Institute
for Medical Research
Johannesburg, South Africa

Dr. Richard L. Zusi
Curator of Birds, United States
National Museum, Smithsonian
Institution
Washington, D.C., U.S.A.

Glossary

Adaptive radiation—Diversification of a species or single ancestral type into several forms that are each adaptively specialized to a specific niche.

Agonistic—Behavioral patterns that are aggressive in context.

Allopatric—Occurring in separate, nonoverlapping geographic areas.

Alpha breeder—The reproductively dominant member of a social unit.

Altricial—An adjective referring to a mammal that is born with little, if any, hair, is unable to feed itself, and initially has poor sensory and thermoregulatory abilities.

Amphibious—Refers to the ability of an animal to move both through water and on land.

Austral—May refer to "southern regions," typically meaning Southern Hemisphere. May also refer to the geographical region included within the Transition, Upper Austral, and Lower Austral Life Zones as defined by C. Hart Merriam in 1892–1898. These zones are often characterized by specific plant and animal communities and were originally defined by temperature gradients especially in the mountains of southwestern North America.

Bergmann's rule—Within a species or among closely related species of mammals, those individuals in colder environments often are larger in body size. Bergmann's rule is a generalization that reflects the ability of endothermic animals to more easily retain body heat (in cold climates) if they have a high body surface to body volume ratio, and to more easily dissipate excess body heat (in hot environments) if they have a low body surface to body volume ratio.

Bioacoustics—The study of biological sounds such as the sounds produced by bats or other mammals.

Biogeographic region—One of several major divisions of the earth defined by a distinctive assemblage of animals and plants. Sometimes referred to as "zoogeographic regions or realms" (for animals) or "phytogeographic regions or realms" (for plants). Such terminology dates from the late nineteenth century and varies considerably. Major biogeographic regions each have a somewhat distinctive flora and fauna. Those generally recognized include Nearctic, Neotropical, Palearctic, Ethiopian, Oriental, and Australian.

Blow—Cloud of vapor and sea water exhaled by cetaceans.

Boreal—Often used as an adjective meaning "northern"; also may refer to the northern climatic zone immediately south of the Arctic; may also include the Arctic, Hudsonian, and Canadian Life Zones described by C. Hart Merriam.

Brachiating ancestor—Ancestor that swung around by the arms.

Breaching—A whale behavior—leaping above the water's surface, then falling back into the water, landing on its back or side.

Cephalopod—Member of the group of mollusks such as squid and octopus.

Cladistic—Evolutionary relationships suggested as "tree" branches to indicate lines of common ancestry.

Cline—A gradient in a measurable characteristic, such as size and color, showing geographic differentiation. Various patterns of geographic variation are reflected as clines or clinal variation, and have been described as "ecogeographic rules."

Cloaca—A common opening for the digestive, urinary, and reproductive tracts found in monotreme mammals.

Colony—A group of mammals living in close proximity, interacting, and usually aiding in early warning of the presence of predators and in group defense.

Commensal—A relationship between species in which one benefits and the other is neither benefited nor harmed.

Congeneric—Descriptive of two or more species that belong to the same genus.

Conspecific—Descriptive of two or more individuals or populations that belong to the same species.

Contact call—Simple vocalization used to maintain communication or physical proximity among members of a social unit.

Convergent evolution—When two evolutionarily unrelated groups of organisms develop similar characteristics due to adaptation to similar aspects of their environment or niche.

Coprophagy—Reingestion of feces to obtain nutrients that were not ingested the first time through the digestive system.

Cosmopolitan—Adjective describing the distribution pattern of an animal found around the world in suitable habitats.

Crepuscular—Active at dawn and at dusk.

Critically Endangered—A technical category used by IUCN for a species that is at an extremely high risk of extinction in the wild in the immediate future.

Cryptic—Hidden or concealed; i.e., well-camouflaged patterning.

Dental formula—A method for describing the number of each type of tooth found in an animal's mouth: incisors (I), canines (C), premolars (P), and molars (M). The formula gives the number of each tooth found in an upper and lower quadrant of the mouth, and the total is multiplied by two for the total number of teeth. For example, the formula for humans is: I2/2 C1/1 P2/2 M3/3 (total, 16, times two is 32 teeth).

Dimorphic—Occurring in two distinct forms (e.g., in reference to the differences in size between males and females of a species).

Disjunct—A distribution pattern characterized by populations that are geographically separated from one another.

Diurnal—Active during the day.

DNA-DNA hybridization—A technique whereby the genetic similarity of different animal groups is determined based on the extent to which short stretches of their DNA, when mixed together in solution in the laboratory, are able to join with each other.

Dominance hierarchy—The social status of individuals in a group; each animal can usually dominate those animals below it in a hierarchy.

Dorso-ventrally—From back to front.

Duetting—Male and female singing and integrating their songs together.

Echolocation—A method of navigation used by some mammals (e.g., bats and marine mammals) to locate objects and investigate surroundings. The animals emit audible "clicks" and determine pathways by using the echo of the sound from structures in the area.

Ecotourism—Travel for the primary purpose of viewing nature. Ecotourism is now "big business" and is used as a non-consumptive but financially rewarding way to protect important areas for conservation.

Ectothermic—Using external energy and behavior to regulate body temperature. "Cold-blooded."

Endangered—A term used by IUCN and also under the Endangered Species Act of 1973 in the United States in reference to a species that is threatened with imminent extinction or extirpation over all or a significant portion of its range.

Endemic—Native to only one specific area.

Endothermic—Maintaining a constant body-temperature using metabolic energy. "Warm-blooded."

Eocene—Geological time period; subdivision of the Tertiary, from about 55.5 to 33.7 million years ago.

Ethology—The study of animal behavior.

Exotic—Not native.

Extant—Still in existence; not destroyed, lost, or extinct.

Extinct—Refers to a species that no longer survives anywhere.

Extirpated—Referring to a local extinction of a species that can still be found elsewhere.

Feral—A population of domesticated animal that lives in the wild.

Flehmen—Lip curling and head raising after sniffing a female's urine.

Forb—Any herb that is not a grass or grass-like.

Fossorial—Adapted for digging.

Frugivorous—Feeds on fruit.

Granivorous—Feeding on seeds.

Gravid—Pregnant.

Gregarious—Occuring in large groups.

Hibernation—A deep state of reduced metabolic activity and lowered body temperature that may last for weeks or months.

Holarctic—The Palearctic and Nearctic bigeographic regions combined.

Hybrid—The offspring resulting from a cross between two different species (or sometimes between distinctive subspecies).

Innate—An inherited characteristic.

Insectivorous—Technically refers to animals that eat insects; generally refers to animals that feed primarily on insects and other arthropods.

Introduced species—An animal or plant that has been introduced to an area where it normally does not occur.

Iteroparous—Breeds in multiple years.

Jacobson's organ—Olfactory organ found in the upper palate that first appeared in amphibians and is most developed in these and in reptiles, but is also found in some birds and mammals.

Kiva—A large chamber wholly or partly underground, and often used for religious ceremonies in Pueblo Indian villages.

Mandible—Technically an animal's lower jaw. The plural, mandibles, is used to refer to both the upper and lower jaw. The upper jaw is technically the maxilla, but often called the "upper mandible."

Marsupial—A mammal whose young complete their embryonic development outside of the mother's body, within a maternal pouch.

Matrilineal—Describing a social unit in which group members are descended from a single female.

Melon—The fat-filled forehead of aquatic mammals of the order Cetacea.

Metabolic rate—The rate of chemical processes in living organisms, resulting in energy expenditure and growth. Metabolic rate decreases when an animal is resting and increases during activity.

Migration—A two-way movement in some mammals, often dramatically seasonal. Typically latitudinal, though in some species is altitudinal or longitudinal. May be short-distance or long-distance.

Miocene—The geological time period that lasted from about 23.8 to 5.6 million years ago.

Molecular phylogenetics—The use of molecular (usually genetic) techniques to study evolutionary relationships between or among different groups of organisms.

Monestrous—Experiencing estrus just once each year or breeding season.

Monogamous—A breeding system in which a male and female mate only with one another.

Monophyletic—A group (or clade) that shares a common ancestor.

Monotypic—A taxonomic category that includes only one form (e.g., a genus that includes only one species; a species that includes no subspecies).

Montane—Of or inhabiting the biogeographic zone of relatively moist, cool upland slopes below timberline dominated by large coniferous trees.

Morphology—The form and structure of animals and plants.

Mutualism—Ecological relationship between two species in which both gain benefit.

Near Threatened—A category defined by the IUCN suggesting possible risk of extinction in the medium term (as opposed to long or short term) future.

Nearctic—The biogeographic region that includes temperate North America. faunal region.

Neotropical—The biogeographic region that includes South and Central America, the West Indies, and tropical Mexico.

New World—A general descriptive term encompassing the Nearctic and Neotropical biogeographic regions.

Niche—The role of an organism in its environment; multidimensional, with habitat and behavioral components.

Nocturnal—Active at night.

Old World—A general term that usually describes a species or group as being from Eurasia or Africa.

Oligocene—The geologic time period occurring from about 33.7 to 23.8 million years ago.

Omnivorous—Feeding on a broad range of foods, both plant and animal matter.

Palearctic—A biogeographic region that includes temperate Eurasia and Africa north of the Sahara.

Paleocene—Geological period, subdivision of the Tertiary, from 65 to 55.5 million years ago.

Pelage—Coat, skin, and hair.

Pelagic—An adjective used to indicate a relationship to the open sea.

Pestiferous—Troublesome or annoying; nuisance.

Phylogeny—A grouping of taxa based on evolutionary history.

Piscivorous—Fish-eating.

Placental—A mammal whose young complete their embryonic development within the mother's uterus, joined to her by a placenta.

Pleistocene—In general, the time of the great ice ages; geological period variously considered to include the last 1 to 1.8 million years.

Pliocene—The geological period preceding the Pleistocene; the last subdivision of what is known as the Tertiary; lasted from 5.5 to 1.8 million years ago.

Polyandry—A breeding system in which one female mates with two or more males.

Polygamy—A breeding system in which either or both male and female may have two or more mates.

Polygyny—A breeding system in which one male mates with two or more females.

Polyphyletic—A taxonomic group that is believed to have originated from more than one group of ancestors.

Post-gastric digestion—Refers to the type of fermentative digestion of vegetative matter found in tapirs and other animals by which microorganisms decompose food in a caecum. This is not as thorough a decomposition as occurs in ruminant digesters.

Precocial—An adjective used to describe animals that are born in an advanced state of development such that they generally can leave their birth area quickly and obtain their own food, although they are often led to food and guarded by a parent.

Proboscis—The prehensile trunk (a muscular hydrostat) found in tapirs, elephants, etc.

Quaternary—The geological period, from 1.8 million years ago to the present, usually including two subdivisions: the Pleistocene, and the Holocene.

Refugium (pl. refugia)—An area relatively unaltered during a time of climatic change, from which dispersion and speciation may occur after the climate readjusts.

Reproductive longevity—The length of an animal's life over which it is capable of reproduction.

Ruminant—An even-toed, hoofed mammal with a four-chambered stomach that eats rapidly to regurgitate its food and chew the cud later.

Scansorial—Specialized for climbing.

Seed dispersal—Refers to how tapirs and other animals transport viable seeds from their source to near or distant, suitable habitats where they can successfully germinate. Such dispersal may occur through the feces, through sputum, or as the seeds are attached and later released from fur, etc.

Semelparity—A short life span, in which a single instance of breeding is followed by death in the first year of life.

Sexual dimorphism—Male and female differ in morphology, such as size, feather size or shape, or bill size or shape.

Sibling species—Two or more species that are very closely related, presumably having differentiated from a common ancestor in the recent past; often difficult to distinguish, often interspecifically territorial.

Sonagram—A graphic representation of sound.

Speciation—The evolution of new species.

Spy-hopping—Positioning the body vertically in the water, with the head raised above the sea surface, sometimes while turning slowly.

Steppe—Arid land with vegetation that can thrive with very little moisture; found usually in regions of extreme temperature range.

Suspensory—Moving around or hanging by the arms.

Sympatric—Inhabiting the same range.

Systematist—A specialist in the classification of organisms; systematists strive to classify organisms on the basis of their evolutionary relationships.

Taxon (pl. taxa)—Any unit of scientific classification (e.g., species, genus, family, order).

Taxonomist—A specialist in the naming and classification of organisms. (See also Systematist. Taxonomy is the older science of naming things; identification of evolutionary relationships has not always been the goal of taxonomists. The modern science of systematics generally incorporates taxonomy with the search for evolutionary relationships.)

Taxonomy—The science of identifying, naming, and classifying organisms into groups.

Territoriality—Refers to an animal's defense of a certain portion of its habitat against other conspecifics. This is often undertaken by males in relation to one another and as a lure to females.

Territory—Any defended area. Territorial defense is typically male against male, female against female, and within a species or between sibling species. Area defended varies greatly among taxa, seasons, and habitats. A territory may include the entire home range, only the area immediately around a nest, or only a feeding area.

Tertiary—The geological period including most of the Cenozoic; from about 65 to 1.8 million years ago.

Thermoregulation—The ability to regulate body temperature; can be either behavioral or physiological.

Tribe—A unit of classification below the subfamily and above the genus.

Truncal erectness—Sitting, hanging, arm-swinging (brachiating), walking bipedally with the backbone held vertical.

Ungulate—A hoofed mammal.

Upper cone—The circle in which the arm can rotate when raised above the head.

Viable population—A population that is capable of maintaining itself over a period of time. One of the major conservation issues of the twenty-first century is determining what is a minimum viable population size. Population geneticists have generally come up with estimates of about 500 breeding pairs.

Vulnerable—A category defined by IUCN as a species that is not Critically Endangered or Endangered, but is still facing a threat of extinction.

GLOSSARY

Mammals species list

Monotremata [Order]

Tachyglossidae [Family]
Tachyglossus [Genus]
T. aculeatus [Species]
Zaglossus [Genus]
Z. bruijni [Species]

Ornithorhynchidae [Family]
Ornithorhynchus [Genus]
O. anatinus [Species]

Didelphimorphia [Order]

Didelphidae [Family]
Caluromys [Genus]
C. derbianus [Species]
C. lanatus
C. philander
Caluromysiops [Genus]
C. irrupta [Species]
Chironectes [Genus]
C. minimus [Species]
Didelphis [Genus]
D. albiventris [Species]
D. aurita
D. marsupialis
D. virginiana
Glironia [Genus]
G. venusta [Species]
Gracilinanus [Genus]
G. aceramarcae [Species]
G. agilis
G. dryas
G. emiliae
G. marica
G. microtarsus
Lestodelphys [Genus]
L. halli [Species]
Lutreolina [Genus]
L. crassicaudata [Species]
Marmosa [Genus]
M. andersoni [Species]
M. canescens
M. lepida

M. mexicana
M. murina
M. robinsoni
M. rubra
M. tyleriana
M. xerophila
Marmosops [Genus]
M. cracens [Species]
M. dorothea
M. fuscatus
M. handleyi
M. impavidus
M. incanus
M. invictus
M. noctivagus
M. parvidens
Metachirus [Genus]
M. nudicaudatus [Species]
Micoureus [Genus]
M. alstoni [Species]
M. constantiae
M. demerarae
M. regina
Monodelphis [Genus]
M. adusta [Species]
M. americana
M. brevicaudata
M. dimidiata
M. domestica
M. emiliae
M. iheringi
M. kunsi
M. maraxina
M. osgoodi
M. rubida
M. scalops
M. sorex
M. theresa
M. unistriata
Philander [Genus]
P. andersoni [Species]
P. opossum
Thylamys [Genus]
T. elegans [Species]

T. macrura
T. pallidior
T. pusilla
T. velutinus

Paucituberculata [Order]

Caenolestidae [Family]
Caenolestes [Genus]
C. caniventer [Species]
C. convelatus
C. fuliginosus
Lestoros [Genus]
L. inca [Species]
Rhyncholestes [Genus]
R. raphanurus [Species]

Microbiotheria [Order]

Microbiotheriidae [Family]
Dromiciops [Genus]
D. gliroides [Species]

Dasyuromorphia [Order]

Dasyuridae [Family]
Antechinus [Genus]
A. bellus [Species]
A. flavipes
A. godmani
A. leo
A. melanurus
A. minimus
A. naso
A. stuartii
A. swainsonii
A. wilhelmina
Dasycercus [Genus]
D. byrnei [Species]
D. cristicauda
Dasykaluta [Genus]
D. rosamondae [Species]
Dasyurus [Genus]
D. albopunctatus [Species]
D. geoffroii
D. hallucatus

D. maculatus
D. spartacus
D. viverrinus
Murexia [Genus]
M. longicaudata [Species]
M. rothschildi
Myoictis [Genus]
M. melas [Species]
Neophascogale [Genus]
N. lorentzi [Species]
Ningaui [Genus]
N. ridei [Species]
N. timealeyi
N. yvonnae
Parantechinus [Genus]
P. apicalis [Species]
P. bilarni
Phascogale [Genus]
P. calura [Species]
P. tapoatafa
Phascolosorex [Genus]
P. doriae [Species]
P. dorsalis
Planigale [Genus]
P. gilesi [Species]
P. ingrami
P. maculata
P. novaeguineae
P. tenuirostris
Pseudantechinus [Genus]
P. macdonnellensis [Species]
P. ningbing
P. woolleyae
Sarcophilus [Genus]
S. laniarius [Species]
Sminthopsis [Genus]
S. aitkeni [Species]
S. archeri
S. butleri
S. crassicaudata
S. dolichura
S. douglasi
S. fuliginosus
S. gilberti
S. granulipes
S. griseoventer
S. hirtipes
S. laniger
S. leucopus
S. longicaudata
S. macroura
S. murina
S. ooldea
S. psammophila
S. virginiae
S. youngsoni

Myrmecobiidae [Family]
Myrmecobius [Genus]
M. fasciatus [Species]

Thylacinidae [Family]
Thylacinus [Genus]
T. cynocephalus [Species]

Peramelemorphia [Order]

Peramelidae [Family]
Chaeropus [Genus]
C. ecaudatus [Species]
Isoodon [Genus]
I. auratus [Species]
I. macrourus
I. obesulus
Macrotis [Genus]
M. lagotis [Species]
M. leucura
Perameles [Genus]
P. bougainville [Species]
P. eremiana
P. gunnii
P. nasuta

Peroryctidae [Family]
Echymipera [Genus]
E. clara [Species]
E. davidi
E. echinista
E. kalubu
E. rufescens
Microperoryctes [Genus]
M. longicauda [Species]
M. murina
M. papuensis
Peroryctes [Genus]
P. broadbenti [Species]
P. raffrayana
Rhynchomeles [Genus]
R. prattorum [Species]

Notoryctemorphia [Order]

Notoryctidae [Family]
Notoryctes [Genus]
N. caurinus [Species]
N. typhlops

Diprotodontia [Order]

Phascolarctidae [Family]
Phascolarctos [Genus]
P. cinereus [Species]

Vombatidae [Family]
Lasiorhinus [Genus]
L. krefftii [Species]
L. latifrons
Vombatus [Genus]
V. ursinus [Species]

Phalangeridae [Family]
Ailurops [Genus]
A. ursinus [Species]

Phalanger [Genus]
P. carmelitae [Species]
P. lullulae
P. matanim
P. orientalis
P. ornatus
P. pelengensis
P. rothschildi
P. sericeus
P. vestitus
Spilocuscus [Genus]
S. maculatus [Species]
S. rufoniger
Strigocuscus [Genus]
S. celebensis [Species]
S. gymnotis
Trichosurus [Genus]
T. arnhemensis [Species]
T. caninus
T. vulpecula
Wyulda [Genus]
W. squamicaudata [Species]

Hypsiprymnodontidae [Family]
Hypsiprymnodon [Genus]
H. moschatus [Species]

Potoroidae [Family]
Aepyprymnus [Genus]
A. rufescens [Species]
Bettongia [Genus]
B. gaimardi [Species]
B. lesueur
B. penicillata
Caloprymnus [Genus]
C. campestris [Species]
Potorous [Genus]
P. longipes [Species]
P. platyops
P. tridactylus

Macropodidae [Family]
Dendrolagus [Genus]
D. bennettianus [Species]
D. dorianus
D. goodfellowi
D. inustus
D. lumholtzi
D. matschiei
D. scottae
D. spadix
D. ursinus
Dorcopsis [Genus]
D. atrata [Species]
D. hageni
D. luctuosa
D. muelleri
Dorcopsulus [Genus]
D. macleayi [Species]
D. vanheurni
Lagorchestes [Genus]

L. asomatus [Species]
L. conspicillatus
L. hirsutus
L. leporides
Lagostrophus [Genus]
  L. fasciatus [Species]
Macropus [Genus]
  M. agilis [Species]
  M. antilopinus
  M. bernardus
  M. dorsalis
  M. eugenii
  M. fuliginosus
  M. giganteus
  M. greyi
  M. irma
  M. parma
  M. parryi
  M. robustus
  M. rufogriseus
  M. rufus
Onychogalea [Genus]
  O. fraenata [Species]
  O. lunata
  O. unguifera
Petrogale [Genus]
  P. assimilis [Species]
  P. brachyotis
  P. burbidgei
  P. concinna
  P. godmani
  P. inornata
  P. lateralis
  P. penicillata
  P. persephone
  P. rothschildi
  P. xanthopus
Setonix [Genus]
  S. brachyurus [Species]
Thylogale [Genus]
  T. billardierii [Species]
  T. brunii
  T. stigmatica
  T. thetis
Wallabia [Genus]
  W. bicolor [Species]

Burramyidae [Family]
Burramys [Genus]
  B. parvus [Species]
Cercartetus [Genus]
  C. caudatus [Species]
  C. concinnus
  C. lepidus
  C. nanus

Pseudocheiridae [Family]
Hemibelideus [Genus]
  H. lemuroides [Species]
Petauroides [Genus]
  P. volans [Species]

Petropseudes [Genus]
  P. dahli [Species]
Pseudocheirus [Genus]
  P. canescens [Species]
  P. caroli
  P. forbesi
  P. herbertensis
  P. mayeri
  P. peregrinus
  P. schlegeli
Pseudochirops [Genus]
  P. albertisii [Species]
  P. archeri
  P. corinnae
  P. cupreus

Petauridae [Family]
Dactylopsila [Genus]
  D. megalura [Species]
  D. palpator
  D. tatei
  D. trivirgata
Gymnobelideus [Genus]
  G. leadbeateri [Species]
Petaurus [Genus]
  P. abidi [Species]
  P. australis
  P. breviceps
  P. gracilis
  P. norfolcensis

Tarsipedidae [Family]
Tarsipes [Genus]
  T. rostratus [Species]

Acrobatidae [Family]
Acrobates [Genus]
  A. pygmaeus [Species]
Distoechurus [Genus]
  D. pennatus [Species]

Xenarthra [Order]

Megalonychidae [Family]
Choloepus [Genus]
  C. didactylus [Species]
  C. hoffmanni

Bradypodidae [Family]
Bradypus [Genus]
  B. torquatus [Species]
  B. tridactylus
  B. variegatus

Myrmecophagidae [Family]
Cyclopes [Genus]
  C. didactylus [Species]
Myrmecophaga [Genus]
  M. tridactyla [Species]
Tamandua [Genus]
  T. mexicana [Species]
  T. tetradactyla

Dasypodidae [Family]
Chlamyphorus [Genus]
  C. retusus [Species]
  C. truncatus
Cabassous [Genus]
  C. centralis [Species]
  C. chacoensis
  C. tatouay
  C. unicinctus
Chaetophractus [Genus]
  C. nationi [Species]
  C. vellerosus
  C. villosus
Dasypus [Genus]
  D. hybridus [Species]
  D. kappleri
  D. novemcinctus
  D. pilosus
  D. sabanicola
  D. septemcinctus
Euphractus [Genus]
  E. sexcinctus [Species]
Priodontes [Genus]
  P. maximus [Species]
Tolypeutes [Genus]
  T. matacus [Species]
  T. tricinctus
Zaedyus [Genus]
  Z. pichiy [Species]

Insectivora [Order]

Erinaceidae [Family]
Atelerix [Genus]
  A. albiventris [Species]
  A. algirus
  A. frontalis
  A. sclateri
Erinaceus [Genus]
  E. amurensis [Species]
  E. concolor
  E. europaeus
Hemiechinus [Genus]
  H. aethiopicus [Species]
  H. auritus
  H. collaris
  H. hypomelas
  H. micropus
  H. nudiventris
Mesechinus [Genus]
  M. dauuricus [Species]
  M. hughi
Echinosorex [Genus]
  E. gymnura [Species]
Hylomys [Genus]
  H. hainanensis [Species]
  H. sinensis
  H. suillus
Podogymnura [Genus]
  P. aureospinula [Species]
  P. truei

Chrysochloridae [Family]
  Amblysomus [Genus]
    A. gunningi [Species]
    A. hottentotus
    A. iris
    A. julianae
  Calcochloris [Genus]
    C. obtusirostris [Species]
  Chlorotalpa [Genus]
    C. arendsi [Species]
    C. duthieae
    C. leucorhina
    C. sclateri
    C. tytonis
  Chrysochloris [Genus]
    C. asiatica [Species]
    C. stuhlmanni
    C. visagiei
  Chrysospalax [Genus]
    C. trevelyani [Species]
    C. villosus
  Cryptochloris [Genus]
    C. wintoni [Species]
    C. zyli
  Eremitalpa [Genus]
    E. granti [Species]

Tenrecidae [Family]
  Echinops [Genus]
    E. telfairi [Species]
  Geogale [Genus]
    G. aurita [Species]
  Hemicentetes [Genus]
    H. semispinosus [Species]
  Limnogale [Genus]
    L. mergulus [Species]
  Microgale [Genus]
    M. brevicaudata [Species]
    M. cowani
    M. dobsoni
    M. dryas
    M. gracilis
    M. longicaudata
    M. parvula
    M. principula
    M. pulla
    M. pusilla
    M. talazaci
    M. thomasi
  Micropotamogale [Genus]
    M. lamottei [Species]
    M. ruwenzorii
  Oryzorictes [Genus]
    O. hova [Species]
    O. talpoides
    O. tetradactylus
  Potamogale [Genus]
    P. velox [Species]
  Setifer [Genus]
    S. setosus [Species]

  Tenrec [Genus]
    T. ecaudatus [Species]

Solenodontidae [Family]
  Solenodon [Genus]
    S. cubanus [Species]
    S. marcanoi
    S. paradoxus

Nesophontidae [Family]
  Nesophontes [Genus]
    N. edithae [Species]
    N. hypomicrus
    N. longirostris
    N. major
    N. micrus
    N. paramicrus
    N. submicrus
    N. zamicrus

Soricidae [Family]
  Anourosorex [Genus]
    A. squamipes [Species]
  Blarina [Genus]
    B. brevicauda [Species]
    B. carolinensis
    B. hylophaga
  Blarinella [Genus]
    B. quadraticauda [Species]
    B. wardi
  Chimarrogale [Genus]
    C. hantu [Species]
    C. himalayica
    C. phaeura
    C. platycephala
    C. styani
    C. sumatrana
  Congosorex [Genus]
    C. polli [Species]
  Crocidura [Genus]
    C. aleksandrisi [Species]
    C. allex
    C. andamanensis
    C. ansellorum
    C. arabica
    C. armenica
    C. attenuata
    C. attila
    C. baileyi
    C. batesi
    C. beatus
    C. beccarii
    C. bottegi
    C. bottegoides
    C. buettikoferi
    C. caliginea
    C. canariensis
    C. cinderella
    C. congobelgica
    C. cossyrensis
    C. crenata

C. crossei
C. cyanea
C. denti
C. desperata
C. dhofarensis
C. dolichura
C. douceti
C. dsinezumi
C. eisentrauti
C. elgonius
C. elongata
C. erica
C. fischeri
C. flavescens
C. floweri
C. foxi
C. fuliginosa
C. fulvastra
C. fumosa
C. fuscomurina
C. glassi
C. goliath
C. gracilipes
C. grandiceps
C. grandis
C. grassei
C. grayi
C. greenwoodi
C. gueldenstaedtii
C. harenna
C. hildegardeae
C. hirta
C. hispida
C. horsfieldii
C. jacksoni
C. jenkinsi
C. kivuana
C. lamottei
C. lanosa
C. lasiura
C. latona
C. lea
C. leucodon
C. levicula
C. littoralis
C. longipes
C. lucina
C. ludia
C. luna
C. lusitania
C. macarthuri
C. macmillani
C. macowi
C. malayana
C. manengubae
C. maquassiensis
C. mariquensis
C. maurisca
C. maxi
C. mindorus

C. minuta
C. miya
C. monax
C. monticola
C. montis
C. muricauda
C. mutesae
C. nana
C. nanilla
C. neglecta
C. negrina
C. nicobarica
C. nigeriae
C. nigricans
C. nigripes
C. nigrofusca
C. nimbae
C. niobe
C. obscurior
C. olivieri
C. orii
C. osorio
C. palawanensis
C. paradoxura
C. parvipes
C. pasha
C. pergrisea
C. phaeura
C. picea
C. pitmani
C. planiceps
C. poensis
C. polia
C. pullata
C. raineyi
C. religiosa
C. rhoditis
C. roosevelti
C. russula
C. selina
C. serezkyensis
C. sibirica
C. sicula
C. silacea
C. smithii
C. somalica
C. stenocephala
C. suaveolens
C. susiana
C. tansaniana
C. tarella
C. tarfayensis
C. telfordi
C. tenuis
C. thalia
C. theresae
C. thomensis
C. turba
C. ultima
C. usambarae

C. viaria
C. voi
C. whitakeri
C. wimmeri
C. xantippe
C. yankariensis
C. zaphiri
C. zarudnyi
C. zimmeri
C. zimmermanni
Cryptotis [Genus]
  C. avia [Species]
  C. endersi
  C. goldmani
  C. goodwini
  C. gracilis
  C. hondurensis
  C. magna
  C. meridensis
  C. mexicana
  C. montivaga
  C. nigrescens
  C. parva
  C. squamipes
  C. thomasi
Diplomesodon [Genus]
  D. pulchellum [Species]
Feroculus [Genus]
  F. feroculus [Species]
Megasorex [Genus]
  M. gigas [Species]
Myosorex [Genus]
  M. babaulti [Species]
  M. blarina
  M. cafer
  M. eisentrauti
  M. geata
  M. longicaudatus
  M. okuensis
  M. rumpii
  M. schalleri
  M. sclateri
  M. tenuis
  M. varius
Nectogale [Genus]
  N. elegans [Species]
Neomys [Genus]
  N. anomalus [Species]
  N. fodiens
  N. schelkovnikovi
Notiosorex [Genus]
  N. crawfordi [Species]
Paracrocidura [Genus]
  P. graueri [Species]
  P. maxima
  P. schoutedeni
Ruwenzorisorex [Genus]
  R. suncoides [Species]
Scutisorex [Genus]
  S. somereni [Species]

Solisorex [Genus]
  S. pearsoni [Species]
Sorex [Genus]
  S. alaskanus [Species]
  S. alpinus
  S. araneus
  S. arcticus
  S. arizonae
  S. asper
  S. bairdii
  S. bedfordiae
  S. bendirii
  S. buchariensis
  S. caecutiens
  S. camtschatica
  S. cansulus
  S. cinereus
  S. coronatus
  S. cylindricauda
  S. daphaenodon
  S. dispar
  S. emarginatus
  S. excelsus
  S. fumeus
  S. gaspensis
  S. gracillimus
  S. granarius
  S. haydeni
  S. hosonoi
  S. hoyi
  S. hydrodromus
  S. isodon
  S. jacksoni
  S. kozlovi
  S. leucogaster
  S. longirostris
  S. lyelli
  S. macrodon
  S. merriami
  S. milleri
  S. minutissimus
  S. minutus
  S. mirabilis
  S. monticolus
  S. nanus
  S. oreopolus
  S. ornatus
  S. pacificus
  S. palustris
  S. planiceps
  S. portenkoi
  S. preblei
  S. raddei
  S. roboratus
  S. sadonis
  S. samniticus
  S. satunini
  S. saussurei
  S. sclateri
  S. shinto

S. sinalis
S. sonomae
S. stizodon
S. tenellus
S. thibetanus
S. trowbridgii
S. tundrensis
S. ugyunak
S. unguiculatus
S. vagrans
S. ventralis
S. veraepacis
S. volnuchini
Soriculus [Genus]
  S. caudatus [Species]
  S. fumidus
  S. hypsibius
  S. lamula
  S. leucops
  S. macrurus
  S. nigrescens
  S. parca
  S. salenskii
  S. smithii
Suncus [Genus]
  S. ater [Species]
  S. dayi
  S. etruscus
  S. fellowesgordoni
  S. hosei
  S. infinitesimus
  S. lixus
  S. madagascariensis
  S. malayanus
  S. mertensi
  S. montanus
  S. murinus
  S. remyi
  S. stoliczkanus
  S. varilla
  S. zeylanicus
Surdisorex [Genus]
  S. norae [Species]
  S. polulus
Sylvisorex [Genus]
  S. granti [Species]
  S. howelli
  S. isabellae
  S. johnstoni
  S. lunaris
  S. megalura
  S. morio
  S. ollula
  S. oriundus
  S. vulcanorum

Talpidae [Family]
Desmana [Genus]
  D. moschata [Species]
Galemys [Genus]
  G. pyrenaicus [Species]

Condylura [Genus]
  C. cristata [Species]
Euroscaptor [Genus]
  E. grandis [Species]
  E. klossi
  E. longirostris
  E. micrura
  E. mizura
  E. parvidens
Mogera [Genus]
  M. etigo [Species]
  M. insularis
  M. kobeae
  M. minor
  M. robusta
  M. tokudae
  M. wogura
Nesoscaptor [Genus]
  N. uchidai [Species]
Neurotrichus [Genus]
  N. gibbsii [Species]
Parascalops [Genus]
  P. breweri [Species]
Parascaptor [Genus]
  P. leucura [Species]
Scalopus [Genus]
  S. aquaticus [Species]
Scapanulus [Genus]
  S. oweni [Species]
Scapanus [Genus]
  S. latimanus [Species]
  S. orarius
  S. townsendii
Scaptochirus [Genus]
  S. moschatus [Species]
Scaptonyx [Genus]
  S. fusicaudus [Species]
Talpa [Genus]
  T. altaica [Species]
  T. caeca
  T. caucasica
  T. europaea
  T. levantis
  T. occidentalis
  T. romana
  T. stankovici
  T. streeti
Urotrichus [Genus]
  U. pilirostris [Species]
  U. talpoides
Uropsilus [Genus]
  U. andersoni [Species]
  U. gracilis
  U. investigator
  U. soricipes

Scandentia [Order]

Tupaiidae [Family]
Anathana [Genus]
  A. ellioti [Species]

Dendrogale [Genus]
  D. melanura [Species]
  D. murina
Ptilocercus [Genus]
  P. lowii [Species]
Tupaia [Genus]
  T. belangeri [Species]
  T. chrysogaster
  T. dorsalis
  T. glis
  T. gracilis
  T. javanica
  T. longipes
  T. minor
  T. montana
  T. nicobarica
  T. palawanensis
  T. picta
  T. splendidula
  T. tana
Urogale [Genus]
  U. everetti [Species]

Dermoptera [Order]

Cynocephalidae [Family]
Cynocephalus [Genus]
  C. variegatus [Species]
  C. volans

Chiroptera [Order]

Pteropodidae [Family]
Acerodon [Genus]
  A. celebensis [Species]
  A. humilis
  A. jubatus
  A. leucotis
  A. lucifer
  A. mackloti
Aethalops [Genus]
  A. alecto [Species]
Alionycteris [Genus]
  A. paucidentata [Species]
Aproteles [Genus]
  A. bulmerae [Species]
Balionycteris [Genus]
  B. maculata [Species]
Boneia [Genus]
  B. bidens [Species]
Casinycteris [Genus]
  C. argynnis [Species]
Chironax [Genus]
  C. melanocephalus [Species]
Cynopterus [Genus]
  C. brachyotis [Species]
  C. horsfieldi
  C. nusatenggara
  C. sphinx
  C. titthaecheileus
Dobsonia [Genus]

MAMMALS SPECIES LIST

D. beauforti [Species]
D. chapmani
D. emersa
D. exoleta
D. inermis
D. minor
D. moluccensis
D. pannietensis
D. peroni
D. praedatrix
D. viridis
Dyacopterus [Genus]
D. spadiceus [Species]
Eidolon [Genus]
E. dupreanum [Species]
E. helvum
Eonycteris [Genus]
E. major [Species]
E. spelaea
Epomophorus [Genus]
E. angolensis [Species]
E. gambianus
E. grandis
E. labiatus
E. minimus
E. wahlbergi
Epomops [Genus]
E. buettikoferi [Species]
E. dobsoni
E. franqueti
Haplonycteris [Genus]
H. fischeri [Species]
Harpyionycteris [Genus]
H. celebensis [Species]
H. whiteheadi
Hypsignathus [Genus]
H. monstrosus [Species]
Latidens [Genus]
L. salimalii [Species]
Macroglossus [Genus]
M. minimus [Species]
M. sobrinus
Megaerops [Genus]
M. ecaudatus [Species]
M. kusnotoi
M. niphanae
M. wetmorei
Megaloglossus [Genus]
M. woermanni [Species]
Melonycteris [Genus]
M. aurantius [Species]
M. melanops
M. woodfordi
Micropteropus [Genus]
M. intermedius [Species]
M. pusillus
Myonycteris [Genus]
M. brachycephala [Species]
M. relicta
M. torquata

Nanonycteris [Genus]
N. veldkampi [Species]
Neopteryx [Genus]
N. frosti [Species]
Notopteris [Genus]
N. macdonaldi [Species]
Nyctimene [Genus]
N. aello [Species]
N. albiventer
N. celaeno
N. cephalotes
N. certans
N. cyclotis
N. draconilla
N. major
N. malaitensis
N. masalai
N. minutus
N. rabori
N. robinsoni
N. sanctacrucis
N. vizcaccia
Otopteropus [Genus]
O. cartilagonodus [Species]
Paranyctimene [Genus]
P. raptor [Species]
Penthetor [Genus]
P. lucasi [Species]
Plerotes [Genus]
P. anchietai [Species]
Ptenochirus [Genus]
P. jagori [Species]
P. minor
Pteralopex [Genus]
P. acrodonta [Species]
P. anceps
P. atrata
P. pulchra
Pteropus [Genus]
P. admiralitatum [Species]
P. aldabrensis
P. alecto
P. anetianus
P. argentatus
P. brunneus
P. caniceps
P. chrysoproctus
P. conspicillatus
P. dasymallus
P. faunulus
P. fundatus
P. giganteus
P. gilliardi
P. griseus
P. howensis
P. hypomelanus
P. insularis
P. leucopterus
P. livingstonei
P. lombocensis

P. lylei
P. macrotis
P. mahaganus
P. mariannus
P. mearnsi
P. melanopogon
P. melanotus
P. molossinus
P. neohibernicus
P. niger
P. nitendiensis
P. ocularis
P. ornatus
P. personatus
P. phaeocephalus
P. pilosus
P. pohlei
P. poliocephalus
P. pselaphon
P. pumilus
P. rayneri
P. rodricensis
P. rufus
P. samoensis
P. sanctacrucis
P. scapulatus
P. seychellensis
P. speciosus
P. subniger
P. temmincki
P. tokudae
P. tonganus
P. tuberculatus
P. vampyrus
P. vetulus
P. voeltzkowi
P. woodfordi
Rousettus [Genus]
R. aegyptiacus [Species]
R. amplexicaudatus
R. angolensis
R. celebensis
R. lanosus
R. leschenaulti
R. madagascariensis
R. obliviosus
R. spinalatus
Scotonycteris [Genus]
S. ophiodon [Species]
S. zenkeri
Sphaerias [Genus]
S. blanfordi [Species]
Styloctenium [Genus]
S. wallacei [Species]
Syconycteris [Genus]
S. australis [Species]
S. carolinae
S. hobbit
Thoopterus [Genus]
T. nigrescens [Species]

Rhinopomatidae [Family]
  Rhinopoma [Genus]
    R. hardwickei [Species]
    R. microphyllum
    R. muscatellum

Emballonuridae [Family]
  Balantiopteryx [Genus]
    B. infusca [Species]
    B. io
    B. plicata
  Centronycteris [Genus]
    C. maximiliani [Species]
  Coleura [Genus]
    C. afra [Species]
    C. seychellensis
  Cormura [Genus]
    C. brevirostris [Species]
  Cyttarops [Genus]
    C. alecto [Species]
  Diclidurus [Genus]
    D. albus [Species]
    D. ingens
    D. isabellus
    D. scutatus
  Emballonura [Genus]
    E. alecto [Species]
    E. atrata
    E. beccarii
    E. dianae
    E. furax
    E. monticola
    E. raffrayana
    E. semicaudata
  Mosia [Genus]
    M. nigrescens [Species]
  Peropteryx [Genus]
    P. kappleri [Species]
    P. leucoptera
    P. macrotis
  Rhynchonycteris [Genus]
    R. naso [Species]
  Saccolaimus [Genus]
    S. flaviventris [Species]
    S. mixtus
    S. peli
    S. pluto
    S. saccolaimus
  Saccopteryx [Genus]
    S. bilineata [Species]
    S. canescens
    S. gymnura
    S. leptura
  Taphozous [Genus]
    T. australis [Species]
    T. georgianus
    T. hamiltoni
    T. hildegardeae
    T. hilli
    T. kapalgensis
    T. longimanus

T. mauritianus
T. melanopogon
T. nudiventris
T. perforatus
T. philippinensis
T. theobaldi

Craseonycteridae [Family]
  Craseonycteris [Genus]
    C. thonglongyai [Species]

Nycteridae [Family]
  Nycteris [Genus]
    N. arge [Species]
    N. gambiensis
    N. grandis
    N. hispida
    N. intermedia
    N. javanica
    N. macrotis
    N. major
    N. nana
    N. thebaica
    N. tragata
    N. woodi

Megadermatidae [Family]
  Cardioderma [Genus]
    C. cor [Species]
  Lavia [Genus]
    L. frons [Species]
  Macroderma [Genus]
    M. gigas [Species]
  Megaderma [Genus]
    M. lyra [Species]
    M. spasma

Rhinolophidae [Family]
  Rhinolophus [Genus]
    R. acuminatus [Species]
    R. adami
    R. affinis
    R. alcyone
    R. anderseni
    R. arcuatus
    R. blasii
    R. borneensis
    R. canuti
    R. capensis
    R. celebensis
    R. clivosus
    R. coelophyllus
    R. cognatus
    R. cornutus
    R. creaghi
    R. darlingi
    R. deckenii
    R. denti
    R. eloquens
    R. euryale
    R. euryotis

R. ferrumequinum
R. fumigatus
R. guineensis
R. hildebrandti
R. hipposideros
R. imaizumii
R. inops
R. keyensis
R. landeri
R. lepidus
R. luctus
R. maclaudi
R. macrotis
R. malayanus
R. marshalli
R. megaphyllus
R. mehelyi
R. mitratus
R. monoceros
R. nereis
R. osgoodi
R. paradoxolophus
R. pearsoni
R. philippinensis
R. pusillus
R. rex
R. robinsoni
R. rouxi
R. rufus
R. sedulus
R. shameli
R. silvestris
R. simplex
R. simulator
R. stheno
R. subbadius
R. subrufus
R. swinnyi
R. thomasi
R. trifoliatus
R. virgo
R. yunanensis

Hipposideridae [Family]
  Anthops [Genus]
    A. ornatus [Species]
  Asellia [Genus]
    A. patrizii [Species]
    A. tridens
  Aselliscus [Genus]
    A. stoliczkanus [Species]
    A. tricuspidatus
  Cloeotis [Genus]
    C. percivali [Species]
  Coelops [Genus]
    C. frithi [Species]
    C. hirsutus
    C. robinsoni
  Hipposideros [Genus]
    H. abae [Species]
    H. armiger

H. ater
H. beatus
H. bicolor
H. breviceps
H. caffer
H. calcaratus
H. camerunensis
H. cervinus
H. cineraceus
H. commersoni
H. coronatus
H. corynophyllus
H. coxi
H. crumeniferus
H. curtus
H. cyclops
H. diadema
H. dinops
H. doriae
H. dyacorum
H. fuliginosus
H. fulvus
H. galeritus
H. halophyllus
H. inexpectatus
H. jonesi
H. lamottei
H. lankadiva
H. larvatus
H. lekaguli
H. lylei
H. macrobullatus
H. maggietaylorae
H. marisae
H. megalotis
H. muscinus
H. nequam
H. obscurus
H. papua
H. pomona
H. pratti
H. pygmaeus
H. ridleyi
H. ruber
H. sabanus
H. schistaceus
H. semoni
H. speoris
H. stenotis
H. turpis
H. wollastoni
Paracoelops [Genus]
P. megalotis [Species]
Rhinonicteris [Genus]
R. aurantia [Species]
Triaenops [Genus]
T. furculus [Species]
T. persicus

Phyllostomidae [Family]
Ametrida [Genus]

A. centurio [Species]
Anoura [Genus]
A. caudifer [Species]
A. cultrata
A. geoffroyi
A. latidens
Ardops [Genus]
A. nichollsi [Species]
Ariteus [Genus]
A. flavescens [Species]
Artibeus [Genus]
A. amplus [Species]
A. anderseni
A. aztecus
A. cinereus
A. concolor
A. fimbriatus
A. fraterculus
A. glaucus
A. hartii
A. hirsutus
A. inopinatus
A. jamaicensis
A. lituratus
A. obscurus
A. phaeotis
A. planirostris
A. toltecus
Brachyphylla [Genus]
B. cavernarum [Species]
B. nana
Carollia [Genus]
C. brevicauda [Species]
C. castanea
C. perspicillata
C. subrufa
Centurio [Genus]
C. senex [Species]
Chiroderma [Genus]
C. doriae [Species]
C. improvisum
C. salvini
C. trinitatum
C. villosum
Choeroniscus [Genus]
C. godmani [Species]
C. intermedius
C. minor
C. periosus
Choeronycteris [Genus]
C. mexicana [Species]
Chrotopterus [Genus]
C. auritus [Species]
Desmodus [Genus]
D. rotundus [Species]
Diaemus [Genus]
D. youngi [Species]
Diphylla [Genus]
D. ecaudata [Species]
Ectophylla [Genus]

E. alba [Species]
Erophylla [Genus]
E. sezekorni [Species]
Glossophaga [Genus]
G. commissarisi [Species]
G. leachii
G. longirostris
G. morenoi
G. soricina
Hylonycteris [Genus]
H. underwoodi [Species]
Leptonycteris [Genus]
L. curasoae [Species]
L. nivalis
Lichonycteris [Genus]
L. obscura [Species]
Lionycteris [Genus]
L. spurrelli [Species]
Lonchophylla [Genus]
L. bokermanni [Species]
L. dekeyseri
L. handleyi
L. hesperia
L. mordax
L. robusta
L. thomasi
Lonchorhina [Genus]
L. aurita [Species]
L. fernandezi
L. marinkellei
L. orinocensis
Macrophyllum [Genus]
M. macrophyllum [Species]
Macrotus [Genus]
M. californicus [Species]
M. waterhousii
Mesophylla [Genus]
M. macconnelli [Species]
Micronycteris [Genus]
M. behnii [Species]
M. brachyotis
M. daviesi
M. hirsuta
M. megalotis
M. minuta
M. nicefori
M. pusilla
M. schmidtorum
M. sylvestris
Mimon [Genus]
M. bennettii [Species]
M. crenulatum
Monophyllus [Genus]
M. plethodon [Species]
M. redmani
Musonycteris [Genus]
M. harrisoni [Species]
Phylloderma [Genus]
P. stenops [Species]
Phyllonycteris [Genus]

P. aphylla [Species]
P. poeyi
Phyllops [Genus]
  P. falcatus [Species]
Phyllostomus [Genus]
  P. discolor [Species]
  P. elongatus
  P. hastatus
  P. latifolius
Platalina [Genus]
  P. genovensium [Species]
Platyrrhinus [Genus]
  P. aurarius [Species]
  P. brachycephalus
  P. chocoensis
  P. dorsalis
  P. helleri
  P. infuscus
  P. lineatus
  P. recifinus
  P. umbratus
  P. vittatus
Pygoderma [Genus]
  P. bilabiatum [Species]
Rhinophylla [Genus]
  R. alethina [Species]
  R. fischerae
  R. pumilio
Scleronycteris [Genus]
  S. ega [Species]
Sphaeronycteris [Genus]
  S. toxophyllum [Species]
Stenoderma [Genus]
  S. rufum [Species]
Sturnira [Genus]
  S. aratathomasi [Species]
  S. bidens
  S. bogotensis
  S. erythromos
  S. lilium
  S. ludovici
  S. luisi
  S. magna
  S. mordax
  S. nana
  S. thomasi
  S. tildae
Tonatia [Genus]
  T. bidens [Species]
  T. brasiliense
  T. carrikeri
  T. evotis
  T. schulzi
  T. silvicola
Trachops [Genus]
  T. cirrhosus [Species]
Uroderma [Genus]
  U. bilobatum [Species]
  U. magnirostrum
Vampyressa [Genus]

V. bidens [Species]
V. brocki
V. melissa
V. nymphaea
V. pusilla
Vampyrodes [Genus]
  V. caraccioli [Species]
Vampyrum [Genus]
  V. spectrum [Species]

Mormoopidae [Family]
  Mormoops [Genus]
    M. blainvillii [Species]
    M. megalophylla
  Pteronotus [Genus]
    P. davyi [Species]
    P. gymnonotus
    P. macleayii
    P. parnellii
    P. personatus
    P. quadridens

Noctilionidae [Family]
  Noctilio [Genus]
    N. albiventris [Species]
    N. leporinus

Mystacinidae [Family]
  Mystacina [Genus]
    M. robusta [Species]
    M. tuberculata

Natalidae [Family]
  Natalus [Genus]
    N. lepidus [Species]
    N. micropus
    N. stramineus
    N. tumidifrons
    N. tumidirostris

Furipteridae [Family]
  Amorphochilus [Genus]
    A. schnablii [Species]
  Furipterus [Genus]
    F. horrens [Species]

Thyropteridae [Family]
  Thyroptera [Genus]
    T. discifera [Species]
    T. tricolor

Myzopodidae [Family]
  Myzopoda [Genus]
    M. aurita [Species]

Molossidae [Family]
  Chaerephon [Genus]
    C. aloysiisabaudiae [Species]
    C. ansorgei
    C. bemmeleni
    C. bivittata
    C. chapini
    C. gallagheri

C. jobensis
C. johorensis
C. major
C. nigeriae
C. plicata
C. pumila
C. russata
Cheiromeles [Genus]
  C. torquatus [Species]
Eumops [Genus]
  E. auripendulus [Species]
  E. bonariensis
  E. dabbenei
  E. glaucinus
  E. hansae
  E. maurus
  E. perotis
  E. underwoodi
Molossops [Genus]
  M. abrasus [Species]
  M. aequatorianus
  M. greenhalli
  M. mattogrossensis
  M. neglectus
  M. planirostris
  M. temminckii
Molossus [Genus]
  M. ater [Species]
  M. bondae
  M. molossus
  M. pretiosus
  M. sinaloae
Mops [Genus]
  M. brachypterus [Species]
  M. condylurus
  M. congicus
  M. demonstrator
  M. midas
  M. mops
  M. nanulus
  M. niangarae
  M. niveiventer
  M. petersoni
  M. sarasinorum
  M. spurrelli
  M. thersites
  M. trevori
Mormopterus [Genus]
  M. acetabulosus [Species]
  M. beccarii
  M. doriae
  M. jugularis
  M. kalinowskii
  M. minutus
  M. norfolkensis
  M. petrophilus
  M. phrudus
  M. planiceps
  M. setiger
Myopterus [Genus]

M. daubentonii [Species]
M. whitleyi
Nyctinomops [Genus]
　N. aurispinosus [Species]
　N. femorosaccus
　N. laticaudatus
　N. macrotis
Otomops [Genus]
　O. formosus [Species]
　O. martiensseni
　O. papuensis
　O. secundus
　O. wroughtoni
Promops [Genus]
　P. centralis [Species]
　P. nasutus
Tadarida [Genus]
　T. aegyptiaca [Species]
　T. australis
　T. brasiliensis
　T. espiritosantensis
　T. fulminans
　T. lobata
　T. teniotis
　T. ventralis

Vespertilionidae [Family]
Antrozous [Genus]
　A. dubiaquercus [Species]
　A. pallidus
Barbastella [Genus]
　B. barbastellus [Species]
　B. leucomelas
Chalinolobus [Genus]
　C. alboguttatus [Species]
　C. argentatus
　C. beatrix
　C. dwyeri
　C. egeria
　C. gleni
　C. gouldii
　C. kenyacola
　C. morio
　C. nigrogriseus
　C. picatus
　C. poensis
　C. superbus
　C. tuberculatus
　C. variegatus
Eptesicus [Genus]
　E. baverstocki [Species]
　E. bobrinskoi
　E. bottae
　E. brasiliensis
　E. brunneus
　E. capensis
　E. demissus
　E. diminutus
　E. douglasorum
　E. flavescens
　E. floweri

E. furinalis
E. fuscus
E. guadeloupensis
E. guineensis
E. hottentotus
E. innoxius
E. kobayashii
E. melckorum
E. nasutus
E. nilssoni
E. pachyotis
E. platyops
E. pumilus
E. regulus
E. rendalli
E. sagittula
E. serotinus
E. somalicus
E. tatei
E. tenuipinnis
E. vulturnus
Euderma [Genus]
　E. maculatum [Species]
Eudiscopus [Genus]
　E. denticulus [Species]
Glischropus [Genus]
　G. javanus [Species]
　G. tylopus
Harpiocephalus [Genus]
　H. harpia [Species]
Hesperoptenus [Genus]
　H. blanfordi [Species]
　H. doriae
　H. gaskelli
　H. tickelli
　H. tomesi
Histiotus [Genus]
　H. alienus [Species]
　H. macrotus
　H. montanus
　H. velatus
Ia [Genus]
　I. io [Species]
Idionycteris [Genus]
　I. phyllotis [Species]
Kerivoula [Genus]
　K. aerosa [Species]
　K. africana
　K. agnella
　K. argentata
　K. atrox
　K. cuprosa
　K. eriophora
　K. flora
　K. hardwickei
　K. intermedia
　K. jagori
　K. lanosa
　K. minuta
　K. muscina

K. myrella
K. papillosa
K. papuensis
K. pellucida
K. phalaena
K. picta
K. smithi
K. whiteheadi
Laephotis [Genus]
　L. angolensis [Species]
　L. botswanae
　L. namibensis
　L. wintoni
Lasionycteris [Genus]
　L. noctivagans [Species]
Lasiurus [Genus]
　L. borealis [Species]
　L. castaneus
　L. cinereus
　L. ega
　L. egregius
　L. intermedius
　L. seminolus
Mimetillus [Genus]
　M. moloneyi [Species]
Miniopterus [Genus]
　M. australis [Species]
　M. fraterculus
　M. fuscus
　M. inflatus
　M. magnater
　M. minor
　M. pusillus
　M. robustior
　M. schreibersi
　M. tristis
Murina [Genus]
　M. aenea [Species]
　M. aurata
　M. cyclotis
　M. florium
　M. fusca
　M. grisea
　M. huttoni
　M. leucogaster
　M. puta
　M. rozendaali
　M. silvatica
　M. suilla
　M. tenebrosa
　M. tubinaris
　M. ussuriensis
Myotis [Genus]
　M. abei [Species]
　M. adversus
　M. aelleni
　M. albescens
　M. altarium
　M. annectans
　M. atacamensis

M. auriculus
M. australis
M. austroriparius
M. bechsteini
M. blythii
M. bocagei
M. bombinus
M. brandti
M. californicus
M. capaccinii
M. chiloensis
M. chinensis
M. cobanensis
M. dasycneme
M. daubentoni
M. dominicensis
M. elegans
M. emarginatus
M. evotis
M. findleyi
M. formosus
M. fortidens
M. frater
M. goudoti
M. grisescens
M. hasseltii
M. horsfieldii
M. hosonoi
M. ikonnikovi
M. insularum
M. keaysi
M. keenii
M. leibii
M. lesueuri
M. levis
M. longipes
M. lucifugus
M. macrodactylus
M. macrotarsus
M. martiniquensis
M. milleri
M. montivagus
M. morrisi
M. muricola
M. myotis
M. mystacinus
M. nattereri
M. nesopolus
M. nigricans
M. oreias
M. oxyotus
M. ozensis
M. peninsularis
M. pequinius
M. planiceps
M. pruinosus
M. ricketti
M. ridleyi
M. riparius
M. rosseti

M. ruber
M. schaubi
M. scotti
M. seabrai
M. sicarius
M. siligorensis
M. simus
M. sodalis
M. stalkeri
M. thysanodes
M. tricolor
M. velifer
M. vivesi
M. volans
M. welwitschii
M. yesoensis
M. yumanensis
Nyctalus [Genus]
N. aviator [Species]
N. azoreum
N. lasiopterus
N. leisleri
N. montanus
N. noctula
Nycticeius [Genus]
N. balstoni [Species]
N. greyii
N. humeralis
N. rueppellii
N. sanborni
N. schlieffeni
Nyctophilus [Genus]
N. arnhemensis [Species]
N. geoffroyi
N. gouldi
N. heran
N. microdon
N. microtis
N. timoriensis
N. walkeri
Otonycteris [Genus]
O. hemprichi [Species]
Pharotis [Genus]
P. imogene [Species]
Philetor [Genus]
P. brachypterus [Species]
Pipistrellus [Genus]
P. aegyptius [Species]
P. aero
P. affinis
P. anchietai
P. anthonyi
P. arabicus
P. ariel
P. babu
P. bodenheimeri
P. cadornae
P. ceylonicus
P. circumdatus
P. coromandra

P. crassulus
P. cuprosus
P. dormeri
P. eisentrauti
P. endoi
P. hesperus
P. imbricatus
P. inexspectatus
P. javanicus
P. joffrei
P. kitcheneri
P. kuhlii
P. lophurus
P. macrotis
P. maderensis
P. mimus
P. minahassae
P. mordax
P. musciculus
P. nanulus
P. nanus
P. nathusii
P. paterculus
P. peguensis
P. permixtus
P. petersi
P. pipistrellus
P. pulveratus
P. rueppelli
P. rusticus
P. savii
P. societatis
P. stenopterus
P. sturdeei
P. subflavus
P. tasmaniensis
P. tenuis
Plecotus [Genus]
P. auritus [Species]
P. austriacus
P. mexicanus
P. rafinesquii
P. taivanus
P. teneriffae
P. townsendii
Rhogeessa [Genus]
R. alleni [Species]
R. genowaysi
R. gracilis
R. minutilla
R. mira
R. parvula
R. tumida
Scotoecus [Genus]
S. albofuscus [Species]
S. hirundo
S. pallidus
Scotomanes [Genus]
S. emarginatus [Species]
S. ornatus

Scotophilus [Genus]
  S. borbonicus [Species]
  S. celebensis
  S. dinganii
  S. heathi
  S. kuhlii
  S. leucogaster
  S. nigrita
  S. nux
  S. robustus
  S. viridis
Tomopeas [Genus]
  T. ravus [Species]
Tylonycteris [Genus]
  T. pachypus [Species]
  T. robustula
Vespertilio [Genus]
  V. murinus [Species]
  V. superans

Primates [Order]

Lorisidae [Family]
  Arctocebus [Genus]
    A. aureus [Species]
    A. calabarensis
  Loris [Genus]
    L. tardigradus [Species]
  Nycticebus [Genus]
    N. coucang [Species]
    N. pygmaeus
  Perodicticus [Genus]
    P. potto [Species]

Galagidae [Family]
  Euoticus [Genus]
    E. elegantulus [Species]
    E. pallidus
  Galago [Genus]
    G. alleni [Species]
    G. gallarum
    G. matschiei
    G. moholi
    G. senegalensis
  Galagoides [Genus]
    G. demidoff [Species]
    G. zanzibaricus
  Otolemur [Genus]
    O. crassicaudatus [Species]
    O. garnettii

Cheirogaleidae [Family]
  Allocebus [Genus]
    A. trichotis [Species]
  Cheirogaleus [Genus]
    C. major [Species]
    C. medius
  Microcebus [Genus]
    Microcebus coquereli [Species]
    Microcebus murinus
    Microcebus rufus

Phaner [Genus]
  P. furcifer [Species]

Lemuridae [Family]
  Eulemur [Genus]
    E. coronatus [Species]
    E. fulvus
    E. macaco
    E. mongoz
    E. rubriventer
  Hapalemur [Genus]
    H. aureus [Species]
    H. griseus
    H. simus
  Lemur [Genus]
    L. catta [Species]
  Varecia [Genus]
    V. variegata [Species]

Indriidae [Family]
  Avahi [Genus]
    A. laniger [Species]
  Indri [Genus]
    I. indri [Species]
  Propithecus [Genus]
    P. diadema [Species]
    P. tattersalli
    P. verreauxi

Lepilemuridae [Family]
  Lepilemur [Genus]
    L. dorsalis [Species]
    L. edwardsi
    L. leucopus
    L. microdon
    L. mustelinus
    L. ruficaudatus
    L. septentrionalis

Daubentoniidae [Family]
  Daubentonia [Genus]
    D. madagascariensis [Species]

Tarsiidae [Family]
  Tarsius [Genus]
    T. bancanus [Species]
    T. dianae
    T. pumilus
    T. spectrum
    T. syrichta

Cebidae [Family]
  Alouatta [Genus]
    A. belzebul [Species]
    A. caraya
    A. coibensis
    A. fusca
    A. palliata
    A. pigra
    A. sara
    A. seniculus
  Callicebus [Genus]

    C. brunneus [Species]
    C. caligatus
    C. cinerascens
    C. cupreus
    C. donacophilus
    C. dubius
    C. hoffmannsi
    C. modestus
    C. moloch
    C. oenanthe
    C. olallae
    C. personatus
    C. torquatus
  Cebus [Genus]
    C. albifrons [Species]
    C. apella
    C. capucinus
    C. olivaceus
  Saimiri [Genus]
    S. boliviensis [Species]
    S. oerstedii
    S. sciureus
    S. ustus
    S. vanzolinii

Callitrichidae [Family]
  Callimico [Genus]
    C. goeldii [Species]
  Callithrix [Genus]
    C. argentata [Species]
    C. aurita
    C. flaviceps
    C. geoffroyi
    C. humeralifer
    C. jacchus
    C. kuhlii
    C. penicillata
    C. pygmaea
  Leontopithecus [Genus]
    L. caissara [Species]
    L. chrysomela
    L. chrysopygus
    L. rosalia
  Saguinus [Genus]
    S. bicolor [Species]
    S. fuscicollis
    S. geoffroyi
    S. imperator
    S. inustus
    S. labiatus
    S. leucopus
    S. midas
    S. mystax
    S. nigricollis
    S. oedipus
    S. tripartitus

Aotidae [Family]
  Aotus [Genus]
    A. azarai [Species]
    A. brumbacki

A. hershkovitzi
A. infulatus
A. lemurinus
A. miconax
A. nancymaae
A. nigriceps
A. trivirgatus
A. vociferans

Pitheciidae [Family]
Cacajao [Genus]
C. calvus [Species]
C. melanocephalus
Chiropotes [Genus]
C. albinasus [Species]
C. satanas
Pithecia [Genus]
P. aequatorialis [Species]
P. albicans
P. irrorata
P. monachus
P. pithecia

Atelidae [Family]
Ateles [Genus]
A. belzebuth [Species]
A. chamek
A. fusciceps
A. geoffroyi
A. marginatus
A. paniscus
Brachyteles [Genus]
B. arachnoides [Species]
Lagothrix [Genus]
L. flavicauda [Species]
L. lagotricha

Cercopithecidae [Family]
Allenopithecus [Genus]
A. nigroviridis [Species]
Cercocebus [Genus]
C. agilis [Species]
C. galeritus
C. torquatus
Cercopithecus [Genus]
C. ascanius [Species]
C. campbelli
C. cephus
C. diana
C. dryas
C. erythrogaster
C. erythrotis
C. hamlyni
C. lhoesti
C. mitis
C. mona
C. neglectus
C. nictitans
C. petaurista
C. pogonias
C. preussi
C. sclateri

C. solatus
C. wolfi
Chlorocebus [Genus]
C. aethiops [Species]
Colobus [Genus]
C. angolensis [Species]
C. guereza
C. polykomos
C. satanas
Erythrocebus [Genus]
E. patas [Species]
Lophocebus [Genus]
L. albigena [Species]
Macaca [Genus]
M. arctoides [Species]
M. assamensis
M. cyclopis
M. fascicularis
M. fuscata
M. maura
M. mulatta
M. nemestrina
M. nigra
M. ochreata
M. radiata
M. silenus
M. sinica
M. sylvanus
M. thibetana
M. tonkeana
Mandrillus [Genus]
M. leucophaeus [Species]
M. sphinx
Miopithecus
M. talapoin
Nasalis [Genus]
N. concolor [Species]
N. larvatus
Papio [Genus]
P. hamadryas [Species]
Presbytis [Genus]
P. comata [Species]
P. femoralis
P. frontata
P. hosei
P. melalophos
P. potenziani
P. rubicunda
P. thomasi
Procolobus [Genus]
P. badius [Species]
P. pennantii
P. preussi
P. rufomitratus
P. verus
Pygathrix [Genus]
P. avunculus [Species]
P. bieti
P. brelichi
P. nemaeus

P. roxellana
Semnopithecus [Genus]
S. entellus [Species]
Theropithecus [Genus]
T. gelada [Species]
Trachypithecus [Genus]
T. auratus [Species]
T. cristatus
T. francoisi
T. geei
T. johnii
T. obscurus
T. phayrei
T. pileatus
T. vetulus

Hylobatidae [Family]
Hylobates [Genus]
H. agilis [Species]
H. concolor
H. gabriellae
H. hoolock
H. klossii
H. lar
H. leucogenys
H. moloch
H. muelleri
H. pileatus
H. syndactylus

Hominidae [Family]
Gorilla [Genus]
G. gorilla [Species]
Homo [Genus]
H. sapiens [Species]
Pan [Genus]
P. paniscus [Species]
P. troglodytes
Pongo [Genus]
P. pygmaeus [Species]

Carnivora [Order]

Canidae [Family]
Alopex [Genus]
A. lagopus [Species]
Atelocynus
A. microtis
Canis [Genus]
C. adustus [Species]
C. aureus
C. latrans
C. lupus
C. mesomelas
C. rufus
C. simensis
Cerdocyon [Genus]
C. thous [Species]
Chrysocyon [Genus]
C. brachyurus [Species]
Cuon [Genus]

C. alpinus [Species]
Dusicyon [Genus]
　D. australis [Species]
Lycaon [Genus]
　L. pictus [Species]
Nyctereutes [Genus]
　N. procyonoides [Species]
Otocyon [Genus]
　O. megalotis [Species]
Pseudalopex [Genus]
　P. culpaeus [Species]
　P. griseus
　P. gymnocercus
　P. sechurae
　P. vetulus
Speothos [Genus]
　S. venaticus [Species]
Urocyon [Genus]
　U. cinereoargenteus [Species]
　U. littoralis
Vulpes [Genus]
　V. bengalensis [Species]
　V. cana
　V. chama
　V. corsac
　V. ferrilata
　V. pallida
　V. rueppelli
　V. velox
　V. vulpes
　V. zerda

Ursidae [Family]
Ailuropoda [Genus]
　A. melanoleuca [Species]
Ailurus [Genus]
　A. fulgens [Species]
Helarctos [Genus]
　H. malayanus [Species]
Melursus [Genus]
　M. ursinus [Species]
Tremarctos [Genus]
　T. ornatus [Species]
Ursus [Genus]
　U. americanus [Species]
　U. arctos
　U. maritimus
　U. thibetanus

Procyonidae [Family]
Bassaricyon [Genus]
　B. alleni [Species]
　B. beddardi
　B. gabbii
　B. lasius
　B. pauli
Potos [Genus]
　P. flavus [Species]
Bassariscus [Genus]
　B. astutus [Species]
　B. sumichrasti

Nasua [Genus]
　N. narica [Species]
　N. nasua
Nasuella [Genus]
　N. olivacea [Species]
Procyon [Genus]
　P. cancrivorus [Species]
　P. gloveralleni
　P. insularis
　P. lotor
　P. maynardi
　P. minor
　P. pygmaeus

Mustelidae [Family]
Amblonyx [Genus]
　A. cinereus [Species]
Aonyx [Genus]
　A. capensis [Species]
　A. congicus
Arctonyx [Genus]
　A. collaris [Species]
Conepatus [Genus]
　C. chinga [Species]
　C. humboldtii
　C. leuconotus
　C. mesoleucus
　C. semistriatus
Eira [Genus]
　E. barbara [Species]
Enhydra [Genus]
　E. lutris [Species]
Galictis [Genus]
　G. cuja [Species]
　G. vittata
Gulo [Genus]
　G. gulo [Species]
Ictonyx [Genus]
　I. libyca [Species]
　I. striatus
Lontra [Genus]
　L. canadensis [Species]
　L. felina
　L. longicaudis
　L. provocax
Lutra [Genus]
　L. lutra [Species]
　L. maculicollis
　L. sumatrana
Lutrogale [Genus]
　L. perspicillata [Species]
Lyncodon [Genus]
　L. patagonicus [Species]
Martes [Genus]
　M. americana [Species]
　M. flavigula
　M. foina
　M. gwatkinsii
　M. martes
　M. melampus
　M. pennanti

　M. zibellina
Meles [Genus]
　M. meles [Species]
Mellivora [Genus]
　M. capensis [Species]
Melogale [Genus]
　M. everetti [Species]
　M. moschata
　M. orientalis
　M. personata
Mephitis [Genus]
　M. macroura [Species]
　M. mephitis
Mustela [Genus]
　M. africana [Species]
　M. altaica
　M. erminea
　M. eversmannii
　M. felipei
　M. frenata
　M. kathiah
　M. lutreola
　M. lutreolina
　M. nigripes
　M. nivalis
　M. nudipes
　M. putorius
　M. sibirica
　M. strigidorsa
　M. vison
Mydaus [Genus]
　M. javanensis [Species]
　M. marchei
Poecilogale [Genus]
　P. albinucha [Species]
Pteronura [Genus]
　P. brasiliensis [Species]
Spilogale [Genus]
　S. putorius [Species]
　S. pygmaea
Taxidea [Genus]
　T. taxus [Species]
Vormela [Genus]
　V. peregusna [Species]

Viverridae [Family]
Arctictis [Genus]
　A. binturong [Species]
Arctogalidia [Genus]
　A. trivirgata [Species]
Chrotogale [Genus]
　C. owstoni [Species]
Civettictis [Genus]
　C. civetta [Species]
Cryptoprocta [Genus]
　C. ferox [Species]
Cynogale [Genus]
　C. bennettii [Species]
Diplogale [Genus]
　D. hosei [Species]
Eupleres [Genus]

E. goudotii [Species]
Fossa [Genus]
   F. fossana [Species]
Genetta [Genus]
   G. abyssinica [Species]
   G. angolensis
   G. genetta
   G. johnstoni
   G. maculata
   G. servalina
   G. thierryi
   G. tigrina
   G. victoriae
Hemigalus [Genus]
   H. derbyanus [Species]
Nandinia [Genus]
   N. binotata [Species]
Macrogalidia [Genus]
   M. musschenbroekii [Species]
Paguma [Genus]
   P. larvata [Species]
Paradoxurus [Genus]
   P. hermaphroditus [Species]
   P. jerdoni
   P. zeylonensis
Osbornictis [Genus]
   O. piscivora [Species]
Poiana [Genus]
   P. richardsonii [Species]
Prionodon [Genus]
   P. linsang [Species]
   P. pardicolor
Viverra [Genus]
   V. civettina [Species]
   V. megaspila
   V. tangalunga
   V. zibetha
Viverricula [Genus]
   V. indica [Species]

Herpestidae [Family]
   Atilax [Genus]
   A. paludinosus [Species]
   Bdeogale [Genus]
   B. crassicauda [Species]
   B. jacksoni
   B. nigripes
   Crossarchus [Genus]
   C. alexandri [Species]
   C. ansorgei
   C. obscurus
   Cynictis [Genus]
   C. penicillata [Species]
   Dologale [Genus]
   D. dybowskii [Species]
   Galerella [Genus]
   G. flavescens [Species]
   G. pulverulenta
   G. sanguinea
   G. swalius
   Galidia [Genus]

G. elegans [Species]
Galidictis [Genus]
   G. fasciata [Species]
   G. grandidieri
Helogale [Genus]
   H. hirtula [Species]
   H. parvula
Herpestes [Genus]
   H. brachyurus [Species]
   H. edwardsii
   H. ichneumon
   H. javanicus
   H. naso
   H. palustris
   H. semitorquatus
   H. smithii
   H. urva
   H. vitticollis
Ichneumia [Genus]
   I. albicauda [Species]
Liberiictis [Genus]
   L. kuhni [Species]
Mungos [Genus]
   M. gambianus [Species]
   M. mungo
Mungotictis [Genus]
   M. decemlineata [Species]
Paracynictis [Genus]
   P. selousi [Species]
Rhynchogale [Genus]
   R. melleri [Species]
Salanoia [Genus]
   S. concolor [Species]
Suricata [Genus]
   S. suricatta [Species]

Hyaenidae [Family]
   Crocuta [Genus]
   C. crocuta [Species]
   Hyaena [Genus]
   H. hyaena [Species]
   Parahyaena [Genus]
   P. brunnea [Species]
   Proteles [Genus]
   P. cristatus [Species]

Felidae [Family]
   Acinonyx [Genus]
   A. jubatus [Species]
   Caracal [Genus]
   C. caracal [Species]
   Catopuma [Genus]
   C. badia [Species]
   C. temminckii
   Felis [Genus]
   F. bieti [Species]
   F. chaus
   F. margarita
   F. nigripes
   F. silvestris

Herpailurus [Genus]
   H. yaguarondi [Species]
Leopardus [Genus]
   L. pardalis [Species]
   L. tigrinus
   L. wiedii
Leptailurus [Genus]
   L. serval [Species]
Lynx [Genus]
   L. canadensis [Species]
   L. lynx
   L. pardinus
   L. rufus
Neofelis [Genus]
   N. nebulosa [Species]
Oncifelis [Genus]
   O. colocolo [Species]
   O. geoffroyi
   O. guigna
Oreailurus [Genus]
   O. jacobita [Species]
Otocolobus [Genus]
   O. manul [Species]
Panthera [Genus]
   P. leo [Species]
   P. onca
   P. pardus
   P. tigris
   Pardofelis
   P. marmorata
Prionailurus [Genus]
   P. bengalensis [Species]
   P. planiceps
   P. rubiginosus
   P. viverrinus
Profelis [Genus]
   P. aurata [Species]
Puma [Genus]
   P. concolor [Species]
Uncia [Genus]
   U. uncia [Species]

Otariidae [Family]
   Arctocephalus [Genus]
   A. australis [Species]
   A. forsteri
   A. galapagoensis
   A. gazella
   A. philippii
   A. pusillus
   A. townsendi
   A. tropicalis
   Callorhinus [Genus]
   C. ursinus [Species]
   Eumetopias [Genus]
   E. jubatus [Species]
   Neophoca [Genus]
   N. cinerea [Species]
   Otaria [Genus]
   O. byronia [Species]

Phocarctos [Genus]
   P. hookeri [Species]
Zalophus [Genus]
   Z. californianus [Species]

Odobenidae [Family]
   Odobenus [Genus]
      O. rosmarus [Species]

Phocidae [Family]
   Cystophora [Genus]
      C. cristata [Species]
   Erignathus [Genus]
      E. barbatus [Species]
   Halichoerus [Genus]
      H. grypus [Species]
   Hydrurga [Genus]
      H. leptonyx [Species]
   Leptonychotes [Genus]
      L. weddellii [Species]
   Lobodon [Genus]
      L. carcinophagus [Species]
   Mirounga [Genus]
      M. angustirostris [Species]
      M. leonina
   Monachus [Genus]
      M. monachus [Species]
      M. schauinslandi
      M. tropicalis
   Ommatophoca [Genus]
      O. rossii [Species]
   Phoca [Genus]
      P. caspica [Species]
      P. fasciata
      P. groenlandica
      P. hispida
      P. largha
      P. sibirica
      P. vitulina

Cetacea [Order]

Platanistidae [Family]
   Platanista [Genus]
      P. gangetica [Species]
      P. minor

Lipotidae [Family]
   Lipotes [Genus]
      L. vexillifer [Species]

Pontoporiidae [Family]
   Pontoporia [Genus]
      P. blainvillei [Species]

Iniidae [Family]
   Inia [Genus]
      I. geoffrensis [Species]

Phocoenidae [Family]
   Australophocaena [Genus]
      A. dioptrica [Species]
   Neophocaena [Genus]

      N. phocaenoides [Species]
   Phocoena [Genus]
      P. phocoena [Species]
      P. sinus
      P. spinipinnis
   Phocoenoides [Genus]
      P. dalli [Species]

Delphinidae [Family]
   Cephalorhynchus [Genus]
      C. commersonii [Species]
      C. eutropia
      C. heavisidii
      C. hectori
   Delphinus [Genus]
      D. delphis [Species]
   Feresa [Genus]
      F. attenuata [Species]
   Globicephala [Genus]
      G. macrorhynchus [Species]
      G. melas
   Grampus [Genus]
      G. griseus [Species]
   Lagenodelphis [Genus]
      L. hosei [Species]
   Lagenorhynchus [Genus]
      L. acutus [Species]
      L. albirostris
      L. australis
      L. cruciger
      L. obliquidens
      L. obscurus
   Lissodelphis [Genus]
      L. borealis [Species]
      L. peronii
   Orcaella [Genus]
      O. brevirostris [Species]
   Orcinus [Genus]
      O. orca [Species]
   Peponocephala [Genus]
      P. electra [Species]
   Pseudorca [Genus]
      P. crassidens [Species]
   Sotalia [Genus]
      S. fluviatilis [Species]
   Sousa [Genus]
      S. chinensis [Species]
      S. teuszii
   Stenella [Genus]
      S. attenuata [Species]
      S. clymene
      S. coeruleoalba
      S. frontalis
      S. longirostris
   Steno [Genus]
      S. bredanensis [Species]
   Tursiops [Genus]
      T. truncatus [Species]

Ziphiidae [Family]
   Berardius [Genus]

      B. arnuxii [Species]
      B. bairdii
   Hyperoodon [Genus]
      H. ampullatus [Species]
      H. planifrons
   Indopacetus [Genus]
      I. pacificus [Species]
   Mesoplodon [Genus]
      M. bidens [Species]
      M. bowdoini
      M. carlhubbsi
      M. densirostris
      M. europaeus
      M. ginkgodens
      M. grayi
      M. hectori
      M. layardii
      M. mirus
      M. peruvianus
      M. stejnegeri
   Tasmacetus [Genus]
      T. shepherdi [Species]
   Ziphius [Genus]
      Z. cavirostris [Species]

Physeteridae [Family]
   Kogia [Genus]
      K. breviceps [Species]
      K. simus
   Physeter [Genus]
      P. catodon [Species]

Monodontidae [Family]
   Delphinapterus [Genus]
      D. leucas [Species]
   Monodon [Genus]
      M. monoceros [Species]

Eschrichtiidae [Family]
   Eschrichtius [Genus]
      E. robustus [Species]

Neobalaenidae [Family]
   Caperea [Genus]
      C. marginata [Species]

Balaenidae [Family]
   Balaena [Genus]
      B. mysticetus [Species]
   Eubalaena [Genus]
      E. australis [Species]
      E. glacialis

Balaenopteridae [Family]
   Balaenoptera [Genus]
      B. acutorostrata [Species]
      B. borealis
      B. edeni
      B. musculus
      B. physalus
   Megaptera [Genus]
      M. novaeangliae [Species]

Tubulidentata [Order]

Orycteropodidae [Family]
Orycteropus [Genus]
O. afer [Species]

Proboscidea [Order]

Elephantidae [Family]
Elephas [Genus]
E. maximus [Species]
Loxodonta [Genus]
L. africana [Species]
L. cyclotis

Hyracoidea [Order]

Procaviidae [Family]
Dendrohyrax [Genus]
D. arboreus [Species]
D. dorsalis
D. validus
Heterohyrax [Genus]
H. antineae [Species]
H. brucei
Procavia [Genus]
P. capensis [Species]

Sirenia [Order]

Dugongidae [Family]
Dugong [Genus]
D. dugon [Species]
Hydrodamalis [Genus]
H. gigas [Species]

Trichechidae [Family]
Trichechus [Genus]
T. inunguis [Species]
T. manatus
T. senegalensis

Perissodactyla [Order]

Equidae [Family]
Equus [Genus]
E. asinus [Species]
E. burchellii
E. caballus
E. grevyi
E. hemionus
E. kiang
E. onager
E. quagga
E. zebra

Tapiridae [Family]
Tapirus [Genus]
T. bairdii [Species]
T. indicus
T. pinchaque
T. terrestris

Rhinocerotidae [Family]
Ceratotherium [Genus]
C. simum [Species]

Dicerorhinus [Genus]
D. sumatrensis [Species]
Diceros [Genus]
D. bicornis [Species]
Rhinoceros [Genus]
R. sondaicus [Species]
R. unicornis

Artiodactyla [Order]

Suidae [Family]
Babyrousa [Genus]
B. babyrussa [Species]
Phacochoerus [Genus]
P. aethiopicus [Species]
P. africanus
Hylochoerus [Genus]
H. meinertzhageni [Species]
Potamochoerus [Genus]
P. larvatus [Species]
P. porcus
Sus [Genus]
S. barbatus [Species]
S. bucculentus
S. cebifrons
S. celebensis
S. heureni
S. philippensis
S. salvanius
S. scrofa
S. timoriensis
S. verrucosus

Tayassuidae [Family]
Catagonus [Genus]
C. wagneri [Species]
Pecari [Genus]
P. tajacu [Species]
Tayassu [Genus]
T. pecari [Species]

Hippopotamidae [Family]
Hexaprotodon [Genus]
H. liberiensis [Species]
H. madagascariensis
Hippopotamus [Genus]
H. amphibius [Species]
H. lemerlei

Camelidae [Family]
Camelus [Genus]
C. bactrianus [Species]
C. dromedarius
Lama [Genus]
L. glama [Species]
L. guanicoe
L. pacos
Vicugna [Genus]
V. vicugna [Species]

Tragulidae [Family]
Hyemoschus [Genus]

H. aquaticus [Species]
Moschiola [Genus]
M. meminna [Species]
Tragulus [Genus]
T. javanicus [Species]
T. napu

Cervidae [Family]
Alces [Genus]
A. alces [Species]
Axis [Genus]
A. axis [Species]
A. calamianensis
A. kuhlii
A. porcinus
Blastocerus [Genus]
B. dichotomus [Species]
Capreolus [Genus]
C. capreolus [Species]
C. pygargus
Cervus [Genus]
C. albirostris [Species]
C. alfredi
C. duvaucelii
C. elaphus
C. eldii
C. mariannus
C. nippon
C. schomburgki
C. timorensis
C. unicolor
Dama [Genus]
D. dama [Species]
D. mesopotamica
Elaphodus [Genus]
E. cephalophus [Species]
Elaphurus [Genus]
E. davidianus [Species]
Hippocamelus [Genus]
H. antisensis [Species]
H. bisulcus
Hydropotes [Genus]
H. inermis [Species]
Mazama [Genus]
M. americana [Species]
M. bricenii
M. chunyi
M. gouazoupira
M. nana
M. rufina
Moschus [Genus]
M. berezovskii [Species]
M. chrysogaster
M. fuscus
M. moschiferus
Muntiacus [Genus]
M. atherodes [Species]
M. crinifrons
M. feae
M. gongshanensis

M. muntjak
M. reevesi
Odocoileus [Genus]
   O. hemionus [Species]
   O. virginianus
Ozotoceros [Genus]
   O. bezoarticus [Species]
Pudu [Genus]
   P. mephistophiles [Species]
   P. puda
Rangifer [Genus]
   R. tarandus [Species]

Giraffidae [Family]
  Giraffa [Genus]
    G. camelopardalis [Species]
  Okapia [Genus]
    O. johnstoni [Species]

Antilocapridae [Family]
  Antilocapra [Genus]
    A. americana [Species]

Bovidae [Family]
  Addax [Genus]
    A. nasomaculatus [Species]
  Aepyceros [Genus]
    A. melampus [Species]
  Alcelaphus [Genus]
    A. buselaphus [Species]
  Ammodorcas [Genus]
    A. clarkei [Species]
  Ammotragus [Genus]
    A. lervia [Species]
  Antidorcas [Genus]
    A. marsupialis [Species]
  Antilope [Genus]
    A. cervicapra [Species]
  Bison [Genus]
    B. bison [Species]
    B. bonasus
  Bos [Genus]
    B. frontalis [Species]
    B. grunniens
    B. javanicus
    B. sauveli
    B. taurus
  Boselaphus [Genus]
    B. tragocamelus [Species]
  Bubalus [Genus]
    B. bubalis [Species]
    B. depressicornis
    B. mephistopheles
    B. mindorensis
    B. quarlesi
  Budorcas [Genus]
    B. taxicolor [Species]
  Capra [Genus]
    C. caucasica [Species]
    C. cylindricornis
    C. falconeri
    C. hircus

C. ibex
C. nubiana
C. pyrenaica
C. sibirica
C. walie
Cephalophus [Genus]
  C. adersi [Species]
  C. callipygus
  C. dorsalis
  C. harveyi
  C. jentinki
  C. leucogaster
  C. maxwellii
  C. monticola
  C. natalensis
  C. niger
  C. nigrifrons
  C. ogilbyi
  C. rubidus
  C. rufilatus
  C. silvicultor
  C. spadix
  C. weynsi
  C. zebra
Connochaetes [Genus]
  C. gnou [Species]
  C. taurinus
Damaliscus [Genus]
  D. hunteri [Species]
  D. lunatus
  D. pygargus
Dorcatragus [Genus]
  D. megalotis [Species]
Gazella [Genus]
  G. arabica [Species]
  G. bennettii
  G. bilkis
  G. cuvieri
  G. dama
  G. dorcas
  G. gazella
  G. granti
  G. leptoceros
  G. rufifrons
  G. rufina
  G. saudiya
  G. soemmerringii
  G. spekei
  G. subgutturosa
  G. thomsonii
Hemitragus [Genus]
  H. hylocrius [Species]
  H. jayakari
  H. jemlahicus
Hippotragus [Genus]
  H. equinus [Species]
  H. leucophaeus
  H. niger
Kobus [Genus]
  K. ellipsiprymnus [Species]

K. kob
K. leche
K. megaceros
K. vardonii
Litocranius [Genus]
  L. walleri [Species]
Madoqua [Genus]
  M. guentheri [Species]
  M. kirkii
  M. piacentinii
  M. saltiana
Naemorhedus [Genus]
  N. baileyi [Species]
  N. caudatus
  N. crispus
  N. goral
  N. sumatraensis
  N. swinhoei
Neotragus [Genus]
  N. batesi [Species]
  N. moschatus
  N. pygmaeus
Oreamnos [Genus]
  O. americanus [Species]
Oreotragus [Genus]
  O. oreotragus [Species]
Oryx [Genus]
  O. dammah [Species]
  O. gazella
  O. leucoryx
Ourebia [Genus]
  O. ourebi [Species]
Ovibos [Genus]
  O. moschatus [Species]
Ovis [Genus]
  O. ammon [Species]
  O. aries
  O. canadensis
  O. dalli
  O. nivicola
  O. vignei
Pantholops [Genus]
  P. hodgsonii [Species]
Pelea [Genus]
  P. capreolus [Species]
Procapra [Genus]
  P. gutturosa [Species]
  P. picticaudata
  P. przewalskii
Pseudois [Genus]
  P. nayaur [Species]
  P. schaeferi
Raphicerus [Genus]
  R. campestris [Species]
  R. melanotis
  R. sharpei
Redunca [Genus]
  R. arundinum [Species]
  R. fulvorufula
  R. redunca

Rupicapra [Genus]
  R. pyrenaica [Species]
  R. rupicapra
Saiga [Genus]
  S. tatarica [Species]
Sigmoceros [Genus]
  S. lichtensteinii [Species]
Sylvicapra [Genus]
  S. grimmia [Species]
Syncerus [Genus]
  S. caffer [Species]
Taurotragus [Genus]
  T. derbianus [Species]
  T. oryx
Tetracerus [Genus]
  T. quadricornis [Species]
Tragelaphus [Genus]
  T. angasii [Species]
  T. buxtoni
  T. eurycerus
  T. imberbis
  T. scriptus
  T. spekii
  T. strepsiceros

Pholidota [Order]

Manidae [Family]
  Manis [Genus]
    M. crassicaudata [Species]
    M. gigantea
    M. javanica
    M. pentadactyla
    M. temminckii
    M. tetradactyla
    M. tricuspis

Rodentia [Order]

Aplodontidae [Family]
  Aplodontia [Genus]
    A. rufa [Species]

Sciuridae [Family]
  Aeretes [Genus]
    A. melanopterus [Species]
  Aeromys [Genus]
    A. tephromelas [Species]
    A. thomasi
  Ammospermophilus [Genus]
    A. harrisii [Species]
    A. insularis
    A. interpres
    A. leucurus
    A. nelsoni
  Atlantoxerus [Genus]
    A. getulus [Species]
  Belomys [Genus]
    B. pearsonii [Species]
  Biswamoyopterus [Genus]
    B. biswasi [Species]

Callosciurus [Genus]
  C. adamsi [Species]
  C. albescens
  C. baluensis
  C. caniceps
  C. erythraeus
  C. finlaysonii
  C. inornatus
  C. melanogaster
  C. nigrovittatus
  C. notatus
  C. orestes
  C. phayrei
  C. prevostii
  C. pygerythrus
  C. quinquestriatus
Cynomys [Genus]
  C. gunnisoni [Species]
  C. leucurus
  C. ludovicianus
  C. mexicanus
  C. parvidens
Dremomys [Genus]
  D. everetti [Species]
  D. lokriah
  D. pernyi
  D. pyrrhomerus
  D. rufigenis
Epixerus [Genus]
  E. ebii [Species]
  E. wilsoni
Eupetaurus [Genus]
  E. cinereus [Species]
Exilisciurus [Genus]
  E. concinnus [Species]
  E. exilis
  E. whiteheadi
Funambulus [Genus]
  F. layardi [Species]
  F. palmarum
  F. pennantii
  F. sublineatus
  F. tristriatus
Funisciurus [Genus]
  F. anerythrus [Species]
  F. bayonii
  F. carruthersi
  F. congicus
  F. isabella
  F. lemniscatus
  F. leucogenys
  F. pyrropus
  F. substriatus
Glaucomys [Genus]
  G. sabrinus [Species]
  G. volans
Glyphotes [Genus]
  G. simus [Species]
Heliosciurus [Genus]
  H. gambianus [Species]

H. mutabilis
H. punctatus
H. rufobrachium
H. ruwenzorii
H. undulatus
Hylopetes [Genus]
  H. alboniger [Species]
  H. baberi
  H. bartelsi
  H. fimbriatus
  H. lepidus
  H. nigripes
  H. phayrei
  H. sipora
  H. spadiceus
  H. winstoni
Hyosciurus [Genus]
  H. heinrichi [Species]
  H. ileile
Iomys [Genus]
  I. horsfieldi [Species]
  I. sipora
Lariscus [Genus]
  L. hosei [Species]
  L. insignis
  L. niobe
  L. obscurus
Marmota [Genus]
  M. baibacina [Species]
  M. bobak
  M. broweri
  M. caligata
  M. camtschatica
  M. caudata
  M. flaviventris
  M. himalayana
  M. marmota
  M. menzbieri
  M. monax
  M. olympus
  M. sibirica
  M. vancouverensis
Menetes [Genus]
  M. berdmorei [Species]
Microsciurus [Genus]
  M. alfari [Species]
  M. flaviventer
  M. mimulus
  M. santanderensis
Myosciurus [Genus]
  M. pumilio [Species]
Nannosciurus [Genus]
  N. melanotis [Species]
Paraxerus [Genus]
  P. alexandri [Species]
  P. boehmi
  P. cepapi
  P. cooperi
  P. flavovittis
  P. lucifer

P. ochraceus
P. palliatus
P. poensis
P. vexillarius
P. vincenti
Petaurillus [Genus]
P. emiliae [Species]
P. hosei
P. kinlochii
Petaurista [Genus]
P. alborufus [Species]
P. elegans
P. leucogenys
P. magnificus
P. nobilis
P. petaurista
P. philippensis
P. xanthotis
Petinomys [Genus]
P. crinitus [Species]
P. fuscocapillus
P. genibarbis
P. hageni
P. lugens
P. sagitta
P. setosus
P. vordermanni
Prosciurillus [Genus]
P. abstrusus [Species]
P. leucomus
P. murinus
P. weberi
Protoxerus [Genus]
P. aubinnii [Species]
P. stangeri
Pteromys [Genus]
P. momonga [Species]
P. volans
Pteromyscus [Genus]
P. pulverulentus [Species]
Ratufa [Genus]
R. affinis [Species]
R. bicolor
R. indica
R. macroura
Rheithrosciurus [Genus]
R. macrotis [Species]
Rhinosciurus [Genus]
R. laticaudatus [Species]
Rubrisciurus [Genus]
R. rubriventer [Species]
Sciurillus [Genus]
S. pusillus [Species]
Sciurotamias [Genus]
S. davidianus [Species]
S. forresti
Sciurus [Genus]
S. aberti [Species]
S. aestuans
S. alleni

S. anomalus
S. arizonensis
S. aureogaster
S. carolinensis
S. colliaei
S. deppei
S. flammifer
S. gilvigularis
S. granatensis
S. griseus
S. ignitus
S. igniventris
S. lis
S. nayaritensis
S. niger
S. oculatus
S. pucheranii
S. pyrrhinus
S. richmondi
S. sanborni
S. spadiceus
S. stramineus
S. variegatoides
S. vulgaris
S. yucatanensis
Spermophilopsis [Genus]
S. leptodactylus [Species]
Spermophilus [Genus]
S. adocetus [Species]
S. alashanicus
S. annulatus
S. armatus
S. atricapillus
S. beecheyi
S. beldingi
S. brunneus
S. canus
S. citellus
S. columbianus
S. dauricus
S. elegans
S. erythrogenys
S. franklinii
S. fulvus
S. lateralis
S. madrensis
S. major
S. mexicanus
S. mohavensis
S. mollis
S. musicus
S. parryii
S. perotensis
S. pygmaeus
S. relictus
S. richardsonii
S. saturatus
S. spilosoma
S. suslicus
S. tereticaudus

S. townsendii
S. tridecemlineatus
S. undulatus
S. variegatus
S. washingtoni
S. xanthoprymnus
Sundasciurus [Genus]
S. brookei [Species]
S. davensis
S. fraterculus
S. hippurus
S. hoogstraali
S. jentinki
S. juvencus
S. lowii
S. mindanensis
S. moellendorffi
S. philippinensis
S. rabori
S. samarensis
S. steerii
S. tenuis
Syntheosciurus [Genus]
S. brochus [Species]
Tamias [Genus]
T. alpinus [Species]
T. amoenus
T. bulleri
T. canipes
T. cinereicollis
T. dorsalis
T. durangae
T. merriami
T. minimus
T. obscurus
T. ochrogenys
T. palmeri
T. panamintinus
T. quadrimaculatus
T. quadrivittatus
T. ruficaudus
T. rufus
T. senex
T. sibiricus
T. siskiyou
T. sonomae
T. speciosus
T. striatus
T. townsendii
T. umbrinus
Tamiasciurus [Genus]
T. douglasii [Species]
T. hudsonicus
T. mearnsi
Tamiops [Genus]
T. macclellandi [Species]
T. maritimus
T. rodolphei
T. swinhoei
Trogopterus [Genus]

T. xanthipes [Species]
Xerus [Genus]
  X. erythropus [Species]
  X. inauris
  X. princeps
  X. rutilus

Castoridae [Family]
Castor [Genus]
  C. canadensis [Species]
  C. fiber

Geomyidae [Family]
Geomys [Genus]
  G. arenarius [Species]
  G. bursarius
  G. personatus
  G. pinetis
  G. tropicalis
Orthogeomys [Genus]
  O. cavator [Species]
  O. cherriei
  O. cuniculus
  O. dariensis
  O. grandis
  O. heterodus
  O. hispidus
  O. lanius
  O. matagalpae
  O. thaeleri
  O. underwoodi
Pappogeomys [Genus]
  P. alcorni [Species]
  P. bulleri
  P. castanops
  P. fumosus
  P. gymnurus
  P. merriami
  P. neglectus
  P. tylorhinus
  P. zinseri
Thomomys [Genus]
  T. bottae [Species]
  T. bulbivorus
  T. clusius
  T. idahoensis
  T. mazama
  T. monticola
  T. talpoides
  T. townsendii
  T. umbrinus
Zygogeomys [Genus]
  Z. trichopus [Species]

Heteromyidae [Family]
Chaetodipus [Genus]
  C. arenarius [Species]
  C. artus
  C. baileyi
  C. californicus

C. fallax
C. formosus
C. goldmani
C. hispidus
C. intermedius
C. lineatus
C. nelsoni
C. penicillatus
C. pernix
C. spinatus
Dipodomys [Genus]
  D. agilis [Species]
  D. californicus
  D. compactus
  D. deserti
  D. elator
  D. elephantinus
  D. gravipes
  D. heermanni
  D. ingens
  D. insularis
  D. margaritae
  D. merriami
  D. microps
  D. nelsoni
  D. nitratoides
  D. ordii
  D. panamintinus
  D. phillipsii
  D. spectabilis
  D. stephensi
  D. venustus
Microdipodops [Genus]
  M. megacephalus [Species]
  M. pallidus
Heteromys [Genus]
  H. anomalus [Species]
  H. australis
  H. desmarestianus
  H. gaumeri
  H. goldmani
  H. nelsoni
  H. oresterus
Liomys [Genus]
  L. adspersus [Species]
  L. irroratus
  L. pictus
  L. salvini
  L. spectabilis
Perognathus [Genus]
  P. alticola [Species]
  P. amplus
  P. fasciatus
  P. flavescens
  P. flavus
  P. inornatus
  P. longimembris
  P. merriami
  P. parvus
  P. xanthanotus

Dipodidae [Family]
Allactaga [Genus]
  A. balikunica [Species]
  A. bullata
  A. elater
  A. euphratica
  A. firouzi
  A. hotsoni
  A. major
  A. severtzovi
  A. sibirica
  A. tetradactyla
  A. vinogradovi
Allactodipus [Genus]
  A. bobrinskii [Species]
Cardiocranius [Genus]
  C. paradoxus [Species]
Dipus [Genus]
  D. sagitta [Species]
Eozapus [Genus]
  E. setchuanus [Species]
Eremodipus [Genus]
  E. lichtensteini [Species]
Euchoreutes [Genus]
  E. naso [Species]
Jaculus [Genus]
  J. blanfordi [Species]
  J. jaculus
  J. orientalis
  J. turcmenicus
Napaeozapus [Genus]
  N. insignis [Species]
Paradipus [Genus]
  P. ctenodactylus [Species]
Pygeretmus [Genus]
  P. platyurus [Species]
  P. pumilio
  P. shitkovi
Salpingotus [Genus]
  S. crassicauda [Species]
  S. heptneri
  S. kozlovi
  S. michaelis
  S. pallidus
  S. thomasi
Sicista [Genus]
  S. armenica [Species]
  S. betulina
  S. caucasica
  S. caudata
  S. concolor
  S. kazbegica
  S. kluchorica
  S. napaea
  S. pseudonapaea
  S. severtzovi
  S. strandi
  S. subtilis
  S. tianshanica
Stylodipus [Genus]

MAMMALS SPECIES LIST

S. *andrewsi* [Species]
S. *sungorus*
S. *telum*
Zapus [Genus]
    Z. *hudsonius* [Species]
    Z. *princeps*
    Z. *trinotatus*

Muridae [Family]
  Abditomys [Genus]
    A. *latidens* [Species]
  Abrawayaomys [Genus]
    A. *ruschii* [Species]
  Acomys [Genus]
    A. *cahirinus* [Species]
    A. *cilicicus*
    A. *cinerasceus*
    A. *ignitus*
    A. *kempi*
    A. *louisae*
    A. *minous*
    A. *mullah*
    A. *nesiotes*
    A. *percivali*
    A. *russatus*
    A. *spinosissimus*
    A. *subspinosus*
    A. *wilsoni*
  Aepeomys [Genus]
    A. *fuscatus* [Species]
    A. *lugens*
  Aethomys [Genus]
    A. *bocagei* [Species]
    A. *chrysophilus*
    A. *granti*
    A. *hindei*
    A. *kaiseri*
    A. *namaquensis*
    A. *nyikae*
    A. *silindensis*
    A. *stannarius*
    A. *thomasi*
  Akodon [Genus]
    A. *aerosus* [Species]
    A. *affinis*
    A. *albiventer*
    A. *azarae*
    A. *bogotensis*
    A. *boliviensis*
    A. *budini*
    A. *cursor*
    A. *dayi*
    A. *dolores*
    A. *fumeus*
    A. *hershkovitzi*
    A. *illuteus*
    A. *iniscatus*
    A. *juninensis*
    A. *kempi*
    A. *kofordi*
    A. *lanosus*

A. *latebricola*
A. *lindberghi*
A. *longipilis*
A. *mansoensis*
A. *markhami*
A. *mimus*
A. *molinae*
A. *mollis*
A. *neocenus*
A. *nigrita*
A. *olivaceus*
A. *orophilus*
A. *puer*
A. *sanborni*
A. *sanctipaulensis*
A. *serrensis*
A. *siberiae*
A. *simulator*
A. *spegazzinii*
A. *subfuscus*
A. *surdus*
A. *sylvanus*
A. *toba*
A. *torques*
A. *urichi*
A. *varius*
A. *xanthorhinus*
Allocricetulus [Genus]
  A. *curtatus* [Species]
  A. *eversmanni*
Alticola [Genus]
  A. *albicauda* [Species]
  A. *argentatus*
  A. *barakshin*
  A. *lemminus*
  A. *macrotis*
  A. *montosa*
  A. *roylei*
  A. *semicanus*
  A. *stoliczkanus*
  A. *stracheyi*
  A. *strelzowi*
  A. *tuvinicus*
Ammodillus [Genus]
  A. *imbellis* [Species]
Andalgalomys [Genus]
  A. *olrogi* [Species]
  A. *pearsoni*
Andinomys [Genus]
  A. *edax* [Species]
Anisomys [Genus]
  A. *imitator* [Species]
Anonymomys [Genus]
  A. *mindorensis* [Species]
Anotomys [Genus]
  A. *leander* [Species]
Apodemus [Genus]
  A. *agrarius* [Species]
  A. *alpicola*
  A. *argenteus*

A. *arianus*
A. *chevrieri*
A. *draco*
A. *flavicollis*
A. *fulvipectus*
A. *gurkha*
A. *hermonensis*
A. *hyrcanicus*
A. *latronum*
A. *mystacinus*
A. *peninsulae*
A. *ponticus*
A. *rusiges*
A. *semotus*
A. *speciosus*
A. *sylvaticus*
A. *uralensis*
A. *wardi*
Apomys [Genus]
  A. *abrae* [Species]
  A. *datae*
  A. *hylocoetes*
  A. *insignis*
  A. *littoralis*
  A. *microdon*
  A. *musculus*
  A. *sacobianus*
Arborimus [Genus]
  A. *albipes* [Species]
  A. *longicaudus*
  A. *pomo*
Archboldomys [Genus]
  A. *luzonensis* [Species]
Arvicanthis [Genus]
  A. *abyssinicus* [Species]
  A. *blicki*
  A. *nairobae*
  A. *niloticus*
  A. *somalicus*
Arvicola [Genus]
  A. *sapidus* [Species]
  A. *terrestris*
Auliscomys [Genus]
  A. *boliviensis* [Species]
  A. *micropus*
  A. *pictus*
  A. *sublimis*
Baiomys [Genus]
  B. *musculus* [Species]
  B. *taylori*
Bandicota [Genus]
  B. *bengalensis* [Species]
  B. *indica*
  B. *savilei*
Batomys [Genus]
  B. *dentatus* [Species]
  B. *granti*
  B. *salomonseni*
Beamys [Genus]
  B. *hindei* [Species]

B. major
Berylmys [Genus]
   B. berdmorei [Species]
   B. bowersi
   B. mackenziei
   B. manipulus
Bibimys [Genus]
   B. chacoensis [Species]
   B. labiosus
   B. torresi
Blanfordimys [Genus]
   B. afghanus [Species]
   B. bucharicus
Blarinomys [Genus]
   B. breviceps [Species]
Bolomys [Genus]
   B. amoenus [Species]
   B. lactens
   B. lasiurus
   B. obscurus
   B. punctulatus
   B. temchuki
Brachiones [Genus]
   B. przewalskii [Species]
Brachytarsomys [Genus]
   B. albicauda [Species]
Brachyuromys [Genus]
   B. betsileoensis [Species]
   B. ramirohitra
Bullimus [Genus]
   B. bagobus [Species]
   B. luzonicus
Bunomys [Genus]
   B. andrewsi [Species]
   B. chrysocomus
   B. coelestis
   B. fratrorum
   B. heinrichi
   B. penitus
   B. prolatus
Calomys [Genus]
   C. boliviae [Species]
   C. callidus
   C. callosus
   C. hummelincki
   C. laucha
   C. lepidus
   C. musculinus
   C. sorellus
   C. tener
Calomyscus [Genus]
   C. bailwardi [Species]
   C. baluchi
   C. hotsoni
   C. mystax
   C. tsolovi
   C. urartensis
Canariomys [Genus]
   C. tamarani [Species]
Cannomys [Genus]

C. badius [Species]
Cansumys [Genus]
   C. canus [Species]
Carpomys [Genus]
   C. melanurus [Species]
   C. phaeurus
Celaenomys [Genus]
   C. silaceus [Species]
Chelemys [Genus]
   C. macronyx [Species]
   C. megalonyx
Chibchanomys [Genus]
   C. trichotis [Species]
Chilomys [Genus]
   C. instans [Species]
Chiromyscus [Genus]
   C. chiropus [Species]
Chinchillula [Genus]
   C. sahamae [Species]
Chionomys [Genus]
   C. gud [Species]
   C. nivalis
   C. roberti
Chiropodomys [Genus]
   C. calamianensis [Species]
   C. gliroides
   C. karlkoopmani
   C. major
   C. muroides
   C. pusillus
Chiruromys [Genus]
   C. forbesi [Species]
   C. lamia
   C. vates
Chroeomys [Genus]
   C. andinus [Species]
   C. jelskii
Chrotomys [Genus]
   C. gonzalesi [Species]
   C. mindorensis
   C. whiteheadi
Clethrionomys [Genus]
   C. californicus [Species]
   C. centralis
   C. gapperi
   C. glareolus
   C. rufocanus
   C. rutilus
   C. sikotanensis
Coccymys [Genus]
   C. albidens [Species]
   C. ruemmleri
Colomys [Genus]
   C. goslingi [Species]
Conilurus [Genus]
   C. albipes [Species]
   C. penicillatus
Coryphomys [Genus]
   C. buhleri [Species]
Crateromys [Genus]

C. australis [Species]
   C. paulus
   C. schadenbergi
Cremnomys [Genus]
   C. blanfordi [Species]
   C. cutchicus
   C. elvira
Cricetomys [Genus]
   C. emini [Species]
   C. gambianus
Cricetulus [Genus]
   C. alticola [Species]
   C. barabensis
   C. kamensis
   C. longicaudatus
   C. migratorius
   C. sokolovi
Cricetus [Genus]
   C. cricetus [Species]
Crossomys [Genus]
   C. moncktoni [Species]
Crunomys [Genus]
   C. celebensis [Species]
   C. fallax
   C. melanius
   C. rabori
Dacnomys [Genus]
   D. millardi [Species]
Dasymys [Genus]
   D. foxi [Species]
   D. incomtus
   D. montanus
   D. nudipes
   D. rufulus
Delanymys [Genus]
   D. brooksi [Species]
Delomys [Genus]
   D. dorsalis [Species]
   D. sublineatus
Dendromus [Genus]
   D. insignis [Species]
   D. kahuziensis
   D. kivu
   D. lovati
   D. melanotis
   D. mesomelas
   D. messorius
   D. mystacalis
   D. nyikae
   D. oreas
   D. vernayi
Dendroprionomys [Genus]
   D. rousseloti [Species]
Deomys [Genus]
   D. ferrugineus [Species]
Dephomys [Genus]
   D. defua [Species]
   D. eburnea
Desmodilliscus [Genus]
   D. braueri [Species]

Desmodillus [Genus]
   D. auricularis [Species]
Dicrostonyx [Genus]
   D. exsul [Species]
   D. groenlandicus
   D. hudsonius
   D. kilangmiutak
   D. nelsoni
   D. nunatakensis
   D. richardsoni
   D. rubricatus
   D. torquatus
   D. unalascensis
   D. vinogradovi
Desmomys [Genus]
   D. harringtoni [Species]
Dinaromys [Genus]
   D. bogdanovi [Species]
Diomys [Genus]
   D. crumpi [Species]
Diplothrix [Genus]
   D. legatus [Species]
Echiothrix [Genus]
   E. leucura [Species]
Eropeplus [Genus]
   E. canus [Species]
Eligmodontia [Genus]
   E. moreni [Species]
   E. morgani
   E. puerulus
   E. typus
Eliurus [Genus]
   E. majori [Species]
   E. minor
   E. myoxinus
   E. penicillatus
   E. tanala
   E. webbi
Ellobius [Genus]
   E. alaicus [Species]
   E. fuscocapillus
   E. lutescens
   E. talpinus
   E. tancrei
Eolagurus [Genus]
   E. luteus [Species]
   E. przewalskii
Eothenomys [Genus]
   E. chinensis [Species]
   E. custos
   E. eva
   E. inez
   E. melanogaster
   E. olitor
   E. proditor
   E. regulus
   E. shanseius
Euneomys [Genus]
   E. chinchilloides [Species]
   E. fossor

   E. mordax
   E. petersoni
Galenomys [Genus]
   G. garleppi [Species]
Geoxus [Genus]
   G. valdivianus [Species]
Gerbillurus [Genus]
   G. paeba [Species]
   G. setzeri
   G. tytonis
   G. vallinus
Gerbillus [Genus]
   G. acticola [Species]
   G. allenbyi
   G. andersoni
   G. bilensis
   G. bottai
   G. burtoni
   G. cheesmani
   G. dalloni
   G. diminutus
   G. dunni
   G. floweri
   G. gerbillus
   G. grobbeni
   G. henleyi
   G. hoogstraali
   G. juliani
   G. lowei
   G. maghrebi
   G. mesopotamiae
   G. nancillus
   G. nigeriae
   G. percivali
   G. poecilops
   G. pulvinatus
   G. pyramidum
   G. riggenbachi
   G. ruberrimus
   G. somalicus
   G. syrticus
   G. vivax
Golunda [Genus]
   G. ellioti [Species]
Grammomys [Genus]
   G. aridulus [Species]
   G. caniceps
   G. dolichurus
   G. gigas
   G. macmillani
   G. rutilans
Graomys [Genus]
   G. domorum [Species]
   G. griseoflavus
Gymnuromys [Genus]
   G. roberti [Species]
Habromys [Genus]
   H. chinanteco [Species]
   H. lepturus
   H. lophurus

   H. simulatus
Hadromys [Genus]
   H. humei [Species]
Haeromys [Genus]
   H. margarettae [Species]
   H. minahassae
   H. pusillus
Hapalomys [Genus]
   H. delacouri [Species]
   H. longicaudatus
Heimyscus [Genus]
   H. fumosus [Species]
Hodomys [Genus]
   H. alleni [Species]
Holochilus [Genus]
   H. brasiliensis [Species]
   H. chacarius
   H. magnus
   H. sciureus
Hybomys [Genus]
   H. basilii [Species]
   H. eisentrauti
   H. lunaris
   H. planifrons
   H. trivirgatus
   H. univittatus
Hydromys [Genus]
   H. chrysogaster [Species]
   H. habbema
   H. hussoni
   H. neobrittanicus
   H. shawmayeri
Hylomyscus [Genus]
   H. aeta [Species]
   H. alleni
   H. baeri
   H. carillus
   H. denniae
   H. parvus
   H. stella
Hyomys [Genus]
   H. dammermani [Species]
   H. goliath
Hyperacrius [Genus]
   H. fertilis [Species]
   H. wynnei
Hypogeomys [Genus]
   H. antimena [Species]
Ichthyomys [Genus]
   I. hydrobates [Species]
   I. pittieri
   I. stolzmanni
   I. tweedii
Irenomys [Genus]
   I. tarsalis [Species]
Isthmomys [Genus]
   I. flavidus [Species]
   I. pirrensis
Juscelinomys [Genus]
   J. candango [Species]

J. vulpinus
Kadarsanomys [Genus]
  K. sodyi [Species]
Komodomys [Genus]
  K. rintjanus [Species]
Kunsia [Genus]
  K. fronto [Species]
  K. tomentosus
Lagurus [Genus]
  L. lagurus [Species]
Lamottemys [Genus]
  L. okuensis [Species]
Lasiopodomys [Genus]
  L. brandtii [Species]
  L. fuscus
  L. mandarinus
Leggadina [Genus]
  L. forresti [Species]
  L. lakedownensis
Leimacomys [Genus]
  L. buettneri [Species]
Lemmiscus [Genus]
  L. curtatus [Species]
Lemmus [Genus]
  L. amurensis [Species]
  L. lemmus
  L. sibiricus
Lemniscomys [Genus]
  L. barbarus [Species]
  L. bellieri
  L. griselda
  L. hoogstraali
  L. linulus
  L. macculus
  L. mittendorfi
  L. rosalia
  L. roseveari
  L. striatus
Lenomys [Genus]
  L. meyeri [Species]
Lenothrix [Genus]
  L. canus [Species]
Lenoxus [Genus]
  L. apicalis [Species]
Leopoldamys [Genus]
  L. edwardsi [Species]
  L. neilli
  L. sabanus
  L. siporanus
Leporillus [Genus]
  L. apicalis [Species]
  L. conditor
Leptomys [Genus]
  L. elegans [Species]
  L. ernstmayri
  L. signatus
Limnomys [Genus]
  L. sibuanus [Species]
Lophiomys [Genus]
  L. imhausi [Species]

Lophuromys [Genus]
  L. cinereus [Species]
  L. flavopunctatus
  L. luteogaster
  L. medicaudatus
  L. melanonyx
  L. nudicaudus
  L. rahmi
  L. sikapusi
  L. woosnami
Lorentzimys [Genus]
  L. nouhuysi [Species]
Macrotarsomys [Genus]
  M. bastardi [Species]
  M. ingens
Macruromys [Genus]
  M. elegans [Species]
  M. major
Malacomys [Genus]
  M. cansdalei [Species]
  M. edwardsi
  M. longipes
  M. lukolelae
  M. verschureni
Malacothrix [Genus]
  M. typica [Species]
Mallomys [Genus]
  M. aroaensis [Species]
  M. gunung
  M. istapantap
  M. rothschildi
Malpaisomys [Genus]
  M. insularis [Species]
Margaretamys [Genus]
  M. beccarii [Species]
  M. elegans
  M. parvus
Mastomys [Genus]
  M. angolensis [Species]
  M. coucha
  M. erythroleucus
  M. hildebrandtii
  M. natalensis
  M. pernanus
  M. shortridgei
  M. verheyeni
Maxomys [Genus]
  M. alticola [Species]
  M. baeodon
  M. bartelsii
  M. dollmani
  M. hellwaldii
  M. hylomyoides
  M. inas
  M. inflatus
  M. moi
  M. musschenbroekii
  M. ochraceiventer
  M. pagensis
  M. panglima

  M. rajah
  M. surifer
  M. wattsi
  M. whiteheadi
Mayermys [Genus]
  M. ellermani [Species]
Megadendromus [Genus]
  M. nikolausi [Species]
Megadontomys [Genus]
  M. cryophilus [Species]
  M. nelsoni
  M. thomasi
Megalomys [Genus]
  M. desmarestii [Species]
  M. luciae
Melanomys [Genus]
  M. caliginosus [Species]
  M. robustulus
  M. zunigae
Melasmothrix [Genus]
  M. naso [Species]
Melomys [Genus]
  M. aerosus [Species]
  M. bougainville
  M. burtoni
  M. capensis
  M. cervinipes
  M. fellowsi
  M. fraterculus
  M. gracilis
  M. lanosus
  M. leucogaster
  M. levipes
  M. lorentzii
  M. mollis
  M. moncktoni
  M. obiensis
  M. platyops
  M. rattoides
  M. rubex
  M. rubicola
  M. rufescens
  M. spechti
Meriones [Genus]
  M. arimalius [Species]
  M. chengi
  M. crassus
  M. dahli
  M. hurrianae
  M. libycus
  M. meridianus
  M. persicus
  M. rex
  M. sacramenti
  M. shawi
  M. tamariscinus
  M. tristrami
  M. unguiculatus
  M. vinogradovi
  M. zarudnyi

Mesembriomys [Genus]
    M. gouldii [Species]
    M. macrurus
Mesocricetus [Genus]
    M. auratus [Species]
    M. brandti
    M. newtoni
    M. raddei
Microdillus [Genus]
    M. peeli [Species]
Microhydromys [Genus]
    M. musseri [Species]
    M. richardsoni
Micromys [Genus]
    M. minutus [Species]
Microryzomys [Genus]
    M. altissimus [Species]
    M. minutus
Microtus [Genus]
    M. abbreviatus [Species]
    M. agrestis
    M. arvalis
    M. bavaricus
    M. breweri
    M. cabrerae
    M. californicus
    M. canicaudus
    M. chrotorrhinus
    M. daghestanicus
    M. duodecimcostatus
    M. evoronensis
    M. felteni
    M. fortis
    M. gerbei
    M. gregalis
    M. guatemalensis
    M. guentheri
    M. hyperboreus
    M. irani
    M. irene
    M. juldaschi
    M. kermanensis
    M. kirgisorum
    M. leucurus
    M. limnophilus
    M. longicaudus
    M. lusitanicus
    M. majori
    M. maximowiczii
    M. mexicanus
    M. middendorffi
    M. miurus
    M. mongolicus
    M. montanus
    M. montebelli
    M. mujanensis
    M. multiplex
    M. nasarovi
    M. oaxacensis
    M. obscurus

    M. ochrogaster
    M. oeconomus
    M. oregoni
    M. pennsylvanicus
    M. pinetorum
    M. quasiater
    M. richardsoni
    M. rossiaemeridionalis
    M. sachalinensis
    M. savii
    M. schelkovnikovi
    M. sikimensis
    M. socialis
    M. subterraneus
    M. tatricus
    M. thomasi
    M. townsendii
    M. transcaspicus
    M. umbrosus
    M. xanthognathus
Millardia [Genus]
    M. gleadowi [Species]
    M. kathleenae
    M. kondana
    M. meltada
Muriculus [Genus]
    M. imberbis [Species]
Mus [Genus]
    M. baoulei [Species]
    M. booduga
    M. bufo
    M. callewaerti
    M. caroli
    M. cervicolor
    M. cookii
    M. crociduroides
    M. famulus
    M. fernandoni
    M. goundae
    M. haussa
    M. indutus
    M. kasaicus
    M. macedonicus
    M. mahomet
    M. mattheyi
    M. mayori
    M. minutoides
    M. musculoides
    M. musculus
    M. neavei
    M. orangiae
    M. oubanguii
    M. pahari
    M. phillipsi
    M. platythrix
    M. saxicola
    M. setulosus
    M. setzeri
    M. shortridgei
    M. sorella

    M. spicilegus
    M. spretus
    M. tenellus
    M. terricolor
    M. triton
    M. vulcani
Mylomys [Genus]
    M. dybowskii [Species]
Myomys [Genus]
    M. albipes [Species]
    M. daltoni
    M. derooi
    M. fumatus
    M. ruppi
    M. verreauxii
    M. yemeni
Myopus [Genus]
    M. schisticolor [Species]
Myospalax [Genus]
    M. aspalax [Species]
    M. epsilanus
    M. fontanierii
    M. myospalax
    M. psilurus
    M. rothschildi
    M. smithii
Mystromys [Genus]
    M. albicaudatus [Species]
Nannospalax [Genus]
    N. ehrenbergi [Species]
    N. leucodon
    N. nehringi
Neacomys [Genus]
    N. guianae [Species]
    N. pictus
    N. spinosus
    N. tenuipes
Nectomys [Genus]
    N. palmipes [Species]
    N. parvipes
    N. squamipes
Nelsonia [Genus]
    N. goldmani [Species]
    N. neotomodon
Neofiber [Genus]
    N. alleni [Species]
Neohydromys [Genus]
    N. fuscus [Species]
Neotoma [Genus]
    N. albigula [Species]
    N. angustapalata
    N. anthonyi
    N. bryanti
    N. bunkeri
    N. chrysomelas
    N. cinerea
    N. devia
    N. floridana
    N. fuscipes
    N. goldmani

N. lepida
N. martinensis
N. mexicana
N. micropus
N. nelsoni
N. palatina
N. phenax
N. stephensi
N. varia
Neotomodon [Genus]
   N. alstoni [Species]
Neotomys [Genus]
   N. ebriosus [Species]
Nesomys [Genus]
   N. rufus [Species]
Nesokia [Genus]
   N. bunnii [Species]
   N. indica
Nesoryzomys [Genus]
   N. darwini [Species]
   N. fernandinae
   N. indefessus
   N. swarthi
Neusticomys [Genus]
   N. monticolus [Species]
   N. mussoi
   N. oyapocki
   N. peruviensis
   N. venezuelae
Niviventer [Genus]
   N. andersoni [Species]
   N. brahma
   N. confucianus
   N. coxingi
   N. cremoriventer
   N. culturatus
   N. eha
   N. excelsior
   N. fulvescens
   N. hinpoon
   N. langbianis
   N. lepturus
   N. niviventer
   N. rapit
   N. tenaster
Notiomys [Genus]
   N. edwardsii [Species]
Notomys [Genus]
   N. alexis [Species]
   N. amplus
   N. aquilo
   N. cervinus
   N. fuscus
   N. longicaudatus
   N. macrotis
   N. mitchellii
   N. mordax
Nyctomys [Genus]
   N. sumichrasti [Species]
Ochrotomys [Genus]

O. nuttalli [Species]
Oecomys [Genus]
   O. bicolor [Species]
   O. cleberi
   O. concolor
   O. flavicans
   O. mamorae
   O. paricola
   O. phaeotis
   O. rex
   O. roberti
   O. rutilus
   O. speciosus
   O. superans
   O. trinitatis
Oenomys [Genus]
   O. hypoxanthus [Species]
   O. ornatus
Oligoryzomys [Genus]
   O. andinus [Species]
   O. arenalis
   O. chacoensis
   O. delticola
   O. destructor
   O. eliurus
   O. flavescens
   O. fulvescens
   O. griseolus
   O. longicaudatus
   O. magellanicus
   O. microtis
   O. nigripes
   O. vegetus
   O. victus
Ondatra [Genus]
   O. zibethicus [Species]
Onychomys [Genus]
   O. arenicola [Species]
   O. leucogaster
   O. torridus
Oryzomys [Genus]
   O. albigularis [Species]
   O. alfaroi
   O. auriventer
   O. balneator
   O. bolivaris
   O. buccinatus
   O. capito
   O. chapmani
   O. couesi
   O. devius
   O. dimidiatus
   O. galapagoensis
   O. gorgasi
   O. hammondi
   O. intectus
   O. intermedius
   O. keaysi
   O. kelloggi
   O. lamia

O. legatus
O. levipes
O. macconnelli
O. melanotis
O. nelsoni
O. nitidus
O. oniscus
O. palustris
O. polius
O. ratticeps
O. rhabdops
O. rostratus
O. saturatior
O. subflavus
O. talamancae
O. xantheolus
O. yunganus
Osgoodomys [Genus]
   O. banderanus [Species]
Otomys [Genus]
   O. anchietae [Species]
   O. angoniensis
   O. denti
   O. irroratus
   O. laminatus
   O. maximus
   O. occidentalis
   O. saundersiae
   O. sloggetti
   O. tropicalis
   O. typus
   O. unisulcatus
Otonyctomys [Genus]
   O. hatti [Species]
Ototylomys [Genus]
   O. phyllotis [Species]
Oxymycterus [Genus]
   O. akodontius [Species]
   O. angularis
   O. delator
   O. hiska
   O. hispidus
   O. hucucha
   O. iheringi
   O. inca
   O. nasutus
   O. paramensis
   O. roberti
   O. rufus
Pachyuromys [Genus]
   P. duprasi [Species]
Palawanomys [Genus]
   P. furvus [Species]
Papagomys [Genus]
   P. armandvillei [Species]
   P. theodorverhoeveni
Parahydromys [Genus]
   P. asper [Species]
Paraleptomys [Genus]
   P. rufilatus [Species]

P. wilhelmina
Parotomys [Genus]
   P. brantsii [Species]
   P. littledalei
Paruromys [Genus]
   P. dominator [Species]
   P. ursinus
Paulamys [Genus]
   P. naso [Species]
Pelomys [Genus]
   P. campanae [Species]
   P. fallax
   P. hopkinsi
   P. isseli
   P. minor
Peromyscus [Genus]
   P. attwateri [Species]
   P. aztecus
   P. boylii
   P. bullatus
   P. californicus
   P. caniceps
   P. crinitus
   P. dickeyi
   P. difficilis
   P. eremicus
   P. eva
   P. furvus
   P. gossypinus
   P. grandis
   P. gratus
   P. guardia
   P. guatemalensis
   P. gymnotis
   P. hooperi
   P. interparietalis
   P. leucopus
   P. levipes
   P. madrensis
   P. maniculatus
   P. mayensis
   P. megalops
   P. mekisturus
   P. melanocarpus
   P. melanophrys
   P. melanotis
   P. melanurus
   P. merriami
   P. mexicanus
   P. nasutus
   P. ochraventer
   P. oreas
   P. pectoralis
   P. pembertoni
   P. perfulvus
   P. polionotus
   P. polius
   P. pseudocrinitus
   P. sejugis
   P. simulus

P. sitkensis
P. slevini
P. spicilegus
P. stephani
P. stirtoni
P. truei
P. winkelmanni
P. yucatanicus
P. zarhynchus
Petromyscus [Genus]
   P. barbouri [Species]
   P. collinus
   P. monticularis
   P. shortridgei
Phaenomys [Genus]
   P. ferrugineus [Species]
Phaulomys [Genus]
   P. andersoni [Species]
   P. smithii
Phenacomys [Genus]
   P. intermedius [Species]
   P. ungava
Phloeomys [Genus]
   P. cumingi [Species]
   P. pallidus
Phyllotis [Genus]
   P. amicus [Species]
   P. andium
   P. bonaeriensis
   P. caprinus
   P. darwini
   P. definitus
   P. gerbillus
   P. haggardi
   P. magister
   P. osgoodi
   P. osilae
   P. wolffsohni
   P. xanthopygus
Pithecheir [Genus]
   P. melanurus [Species]
   P. parvus
Phodopus [Genus]
   P. campbelli [Species]
   P. roborovskii
   P. sungorus
Platacanthomys [Genus]
   P. lasiurus [Species]
Podomys [Genus]
   P. floridanus [Species]
Podoxymys [Genus]
   P. roraimae [Species]
Pogonomelomys [Genus]
   P. bruijni [Species]
   P. mayeri
   P. sevia
Pogonomys [Genus]
   P. championi [Species]
   P. loriae
   P. macrourus

P. sylvestris
Praomys [Genus]
   P. delectorum [Species]
   P. hartwigi
   P. jacksoni
   P. minor
   P. misonnei
   P. morio
   P. mutoni
   P. rostratus
   P. tullbergi
Prionomys [Genus]
   P. batesi [Species]
Proedromys [Genus]
   P. bedfordi [Species]
Prometheomys [Genus]
   P. schaposchnikowi [Species]
Psammomys [Genus]
   P. obesus [Species]
   P. vexillaris
Pseudohydromys [Genus]
   P. murinus [Species]
   P. occidentalis
Pseudomys [Genus]
   P. albocinereus [Species]
   P. apodemoides
   P. australis
   P. bolami
   P. chapmani
   P. delicatulus
   P. desertor
   P. fieldi
   P. fumeus
   P. fuscus
   P. glaucus
   P. gouldii
   P. gracilicaudatus
   P. hermannsburgensis
   P. higginsi
   P. johnsoni
   P. laborifex
   P. nanus
   P. novaehollandiae
   P. occidentalis
   P. oralis
   P. patrius
   P. pilligaensis
   P. praeconis
   P. shortridgei
Pseudoryzomys [Genus]
   P. simplex [Species]
Punomys [Genus]
   P. lemminus [Species]
Rattus [Genus]
   R. adustus [Species]
   R. annandalei
   R. argentiventer
   R. baluensis
   R. bontanus
   R. burrus

R. colletti
R. elaphinus
R. enganus
R. everetti
R. exulans
R. feliceus
R. foramineus
R. fuscipes
R. giluwensis
R. hainaldi
R. hoffmanni
R. hoogerwerfi
R. jobiensis
R. koopmani
R. korinchi
R. leucopus
R. losea
R. lugens
R. lutreolus
R. macleari
R. marmosurus
R. mindorensis
R. mollicomulus
R. montanus
R. mordax
R. morotaiensis
R. nativitatis
R. nitidus
R. norvegicus
R. novaeguineae
R. osgoodi
R. palmarum
R. pelurus
R. praetor
R. ranjiniae
R. rattus
R. sanila
R. sikkimensis
R. simalurensis
R. sordidus
R. steini
R. stoicus
R. tanezumi
R. tawitawiensis
R. timorensis
R. tiomanicus
R. tunneyi
R. turkestanicus
R. villosissimus
R. xanthurus
Reithrodon [Genus]
R. auritus [Species]
Reithrodontomys [Genus]
R. brevirostris [Species]
R. burti
R. chrysopsis
R. creper
R. darienensis
R. fulvescens
R. gracilis

R. hirsutus
R. humulis
R. megalotis
R. mexicanus
R. microdon
R. montanus
R. paradoxus
R. raviventris
R. rodriguezi
R. spectabilis
R. sumichrasti
R. tenuirostris
R. zacatecae
Rhabdomys [Genus]
R. pumilio [Species]
Rhagomys [Genus]
R. rufescens [Species]
Rheomys [Genus]
R. mexicanus [Species]
R. raptor
R. thomasi
R. underwoodi
Rhipidomys [Genus]
R. austrinus [Species]
R. caucensis
R. couesi
R. fulviventer
R. latimanus
R. leucodactylus
R. macconnelli
R. mastacalis
R. nitela
R. ochrogaster
R. scandens
R. venezuelae
R. venustus
R. wetzeli
Rhizomys [Genus]
R. pruinosus [Species]
R. sinensis
R. sumatrensis
Rhombomys [Genus]
R. opimus [Species]
Rhynchomys [Genus]
R. isarogensis [Species]
R. soricoides
Saccostomus [Genus]
S. campestris [Species]
S. mearnsi
Scapteromys [Genus]
S. tumidus [Species]
Scolomys [Genus]
S. melanops [Species]
S. ucayalensis
Scotinomys [Genus]
S. teguina [Species]
S. xerampelinus
Sekeetamys [Genus]
S. calurus [Species]
Sigmodon [Genus]

S. alleni [Species]
S. alstoni
S. arizonae
S. fulviventer
S. hispidus
S. inopinatus
S. leucotis
S. mascotensis
S. ochrognathus
Sigmodontomys [Genus]
S. alfari [Species]
S. aphrastus
Solomys [Genus]
S. ponceleti [Species]
S. salamonis
S. salebrosus
S. sapientis
S. spriggsarum
Spalax [Genus]
S. arenarius [Species]
S. giganteus
S. graecus
S. microphthalmus
S. zemni
Spelaeomys [Genus]
S. florensis [Species]
Srilankamys [Genus]
S. ohiensis [Species]
Stenocephalemys [Genus]
S. albocaudata [Species]
S. griseicauda
Steatomys [Genus]
S. caurinus [Species]
S. cuppedius
S. jacksoni
S. krebsii
S. parvus
S. pratensis
Stenomys [Genus]
S. ceramicus [Species]
S. niobe
S. richardsoni
S. vandeuseni
S. verecundus
Stochomys [Genus]
S. longicaudatus [Species]
Sundamys [Genus]
S. infraluteus [Species]
S. maxi
S. muelleri
Synaptomys [Genus]
S. borealis [Species]
S. cooperi
Tachyoryctes [Genus]
T. ankoliae [Species]
T. annectens
T. audax
T. daemon
T. macrocephalus
T. naivashae

T. rex
T. ruandae
T. ruddi
T. spalacinus
T. splendens
Taeromys [Genus]
  T. arcuatus [Species]
  T. callitrichus
  T. celebensis
  T. hamatus
  T. punicans
  T. taerae
Tarsomys [Genus]
  T. apoensis [Species]
  T. echinatus
Tateomys [Genus]
  T. macrocercus [Species]
  T. rhinogradoides
Tatera [Genus]
  T. afra [Species]
  T. boehmi
  T. brantsii
  T. guineae
  T. inclusa
  T. indica
  T. kempi
  T. leucogaster
  T. nigricauda
  T. phillipsi
  T. robusta
  T. valida
Taterillus [Genus]
  T. arenarius [Species]
  T. congicus
  T. emini
  T. gracilis
  T. harringtoni
  T. lacustris
  T. petteri
  T. pygargus
Tscherskia [Genus]
  T. triton [Species]
Thallomys [Genus]
  T. loringi [Species]
  T. nigricauda
  T. paedulcus
  T. shortridgei
Thalpomys [Genus]
  T. cerradensis [Species]
  T. lasiotis
Thamnomys [Genus]
  T. kempi [Species]
  T. venustus
Thomasomys [Genus]
  T. aureus [Species]
  T. baeops
  T. bombycinus
  T. cinereiventer
  T. cinereus
  T. daphne

T. eleusis
T. gracilis
T. hylophilus
T. incanus
T. ischyurus
T. kalinowskii
T. ladewi
T. laniger
T. monochromos
T. niveipes
T. notatus
T. oreas
T. paramorum
T. pyrrhonotus
T. rhoadsi
T. rosalinda
T. silvestris
T. taczanowskii
T. vestitus
Tokudaia [Genus]
  T. muenninki [Species]
  T. osimensis
Tryphomys [Genus]
  T. adustus [Species]
Tylomys [Genus]
  T. bullaris [Species]
  T. fulviventer
  T. mirae
  T. nudicaudus
  T. panamensis
  T. tumbalensis
  T. watsoni
Typhlomys [Genus]
  T. chapensis [Species]
  T. cinereus
Uranomys [Genus]
  U. ruddi [Species]
Uromys [Genus]
  U. anak [Species]
  U. caudimaculatus
  U. hadrourus
  U. imperator
  U. neobritanicus
  U. porculus
  U. rex
Vandeleuria [Genus]
  V. nolthenii [Species]
  V. oleracea
Vernaya [Genus]
  V. fulva [Species]
Volemys [Genus]
  V. clarkei [Species]
  V. kikuchii
  V. millicens
  V. musseri
Wiedomys [Genus]
  W. pyrrhorhinos [Species]
Wilfredomys [Genus]
  W. oenax [Species]
  W. pictipes

Xenomys [Genus]
  X. nelsoni [Species]
Xenuromys [Genus]
  X. barbatus [Species]
Xeromys [Genus]
  X. myoides [Species]
Zelotomys [Genus]
  Z. hildegardeae [Species]
  Z. woosnami
Zygodontomys [Genus]
  Z. brevicauda [Species]
  Z. brunneus
Zyzomys [Genus]
  Z. argurus [Species]
  Z. maini
  Z. palatilis
  Z. pedunculatus
  Z. woodwardi

Anomaluridae [Family]
  Anomalurus [Genus]
    A. beecrofti [Species]
    A. derbianus
    A. pelii
    A. pusillus
  Idiurus [Genus]
    I. macrotis [Species]
    I. zenkeri
  Zenkerella [Genus]
    Z. insignis [Species]

Pedetidae [Family]
  Pedetes [Genus]
    P. capensis [Species]

Ctenodactylidae [Family]
  Ctenodactylus [Genus]
    C. gundi [Species]
    C. vali
  Felovia [Genus]
    F. vae [Species]
  Massoutiera [Genus]
    M. mzabi [Species]
  Pectinator [Genus]
    P. spekei [Species]

Myoxidae [Family]
  Dryomys [Genus]
    D. laniger [Species]
    D. nitedula
    D. sichuanensis
  Eliomys [Genus]
    E. melanurus [Species]
    E. quercinus
  Glirulus [Genus]
    G. japonicus [Species]
  Graphiurus [Genus]
    G. christyi [Species]
    G. hueti
    G. lorraineus
    G. monardi
    G. ocularis

G. parvus
G. rupicola
Muscardinus [Genus]
   M. avellanarius [Species]
Myomimus [Genus]
   M. personatus [Species]
   M. roachi
   M. setzeri
Myoxus [Genus]
   M. glis [Species]
Selevinia [Genus]
   S. betpakdalaensis [Species]

Petromuridae [Family]
Petromus [Genus]
   P. typicus [Species]

Thryonomyidae [Family]
Thryonomys [Genus]
   T. gregorianus [Species]
   T. swinderianus

Bathyergidae [Family]
Bathyergus [Genus]
   B. janetta [Species]
   B. suillus
Cryptomys [Genus]
   C. bocagei [Species]
   C. damarensis
   C. foxi
   C. hottentotus
   C. mechowi
   C. ochraceocinereus
   C. zechi
Georychus [Genus]
   G. capensis [Species]
Heliophobius [Genus]
   H. argenteocinereus
Heterocephalus [Genus]
   H. glaber [Species]

Hystricidae [Family]
Atherurus [Genus]
   A. africanus [Species]
   A. macrourus
Hystrix [Genus]
   H. africaeaustralis [Species]
   H. brachyura
   H. crassispinis
   H. cristata
   H. indica
   H. javanica
   H. pumila
   H. sumatrae
Trichys [Genus]
   T. fasciculata [Species]

Erethizontidae [Family]
Coendou [Genus]
   C. bicolor [Species]
   C. koopmani
   C. prehensilis

C. rothschildi
Echinoprocta [Genus]
   E. rufescens [Species]
Erethizon [Genus]
   E. dorsatum [Species]
Sphiggurus [Genus]
   S. insidiosus [Species]
   S. mexicanus
   S. pallidus
   S. spinosus
   S. vestitus
   S. villosus

Chinchillidae [Family]
Chinchilla [Genus]
   C. brevicaudata [Species]
   C. lanigera
Lagidium [Genus]
   L. peruanum [Species]
   L. viscacia
   L. wolffsohni
Lagostomus [Genus]
   L. maximus [Species]

Dinomyidae [Family]
Dinomys [Genus]
   D. branickii [Species]

Caviidae [Family]
Cavia [Genus]
   C. aperea [Species]
   C. fulgida
   C. magna
   C. porcellus
   C. tschudii
Dolichotis [Genus]
   D. patagonum [Species]
   D. salinicola
Galea [Genus]
   G. flavidens [Species]
   G. spixii
Kerodon [Genus]
   K. rupestris [Species]
Microcavia [Genus]
   M. australis [Species]
   M. niata
   M. shiptoni

Hydrochaeridae [Family]
Hydrochaeris [Genus]
   H. hydrochaeris [Species]

Dasyproctidae [Family]
Dasyprocta [Genus]
   D. azarae [Species]
   D. coibae
   D. cristata
   D. fuliginosa
   D. guamara
   D. kalinowskii
   D. leporina
   D. mexicana

D. prymnolopha
D. punctata
D. ruatanica
Myoprocta [Genus]
   M. acouchy [Species]
   M. exilis

Agoutidae [Family]
Agouti [Genus]
   A. paca [Species]
   A. taczanowskii

Ctenomyidae [Family]
Ctenomys [Genus]
   C. argentinus [Species]
   C. australis
   C. azarae
   C. boliviensis
   C. bonettoi
   C. brasiliensis
   C. colburni
   C. conoveri
   C. dorsalis
   C. emilianus
   C. frater
   C. fulvus
   C. haigi
   C. knighti
   C. latro
   C. leucodon
   C. lewisi
   C. magellanicus
   C. maulinus
   C. mendocinus
   C. minutus
   C. nattereri
   C. occultus
   C. opimus
   C. pearsoni
   C. perrensis
   C. peruanus
   C. pontifex
   C. porteousi
   C. saltarius
   C. sericeus
   C. sociabilis
   C. steinbachi
   C. talarum
   C. torquatus
   C. tuconax
   C. tucumanus
   C. validus

Octodontidae [Family]
Aconaemys [Genus]
   A. fuscus [Species]
   A. sagei
Octodon [Genus]
   O. bridgesi [Species]
   O. degus
   O. lunatus

Octodontomys [Genus]
  O. gliroides [Species]
Octomys [Genus]
  O. mimax [Species]
Spalacopus [Genus]
  S. cyanus [Species]
Tympanoctomys [Genus]
  T. barrerae [Species]

Abrocomidae [Family]
  Abrocoma [Genus]
    A. bennetti [Species]
    A. boliviensis
    A. cinerea

Echimyidae [Family]
  Boromys [Genus]
    B. offella [Species]
    B. torrei
  Brotomys [Genus]
    B. contractus [Species]
    B. voratus
  Carterodon [Genus]
    C. sulcidens [Species]
  Clyomys [Genus]
    C. bishopi [Species]
    C. laticeps
  Chaetomys [Genus]
    C. subspinosus [Species]
  Dactylomys [Genus]
    D. boliviensis [Species]
    D. dactylinus
    D. peruanus
  Diplomys [Genus]
    D. caniceps [Species]
    D. labilis
    D. rufodorsalis
  Echimys [Genus]
    E. blainvillei [Species]
    E. braziliensis
    E. chrysurus
    E. dasythrix
    E. grandis
    E. lamarum
    E. macrurus
    E. nigrispinus
    E. pictus
    E. rhipidurus
    E. saturnus
    E. semivillosus
    E. thomasi
    E. unicolor
  Euryzygomatomys [Genus]
    E. spinosus [Species]
  Heteropsomys [Genus]
    H. antillensis [Species]
    H. insulans
  Hoplomys [Genus]
    H. gymnurus [Species]
  Isothrix [Genus]
    I. bistriata [Species]

I. pagurus
Kannabateomys [Genus]
  K. amblyonyx [Species]
Lonchothrix [Genus]
  L. emiliae [Species]
Makalata [Genus]
  M. armata [Species]
Mesomys [Genus]
  M. didelphoides [Species]
  M. hispidus
  M. leniceps
  M. obscurus
  M. stimulax
Olallamys [Genus]
  O. albicauda [Species]
  O. edax
Proechimys [Genus]
  P. albispinus [Species]
  P. amphichoricus
  P. bolivianus
  P. brevicauda
  P. canicollis
  P. cayennensis
  P. chrysaeolus
  P. cuvieri
  P. decumanus
  P. dimidiatus
  P. goeldii
  P. gorgonae
  P. guairae
  P. gularis
  P. hendeei
  P. hoplomyoides
  P. iheringi
  P. longicaudatus
  P. magdalenae
  P. mincae
  P. myosuros
  P. oconnelli
  P. oris
  P. poliopus
  P. quadruplicatus
  P. semispinosus
  P. setosus
  P. simonsi
  P. steerei
  P. trinitatis
  P. urichi
  P. warreni
Puertoricomys [Genus]
  P. corozalus [Species]
Thrichomys [Genus]
  T. apereoides [Species]

Capromyidae [Family]
  Capromys [Genus]
    C. pilorides [Species]
  Geocapromys [Genus]
    G. brownii [Species]
    G. thoracatus
  Hexolobodon [Genus]

H. phenax [Species]
Isolobodon [Genus]
  I. montanus [Species]
  I. portoricensis
Mesocapromys [Genus]
  M. angelcabrerai [Species]
  M. auritus
  M. nanus
  M. sanfelipensis
Mysateles [Genus]
  M. garridoi [Species]
  M. gundlachi
  M. melanurus
  M. meridionalis
  M. prehensilis
Plagiodontia [Genus]
  P. aedium [Species]
  P. araeum
  P. ipnaeum
Rhizoplagiodontia [Genus]
  R. lemkei [Species]

Heptaxodontidae [Family]
  Amblyrhiza [Genus]
    A. inundata [Species]
  Clidomys [Genus]
    C. osborni [Species]
    C. parvus
  Elasmodontomys [Genus]
    E. obliquus [Species]
  Quemisia [Genus]
    Quemisia gravis [Species]

Myocastoridae [Family]
  Myocastor [Genus]
    M. coypus [Species]

Lagomorpha [Order]

Ochotonidae [Family]
  Ochotona [Genus]
    O. alpina [Species]
    O. cansus
    O. collaris
    O. curzoniae
    O. dauurica
    O. erythrotis
    O. forresti
    O. gaoligongensis
    O. gloveri
    O. himalayana
    O. hyperborea
    O. iliensis
    O. koslowi
    O. ladacensis
    O. macrotis
    O. muliensis
    O. nubrica
    O. pallasi
    O. princeps
    O. pusilla
    O. roylei

O. rufescens
O. rutila
O. thibetana
O. thomasi
Prolagus [Genus]
   P. sardus [Species]

Leporidae [Family]
  Brachylagus [Genus]
    B. idahoensis [Species]
  Bunolagus [Genus]
    B. monticularis [Species]
  Caprolagus [Genus]
    C. hispidus [Species]
  Lepus [Genus]
    L. alleni [Species]
    L. americanus
    L. arcticus
    L. brachyurus
    L. californicus
    L. callotis
    L. capensis
    L. castroviejoi
    L. comus
    L. coreanus
    L. corsicanus
    L. europaeus
    L. fagani
    L. flavigularis
    L. granatensis
    L. hainanus
    L. insularis

L. mandshuricus
L. nigricollis
L. oiostolus
L. othus
L. pequensis
L. saxatilis
L. sinensis
L. starcki
L. timidus
L. tolai
L. townsendii
L. victoriae
L. yarkandensis
Nesolagus [Genus]
  N. netscheri [Species]
Oryctolagus [Genus]
  O. cuniculus [Species]
Pentalagus [Genus]
  P. furnessi [Species]
Poelagus [Genus]
  P. marjorita [Species]
Pronolagus [Genus]
  P. crassicaudatus [Species]
  P. randensis
  P. rupestris
Romerolagus [Genus]
  R. diazi [Species]
Sylvilagus [Genus]
  S. aquaticus [Species]
  S. audubonii
  S. bachmani
  S. brasiliensis

S. cunicularius
S. dicei
S. floridanus
S. graysoni
S. insonus
S. mansuetus
S. nuttallii
S. palustris
S. transitionalis

Macroscelidea [Order]

Macroscelididae [Family]
  Elephantulus [Genus]
    E. brachyrhynchus [Species]
    E. edwardii
    E. fuscipes
    E. fuscus
    E. intufi
    E. myurus
    E. revoili
    E. rozeti
    E. rufescens
    E. rupestris
  Macroscelides [Genus]
    M. proboscideus [Species]
  Petrodromus [Genus]
    P. tetradactylus [Species]
  Rhynchocyon [Genus]
    R. chrysopygus [Species]
    R. cirnei
    R. petersi

• • • • •

A brief geologic history of animal life

A note about geologic time scales: A cursory look will reveal that the timing of various geological periods differs among textbooks. Is one right and the others wrong? Not necessarily. Scientists use different methods to estimate geological time—methods with a precision sometimes measured in tens of millions of years. There is, however, a general agreement on the magnitude and relative timing associated with modern time scales. The closer in geological time one comes to the present, the more accurate science can be—and sometimes the more disagreement there seems to be. The following account was compiled using the more widely accepted boundaries from a diverse selection of reputable scientific resources.

Geologic time scale

Era	Period	Epoch	Dates	Life forms
Proterozoic			2,500-544 mya*	First single-celled organisms, simple plants, and invertebrates (such as algae, amoebas, and jellyfish)
Paleozoic	Cambrian		544-490 mya	First crustaceans, mollusks, sponges, nautiloids, and annelids (worms)
	Ordovician		490-438 mya	Trilobites dominant. Also first fungi, jawless vertebrates, starfishes, sea scorpions, and urchins
	Silurian		438-408 mya	First terrestrial plants, sharks, and bony fishes
	Devonian		408-360 mya	First insects, arachnids (scorpions), and tetrapods
	Carboniferous	Mississippian	360-325 mya	Amphibians abundant. Also first spiders, land snails
		Pennsylvanian	325-286 mya	First reptiles and synapsids
	Permian		286-248 mya	Reptiles abundant. Extinction of trilobytes. Most modern insect orders
Mesozoic	Triassic		248-205 mya	Diversification of reptiles: turtles, crocodiles, therapsids (mammal-like reptiles), first dinosaurs, first flies
	Jurassic		205-145 mya	Insects abundant, dinosaurs dominant in later stage. First mammals, lizards, frogs, and birds
	Cretaceous		145-65 mya	First snakes and modern fish. Extinction of dinosaurs and ammonites, rise and fall of toothed birds
Cenozoic	Tertiary	Paleocene	65-55.5 mya	Diversification of mammals
		Eocene	55.5-33.7 mya	First horses, whales, monkeys, and leafminer insects
		Oligocene	33.7-23.8 mya	Diversification of birds. First anthropoids (higher primates)
		Miocene	23.8-5.6 mya	First hominids
		Pliocene	5.6-1.8 mya	First australopithecines
	Quaternary	Pleistocene	1.8 mya-8,000 ya	Mammoths, mastodons, and Neanderthals
		Holocene	8,000 ya-present	First modern humans

*Millions of years ago (mya)

Index

Bold page numbers indicate the primary discussion of a topic; page numbers in italics indicate illustrations; "t" indicates a table.

INDEX

INDEX

INDEX

INDEX

F

INDEX

INDEX

Humbback whales *(continued)*
16:42t
Hunting, 12:24–25, 14:248
See also Humans
Hunting dogs, 12:146, 14:288, 14:292
Huon tree kangaroos. *See* Matschie's tree
kangaroos
Huskies, Siberian, 14:289, 14:292, 14:293
Hussar monkeys. *See* Patas monkeys
Hutias, 16:123, **16:461–467**, 16:464,
16:467t
See also Giant hutias
Hutterer, R., 16:433
Hutton's tube-nosed bats, 13:526t
Hyaena brunnea. See Brown hyenas
Hyaena hyaena. See Striped hyenas
Hyaena hyaena barbara, 14:360
Hyaena hyaena dubbah, 14:360
Hyaena hyaena hyaena, 14:360
Hyaena hyaena sultana, 14:360
Hyaena hyaena syriaca, 14:360
Hyaenidae. *See* Aardwolves; Hyenas
Hybrid gibbons, 14:208–209, 14:210
Hydrochaeridae. *See* Capybaras
Hydrochaeris hydrochaeris. See Capybaras
Hydrodamalinae. *See* Steller's sea cows
Hydrodamalis gigas. See Steller's sea cows
Hydrodamalis stelleri. See Giant sea cows
Hydromys chrysogaster. See Golden-bellied
water rats
Hydropotes spp. *See* Water deer
Hydropotes inermis. See Chinese water deer
Hydropotes inermis argyropus. See Siberian
water deer
Hydropotes inermis inermis, 15:373
Hydropotinae. *See* Chinese water deer
Hydrurga leptonyx. See Leopard seals
Hyemoschus spp. *See* Water chevrotains
Hyemoschus aquaticus. See Water chevrotains
Hyenas, 12:142, 14:256, 14:259, 14:260,
14:359–367, 14:361, 14:363
Hylobates spp., 14:207–208, 14:209, 14:211
Hylobates agilis. See Agile gibbons
Hylobates agilis albibarbis, 14:208
Hylobates concolor, 14:207
Hylobates gabriellae, 14:207
Hylobates hoolock. See Hoolock gibbons
Hylobates klossi. See Kloss gibbons
Hylobates lar. See Lar gibbons
Hylobates leucogenys, 14:207
Hylobates moloch. See Moloch gibbons
Hylobates muelleri. See Mueller's gibbons
Hylobates pileatus. See Pileated gibbons
Hylobates syndactylus. See Siamangs
Hylobatidae. *See* Gibbons
Hylochoerus spp., 15:275, 15:276
Hylochoerus meinertzhageni. See Forest hogs
Hylomyinae, 13:203, 13:204
Hylomys spp., 13:203
Hylomys hainanensis. See Hainan gymnures
Hylomys megalotis. See Long-eared lesser
gymnures
Hylomys sinens, 13:207
Hylomys sinensis. See Shrew gymnures
Hylomys suillus. See Lesser gymnures
Hylomys suillus parvus, 13:207
Hylopetes spp., 16:136
Hylopetes lepidus. See Gray-cheeked flying
squirrels
Hyotherium spp., 15:265

Hypercapnia, 12:73
Hyperoodon spp. *See* Bottlenosed whales
Hyperoodon ampullatus. See Northern
bottlenosed whales
Hyperoodon planifrons. See Southern
bottlenosed whales
Hypertragulidae, 15:265
Hypnomys spp., 16:317–318
Hypogeomys antimena. See Malagasy giant rats
Hypoxia, 12:73, 12:114
Hypsignathus monstrosus. See Hammer-headed
fruit bats
Hypsiprimnodontidae. *See* Musky rat-
kangaroos
Hypsiprymnodon moschatus. See Musky rat-
kangaroos
Hyracoidea. *See* Hyraxes
Hyracotherium spp., 15:132
Hyraxes, **15:177–190**, 15:178, 15:185
behavior, 15:180–181
conservation status, 15:184
digestive system, 12:14–15
distribution, 15:177, 15:179
evolution, 15:131, 15:133, 15:177
feeding ecology, 15:182
habitats, 15:179–180
humans and, 15:184
physical characteristics, 15:178–179
reproduction, 12:94, 12:95, 15:182–183
species of, 15:186–188, 15:189t
taxonomy, 15:134, 15:177–178
Hyspiprymnodon bartholomai, 13:69–70
Hystricidae. *See* Old World porcupines
Hystricognathi, 12:11, 12:71, 16:121, 16:122,
16:125
Hystricomorph rodents, 12:92, 12:94
Hystrix spp., 16:351
Hystrix africaeaustralis. See South African
porcupines
Hystrix brachyura. See Common porcupines
Hystrix crassispinis. See Thick-spined
porcupines
Hystrix cristata. See North African porcupines
Hystrix indica. See Indian crested porcupines
Hystrix javanica. See Javan short-tailed
porcupines
Hystrix macroura, 16:358
Hystrix pumila. See Indonesian porcupines
Hystrix sumatrae. See Sumatran porcupines
Hysudricus maximus, 15:164

I

Iberian desmans. *See* Pyrenean desmans
Iberian lynx, 14:262, 14:370–371, 14:374,
14:378, 14:387, 14:389
Ibex, 16:4, 16:8, 16:9, 16:90, 16:92–93, 16:95
Ice Age, **12:17–25**
Ice rats, 16:283
Ichneumia spp., 14:347
Ichneumia albicauda. See White-tailed
mongooses
Ichnumeons. *See* Egyptian mongooses
Ichthyomyines, 16:264, 16:268, 16:269
Ichthyomys spp., 16:266, 16:269
Ichthyomys pittieri, 16:269
Ictonyx spp. *See* Zorillas
Ictonyx striatus. See Zorillas

Idaho ground squirrels, 16:147, 16:150,
16:153, 16:157–158
Idionycteris phyllotis. See Allen's big-eared bats
Idiurus macrotis. See Big-eared flying mice
Idiurus zenkeri. See Zenker's flying mice
Ili pikas, 16:487, 16:495, 16:502t
Imitation, 12:158
See also Behavior
Immersion exhibits, 12:205
Immunocontraception, 12:188
See also Reproduction
Immunoglobins, neonatal, 12:126
Impalas, 16:36, 16:41–42
behavior, 16:5, 16:31, 16:39, 16:41–42
black-faced, 16:34
conservation status, 16:31
distribution, 16:29, 16:39
evolution, 16:27
habitats, 16:30
physical characteristics, 16:28
predation and, 12:116
reproduction, 16:32
smell, 12:80
taxonomy, 16:27
Implantation, delayed, 12:109–110
See also Reproduction
Imposter hutias, 16:467t
Inbreeding, 12:110, 12:222
See also Captive breeding; Domestication
Incisors, 12:46
See also Teeth
Incus, 12:36
Indexes
indirect observations and, 12:199–202
population, 12:195–196
Indian blackbuck antelopes, 15:271
Indian civets, 14:336, 14:345t
Indian crested porcupines, 16:352, 16:356,
16:358, 16:359–360
Indian desert hedgehogs, 13:204, 13:206,
13:208, 13:210
Indian elephants. *See* Asian elephants
Indian flying foxes, 13:307, 13:320, 13:322,
13:324, 13:327, 13:328
Indian fruit bats, 13:305, 13:338, 13:344
Indian giant squirrels, 16:167, 16:174t
Indian gray mongooses, 14:351–352
Indian mongooses, 12:190–191, 14:350,
14:352
Indian muntjacs, 15:343, 15:344, 15:346,
15:347, 15:350, 15:351–352
Indian pangolins, 16:110, 16:113, 16:114,
16:115, 16:116
Indian porcupines. *See* Indian crested
porcupines
Indian rabbits. *See* Central American agoutis
Indian rhinoceroses, 15:223, 15:251, 15:254,
15:256, 15:258–259
behavior, 15:218
conservation status, 12:224, 15:222, 15:254
habitats, 15:252–253
humans and, 15:254–255
Indian spotted chevrotains. *See* Spotted
mouse deer
Indian tapirs. *See* Malayan tapirs
Indian tree shrews, 13:197, 13:292, 13:293,
13:294
Indian wild asses. *See* Khurs
Indian Wildlife Protection Act of 1972, 15:280
Indian wolves, 12:180

INDEX

North American black bears. *See* American black bears
North American kangaroo rats, 16:124
North American mountain goats, 12:131
North American plains bison. *See* American bison
North American pocket gophers, 16:126
North American porcupines, 16:*365*, 16:*370*, 16:*371–372*
 behavior, 16:367–368
 distribution, 16:123, 16:367
 feeding ecology, 12:*123*
 humans and, 16:127
 physical characteristics, 12:38
 reproduction, 16:*366*
North American pronghorns, 12:40
North American red squirrels, 16:*127*, 16:*168*, 16:*173*
 See also Red squirrels
North American sheep, 16:92
 See also Sheep
North Atlantic right whales, 15:10, 15:107, 15:*108*, 15:*109*, 15:*111*, 15:*114*, 15:*115*, 15:116
North Chinese flying squirrels, 16:141*t*
 See also Flying squirrels
North Indian muntjacs, 15:*345*
 See also Muntjacs
North Moluccan flying foxes, 13:331*t*
North Pacific right whales, 15:107, 15:109, 15:111–112, 15:*114*, 15:116–*117*
 See also Right whales
North Sea beaked whales. *See* Sowerby's beaked whales
Northern Bahian blond titis, 14:11, 14:148
Northern bats, 13:500
Northern bettongs, 13:74, 13:76, 13:77, 13:*79*
Northern bog lemmings, 16:227, 16:238*t*
Northern bottlenosed whales, 15:*60*, 15:*61*, 15:62, 15:*63*, 15:*64–65*
Northern brown bandicoots, 13:2, 13:4, 13:6, 13:*10*, 13:*11*, 13:*13*, 13:*14–15*
Northern brushtail possums, 13:66*t*
Northern cave-lions, 12:24
Northern chamois, 16:6, 16:87, 16:*96*, 16:*100*
Northern collared lemmings, 16:226, 16:227, 16:237*t*
Northern common cuscuses. *See* Common cuscuses
Northern elephant seals, 14:*425*, 14:431
 behavior, 14:*421*, 14:*423*
 distribution, 14:*429*
 feeding ecology, 14:419
 humans and, 14:424
 lactation, 12:127
 physical characteristics, 12:67
 reproduction, 14:421–422
Northern flying squirrels, 16:135
 See also Flying squirrels
Northern four-toothed whales. *See* Baird's beaked whales
Northern fur seals, 14:*401*, 14:*402*
 conservation status, 12:*219*
 distribution, 14:394, 14:395
 evolution, 14:393
 reproduction, 12:110, 14:396, 14:398, 14:399
Northern ghost bats, 13:*359*, 13:*360*, 13:362
Northern gliders, 13:133*t*
Northern gracile mouse opossums, 12:264*t*

Northern grasshopper mice, 16:267
Northern greater bushbabies, 14:25, 14:*25*, 14:*27*, 14:*29*, 14:31–32
Northern hairy-nosed wombats, 13:39, 13:40, 13:51, 13:52, 13:53, 13:*54*, 13:*55*, 13:56
Northern leaf-nosed bats, 13:*403*, 13:*404*
Northern lesser bushbabies. *See* Senegal bushbabies
Northern long-nosed armadillos. *See* Llanos long-nosed armadillos
Northern marsupial moles, 13:26, 13:*28*
Northern mastiff bats, 13:493*t*
Northern minke whales, 12:*200*, 15:120, 15:*122*, 15:125, 15:*126*, 15:128–129
Northern muriquis, 14:11, 14:156, 14:159, 14:*161*, 14:*164*, 14:165
Northern naked-tailed armadillos, 13:183, 13:190*t*
Northern needle-clawed bushbabies, 14:26, 14:32*t*
Northern pikas, 16:*493*, 16:*496*, 16:*498–499*
Northern plains gray langurs, 14:10, 14:*173*, 14:175, 14:*176*, 14:*180*
Northern planigales. *See* Long-tailed planigales
Northern pocket gophers, 16:*187*, 16:*189*, 16:195*t*
Northern pudus, 15:379, 15:396*t*
Northern quolls, 12:281, 12:282, 12:299*t*
Northern raccoons, 12:*189*, 14:*309*, 14:310, 14:*311*, 14:*312*, 14:*313*
Northern red-legged pademelons. *See* Red-legged pademelons
Northern right whale dolphins, 15:42
Northern river otters, 12:109, 14:*322*, 14:*323*
Northern sea elephants. *See* Northern elephant seals
Northern sea lions. *See* Steller sea lions
Northern short-tailed bats. *See* Lesser New Zealand short-tailed bats
Northern short-tailed shrews, 13:*195*, 13:*250*, 13:*253*, 13:256–257
Northern smooth-tailed tree shrews, 13:297*t*
Northern sportive lemurs, 14:78, 14:*83*
Northern tamanduas, 13:171, 13:*176*, 13:*178*
 See also Anteaters
Northern three-toed jerboas. *See* Hairy-footed jerboas
Northern water shrews. *See* Eurasian water shrews
Northern water voles, 12:70–71, 12:182–183, 16:229, 16:*231*, 16:*233–234*
Norway lemmings, 16:227, 16:*231*, 16:233, 16:235
Norway rats. *See* Brown rats
Noses, 12:8, 12:80, 12:*81*
 See also Physical characteristics
Nostrils, 12:8
 See also Physical characteristics
Notharctus spp., 14:1
Nothrotheriops spp., 13:151
Notiomys edwardsii, 16:266
Notiosorex spp., 13:248, 13:250
Notiosorex crawfordi. *See* Desert shrews
Notiosoricini, 13:247
Notomys alexis. *See* Australian jumping mice
Notomys fuscus. *See* Dusky hopping mice
Notonycteris spp., 13:413
Notopteris macdonaldi. *See* Long-tailed fruit bats

Notoryctemorphia. *See* Marsupial moles
Notoryctes spp., 13:25, 13:26
Notoryctes caurinus. *See* Northern marsupial moles
Notoryctes typhlops. *See* Marsupial moles
Notoryctidae. *See* Marsupial moles
Notoungulata, 12:11, 12:132, 15:131, 15:133, 15:134, 15:135, 15:136, 15:137
Nourishment. *See* Feeding ecology
Novae's bald uakaris, 14:148, 14:151
NST (Non-shivering thermogenesis), 12:113
Nubian ibex, 15:146, 16:87, 16:91, 16:92, 16:94, 16:95
Nubian wild asses, 12:176–177
 See also African wild asses
Nubra pikas, 16:502*t*
Nuciruptor spp., 14:143
Nuclear DNA (nDNA), 12:26–27, 12:29–30, 12:31, 12:33
Nucleotide bases, 12:26–27
Numbats, 12:137, 12:277–284, **12:303–306**, 12:*304*, 12:*305*, 12:*306*
Numbers, understanding, 12:155–156, 12:162
 See also Cognition
Nutrias. *See* Coypus
Nutritional adaptations, **12:120–128**
Nutritional ecology. *See* Feeding ecology
Nyalas, 16:6, 16:11, 16:25*t*
Nyctalus spp. *See* Noctules
Nyctalus lasiopterus. *See* Greater noctules
Nyctalus leisleri. *See* Leisler's bats
Nyctalus noctula. *See* Noctules
Nyctereutes procyonoides. *See* Raccoon dogs
Nycteridae. *See* Slit-faced bats
Nycteris spp. *See* Slit-faced bats
Nycteris arge. *See* Bate's slit-faced bats
Nycteris avrita, 13:373
Nycteris gambiensis. *See* Gambian slit-faced bats
Nycteris grandis. *See* Large slit-faced bats
Nycteris hispida. *See* Hairy slit-faced bats
Nycteris intermedia. *See* Intermediate slit-faced bats
Nycteris javanica. *See* Javan slit-faced bats
Nycteris macrotis. *See* Large-eared slit-faced bats
Nycteris madagascarensis. *See* Madagascar slit-faced bats
Nycteris major. *See* Ja slit-faced bats
Nycteris thebaica. *See* Egyptian slit-faced bats
Nycteris woodi. *See* Wood's slit-faced bats
Nycticebus bengalensis. *See* Bengal slow lorises
Nycticebus coucang. *See* Sunda slow lorises
Nycticebus pygmaeus. *See* Pygmy slow lorises
Nycticeius humeralis. *See* Evening bats
Nyctimene spp., 13:335
Nyctimene rabori. *See* Philippine tube-nosed fruit bats
Nyctimene robinsoni. *See* Queensland tube-nosed bats
Nyctinomops femorosaccus. *See* Pocketed free-tailed bats

O

Obdurodon spp., 12:228
Obligatory vivipary, 12:6
 See also Reproduction

INDEX

INDEX

INDEX

INDEX

INDEX